Western Plainchant

London, British Library, Harley 4951, fo. 298ᵛ, showing the start of the section for mode 3 in the tonary (cf. Ex. III.14.1–2). By permission of the British Library.

Western Plainchant

A Handbook

DAVID HILEY

CLARENDON PRESS · OXFORD

1993

Oxford University Press, Walton Street, Oxford OX2 6DP

Oxford New York Toronto
Delhi Bombay Calcutta Madras Karachi
Kuala Lumpur Singapore Hong Kong Tokyo
Nairobi Dar es Salaam Cape Town
Melbourne Auckland Madrid
and associated companies in
Berlin Ibadan

Oxford is a trade mark of Oxford University Press

Published in the United States by
Oxford University Press Inc., New York

First published 1993
Reprinted 1993

British Library Cataloguing in Publication Data
Data available
ISBN 0-19-816289-8

Library of Congress Cataloging in Publication Data
Western plainchant: a handbook / David Hiley.
Includes bibliographical references and index.
1. Chants (Plain, Gregorian, etc.)—History and criticism.
I. Title.
ML3082.H54   1992   782.32'22—dc20   92-13020
ISBN 0-19-816289-8

Typeset by Hope Services (Abingdon) Ltd.
Printed in Great Britain
on acid-free paper by
St. Edmundsbury Press, Bury St. Edmunds

for Marjorie and Henry

PREFACE

I have been up to London to get the book I am writing, out of the British Museum.
I have got a lot of it out, and I shall go again presently to get some more; and
when I have got it all, there will be another book . . . So many people were there,
getting out their books. It doesn't seem to matter everything's being in books
already: I don't mind it at all. There are attendants there on purpose to bring it to
you. That is how books are made, and it is difficult to think of any other way. I
mean the kind called serious . . .

No doubt the author of any reference book such as the present one could echo the
sentiments of Miss Charity Marcon, in Ivy Compton-Burnett's *Daughters and Sons*.
A great deal of my book is indeed got out of others, as the bibliography and references
in the text make clear. What justification is there for this, and what is the purpose of
the book?

In the first place I wished to provide a book of reference both for those coming new
to plainchant and for those needing guidance in the specialist literature. The book
starts with the assumption, reasonable in this secular age, that many things about the
liturgy and its plainchant, even quite basic matters, are unfamiliar to the reader. At
every stage in the encounter with plainchant one comes up against specialist
terminology and concepts which constitute a real obstacle. That is in the nature of the
subject, for ecclesiastical ritual is essentially exclusive, remote from everyday
experience, reserved for specially trained personnel. To start with essentials does not
mean, however, that difficulties have been avoided. I have not, I hope, confused
inexperience with lack of intelligence. The reader will encounter here many complex
problems, both those for which scholars have found solutions and others which
remain obscure. I have also illustrated techniques of research and given examples, not
just of the music in plainchant sources, but also of their make-up, the way they
deploy their material, and their notation.

Those with access to well-equipped libraries will find here sufficient references to
further specialist literature. But I have also tried to make the book self-explanatory,
and well enough illustrated, so that it will be useful also to anyone interested in and
able to read music.

Such a book fulfils a need primarily because of the enormous expansion of
plainchant studies in the last few decades. The last major work of synthesis in the
English language, Apel's *Gregorian Chant*, is now over thirty years old. It has not, of
course, been my intention to try and replace Apel, let alone the more comprehensive
Einführung in die gregorianischen Melodien by Peter Wagner from before the First
World War, both of which remain indispensable. But much has been explored and
discovered since their day. The best modern survey, the article 'Plainchant' by

Kenneth Levy and John Emerson in *The New Grove Dictionary*, is necessarily brief (but with an excellent bibliography). The writings of scholars such as the late Bruno Stäblein—his *Schriftbild der einstimmigen Musik* and articles in *Die Musik in Geschichte und Gegenwart*—together practically constitute a textbook on chant. But an up-to-date one-volume work is clearly required.

The compression and omissions entailed in such a work are the least pleasant things facing an author. For my own undoubted sins of omission I have tried to make amends in the bibliography, by citing literature to which I could not do justice in the main text. The bibliography should go some way towards being a reference tool in itself. In writing the book I tried to take into account literature up to 1990. I take this opportunity to mention the recent appearance of a new chant bibliography with over 4,200 items by Thomas Kohlhase and Günther Michael Paucker, in *Beiträge zur Gregorianik*, 9–10 (1990). To one work which appeared when the main text was all but complete, but whose contents I knew intimately, I have made no reference. I acted as co-editor with Richard Crocker of the new edition of *The New Oxford History of Music*, ii: *The Early Middle Ages to 1300* (Oxford, 1989), and it was not always easy to avoid unconscious borrowing from it. At any rate, that volume now takes its place beside Wagner, Apel, and Stäblein's *Schriftbild* as an essential part of the chant scholar's library.

Historical writing about plainchant is a relatively young phenemonon. Prince-Abbot Martin Gerbert's *De cantu et musica sacra a prima ecclesiae aetate usque ad praesens tempus* of 1774, the ancestor of all musicological writing on plainchant, was motivated by the desire to stimulate reforms in the church music of his own time. Knowledge of the past would make men conscious of the malpractices of the present: 'I desire . . . to lay open to view the astounding abuses in these matters, which in my opinion are the gravest in the discipline of our church' (quoted in p. lviii* of Othmar Wessely's introduction to the 1968 reprint of Gerbert's work). The work of the great restorers of the late nineteenth and early twentieth centuries was imbued with the spirit of reform, to be achieved, as with Gerbert, through recovery of a former ideal state. Work which is of incalculable value for musicologists sprang from these reforming efforts: the series of facsimiles *Paléographie musicale*, the volumes of *Le Graduel romain*, the studies of notational practices made by Dom Eugène Cardine and his pupils (I cite here only a few examples from the work of monks of the French Benedictine monastery of Solesmes). However, the fact that more than musicology was involved inevitably affected the selection of information and the use to which it was put. This type of counterpoint between scholars and their material is of course common to all scholarship. In the case of chant studies the consequences of different viewpoint and purposes can be appreciated by comparing Cardine's 'Vue d'ensemble sur le chant grégorien' and Levy and Emerson's 'Plainchant'.

The researches of the Benedictine fathers also had a definite practical purpose: the provision of model, definitive chant-books for use in the Roman Church. Consequently, their work was concerned with those chants sung in the Roman liturgy of their own time, but not with parts of the repertory, such as sequences and tropes, which were no longer admitted, though important in the Middle Ages. In recent years mostly lay scholars have devoted considerable energy to those repertories, and it is perhaps not fanciful to see this partly as a reaction against the bias of earlier chant studies. In my own case, it led among other things to a decision to transcribe all musical examples from original sources, rather than rely on modern service-books, slight though the differences between them may sometimes be.

A great deal of the satisfaction gained from writing a book such as this comes from sorting out problems in one's own mind, for, as the renowned author of a monograph on Lassus succinctly expressed it: 'Nothing clears up a case so much as stating it to another person.' Yet even more satisfying has been the ever-increasing wonderment at the variety and richness of the chant repertory, a delight to anyone working in the area. I hope that the musical examples, at least, will help further the appreciation of these treasures.

Like Miss Marcon, I owe a considerable debt to the British Library, but also to the Anselm Hughes Library of Royal Holloway and Bedford New College, University of London, where I once taught; to the University of London Library at Senate House, and more recently to the Universitätsbibliothek of Regensburg University, where I now teach. Many of my musical examples were transcribed from microfilms, work with which has in many cases been made possible by grants from the Central Research Fund of the University of London. My principal debt is to the numerous fellow chant scholars from whom I have learned over the years, many of whom could have written a better book than this. It is the fate of textbooks to be used for target-practice, as it were, by better-informed teachers, but I hope my colleagues will find here something they can set before their students with reasonable confidence. Above all, I hope the book will straighten the path of those who, like myself a couple of decades ago, are trying to find their way in an initially foreign, often bewildering, but always fascinating and rewarding musical world.

I am most grateful to the libraries which kindly supplied photographs for the plates. Plates 1, 4–6, 8–9, 11–14, and 18 are published by permission of the British Library; plates 2, 7, and 16 by permission of the Bodleian Library; plate 10 by permission of the Syndics of Cambridge University Library; plate 15 by permission of the Master and Fellows of University College, Oxford; and plate 17 by permission of Edinburgh University Library.

In conclusion I wish to thank those who have played an especially important part in the production of the book: Malcolm Gerratt, who launched the project and fanned it along for several years; Bruce Phillips, who brought it into the safe haven of Oxford University Press, and the staff of OUP, especially Leofranc Holford-Strevens and Bonnie Blackburn, who did far more for the book than an author has a right to expect from his copy-editor. Greatest of all is my debt to my wife Ann, without whose patience and encouragement I should never have reached the end of the long voyage.

D.H.

CONTENTS

ANALYTICAL TABLE

I. PLAINCHANT IN THE LITURGY

II. CHANT GENRES

III. LITURGICAL BOOKS AND PLAINCHANT SOURCES

IV. NOTATION

PLATES

V. PLAINCHANT AND EARLY MUSIC THEORY

VI. PLAINCHANT UP TO THE EIGHTH CENTURY

VII. THE CAROLINGIAN CENTURY

VIII. GREGORIAN CHANT AND OTHER CHANT REPERTORIES

IX. PERSONS AND PLACES

X. REFORMATIONS OF GREGORIAN CHANT

XI. THE RESTORATION OF MEDIEVAL CHANT

ILLUSTRATIONS

Plates

For list see above, p.xvii.

Maps

Figure

TABLES

MUSIC EXAMPLES

NOTE ON MUSIC EXAMPLES

Slurs are used only for notes joined in the same sign in the original notation.

ꞁ = oriscus   ᴎ = neume at the semitone step ('mi-neume')

ɯ = quilisma   ᾽ = apostropha

⌐° ⌐ = liquescent note

Pitch-letters follow Guidonian, not modern practice.

Γ A B C D E F G a b c d e f g $\overset{a}{a}$ $\overset{b}{b}$ $\overset{c}{c}$ etc.

ABBREVIATIONS

AcM	*Acta musicologica*
AfMw	*Archiv für Musikwissenschaft*
AH	*Analecta hymnica medii aevi*, ed. Guido Maria Dreves, Clemens Blume, and Henry Marriott Bannister, 55 vols. (Leipzig, 1886–1922); *Register*, ed. Max Lütolf, 3 vols. (Berne, 1978). Text editions of hymns: 2, 4, 11–12, 14, 16, 19, 22–3, 27, 43, 48, 50–2; of rhymed offices: 5, 13, 17–18, 24–6, 28, 45; of sequences: 7–10, 34, 37, 39–40, 42, 44, 53–5; of tropes: 47, 49. For individual volumes see Bibliography
AM	*Antiphonale monasticum pro diurnis horis* (Tournai, 1934)
AMS	René-Jean Hesbert, *Antiphonale missarum sextuplex* (Brussels, 1935) [Monza, Basilica S. Giovanni, CIX: 'Modoetiensis'; Zürich, Zentralbibliothek, Rheinau 30: 'Rhenaugiensis'; Brussels, Bibliothèque Royale, 10127–10144: 'Blandiniensis'; Paris, Bibliothèque Nationale, lat. 17436: 'Compendiensis'; Paris, Bibliothèque Nationale, lat. 12050: 'Corbiensis'; Paris, Bibliothèque Sainte-Geneviève, 111: 'Silvanectensis']
AR	*Antiphonale sacrosanctae Romanae ecclesiae* (Rome, 1912)
AS	*Antiphonale Sarisburiense: A Reproduction in Facsimile of a Manuscript of the Thirteenth Century*, ed. Walter Howard Frere (London, 1901–24)
BzG	*Beiträge zur Gregorianik*
CAO	*Corpus antiphonalium officii*, ed. René-Jean Hesbert (Rerum ecclesiasticarum documenta, Series maior, Fontes, 7–12; Rome, 1963–79). For individual volumes see Bibliography
CCM	Corpus consuetudinum monasticarum, ed. Kassius Hallinger (Siegburg, 1963–). For individual volumes see Bibliography
CM	Consuetudines monasticae, ed. Bruno Albers. For individual volumes see Bibliography
CS	Edmond de Coussemaker, *Scriptorum de musica medii aevi novam seriem . . .*, 4 vols. (Paris, 1864, 1867, 1869, 1876)
CSM	Corpus scriptorum de musica (Rome, 1950–). For individual volumes see Bibliography
CT	Corpus troporum (Studia Latina Stockholmiensia; Stockholm, 1975–). For individual volumes see Bibliography
DACL	Fernand Cabrol, Henri Leclercq, and Henri Marrou (eds.), *Dictionnaire d'archéologie chrétienne et de liturgie*, 15 vols. in 30 (Paris, 1907–53)
DMA	Divitiae musicae artis, ed. Joseph Smits van Waesberghe (Buren, 1975–). For individual volumes see Bibliography
EG	*Études grégoriennes*
EL	*Ephemerides liturgicae*
EMH	*Early Music History*
GR	*Graduale sacrosanctae Romanae ecclesiae* (Rome, 1908)

GS   Martin Gerbert, *Scriptores ecclesiastici de musica sacra potissimum*, 3 vols. (Sankt-Blasien, 1784) [to be used in conjunction with Bernhard 1989]

GT   *Graduale triplex*. ed. Marie-Claire Billecocq and Rupert Fischer (Solesmes, 1979) [*Graduale Romanum* of 1974 with neumes of Laon 239, Einsiedeln 121, St Gall 359 and 339, etc.]

HMT   Hans-Heinrich Eggebrecht (ed.), *Handwörterbuch der musikalischen Terminologie* (Mainz, 1971–)

JAMS   *Journal of the American Musicological Society*

JPMMS   *Journal of the Plainsong & Mediaeval Music Society*

KmJb   *Kirchenmusikalisches Jahrbuch*

LR   *Liber responsorialis pro festis I. classis et communi sanctorum juxta ritum monasticum* (Solesmes, 1894)

LU   *Liber usualis missae et officii pro dominicis et festis I. vel II. classis* (Rome, 1921)

MD   *Musica disciplina*

Mf   *Die Musikforschung*

MGG   Friedrich Blume (ed.), *Die Musik in Geschichte und Gegenwart: Allgemeine Enzyklopädie der Musik*, 17 vols. (Kassel, 1949–86)

MGH   Monumenta Germaniae historica. For individual volumes see Bibliography

MMMA   Monumenta monodica medii aevi (Kassel, 1956–). For individual volumes see Bibliography

MMS   Monumenta musicae sacrae, ed. René-Jean Hesbert. For individual volumes see Bibliography

MQ   *Musical Quarterly*

NCE   *The New Catholic Encyclopedia*, 15 vols. (New York, 1967; supplementary vols. 1974, 1979)

NG   Stanley Sadie (ed.), *The New Grove Dictionary of Music and Musicians*, 20 vols. (London, 1980)

PalMus   Paléographie musicale: Les principaux manuscrits de chant grégorien, ambrosien, mozarabe, gallican [premier série, deuxième série]. For individual volumes see Bibliography

PL   Patrologiae cursus completus, series latina, ed. J.-P. Migne, 221 vols. (Paris, 1844–64)

RB   *Revue bénédictine*

RCG   *Revue du chant grégorien*

RG   *Revue grégorienne*

RIM   *Rivista italiana di musicologia*

RISM   Répertoire international des sources musicales (Munich and Duisburg). For individual volumes see Bibliography

SMH   *Studia musicologica Academiae scientiarum Hungaricae*

VGA   Veröffentlichungen der Gregorianischen Akademie zu Freiburg in der Schweiz

BIBLIOGRAPHY

ABATE, G., 'Il primitivo breviario francescano', *Miscellanea francescana*, 60 (1960), 47–240.

ABERT, HERMANN, *Die aesthetischen Grundsätze der mittelalterlichen Melodiebildung* (Halle, 1905).

—— *Die Musikanschauung des Mittelalters und ihre Grundlagen* (Halle, 1905).

ABRAHAMSEN, ERIK, *Éléments romans et allemands dans le chant grégorien et la chanson populaire en Danemark* (Publications de l'Académie grégorienne de Fribourg (Suisse), 11; Copenhagen, 1923).

Acta sanctorum collecta . . . a Sociis Bollandianis (Antwerp, 1643– ; 3rd edn., Paris and Brussels, 1863–).

Acta sanctorum ordinis sancti Benedicti, ed. Jean Mabillon, Luc d'Achery, *et al.*, 9 vols. (Paris, 1668–1701).

ADLER Festschrift 1930 = *Studien zur Musikgeschichte: Festschrift für Guido Adler* (Vienna, 1930).

AGUSTONI, LUIGI, *Gregorianischer Choral: Elemente und Vortragslehre mit besonderer Berücksichtigung der Neumenkunde* (Fribourg, 1963).

—— 'Die Frage der Tonstufen SI und MI', *BzG* 4 (1987), 47–101.

—— and GÖSCHL, Johannes Berchmans, *Einführung in die Interpretation des Gregorianischen Chorals, 1: Grundlagen* (Regensburg, 1987).

AH 2 (1887) = *Hymnarius Moissiacensis: Das Hymnar der Abtei Moissac im 10. Jahrhundert nach einer Handschrift der Rossiana*, ed. Guido Maria Dreves.

AH 4 (1888) = *Hymni inediti: Liturgische Hymnen des Mittelalters aus handschriftlichen Breviarien, Antiphonalien und Processionalien*, ed. Dreves.

AH 5 (1889), 13 (1892), 18 (1894), 24 (1896), 25–6 (1897), 28 (1898), 45 (1904) = *Historiae rhythmicae: Liturgische Reimoffizien des Mittelalters aus Handschriften und Wiegendrucken*, ed. Dreves.

AH 7 (1889) = *Prosarium Lemovicense: Die Prosen der Abtei St. Martial zu Limoges, aus Troparien des 10., 11. und 12. Jahrhunderts*, ed. Dreves.

AH 8–9 (1890), 10 (1891) = *Sequentiae ineditae: Liturgische Prosen des Mittelalters aus Handschriften und Frühdrucken*, ed. Dreves.

AH 11 (1891), 12 (1892), 19 (1895), 22 (1895), 23 (1896), 43 (1903) = *Hymni inediti: Liturgische Hymnen des Mittelalters aus Handschriften und Wiegendrucken*, ed. Dreves.

AH 13 see AH 5.

AH 14 (1893) = *Hymnarius Severinianus: Das Hymnar der Abtei S. Severin in Neapel nach den Codices Vaticanus 7172 und Parisinus 1092*, ed. Dreves.

AH 16 (1894) = *Hymnodia Hiberica: Spanische Hymnen des Mittelalters aus liturgischen Handschriften und Druckwerken römischen Ordos*, ed. Dreves.

AH 17 (1894) = *Hymnodia Hiberica: Liturgische Reimoffizien aus spanischen Brevieren. Im Anhange: Carmina Compostellana, die Lieder des s.g. Codex Calixtinus*, ed. Dreves.

AH 18 see AH 5.

AH 19, 22, 23 see AH 11.

AH 24–6, 28 *see* AH 5.

AH 27 (1897) = *Hymnodia Gotica: Die mozarabischen Hymnen des alt-spanischen Ritus aus handschriftlichen und gedruckten Quellen*, ed. Clemens Blume.

AH 34 (1900), 37 (1901), 39 (1902), 42 (1903), 44 (1904) = *Sequentiae ineditae: Liturgische Prosen des Mittelalters aus Handschriften und Frühdrucken*, ed. Blume.

AH 40 (1902) = *Sequentiae ineditae: Liturgische Prosen des Mittelalters aus Handschriften und Frühdrucken*, ed. Henry Marriott Bannister.

AH 42 *see* AH 34.

AH 43 *see* AH 11.

AH 44 *see* AH 34.

AH 45 *see* AH 5.

AH 47 (1905) = *Tropi graduales: Tropen des Missale im Mittelalter. I. Tropen zum Ordinarium Missae. Aus handschriftlichen Quellen*, ed. Blume and Bannister.

AH 48 (1905), 50 (1907) = *Hymnographi Latini: Lateinische Hymnendichter des Mittelalters aus gedruckten und ungedruckten Quellen*, ed. Dreves.

AH 49 (1906) = *Tropi graduales: Tropen des Missale im Mittelalter. II. Tropen zum Proprium Missae. Aus handschriftlichen Quellen*, ed. Blume.

AH 50 *see* AH 48.

AH 51 (1908) = *Thesauri hymnologici hymnarium: Die Hymnen des Thesaurus Hymnologicus H. A. Daniels und anderer Hymnen-Ausgaben. I. Die Hymnen des 5.–11. Jahrhunderts und die irisch-keltische Hymnodie aus den ältesten Quellen*, ed. Blume.

AH 52 (1909) = *Thesauri hymnologici hymnarium . . . II. Die Hymnen des 12.–16. Jahrhunderts aus den ältesten Quellen*, ed. Blume.

AH 53 (1911) = *Thesauri hymnologici prosarium: Liturgische Prosen erster Epoche aus den Sequenzenschulen des Abendlandes, insbesondere die dem Notkerus Balbulus zugeschriebenen, nebst Skizze über den Ursprung der Sequenz, aufgrund der Melodien aus den Quellen des 10.–16. Jahrhunderts*, ed. Blume and Bannister.

AH 54 (1915) = *Thesauri hymnologici prosarium: Liturgische Prosen des Übergangsstiles und der zweiten Epoche, insbesondere die dem Adam von Sanct Victor zugeschriebenen, aus Handschriften und Frühdrucken*, ed. Blume and Bannister.

AH 55 (1922) = *Thesauri hymnologici prosarium: Liturgische Prosen zweiter Epoche auf Feste der Heiligen nebst einem Anhange: Hymnologie des Gelderlandes und des Haarlemer Gebietes aus Handschriften und Frühdrucken*, ed. Blume.

ÄIKÄÄ, ERMO, 'Ein marianischer Gloria-Tropus', paper read at the First European Science Foundation International Workshop on Tropes, Munich, 1983.

ALBAROSA, NINO, 'La scuola gregoriana di Eugène Cardine', *RIM* 9 (1974), 269–97; 12 (1977) 136–52.

—— 'The Pontificio Istituto di Musica Sacra in Rome and the Semiological School of Dom Eugène Cardine', *JPMMS* 6 (1983), 26–33.

—— 'Paleografi non semiologi?', in Huglo 1987, 101–5.

Albers *see* CM.

ALBRECHT (HANS) Gedenkschrift 1962 = Wilfried Brennecke and Hans Haase (eds.), *Hans Albrecht in Memoriam* (Kassel, 1962).

ALBRECHT (OTTO) Festschrift 1980 = John Walter Hill (ed.), *Studies in Musicology in Honor of Otto E. Albrecht* (Kassel, 1980).

ALFONZO, PIO, *L'antifonario dell'Uffizio Romano: Note sulle origini della composizione dei testi* (Monografie liturgiche, 3; Subiaco, 1935).

—— *I responsori biblici dell'Uffizio Romano* (Lateranum, NS, 2/1; Rome, 1936).

ALLWORTH, CHRISTOPHER, 'The Medieval Processional: Donaueschingen MS 882', *EL* 84 (1970), 169–86.

ALMEIDA, FORTUNATO DE, *História da igreja em Portugal*, 8 vols. in 4 (Coimbra, 1910–22; 2nd edn., 1930).

AMELLI, AMBROGIO M., 'L'epigramma di Paolo Diacono intorno al canto Gregoriano e Ambrosiano'. *Memorie storiche forogiuliesi*, 9 (1913), 153–75.

AMIET, ROBERT, 'Un bréviaire noté du XIVe siècle de la primatiale Saint-Jean de Lyon', *EG* 15 (1975), 201–6.

—— 'Le Liber misticus de Combret, au diocèse d'Elne', *EG* 20 (1981), 5–68.

ANDERSEN, M. GEERT, and RAASTED, JØRGEN, *Inventar over Det kongelige Biblioteks fragmentsamling* [Inventory of the Fragment Collection in the Royal Library] (Hjælpemidler, 6; Copenhagen, 1983).

ANDERSON, GORDON A. (ed.), *The Las Huelgas Manuscript, Burgos, Monasterio de las Huelgas*, 2 vols. (Corpus mensurabilis musicae, 79; Rome, 1982).

ANDERSON (GORDON) Gedenkschrift 1984 = *Gordon Athol Anderson (1929–1981) in memoriam*, 2 vols. (Henryville–Ottawa–Binningen, 1984).

ANDERSON, WARREN, 'Alypius', *NG*.

ANDOYER, RAPHAEL, 'Le Chant romain antégrégorien', *RCG* 20 (1911–12), 69–75, 107–14.

—— 'L'Ancienne Liturgie de Bénévent', *RCG* 20 (1911–12), 176–83; 21 (1912–13), 14–20, 44–51, 81–5, 112–15, 144–8, 169–74; 22 (1913–14), 8–11, 41–4, 80–3, 106–11, 141–5, 170–2; 23 (1919–20), 44–50, 116–18, 151–3, 182–3; 24 (1920–1), 48–50, 87–9, 146–8, 182–5.

ANDRIEU, MICHEL, 'Règlement d'Angilramne de Metz (768–791) fixant les honoraires de quelques fonctions liturgiques', *Revue des sciences religieuses*, 10 (1930), 349–69.

—— *Les Ordines Romani du haut moyen âge* (Spicilegium sacrum Lovaniense, 11, 23–4, 28–9; Louvain, 1931–61).

—— *Le Pontifical romain au moyen âge* (Studi e testi, 86–8, 99; Rome, 1938–41).

ANDRIEU Festschrift 1956 = *Mélanges en l'honneur de M. Michel Andrieu* (Strasburg, 1956).

ANGERER, JOACHIM FRIDOLIN, *Die liturgisch-musikalische Erneuerung der Melker Reform: Studien zur Erforschung der Musikpraxis in den Benediktinerklöstern des 15. Jahrhunderts* (Österreichische Akademie der Wissenschaften, Philologisch-historische Klasse, 287; Vienna, 1974).

—— 'Die Consuetudines Monasticae als Quelle für die Musikwissenschaft', in Haberl Festschrift 1977, 237.

—— *Lateinische und deutsche Gesänge aus der Zeit der Melker Reform* (Forschungen zur älteren Musikgeschichte, 2; Vienna, 1979).

ANGLÈS, HIGINI, 'Epístola farcida del martiri de Sant Esteve', *Vida cristiana*, 9 (1922), 69–75.

—— (ed.), *El còdex musical de Las Huelgas (Música a veus dels segles XIII–XIV)*, 3 vols. (Biblioteca de Catalunya, Publicacions del Departament de Mùsica, 6; Barcelona, 1931).

—— *La música a Catalunya fins al segle XIII* (Biblioteca de Catalunya, Publicacions del Departament de Música, 10; Barcelona, 1935).

—— 'La mùsica medieval en Toledo hasta el siglo XI', *Gesammelte Aufsätze zur Kulturgeschichte Spaniens*, 7 (Spanische Forschungen der Görresgesellschaft, 1st ser.; Münster, 1938), 1–68.

—— *Historia de la música medieval en Navarra* (Pamplona, 1970).

ANGLÈS, HIGINI and SUBIRÁ, JOSÉ, *Catálogo musical de la Biblioteca nacional de Madrid, 1: Manuscritos* (Barcelona, 1946).

ANGLÈS Festschrift 1958–61 = *Miscelánea en homenaje a Monseñor Higinio Anglès*, 2 vols. (Barcelona, 1958, 1961).

Antifonario visigótico mozárabe de la catedral de León (Monumenta Hispaniae sacra, serie litúrgica, v/2; Madrid, 1953) (see also Brou and Vives 1959).

Antiphonale Hispaniae vetus (s. X–XI). Biblioteca de la Universidad de Zaragoza (Saragossa, 1986).

Antiphonale missarum juxta ritum sanctae ecclesiae Mediolanensis [ed. G. M. Suñol] (Rome, 1935).

Antiphonale monasticum secundum traditionem Helveticae Congregationis Benedictinae (1943).

Antiphonale Pataviense (Wien 1519), ed. Karlheinz Schlager (Das Erbe deutscher Musik, 88; Kassel, 1985).

Antiphonale Silense: British Library Mss. Add. 30.850, ed. Ismael Fernández de la Cuesta (Madrid, 1985).

Antiphonar von St. Peter [Vienna, Nationalbibliothek, series nova 2700], 2 vols. (Codices selecti, 21; Graz, 1969–73, 1974).

APEL, WILLI, *Gregorian Chant* (Bloomington, Ind., 1958).

ARCHER, PETER, *The Christian Calendar and the Gregorian Reform* (New York, 1941).

ARLT, WULF, *Ein Festoffizium des Mittelalters aus Beauvais in seiner liturgischen und musikalischen Bedeutung*, 2 vols. (Cologne, 1970).

—— 'Einstimmige Lieder des 12. Jahrhunderts und Mehrstimmiges in französischen Handschriften des 16. Jahrhunderts aus Le Puy', *Schweizer Beiträge zur Musikwissenschaft*, 3 (1978), 7–47.

—— 'Zur Interpretation der Tropen', *Forum musicologicum*, 3 (1982), 61–90.

—— and STAUFFACHER, MATTHIAS, *Engelberg Stiftsbibliothek Codex 314* (Winterthur, 1986).

ARNESE, RAFFAELE, *I codici notati della Biblioteca nazionale di Napoli* (Florence, 1967).

ARNOUX, G., *Musique platonicienne—âme du monde* (Paris, 1960).

ASHWORTH, HENRY, 'Did St. Augustine Bring the Gregorianum to England?' *EL* 72 (1958), 39–43.

—— 'The Liturgical Prayers of St. Gregory the Great', *Traditio*, 15 (1959), 107–61.

—— 'Further Parallels to the "Hadrianum" from St. Gregory the Great's Commentary on the First Book of Kings', *Traditio*, 16 (1960), 364–73.

ASKETORP, BODIL, 'The Musical Contents of Two Danish Pontificals from the Late Middle Ages', *JPMMS* 7 (1984), 28–46.

ATCHLEY, E. G. C. F., *Ordo Romanus primus* (London, 1905).

ATKINSON, CHARLES M., 'The Earliest Agnus Dei Melody and its Tropes', *JAMS* 30 (1977), 1–19.

—— '*O amnos tu theu*: The Greek Agnus Dei in the Roman Liturgy from the Eighth to the Eleventh Century', *KmJb* 65 (1981), 7–30.

—— 'Zur Entstehung und Überlieferung der "Missa Graeca"', *AfMw* 39 (1982), 113–45.

—— 'The *Parapteres*: *Nothi* or Not?', *MQ* 68 (1982), 32–59.

—— 'On the Interpretation of *Modi, quos abusive tonos dicimus*', in Gallacher and Damico 1989, 147–61.

—— 'The *Doxa*, the *Pisteuo*, and the *Ellinici Fratres*: Some Anomalies in the Transmission of the Chants of the "Missa Graeca"', *Journal of Musicology*, 7 (1989), 81–106.

——'From "Vitium" to "Tonus acquisitus": On the Evolution of the Notational Matrix of Medieval Chant', in Cantus Planus 1990, 181–97.

——'Parapter', *Handwörterbuch der musikalischen Terminologie*, ed. Hans-Heinrich Eggebrecht (Mainz, 1971–).

AUDA, ANTOINE, *L'École musicale liégeoise au Xe siècle: Étienne de Liège* (Brussels, 1923).

——*Les Modes et les tons de la musique et spécialement de la musique médiévale* (Académie royale de Belgique. Classe des beaux arts. Mémoires, iii, pt. 1; Brussels, 1930).

——*Contribution à l'histoire de l'origine des modes et des tons grégoriens* (Grenoble, 1932) = RCG 36 (1932), 33–9, 72–7, 105–11, 130–2.

AUGUSTINE, ST, *De musica, a Synopsis*, ed. W. F. J. Knight (London, 1949).

——*On Music: De musica. The Fathers of the Church, a New Translation*, trans. R. C. Taliaferro (Writings of Saint Augustine, 2; New York, 1947), 153–379.

AVENARY, HANOCH, 'The Northern and Southern Idioms of Early European Music: A New Approach to an Old Problem', AcM 49 (1977), 27–49.

AVERY, MYRTILLA, *The Exultet Rolls of South Italy* (Princeton, NJ, 1936).

BABB, WARREN, *Hucbald, Guido, and John on Music*, ed., with introductions, by Claude V. Palisca, index of chants by Alejandro Enrique Planchart (New Haven, Conn., 1978).

Bagnall *see* Yardley.

BAILEY, TERENCE, *The Processions of Sarum and the Western Church* (Toronto, 1971).

——*The Intonation Formulas of Western Chant* (Toronto, 1974).

——'Accentual and Cursive Cadences in Gregorian Psalmody', *JAMS* 29 (1976), 463–71.

——'Modes and Myth', *Studies in Music from the University of Western Ontario*, 1 (1976), 43–54.

——'Ambrosian Psalmody: An Introduction', *Studies in Music from the University of Western Ontario*, 2 (1977), 65–78.

——'De modis musicis: A New Edition and Explanation', *KmJb* 61–2 (1977–8), 47–60.

——'Ambrosian Psalmody: The Formulae', *Studies in Music from the University of Western Ontario*, 3 (1978), 72–96.

——*Commemoratio brevis de tonis et psalmis modulandis: Introduction, Critical Edition, Translation* (Ottawa, 1979).

——*The Ambrosian Alleluias* (Englefield Green, 1983).

——'Ambrosian Chant in Southern Italy', *JPMMS* 6 (1983), 1–7.

——*The Ambrosian Cantus* (Ottawa, 1987).

——'Milanese Melodic Tropes', *JPMMS* 11 (1988), 1–12.

——and MERKLEY, PAUL, *The Antiphons of the Ambrosian Office* (Ottawa, 1989).

——— ——*The Melodic Tradition of the Ambrosian Office-Antiphons* (Ottawa, 1990).

BALDOVIN, JOHN F., *The Urban Character of Christian Worship: The Origins, Development, and Meaning of Stational Liturgy* (Orientalia Christiana analecta, 228; Rome, 1987).

BALTZER, REBECCA A., 'Johannes de Garlandia', *NG*.

Bamberg 6 = *Die Handschrift Bamberg Staatsbibliothek Lit. 6* (Monumenta palaeographica Gregoriana, 2; Münsterschwarzach, [1986]).

BANNISTER, HENRY MARRIOTT, *Monumenti vaticani di paleografia musicale latina*, 2 vols. (Codices e vaticanis selecti, phototypice expressi, 11; Leipzig, 1913).

——*see also* AH.

BARALLI, RAFFAELLO, 'A proposito di un piccolo trattato sul canto ecclesiastico in un manoscritto dei sec. X–XI', *Rassegna gregoriana*, 4 (1905), 59–66.

BÁRDOS, KORNÉL, *Volksmusikartige Variierungstechnik in den ungarischen Passionen, 15. bis 18. Jahrhundert* (Musicologia Hungarica, NS, 5; Budapest, 1975).

BAREZZANI, MARIA TERESA ROSA, *La notazione neumatica di un codice bresciano (secolo XI)* (Cremona, 1981).

BAROFFIO, BONIFAZIO, *Die Offertorien der ambrosianischen Kirche: Vorstudie zur kritischen Ausgabe der mailändischen Gesänge* (Diss., Cologne University, 1964).

——'Die mailändische Überlieferung des Offertoriums *Sanctificavit*', in Stäblein Festschrift 1967, 1–8.

——'Ambrosianische Liturgie', in Fellerer 1972, 191–204.

——'Le origini del canto liturgico nella Chiesa latina e la formazione dei repertori italici', *Renovatio*, 13 (1978), 26–52.

——'Benevent', *MGG* (supplement).

——'Ambrosian Rite, Music of the', *NG*.

——and STEINER, RUTH, 'Offertory', *NG*.

BATIFFOL, PIERRE, *Histoire du bréviaire roman* (3rd edn., Paris, 1911); trans.: *History of the Roman Breviary* (London, 1913).

BAUDOT, JULES, 'Bénédictionnaire', *DACL*.

BÄUMER, SUITBERT, *Geschichte des Breviers*, 2 vols. (Freiburg, 1895); trans. and rev. R. Biron, *Histoire du bréviaire*, 2 vols. (Paris, 1905).

BAUMSTARK, ANTON, *Comparative Liturgy* (London, 1958).

BAUTIER, ROBERT-HENRI, GILLES, MONIQUE, DUCHEZ, MARIE-ELISABETH, and HUGLO, MICHEL, *Odorannus de Sens: Opera omnia* (Paris, 1972).

BAUTIER-REGNIER, ANNE-MARIE, 'À propos du sens de *neuma* et de *nota* en latin médiéval', *Revue belge de musicologie*, 18 (1964), 1–9.

BAYART, PAUL, *Les Offices de Saint Winnoc et de Saint Oswald d'après le manuscrit 14 de la Bibliothèque de Bergues*, Annales du Comité Flamand de France 35 (Lille, 1926).

BEARE, WILLIAM, *Latin Verse and European Song: A Study in Accent and Rhythm* (London, 1957).

BECKER, GUSTAV HEINRICH, *Catalogi bibliothecarum antiqui* (Bonn, 1885).

BECKER, HANSJAKOB, *Die Responsorien des Kartäuserbreviers* (Münchener theologische Studien, II. Systematische Abteilung, 39; Munich, 1971).

——*Das Tonale Guigos I: Ein Beitrag zur Geschichte des liturgischen Gesangs und der Ars Musica im Mittelalter* (Münchener Beiträge zur Mediävistik und Renaissance-Forschung, 23; Munich, 1975).

BEDE, *A History of the English Church and People*, ed. Leo Sherley-Price (Harmondsworth, 1955; 2nd edn., 1968).

——*Opera historica*, ed. C. Plummer, 2 vols. (Oxford, 1896).

BENOÎT-CASTELLI, GEORGES, 'Le "Praeconium paschale"', *EL* 67 (1953), 309–34.

——'Un processional anglais du XIVème siècle: Le processional dit "de Rollington"', *EL* 75 (1961), 281–326.

BENSON, ROBERT L., CONSTABLE, GILES, and LANHAM, CAROL D. (eds.), *Renaissance and Renewal in the Twelfth Century* (Cambridge, Mass., 1982).

BENTON, JOHN F., 'Nicolas of Clairvaux and the Twelfth-Century Sequence, with Special Reference to Adam of St. Victor', *Traditio*, 18 (1962), 149–79.

BERGER, HUGO, *Untersuchungen zu den Psalmdifferenzen* (Kölner Beiträge zur Musikforschung, 37; Regensburg, 1966).

BERNARD, MADELEINE, 'L'Officium Stellae Nivernais', *Revue de musicologie*, 51 (1965), 52–65.

——*Bibliothèque Sainte-Geneviève* (Répertoire de manuscrits médiévaux contenant des notations musicales, 1; Paris, 1965).

——*Bibliothèque Mazarine* (Répertoire de manuscrits médiévaux contenant des notations musicales, 2; Paris, 1966).

——*Bibliothèque de l'Arsenal, Bibliothèque nationale (musique), des beaux-arts, de l'Université et petits fonds* (Répertoire de manuscrits médiévaux contenant des notations musicales, 3; Paris, 1974).

——'Un recueil inédit du XIIe siècle et la copie aquitaine de l'office versifié de saint Bernard', *EG* 16 (1977), 145–59.

——'Les Offices versifiés attribués à Léon IX (1002–1054)', *EG* 19 (1980), 89–164.

BERNHARD, MICHAEL, *Wortkonkordanz zu Anicius Manlius Severinus Boethius De Institutione Musica* (Bayerische Akademie der Wissenschaften, Veröffentlichungen der Musikhistorischen Kommission, 4; Munich, 1979).

——*Studien zur Epistola de harmonica institutione des Regino von Prüm* (Bayerische Akademie der Wissenschaften, Veröffentlichungen der Musikhistorischen Kommission, 5; Munich, 1979).

——*Clavis Gerberti: Eine Revision von Martin Gerberts Scriptores ecclesiastici de musica sacra potissimum (St. Blasien 1784) 1* (Bayerische Akademie der Wissenschaften, Veröffentlichungen der Musikhistorischen Kommission, 7; Munich, 1989).

[BERRY, MARY] Mother Thomas More, 'The Performance of Plainsong in the Later Middle Ages and the Sixteenth Century', *Proceedings of the Royal Musical Association*, 92 (1965–6), 121–34.

——'Gregorian Chant. The Restoration of the Chant and Seventy-five Years of Recording', *Early Music*, 7 (1979), 197–217.

——arts. 'Cistercian Monks, 'Dominican Friars', 'Sarum Rite, Music of the', *NG*.

BESCOND, ALBERT, *Le Chant grégorien* (Paris, 1972).

BESSELER Festschrift 1961 = *Festschrift Heinrich Besseler zum sechzigsten Geburtstag* (Leipzig, 1961).

BEWERUNGE, H., 'The Metrical Cursus in the Antiphon Melodies', *Zeitschrift der internationalen Musikgesellschaft*, 13 (1910–11), 227.

Biblioteca sanctorum, 13 vols. (Rome, 1961–70).

BILLECOCQ, MARIE-CLAIRE, 'Lettres ajoutées à la notation neumatique du codex 239 de Laon', *EG* 17 (1978), 71–44.

BINFORD-WALSH, HILDE M., 'The Ordering of Melody in Aquitanian Chant: A Study of Mode One Introit Tropes', in Cantus Planus 1990, 327–39.

BIRKNER, GÜNTHER, 'Eine "Sequentia sancti Johannis confessoris" in Trogir (Dalmatien)', *Musik des Ostens*, 2 (1963), 91.

BISCHOFF, BERNHARD, 'Otloh', in Karl Langosch (ed.), *Die deutsche Literatur des Mittelalters. Verfasserlexicon, 5: Nachtragband* (Berlin, 1955), 658–70.

——'Panorama der Handschriftenüberlieferung aus der Zeit Karls des Großen', in Braunfels, ii (1965), 233–54 = Bischoff 1966, 5–38.

——*Mittelalterliche Studien: Ausgewählte Aufsätze zur Schriftkunde und Literaturgeschichte*, 3 vols. (Stuttgart, 1966, 1967, 1981).

——'Frühkarolingische Handschriften und ihre Heimat', *Scriptorium*, 22 (1968), 306–14 [index to Bischoff 1965].

BISCHOFF, BERNHARD, *Die südostdeutschen Schreibschulen und Bibliotheken in der Karolingerzeit*, 2 vols. (3rd edn., Wiesbaden, 1974, 1980).

——*Paläographie des römischen Altertums und des abendländischen Mittelalters* (Grundlagen der Germanistik, 24; Berlin, 1979; 2nd edn., 1986).

——*Kalligraphie in Bayern: 8.–12. Jahrhundert* (Wiesbaden, 1981).

——*see also Carmina Burana*.

BISHOP, EDMUND, *Liturgica historica* (Oxford, 1918).

BISHOP, WILLIAM CHATTERLY, *The Mozarabic and Ambrosian Rites: Four Essays in Comparative Liturgiology* (Alcuin Club Tracts, 15; London, 1924).

BJORK, DAVID, 'Early Repertories of the *Kyrie Eleison*', *KmJb* 63–4 (1979–80), 9–43.

——'On the Dissemination of *Quem quaeritis* and the *Visitatio Sepulchri* and the Chronology of their Early Sources', *Comparative Drama*, 14 (1980), 46–69.

——'Early Settings of the Kyrie Eleison and the Problem of Genre Definition', *JPMMS* 3 (1980), 40–8.

——'The Kyrie Trope', *JAMS* 33 (1980), 1–41.

——'The Early Frankish Kyrie Text: A Reappraisal', *Viator*, 12 (1981), 9–35.

BJÖRKVALL, GUNILLA, 'Offertory Prosulas for Advent in Italian and Aquitanian Manuscripts', in Cantus Planus 1990, 377–400.

——and STEINER, RUTH, 'Some Prosulas for Offertory Antiphons', *JPMMS* 5 (1982), 17–35.

——*see also* CT.

BLACHLEY, ALEX, 'Some Observations on the "Germanic" Plainchant Tradition', in Sanders Festschrift 1990, 85–117.

BLANCHARD, P., 'La Correspondance apocryphe du pape S. Damase et de S. Jérôme sur le psautier et le chant de l'"Alleluia"', *EL* 63 (1949), 376–88.

BLEZZARD, JUDITH, RYLE, STEPHEN, and ALEXANDER, JONATHAN, 'New Perspectives on the Feast of the Crown of Thorns', *JPMMS* 10 (1987), 23–47.

BLOCH, HERBERT, *Monte Cassino in the Middle Ages*, 3 vols. (Rome, 1986).

Blume *see* AH.

BOE, JOHN, 'A New Source for Old Beneventan Chant: The Santa Sophia Maundy in MS Ottoboni lat. 145', *AcM* 52 (1980), 122–33.

——'Gloria A and the Roman Easter Vigil', *MD* 36 (1982), 5–37.

——'The Beneventan Apostrophus in South Italian Notation, A.D. 1000–1100', *EMH* 3 (1983), 43–66.

——'The "Lost" Palimpsest Kyries in the Vatican Manuscript Urbinas latinus 602', *JPMMS* 8 (1985), 1–24.

——'Hymns and Poems at Mass in Eleventh-Century Southern Italy (other than Sequences)', in Congress Bologna 1987, iii. 515–41.

——*Beneventanum Troporum Corpus II: Ordinary Chants and Tropes for the Mass from Southern Italy, A.D. 1000–1250. Part 1: Kyrie eleison* (Madison, Wisc., 1989); *Part 2: Gloria in excelsis* (Madison, Wisc., 1990).

——'Italian and Roman Verses for Kyrie leyson in the MSS Cologne-Genève, Bibliotheca Bodmeriana 74 and Vaticanus latinus 5319', in Leonardi and Menesto 1990, 337–84.

BOETHIUS, *De institutione arithmetica libri duo. De institutione musica libri quinque*, ed. G. Friedlein (Leipzig, 1867).

——*'Fundamentals of Music' by Anicius Manlius Severinus Boethius*, trans. with an introduction and notes by C. M. Bower, ed. by Claude V. Palisca (New Haven, Conn., 1989).

BOMM, URBANUS, *Der Wechsel der Modalitätbestimmung in der Tradition der Meßgesänge im IX. bis XIII. Jahrhundert und sein Einfluß auf die Tradition ihrer Melodien* (Diss., Göttingen, 1928; Einsiedeln, 1929).

BONA, J., *Rerum liturgicarum libri duo* (Rome, 1671; Paris, 1672).

BONASTRE, F., 'Estudis sobre la verbeta: la verbeta a Catalunya durant els segles XI–XVI' (Diss., Tarragona, 1982).

BONNIWELL, WILLIAM R., *A History of the Dominican Liturgy* (New York, 1944).

BOOR, HELMUT DE, *Die Textgeschichte der lateinischen Osterfeiern* (Tübingen, 1967).

BORELLA, PIETRO, *Il rito ambrosiano* (Brescia, 1964).

VAN DEN BORREN Festschrift 1945 = S. Clercx and A. van der Linden (eds.), *Hommage à Charles van den Borren* (Antwerp, 1945).

BORSAI, ILONA, 'Coptic Rite, Music of the', *NG*.

BOSSE, DETLEV, *Untersuchung einstimmiger mittelalterlicher Melodien zum 'Gloria in excelsis deo'* (Regensburg, 1955).

BOTTE, BERNARD, *Les Origines de la Noël et de l'Épiphanie* (Louvain, 1932).

BOUTEMY Festschrift 1976 = Guy Cambier (ed.), *Hommages à André Boutemy* (Collection Latomus, 145; Brussels, 1976).

BOUYER, LOUIS, *La Vie de la liturgie* (Paris, 1956); trans.: *Life and Liturgy* (London, 1956).

—— *Le Rite et l'homme, sacralité naturelle et liturgie* (Paris, 1962); trans.: *Rite and Man* (London, 1963).

BOWER, CALVIN M., 'Boethius' "The Principles of Music", an Introduction, Translation, and Commentary' (Diss., George Peabody College, 1967; UMI 67–15005).

—— 'Natural and Artificial Music: The Origins and Development of an Aesthetic Concept', *MD* 25 (1971), 17–33.

—— 'Boethius and Nicomachus: An Essay Concerning the Sources of *De Institutione Musica*', *Vivarium*, 16 (1978), 1–45.

—— 'An Alleluia for Mater', in Herz Festschrift 1982, 98–116.

—— 'The Modes of Boethius', *Journal of Musicology*, 3 (1984), 252–63.

—— 'Boethius' "De institutione musica". A Hand-List of Manuscripts', *Scriptorium*, 42 (1988), 205–51.

—— 'The Grammatical Model of Musical Understanding in the Middle Ages', in Gallacher and Damico 1989, 133–45.

BOYCE, JAMES JOHN, 'The Office of St Mary of Salome', *JPMMS* 11 (1988), 25–47.

—— 'The Office of the Three Marys in the Carmelite Liturgy', *JPMMS* 12 (1989), 1–38.

—— 'The Medieval Carmelite Office Tradition', *AcM* 62 (1990), 119–51.

BOYLE, LEONARD E., *Medieval Latin Palaeography: A Bibliographical Introduction* (Toronto Medieval Bibliographies, 8; Toronto, 1984).

BRADSHAW, PAUL, *Daily Prayer in the Early Church* (Alcuin Club Collections, 63; London, 1981).

BRAGANÇA, JOAQUIM O., *Missal de Mateus: Manuscrito 1000 da Biblioteca Pública e Arquivo Distrital de Braga* (Lisbon, 1975).

—— *Processionale-Tropario de Alcobaça* (Lisbon, 1985).

Bragard *see* CSM 3.

BRANNER, ROBERT, 'Two Parisian Capella Books in Bari', *Gesta*, 8 (1969), 14–19.

—— *Manuscript Painting in Paris during the Reign of Saint Louis* (Berkeley, Calif., 1977).

BRAUNFELS, WOLFGANG (ed.), *Karl der Große: Lebenswerk und Nachleben*, 5 vols. (Düsseldorf, 1965–8).

Breviarium Nidrosiense (Parisius 1519) (fac. edn., Oslo, 1964).

Briggs, Henry B., *The Musical Notation of the Middle Ages Exemplified by Facsimiles of Manuscripts Written between the 10th and 16th Centuries Inclusive* (London, 1890).

Brockett, Clyde Waring, *Antiphons, Responsories, and other Chants of the Mozarabic Rite* (Brooklyn, 1968).

—— 'Unpublished Antiphons and Antiphon Series Found in the Gradual of St. Yrieix', *MD* 26 (1972), 67–94.

—— 'Easter Monday Antiphons and the Peregrinus Play', *KmJb* 61–2 (1977–8). 29–46.

Bronarski, L., 'Die Quadripartita figura in der mittelalterlichen Musiktheorie', in Wagner Festschrift 1926, 27–43.

Brou, Louis, 'Les Impropères du Vendredi-Saint', *RG* 20 (1935), 161–79; 21 (1936), 8–16; 22 (1937), 19, 44–51.

—— 'L'Alléluia gréco-latin "Dies sanctificatus" de la messe du jour de Noël: Origine et évolution d'un chant bilingue et protéiforme', *RG* 23 (1938), 170–5; 24 (1939), 1–8, 81–9, 202–13.

—— 'Le Répons "Ecce quomodo moritur" dans les traditions romaine et espagnole', *RB* 41 (1939), 144–68.

—— 'Le Psallendum de la messe et les chants connexes d'après les sources manuscrites', *EL* 61 (1947), 13–54.

—— 'Études sur la liturgie mozarabe: Le Trisagion de la messe d'après les sources manuscrites', *EL* 61 (1947), 309–34.

—— 'Les "Benedictiones" ou cantique des trois enfants dans l'ancienne messe espagnole', *Hispania sacra*, 1 (1948), 21–33.

—— 'Les Chants en langue grecque dans les liturgies latines', *Sacris erudiri*, 1 (1948), 165–80; 4 (1952), 226–38.

—— 'L'Antiphonaire wisigothique et l'antiphonaire grégorien au début du viiie siècle. Essai de musicologie comparée', *Anuario musical*, 5 (1950), 3–10.

—— 'L'Alléluia dans la liturgie mozarabe. Étude liturgico-musicale d'après les manuscrits', *Anuario musical*, 6 (1951), 3–90.

—— 'Séquences et tropes dans la liturgie mozarabe', *Hispania sacra*, 4 (1951), 1–15.

—— 'Notes de paléographie musicale mozarabe', *Anuario musical*, 7 (1952), 51–76; 10 (1955), 23–44.

—— 'Études sur le missel et le bréviaire "mozarabes" imprimés', *Hispania sacra*, 11 (1958), 1–50.

—— 'L'Ancien Office de saint Vaast, évêque d'Arras', *EG* 4 (1961), 7–42.

—— and Vives, José, *Antifonario visigótico mozárabe de la Catedral de León: Edición del texto, notas e índices* (Monumenta hispaniae sacra. Serie litúrgica, v/1; Barcelona and Madrid, 1959).

Brovelli, F., 'Per uno studio dei messali francesi del XVIII° secolo, saggio di analisi', *EL* 96 (1982), 279–406.

Brown, Julian, Patterson, Sonia, and Hiley, David, 'Further Observations on W1', *JPMMS* 4 (1981), 53–80.

Browne, Alma Colk, 'The a–p System of Letter Notation', *MD* 35 (1981), 5–54.

Brunhölzl, Franz, 'Zur Antiphon "Alma redemptoris mater"', *Studien und Mitteilungen zur Geschichte des Benediktinischen Ordens und seiner Zweige*, 78 (1967), 321–4.

—— *Geschichte der lateinischen Literatur des Mittelalters*, i (Munich, 1975).

BRUNNER, LANCE W., 'A Perspective on the Southern Italian Sequence: The Second Tonary of the Manuscript Monte Cassino 318', *EMH* 1 (1981), 117–64.

—— 'The Performance of Plainchant: Some Preliminary Observations of the New Era', *Early Music*, 10 (1982), 317–28.

—— 'Catalogo delle sequenze in manoscritti di origine italiana', *RIM* 20 (1985), 191–276.

BRYANT, DAVID, 'Aquileia', *NG*.

BUCHNER, M., 'Die "Vita Chrodegangi": Eine kirchenpolitische Tendenzschrift aus der Mitte des 9. Jahrhunderts, zugleich eine Untersuchung zur Entwicklung der Primatial- und Vikariatsidee', *Zeitschrift der Savigny-Stiftung für Rechtsgeschichte, Kanonische Abteilung*, 16 (1927), 1–36.

BUJIĆ, BOJAN, 'Zadarski neumatski fragmenti v Oxfordu/Neumatic Fragments of Zadar in Oxford', *Muzikološki zbornik*, 4 (1968), 28.

BUKOFZER, MANFRED, 'Speculative Thinking in Medieval Music', *Speculum*, 17 (1942), 165–80.

BULST, NEITHART, *Untersuchungen zu den Klosterreformen Wilhelms von Dijon (962–1031)* (Pariser historische Studien, 11; Bonn, 1973).

—— 'Rodulfus Glabers Vita domni Willelmi abbatis. Neue Edition nach einer Handschrift des 11. Jahrhunderts (Paris, Bibl. nat., lat. 5390)', *Deutsches Archiv für Erforschung des Mittelalters*, 30 (1974), 450–87.

BULST Festschrift 1960 = Hans Robert Jauss (ed.), *Medium aevum vivum: Festschrift für Walter Bulst* (Heidelberg, 1960).

BUSCHINGER, D., and CRÉPIN, A. (eds.), *Musique, littérature et société au Moyen Âge* (Paris, 1980).

BUŽGA, JAROSLAV, 'Kantional (tschechisch)', *MGG*.

CABANISS, A., *Amalarius of Metz* (Amsterdam, 1954).

CABROL, FERNAND, *Les Livres de la liturgie latine* (Paris, 1930); trans.: *The Books of the Latin Liturgy* (London, 1932).

—— 'Mozarabe (La liturgie)', *DACL*.

CALCIDIUS, *Platon, Timaeus. A Calcidio translatus comentarioque instructus*, ed. J. Waszink (London and Leiden, 1962).

CALDWELL, JOHN, 'The *De Institutione Arithmetica* and the *De Institutione Musica*', in Gibson 1981, 135–54.

CALLEWAERT, CAMILLE, 'S. Grégoire, les scrutins et quelques messes quadragésimales', *EL* 53 (1939), 191–203 = Callewaert 1940, 659–71.

—— *Sacris erudiri* (Steenbrugge, 1940).

CAMPBELL, THOMAS P., and DAVIDSON, CLIFFORD (eds.), *The Fleury Playbook: Essays and Studies* (Kalamazoo, Mich., 1985).

CANAL, J. M., 'Sobre el autor del antifonario cisterciense', *EL* 74 (1960), 36–47.

Cantus Planus 1990 = *International Musicological Society Study Group 'Cantus Planus'. Papers Read at the Third Meeting, Tihany, Hungary, 19–24 September 1988*, ed. László Dobszay, Péter Halász, János Mezei, and Gábor Prószéky (Budapest, 1990).

Cantus selecti ex libris Vaticanis et Solesmensibus excerpti (Solesmes, 1949).

CAO 1 (1963) = *Manuscripti 'cursus romanus'* [Bamberg, Staatsbibliothek, lit. 23; Paris, Bibliothèque Nationale, lat. 17436; Durham, Cathedral Chapter Library, B. iii. 11; Ivrea, Biblioteca Capitolare, 106; Monza, Basilica S. Giovanni, C. 12/75; Verona, Biblioteca Capitolare, 98].

CAO 2 (1965) = *Manuscripti 'cursus monasticus'* [St. Gall, Stifsbibliothek, 390–391; Zürich, Zentralbibliothek, Rheinau 28; Paris, Bibliothèque Nationale, lat. 17296; Paris, Bibliothèque Nationale, lat. 12584; London, British Library, Add. 30850; Benevento, Archivio Capitolare, 21].

CAO 3 (1968) = *Invitatoria et antiphonae.*

CAO 4 (1970) = *Responsoria, versus, hymni, varia.*

CAO 5 (1975) = *Fontes earumque prima ordinatio.*

CAO 6 (1979) = *Secunda et tertia ordinationes.*

CAO–ECE = *Corpus Antiphonalium Officii—Ecclesiarum Centralis Europae: A Preliminary Report*, ed. László Dobszay and Gábor Prószéky (Budapest, 1988).

CAO–ECE I/A Salzburg (Temporale), ed. László Dobszay (Budapest, 1990).

CAPELLE, B., 'Le Kyrie de la messe et le pape Gélase', *RB* 46 (1934), 126–44.

CARDINE, EUGÈNE, 'La Psalmodie des introïts', *RG* 26 (1947), 172–7, 229–36; 27 (1948), 16–25.

——— 'La Corde récitative du 3ᵉ ton psalmodique dans l'ancienne tradition sangallienne', *EG* 1 (1954), 47–52.

——— 'Le Chant grégorien est-il mesuré?', *EG* 6 (1963), 7–38; trans. A. Dean: *Is Gregorian Chant Measured Music?* (Solesmes, 1964).

——— *Graduel neumé* (Solesmes, 1966).

——— *Semiologia gregoriana* (Rome, 1968); French trans.: 'Sémiologie grégorienne', *EG* 11 (1970), 1158; trans. Robert M. Fowels: *Gregorian Semiology* (Solesmes, 1982).

——— 'Vue d'ensemble sur le chant grégorien', *EG* 16 (1977), 173–92.

CARDINE Festschrift 1980 = Johannes Berchmans Göschl (ed.), *Ut mens concordet voci: Festschrift Eugène Cardine zum 75. Geburtstag* (St. Ottilien, 1980).

Carmina Burana: Faksimile-Ausgabe der Hs. Clm 4660 + 4660a, ed. Bernhard Bischoff (Brooklyn, 1967).

CASSIODORUS, *Institutiones divinarum et humanarum rerum*, ed. R. A. B. Mynors (Oxford, 1937).

CASTIGLIONI Festschrift 1957 = *Studi in onore di Mons. Carlo Castiglioni* (Milan, 1957).

CASTRO, EVA, 'Le Long Chemin de Moissac à S. Millan (Le troparium de la Real Acad. Hist., Aemil 51)', in Leonardi and Menesto 1990, 243–63.

CATTANEO, ENRICO, 'I canti della frazione e comunione nella liturgia ambrosiana', in Mohlberg Festschrift 1948, i. 147–74.

——— *Note storiche sul canto ambrosiano* (Milan, 1950).

CAVALLO, GUGLIELMO, *Rotoli di Exultet dell'Italia meridionale* (Bari, 1973).

CCM 9 (1976) = *Consuetudines Floriacenses saeculi tertii decimi*, ed. Anselmo Davril.

CCM 10 (1980) = *Liber Tramitis aevi Odilonis Abbatis*, ed. Peter Dinter.

CCM 12 (2 vols., 1975, 1987) = *Consuetudines Fructuarienses–Sanblasianae*, ed. Luchesius G. Spätling and Peter Dinter.

CENSORINUS, *De die natali liber ad Q. Caerellium, accedit Anonymi cujusdam Epitoma disciplinarum (Fragmentum Censorini)*, ed. Nicolas Sallmann (Leipzig, 1983).

——— *Le jour natal, traduction annotée*, trans. Guillaume Rocca-Serra (Paris, 1980).

CHADWICK, HENRY, *Boethius: The Consolations of Music, Logic, Theology, and Philosophy* (Oxford, 1981).

CHAILLEY, JACQUES, 'Un document sur la danse ecclésiastique', *AcM* 21 (1949), 18–24.

——— 'Le Mythe des modes grecs', *AcM* 28 (1956), 137–63.

—— 'Les Anciens Tropaires et séquentiaires de l'École de Saint-Martial de Limoges (x^e–xie siècle)', *EG* 2 (1957), 163–88.

—— *L'Imbroglio des modes* (Paris, 1960).

—— *L'École musicale de Saint-Martial de Limoges jusqu'à la fin du XIe siècle* (Paris, 1960).

—— *Alia Musica (Traité de musique du IXe siècle). Édition critique commentée avec une introduction sur l'origine de la nomenclature modale pseudo-grecque au Moyen-Âge* (Paris, 1965).

CHAMBERS, E. K., *The Mediaeval Stage*, 2 vols. (Oxford, 1903).

CHARLESWORTH, K., *The Odes of Solomon* (2nd edn., Missoula, Mont., 1977).

CHARTIER, YVES, 'L'Epistola de armonica institutione de Réginon de Prüm' (Diss., Ottawa, 1965).

—— 'La Musica d'Hucbald de Saint-Amand (traité de musique du xie siècle). Introduction, établissement du texte, traduction et commentaire' (Diss., Sorbonne, 1973).

—— 'Hucbald de Saint-Amand et la notation musicale', in Huglo 1987, 145–55.

CHAVANON, JULES, *Adémar de Chabannes: Chronique* (Collection des Textes; Paris, 1897).

CHAVASSE, ANTOINE, 'Le Carême romain et les scrutins prébaptismaux avant le ixe siècle', *Recherches de science religieuse*, 35 (1948), 325–81.

—— 'Les Plus Anciens Types du lectionnaire et de l'antiphonaire romain de la messe', *RB* 62 (1952), 3–94.

—— 'Cantatorium et Antiphonale missarum. Quelques procédés de confection. L'Antiphonaire des dimanches après la Pentecôte. Les graduels du Sanctoral', *Ecclesia orans*, 1 (1984), 15–55.

CHEVALIER, ULYSSE, *Repertorium hymnologicum. Catalogue des chants, hymnes, proses, séquences, tropes en usage dans l'église latine depuis les origines jusqu'à nos jours*, 6 vols. (Louvain and Brussels, 1892–1920).

—— *Prosolarium Anicensis: Office en vers de la Circoncision en usage dans l'église du Puy* (Bibliothèque liturgique, 5; Paris, 1894).

—— *Sacramentaire et martyrologe de l'abbaye de Saint-Rémy: Martyrologe, calendrier, ordinaires et prosaire de la métropole de Reims (VIIIe–XIIIe siècle), publiés d'après les manuscrits de Paris, Londres, Reims et Assise* (Bibliothèque liturgique, 7; Paris, 1900).

—— *Ordinaire et coutumier de l'église cathédrale de Bayeux (XIIIe siècle), publiés d'après les manuscrits originaux* (Bibliothèque liturgique, 8; Paris, 1902).

CHIBNALL, MARJORIE, *The Ecclesiastical History of Orderic Vitalis*, 6 vols. (Oxford, 1969–80).

CHOISSELET, DANIELE, and VERNET, PLACIDE (eds.), *Les Ecclesiastica Officia cisterciens du XIIe siècle* (Oelenberg, 1989).

CHOMTON, L., *Histoire de l'église Saint-Bénigne de Dijon* (Dijon, 1900).

CLAIRE, JEAN, 'L'Évolution modale dans les répertoires liturgiques occidentaux', *RG* 40 (1962), 196–211, 229–45.

—— 'La Psalmodie responsoriale antique', *RG* 41 (1963), 8–29, 49–62, 77–102.

—— 'L'Évolution modale dans les récitatifs liturgiques', *RG* 41 (1963), 127–51.

—— 'Les Répertoires liturgiques latins avant l'octoéchos. I. L'office férial romano-franc', *EG* 15 (1975), 5–192.

—— 'The Tonus peregrinus—A Question Well Put?', *Orbis musicae: Studies in Musicology*, 7 (1979–80), 3–14.

—— 'Les Psaumes graduels au cœur de la liturgie quadragésimale', *EG* 21 (1986), 5–12.

CLAIRE, JEAN, 'Le Rituel quadragésimal des catechumènes à Milan', in Gy Festschrift 1990, 131–51.

CLARK, J. M., *The Abbey of St. Gall as a Centre of Literature and Art* (Cambridge, 1926).

CLEMOES Festschrift = Michael Lapidge and Helmut Gneuss (eds.), *Learning and Literature in Anglo-Saxon England* (Cambridge, 1985).

CM 1 (Stuttgart and Vienna, 1900) = *Consuetudines Farfenses*.

CM 2 (Montecassino, 1905) = *Consuetudines Cluniacenses antiquiores*.

CM 3 (Montecassino, 1907) = *Statuta Murbacensia, Capitula Aquisgranensia*.

CM 4 (Montecassino, 1911) = *Consuetudines Fructuarienses* (etc.).

CM 5 (Montecassino, 1912) = *Redactio consuetudinem saeculi duodecimi* (etc.).

COCHERIL, MAUR, *L'Évolution historique du kyriale cistercien* (Port-du-Salut, 1956).

—— 'Le "Tonale Sancti Bernardi" et la définition du "ton"', *Cîteaux: Commentarii cistercienses*, 13 (1962), 35–66.

Codex Albensis, ein Antiphonar aus dem 12. Jahrhundert, ed. Zoltán Falvy and L. Mezey (Monumenta Hungariae musicae, 1; Budapest, 1963) [fac. of Graz, Universitätsbibliothek, 211].

CODY, AELRED, 'The Early History of the Octoechos in Syria', in Garsoian, Mathews, and Thomson 1982, 89–113.

COLETTE, MARIE-NOËL, 'La Notation du demi-ton dans le manuscrit Paris, B. N. Lat. 1139 et dans quelques manuscrits du Sud de la France', in Leonardi and Menesto 1990, 297–311.

COLGRAVE, BERTRAM, *The Earliest Life of Gregory the Great by an Anonymous Monk of Whitby* (Lawrence, Kansas, 1968).

COLLINS, A. JEFFERIES, *Manuale ad usum percelebris ecclesie Sarisburiensis* (Henry Bradshaw Society, 91; Chichester, 1960).

—— *The Bridgettine Breviary of Syon Abbey: From the MS. with English Rubrics F. 4. 11 at Magdalene College, Cambridge* (Henry Bradshaw Society, 96; Worcester, 1969).

COMBE, PIERRE, *Histoire de la restauration du chant grégorien d'après des documents inédits: Solesmes et l'édition vaticane* (Solesmes, 1969).

Congress Berkeley 1977 = Daniel Heartz and Bonnie Wade (eds.), *International Musicological Society: Report of the Twelfth Congress Berkeley 1977* (Kassel, 1981).

Congress Berlin 1974 = Helmut Kühn and Peter Nitsche (eds.), *Gesellschaft für Musikforschung: Bericht über den internationalen musikwissenschaftlichen Kongreß Berlin 1974* (Kassel, 1977).

Congress Bologna 1987 = *Atti del XIV congresso della Società Internazionale di Musicologia: Trasmissione e recezione delle forme di cultura musicale, Bologna, 27 agosto-1 settembre 1987*, ed. Angelo Pompilio, Donatella Restani, Lorenzo Bianconi, and F. Alberto Gallo, 3 vols. (Rome, 1990).

Congress Cluny 1949 = *À Cluny: Congrès scientifique: Fêtes et cérémonies liturgiques en l'honneur des saints abbés Odon et Odilon 9–11 juillet 1949* (Dijon, 1950).

Congress Cologne 1958 = G. Abraham, S. Clercx-Lejeune, H. Federhofer, and W. Pfannkuch (eds.), *International Musicological Society: Bericht über den siebenten internationalen musikwissenschaftlichen Kongress Köln* (Kassel, 1959).

Congress Copenhagen 1972 = *International Musicological Society: Report of the Eleventh Congress Copenhagen 1972* (Copenhagen, 1974).

Congress Hamburg 1956 = W. Gerstenberg, H. Husmann, and H. Heckmann (eds.), *Gesellschaft für Musikforschung: Bericht über den internationalen musikwissenschaftlichen Kongreß Hamburg 1956* (Kassel, 1957).

Congress Kassel 1962 = G. Reichert and M. Just (eds.), *Gesellschaft für Musikforschung: Bericht über den internationalen musikwissenschaftlichen Kongreß Kassel 1962* (Kassel, 1962).

Congress Leipzig 1925 = *Bericht üben den I. musikwissenschaftlichen Kongreß der deutschen Musikgesellschaft in Leipzig vom 4. bis 8. Juni 1925* (Leipzig, 1926).

Congress New York 1961 = Jan LaRue (ed.), *International Musicological Society: Report of the Eighth Congress New York 1961* (Kassel, 1962).

Congress Salamanca 1985 = *España en la Música de Occidente: Actas del Congreso Internacional celebrado en Salamanca (29 de octubre–5 de noviembre de 1985)*, ed. Emilio Casares Rodicio, Ismael Fernández de la Cuesta, and José López-Calo, 2 vols. (Madrid, 1987).

Congress Strasburg 1982 = Marc Honegger, Christian Meyer, and Paul Prévost (eds.), *La Musique et le rite sacré et profane: Actes du XIIIᵉ Congrès de la Société Internationale de Musicologie Strasbourg 1982*, 2 vols. (Strasburg, 1986).

Congress Todi 1958 = *Spiritualità cluniacense, 12–15 ottobre 1958* (Convegni del Centro di studi sulla spiritualità medievale, 2; Todi, 1960).

Congress Utrecht 1952 = International Musicological Society, *Report of the Fifth Congress Utrecht 1952* (Amsterdam, 1953).

Congress Vienna 1954 = *Zweiter internationaler Kongreß für katholische Kirchenmusik Wien 4.–10. Oktober 1954: Bericht* (Vienna, 1956).

Congress Vienna 1956 = Erich Schenk (ed.), *Gesellschaft für Musikforschung: Bericht über den internationalen musikwissenschaftlichen Kongreß Wien, Mozartjahr 1956* (Graz and Cologne, 1958).

CONNOLLY, THOMAS H., 'The Introits and Communions of the Old Roman Chant' (Diss., Harvard University, 1972).

—— 'Introits and Archetypes: Some Archaisms of the Old Roman Chant', *JAMS* 25 (1972), 157–74.

—— 'The Graduale of S. Cecilia in Trastevere and the Old Roman Tradition', *JAMS* 28 (1975), 413–58.

—— 'Musical Observance of Time in Early Roman Chant', in Otto Albrecht Festschrift 1980, 3–18.

—— 'Psalm II', *NG*.

CONSTABLE, GILES, *Medieval Monasticism: A Select Bibliography* (Toronto Medieval Bibliographies, 6; Toronto, 1976).

Constitution on the Sacred Liturgy of the Second Vatican Council, in Austin P. Flannery (ed.), *Documents of Vatican II* (Grand Rapids, Mich., 1975; 2nd edn., 1984), 1–282.

CONTRENI, JOHN, *The Cathedral School of Laon from 850–930: Its Manuscripts and Masters* (Munich, 1978).

CONYBEARE, FREDERICK C. (ed.), *Philo: About the Contemplative Life or the Fourth Book of the Treatise Concerning Virtues* (London, 1895).

CORBIN, SOLANGE, *La Musique religieuse portugaise au Moyen-Âge* (Paris, 1952).

—— 'Le *Cantus Sibyllae*: Origines et premiers textes', *Revue de musicologie*, 34 (1952), 1–10.

—— 'Le Manuscrit 201 d'Orléans: Drames liturgiques dits de Fleury', *Romania*, 74 (1953), 1–43.

—— 'Valeur et sens de la notation alphabétique à Jumièges et en Normandie', in *Jumièges: Congrès scientifique du XIIIᵉ centenaire*, ii (Rouen, 1955), 913–24.

Corbin, Solange, *La Déposition liturgique du Christ au Vendredi Saint: Sa Place dans l'histoire des rites et du théâtre religieux. Analyse des documents portugais* (Paris, 1960).

——*L'Eglise à la conquête de sa musique* (Paris, 1960).

——*Die Neumen* (Palaeographie der Musik, i/3; Cologne, 1977).

——'Neumatic Notations, I–IV', *NG*.

Cornford, Frances Macdonald, *Plato's Cosmology: The Timaeus of Plato Translated with a Running Commentary* (London, 1937).

Coussemaker, Edmond de, *Histoire de l'harmonie au moyen-âge* (Paris, 1852).

——*Drames liturgiques du moyen-âge* (Paris, 1861).

——*see also* CS.

Cowdrey, Herbert E. J., *The Cluniacs and the Gregorian Reform* (Oxford, 1970).

——*The Age of Abbot Desiderius: Montecassino, the Papacy and the Normans in the Eleventh and Early Twelfth Centuries* (Oxford, 1983).

Crocker, Richard L., 'The Repertoire of Proses at Saint Martial de Limoges (Tenth and Eleventh Centuries)' (Diss., Yale University, 1957; UMI 68–4915).

——'The Repertory of Proses at Saint Martial de Limoges in the 10th Century', *JAMS* 11 (1958), 149–64.

——'*Musica rhythmica* and *Musica metrica* in Antique and Medieval Theory', *Journal of Music Theory*, 2 (1958), 2–23.

——'Discant, Counterpoint and Harmony', *JAMS* 15 (1962), 1–21.

——'Pythagorean Mathematics and Music', *Journal of Aesthetics and Art Criticism*, 22 (1963), 189–98, 325–35.

——'The Troping Hypothesis', *MQ* 52 (1966), 183–203.

——'Aristoxenus and Greek Mathematics', in Reese Festschrift 1967, 96–110.

——'A New Source for Medieval Music Theory', *AcM* 39 (1967), 161–71.

——'Some Ninth-Century Sequences', *JAMS* 20 (1967), 367–402.

——'Hermann's Major Sixth', *JAMS* 25 (1972), 19–37.

——'The Sequence', in Schrade Gedenkschrift 1973, 269–322.

——*The Early Medieval Sequence* (Berkeley, Calif., 1977).

——'Alphabet Notations for Early Medieval Music', in Charles Jones Festschrift 1979, ii. 79–104.

——'Matins Antiphons at St. Denis', *JAMS* 39 (1986), 441–90.

——arts. 'Agnus Dei', 'Credo', 'Gloria in excelsis Deo', 'Kyrie eleison', 'Sanctus', 'Sequence (i)', 'Versus', *NG*.

——and Hiley, David (eds.), *New Oxford History of Music*, ii. *The Early Middle Ages to 1300* (2nd edn., Oxford, 1989).

Cserba, Simon M. (ed.), *Hieronymus de Moravia O. P., Tractatus de musica* (Freiburger Studien zur Musikwissenschaft, 2; Regensburg, 1935).

CSM 1 (1950) = *Johannis Affligemensis De musica cum tonario*, ed. Joseph Smits van Waesberghe.

CSM 2 (1951) = *Aribonis De musica*, ed. Joseph Smits van Waesberghe.

CSM 3 (7 vols., 1955–73) = *Jacobi Leodiensis Speculum musicae*, ed. Roger Bragard.

CSM 4 (1955) = *Guidonis Aretini Micrologus*, ed. Joseph Smits van Waesberghe.

CSM 14 (1970) = *Walteri Odington Summa de speculatione musicae*, ed. Frederick F. Hammond.

CSM 21 (1975) = *Aureliani Reomensis Musica disciplina*, ed. Lawrence Gushee.

CSM 23 (1975) = *Willehelmi Hirsaugensis Musica*, ed. Denis Harbinson.

CSM 24 (1974) = *Epistola S. Bernardi de revisione cantus Cisterciensis et Tractatus scriptus ab auctore incerto Cisterciense, Cantum quem Cisterciensis ordinis ecclesiae cantare consueverant*, ed. Francis J. Guentner.

CSM 25 (1975) = *Ameri Practica artis musice*, ed. Cesarino Ruini.

CSM 29 (1976) = *Petrus de Cruce Ambianensi[s]: Tractatus de Tonis*, ed. Denis Harbinson.

CT 1 (SLS 21, 1975) = *Tropes du propre de la messe 1: Cycle de Noël*, ed. Ritva Jonsson.

CT 2 (SLS 22, 1976) = *Prosules de la messe 1: Tropes de l'alléluia*, ed. Olof Marcusson.

CT 3 (SLS 25, 1982) = *Tropes du propre de la messe 2: Cycle de Pâques*, ed. Gunilla Björkvall, Gunilla Iversen, and Ritva Jonsson.

CT 4 (SLS 26, 1980) = *Tropes de l'Agnus Dei*, ed. Gunilla Iversen.

CT 5 (SLS 32, 1986) = *Les Deux Tropaires d'Apt, mss. 17 et 18*, ed. Gunilla Björkvall.

CT 6 (SLS 31, 1986) = *Prosules de la messe 2: Les prosules limousines de Wolfenbüttel*, ed. Eva Odelman.

CT 7 (SLS 34, 1990) = *Tropes du Sanctus*, ed. Gunilla Iversen.

CURRAN, M., *The Antiphonary of Bangor and the Early Irish Monastic Liturgy* (Dublin, 1984).

CUTTER, PAUL F., 'The Question of the "Old-Roman" Chant: A Reappraisal', *AcM* 39 (1967), 2–20.

—— 'The Old-Roman Chant Tradition: Oral or Written?', *JAMS* 20 (1967), 167–81.

—— 'The Old-Roman Responsories of Mode 2' (Diss., Princeton University, 1969; UMI 70–08358).

—— 'Die altrömischen und gregorianischen Responsorien im zweiten Modus', *KmJb* 54 (1970), 33–40.

—— 'Oral Transmission of the Old-Roman Responsories?', *MQ* 62 (1976), 182–94.

—— *Musical Sources of the Old-Roman Mass* (Musicological Studies and Documents, 36; Neuhausen-Stuttgart, 1979).

[——] 'Responsory', *NG*.

DADELSEN Festschrift 1978 = Thomas Kohlhase and Volker Scherliess (eds.), *Festschrift Georg von Dadelsen zum 60. Geburtstag* (Neuhausen-Stuttgart, 1978).

DAHLHAUS Festschrift 1988 = H. Danuser, H. de la Motte-Haber, S. Leopold, and N. Miller (eds.), *Das musikalische Kunstwerk: Geschichte–Ästhetik–Theorie. Festschrift Carl Dahlhaus zum 60. Geburtstag* (Laaber, 1988).

DALTON, J. N. and (vol. iv) DOBLE, G., *Ordinale Exon* (Henry Bradshaw Society, 37, 38, 63, and 79; London, 1909, 1909, 1926, and 1941).

DAMILANO, PIERO, 'Laudi latine in un antifonario bobbiese del trecento', *Collectanea historiae musicae*, 3 (1963), 15–57.

—— 'Sequenze bobbiese', *RIM* 2 (1967), 3–35.

D'ANGERS, O., 'Le Chant liturgique dans l'Ordre de Saint-François aux origines', *Études franciscaines*, 25 (1975), 157–306.

DAVID, LUCIEN, and HANDSCHIN, JACQUES, 'Un point d'histoire grégorienne. Guillaume de Fécamp', *RCG* 39 (1935–6), 180–3; 40 (1936–7), 11–17 (see also Handschin 1936–8).

DAVIES, JOHN G. (ed.), *A Dictionary of Liturgy and Worship* (London, 1972).

—— *A New Dictionary of Liturgy and Worship* (London, 1986).

DAVRIL, ANSELME, 'À propos d'un bréviaire manuscrit de Cluny conservé à Saint-Victor-sur-Rhins', *RB* 93 (1983), 108–22.

—— 'Johann Drumbl and the Origin of the *Quem quaeritis*: A Review Article', *Comparative Drama*, 20 (1986), 65–75.

DEANESLY, MARGARET, *Augustine of Canterbury* (London, 1964).

DECHEVRENS, ANTOINE, *Études de science musicale*, 3 vols. (Paris, 1898).

—— *Composition musicale et composition littéraire à propos du chant grégorien*, 2 vols. (Paris, 1910).

DELALANDE, DOMINIQUE, *Vers la version authentique du Graduel grégorien: Le Graduel des Prêcheurs* (Bibliothèque d'histoire Dominicaine, 2; Paris, 1949).

DELAPORTE, YVES, 'Fulbert de Chartres et l'école chartraine de chant liturgique au XIe siècle', *EG* 2 (1957), 51–81.

—— 'L'Office fécampois de saint Taurin', *L'Abbaye bénédictine de Fécamp, ouvrage scientifique du XIIIe centenaire, 658–1958* (Fécamp, 1959–60), ii. 171–89, 377.

DELÉGLISE, FRANÇOIS, '"Illustris civitas", office rimé de saint Théodule (XIIIe siècle)', *Vallesia*, 38 (Sion, 1983), 173–308.

DESHUSSES, JEAN, *Le Sacramentaire grégorien: Ses principales formes d'après les plus anciens manuscrits* (Spicilegium Friburgense, 16; Fribourg, 1971; 2nd edn., 1979), and 24 (1979).

—— and DARRAGON, B., *Concordances et tableaux pour l'étude des grands sacramentaires* (Spicilegii Friburgensis Subsidia, 9–14; Fribourg, 1982–3).

DEWICK, E. S., and (vol. ii) FRERE, W. H., *The Leofric Collectar* (Henry Bradshaw Society, 45 and 56; London, 1914, 1921).

DICKINSON, FRANCIS HENRY (ed.), *Missale ad usum insignis et praeclarae ecclesiae Sarum* (Burntisland, 1861–83).

VAN DIJK, S. J. P., 'St Bernard and the *Instituta Patrum* of St Gall', *MD* 4 (1950), 99–109.

—— 'The Legend of "the Missal of the Papal Chapel" and the Fact of Cardinal Orsini's Reform', *Sacris erudiri*, 8 (1956), 76–142.

—— 'The Authentic Missal of the Papal Chapel', *Scriptorium*, 14 (1960), 257–314.

—— 'The Urban and Papal Rites in Seventh- and Eighth-Century Rome', *Sacris erudiri*, 12 (1961), 411–87.

—— 'The Old-Roman Rite', *Studia patristica*, 5, Texte und Untersuchungen, 80 (1962), 185–205.

—— 'Papal Schola "versus" Charlemagne', in Smits van Waesberghe Festschrift 1963, 21–30.

—— 'Gregory the Great Founder of the Urban "Schola cantorum"', *EL* 77 (1963), 335–56.

—— *Sources of the Modern Roman Liturgy: The Ordinals by Haymo of Faversham and Related Documents (1243–1307)*, 2 vols. (Leiden, 1963).

—— 'Recent Developments in the Study of the Old-Roman Rite', *Studia patristica*, 8, Texte und Untersuchungen, 93 (1966), 299–319.

—— 'The Medieval Easter Vespers of the Roman Clergy', *Sacris erudiri*, 19 (1969–70), 261–363.

—— (completed by Joan Hazelden Walker), *The Ordinal of the Papal Court from Innocent III to Boniface VIII, and Related Documents* (Spicilegium Friburgense, 22; Fribourg, 1975).

—— and WALKER, JOAN HAZELDEN, *The Origins of the Modern Roman Liturgy: The Liturgy of the Papal Court and the Franciscan Order in the Thirteenth Century* (London, 1960).

DIX, GREGORY, *The Shape of the Liturgy* (London, 1945).

DMA A.I (1975) = *De numero tonorum litterae episcopi A. ad coepiscopum E. missae ac Commentum super tonos episcopi E. (ad 1000)*.

DMA A.II (1981) = *Adaboldi Episcopi Ultraiectensis: Per musicae instrumentalis artem ad harmoniam musicae mundanae (ad 1025)*.

DMA A.III (1975) = *Tres tractatuli Guidonis Aretini. [1] Guidonis Prologus in antiphonarium*.

DMA A.IV (1985) = *Tres tractatuli Guidonis Aretini. [2] Guidonis Aretini Regulae rhythmicae*.

[DMA A.V = Tres tractatuli Guidonis Aretini. [3] Guidonis Aretini Epistola ad Michaelem—did not appear].

DMA A.VIa (1978) = *Bernonis Augiensis abbatis de arte musica disputationes traditae, pars A: Bernonis Augiensis De mensurando monochordo*.

DMA A.VIb (1979) = *Bernonis Augiensis abbatis de arte musica disputationes traditae, pars B: Quae ratio est inter tria opera de arte musica Bernonis Augiensis*.

DMA A.VII (1977) = *Musica Domni Heinrici Augustensis magistri*.

DMA A.VIIIa (1988) = *Summula: Tractatus metricus de musica*.

DMA A.Xa (1979) = *Codex Oxoniensis Bibliothecae Bodleianae, Rawlinson C 270, pars A: De vocum consonantiis ac De re musica (Osberni Cantuariensis?)*.

DMA A.Xb (1979) = *Codex Oxoniensis Bibliothecae Bodleianae, Rawlinson C 270, pars B: XVII tractatuli a quodam studioso peregrino ad annum MC collecti*.

DOBSON, ERIC J., and HARRISON, FRANK LL., *Medieval English Songs* (London, 1979).

DOBSZAY, LÁSZLÓ, 'The System of the Hungarian Plainsong Sources', *SMH* 27 (1985), 37–65.

—— 'The Program "CAO–ECE"', *SMH* 30 (1988), 355–60.

—— 'Experiences in the Musical Classification of Antiphons', in Cantus Planus 1990, 143–56.

—— 'Plainchant in Medieval Hungary', *JPMMS* 13 (1990), 49–78.

—— and PRÓSZÉKY, GÁBOR, *Corpus Antiphonalium Officii—Ecclesiarum Centralis Europae: A Preliminary Report* (Budapest, 1988).

DOLD, ALBAN, *Das älteste Liturgiebuch der lateinischen Kirche: Ein altgallikanisches Lektionar des 5.–6. Jhs. aus dem Wolfenbüttler Palimpsest-Codex Weissenburgensis 76* (Texte und Arbeiten, 26–8; Beuron, 1936).

—— *Neue St. Galler vorhieronymianische Propheten-Fragmente der St. Galler Sammelhandschrift 1398b zugehörig: Anhang 1. Ein neues Winitharfragment mit liturgischen Texten, 2. Irische Isidorfragmente des 7. Jahrhunderts* (Texte und Arbeiten, 31; Beuron, 1940).

DONATO, GIUSEPPE, *Gli elementi costituitivi dei tonari* (Messina, 1978).

DONOVAN, RICHARD B., *The Liturgical Drama in Medieval Spain* (Toronto, 1958).

DOREN, ROMBAUT VAN, *Étude sur l'influence musicale de l'abbaye de Saint-Gall (VIIIᵉ au XIᵉ siècle)* (Brussels, 1925).

Dreves *see* AH.

DRINKWELDER, OTTO, *Ein deutsches Sequentiar aus dem Ende des 12. Jahrhunderts* (VGA 8; Graz and Vienna, 1914) [Berlin, Staatsbibliothek Preußischer Kulturbesitz, Mus. 40048].

DRONKE, PETER, 'The Beginnings of the Sequence', *Beiträge zur Geschichte der deutschen Sprache und Literatur*, 87 (1965), 43–73.

—— 'Types of Poetic Art in Tropes', in Silagi 1985, 1–24.

DROSTE, DIANE LYNNE, 'The Musical Notation and Transmission of the Music of the Sarum Use, 1225–1500' (Diss., Toronto, 1983; Canadian theses 59826).

DRUMBL, JOHANN, 'Zweisprachige Antiphonen zur Kreuzverehrung', *Italia medioevale e umanistica*, 19 (1976), 41–55.

—— 'Ursprung des liturgischen Spiels', *Italia medioevale e umanistica*, 22 (1979), 45–96.

—— *Quem queritis: Teatro sacro dell'alto medioevo* (Rome, 1981).

DUBOIS, JACQUES, *Le Martyrologe d'Usuard* (Subsidia hagiographica, 40; Brussels, 1965).

DUCHESNE, LOUIS, *Origines du culte chrétien: Étude sur la liturgie latine avant Charlemagne* (5th edn., Paris, 1919); trans.: *Christian Worship: Its Origin and Evolution. A Study of the Latin Liturgy up to the Time of Charlemagne* (London, 1919).

DUCHESNE, LOUIS, *Le Liber pontificalis*, 2 vols. (Paris, 1886 and 1892); rev. edn. by Cyrille Vogel, 3 vols. (Paris, 1955, 1955, 1957).

DUCHEZ, MARIE-ELISABETH, *Imago mundi: La naissance de la théorie musicale occidentale dans les commentaires carolingiens de Martianus Capella* (Paris, 1979).

——'La Représentation spatio-verticale du caractère musical grave–aigu et l'élaboration de la notion de hauteur de son dans la conscience occidentale', *AcM* 51 (1979), 54–73.

——'Description grammaticale et description arithmétique des phénomènes musicaux: Le tournant du IXᵉ siècle', *Sprache und Erkenntnis im Mittelalter: Akten des VI. Internationalen Kongresses für mittelalterliche Philosophie Bonn 1977* (Miscellanea medievalia, 13/2; Berlin and New York, 1981), 561–79.

——'Des neumes à la portée', in Huglo 1987, 57–60.

DUMAS, A., and DESHUSSES, JEAN, *Liber sacramentorum Gellonensis* (Corpus Christianorum, 159–159A; Turnhout, 1981).

DYER, JOSEPH, 'The Offertories of Old Roman Chant: A Musico-liturgical Investigation' (Diss., Boston University, 1971; UMI 71–26401).

——'Singing with Proper Refinement from *De modo bene cantandi* (1474) by Conrad von Zabern', *Early Music*, 6 (1978), 207–27.

——'Augustine and the "Hymni ante oblatium": The Earliest Offertory Chants?', *Revue des études augustiniennes*, 27 (1981), 85–99.

——'The Offertory Chant of the Roman Liturgy and its Musical Form', *Studi musicali*, 11 (1982), 3–30.

——'Latin Psalters, Old Roman and Gregorian Chants', *KmJb* 68 (1984), 11–30.

——'Monastic Psalmody of the Middle Ages', *RB* 99 (1989), 41–74.

——'The Singing of Psalms in the Early-Medieval Office', *Speculum*, 64 (1989), 535–78.

——'On the Monastic Origins of Western Music Theory', in Cantus Planus 1990, 199–225.

EBEL, BASILIUS, *Das älteste alemannische Hymnar mit Noten, Kodex 366 (472) Einsiedeln (XII. Jahrhundert)* (VGA 17; Einsiedeln, 1930).

EBERLE, L., *The Rule of the Master* (Cistercian Studies Series, 6; Kalamazoo, Mich., 1977).

Echternacher Sakramentar und Antiphonar: Vollständige Facsimile-Ausgabe im Originalformat der Handschrift 1946 . . . aus dem Besitz der Hessischen Landes- und Hochschulbibliothek Darmstadt. With Essays by Kurt Hans Staub, Paul Ulveling and Franz Unterkircher (Codices Selecti, 74; Graz, 1982).

EDWARDS, KATHLEEN, *The English Secular Cathedrals in the Middle Ages* (Manchester, 1949; 2nd edn., 1967).

EDWARDS, OWAIN TUDOR, 'A Fourteenth-Century Welsh Sarum Antiphonal: National Library of Wales ms. 20541', *JPMMS* 10 (1987), 15–21.

——*Matins, Lauds and Vespers for St David's Day: The Medieval Office of the Welsh Patron Saint in National Library of Wales MS 20541 E* (Woodbridge and Wolfeboro, 1990).

EELES, FRANCIS CHARLES, *The Holyrood Ordinale: A Scottish Version of a Directory of English Augustinian Canons with Manual and Other Liturgical Forms* (The Book of the Old Edinburgh Club, 7; Edinburgh, 1916).

EGGEN, ERIK, *The Sequences of the Archbishopric of Nidaros* (Bibliotheca Arnamagnæana, 212; Copenhagen, 1968).

EHRENSBERGER, HUGO, *Bibliotheca liturgica manuscripta: Nach Handschriften der großherzoglich badischen Hof- und Landesbibliothek* (Karlsruhe, 1887).

EKBERG, GUDRUN, 'Ekphonetic Notation', *NG*.

EKENBERG, ANDERS, *Cur cantatur? Die Funktionen des liturgischen Gesangs nach den Autoren der Karolingerzeit* (Stockholm, 1987).

ELLINWOOD, LEONARD, *Musica Hermanni Contracti* (Eastman School of Music Studies, 2; Rochester, NY, 1936).

EMERSON, JOHN A., 'The Recovery of the Wolffheim Antiphonal', *Annales musicologiques*, 6 (1958–63), 69–97.

—— 'Fragments of a Troper from Saint-Martial de Limoges', *Scriptorium*, 16 (1962), 369–72.

—— 'Über Entstehung und Inhalt von MüD', *KmJb* 48 (1964), 33–60.

—— 'Two Newly Identified Offices for Saints Valeria and Austriclinianus by Adémar de Chabannes (Ms Paris, Bibl. Nat., Latin 909, fols. 79–85v)', *Speculum*, 40 (1965), 31–46.

—— 'Sources, MS, II: Western Plainchant', *NG*.

ENGEL Festschrift 1964 = *Festschrift Hans Engel zum siebzigsten Geburtstag* (Kassel, 1964).

EPSTEIN, MARCY J., '*Ludovicus decus regnantium*: Perspectives on the Rhymed Office', *Speculum*, 53 (1978), 283–333.

ERBACHER, RHABANUS, *Tonus peregrinus: Aus der Geschichte eines Psalmtons* (Münster-schwarzach, 1971).

ESCUDIER, DENIS, 'La Notation musicale de St. Vaast: Étude d'une particularité graphique', in Huglo 1987, 107–18.

EVANS, JOAN, *Monastic Life at Cluny, 910–1157* (Oxford, 1931).

EVANS, PAUL, 'Some Reflections on the Origins of the Trope', *JAMS* 14 (1961), 119–30.

—— 'The *Tropi ad sequentiam*', in Strunk Festschrift 1968, 73–82.

—— *The Early Trope Repertory of Saint Martial de Limoges* (Princeton, NJ, 1970).

—— 'Northern French Elements in an Early Aquitanian Troper', in Husmann Festschrift 1970, 103–10.

EVANS, ROGER WILLIAM, 'Amalarius of Metz and the Singing of the Carolingian Offices' (Diss., New York University, 1977; UMI 77–19546).

FALCONER, KEITH, 'Early Versions of the Gloria Trope *Pax sempiterna Christus*', *JPMMS* 7 (1984), 18–27.

—— 'A Kyrie and Three Gloria Tropes in a Norwegian Manuscript Fragment', *Svensk Tidskrift för Musikforskning*, 67 (1985), 77–88.

—— 'Some Early Tropes to the Gloria' (Diss., Princeton University, 1989; UMI 89–20341).

FALLOWS, DAVID, 'Sources, MS, III, 2: Secular Monophony, Latin', *NG*.

FALVY, ZOLTÁN, *Drei Reimoffizien aus Ungarn und ihre Musik* (Musicologia Hungarica, NS, 2; Budapest, 1968).

—— *see also* Codex Albensis.

FARMER, DAVID HUGH, *The Oxford Dictionary of Saints* (Oxford, 1978).

FASSLER, MARGOT, 'Musical Exegesis in the Sequences of Adam and the Canons of St. Victor' (Diss., Cornell University, 1983; UMI 83–28706).

—— 'Who was Adam of St. Victor? The Evidence of the Sequence Manuscripts', *JAMS* 37 (1984), 233–69.

—— 'The Office of the Cantor in Early Western Monastic Rules and Customaries: A Preliminary Investigation', *EMH* 5 (1985), 29–51.

—— 'Accent, Meter, and Rhythm in Medieval Treatises "De rithmis"', *Journal of Musicology*, 5 (1987), 164–90.

FEICHT Festschrift 1967 = Zofia Lissa (ed.), *Studia Hieronymo Feicht septuagenario dedicata* (Warsaw, 1967).

FELDER, HILARIN, *Die liturgischen Reimoffizien auf den heiligen Franciscus und Antonius. Gedichtet und componiert durch Julian von Speier* (Fribourg, 1901).

FELLERER, KARL GUSTAV, 'Zur Choralpflege und Chorallehre im 17./18. Jahrhundert', *KmJb* 56 (1972), 51–72.

—— (ed.), *Geschichte der katholischen Kirchenmusik*, i (Kassel, 1972).

—— 'Zur Choral-Restauration in Frankreich um die Mitte des 19. Jahrhunderts', *KmJb* 58–9 (1974–5), 135–47.

——*Kirchenmusik im 19. Jahrhundert* (Studien zur Musik des 19. Jahrhunderts, 2; Regensburg, 1985).

—— 'Gregorianik im 19. Jahrhundert', in Fellerer 1985, 9–95.

—— arts. 'Caecilianismus', 'Choralreform', *MGG*.

FELLERER Festschrift 1962 = Heinrich Hüschen (ed.), *Festschrift Karl Gustav Fellerer zum sechzigsten Geburtstag am 7. Juli 1962* (Regensburg, 1962).

FENLON, IAIN (ed.), *Cambridge Music Manuscripts 900–1700* (Cambridge, 1982).

FERNÁNDEZ DE LA CUESTA, ISMAEL, *Manuscritos y fuentes musicales en España: Edad media* (Madrid, 1980).

—— 'La irrupción del canto gregoriano en España', *Revista de musicología*, 8 (1985), 239–48.

—— *see also Antiphonale Silense*.

FÉROTIN, M., *Le Liber ordinum en usage dans l'église wisigothique et mozarabe d'Espagne du cinquième à l'onzième siècle* (Monumenta ecclesiae liturgica, 6; Paris, 1904).

FERRETTI, PAOLO, *Principi teoretici e practici di canto gregoriano* (Rome, 1905; 3rd edn., 1913).

——*Il cursu metrico e il ritmo delle melodie gregoriane* (Rome, 1913).

—— 'Étude sur la notation acquitaine d'après le Graduel de Saint-Yrieix', PalMus 13, 54–211.

—— 'I manoscritti gregoriani dell'Archivio di Montecassino', in *Casinensia: Miscellanea di studi cassinensi pubblicati in occasione del XIV centenario della fondazione della badia di Montecassino*, i (Montecassino, 1929), 187–203.

——*Estetica gregoriana: Trattato delle forme musicali del canto gregoriano*, i (Rome, 1934); trans. A. Agaësse: *Esthétique grégorienne, ou traité des formes musicales du chant grégorien*, i (Solesmes, 1938).

——*Estetica gregoriana dei recitativi liturgici*, ed. P. Ernetti (Venice, 1964).

FIALA, V., and IRTENKAUF, W., 'Versuch einer liturgischen Nomenklatur', *Zeitschrift für Bibliothekswesen und Bibliographie. Sonderheft 1: Zur Katalogisierung mittelalterlicher und neuerer Handschriften* (Frankfurt, 1963), 105–37.

FICKETT, MARTHA VAN ZANDT, 'Chants for the Feast of St. Martin of Tours' (Diss., Catholic University of America, 1983; UMI 84–01475).

FISHER (KURT VON) Festschrift 1973 = Hans-Heinrich Eggebrecht and Max Lütolf (eds.), *Studien zur Tradition in der Musik: Kurt von Fischer zum 60. Geburtstag* (Munich, 1973).

FISHER (KURT VON) Festschrift 1973 = Hans-Heinrich Eggebrecht and Max Lütolf (eds.), *Studien zur Tradition in der Musik: Kurt von Fischer zum 60. Geburtstag* (Munich, 1973).

FISCHER, LUDWIG (ed.), *Bernhardi cardinalis et Lateranensis ecclesiae prioris, ordo officiorum ecclesiae Lateranensis* (Munich, 1916).

Fischer, Pieter *see* RISM.

FISCHER (WILHELM) Festschrift 1956 = *Festschrift Wilhelm Fischer zum 70. Geburtstag, überreicht im Mozartjahr 1956* (Innsbrücker Beiträge zur Kulturwissenschaft, Sonderheft 3; Innsbruck, 1956).

FLEMING, KEITH, 'The Editing of Some Communion Melodies in Medieval Chant Manuscripts' (Diss., Catholic University of America, 1979; UMI 79–20544).

FLINT, VALERIE I. J., 'Are Heinricus of Augsburg and Honorius Augustodunensis the Same Person?' *RB*, 92 (1982), 148–68.

FLOROS, CONSTANTIN, *Universale Neumenkunde*, 3 vols. (Kassel, 1970).

—— *Einführung in die Neumenkunde* (Wilhelmshafen, 1980).

FLOTZINGER, RUDOLF, 'Zu Herkunft und Datierung der Gradualien Graz 807 und Wien 13314', *SMH* 31 (1989), 57–80.

FLOYD, MALCOLM, 'Processional Chants in English Sources', *JPMMS* 13 (1990), 1–48.

FOLEY, E., 'The Cantor in Historical Perspective', *Worship*, 56 (1982), 194–213.

FONTAINE, G., 'Présentation des missels diocésains français du XVIIᵉ au XIXᵉ siècles', *La Maison-Dieu*, 141 (1980), 97–166.

FORTESCUE, ADRIAN, *The Ceremonies of the Roman Rite Described* (London, 1917; rev. J. B. O'Connell, 12th edn., 1962).

FOURNIER, DOMINIQUE M., 'Sources scripturaires et provenance liturgique des pièces de chant du graduel de Paul VI. (I. Ancien Testament)', *EG* 21 (1986), 97–114.

—— 'Sources scripturaires et provenance liturgique des pièces de chant du graduel de Paul VI. (II. Les Psaumes)', *EG* 22 (1988), 109–75.

FRANCA, UMBERTO, *Le antifone bibliche dopo Pentecoste* (Studia anselmiana, 73; Rome, 1977).

FREISTEDT, HEINRICH, *Die liqueszierenden Noten des gregorianischen Chorals* (Fribourg, 1929).

FRÉNAUD, GEORGES, 'Les Témoins indirects du chant liturgique en usage à Rome aux IXᵉ et Xᵉ siècles', *EG* 3 (1959), 41–74.

FRERE, WALTER HOWARD, *The Winchester Troper from MSS of the Xth and XIth Centuries* (Henry Bradshaw Society, 8; London, 1894).

—— *Bibliotheca musico-liturgica: A Descriptive Handlist of the Musical and Latin-liturgical MSS of the Middle Ages Preserved in the Libraries of Great Britain and Northern Ireland*, i (London, 1894 and 1901), ii (1932).

—— *The Use of Sarum: I. The Sarum Customs as Set Forth in the Consuetudinary and Customary* (Cambridge, 1898); *II. The Ordinal and Tonal* (Cambridge, 1901).

—— *Pontifical Services* (Alcuin Club Collections, 3–4; London, 1901).

—— *The Principles of Religious Ceremonial* (London, 1906: 2nd edn., 1928).

—— (ed.), *Pars antiphonarii* (London, 1923) [fac. of Durham Cathedral Library B. iii. 11].

—— *Studies in Early Roman Liturgy, i: The Calendar* (Alcuin Club Collections, 28; London, 1930).

—— *Studies in Early Roman Liturgy, ii: The Roman Gospel-Lectionary* (Alcuin Club Collections, 30; London, 1934).

—— *Studies in Early Roman Liturgy, iii: The Roman Epistle-Lectionary* (Alcuin Club Collections, 32; London, 1935).

—— and BROWN, LANGTON E. G., *The Hereford Breviary* (Henry Bradshaw Society, 26, 40, and 46; London, 1904, 1911, and 1915).

—— *see also AS; Graduale Sarisburiense*.

FROGER, JACQUES, 'L'Alléluia dans l'usage romain et la réforme de saint Grégoire', *EL* 62 (1948), 6–48.

FROGER, JACQUES, 'Le Chant de l'introït', *EL* 62 (1948), 248–55.

—— *Les Chants de la messe au VIIIe et IXe siècles* (Tournai, 1950) = *RG* 26 (1947), 165–72, 218–28; 27 (1948), 56–62; 98–107; 28 (1949), 58–65, 94–102.

—— 'L'Édition critique de l'Antiphonale Missarum romain par les moines de Solesmes', *EG* 1 (1954), 151–7.

—— 'L'Épître de Notker sur les "lettres significatives"', *EG* 5 (1962), 23–72.

—— *La Critique des textes et son automatisation* (Paris, 1968).

—— 'Les Prétendus Quarts de ton dans le chant grégorien et les symboles du ms. H. 159 de Montpellier', *EG* 17 (1978), 145–79.

—— 'The Critical Edition of the Roman Gradual by the Monks of Solesmes', *JPMMS* 1 (1978), 81–97.

—— 'La Méthode de Dom Hesbert dans le volume V du *Corpus Antiphonalium Officii*', *EG* 18 (1979), 97–143.

—— 'Le Fragment de Lucques (fin du VIIIe siècle)', *EG* 18 (1979), 145–55.

—— 'Le Lieu de destination et de provenance du "Compendiensis"', in Cardine Festschrift 1980, 338–53.

—— 'La Méthode de Dom Hesbert dans le volume VI du *Corpus Antiphonalium Officii*', *EG* 19 (1980), 185–96.

FULLER, DAVID, 'Plainchant musical', *NG*.

FULLER, SARAH, 'Aquitanian Polyphony of the Eleventh and Twelfth Centuries' (Diss., University of California at Berkeley, 1969; UMI 70–13051).

—— 'An Anonymous Treatise *dictus de Sancto Martiale*: A New Source for Cistercian Music Theory', *MD* 31 (1977), 5–30.

—— 'Theoretical Foundations of Early Organum Theory', *AcM* 53 (1981), 52–84.

GAJARD, JOSEPH, *La Méthode de Solesmes, ses principes constitutifs, ses règles pratiques d'interpretation* (Tournai, 1951).

—— 'Les Récitations modales des 3e et 4e modes dans les manuscrits bénéventains et aquitains', *EG* 1 (1954), 9–45.

GALLACHER, PATRICK J., and DAMICO, HELEN (eds.), *Hermeneutics and Medieval Culture* (Albany, NY, 1989).

GALLO, F. ALBERTO, 'Philological Works on Musical Treatises of the Middle Ages. A Bibliographical Report', *AcM* 44 (1972), 78–101.

GAMBER, KLAUS, *Sakramentartypen: Versuch einer Gruppierung der Handschriften und Fragmente bis zur Jahrtausendwende* (Texte und Arbeiten, 49–50; Beuron, 1958).

—— 'Das Lectionar und Sakramentar des Musäus von Massilia', *RB* 69 (1959), 118–215.

—— 'Fragment eines mittelitalienischen Plenarmissale aus dem 8. Jh.', *EL* 76 (1962), 335–41.

—— *Codices liturgici Latini antiquiores* (Spicilegii Friburgensis, Subsidia, 1; Freiburg, 1963; 2 vols., 2nd edn., 1968; Subsidia [supplement and indices], 1988).

GARAND, M., 'Le Missel clunisien de Nogent-le-Rotrou', in Boutemy Festschrift 1976, 129–51.

GARSOIAN, NINA G., MATHEWS, THOMAS F., and THOMSON, ROBERT W. (eds.), *East of Byzantium: Syria and Armenia in the Formative Period* (Dumbarton Oaks, 1982).

GASTOUÉ, AMEDÉE, 'Le Chant gallican', *RCG* 41 (1937), 101–6, 131–3, 167–76; 42 (1938), 5–12, 57–62, 76–80, 107–12, 146–51, 171–6; 43 (1939), 7–12, 44–6; also published as a book (Grenoble, 1939).

GATARD, A., 'Ambrosien (chant)', *DACL*.

GAUTIER, LÉON, *Histoire de la poésie liturgique au moyen âge*, i: *Les Tropes* (Paris, 1886).

GAVEL, M. H., 'À propos des erreurs d'accentuation latine dans les livres liturgiques', *EG* 1 (1954), 83–148.

GELINEAU, JOSEPH, *Chant et musique dans le culte chrétien, principes, lois et applications* (Paris, 1962); trans. C. Howell: *Voices and Instruments in Christian Worship* (Collegeville, Minn., 1964).

GENNRICH, FRIEDRICH, *Grundriß einer Formenlehre des mittelalterlichen Liedes als Grundlage einer musikalischen Formenlehre des Liedes* (Halle, 1932).

GERBERT, MARTIN, *De cantu et musica sacra a prima ecclesiae aetate usque ad praesens tempus*, 2 vols. (Sankt Blasien, 1774; repr. Graz, 1968 with introduction and index by Othmar Wessely).

—— *see also* GS.

GEVAERT, FRANÇOIS-AUGUSTE, *Les Origines du chant liturgique de l'église latine* (Ghent, 1890).

—— *La Melopée antique dans le chant de l'église latine* (Ghent, 1895).

GIBSON, MARGARET (ed.), *Boethius* (Oxford, 1981).

GILISSEN Festschrift 1985 = Jacques Lemaire and Émile van Balberghe (eds.), *Calames et cahiers: Mélanges de codicologie et de paléographie offerts à Léon Gilissen* (Brussels, 1985).

GILLINGHAM, BRYAN (ed.), *Paris, Bibliothèque Nationale, fonds latin 1139* (Publications of Mediaeval Manuscripts, 14; Ottawa, 1987).

—— *Paris, Bibliothèque Nationale, fonds latin 3719* (Publications of Mediaeval Manuscripts, 15; Ottawa, 1987).

—— *Paris, Bibliothèque Nationale, fonds latin 3549 and London, B. L., Add. 36,881* (Publications of Mediaeval Manuscripts, 16; Ottawa, 1987).

—— *Cambridge, University Library, Ff. i. 17 (1)* (Publications of Mediaeval Manuscripts, 17; Ottawa, 1989).

GINDELE, CORBINIAN, 'Chordirektion des gregorianischen Gesangs im Mittelalter', *Studien und Mitteilungen zur Geschichte des Benediktiner Ordens und seiner Zweige*, 63 (1951), 31–44.

—— 'Doppelchor und Psalmvortrag im Frühmittelalter', *Mf* 6 (1953), 296–300.

GINGRAS, G., *Egeria: Diary of a Pilgrimage* (Ancient Christian Writers, 38; New York, 1970).

GJERLØW, LILLI, *Adoratio Crucis, the Regularis Concordia and the Decreta Lanfranci: Manuscript Studies in the Early Medieval Church of Norway* (Oslo, 1961).

—— *Ordo Nidrosiensis ecclesiae* (Libri liturgici provinciae Nidrosis medii aevi, 2; Oslo, 1968).

—— 'Votive Masses Found in Oslo', *EL* 84 (1970), 113–28.

—— *Antiphonarium Nidrosiensis ecclesiae* (Libri liturgici provinciae Nidrosis medii aevi, 3; Oslo, 1979).

—— *Liturgica Islandica* (Bibliotheca Arnamagnaeana, 35–6; Copenhagen, 1980).

GLEESON, PHILIP, 'Dominican Liturgical Manuscripts before 1254', *Archivum fratrum praedicatorum*, 42 (1972), 81–135.

GMELCH, JOSEPH, *Die Vierteltonstufen im Meßtonale von Montpellier* (Eichstätt, 1911).

GNEUSS, HELMUT, *Hymnar und Hymnarien im englischen Mittelalter* (Tübingen, 1968).

GOEDE, N. DE, *The Utrecht Prosarium* (Monumenta musica Neerlandica, 6; Amsterdam, 1965) [Utrecht, Bibliotheek der Rijksuniversiteit, 417].

GÖLLNER, THEODOR, 'Unknown Spanish Passion Tones in Sixteenth-Century Hispanic Sources', *JAMS* 28 (1975), 46–71.

GOMBOSI, OTTO, 'Studien zur Tonartlehre des frühen Mittelalters', *AcM* 10 (1938), 174–94; 11 (1939), 28–39, 128–35; 12 (1940), 21–52.

——'Key, Mode, Species', *JAMS* 4 (1951), 20–6.

GÖSCHL, JOHANNES BERCHMANS, *Semiologische Untersuchungen zum Phänomen der gregorianischen Liqueszenz: Der isolierte dreistufige Epiphonus praepunctis, ein Sonderproblem der Liqueszenzforschung*, 2 vols. (Forschungen zur älteren Musikgeschichte, 3; Vienna, 1980).

——'Der gegenwärtige Stand der semiologischen Forschung', *BzG* 1 (1985), 43–102.

Gottesdienst der Kirche: Handbuch der Liturgiewissenschaft, ed. Hans B. Meyer, Hansjörg auf der Mauer, Balthasar Fischer, Angelus A. Häußling, and Bruno Kleinheyer. Vol. iii: *Gestalt des Gottesdienstes: Sprachlich und nichtsprachliche Ausdrucksformen* (Regensburg, 1987).

Graduale Arosiense impressum, ed. Toni Schmid (Laurentius Petri Sällskapets Urkundsserie, 7; Malmö and Lund, 1959–65).

Graduale de tempore et de sanctis juxta ritum Sanctae Romanae Ecclesiae cum cantu Pauli V. Pont. Max. jussu reformato (Rome, 1898).

Graduale Pataviense (Wien 1511): Faksimile, ed. Christian Väterlein (Das Erbe deutscher Musik, 87; Kassel, 1982).

Graduel romain . . . Chant restauré par la Commission de Reims et de Cambrai d'après les anciens manuscrits (Paris, 1874).

Graduel romain: Édition critique par les moines de Solesmes. II: *Les Sources* (Solesmes, 1957); IV: *Le Texte neumatique, i: Le Groupement des manuscrits* (Solesmes, 1960); *ii: Les Relations généalogiques des manuscrits* (Solesmes, 1962).

Graduel romain . . . publiée par la Commission Ecclésiastique de Digne (Marseilles, 1872).

Graduale Romanum . . . restitutum et editum Pauli VI (Solesmes, 1974).

Graduale Sarisburiense = Walter Howard Frere (ed.), *Graduale Sarisburiense: A Reproduction in Facsimile of a Manuscript of the Thirteenth Century, with a Dissertation and Historical Index Illustrating its Development from the Gregorian Antiphonale missarum* (London, 1894).

GRÉGOIRE, RÉGINALD, *Les Homéliaires du moyen âge: Inventaire et analyse des manuscrits* (Rerum ecclesiasticarum documenta, Series maior, 6; Rome, 1966).

——'Repertorium liturgicum italicum', *Studi medievali*, 3rd ser., 9 (1968), 463–592; 11 (1970), 537–56; 14 (1973), 1123–32.

——*Homéliaires liturgiques medievaux* (Biblioteca degli 'Studi medievali', 12; Spoleto, 1980).

GRISAR, HARTMANN, 'Die Gregorbiographie des Paulus Diakonus in ihrer ursprünglichen Gestalt, nach italienischen Handschriften', *Zeitschrift für katholische Theologie*, 11 (1887), 158–73.

GROS, MIQUEL S., 'El Processoner de la Catedral de Vic—Vic, Mus. Episc., MS 117 (CXXIV)', *Miscel·lània litúrgica catalana*, 2 (1983), 73–130.

GROTEFEND, HERMANN, *Zeitrechnung des deutschen Mittelalters und der Neuzeit*, 2 vols. (Hannover, 1891–8).

——*Taschenbuch der Zeitrechnung des deutschen Mittelalters und der Neuzeit* (Hannover and Leipzig, 1898; 10th edn., 1960).

Guentner *see* CSM 24.

GUIDETTI, GIOVANNI DOMENICO, *Directorium chori ad usum sacrosanctae basilicae vaticanae et aliarum cattedralium* (Rome, 1582).

GÜMPEL, KARL-WERNER, 'Zur Interpretation der Tonus-Definition des Tonale Sancti Bernardi', *Akademie der Wissenschaften und der Literatur, Jahrgang 1959/2* (Wiesbaden, 1959), 25–51.

GUSHEE, LAWRENCE A., 'The *Musica Disciplina* of Aurelian of Réôme, a Critical Text and Commentary' (Diss., Yale University, 1963; UMI 64–11873).

—— 'Questions of Genre in Medieval Treatises on Music', in Schrade Gedenkschrift 1973, 365–433.

—— *see also* CSM 21.

GY, PIERRE-MARIE, 'Collectaire, rituel, processional', *Revue des sciences philosophiques et théologiques*, 44 (1960), 441–69.

—— 'Typologie et ecclésiologie des livres liturgiques médiévaux', *La Maison-Dieu*, 121 (1975), 7–21.

—— 'Les Tropes dans l'histoire de la liturgie et de la théologie', in Iversen 1983, 7–16.

—— 'L'Influence des chanoines de Lucques sur la liturgie du Latran', *Revue des sciences religieuses*, 58 (1984),' 31–41.

—— 'Les Répons de l'office nocturne pour la fête de S. Martin', in Giustino Farnedi (ed.), *Traditio et progressio: Studi liturgici in onore del Prof. Adrien Nocent, OSB* (= *Studia anselmiana*, 95 (1988)), 215–23.

GY Festschrift 1990 = *Rituels: Mélanges offerts au Père Gy OP*, ed. Paul de Clerck and Eric Palazzo (Paris, 1990).

GYUG, RICHARD F., 'Tropes and Prosulas in Dalmatian Sources of the Twelfth and Thirteenth Centuries', in Leonardi and Menesto 1990, 409–38.

HAAPANEN, T., *Verzeichnis der mittelalterlichen Handschriftenfragmente in der Universitäts-bibliothek zu Helsingfors* (Helsinki, 1922–32).

—— *Die Neumenfragmente der Universitätsbibliothek Helsingfors: Eine Studie zur ältesten nordischen Musikgeschichte* (Helsinki, 1924).

HAAS, MAX, *Byzantinische und slavische Notationen* (Palaeographie der Musik, 1/2; Cologne, 1973).

—— 'Probleme einer "Universale Neumenkunde"', *Forum musicologicum*, 1 (1975), 305–22.

—— 'Studien zur mittelalterlichen Musiklehre. I. Eine Übersicht über die Musiklehre im Kontext der Philosophie des 13. und frühen 14. Jahrhunderts', *Forum musicologicum*, 3 (1982), 223–456.

HABERL (FERDINAND) Festschrift 1977 = Franz A. Stein (ed.), *Festschrift Ferdinand Haberl zum 70. Geburtstag: Sacerdos et cantus gregoriani magister* (Regensburg, 1977).

HABERL, FRANZ XAVER, 'Geschichte und Wert der offiziellen Chorbücher', *KmJb* 27 (1902), 134–92.

HADDAN, ARTHUR WEST, and STUBBS, WILLIAM, *Councils and Ecclesiastical Documents Relating to Great Britain and Ireland*, 3 vols. (Oxford, 1869–71).

HAIN, KARL, *Ein musikalischer Palimpsest* (VGA 12; Fribourg, 1925).

HALLINGER, KASSIUS, *Gorze-Kluny: Studien zu den monastischen Lebensformen und Gegensätzen im Hochmittelalter* (Studia anselmiana, 22/2324/25; Rome, 1950–1).

—— *see also* CCM.

HALPERIN, DAVID, 'Contributions to a Morphology of Ambrosian Chant' (Diss., Tel Aviv University, 1986).

HÄMEL, ADALBERT, 'Überlieferung und Bedeutung des Liber Sancti Jacobi und des Pseudo-Turpin', *Sitzungsberichte der philosophisch-historischen Klasse der Bayerischen Akademie der Wissenschaften* (1950), no. 2, 1–75.

HAMMERSTEIN, REINHOLD, *Die Musik der Engel: Untersuchungen zur Musikanschauung des Mittelalters* (Berne, 1962).

Hammond *see* CSM 14.

HANDSCHIN, JACQUES, 'Ein mittelalterlicher Beitrag zur Lehre von der Sphärenharmonie', *Zeitschrift für Musikwissenschaft*, 9 (1926–7), 193–208.

——'Über Estampie und Sequenz', *Zeitschrift für Musikwissenschaft*, 12 (1929), 1–20; 13 (1930–1), 113–32.

——'Sequenzprobleme', *Zeitschrift für Musikwissenschaft*, 17 (1935), 242–50.

——'L'Organum à l'église et les exploits de l'abbé Turstin', *RCG* 40 (1936–7), 179–82; 41 (1937–8), 14–19, 41–8 (see also David and Handschin 1935–7).

——'Eine alte Neumenschrift', *AcM* 22 (1950), 69–97; 25 (1953), 87–8.

——'Trope, Sequence, and Conductus', *The New Oxford History of Music*, ii, ed. Anselm Hughes (London, 1952), 128–74.

——'Sur quelques tropaires grecs traduits en latin', *Annales musicologiques*, 2 (1954), 27–60.

——'The Timaeus Scale', *MD* 4 (1950), 3–42.

HANDSCHIN Gedenkschrift 1962 = H. Anglès, G. Birkner, Ch. van den Borren, Fr. Benn, A. Carapetyan, and H. Husmann (eds.), *In memoriam Jacques Handschin* (Strasburg, 1962).

HÄNGGI, A., *Der Rheinauer Liber ordinarius (Zürich Rh 80, Anfang 12. Jh.)* (Spicilegium Friburgense, 1; 1957).

HANNICK, CHRISTIAN, arts. 'Armenian Rite, Music of the', 'Christian Church, Music of the Early', 'Ethiopian Rite, Music of the', 'Georgian Rite, Music of the', *NG*.

HANSEN, FINN EGELAND, *H 159 Montpellier* (Copenhagen, 1974).

——'Editorial Problems Connected with the Transcription of H 159, Montpellier: Tonary of St. Bénigne of Dijon', *EG* 16 (1977), 161–72.

——*The Grammar of Gregorian Tonality*, 2 vols. (Copenhagen, 1979).

HANSLIK, R., *Benedicti Regula* (Corpus scriptorum ecclesiasticorum latinorum, 75; Vienna, 1960; rev. edn., 1977).

HANSSENS, JEAN-MICHAEL, *Amalarii episcopi opera liturgica omnia* (Studi e testi, 13840; Rome, 1948–50).

Harbinson *see* CSM 23.

HARDISON, O. B., *Christian Rite and Christian Drama in the Middle Ages* (Baltimore, Md., 1965).

HARRISON, FRANK LL., *Music in Medieval Britain* (London, 1958; 2nd edn., 1963).

——'Benedicamus, Conductus, Carol: A Newly-Discovered Source', *AcM* 37 (1965), 35–48.

HARTZELL, K. DREW, 'An Unknown English Benedictine Gradual of the Eleventh Century', *Anglo-Saxon England*, 4 (1975), 131–44.

——'An Eleventh-Century English Missal Fragment in the British Library', *Anglo-Saxon England*, 18 (1989), 45–97.

HAUG, ANDREAS, *Gesungene und schriftlich dargestellte Sequenz: Beobachtungen zum Schriftbild der ältesten ostfränkischen Sequenzenhandschriften* (Neuhausen-Stuttgart, 1987).

HAYBURN, ROBERT F., *Papal Legislation on Sacred Music, 95 A.D. to 1977 A.D.* (Collegeville, Minn., 1979).

HEARD, EDMUND BROOKS, 'Alia musica: A Chapter in the History of Medieval Music Theory' (Diss., University of Wisconsin, 1966; UMI 66–13798).

HEIMING, ODILO, 'Zum monastischen Offizium von Kassianus bis Kolumbanus', *Archiv für Liturgiewissenschaft*, 7 (1961), 89–156.

—— 'Das Corpus Ambrosiano-Liturgicum. Ein Bericht', *EL* 92 (1978), 477–80.

HEISLER, MARIA-ELISABETH, 'Die Problematik des "germanischen" oder "deutschen" Choraldialekts', *SMH* 27 (1985), 67–82.

—— 'Studien zum ostfrankischen Choraldialekt' (Diss., Frankfurt, 1987).

HEITZ, CAROL, *Recherches sur les rapports entre architecture et liturgie à l'époque carolingienne* (Paris, 1963).

HELANDER, S., *Ordinarius Lincopensis och dess liturgiska förebilder* [The Ordinal of Linköping and its liturgical models] (Bibliotheca theologica practicae, 4; Lund, 1957).

HENDERSON, ISOBEL, 'Ancient Greek Music', *New Oxford History of Music*, i, ed. Egon Wellesz (Oxford, 1957), 336–403.

HENDERSON, WILLIAM G., *Missale ad usum insignis ecclesiae Eboracensis* (Surtees Society, 59–60; London, 1874).

—— *Missale ad usum percelebris ecclesiae Herfordensis* (Leeds, 1874).

—— (ed.), *Manuale et processionale ad usum insignis ecclesiae Eboracensis* (Surtees Society, 63; London, 1875).

—— (ed.), *Processionale ad usum insignis ac praeclarae ecclesiae Sarum* (Leeds, 1882).

HERLINGER, JAN, *The Lucidarium of Marchetto of Padua: A Critical Edition, Translation, and Commentary* (Chicago, 1985).

HERMESDORFF, MICHAEL, *Graduale juxta usum ecclesiae cathedralis Trevirensis dispositum. Quod ex veteribus codicibus originalibus accuratissime conscriptum et novis interim ordinatis seu indultis festis auctum* (Trier, 1863).

—— *Graduale ad normam cantus S. Gregorii, auf Grund der Forschungs-Resultate und unter Beihilfe der Mitglieder des Vereins zur Erforschung alter Choralhandschriften nach den ältesten und zuverlässigsten Quellen* (Trier, 1876–82).

HERZ Festschrift 1982 = Robert L. Weaver (ed.), *Essays on the Music of J. S. Bach and Other Divers Subjects: A Tribute to Gerhard Herz* (Louisville, Ky., 1982).

HERZO, ANTHONY MARIE, 'Five Aquitanian Graduals: Their Mass Propers and Alleluia Cycles' (Diss., University of Southern California, 1967; UMI 67–10762).

HESBERT, RENÉ-JEAN, 'La Messe "Omnes gentes" du VIIe dimanche après la Pentecôte et l'"Antiphonale Missarum" romain', *RG* 17 (1932), 81–9, 170–9; 18 (1933), 1–14.

—— 'Les Dimanches de Carême dans les manuscrits romano-bénéventains', *EL* 48 (1934), 198–222.

—— 'L'"Antiphonale Missarum" de l'ancien rit bénéventain', *EL* 52 (1938), 28–66, 141–58; 53 (1939), 168–90; 59 (1945), 69–95; 60 (1946), 103–41; 61 (1947), 153–210.

—— 'Un curieux antiphonaire palimpseste de l'office. Rouen, A. 292 (IXe s.)', *RB* 64 (1954), 28–45.

—— 'L'Antiphonaire d'Amalar', *EL* 94 (1980), 176–94.

—— 'L'Antiphonaire de la Curie', *EL* 94 (1980), 431–59.

—— 'The Sarum Antiphoner—Its Sources and Influence', *JPMMS* 3 (1980), 49–55.

—— 'Les Antiphonaires monastiques insulaires', *RB* 112 (1982), 358–75.

—— 'Les Matines de Pâques dans la tradition monastique', *Studia monastica*, 24 (1982), 311–48.

—— *see also* AMS, CAO, and MMS.

HILEY, DAVID (with Julian Brown and Sonia Patterson), 'Further Observations on W1', *JPMMS* 3 (1980), 53–80.

HILEY, DAVID 'The Norman Chant Traditions—Normandy, Britain, Sicily', *Proceedings of the Royal Musical Association*, 107 (1980–1), 1–33.

—— 'The Liturgical Music of Norman Sicily: A Study Centred on Manuscripts 288, 289, 19421 and Vitrina 20–4 of the Biblioteca Nacional, Madrid' (Diss., London, 1981; British Theses D70780/82).

—— 'Some Observations on the Relationships between Trope Repertories', in Iversen 1983, 29–38.

—— 'Quanto c'è di normanno nei tropari siculo-normanni?', *RIM* 18 (1983), 3–28.

—— 'The Plica and Liquescence', in Anderson Gedenkschrift 1984, 379–91.

—— 'Ordinary of Mass Chants in English, North French and Sicilian Manuscripts', *JPMMS* 9 (1986), 1–128.

—— 'Thurstan of Caen and Plainchant at Glastonbury: Musicological Reflections on the Norman Conquest', *Proceedings of the British Academy*, 72 (1986), 57–90.

—— 'The Rhymed Sequence in England: A Preliminary Survey', in Huglo 1987, 227–46.

—— 'The Chant of Norman Sicily: Interaction between the Norman and Italian Traditions', *SMH* 30 (1988), 379–91.

—— 'Cluny, Sequences and Tropes', in Leonardi and Menesto 1990, 125–38.

—— 'Editing the Winchester Sequence Repertory of ca. 1000', in Cantus Planus 1990, 99–113.

—— arts. 'Neo-Gallican Chant', 'Neuma', 'Notation, III, 1: Western, Plainchant', 'Pontifical', *NG*.

HÖFLER, J., 'Rekonstrukcija srednjeveškega sekvenciarija v osrednji Sloveniji/Reconstruction of the medieval sequencer in Central Slovenia', *Muzikološki zbornik*, 3 (1967), 5.

HOFMANN-BRANDT, HELMA, *Die Tropen zu den Responsorien des Officium*, 2 vols. (Diss., Erlangen-Nürnberg University, 1971).

HOHLER, CHRISTOPHER, 'The Proper Office of St. Nicholas and Related Matters with Reference to a Recent Book', *Medium aevum*, 36 (1967), 40–8.

—— 'A Note on Jacobus', *Journal of the Warburg and Courtauld Institutes*, 35 (1972), 31–80.

—— 'Reflections on Some Manuscripts Containing 13th-Century Polyphony', *JPMMS* 1 (1978), 2–38.

—— review of Babb, *Hucbald, Guido, and John on Music*, in *JPMMS* 3 (1980), 57–8.

HOLDER, STEPHEN, 'The Noted Cluniac Breviary-Missal of Lewes: Fitzwilliam Museum Manuscript 369', *JPMMS* 8 (1985), 25–32.

HOLLAENDER, ALBERT, 'The Sarum Illuminator and his School', *Wiltshire Archaeological and Natural History Magazine*, 1 (1942–4), 230–62.

HOLMAN, HANS-JORGEN, 'The Responsoria prolixa of the Codex Worcester F 160' (Diss., Indiana University, 1961; UMI 61–04447).

—— 'Melismatic Tropes in the Responsories for Matins', *JAMS* 16 (1963), 36–46.

HOLSCHNEIDER, ANDREAS, *Die Organa von Winchester: Studien zum ältesten Repertoire polyphoner Musik* (Hildesheim, 1968).

—— 'Instrumental Titles to the Sequentiae of the Winchester Tropers', in Westrup Festschrift 1975, 8–18.

—— 'Die instrumentalen Tonbuchstaben im Winchester Troper', in Dadelsen Festschrift 1978, 155–66.

HOLTZ, LOUIS, 'Quelques aspects de la tradition et de la diffusion des "Institutiones"', in *Flavio Magno Aurelio Cassiodoro: Atti della Settimana di Studi su Cassiodoro* (Cosenza–Rossano–Squillace, 1984), 281–312.

HOMAN, FREDERIC W., 'Final and Internal Cadential Patterns in Gregorian Chant', *JAMS* 17 (1964), 66–77.

HOPPIN, RICHARD H., *Cypriot Plainchant of the Manuscript Torino, Biblioteca Nazionale J. II. 9* (Musicological Studies and Documents, 19; Rome, 1968).

HOURLIER, JACQUES, 'Le Domaine de la notation messine', *RG* 30 (1951), 96–113, 150–8.

—— 'Remarques sur la notation clunisienne', *RG* 30 (1951), 231–40.

—— 'Le Bréviaire de Saint-Taurin: Un livre liturgique clunisien à l'usage de l'Échelle-Saint-Aurin (Paris B. N. lat. 12601)', *EG* 3 (1959), 163–73.

—— 'Notes sur l'antiphonie', in Schrade Gedenkschrift 1973, 116–43.

—— 'L'Origine des neumes', in Cardine Festschrift 1980, 354–61.

—— and HUGLO, MICHEL, 'La Notation paléofranque', *EG* 2 (1957), 212–19.

—— —— 'Étude sur la notation bénéventaine', PalMus 15, 71–161.

HRABANUS MAURUS, *Artium liberalium ordo et natura* (Book 3 of *De clericorum institutione*), PL 107, 377–420.

—— *De computo*, PL 107, 669–728.

HUCKE, HELMUT, 'Untersuchungen zum Begriff "Antiphon" und zur Melodik der Offiziums-antiphonen' (Diss., Freiburg im Br., 1951).

—— 'Musikalische Formen der Offiziumsantiphonen', *KmJb* 37 (1953), 7–33.

—— 'Die Entwicklung des christlichen Kultgesangs zum Gregorianischen Gesang' *Römische Quartalschrift*, 48 (1953), 147–94.

—— 'Die Einführung des Gregorianischen Gesangs im Frankenreich', *Römische Quartalschrift*, 49 (1954), 172–87.

—— 'Improvisation im Gregorianischen Gesang', *KmJb* 38 (1954), 5–8.

—— 'Die Tradition des Gregorianischen Gesanges in der römischen Schola cantorum', in Congress Vienna 1954, 120–3.

—— 'Graduale', *EL* 69 (1955), 262–4.

—— 'Gregorianischer Gesang in altrömischer und fränkischer Überlieferung', *AfMw* 12 (1955), 74–87.

—— 'Die Entstehung der Überlieferung von einer musikalischen Tätigkeit Gregors des Großen', *Mf* 8 (1955), 259–64.

—— 'Die Gregorianische Gradualeweise des 2. Tons und ihre ambrosianischen Parallelen. Ein Beitrag zur Erforschung des Ambrosianischen Gesangs', *AfMw* 13 (1956), 285–314.

—— 'Cantus gregorianus', in Hermann Schmidt 1956–7, ii. 901–50.

—— 'Eine unbekannte Melodie zu den Laudes regiae', *KmJb* 42 (1958), 32–8.

—— 'Zu einigen Problemen der Choralforschung', *Mf* 11 (1958), 385–414.

—— 'Zum Problem des Rhythmus im Gregorianischen Gesang', in Congress Cologne 1958, 141–3.

—— 'War Gregor der Große doch Musiker?', *Mf* 18 (1965), 390–3.

—— 'Tractusstudien', in Stäblein Festschrift 1967, 116–20.

—— 'Die Texte der Offertorien', in Husmann Festschrift 1970, 193–203.

—— 'Le Problème de la musique religieuse', *La Maison-Dieu*, 108 (1971), 7–20.

—— 'Das Responsorium', in Schrade Gedenkschrift 1973, 144–91.

—— 'Die Herkunft der Kirchentonarten und die fränkische Überlieferung des Gregorianischen Gesangs', in Congress Berlin 1974, 257–60.

—— 'Karolingische Renaissance und Gregorianischer Gesang', *Mf* 28 (1975), 4–18.

—— 'Der Übergang von mündlicher zu schriftlicher Musiküberlieferung im Mittelalter', in Congress Berkeley 1977, 180–91.

HUCKE, HELMUT, 'Die Cheironomie und die Entstehung der Newumenschrift', *Mf* 32 (1979), 1–6.

—— 'Towards a New Historical View of Gregorian Chant', *JAMS* 33 (1980), 437–67.

—— 'Zur Aufzeichnung der altrömischen Offertorien', in Cardine Festschrift 1980, 296–313.

—— 'Die Anfänge der Bearbeitung', *Schweizer Jahrbuch für Musikwissenschaft*, NS, 3 (1983), 15–20.

—— 'Die Anfänge der abendländischen Notenschrift', in Elvers Festschrift 1985, 271–88.

—— 'Gregorianische Paläographie als Überlieferungsgeschichte', in Huglo 1987, 61–5.

—— 'Choralforschung und Musikwissenschaft', in Dahlhaus Festschrift 1988, 131–41.

—— 'Gregorianische Fragen', *Mf* 41 (1988), 304–30.

—— 'Responsorium', *MGG*.

—— arts. 'Gradual (i)', 'Gregorian and Old Roman Chant', 'Gregory the Great', 'Tract', *NG*.

—— and HUGLO, MICHEL, 'Communion', *NG*.

HUDOVSKÝ, ZORAN, 'Missale beneventanum MR 166 della Biblioteca metropolitana a Zagrabia', *Jucunda laudatio*, 3 (1965), 306.

—— 'Benedictionale MR 89 of the Metropolitan Library in Zagreb', *SMH* 9 (1967), 55–75.

—— 'Neumatski rukopis Agenda Pontificalis MR 165 Metropolitanske knjižnice u Zagrebu', *Arti musices*, 2 (1971), 17–30.

HUGHES, ANDREW, *Fifteenth Century Liturgical Music: I, Antiphons and Music for Holy Week and Easter* (Early English Church Music, 8; London, 1968).

—— *Medieval Music: The Sixth Liberal Art* (Toronto, 1974; 2nd edn., 1980).

—— *Medieval Manuscripts for Mass and Office: A Guide to their Organization and Terminology* (Toronto, 1982).

—— 'Modal Order and Disorder in the Rhymed Office', *MD* 37 (1983), 29–52.

—— 'Late Medieval Rhymed Offices', *JPMMS* 8 (1985), 33–49.

—— 'Rhymed Office', *NG*.

HUGHES, ANSELM, *Anglo-French Sequelae, Edited from the Papers of the Late Dr. Henry Marriott Bannister* (Burnham, 1934).

—— *The Portiforium of Saint Wulstan (Corpus Christi College, Cambridge, Ms. 391)* (Henry Bradshaw Society, 89 and 90; Leighton Buzzard, 1958, 1960).

—— *The Bec Missal* (Henry Bradshaw Society, 94; Leighton Buzzard, 1963).

HUGHES, DAVID G., 'Further Notes on the Grouping of the Aquitanian Tropers', *JAMS* 19 (1966), 3–12.

—— 'Music for St. Stephen at Laon', in Merritt Festschrift 1972, 137–59.

—— 'Variants in Antiphon Families: Notation and Tradition', in Congress Strasbourg 1982, ii. 29–48.

—— 'Evidence for the Traditional View of the Transmission of Gregorian Chant', *JAMS* 40 (1987), 377–404.

HUGLO, MICHEL, 'La Tradition occidentale des mélodies byzantines du Sanctus', in Johner Festschrift 1950, 40–6.

—— 'La Mélodie grecque du "Gloria in excelsis" et son utilisation dans le Gloria XIV', *RG* 29 (1950), 30–40.

—— 'Origine de la mélodie du Credo authentique de l'Édition Vaticane', *RG* 30 (1951), 68–78.

—— 'Les Antiennes de la procession des reliques: Vestiges du chant "Vieux Roman" dans le pontifical', *RG* 31 (1952), 136–9.

—— 'Un tonaire du graduel de la fin du VIIIe siècle (Paris, B. N. lat. 15139)', *RG* 31 (1952), 176–86, 224–33.

—— 'Les Noms des neumes et leur origine', *EG* 1 (1954), 53–67.

—— 'Le Chant vieux-romain: Manuscrits et témoins indirects', *Sacris erudiri*, 6 (1954), 96–124.

—— 'Vestiges d'un ancien répertoire musical de Haute-Italie', in Congress Vienna 1954, 142–5.

—— 'Antifone antiche per la "fractio panis"', *Ambrosius. Rivista di pastorale ambrosiana*, 31 (1955), 85–95.

—— 'Les Preces des graduels aquitains empruntés à la liturgie hispanique', *Hispania sacra*, 8 (1955), 361–83.

—— 'Le Tonaire de Saint-Bénigne de Dijon', *Annales musicologiques*, 4 (1956), 7–18.

—— 'Trois anciens manuscrits liturgiques d'Auvergne, ii: Le Graduel d'un prieuré clunisien d'Auvergne', *Bulletin historique et scientifique de l'Auvergne*, 77 (1957), 81–104 at 92–100.

—— 'Origine et diffusion des Kyrie', *RG* 37 (1958), 85–7.

—— 'Le Domaine de la notation bretonne', *AcM* 35 (1963), 53–84.

—— 'La Chironomie médiévale', *Revue de musicologie*, 49 (1963), 155–71.

—— 'Le Chant des Béatitudes dans la liturgie hispanique', *Hispania sacra*, 17 (1964), 135–40.

—— 'Les Chants de la Missa greca de Saint-Denis', in Wellesz Festschrift 1966, 74–83.

—— 'Règlement du XIIIe siècle pour la transcription de livres notés', in Stäblein Festschrift 1967, 121–33.

—— 'Un théoricien du XIe siècle: Henri d'Augsbourg', *Revue de musicologie*, 53 (1967), 57–9.

—— 'Un troisième témoin du tonaire carolingien', *AcM* 40 (1968), 22–8.

—— 'L'Auteur du Dialogue sur la Musique attribuée à Odon', *Revue de musicologie*, 55 (1969), 119–71.

—— 'Les Listes alléluiatiques dans les témoins du graduel grégorien', in Husmann Festschrift 1970, 219–27.

—— *Les Tonaires: Inventaire, analyse, comparaison* (Paris, 1971).

—— 'Der Prolog des Odo zugeschriebenen "Dialogus de Musica"', *AfMw* 28 (1971), 134–46.

—— 'Comparaison de la terminologie modale en orient et en occident', in Congress Copenhagen 1972, ii. 758–61.

—— 'L'Introduction en occident des formules byzantines d'intonation', *Studies in Eastern Chant*, 3 (1973), 81–90.

—— 'Tradition orale et tradition écrite dans la transmission des mélodies grégoriennes', in Kurt von Fischer Festschrift 1973, 31–42.

—— 'Le Graduel palimpseste de Plaisance (Paris, B. N., lat. 7102)', *Scriptorium*, 28 (1974), 3–31.

—— 'Le Développement du vocabulaire de l'Ars Musica à l'époque carolingienne', *Latomus*, 34 (1975), 131—51.

—— 'Liturgia e musica sacra aquileiese', in G. Folena (ed.), *Storia della cultura veneta dalle origini al trecento* (Vicenza, 1976), 312–25.

—— 'L'Auteur du traité de musique dedié à Fulgence d'Affligem', *Revue belge de musicologie*, 31 (1977), 5–20.

—— 'Les Livres liturgiques de la Chaise-Dieu', *RB* 87 (1977), 62–96, 289–348.

—— 'Aux origines des tropes d'interpolation: Le trope méloform d'introït', *Revue de musicologie*, 64 (1978), 5–54.

—— 'Les Remaniements de l'Antiphonaire grégorien au IXe siècle: Hélisachar, Agobard,

Amalaire', in *Culto cristiano, politica imperiale carolinga. XVIII Convegno internazionale di studi sulla spiritualità medievale* (Todi, 1979), 87–120.

—— 'Abelard, poète et musicien', *Cahiers de civilisation médiévale X^e–XIIe siècles*, 22 (1979), 349–61.

—— 'On the Origins of the Troper-Proser', *JPMMS* 2 (1979), 11–18.

—— 'Les Débuts de la polyphonie à Paris: Les premiers organa parisiens', *Forum musicologicum*, 3 (1982), 93–164.

—— 'Le Répons-Graduel de la messe: Évolution de la forme. Permanence de la fonction', *Schweizer Jahrbuch für Musikwissenschaft*, NS, 2 (1982), 53–73.

—— 'Remarques sur la notation musicale du bréviaire de Saint-Victor-sur-Rhins', *RB* 93 (1983), 132–6.

—— 'L'Ancien Chant bénéventain', *Ecclesia orans*, 2 (1985), 265–93.

—— 'Analyse codicologique des drames liturgiques de Fleury', in Gilissen Festschrift 1985, 61–78.

—— 'La Notation wisigothique est-elle plus ancienne que les autres notations européennes?', in Congress Salamanca 1985, 19–26.

—— 'La Pénétration des manuscrits aquitains en Espagne', *Revista de musicología*, 8 (1985), 249–56.

—— (ed.), *Musicologie médiévale: Notations et sequences, Table ronde du CNRS à l'IRHT d'Orléans–La Source, 10–12 Septembre 1982* (Paris and Geneva, 1987).

—— *Les Livres de chant liturgique* (Typologie des sources du Moyen Âge occidental, 52; Turnhout, 1988).

—— 'Bibliographie des éditions et études relatives à la théorie musicale du Moyen Âge (1972–1987)', *AcM* 60 (1988), 229–72.

—— 'Bilan de 50 années de recherches (1939–1989) sur les notations musicales de 850 à 1300', *AcM* 62 (1990), 224–59.

—— arts. 'O Antiphons', 'Offertory Chant', *NCE*.

—— arts. 'Antiphon', 'Antiphoner', 'Breviary', 'Cluniac Monks', 'Epistle', 'Exultet', 'Farse', 'Gallican Rite, Music of the' [trans. and rev. from 'Altgallikanische Liturgie' in Fellerer 1972, 21933], 'Gerbert d'Aurillac', 'Gospel', 'Gradual (ii)', 'Litany', 'Missal', 'Odo', 'Processional', 'Tonary', *NG*.

—— et al., *Fonti e paleografia del canto ambrosiano* (Milan, 1956).

—— *see also* RISM B/III/3.

HUNT, NOREEN, *Cluny under St. Hugh, 1049–1109* (London, 1967).

HÜSCHEN, HEINRICH, 'Regino von Prüm, Historiker, Kirchenrechtler und Musiktheoretiker', in Fellerer Festschrift 1962, 205–23.

—— arts. 'Ars musica', 'Artes liberales', 'Augustiner', 'Benediktiner', 'Dominikaner', 'Franziskaner', 'Harmonie', 'Kartäuser', 'Prämonstratenser', 'Zisterzienser', *MGG*.

HÜSCHEN FESTSCHRIFT 1980 = Detlef Altenburg (ed.), *Ars musica, musica scientia: Festschrift Heinrich Hüschen zum 65. Geburtstag* (Beiträge zur rheinischen Musikgeschichte, 126; Kassel, 1980).

HUSMANN, HEINRICH, 'Die St. Galler Sequenztradition bei Notker und Ekkehard', *AcM* 26 (1954), 6–18.

—— 'Sequenz und Prosa', *Annales musicologiques*, 2 (1954), 61–91.

—— 'Das Alleluia Multifarie und die vorgregorianische Stufe des Sequenzengesanges', in Schneider Festschrift 1955, 17–23.

—— 'Alleluia, Vers und Sequenz', *Annales musicologiques*, 4 (1956), 19–53.

—— 'Iustus ut palma. Alleluia und Sequenzen in St. Gallen und St. Martial', *Revue belge de musicologie*, 10 (1956), 112–28.

—— 'Die Alleluia und Sequenzen der Mater-Gruppe', in Congress Vienna 1956, 276–84.

—— 'Zum Großaufbau der Ambrosianischen Alleluia', *Anuario musical*, 12 (1957), 17–33.

—— 'Alleluia, Sequenz und Prosa im altspanischen Choral', in Anglès Festschrift 1958–61, 407–15.

—— 'Sinn und Wesen der Tropen', *AfMw* 16 (1959), 135–47.

—— 'Ecce puerpera genuit. Zur Geschichte der teiltextierten Sequenzen', in Besseler Festschrift 1961, 59–65.

—— *Grundlagen der antiken und orientalischen Musikkultur* (Berlin, 1961).

—— 'Das Graduale von Ediger. Eine neue Quelle der rheinischen Augustinerliturgie', in Fellerer Festschrift 1962, 224–34.

—— 'Die Sequenz Duo tres. Zur Geschichte der Sequenzen in St. Gallen und St. Martial', in Handschin Gedenkschrift 1962, 66–72.

—— 'Zur Stellung des Meßpropriums der österreichischen Augustinerchorherren', *Studien zur Musikwissenschaft*, 25 (1962) (Festschrift für Erich Schenk), 261–75.

—— 'Studien zur geschichtlichen Stellung der Liturgie Kopenhagens. Die Oster- und Pfingstalleluia der Kopenhagener Liturgie und ihre historischen Beziehungen', *Dansk Aarbog for Musikforskning*, 1962, 3–58, and 1964–5, 3–62.

—— 'Zur Überlieferung der Thomas-Offizien', in Smits van Waesberghe Festschrift 1963, 87–8.

—— 'Notre-Dame und Saint-Victor. Repertoire-Studien zur Geschichte der gereimten Prosen', *AcM* 36 (1964), 98–123, 191–221.

—— 'Zur Geschichte der Meßliturgie von Sitten und über ihren Zusammenhang mit den Liturgien von Einsiedeln, Lausanne und Genf', *AfMw* 22 (1965), 217–47.

—— 'Die Handschrift Rheinau 71 der Zentralbibliothek Zürich und die Frage nach Echtheit und Entstehung der St. Galler Sequenzen und Notkerschen Prosen', *AcM* 38 (1966), 118–49.

—— 'Das Einsiedelner Gradual-Sakramentar St. Paul/Kärnten 25.2.25 (Seine Stellung und Bedeutung)', in Feicht Festschrift 1967, 89–95.

—— 'Hymnus und Troparion. Studien zur Geschichte der musikalischen Gattungen von Horologion und Tropologion', *Jahrbuch des Staatlichen Instituts für Musikforschung, Preußischer Kulturbesitz* (Berlin, 1971), 7–86.

—— 'Das Brevier der hl. Klara und seine Bedeutung in der Geschichte des römischen Chorals', *Studi musicali*, 2 (1973), 217–33.

—— 'Ein Missale von Assisi: Baltimore, Walters Art Gallery W. 75', in Hüschen Festschrift 1980, 255–62.

—— 'Zur Herkunft des "Andechser Missale" Clm 3005', *AfMw* 37 (1980), 155–65.

—— 'Notre-Dame-Epoche', *MGG*.

—— 'Syrian Church Music', *NG*.

—— *see also* RISM B/V/1.

HUSMANN Festschrift 1970 = Heinz Becker and Reinhard Gerlach (eds.), *Speculum musicae artis: Festgabe für Heinrich Husmann zum 60. Geburtstag* (Munich, 1970).

HUTTER, JOSEF, *Česká notace I. Neumy* (Prague, 1926), *II. Nota choralis* (Prague, 1930).

—— *Notationis Bohemicae antiquae specimina selecta e codicibus Bohemicis, 1. Neumae, 2. Nota choralis* (Prague, 1931).

Hymnarium Oscense (s. XI), I. Edición facsimil. II. Estudios, ed. Antonio Durán, Ramón Moragas, Juan Villareal (Saragossa, 1987).

IRTENKAUF, WOLFGANG, 'Das Seckauer Cantionarium vom Jahre 1345 (Hs. Graz 756)', *AfMw* 13 (1956), 116–41.

——'Reimoffizium', *MGG*.

ISIDORE, *Etymologiarum sive originum libri XX*, ed. W. M. Lindsay (Oxford, 1911).

Iter Helveticum, ed. Pascal Ladner (Spicilegii Friburgensis Subsidia; Fribourg, 1976–).

Iter Helveticum 1 (SFS 15, 1976) = Josef Leisibach, *Die liturgischen Handschriften der Kantons- und Universitätsbibliothek Freiburg*.

Iter Helveticum 2 (SFS 16, 1977) = Josef Leisibach, *Die liturgischen Handschriften des Kantons Freiburg (ohne Kantonsbibliothek)*.

Iter Helveticum 3 (SFS 17, 1979) = Josef Leisibach, *Die liturgischen Handschriften des Kapitelsarchiv in Sitten*.

Iter Helveticum 4 (SFS 18, 1984) = Josef Leisibach, *Die liturgischen Handschriften des Kantons Wallis (ohne Kapitelsarchiv Sitten)*.

IVERSEN, GUNILLA (ed.), *Research on Tropes* (Stockholm, 1983).

——*see also* CT.

JACOB, A., 'À propos de l'édition de l'Ordinaire de Tongres', *Revue d'histoire ecclésiastique*, 65 (1970), 789–97.

JACOBSTHAL, GUSTAV, *Die chromatische Alteration im liturgischen Gesang der abendländischen Kirche* (Berlin, 1897).

JAKOBS, HERMANN, *Die Hirsauer: Ihre Ausbreitung und Rechtsstellung im Zeitalter des Investiturstreits* (Kölner historische Abhandlungen, 4; Cologne and Graz, 1961).

JAMMERS, EWALD, 'Die Antiphonen der rheinischen Reimoffizien', *EL* 43 (1929), 199–219, 425–51; 44 (1930), 84–99, 342–68.

——*Das Karlsoffizium 'Regali natus'* (Sammlung musikwissenschaftlicher Abhandlungen, 14; Strasburg, 1934).

——*Der gregorianischen Rhythmus: Antiphonale Studien* (Strasburg, 1937).

——*Die Essener Neumenhandschriften der Landes- und Stadtbibliothek Düsseldorf* (Ratingen, 1952).

——'Die paläofrankische Neumenschrift', *Scriptorium*, 7 (1953), 235–59.

——*Musik in Byzanz, im päpstlichen Rom und im Frankenreich: Der Choral als Musik der Textaussprache* (Abhandlungen der Heidelberger Akademie der Wissenschaften, Philosophisch-historische Klasse, 1962/1; Heidelberg, 1962).

——'Einige Anmerkungen zur Tonalität des gregorianischen Gesanges', in Fellerer Festschrift 1962, 235–44.

——*Tafeln zur Neumenkunde* (Tutzing, 1965).

——'Der Choral als Rezitativ', *AfMw* 22 (1965), 143–68.

——'Abendland und Byzanz, II: Kirchenmusik', [1969] *Reallexikon der Byzantinistik*, ed. Peter Wirth.

——*Das Alleluia in der Gregorianischen Messe* (Liturgiewissenschaftliche Quellen und Forschungen, 55; Münster, 1973).

——'Cantio', *MGG*.

——'Cantio', *NG*.

JAN, K. VON (ed.), *Musici scriptores graeci*, 2 vols. (Leipzig, 1895–9).

JANSON, TORE, *Prose Rhythm in Medieval Latin from the 9th to the 13th Century* (Studia Latina Stockholmiensia, 20; 1975).

JEANNETEAU, JEAN, *Los modos gregorianos: Historia—análisis—estética* (Studia Silensia, 11; Abadia de Silos, 1985).

JEFFERY, PETER, 'An Early Cantatorium Fragment Related to MS. Laon 239', *Scriptorium*, 36 (1982), 245–52.

——'The Oldest Sources of the *Graduale*: A Preliminary Checklist of MSS Copied before about 900', *Journal of Musicology*, 2 (1983), 316–21.

——'The Introduction of Psalmody into the Roman Mass by Pope Celestine I (422–432): Reinterpreting a Passage in the *Liber Pontificalis*', *Archiv für Liturgiewissenschaft*, 26 (1984), 147–55.

JESSON, ROY, 'Ambrosian Chant: The Music of the Mass' (Diss., Indiana University, 1955; UMI 00–12833).

——'Ambrosian Chant', in Apel 1958, 465–83.

JOHNER, DOMINICUS, *Wort und Ton im Choral: Ein Beitrag zur Aesthetik des gregorianischen Gesanges* (2nd edn., Leipzig, 1953).

JOHNER Festschrift 1950 = Franz Tack (ed.), *Der kultische Gesang der abendländischen Kirche* (Cologne, 1950).

JOHNSTONE, JOHN G., 'Beyond a Chant: "Tui sunt caeli" and its Tropes', *Studies in the History of Music, 1: Music and Language* (New York, 1983), 24–37.

JONES, CHARLES W., *The Saint Nicholas Liturgy and its Literary Relationships* (Berkeley, 1963).

JONES (CHARLES) Festschrift 1979 = Margot H. King and Wesley M. Stevens (eds.), *Saints, Scholars, and Heroes: Studies in Medieval Culture in Honor of Charles W. Jones*, 2 vols. (Collegeville, Minn., 1979).

JONES, CHESLYN, WAINWRIGHT, GEOFFREY, and YARNOLD, EDWARD (eds.), *The Study of Liturgy* (London, 1978).

JONSSON, RITVA, *Historia: Études sur la genèse des offices versifiés* (Studia Latina Stockholmiensia, 15; 1968).

——'Corpus Troporum', *JPMMS* 1 (1978), 98–115.

——and TREITLER, LEO, 'Medieval Music and Language: A Reconsideration of the Relationship', *Studies in the History of Music, 1: Music and Language* (New York, 1983), 1–23.

——*see also* CT.

JOUNEL, PIERRE, *Le Culte des saints dans les basiliques du Latran et du Vatican au douzième siècle* (Rome, 1977).

JUNGMANN, JOSEF A., *Missarum sollemnia: Eine genetische Erklärung der römischen Messe*, 2 vols. (Vienna, 1948; 5th edn., 1962, 6th edn., 1966); trans. Francis A. Brunner from 2nd edn., 1951: *The Mass of the Roman Rite*, 2 vols. (New York, 1951, 1955); abridged version by Charles K. Riepe, 1 vol. (New York, 1959).

——*Liturgisches Erbe und pastorale Gegenwart: Studien und Vorträge* (Innsbruck, 1960); trans.: *Pastoral Liturgy* (London, 1962).

JUNGMANN Festschrift 1959 = Balthasar Fischer and J. Wagner (eds.), *Paschatis sollemnia: Studien zu Osterfeier und Osterfrömmigkeit. Festschrift J. A. Jungmann* (Basel, 1959).

KÄHMER, INGE, *Die Offertoriums-Überlieferung in Rom Vat. lat. 5319* (Diss., University of Cologne, 1971).

KAINZBAUER, XAVER, 'Der Tractus Tetrardus: Eine centologische Untersuchung', *BzG* 11 (1991), 1–132.

KANTOROWICZ, ERNST H., *Laudes regiae: A Study in Liturgical Acclamations and Mediaeval Ruler Worship. With a Study of the Music of the Laudes and Musical Transcriptions* by Manfred F. Bukofzer (Berkeley and Los Angeles, 1946).

KARNOWKA, G.-H., *Breviarium Passaviense: Das Passauer Brevier im Mittelalter und die Breviere der altbayrischen Kirchenprovinz* (Münchener Theologische Studien, II/44; St. Ottilien, 1983).

Karolingisches Sakramentar: Codex Vindobonensis 958 (Codices selecti, 25; Graz, 1971).

KEIL, HEINRICH, *Grammatici latini*, 8 vols. (Leipzig, 1857–80).

KELLY, THOMAS FORREST, 'Melodic Elaboration in Responsory Melismas', *JAMS* 27 (1974), 461–74.

—— 'New Music from Old: The Structuring of Responsory Proses', *JAMS* 30 (1977), 366–90.

—— 'Introducing the *Gloria in excelsis*', *JAMS* 37 (1984), 479–506.

—— 'Melisma and Prosula: The Performance of Responsory Tropes', in Silagi 1985, 163–80.

—— 'Montecassino and Old Beneventan Chant', *EMH* 5 (1985), 53–83.

—— 'Beneventan and Milanese Chant', *Journal of the Royal Musical Association*, 112 (1987), 173–95.

—— 'Neuma triplex', *AcM* 60 (1988), 1–30.

—— *The Beneventan Chant* (Cambridge, 1989).

KENNEDY, V. L., 'For a New Edition of the *Micrologus* of Bernold of Constance', in Andrieu Festschrift 1956, 229–41.

KER, NEIL R., *Medieval Libraries of Great Britain: A List of Surviving Books* (2nd edn., London, 1964); supplement ed. Andrew G. Watson (London, 1987).

KIENLE, AMBROSIUS, 'Notizen über das Dirigieren mittelalterlicher Gesangschöre', *Vierteljahrsschrift für Musikwissenschaft*, 1 (1885), 158–69.

KING, ALEC HYATT, *Four Hundred Years of Music Printing* (London, 1964; 2nd edn., 1968).

KING, ARCHDALE A., *The Rites of Eastern Christendom* (Rome, 1947–8).

—— *Liturgies of the Religious Orders* (Milwaukee, Wisc., 1955, London, 1956).

—— *Liturgies of the Primatial Sees* (Milwaukee, Wisc., and London, 1957).

—— *Liturgy of the Roman Church* (Milwaukee, Wisc., and London, 1957).

—— *Liturgies of the Past* (Milwaukee, Wisc., and London, 1959).

KLAUSER, THEODOR, 'Eine Stationsliste der Metzer Kirche aus dem 8. Jhd. wahrscheinlich ein Werk Chrodegangs', *EL* 44 (1930), 162–93.

—— 'Die liturgischen Austauschbeziehungen zwischen der römischen und der fränkisch-deutschen Kirche vom achten bis zum elften Jahrhundert', *Historisches Jahrbuch der Görres-Gesellschaft*, 53 (1933), 169–89.

—— *Das römische Capitulare Evangeliorum*, i: *Typen* (Liturgiewissenschaftliche Quellen und Forschungen, 28; Münster, 1935).

—— 'Der Übergang der römischen Kirche von der griechischen zur lateinischen Liturgiesprache', in *Miscellanea G. Mercati* (Studi e testi, 121; Rome, 1946), 467–82.

—— *Kleine abendländische Liturgiegeschichte* (Bonn, 1965); trans.: *A Short History of the Western Liturgy* (London, 1969).

KLÖCKNER, STEFAN, 'Analytische Untersuchungen an 16 Introiten im I. Ton des altrömischen und des fränkisch-gregorianischen Repertoires hinsichtlich einer bewußten melodischen Abhängigkeit', *BzG* 5 (1988), 3–95.

KNOWLES, DAVID, *The Monastic Constitutions of Lanfranc* (London, 1951).

—— *The Monastic Order in England* (Cambridge, 1940; 2nd edn., 1963).

—— *Christian Monasticism* (New York, 1969).

KOHRS, KLAUS HEINRICH, *Die aparallelen Sequenzen: Repertoire, liturgische Ordnung, musikalischer Stil* (Beiträge zur Musikforschung, 6; Salzburg, 1978).

KONRAD, KAREL, *Geschichte des alttschechischen Kirchengesanges* (Prague, 1881).

KOWALEWICZ, H., *Cantica medii aevi polono-latina*, i: *Sequentiae* (Bibliotheca latina medii et recentioris aevi, 15; Warsaw, 1964).

KRÄMER, SIGRID, *Handschriftenerbe des deutschen Mittelalters* (Mittelalterliche Bibliothekskataloge Deutschlands und der Schweiz, 5 (Ergänzungsband), 1. Aachen–Kochel, 2. Köln–Zyfflich; Munich, 1989).

KRIEG, EDUARD, *Das lateinische Osterspiel von Tours* (Würzburg, 1956).

KROON, SIGURD, *Ordinarium missae* (Lunds universitets årsskrift, NS, sect. 1, vol. 39, no. 6; Lund, 1953).

KURZEJA, ADALBERT, *Der älteste Liber Ordinarius der Trierer Domkirche: London, Brit. Mus., Harley 2958, Anfang 14. Jh.* (Liturgiewissenschaftliche Quellen und Forschungen, 52; Münster, 1970).

Kyriale seu ordinarium missae (Rome, 1905).

LABHARDT, FRANK, *Das Sequentiar Cod. 546 der Stiftsbibliothek von St. Gallen und seine Quellen* (Publikationen der Schweizerischen Musikforschenden Gesellschaft, Ser. 2, 8; 2 vols., Berne, 1959, 1963).

LA FAGE, ADRIEN DE, *Essais de diphthérographie musicale ou notices, descriptions, analyses, extraits et reproductions de manuscrits relatifs à la pratique, à la théorie et à l'historie de la musique*, 2 vols. (Paris, 1864) [includes edition of Paris 7211].

LAISTNER, MAX LUDWIG W., *Thought and Letters in Western Europe A.D. 500–900* (London, 1931).

LAMBILOTTE, LOUIS, *Antiphonaire de saint Grégoire: facsimilé du manuscrit de Saint-Gall* (Brussels, 1851, 2/1867).

LAMBRES, BENOÎT, 'Le Chant des Chartreux', *Revue belge de musicologie*, 24 (1970), 17–42.

—— 'L'Antiphonaire des Chartreux', *EG* 14 (1973), 213–18.

LAMOTHE, DONAT R., and CONSTANTINE, CYPRIAN G. (eds.), *Matins at Cluny for the Feast of St. Peter's Chains* (London, 1986).

LANDWEHR-MELNICKI, MARGARETHA, *Das einstimmige Kyrie des lateinischen Mittelalters* (Regensburg, 1955).

—— *see also* MMMA and Melnicki.

LAUM, BERNHARD, *Das alexandrinische Akzentuationssystem unter Zugrundelegung der theoretischen Lehren der Grammatiker und mit Heranziehung der praktischen Verwendung in den Papyri* (Studien zur Geschichte und Kultur des Altertums, 4. Ergänzungsband; Paderborn, 1928).

LAUNAY, DENISE, 'Un esprit critique au temps de Jumilhac: Dom Jacques Le Clerc, bénédictin de la Congrégation de Saint-Maur', *EG* 19 (1980), 197–219.

LAWLEY, STEPHEN W. (ed.), *Breviarium ad usum insignis ecclesiae Eboracensis* (Surtees Society, 71, 75; London, 1880, 1883).

LECHNER, JOSEPH, and EISENHOFER, LUDWIG, *Liturgik des römischen Ritus* (Freiburg, 1953); trans.: *The Liturgy of the Roman Rite* (Freiburg and Edinburgh, 1961).

LECLERCQ, HENRI, arts. 'Gallicane (liturgie)', 'Liturgies néo-gallicanes', *DACL*.

LECLERCQ, JEAN, *L'Amour des lettres et le désir de Dieu: Initiation aux auteurs monastiques du moyen âge* (Paris, 1957); trans. Catharine Misrahi: *The Love of Learning and the Desire for God* (New York, 1961; 2nd edn., 1974).

LEDWON, JACOB CARL, 'The Winter Office of Sant'Eutizio di Norcia: A Study of the Contents and Construction of Biblioteca Vallicelliana Manuscripts C 13 and C 5' (Diss., State University of New York at Buffalo, 1986; UMI 86–19339).

LEEB, HELMUT, *Die Psalmodie bei Ambrosius* (Wiener Beiträge zur Theologie, 17; Vienna, 1967).

—— *Die Gesänge im Gemeindegottesdienst von Jerusalem (vom 5. bis 8. Jahrhundert)* (Wiener Beiträge zur Theologie, 28; Vienna, 1979).

LEFÈVRE, PLACIDE F., 'La Nouvelle Édition du Processionale Praemonstratense de 1932', *Analecta praemonstratensia*, 9 (1933), 170–5.

—— 'Le Nouvel Antiphonaire de Prémontré', *Analecta praemonstratensia*, 13 (1937), 63–9.

—— *L'Ordinaire de Prémontré d'après des manuscrits du XIIe et du XIIIe siècle* (Bibliothèque de la 'Revue d'histoire ecclésiastique', 22; Louvain, 1941).

—— *Coutumiers liturgiques de Prémontré du XIIIe et du XIVe siècles* (Bibliothèque de la 'Revue d'histoire ecclésiastique', 27; Louvain, 1953).

—— *La Liturgie de Prémontré: Histoire, formulaire, chant et cérémonial* (Bibliotheca analectorum praemonstratensium, 1; Louvain, 1957).

—— *Les Ordinaires des collégiales Saint-Pierre à Louvain et Saints-Pierre-et-Paul à Anderlecht d'après des manuscrits du XIVe siècle* (Bibliothèque de la 'Revue d'histoire ecclésiastique', 36; Louvain, 1960).

—— *L'Ordinaire de la collégiale autrefois cathédrale de Tongres, d'après un manuscrit du XVe siècle* (Spicilegium sacrum Lovaniense, Études et documents, 345; Louvain, 1967–8).

—— 'L'Antiphonale psalterii d'après le rite de Prémontré', *Analecta praemonstratensia*, 44 (1968), 247–74.

—— 'Les Répons prolixes aux heures diurnes du Triduum sacrum dans la liturgie canoniale', *Analecta praemonstratensia*, 48 (1972), 5–19.

LEFFERTS, PETER M., 'Cantilena and Antiphon: Music for Marian Services in Late Medieval England', in Sanders Festschrift 1990, 247–82.

LEGG, JOHN WICKHAM, *Missale ad usum ecclesiae Wertmonasteriensis* (Henry Bradshaw Society, 1; London, 1891; 5 (1893); and 12 (1897).

—— *The Processional of the Nuns of Chester* (Henry Bradshaw Society, 18; London, 1899).

—— *The Sarum Missal Edited from Three Early Manuscripts* (Oxford, 1916).

LE HOLLADAY, RICHARD, 'The Musica Enchiriadis and Scholia Enchiriadis: A Translation and Commentary' (Diss., Ohio State University, 1977).

LEJAY, PAUL, 'Ambrosien (rit)', *DACL*.

LEMARIÉ, J., 'Les Antiennes "Veterem hominem" du jour Octave de l'Épiphanie et les antiennes d'origine grecque de l'Épiphanie', *EL* 72 (1958), 3–38.

—— *Le Bréviaire de Ripoll: Paris B. N. lat. 742* (Scripta et documenta, 14; Montserrat, 1965).

LEÓN TELLO, FRANCISCO JOSÉ, *Estudios de historia de la teoría musical* (Madrid, 1962).

LEONARDI, CLAUDIO, and MENESTO, ENRICO (eds.), *La tradizione dei tropi liturgici* (Spoleto, 1990).

LEROQUAIS, VICTOR, *Les Sacramentaires et les missels manuscrits des bibliothèques publiques de France*, 3 vols. (Paris, 1924).

—— *Les Bréviaires manuscrits des bibliothèques publiques de France*, 5 vols. (Paris, 1934).

—— *Le Bréviaire-missel du prieuré clunisien de Lewes* (Paris, 1935).

—— *Les Pontificaux manuscrits des bibliothèques publiques de France*, 3 vols. (Paris, 1937).

LEROQUAIS, VICTOR, *Les Psautiers manuscrits latins des bibliothèques de France*, 2 vols. (Mâcon, 1940–1).

LE ROUX, MARY PROTHASE, 'The "De harmonica institutione" and "Tonarius" of Regino of Prüm' (Diss., Catholic University of America, 1965; UMI 66–00318).

LE ROUX, RAYMOND, 'Aux origines de l'office festif: Les antiennes et les psaumes de Matines et de Laudes pour Noël et le 1ᵉʳ janvier', *EG* 4 (1961), 65–170.

—— 'Les Graduels des dimanches après la Pentecôte', *EG* 5 (1962), 119–30.

—— 'Les Répons de Psalmis pour le Matines de l'Épiphanie à la Septuagésime', *EG* 6 (1963), 39–148.

—— 'Guillaume de Volpiano. Son cursus liturgique au Mont-Saint-Michel et dans les abbayes normandes', *Millenaire monastique du Mont-Saint-Michel*, 1 (Paris, 1967), 417–72.

—— 'Repons du Triduo sacro et de Pâques', *EG* 18 (1979), 157–76.

LEVY, KENNETH J., 'The Byzantine Sanctus and its Modal Tradition in East and West', *Annales musicologiques*, 6 (1958–63), 7–67.

—— 'The Italian Neophytes' Chants', *JAMS* 23 (1970), 181–227.

—— '"Lux de luce": The Origin of an Italian Sequence', *MQ* 57 (1971), 40–61.

—— 'The Trisagion in Byzantium and the West', in Congress Copenhagen 1972, 761–5.

—— 'A Gregorian Processional Antiphon', *Schweizer Jahrbuch für Musikwissenschaft*, NS, 2 (1982), 91–102.

—— 'Toledo, Rome and the Legacy of Gaul', *EMH* 4 (1984), 49–99.

—— 'Old-Hispanic Chant in its European Context', in Congress Salamanca 1985, i. 3–14.

—— 'Charlemagne's Archetype of Gregorian Chant', *JAMS* 40 (1987), 1–30.

—— 'On the Origin of Neumes', *EMH* 7 (1987), 59–90.

—— arts. 'Byzantine Rite, Music of the', 'Ravenna Rite, Music of the', 'Trisagion', *NG*.

—— and EMERSON, JOHN, 'Plainchant', *NG*.

Liber gradualis (Tournai, 1883; 2nd edn., 1895).

Liber hymnarius cum invitatoriis & aliquibus responsoriis (Solesmes, 1982).

Liber vesperalis juxta ritum sanctae ecclesiae Mediolanensis [ed. G. M. Suñol] (Rome, 1939).

LICKLEDER, CHRISTOPH, *Choral und figurierte Kirchenmusik in der Sicht Franz Xaver Witts anhand der Fliegenden Blätter und der Musica Sacra* (Documenta Caeciliana, 3; Regensburg, 1988).

LIPPHARDT, WALTER, 'Studien zur Rhythmik der Antiphonen', *Mf* 3 (1950), 47–60, 224–34.

—— 'Das Moosburger Cantionale', *Jahrbuch für Liturgik und Hymnologie*, 3 (1957), 113–17.

—— 'Das Herodesspiel von Le Mans nach den Handschriften Madrid, Bibl. Nac. 288 und 289 (11. und 12. Jhd.)', in Smits van Waesberghe Festschrift 1963, 107–22.

—— *Der karolingische Tonar von Metz* (Liturgiewissenschaftliche Quellen und Forschungen, 43; Münster-Westfalen, 1965).

—— *Lateinische Osterfeiern und Osterspiele*, 6 vols. (Berlin, 1975–81).

—— 'Mensurale Hymnenaufzeichnungen in einem Hymnar des 15. Jahrhunderts aus St. Peter, Salzburg (Michaelbeuern Ms. Cart. 1)', in Cardine Festschrift 1980, 458–87.

—— 'Liturgische Dramen', *MGG*.

LLOYD, RICHARD WINGATE, 'Cluny Epigraphy', *Speculum*, 7 (1932), 336–49.

LOEW, ELIAS AVERY, *The Beneventan Script: A History of the South Italian Minuscule*. Second edition prepared and enlarged by Virginia Brown, 2 vols. (Rome, 1980).

LOHR, I., *Solmisation und Kirchentonarten* (Basel, 1943).

LORIQUET, HENRI, POTHIER, JOSEPH, and COLETTE, ARMAND K., *Le Graduel de l'église cathédrale de Rouen au XIIIᵉ siècle*, 2 vols. (Rouen, 1907) [Paris, Bibliothèque Nationale, lat. 904].

LOT, FERDINAND (ed.), *Hariulf: Chronique de l'Abbaye de Saint-Riquier (Vᵉ siècle–1104)* (Collection de Textes; Paris, 1894).

LOWE, ELIAS AVERY [= Loew], 'Two New Latin Liturgical Fragments on Mount Sinai', *RB* 74 (1964), 252–83.

LUNDÉN, T., *Den heliga Birgitta och den helige Petrus av Skänninge: Officium parvum beate Marie Virginis [The Lady-offices of St Bridget and the Venerable Peter of Skänninge]* (Acta universitatis Upsaliensis, Studia historico-ecclesiastica Upsaliensia, 27; 1976).

LÜTOLF, MAX (ed.), *Das Graduale von Santa Cecilia in Trastevere (Cod. Bodmer 74)*, 2 vols. (Cologne-Geneva, 1987).

LUTZ, CORA E., *Joannis Scotti Annotationes in Marcianus* (Cambridge, Mass., 1939).

—— 'Remigius' Ideas on the Classification of the Seven Liberal Arts', *Traditio*, 12 (1956), 65–86.

—— 'The Commentary of Remigius of Auxerre on Martianus Capella', *Medieval Studies*, 19 (1957), 138–56.

—— (ed.), *Remigii Autissiodorensis commentum in Martianum Capellam*, 2 vols. (Leiden, 1962–5).

—— *Schoolmasters of the Tenth Century* (Hamden, Conn., 1977).

Maas *see* RISM B/III/1.

McARTHUR, A. A., *The Evolution of the Christian Year* (London, 1953).

McCARTHY, M. C., *The Rule for Nuns of St. Caesarius of Arles: A Translation with a Critical Introduction* (The Catholic University of America Studies in Medieval History, NS, 16; Washington, DC, 1960).

McGEE, TIMOTHY J., 'The Liturgical Placements of the *Quem quaeritis* Dialogue', *JAMS* 29 (1976), 1–29.

MACHABEY, ARMAND, *Genèse de la tonalité musicale classique des origines au XVᵉ siècle* (Paris, 1955).

MACIEJEWSKI, T., 'Graduałz Chełmna [The Chelmno Gradual]', *Musica medii aevi*, 4 (1973), 164–245.

McKINNON, JAMES W., 'The Meaning of the Patristic Polemic against Musical Instruments', *Current Musicology*, 1 (1965), 69–82.

—— 'The Tenth Century Organ at Winchester', *The Organ Yearbook*, 5 (1974), 4–19.

—— 'The Fifteen Temple Steps and the Gradual Psalms', *Imago musicae*, 1 (1984), 29–49.

—— 'On the Question of Psalmody in the Ancient Synagogue', *EMH* 6 (1986), 159–91.

—— 'The Fourth-Century Origin of the Gradual', *EMH* 7 (1987), 91–106.

—— *Music in Early Christian Literature* (Cambridge, 1987).

—— 'The Patristic Jubilus and the Alleluia of the Mass', in Cantus Planus 1990, 61–70.

McKITTERICK, ROSAMUND, *The Frankish Church and the Carolingian Reforms, 789–895* (London, 1977).

—— *The Frankish Kingdoms under the Carolingians, 751–987* (London, 1983).

McROBERTS, DAVID, 'Some 16th-Century Scottish Breviaries and their Place in the History of the Scottish Liturgy', *The Innes Review*, 3 (1952), 33–48.

—— 'Catalogue of Scottish Medieval Liturgical Books and Fragments', *The Innes Review*, 3 (1952), 49–63 (published separately, Glasgow, 1953).

—— 'The Medieval Scottish Liturgy Illustrated by Surviving Documents', *Transactions of the Scottish Ecclesiological Society*, 15 (1957), 22–40.

MACROBIUS, *Commentarius in Somnium Scipionis*, ed. James Willis (Leipzig, 1970); *Commentary on the Dream of Scipio*, trans. William Harris Stahl (New York, 1952).

MADRIGNAC, ANDRÉ, 'Les Formules centons des Alléluia anciens', *EG* 20 (1981), 3–4; 21 (1986), 27–45.

MAGISTRETTI, MARCO (ed.), *Beroldus sive ecclesiae Ambrosianae Mediolanensis kalendarium et ordines saec. xii* (Milan, 1894).

—— (ed.)', *Manuale Ambrosianum ex codice saec. xi olim in usum canonicae Vallis Travaliae*, 2 vols. (Milan, 1904, 1905).

MALLET, JEAN, and THIBAUT, ANDRÉ, *Les Manuscrits en écriture bénéventaine de la Bibliothèque capitulaire de Bénévent*, i: *Manuscrits 1–18* (Paris, 1984).

MANITIUS, MAXIMILIANUS, *Geschichte der lateinischen Literatur des Mittelalters*, 3 vols. (Handbücher der klassischen Altertumswissenschaft; Munich, 1911–31).

MANSI, GIOVANNI DOMENICO, *Sacrorum conciliorum nova et amplissima collectio*, 31 vols. (Florence and Venice, 1757–98); 53 vols. (Paris, Leipzig, and Arnheim, 1901–27).

Marcusson *see* CT.

MARKOVITS, M., *Das Tonsystem der abendländischen Musik im frühen Mittelalter* (Publikationen der Schweizerischen Musikforschenden Gesellschaft, ser. 2, 30; Berne, 1977).

MAROSSZÉKI, SOLUTOR RODOLPHE, *Les Origins du chant cistercien: Recherches sur les réformes du plain-chant cistercien au XIIe siècle* (Analecta Sacri Ordinis Cisterciensis, 8; Rome, 1952).

MARTÈNE, EDMOND, *De antiquis ecclesiae ritibus*, 3 vols. (Rouen, 1700–2).

MARTIANUS CAPELLA, *De nuptiis Philologiae et Mercurii*, ed. James Willis (Leipzig, 1983); *The Marriage of Philology and Mercury: Martianus Capella*, trans. William Harris Stahl (New York, 1977).

MARTIMORT, AIMÉ-GEORGES (ed.), *L'Église en prière: Introduction à la liturgie* (Paris, 1961; 3rd edn., 1965); first two of the four parts trans.: *The Church at Prayer: Introduction to the Liturgy* (New York, 1968) and *The Church at Prayer: The Eucharist* (New York, 1973).

—— 'Origine et signification de l'alléluia de la messe romaine', in Quasten Festschrift 1970, 811–34.

—— (ed.), *L'Église en prière, édition nouvelle*, 4 vols. (Paris, 1983), trans. Matthew J. O'Connell: *The Church at Prayer, New Edition*, 4 vols. (Collegeville, Minn., and London, 1987, 1987, 1988, and 1986).

—— 'A propos du nombre des lectures à la messe', *Revue des sciences religieuses*, 58 (1984), 42–51.

MARXER, OTTO, *Zur spätmittelalterlichen Choralgeschichte St. Gallens: Der Codex 546 der St. Galler Stiftsbibliothek* (VGA 3; St. Gallen, 1908).

MAS, JOSIANE, 'La Notation catalane', *Revista de musicología*, 11 (1988), 11–30.

MATHIAS, R. X., *Die Tonarien* (Diss., Graz, 1903).

MEIER, BERNHARD, 'Modale Korrektur und Wortausdeutung im Choral der Editio Medicaea', *KmJb* 53 (1969), 101–32.

MELNICKI, MARGARETA, and STÄBLEIN, BRUNO, 'Graduale (Buch)', *MGG*.

—— *see also* MMMA and Landwehr-Melnicki.

MÉNAGER, ARMAND, 'Aperçu sur la notation du manuscrit 239 de Laon. Sa concordance avec les codices rythmiques sangalliens', PalMus 10, 177–211.

—— 'Étude sur la notation du manuscrit 47 de Chartres', PalMus 11, 41–131.

MERCENIER, F., 'La Plus Ancienne Prière à la Vierge "Sub tuum praesidium"', *Questions liturgiques et paroissiales*, 25 (1940), 33.

MERLETTE, B., 'Écoles et bibliothèques à Laon du déclin de l'antiquité au developpement de l'université', *Enseignement et vie intellectuelle, Actes du 95ᵉ Congrès des Sociétés Savantes* (Paris, 1975), i. 21–53.

MERRITT Festschrift = Laurence Berman (ed.), *Words and Music: The Scholar's View. A Medley of Problems and Solutions Compiled in Honor of A. Tillman Merritt by Sundry Hands* (Cambridge, Mass., 1972).

METTENLEITER, DOMINICUS, *Musikgeschichte der Stadt Regensburg* (Regensburg, 1865).

Meyer, Christian *see* RISM B/III/3.

MEYER, KATHI, 'The Eight Gregorian Modes on the Cluny Capitals', *Art Bulletin*, 34 (1952), 75–94.

MGH (Antiquitates) *Poetae Latini aevi Carolini*, i, ed. Ernst Dümmler (1881).

MGH (Antiquitates) *Poetae Latini aevi Carolini*, iv/1, ed. Paul von Winterfeld (1899).

MGH (Epistolae) *Epistolae Karolini aevi*, ii, ed. Ernst Dümmler (1895).

MGH (Epistolae) *Epistolae Karolini aevi*, iii, ed. Ernst Dümmler, Karl Hampe, *et al.* (1898–9).

MGH (Leges) *Capitularia regum Francorum*, i, ed. Alfred Boretius (1883).

MGH (Leges) *Concilia aevi Karolini*, i/2, ed. Albert Werminghoff (1906).

MGH (Scriptores) *Annales et chronica aevi Salici*, ed. Georg Heinrich Pertz (1844).

MGH (Scriptores) *Die Chronik von Montecassino (Chronica monasterii Casinensis)*, ed. Hartmut Hoffmann (1980).

MGH (Scriptores) *Liber Pontificalis*, i, ed. Theodor Mommsen (1898).

MGH (Scriptores) *Passiones vitaeque sanctorum aevi Merovingici et antiquiorum aliquot*, i, ed. Bruno Krusch (1896).

MGH (Scriptores) *Scriptores rerum Sangallensium: Annales, chronica et historiae aevi Carolini*, ed. Georg Heinrich Pertz (1829).

MIAZGA, TADEUSZ, *Die Melodien des einstimmigen Credo der römisch-katholischen lateinischen Kirche: Eine Untersuchung der Melodien in den handschriftlichen Überlieferungen mit besonderer Berücksichtigung der polnischen Handschriften* (Graz, 1976).

—— *Pontyfikały Polskie* (Graz, 1981).

MILLER, JOHN H., *Fundamentals of the Liturgy* (Notre Dame, Ind., 1959).

—— 'Liturgy', *NCE*.

MILVEDEN, INGMAR, 'Manuskript, Mönch und Mond. Ein Hauptteil des Cod. Upsal. C 23 in quellenkritischer Beleuchtung', *Svensk tidskrift för musikforskning*, 46 (1964), 9.

—— 'Neue Funde zur Brynolphus-Kritik', *Svensk tidskrift för musikforskning*, 54 (1972), 5–51.

Missale notatum Strigoniense ante 1341 in Posonio, ed. Janka Szendrei and Richard Rybarič (Musicalia Danubiana, 1; Budapest, 1982).

Missale plenarium Bib. Capit. Gnesnensis Ms. 149, ed. Krzysztof Biegański and Jerzy Woronczak (Antiquitates musicae in Polonia, 1112; Warsaw and Graz, 1970, 1972).

MISSET, E., and WEALE, W. H. J., *Analecta liturgica II: Thesaurus hymnologicus 1–2* (Bruges and Lille, 1888, 1892).

—— and AUBRY, PIERRE, *Les Proses d'Adam de Saint Victor* (Paris, 1900).

Mittelalterliche Bibliotheks-Kataloge Deutschlands und der Schweiz (Munich, 1918–) (see also Krämer 1989).

Mittelalterliche Bibliotheks-Kataloge Oesterreichs (Vienna, 1915–).

MMMA 1 (1956) = *Hymnen I. Die mittelalterlichen Hymnenmelodien des Abendlandes*, ed. Bruno Stäblein.

MMMA 2 (1970) = Bruno Stäblein (introduction) and Margareta Landwehr-Melnicki (edition), *Die Gesänge des altrömischen Graduale Vat. lat. 5319.*

MMMA 3 (1970) = *Introitus-Tropen I. Das Repertoire der südfranzösischen Tropare des 10. und 11. Jahrhunderts*, ed. Günther Weiß.

MMMA 7 (1968) = *Alleluia-Melodien I, bis 1100*, ed. Karlheinz Schlager.

MMMA 8 (1987) = *Alleluia-Melodien II, ab 1100*, ed. Karlheinz Schlager.

MMS 1 (Mâcon, 1952) = *Le Prosaire de la Sainte-Chapelle* [part of Bari, Biblioteca Capitolare, 1].

MMS 2 (Mâcon, 1954) = *Les Manuscrits musicaux de Jumièges.*

MMS 3 (Touen, 1960) = *Le Prosaire d'Aix-la-Chapelle* [part of Aachen, Bischöfliche Diözesanbibliothek, 13 (XII)].

MMS 3 (Rouen, 1960) = *Le Tropaire-prosaire de Dublin* [part of Cambridge, University Library, Add. 710].

MMS 5 (Paris, 1981) = *Le Graduel de St. Denis* [Paris, Bibliothèque Mazarine, 384].

MOBERG, CARL ALLAN, *Über die schwedischen Sequenzen: Eine musikgeschichtliche Studie*, 2 vols. (VGA 13; Uppsala, 1927).

—— *Die liturgischen Hymnen in Schweden* (Uppsala, 1947).

MOCQUEREAU, ANDRÉ, 'Origine et classement de differentes écritures neumatiques: 1. Notation oratoire ou chironomique, 2. Notation musicale ou diastématique', PalMus 1, 96–160.

—— 'Neumes-accents liquescents ou sémi-vocaux', PalMus 2, 37–86.

—— *Le Nombre musical grégorien ou rythmique grégorienne*, 2 vols. (Rome, 1908–1927).

—— 'La Pensée pontificale et la restauration grégorienne', *RG* 5 (1920), 181–9; 6 (1921), 9–18, 46–53.

MODERINI, AVE, *La notazione neumatica di Nonantola*, 2 vols. (Cremona, 1970).

MOELLER, E., *Corpus benedictionum pontificalium*, 4 vols. (Corpus Christianorum, 162; Turnhout, 197–19).

MOHLBERG, L. CUNIBERT (ed.), *Missale Gothicum* (Rerum ecclesiasticarum documenta, Series maior, Fontes, 5; Rome, 1961).

—— EIZENHÖFER, L., and SIFFRIN, P. (eds.), *Sacramentarium Veronense* (Rerum ecclesiasticarum documenta, Series maior, Fontes, 1; Rome, 1956).

—— —— —— (eds.), *Missale Gallicanum vetus* (Rerum ecclesiasticarum documenta, Series maior, Fontes, 3; Rome, 1958).

—— —— —— (eds.), *Liber sacramentorum Romanae aecclesiae ordinis anni circuli (Sacramentarium Gelasianum)* (Rerum ecclesiasticarum documenta, Series maior, Fontes, 4; Rome, 1960).

MOHLBERG Festschrift 1948 = *Miscellanea liturgica in honorem L. Cunibert Mohlberg*, 2 vols. (Bibliotheca ephemerides liturgicae, 23; Rome, 1948).

MOLITOR, RAPHAEL, *Die nach-Tridentinische Choralreform zu Rom*, 2 vols. (Leipzig, 1901–2).

—— *Deutsche Choral-Wiegendrucke: Ein Beitrag zur Geschichte des Chorals und des Notendruckes in Deutschland* (Regensburg, 1904).

MÖLLER, HARTMUT, 'Research on the Antiphoner—Problems and Perspectives', *JPMMS* 10 (1987), 1–14.

—— 'Zur Reichenauer Offiziumstradition der Jahrtausendwende', *SMH* 29 (1987), 35–61.

Möller, Hartmut, 'Deutsche Neumen—St. Galler Neumen. Zur Einordnung der Echternacher Neumenschrift', *SMH* 30 (1988), 415–30.

—— (ed.), *Das Quedlinburger Antiphonar (Berlin, Staatsbibliothek Preußischer Kulturbesitz Mus. ms. 40047)*, 3 vols. (Tutzing, 1990).

—— 'Die Prosula "Psalle modulamina" (Mü 9543) und ihre musikhistorische Bedeutung', in Leonardi and Menesto 1990, 279–96.

Mone, F. J., *Lateinische und griechische Messen aus dem zweiten bis sechsten Jahrhundert* (Frankfurt, 1850; PL 138, 862–82).

Moneta Caglio, Ernesto, 'I responsori "cum infantibus" nella liturgia ambrosiana', in Castiglioni Festschrift 1957, 481–574.

Morawski, Jerzy, *Polska liryka muzycna w średniowieczu: Repertuar sekwencyjny cystersów (XIII–XVI w.)* [résumé: 'Polish musical lyric in the Middle Ages: The Cistercians' sequence repertory (13th–16th c.)'] (Warsaw, 1973).

—— 'Recherches sur les variantes régionales dans le chant grégorien', *SMH* 30 (1988), 403–14.

Morin, Germain, 'Fragments inédits et jusqu'à présent uniques d'Antiphonaire Gallican', *RB* 22 (1905), 329–56.

—— 'Les Plus Anciens Comes ou lectionnaire de l'Église romaine', *RB* 27 (1910), 41–74.

—— 'Liturgie et basiliques de Rome au milieu du viiᵉ siècle', *RB* 28 (1911), 296–330.

—— 'Le Plus Ancien Monument qui existe de la liturgie gallicane : Le lectionnaire palimpseste de Wolfenbüttel', *EL* 51 (1937), 3–12.

—— (ed.), *Sancti Caesarii Arelatensis opera omnia* (Maredsous, 1942).

Mühlmann, Wilhelm, *Die Alia Musica (Gerbert, Scriptores 1): Quellenfrage, Umfang, Inhalt und Stammbaum* (Leipzig, 1914).

Müller, Herman, *Die Musik Wilhelms von Hirschau. Widerherstellung, Übersetzung und Erklärung seines musik-theoretischen Werkes* (Frankfurt, 1883).

—— *Hucbalds echte und unechte Schriften über Musik* (Leipzig, 1884).

Muller, H. F., 'Pre-history of the Medieval Drama: The Antecedents of the Tropes and the Conditions of their Appearance', *Zeitschrift für romanische Philologie*, 44 (1924–5), 544–75.

Müller-Blattau Festschrift 1966 = *Zum 70. Geburtstag von Joseph Müller-Blattau* (Saarbrücker Studien zur Musikwissenschaft, 1; Kassel, 1966).

Murphy, Joseph Michael, 'The Communions of the Old Roman Chant' (Diss., University of Pennsylvania, 1977; UMI 78–06625).

Murray, Gregory, *Gregorian Chant according to the Manuscripts* (London, 1963).

La Musique dans la liturgie (La Maison-Dieu, 108; 1971).

Nejedlý, Zdeněk, *Zpěv předhusitský/Geschichte des vorhussitischen Gesanges in Böhmen* (Prague, 1904; 2nd edn., 1954) [vol. i of *Djiny husitského zpěvu*, 3 vols. (Prague, 1904–13; 2nd edn., 1954–6)].

Netzer, H., *L'Introduction de la messe romaine en France sous les Carolingiens* (Paris, 1910).

Neufville, J. *see* Vogüé, Adalbert de.

Nicholson, Edward Williams Byron, *Early Bodleian Music 3: Introduction to the Study of Some of the Oldest Latin Manuscripts in the Bodleian Library Oxford* (London, 1913).

Nicolau, M. G., *L'Origine du 'cursus' rythmique et les débuts de l'accent d'intensité en latin* (Paris, 1930).

NIEMÖLLER, KLAUS WOLFGANG, *Die Musica gregoriana des Nicolaus Wollick: Opus aureum, Köln 1501, pars I/II* (Beiträge zur rheinischen Musikgeschichte, 11; Cologne, 1955).

NORBERG, DAG, *Introduction à l'étude de la versification latine médiévale* (Studia Latina Stockholmiensia, 5; 1958).

NORDEN, EDUARD, *Die antike Kunstprosa* (4th edn., Leipzig, 1923).

NORTIER, GENEVIÈVE, 'Les Bibliothèques médiévales des abbayes bénédictines de Normandie', *Revue Mabillon*, 187 (1957), 1–34, 57–83, 135–71, 214–44; 188 (1958), 1–19, 99–127, 165–75, 249–57; new edition (Paris, 1971).

NORTON, MICHAEL LEE, 'The Type II Visitatio Sepulchri: A Repertorial Study' (Diss., Ohio State University, 1983; UMI 83–11783).

—— 'Of "Stages" and "Types" in Visitatione Sepulchri', *Comparative Drama*, 21 (1987), 34–61, 127–44.

NOWACKI, EDWARD, 'The Syntactical Analysis of Plainchant', in Congress Berkeley 1977, 191–201.

—— 'Studies on the Office Antiphons of the Old Roman Manuscripts' (Diss., Brandeis University, 1980; UMI 80–24546).

—— 'The Gregorian Office Antiphons and the Comparative Method', *Journal of Musicology*, 4 (1985), 243–75.

—— 'Text Declamation as a Determinant of Melodic Form in the Old Roman Eight-Mode Tracts', *EMH* 6 (1986), 193–225.

—— 'The Performance of Office Antiphons in Twelfth-Century Rome', in Cantus Planus 1990, 79–92.

ODELMAN, EVA, 'Comment a-t-on appelé les tropes? Observations sur les rubriques des tropes des Xᵉ et XIᵉ siècles', *Cahiers de civilisation médiévale*, 18 (1975), 15–36.

—— *see also* CT.

OESCH, HANS, *Guido von Arezzo: Biographisches und Theoretisches unter besonderer Berücksichtigung der sogennanten odonischen Traktate* (Publikationen der Schweizerischen Musikforschenden Gesellschaft, ser. 2, 4; Berne, 1954).

—— *Berno und Hermann von Reichenau als Musiktheoretiker* (Publikationen der Schweizerischen Musikforschenden Gesellschaft, ser. 2, 9; Berne, 1961).

Offertoriale triplex, ed. Rupert Fischer (Solesmes, 1985) [Ott 1935 with neumes of Laon 239 and Einsiedeln 121 or St Gall 339].

OMLIN, EPHREM, *Die Sankt Gallischen Tonarbuchstaben: Ein Beitrag zur Entwicklungsgeschichte der Offiziumsantiphonen in Bezug auf ihre Tonarten und Psalmkadenzen* (VGA 18; Regensburg, 1934).

ÖNNERFORS, A., 'Zur Offiziendichtung im schwedischen Mittelalter—mit einer Edition des Birger Gregersson zugeschriebenen "Officium S. Botuida"', *Mittellateinisches Jahrbuch*, 3 (1966), 55–93.

OREL, DOBROSLAV, *Kancionál Franusův z roku 1505: Příspěvek k vývoji notace a k dějinám zpěvu jednohlasého i vícehlasého doby Jagellovců v Čechách* (Prague, 1922).

—— HORNOF, VLADIMIR, and VOSYKA, VÁCLAV, *Český Kancionál* (Prague, 1921).

OSSING, H., *Untersuchungen zum Antiphonale Monasteriense (Alopecius Druck 1537): Ein Vergleich mit den Handschriften des Münsterlandes* (Kölner Beiträge zur Musikwissenschaft, 39; Regensburg, 1966).

OTT, KARL, *Offertoriale sive versus offertoriorum* (Tournai, 1935).

OTTOSEN, KNUD, *L'Antiphonaire latin au Moyen-Âge: Réorganisation des séries de répons de l'Avent classés par R.-J. Hesbert* (Rerum ecclesiasticarum documenta, extra seriem; Rome, 1986).

OTTÓSSON, ROBERT A., *Sancti Thorlaci Episcopi officia rhythmica et proprium missae in AM 241a folio* (Bibliotheca Arnamagnæana Supplementum, 3; Copenhagen, 1959).

Oxford Classical Dictionary, ed. Nicholas G. Hammond (2nd edn., Oxford, 1970).

Oxford Dictionary of the Christian Church, ed. Frank L. Cross and Elizabeth A. Livingstone (2nd edn., Oxford, 1974).

PAGE, CHRISTOPHER, 'The Earliest English Keyboard', *Early Music*, 7 (1979), 309–14.

PALISCA, CLAUDE V., arts. 'Anonymous Theoretical Writings', 'Theory, Theorists', *NG*.

PalMus 1 (Solesmes, 1889) = *Le Codex 339 de la Bibliothèque de Saint-Gall (Xe siècle): Antiphonale missarum sancti Gregorii.*

PalMus 2 (Solesmes, 1891) = *Le Répons-graduel Justus ut palma, réproduit en fac-similé d'après plus de deux cents antiphonaires manuscrits du IXe au XVIIe siècle.*

PalMus 3 (Solesmes, 1892) = *Le Répons-graduel Justus ut palma: Deuxième partie.*

PalMus 4 (Solesmes, 1894) = *Le Codex 121 de la Bibliothèque d'Einsiedeln (IXe–XIe siècle): Antiphonale missarum sancti Gregorii.*

PalMus 5 (Solesmes, 1896) = *Antiphonarium Ambrosianum du Musée Britannique (XIIe siècle), Codex Additional 34209.*

PalMus 6 (Solesmes, 1900) = *Antiphonarium Ambrosianum du Musée Britannique (XIIe siècle), Codex Additional 34209: Transcription.*

PalMus 7 (Solesmes, 1901) = *Antiphonarium tonale missarum, XIe siècle: Codex H. 159 de la Bibliothèque de l'École de Médecine de Montpellier.*

PalMus 8 (Solesmes, 1901–5) = *Antiphonarium tonale missarum, XIe siècle: Codex H. 159 de la Bibliothèque de l'École de Médecine de Montpellier. Phototypies.*

PalMus 9 (Solesmes, 1906) = *Antiphonaire monastique, XIIe siècle: Codex 601 de la Bibliothèque Capitulaire de Lucques.*

PalMus 10 (Solesmes, 1909) = *Antiphonale missarum sancti Gregorii, IXe–Xe siècle: Codex 239 de la Bibliothèque de Laon.*

PalMus 11 (Solesmes, 1912) = *Antiphonale missarum sancti Gregorii, Xe siècle: Codex 47 de la Bibliothèque de Chartres.*

PalMus 12 (Solesmes, 1922) = *Antiphonaire monastique, XIIIe siècle: Codex F. 160 de la Bibliothèque de la Cathédrale de Worcester.*

PalMus 13 (Solesmes, 1925) = *Le Codex 903 de la Bibliothèque Nationale de Paris (XIe siècle): Graduel de Saint-Yrieix.*

PalMus 14 (Solesmes, 1931) = *Le Codex 10673 de la Bibliothèque Vaticane fonds latin (XIe siècle): Graduel Bénéventain.*

PalMus 15 (Solesmes, 193757) = *Le Codex VI. 34 de la Bibliothèque Capitulaire de Bénévent (XIe–XIIe siècle): Graduel de Bénévent avec prosaire et tropaire.*

PalMus 16 (Solesmes, 1955) = *Le Manuscrit du Mont-Renaud, Xe siècle: Graduel et antiphonaire de Noyon.*

PalMus 17 (Solesmes, 1958) = *Fragments des manuscrits de Chartres, réproduction phototypique: Présentation par le Chanoine Yves Delaporte.*

PalMus 18 (Berne, 1969) = *Le Codex 123 de la Bibliothèque Angelica de Rome (XIe siècle): Graduel et tropaire de Bologne.*

PalMus 19 (Berne, 1974) = *Le Manuscrit 807, Universitätsbibliothek Graz (XIIe siècle): Graduel de Klosterneuburg.*

PalMus 20 (Berne, 1983) = *Le Manuscrit VI-33, Archivio Arcivescovile Benevento: Missel de Bénévent (début du XIe siècle)*.

PalMus II/1 (Solesmes, 1900) = *Antiphonaire de l'office monastique transcrit par Hartker: MSS. Saint-Gall 390–391 (980–1011)*.

PalMus II/2 (Solesmes, 1924) = *Cantatorium, IXe siècle: N° 359 de la Bibliothèque de Saint-Gall*.

PANNAIN, GUIDO, 'Liber musicae. Un teorico anonimo del XIV secolo', *Rivista musicale italiana*, 27 (1920), 407–40.

PAREDI, A., 'Milanese Rite', *NCE*.

PARVIO, M., *Missale Aboense secundum ordinem Fratrum Praedicatorum* (Helsinki, 1971).

PASCHER, JOSEPH, *Das liturgische Jahr* (Munich, 1963).

PATIER, DOMINIQUE, 'Un office rythmique tchèque du xivème siècle: Étude comparative avec quelques offices hongrois', *SMH* 12 (1970), 41–129.

—— 'L'Office rythmique de sainte Ludmila', *EG* 21 (1986), 49–96.

PEREGO, CAMILLO, *La regola del canto fermo ambrosiano* (Milan, 1622).

PESCE, DOLORES, 'B-Flat: Transposition or Transformation?', *Journal of Musicology*, 4 (1985–6), 330–49.

—— *The Affinities and Medieval Transposition* (Bloomington, Ind., 1987).

PFAFF, RICHARD W., *New Liturgical Feasts in Later Medieval England* (Oxford, 1970).

—— *Medieval Latin Liturgy: A Select Bibliography* (Toronto Medieval Bibliographies, 9; Toronto, 1982).

PHILLIPS, NANCY, 'Musica and Scolica enchiriadis: Its Literary, Theoretical, and Musical Sources' (Diss., New York University, 1984; UMI 85–05525).

—— 'The Dasia Notation and its Manuscript Tradition', in Huglo 1987, 157–73.

—— and HUGLO, MICHEL, 'The Versus Rex caeli—Another Look at the So-called Archaic Sequence', *JPMMS* 5 (1982), 36–43.

—— —— 'Le "De musica" de saint Augustin et l'organisation de la durée musicale du ixe au xiie siècles', *Recherches augustiniennes*, 20 (1985), 117–31.

PICONE, CARMELO, 'Il "salicus" con lettere espressive nel codice di Laon 239', *EG* 16 (1977), 7–143.

PIKULIK, JERZY (ed.), *État des recherches sur la musique religieuse dans la culture polonaise* (Warsaw, 1973).

—— 'Les Offices polonais de saint Adalbert', in Pikulik 1973, 306–72.

—— 'Sekwencje Polskie/Polish sequences', *Musica medii aevi*, 4 (1973), 7–128; 5 (1976), 6–194.

—— *Indeks sekwencji w polskich rękopisach muzycznych* [résumé: 'Sequence index in the Polish musical manuscripts']. *Sekwencje zespołu rękopisów tarnowskich* [résumé: 'Sequences from the Tarnow manuscript set'] (Warsaw, 1974).

—— 'Indeks śpiewów ordinarium missae w graduałach polskich do 1600 R.' [résumé: 'Index des chants polonais de l'ordinaire de la messe au moyen âge'], *Muzyka religijna w Polsce: Materiały i studia*, 2 (1978), 139–272.

—— *Śpiewy alleluia o najświętszej Maryi Pannie w polskich graduałach przedtrydenckich* [résumé: 'Les chants alleluia en l'honneur de la Vierge Marie dans les graduels polonais antétridentins'], *Muzyka religijna w Polsce: Materiały i studia*, 6 (1984) [complete volume].

—— 'Les Tropes du Kyrie et du Sanctus dans les graduels polonais médiévaux', in Leonardi and Menesto 1990, 325–35.

PINEAU, C., *Le Plain-chant musical en France au XVIIe siècle* (n.p., 1955).

PINELL, J. M., 'Los textos de la antigua liturgia hispánica: Fuentes para su estudio', in Rivera Recio 1965, 109–64.

PITMAN, GROVER ALLEN, 'The Lenten Offertories of the Aquitanian Manuscripts' (Diss., Catholic University of America, 1973; UMI 73–12885).

PLANCHART, ALEJANDRO ENRIQUE, *The Repertory of Tropes at Winchester*, 2 vols. (Princeton, 1977).

—— 'The Transmission of Medieval Chant', in Iain Fenlon (ed.), *Music in Medieval and Early Modern Europe: Patronage, Sources and Texts* (Cambridge, 1981), 347–63.

—— 'About Tropes', *Schweizer Jahrbuch für Musikwissenschaft*, NS, 2 (1982), 125–35.

—— 'Italian Tropes', *Mosaic*, 18/4 (1985), 11–31.

—— 'On the Nature of Transmission and Change in Trope Repertories', *JAMS* 41 (1988), 215–49.

—— 'The Interaction between Montecassino and Benevento', in Leonardi and Menesto 1990, 385–407.

—— and FULLER, SARAH, 'St Martial', *NG*.

PLANER, JOHN HARRIS, 'The Ecclesiastical Modes in the Late Eighth Century' (Diss., University of Michigan, 1970; UMI 71–15270).

PLOCEK, VÁCLAV, *Catalogus codicum notis musicis instructorum qui in Bibliotheca publica rei publicae Bohemicae socialisticae in Bibliotheca universitatis Pragensis servantur*, 2 vols. (Prague, 1973).

POCKNEE, CYRIL E., *The French Diocesan Hymns and their Melodies* (London, 1954).

PODLAHA, ADOLF PATERA-ANTONIN, *Handschriften-Katalog der Bibliothek des Prager Metropolitankapitels* (Prague, 1910).

POLHEIM, KARL, *Die lateinische Reimprosa* (Berlin, 1925; 2nd edn., 1963).

PONCHELET, RENÉ, 'Le Salicus en composition dans le codex St. Gall 359', *EG* 14 (1973), 7–125.

PONTE, JOSEPH P., 'Aureliani Reomensis Musica disciplina: A Revised Text, Translation and Commentary' (Diss., Brandeis University, 1961; UMI 62–01207).

—— *Aurelian of Réôme (ca. 843): The Discipline of Music (Musica disciplina)* (Colorado College Music Press, Translations, 3; 1968).

PORAS, GRÉGOIRE, 'Recherches sur les conditions du chant liturgique pendant le haut moyen âge', *EG* 21 (1986), 23–5.

PORTER, W. S., *The Gallican Rite* (London, 1958).

POTHIER, JOSEPH, *Les Mélodies grégoriennes d'après la tradition* (Tournai, 1880).

POTIRON, HENRI, *La Théorie harmonique de trois groupes modaux et l'accord final des 3e et 4e modes* (Monographies grégoriennes, 6; Tournai, 1925).

—— *La Modalité grégorienne* (Monographies grégoriennes, 9; Tournai, 1928).

—— *L'Origine des modes grégoriennes* (Tournai, 1948).

—— *La Notation grecque et Boèce* (Tournai, 1951).

—— *Boèce, théoricien de la musique grecque* (Paris, 1954).

—— 'La Modalité de la Messe I', *RG* 33 (1954), 45–8.

—— 'La Notation grecque dans l'Institution harmonique d'Hucbald', *EG* 2 (1957), 37–50.

—— 'La Terminologie grecque des modes', *EG* 7 (1967), 57–61.

—— 'Valeur et traduction de la notation grecque', *EG* 15 (1975), 193–9.

POWERS, HAROLD S., 'Mode', *NG*.

PRADO, GERMAN, *Manual de liturgia hispano-visigótica o mozárabe* (Madrid, 1927).

PRADO, GERMAN, *Historia del rito mozárabe y toledano* (Santo Domingo de Silos, 1928).

—— *Cantus lamentationum pro ultimo triduo hebdomadae majoris juxta hispanos codices* (Paris, 1934).

PRIM, J., '"Chant sur le livre" in French Churches in the 18th Century', *JAMS* 14 (1961), 37–49.

Processionale ad usum Sarum (London: Richard Pynson, 1502) (fac., Clarabricken, 1980).

Processionale monasticum ad usum congregationis gallicae ordinis Sancti Benedicti (Solesmes, 1893).

PROCTER, FRANCIS, and WORDSWORTH, CHRISTOPHER (eds.), *Breviarium ad usum insignis ecclesiae Sarum*, 3 vols. (Cambridge, 1882, 1879, 1886).

Psalterium monasticum (Solesmes, 1981).

QUASTEN Festschrift 1970 = P. Granfield and Josef Jungmann (eds.), *Kyriakon: Festschrift Johannes Quasten*, 2 vols. (Münster, 1970).

QUENTIN, HENRI, *Les Martyrologes historiques du moyen âge* (Paris, 1908).

RAASTED, JØRGEN, *Intonation Formulas and Modal Signatures in Byzantine Musical Manuscripts* (Monumenta musicae Byzantinae, Subsidia, 7; Copenhagen, 1966).

RABY, F. J. E., *A History of Christian-Latin Poetry from the Beginnings to the Close of the Middle Ages* (Oxford, 1927; 2nd edn., 1953).

RADÓ, POLYCARP, *Libri liturgici manuscripti bibliothecarum Hungariae et limitropharum regionum*'(Budapest, 1973).

RAJECKY, BENJAMIN, and RADÓ, POLYCARP, *Hymnen und Sequenzen* (Melodiarum Hungariae medii aevi, 1; Budapest, 1956; rev. edn. 1982 with new supplementary vol.).

RAND, EDWARD KENNARD, *Founders of the Middle Ages* (Cambridge, Mass., 1929).

RANDEL, DON M., *The Responsorial Psalm Tones for the Mozarabic Office* (Princeton, 1969).

—— 'Responsorial Psalmody in the Mozarabic Rite', *EG* 10 (1969), 87–116.

—— *An Index to the Chant of the Mozarabic Rite* (Princeton, 1973).

—— 'Antiphonal Psalmody in the Mozarabic Rite', in Congress Berkeley 1977, 414–22.

—— 'El antiguo rito hispánico y la salmodía primitiva en occidente', *Revista de musicología*, 8 (1985), 229–38.

—— 'Mozarabic Rite, Music of the', *NG*.

RANKE, E., *Codex Fuldensis NT latine, interprete Hieronymo* (Marburg and Leipzig, 1868).

RANKIN, SUSAN K., 'The Music of the Medieval Liturgical Drama in France and in England' (Diss., University of Cambridge, 1981).

—— 'The Mary Magdalene Scene in the *Visitatio Sepulchri* Ceremonies', *EMH* 1 (1981), 227–55.

—— 'A New English Source of the *Visitatio Sepulchri*', *JPMMS* 4 (1981), 1–11.

—— 'From Memory to Record: Musical Notations in Manuscripts from Exeter', *Anglo-Saxon England*, 13 (1984), 97–112.

—— 'The Liturgical Background of the Old English Advent Lyrics: A Reappraisal', in Clemoes Festschrift 1985, 317–40.

—— 'Musical and Ritual Aspects of *Quem queritis*', in Silagi 1985, 181–92.

—— 'Neumatic Notations in Anglo-Saxon England', in Huglo 1987, 129–44.

RASMUSSEN, NIELS KROGH, *Les Pontificaux du haut moyen-âge* ([Diss., Paris]; Aarhus, 1977).

RATCLIFF, E. C., *Expositio antiquae liturgiae Gallicanae* (Henry Bradshaw Society, 98; London, 1971).

RECKOW, FRITZ, *Der Musiktraktat des Anonymus 4: Edition und Interpretation der organum purum-Lehre* (Beihefte zum Archiv für Musikwissenschaft, 4–5; Wiesbaden, 1967).

REESE Festschrift 1967 = Jan LaRue (ed.), *Aspects of Medieval and Renaissance Music, a Birthday Offering to Gustave Reese* (New York, 1967).

REHLE, SIEGHILD, 'Missale Beneventanum (Codex VI 33 des Erzbischöflichen Archivs von Benevent)', *Sacris erudiri*, 21 (1972–3), 323–405.

—— and GAMBER, KLAUS, *Sacramentarium Arnonis: Die Fragmente des Salzburger Exemplars* (Textus patristici et liturgici, 8 and 10; Regensburg, 1970, 1973).

REICHERT, GEORG, 'Strukturprobleme der älteren Sequenz', *Deutsche Vierteljahrsschrift für Literaturwissenschaft und Geistesgeschichte*, 23 (1949), 227–51.

REIER, ELLEN JANE, 'The Introit Trope Repertory at Nevers: MSS Paris B. N. lat. 9449 and Paris B. N. n. a. lat. 1235' (Diss., University of California at Berkeley, 1981; UMI 82–12077).

REIMER, ERICH (ed.), *Johannes de Garlandia: De mensurabili musica. Kritische Edition mit Kommentar und Interpretation der Notationslehre* (Beihefte zum Archiv für Musikwissenschaft, 10–11; Wiesbaden, 1972).

—— 'Musicus und Cantor. Zur Sozialgeschichte eines musikalischen Lehrstücks', *AfMw* 35 (1978), 1–32.

RENAUDIN, ANDRÉ, 'Deux antiphonaires de Saint-Maur-des-Fossés: B. N. lat. 12584 et 12044', *EG* 13 (1972), 53–119.

RENOUX, ATHANASE, 'Un manuscrit du Lectionnaire arménien de Jérusalem (cod. Jerus. arm. 121)', *Le Muséon*, 74 (1961), 361–85; 75 (1962), 385–97.

—— *Le Codex arménien Jérusalem 121* (Patrologia orientalis, 35/1 and 36/2; Turnhout, 1969, 1971).

RIBAY, BERNARD, 'Les Graduels en IIA', *EG* 22 (1988), 43–107.

RIEMANN, HUGO, *Notenschrift und Notendruck* (Leipzig, 1896).

—— *Geschichte der Musiktheorie im IX.–XIX. Jahrhundert* (2nd edn., Berlin, 1920); trans. Raymond H. Haggh: *History of Music Theory, Books I and II* (Lincoln, Nebr., 1962).

RIGHETTI, MARIO, *Manuale di storia liturgica*, 4 vols. (Milan, 1945–53; 3rd edn., 1964–9).

RISM B/III = *The Theory of Music from the Carolingian Era up to 1400: Descriptive Catalogue of Manuscripts*.

RISM B/III/1 (1961) = Joseph Smits van Waesberghe, Pieter Fischer, and Christian Maas, *Austria, Belgium, Switzerland, Denmark, France, Luxembourg, Netherlands*.

RISM B/III/2 (1968) = Pieter Fischer, *Italy*.

RISM B/III/3 (1987) = Michel Huglo and Christian Meyer, *Federal Republic of Germany*.

RISM B/V/1 (1964) = Heinrich Husmann, *Tropen- und Sequenzenhandschriften*.

RIVERA RECIO, J. F. (ed.), *Estudios sobre la liturgia mozárabe* (Toledo, 1965).

ROBERT, LÉON, 'Les Chants du célébrant', *RG* 41 (1963), 113–26.

ROBERT, MICHEL, 'Le Graduel du Mont Saint-Michel', *Millenaire monastique du Mont-Saint-Michel*, 1 (Paris, 1967), 379–82.

ROBERTSON, ANNE WALTERS, '*Benedicamus Domino*: The Unwritten Tradition', *JAMS* 41 (1988), 1–62.

—— *The Service Books of the Royal Abbey of Saint-Denis: Images of Ritual and Music in the Middle Ages* (Oxford, 1991).

ROEDERER, CHARLOTTE D., 'Can We Identify an Aquitanian Chant Style?', *JAMS* 27 (1974), 75–99.

Rojo, C., and Prado, G., *El canto mozárabe: Estudio histórico-crítico de su antigüedad y estado actual* (Barcelona, 1929).

Rönnau, Klaus, *Die Tropen zum Gloria in excelsis Deo, unter besonderer Berücksichtigung des Repertoires der St. Martial-Handschriften* (Wiesbaden, 1967).

—— 'Regnum tuum solidum', in Stäblein Festschrift 1967, 195–205.

Rousseau, O., *Histoire du mouvement liturgique: Esquisse historique depuis le début du XIXe siècle jusqu'au pontificat de Pie X* (Paris, 1945).

Ruini *see* CSM 25.

Russell, Carlton Thrasher, 'The Southern French Tonary in the Tenth and Eleventh Centuries' (Diss., Princeton University, 1966; UMI 66–07182).

Russell, Tilden A., 'A Poetic Key to a Pre-Guidonian Psalm and the *Echemata*', *JAMS* 34 (1981), 109–18.

Rutter, Philip, 'The Epiphany Trope Cycle in Paris, Bibliothèque Nationale, fonds latin 1240', in Leonardi and Menesto 1990, 313–24.

Sablayrolles, Maur, 'À la recherche des manuscrits grégoriens espagnoles: Iter Hispanicum', *Sammelbände der Internationalen Musik-Gesellschaft*, 13 (1911–12), 205–47, 401–32, 509–31.

Sachs, Klaus-Jürgen, 'Gerbertus cognomento musicus: Zur musikgeschichtlichen Stellung des Gerbert von Rheims', *AfMw* 29 (1972), 257–74.

Sackur, Ernst, *Die Cluniacenser in ihrer kirchlichen und allgemeingeschichtlichen Wirksamkeit bis zur Mitte des elften Jahrhunderts*, 2 vols. (Halle, 1892, 1894).

Sacred Music and Liturgy [De sacra musica et sacra liturgia]. *Instruction of the Sacred Congregation of Rites September 3rd, 1958*, trans. J. B. O'Connell (London, 1959).

Sallmann, Nicolas, 'Censorinus' "De die natali". Zwischen Rhetorik und Wissenschaft', *Hermes*, 111 (1983), 233–58.

Salmon, Pierre, *L'Office divin: Histoire de la formation du bréviaire* (Lex orandi, 27; Paris, 1959); trans.: *The Breviary through the Centuries* (Collegeville, Minn., 1962).

—— 'Un bréviaire-missel du xie siècle: Le manuscrit vatican lat. 7018', in *Mélanges Eugène Tisserant*, vii (Studi e testi, 237; Rome, 1964), 327–43.

—— *L'Office divin au moyen âge* (Lex orandi, 43; Paris, 1967).

—— *Les Manuscrits liturgiques latins de la Bibliothèque Vaticane* (Studi e testi, 251, 253, 260, 267, and 270; Rome, 1968–72).

Salvat, M., 'Un traité de musique du xiiie siècle: Le De musica de Barthélemi L'Anglais', in Buschinger and Crépin 1980, 345–60.

Sanders Festschrift 1990 = Peter M. Lefferts and Brian Seirup (eds.), *Studies in Medieval Music: Festschrift for Ernest H. Sanders* = *Current Musicology*, 45–7 (1990).

Sandon, Nick, *The Use of Salisbury: The Ordinary of the Mass* (Lustleigh, 1984).

—— *The Use of Salisbury, 2: The Proper of the Mass in Advent* (Lustleigh, 1986).

—— *The Use of Salisbury, 3: The Proper of the Mass from Septuagesima to Palm Sunday* (Lustleigh, 1991).

Scharnagl, August, 'Offertorium', *MGG*.

Scheiwiler, A., *Das Kloster St. Gallen: Die Geschichte eines Kulturzentrums* (Einsiedeln and Cologne, 1938).

Schildbach, Martin, *Das einstimmige Agnus Dei und seine handschriftliche Überlieferung vom 10. bis zum 16. Jahrhundert* (Diss., Erlangen-Nürnberg University, 1967).

Schlager, Karlheinz, *Thematischer Katalog der ältesten Alleluia-Melodien aus Handschriften des 10. und 11. Jahrhunderts, ausgenommen das ambrosianische, alt-spanische und alt-römische Repertoire* (Erlanger Arbeiten zur Musikwissenschaft, 2; Munich, 1965).

—— 'Ein beneventanisches Alleluia und seine Prosula', in Stäblein Festschrift 1967, 217–25.

—— 'Anmerkungen zu den zweiten Alleluja-Versen', *AfMw* 24 (1967), 199–219.

—— 'The Microfilm Archive of Medieval Music Manuscripts at the Institut für Musikwissenschaft of Erlangen-Nürnberg University', *JPMMS* 2 (1970), 61–4.

—— 'Choraltextierung und Melodieverständnis im frühen und späten Mittelalter', in Cardine Festschrift 1980, 314–37.

—— 'Trinitas, Deitas, Unitas—A Trope for the Sanctus of Mass', *JPMMS* 6 (1983), 8–14.

—— 'Tropen als Forschungsbereich der Musikwissenschaft—Vom Lebenslauf eines Melismas', in Iversen 1983, 17–28.

—— 'Te Deum', *MGG*.

—— 'Alleluia. I', *NG*.

—— *see also Antiphonale Pataviense* and MMMA 7–8.

SCHMID, HANS (ed.), *Musica et Scolica Enchiriadis una cum aliquibus tractatulis adiunctis* (Bayerische Akademie der Wissenschaften, Veröffentlichungen der Musikhistorischen Kommission, 3; Munich, 1981).

SCHMID, TONI, 'Smärre liturgiska bidrag VIII. Om Sankt Swithunusmässen i Sverige', *Nordisk Tidskrift för Bok- och Biblioteksväsen*, 31 (Uppsala, 1944), 25–34.

—— *see also Graduale Arosiense.*

SCHMIDT, HANS, *Untersuchungen zu den Tractus des zweiten Tons aus dem Codex St. Gallen 359* (Diss., Bonn, 1955).

—— 'Die Tractus des zweiten Tons in Gregorianischer und stadtrömischer Überlieferung', in Schmidt-Görg Festschrift 1957, 283–302.

—— 'Untersuchungen zu den Tractus des zweiten Tones', *KmJb* 42 (1958), 1–25.

—— 'Gregorianik—Legende oder Wahrheit?', in Hüschen Festschrift 1980, 400–11.

SCHMIDT, HERMANN, *Hebdomada sancta*, 2 vols. (Rome, 1956, 1957).

SCHMIDT-GÖRG Festschrift 1957 = Dagmar Weise (ed.), *Festschrift Joseph Schmidt-Görg zum 60. Geburtstag* (Bonn, 1957).

SCHMITZ, ARNOLD, 'Ein schlesisches Cantional aus dem 15. Jahrhundert', *Archiv für Musikforschung*, 1 (1936), 385–423.

SCHNEIDER Festschrift 1955 = Walther Vetter (ed.), *Festschrift Max Schneider zum achzigsten Geburtstage* (Leipzig, 1955).

SCHOENBAUM, CAMILLO, 'Hymnologische Forschung in der Tschechoslowakei', *Jahrbuch für Liturgik und Hymnologie*, 5 (1960), 157–65.

SCHRADE, LEO, 'Die Darstellung der Töne an den Kapitellen der Abteikirche zu Cluny (Ein Beitrag zum Symbolismus in mittelalterlicher Kunst)', *Deutsche Vierteljahrsschrift für Literatur- und Geisteswissenschaft*, 7 (1929), 229–66 = Schrade 1967, 113–51.

—— *De scientia musicae studia atque orationes*, ed. Ernst Lichtenhahn (Berne, 1967).

SCHRADE Gedenkschrift 1973 = Wulf Arlt, Ernst Lichtenhahn, and Hans Oesch (eds.), *Gattungen der Musik in Einzeldarstellungen: Gedenkschrift für Leo Schrade* (Berne, 1973).

SCHUBIGER, ANSELM, *Die Sängerschule St. Gallens vom 8. bis 12. Jahrhundert: Ein Beitrag zur Gesangsgeschichte des Mittelalters* (Einsiedeln, 1858).

SCHUELLER, HERBERT M., *The Idea of Music: An Introduction to Musical Aesthetics in Antiquity and the Middle Ages* (Kalamazoo, Mich., 1988).

SCHULER, ERNST AUGUST, *Die Musik der Osterfeiern, Osterspiele und Passionen des Mittelalters* (Kassel, 1951).

SCHULER, MANFRED, 'Die Musik an den Höfen der Karolinger', *AfMw* 27 (1970), 23–40.

SCHWYZER, EDUARD, *Griechische Grammatik* (Handbuch der Altertumswissenschaft, II/1/1; Munich, 1939).

SEAY, ALBERT, 'An Anonymous Treatise from St. Martial', *Annales musicologiques*, 5 (1957), 7–42.

SEEBASS, TILMAN, 'Musik und Musikinstrumente in der Tonarillustration: Organologische, stilistische und ikonographische Studie anhand der Handschrift Paris fonds latin 1118' (Diss., Basel University, 1970).

SERVATIUS, VIVECA, *Cantus Sororum: Musik- und liturgiegeschichtliche Studien zu den Antiphonen des birgittinischen Eigenrepertoires, nebst 91 Transkriptionen* (Uppsala, 1990).

SESINI, UGO, *La notazione comasca nel cod. Ambrosiano E. 68 supra* (Milan, 1932).

SEVESTRE, NICOLE, 'The Aquitanian Tropes of the Easter Introit—A Musical Analysis', *JPMMS* 3 (1980), 26–39.

SIDLER, HUBERT, 'Studien zu den alten Offertorien mit ihren Versen' (Diss., Fribourg, 1939).

SIGL, MAXIMILIAN, *Zur Geschichte des Ordinarium Missae in der deutschen Choralüberlieferung*, 2 vols. (Regensburg, 1911).

SILAGI, GABRIEL (ed.), *Liturgische Tropen: Referate zweier Colloquien des Corpus Troporum in München (1983) und Canterbury (1984)* (Münchener Beiträge zur Mediävistik und Renaissance-Forschung, 36; Munich, 1985).

SKEAT, W. W., *The Book of Lindisfarne: The Holy Gospels in Anglo-Saxon, Northumbrian, and Old Mercian Versions*, 4 vols. (Cambridge, 1871–87).

SMITH, JOHN A., 'The Ancient Synagogue, the Early Church and Singing', *Music and Letters*, 65 (1984), 1–16.

SMITS VAN WAESBERGHE, JOSEPH, *Muziekgeschiedenis der Middeleeuwen*, 2 vols. (Tilburg, 1938–42).

—— 'Some Music Treatises and Their Interrelation: A School of Liège c. 1050–1200?', *MD* 3 (1949), 25–32, 95–118.

—— 'The Musical Notation of Guido of Arezzo', *MD* 5 (1951), 15–53.

—— 'Guido of Arezzo and Musical Improvisation', *MD* 5 (1951), 55–64.

—— *Cymbala (Bells in the Middle Ages): Edition of Texts and Introduction* (Rome, 1951).

—— 'La Place exceptionelle de l'Ars Musica dans le développement des sciences au siècle des Carolingiens', *RG* 31 (1952), 81–104; trans. H. Schulze: 'Die besonderer Stellung der Ars Musica im Zeitalter der Karolinger', in Smits van Waesberghe 1976, 48–70.

—— *De musico-paedagogico et theoretico Guidone Aretino eiusque vita et moribus* (Florence, 1953).

—— 'Neues über die Schola Cantorum zu Rom', in Congress Vienna 1954, 111–19.

—— 'Musikalische Beziehungen zwischen Aachen, Köln, Lüttich und Maastricht vom 11. bis zum 13. Jahrhundert', in *Beiträge zur Musikgeschichte der Stadt Aachen* (Cologne, 1954), 5–13.

—— *A Textbook of Melody: A Course in Functional Melodic Analysis* (Rome, 1955).

—— *Expositiones in Micrologum Guidonis Aretini* (Amsterdam, 1957) [Liber argumentorum, Liber specierum, Metrologus, Commentarius in Micrologum Guidonis Aretini].

—— 'Les Origines de la notation alphabétique au moyen-âge', *Anuario musical*, 12 (1957), 3–14.

—— 'Over het onstaan van Sequens en Prosula en beider oorspronkelijke uitvoeringswijze', *Feestaflevering ter gelegenheid van de zestige verjaardag van Prof. Dr. K. Ph. Bernet*

Kempers = Orgaan K. N. T. V., Officeel Maandblad van de Koninklijke Nederlandsche Toonkunstenaars Vereeniging (Amsterdam, 1957), 41–57; shorter version: 'Zur ursprünglichen Vortragsweise der Prosulen, Sequenzen und Organa', in Congress Cologne 1958, 251–4.

—— 'L'État actuel des recherches scientifiques dans le domaine du chant grégorien', in *Actes du troisième congrès international de musique sacrée, Paris 1ᵉʳ–8ᵐᵉ juillet 1957* (Paris, 1959), 206–17.

—— 'Die Imitation der Sequenztechnik in den Hosanna-Prosulen', in Fellerer Festschrift 1962, 485–90.

—— '"De glorioso officio . . . dignitate apostolica . . ." (Amalar); zum Aufbau der Groß-Alleluia in den päpstlichen Osterverspern', in Wellesz Festschrift 1966, 4873 = Smits van Waesberghe 1976, 117–46.

—— *Musikerziehung* (Musikgeschichte in Bildern, 3, pt. 3; Leipzig, 1969).

—— 'Neue Kompositionen des Johannes von Metz (um 975), Hucbalds von St. Amand und Sigeberts von Gembloux?', in Husmann Festschrift 1970, 285–303.

—— 'Studien über das Lesen (pronunciare), das Zitieren und über die Herausgabe lateinischer musiktheoretischer Traktate', *AfMw* 28 (1971), 155–200; 29 (1972), 64–86.

—— 'Gedanken über den inneren Traditionsprozeß in der Geschichte der Musik des Mittelalters', in Kurt von Fischer Festschrift 1973, 7–30.

—— 'Wie Wortwahl und Terminologie bei Guido von Arezzo entstanden und überliefert wurden', *AfMw* 31 (1974), 73–86.

—— *Dia-pason, de omnibus: Ausgewählte Aufsätze von Josef Smits van Waesberghe: Festgabe zu seinem 75. Geburtstag* (Buren, 1976).

—— *see also* CSM 1, 2, and 4; DMA, RISM B/III/1.

Smits van Waesberghe Festschrift = Pieter Fischer (ed.), *Organicae voces: Festschrift J. Smits van Waesberghe* (Amsterdam, 1963).

Snow, Robert, 'The Old-Roman Chant', in Apel 1958, 484–505.

Söhner, L., *Die Geschichte der Begleitung des gregorianischen Chorals in Deutschland vornehmlich im 18. Jahrhundert* (Augsburg, 1931).

Sowa, Heinrich, 'Zur Handschrift Clm 9921', *AcM* 5 (1933), 60–5, 107–20.

—— *Quellen zur Transformation der Antiphonen: Tonar- und Rhythmus-Studien* (Kassel, 1935).

Sowulewska, Halina, 'Les Relations de proximité entre les graduels polonais des Prémonstrés et les manuscrits européens', *SMH* 27 (1985), 123–9.

Spanke, Hans, 'Das Moosburger Graduale', *Zeitschrift für romanische Philologie*, 50 (1930), 582–95.

—— 'St. Martial-Studien. Ein Beitrag zur frühromanischen Metrik', *Zeitschrift für französische Sprache und Literatur*, 54 (1930), 282–317, 385–422; 56 (1931), 450–78.

—— 'Rhythmen- und Sequenzstudien', *Studi medievali*, ns, 4 (1931), 286–320.

—— 'Aus der Vorgeschichte und Frühgeschichte der Sequenz', *Zeitschrift für deutsches Altertum und deutsche Literatur*, 71 (1934), 1–39.

—— *Beziehungen zwischen romanischer und mittellateinischer Lyrik mit besonderer Berücksichtigung der Metrik und Musik* (Abhandlungen der Gesellschaft der Wissenschaften zu Göttingen, Philologisch-historische Klasse, 3rd ser., 18; Berlin, 1936).

—— 'Sequenz und Lai', *Studi medievali*, ns, 11 (1938), 12–68.

—— 'Die Kompositionskunst der Sequenzen Adams von St. Victor', *Studi medievali*, ns, 14 (1951), 1–30.

SPUNAR, PAVEL, 'Das Troparium des Prager Dekans Vit (Prag Kapitelbibliothek, Cim 4)', *Scriptorium*, 9 (1957), 50–62.

SRAWLEY, J. H., *The Early History of the Liturgy* (Cambridge, 1913; 2nd edn., 1947).

STÄBLEIN, BRUNO, 'Zur Geschichte der choralen Pange lingua-Melodie', in Johner Festschrift 1950, 72–5.

—— 'Von der Sequenz zum Strophenlied. Eine neue Sequenzemelodie "archaischen" Stiles', *Mf* 7 (1954), 257–68, 511.

—— 'Zur Frühgeschichte der Sequenz', *AfMw* 18 (1961), 1–33.

—— 'Die Unterlegung von Texten unter Melismen. Tropus, Sequenz und andere Formen', in Congress New York 1961, 12–29.

—— 'Das sogenannte aquitanische "Alleluia Dies sanctificatus" und seine Sequenz', in Hans Albrecht Gedenkschrift 1962, 22–6.

—— 'Die Schwanenklage. Zum Problem Lai—Planctus—Sequenz', in Fellerer Festschrift 1962, 491–502.

—— 'Der Tropus "Dies sanctificatus" zum Alleluia "Dies sanctificatus"', *Studien zur Musikwissenschaft*, 25 (1962) (Festschrift für Erich Schenk), 504–15.

—— 'Notkeriana', *AfMw* 19–20 (1962–3), 84–99.

—— 'Zum Verständnis des "klassischen" Tropus', *AcM* 35 (1963), 84–95.

—— 'Modale Rhythmen im Saint-Martial-Repertoire?', in Blume Festschrift 1963, 340–62.

—— 'Zwei Textierungen des Alleluia Christus Resurgens in St. Emmeram Regensburg', in Smits van Waesberghe Festschrift 1963, 157–67.

—— 'Die Sequenzmelodie "Concordia" und ihr geschichtlicher Hintergrund', in Engel Festschrift 1964, 364–92.

—— 'Zur Musik des Ludus de Antichristo', in Müller-Blattau Festschrift 1966, 312–27.

—— 'Der "altrömische" Choral in Oberitalien und im deutschen Süden', *Mf* 19 (1966), 3–9.

—— '"Gregorius Presul", der Prolog zum römischen Antiphonale', in Vötterle Festschrift 1968, 537–61.

—— 'Thèses équalistes et mensuralistes', *Encyclopédie des musiques sacrés*, 2 (Paris, 1969), 80–98.

—— 'Zwei Melodien der altirischen Liturgie', in Fellerer Festschrift 1973, 590–7.

—— *Schriftbild der einstimmigen Musik* (Musikgeschichte in Bildern, 3, pt. 4; Leipzig, 1975).

—— 'Pater noster-Tropen', in Haberl Festschrift 1977, 247–78.

—— 'Einiges Neue zum Thema "archaische Sequenz"', in Festschrift Dadelsen 1978, 352–83.

—— arts. 'Agnus Dei', 'Alleluja', 'Antiphon', 'Antiphonar', 'Brevier', 'Cantatorium', 'Canticum', 'Choral', 'Communio', 'Credo', 'Deutschland. B. Mittelalter. I. Der römische Choral im Norden', 'Epistel, A. Katholisch', 'Evangelium, A. Katholisch', 'Exultet', 'Frühchristliche Musik', 'Gallikanische Liturgie', 'Gemeindegesang, A. Mittelalter', 'Gloria in excelsis Deo', 'Graduale (Gesang)', 'Gregor I.', 'Gregorianik', 'Hymnar', 'Hymnus, B. Der lateinischen Hymnus', 'Improperien', 'Introitus', 'Invitatorium', 'Kyriale', 'Kyrie', 'Litanei', 'Messe, A. Die lateinische Messe', 'Missale', 'Passion, A. Die einstimmige lateinische Passion', 'Pater noster', 'Präfation', 'Psalm, B', 'Saint-Martial', 'Sequenz (Gesang)', 'Tropus', 'Versus', *MGG*.

—— *see also* MMMA.

STÄBLEIN Festschrift 1967 = Martin Ruhnke (ed.), *Festschrift Bruno Stäblein zum 70. Geburtstag* (Kassel, 1967).

STAHL, WILLIAM HARRIS, *Martianus Capella and the Seven Liberal Arts* (New York, 1971).

Steer, Georg, '"Carmina Burana" in Südtirol: Zur Herkunft des clm 4660', *Zeitschrift für deutsches Altertum und deutsche Literatur*, 112 (1983), 1–37.

Steglich, Rudolf, *Die Quaestiones in musica; ein Choraltraktat des zentralen Mittelalters und ihr mutmaßlicher Verfasser Rudolf von St. Trond (1070–1138)* (Publikationen der Internationalen Musikgesellschaft, Beihefte II/10; Leipzig, 1911).

Stein, Franz A., *Das Moosburger Graduale* (Diss., Freiburg im Br., 1956).

Steinen, Wolfram von den, 'Die Anfänge der Sequenzendichtung', *Zeitschrift für schweizerische Kirchengeschichte*, 40 (1946), 190–212, 241–68; 41 (1947), 19–48, 122–62.

—— *Notker der Dichter und seine geistige Welt*, 2 vols. (Berne, 1948).

Steiner, Ruth, 'Some Questions about the Gregorian Offertories and their Verses', *JAMS* 19 (1966), 162–81.

—— 'The Prosulae of the MS Paris BN lat. 1118', *JAMS* 22 (1969), 367–93.

—— 'The Responsories and Prosa for St. Stephen's Day at Salisbury', *MQ* 56 (1970), 162–82.

—— 'Some Melismas for Office Responsories', *JAMS* 26 (1973), 108–31.

—— 'The Gregorian Chant Melismas of Christmas Matins', in *Essays in Honor of Charles Warren Fox* (Rochester, NY, 1979), 241–53.

—— 'The Canticle of the Three Children as a Chant of the Roman Mass', *Schweizer Jahrbuch für Musikwissenschaft*, NS, 2 (1982), 81–90.

—— 'Antiphons for the Benedicite at Lauds', *JPMMS* 7 (1984), 1–17.

—— 'The Music for a Cluny Office of St. Benedict', in Verdon and Dally 1984, 81–113.

—— 'Reconstructing the Repertory of Invitatory Tones and their Uses at Cluny in the Late 11th Century', in Huglo 1987, 175–82.

—— arts. 'Cantatorium', 'Compline', 'Cursus', 'Gregorian Chant', 'Hymn, II. Monophonic Latin', 'Introit', 'Invitatory', 'Lord's Prayer', 'Prosula', 'Psalter', 'Te Deum. 1.2.', 'Trope', *NG*.

—— and Levy, Kenneth, 'Liturgy and Liturgical Books', *NG*.

Stenzl, Jörg, *Repertorium der liturgischen Musikhandschriften der Diözesen Sitten, Lausanne und Genf*, 1 (VGA, NS, 1; Fribourg, 1972).

Stephan, Rudolf, 'Lied, Tropus und Tanz im Mittelalter', *Zeitschrift für deutsches Altertum und deutsche Literatur*, 87 (1956), 147–62.

Stevens, John, *Words and Music in the Middle Ages: Song, Narrative, Dance and Drama, 1050–1350* (Cambridge, 1986).

—— 'Medieval Drama, II. Liturgical Drama', *NG*.

Stevenson, Joseph (ed.), *Chronicon monasterii de Abingdon*, 2 vols. (Rerum Britannicarum medii aevi scriptores (Rolls Series), 2; London, 1858).

Stotz, Peter, *Sonderformen der sapphischen Dichtung* (Munich, 1982).

Stratman, C. J., *Bibliography of Medieval Drama* (Berkeley, Cal., 1954; 2nd edn., 1972).

Strecker, Karl, *Introduction to Medieval Latin*, trans. and rev. Robert B. Palmer (Dublin and Zürich, 1957).

Strehl, Reinhard, 'Zum Zusammenhang von Tropus und Prosa "Ecce iam Christus"', *Mf* 17 (1964), 269–71.

Strömberg, Bengt, *Den pontifikala liturgin i Lund och Roskilde under medeltiden* [The pontifical liturgy in Lund and Roskilde during the Middle Ages] (Studia theologica Lundensia, 9; Lund, 1955).

Strunk, Oliver, 'Intonations and Signatures of the Byzantine Modes', *MQ* 31 (1945), 339–55 = Strunk 1977, 19–36.

—— *Source Readings in Music History* (New York, 1950).

—— 'The Antiphons of the Oktoechos', *JAMS* 13 (1960), 50–67 = Strunk 1977, 165–90.

—— 'The Latin Antiphons for the Octave of the Epiphany', in *Mélanges Georges Ostrogorsky*, ii (Recueil de travaux de l'Institut d'Études byzantines, 8; Belgrade, 1964), 417–26 = Strunk 1977, 208–19.

—— 'Tropus and Troparion', in Husmann Festschrift 1970, 305–11 = Strunk 1977, 268–76.

—— 'Die Gesänge der byzantinisch-griechischen Liturgie', in Fellerer 1972, 12847; trans.: 'The Chants of the Byzantine-Greek Liturgy' in Strunk 1977, 297–330.

—— *Essays on Music in the Byzantine World* (New York, 1977).

STRUNK Festschrift 1968 = Harold S. Powers (ed.), *Studies in Music History: Esays for Oliver Strunk* (Princeton, NJ, 1968).

STUART, NICHOLAS, 'Melodic "Corrections" in an Eleventh-Century Gradual (Paris, B. N., lat. 903)', *JPMMS* 2 (1979), 2–10.

STUBBS, WILLIAM, *Memorial of Saint Dunstan* (Rerum Britannicarum medii aevi scriptores (Rolls Series), 63; London, 1874).

SUCHIER, W., 'Die Entstehung des mittellateinischen und romanischen Verssystems', *Romanistisches Jahrbuch*, 3 (1950), 529–63.

SUÑOL, GRÉGOIRE MARIE, *Introduction à la paléographie musicale grégorienne* (Tournai, 1935); rev. and enlarged trans. of Gregori Maria Sunyol, *Introducció a la paleografia musical gregoriana* (Montserrat, 1925).

—— *see also Antiphonale missarum* and *Liber vesperalis*.

SYMONS, THOMAS, *Regularis Concordia: The Monastic Agreement of the Monks and Nuns of the English Nation* (London, 1953).

SZENDREI, JANKA, 'Zur Notations- und Vortragsweise der Prosulen nach den ungarischen Handschriften', in *Magyar Könyvszemle* (Budapest, 1972), 157–65.

—— *A magyar középkor hangjegyes forrásai* [résumé: 'Notierte Quellen des ungarischen Mittelalters'] (Budapest, 1981).

—— *Középkori hangjegyírások Magyarországon* [résumé: 'Mittelalterliche Choralnotationen in Ungarn'] (Budapest, 1983).

—— 'Beobachtungen an der Notation des Zisterzienser-Antiphonars Cod. 1799** in der Österreichischen Nationalbibliothek', *SMH* 27 (1985), 273–90.

—— 'The Introduction of Staff Notation into Middle Europe', *SMH* 28 (1986), 303–19.

—— 'Die Geschichte der Graner Choralnotation', *SMH* 30 (1988), 5–234.

—— 'Choralnotationen in Mitteleuropa', *SMH* 30 (1988), 437–46.

—— *see also Missale notatum Strigoniense*.

SZIGETI, K., 'Denkmäler des gregorianischen Chorals aus dem ungarischen Mittelalter', *SMH* 4 (1963), 129–72.

SZÖVERFFY, JOSEF, *Die Annalen der lateinischen Hymnendichtung*, 2 vols. (Berlin, 1964–5).

—— *Repertorium hymnologicum novum* (Berlin, 1983).

TACK, FRANZ, *Der gregorianische Choral* (Das Musikwerk, 18; trans. Everett Helm: *Gregorian Chant*, Anthology of Music, 18; Cologne, 1960).

TAFT, ROBERT, *The Liturgy of the Hours in East and West: The Origins of the Divine Office and its Meaning for Today* (Collegeville, Minn., 1985).

TARCHNISCHVILI, MICHEL, *Le Grand Lectionnaire de l'église de Jérusalem (V^e–VIIIe siècles)* (Corpus scriptorum Christianorum orientalium, 188–9, 204–5; Louvain, 1959, 1960).

TERRIZZI, FRANCESCO (ed.), *Missale antiquum S. Panormitanae ecclesiae* (Rerum ecclesiasticarum documenta, series maior, 13; Rome, 1970).

THANNABAUR, PETER JOSEF, *Das einstimmige Sanctus der römischen Messe in der handschriftlichen Überlieferung des 11.–16. Jahrhunderts* (Erlanger Arbeiten zur Musikwissenschaft, 1; Munich, 1962).

—— 'Anmerkung zur Verbreitung und Struktur der Hosanna-Tropen im deutschsprachigen Raum und den Ostländern', in Stäblein Festschrift 1967, 250–9.

—— 'Sanctus', *MGG*.

THIBAUT, JEAN BAPTISTE, *Origine byzantine de la notation neumatique de l'église latine* (Paris, 1907).

—— *Monuments de la notation ekphonétique et neumatique de l'église latine* (St Petersburg, 1912).

—— *Monuments de la notation ekphonétique et hagiopolite de l'église grecque* (St Petersburg, 1913).

THODBERG, CHRISTIAN, *Der byzantinische Alleluiarionzyklus* (Monumenta musicae Byzantinae, Subsidia, 8; Copenhagen, 1966).

THOMPSON, EDWARD MAUNDE, *An Introduction to Greek and Latin Palaeography* (Oxford, 1912).

TILLYARD, H. J. W., *Handbook of the Middle Byzantine Notation* (Monumenta musicae Byzantinae, Subsidia, 1/1; Copenhagen, 1935).

TOLHURST, J. B. L., *The Monastic Breviary of Hyde Abbey, Winchester (mss. Rawlinson Liturg. e. 1*, and Gough Liturg. 8, in the Bodleian Library, Oxford)* (Henry Bradshaw Society, 69; London, 1932; 70 (1933), 71 (1934), 76 (1938), 78 (1940), and 80 (1943)).

TRAUB, ANDREAS, 'Hucbald von Saint-Amand. De harmonica institutione', *BzG* 7 (1989), 3–101.

TREITLER, LEO, 'Musical Syntax in the Middle Ages: Background to an Aesthetic Problem', *Perspectives of New Music*, 4 (1965–6), 75–85.

—— 'The Aquitanian Repertories of Sacred Monody in the Eleventh and Twelfth Centuries' (Diss., Princeton University, 1967; UMI 67–9613).

—— 'On the Structure of the Alleluia Melisma: A Western Tendency in Western Chant', in Strunk Festschrift 1968, 59–72.

—— 'Homer and Gregory: The Transmission of Epic Poetry and Plainchant', *MQ* 60 (1974), 333–72.

—— contribution to symposium '"Peripherie" und "Zentrum"', in Congress Berlin 1974, 58–74.

—— '"Centonate" Chant: Übles Flickwerk or *E pluribus unus?*', *JAMS* 28 (1975), 1–23.

—— 'Transmission and the Study of Music History', in Congress Berkeley 1977, 202–11.

—— 'Oral, Written, and Literate Process in the Transmission of Medieval Music', *Speculum*, 56 (1981), 471–91.

—— 'Observations on the Transmission of Some Aquitanian Tropes', *Forum musicologicum*, 3 (1982), 11–60.

—— 'The Early History of Music Writing in the West', *JAMS* 35 (1982), 237–79.

—— 'From Ritual through Language to Music', *Schweizer Jahrbuch für Musikwissenschaft*, NS, 2 (1982), 109–24.

—— 'Reading and Singing: On the Genesis of Occidental Music-Writing', *EMH* 4 (1984), 135–208.

—— 'Paleography and Semiotics', in Huglo 1987, 17–27.

TURCO, ALBERTO, *Tracce della modalità arcaica nella salmodia del Temporale e del Sanctorale* (Milan, 1972).

—— 'Les Répertoires liturgiques latins en marche vers l'octoéchos. La psalmodie grégorienne des fêtes du Temporal et du Sanctoral', *EG* 18 (1979), 177–233.

TURCO, ALBERTO, 'Introito "Tibi dixit" e "Alleluia. Dies sanctificatus"', in Cardine Festschrift 1980, 257–67.

—— 'Melodie-tipo e timbri modali nell'Antiphonale romanum', *Studi gregoriani*, 3 (1987), 191–241.

TURNER, DEREK HOWARD, *The Missal of the New Minster, Winchester (Le Havre, Bibliothèque Municipale, MS 330)* (Henry Bradshaw Society, 93; Leighton Buzzard, 1962).

TYRER, JOHN WALTON, *Historical Survey of Holy Week: Its Services and Ceremonial* (Alcuin Club Collections, 29; London, 1932).

UDOVICH, JOANN, 'The Magnificat Antiphons for the Ferial Office', *JPMMS* 3 (1980), 1–25.

—— 'Modality, Office Antiphons, and Psalmody: The Musical Authority of the Twelfth-Century Antiphonal from St.-Denis' (Diss., University of North Carolina at Chapel Hill, 1985; UMI 85–27331).

UHLFELDER, MYRA L. (trans.), *Joannes Scotus, Erigena: Periphyseon. On the Division of Nature*, with summaries by Jean A. Potter (Library of Liberal Arts, 157; Indianapolis, Ind., 1976).

ULLMANN, PÉTER, 'Bericht über die vergleichende Repertoire-Analyse der Breviere aus Ungarn', *SMH* 27 (1985), 185–92.

UNDERWOOD, PETER, 'Melodic Traditions in Medieval English Antiphoners', *JPMMS* 5 (1982), 1–12.

UNDHAGEN, CARL-GUSTAV (ed.), *Birger Gregerssons 'Birgitta-officium'* (Svenska Fornskrifts-ällskapet, ser. 2; Latinska skrifter, 6; Stockholm, 1960).

VAGAGGINI, CIPRIANO, *Il senso teologico della liturgia: Saggio di liturgia teologica generale*, 2 vols. (Rome, 1957; 4th edn., 1965); trans.: *Theological Dimensions of the Liturgy*, 2 vols. (Collegeville, Minn., 1959).

VALOUS, GUY de, *Le Monachisme clunisien des origines au XVᵉ siècle: Vie intérieure des monastères et organisation de l'ordre*, 2 vols. (Ligugé and Paris, 1935; 2nd edn., 1970).

VAN DER WERF, HENDRIK, *The Emergence of Gregorian Chant*, i, 2 vols. (Rochester, NY, 1983).

VAN DEUSEN, NANCY M., 'An Historical and Stylistic Comparison of the Graduals of Gregorian and Old Roman Chant' (Diss., Indiana University, 1972; UMI 73–09783).

—— *Music at Nevers Cathedral—Principle Sources of Medieval Chant*, 2 vols. (Henryville, 1980).

—— 'The Sequence Repertory at Nevers Cathedral', *Forum musicologicum*, 2 (1980), 44–59.

—— 'Style, Nationality and the Sequence in the Middle Ages', *JPMMS* 5 (1982), 44–55.

VANICKÝ, JAROSLAV, 'Frater Domaslev (Domaslaus), der älteste bekannte Sequenzdichter Böhmens', *Jahrbuch für Liturgik und Hymnologie*, 5 (1960), 118–22.

Variae preces ex liturgia, tum hodierna tum antiqua (Solesmes, 1896).

Väterlein *see Graduale Pataviense*.

VECCHI, GIUSEPPE, *Uffici drammatici padovani* (Florence, 1954).

—— (ed.), *Troparium sequentiarium Nonantulanum, Cod. Casanat. 1741* (Monumenta lyrica medii aevi italica, I. Latina, i; Modena, 1955).

VELIMIROVIĆ, MILOŠ, arts. 'Echos', 'Russian and Slavonic church music', *NG*.

VERDON, TIMOTHY GREGORY, and DALLEY, JOHN (eds.), *Monasticism and the Arts* (Syracuse, NY, 1984).

VERHEIJEN, LUC, *La Règle de Saint Augustin*, 2 vols. (Paris, 1967).

VIDAKOVIĆ, A., 'I nuovi confini della scrittura neumatica musicale nell'Europa sud-est', *Studien zur Musikwissenschaft*, 24 (1960), 5–12.

VILLETARD, HENRI, *Office de Pierre de Corbeil (Office de la Circoncision) improprement appelé 'Office des Fous'* (Paris, 1907).

—— *La Danse ecclésiastique à la metropole de Sens* (Paris, 1911).

—— *Odorannus de Sens et son uvre musicale* (Paris, 1912).

—— *Office de St Savinien et de St Potentien, premiers évêques de Sens: Catalogue sommaire des livres liturgiques de l'ancien diocèse de Sens* (Paris, 1956).

VINCENT, ALEXANDRE-JOSEPH-HYDULPHE, 'Emploi des quarts de ton dans le chant grégorien, constaté dans l'Antiphonaire de Montpellier', *Revue archéologique*, 11 (1854), 262–72.

VIVELL, COELESTIN, *Initia tractatuum musices ex codicibus editorum* (Graz, 1912).

—— *Frutolfi Breviarium de musica et tonarius* (Akademie der Wissenschaften in Wien, Philosophisch-historische Klasse. Sitzungsberichte 188, Abhandlung 2; Vienna, 1919).

VIVES, J., *Oracional visigótico* (Monumenta Hispaniae sacra, Serie liturgica, 1; Barcelona, 1946).

VOGEL, CYRILLE, 'Les Échanges liturgiques entre Rome et les pays francs jusqu'à l'époque de Charlemagne', in *Le chiese nei regni dell'Europa occidentale e i loro rapporti con Roma fino all' 800* (Settimane di studi del Centro italiano di studi sull'alto medioevo, 7; Spoleto, 1960), 185–95.

—— 'La Réforme cultuelle sous Pépin le Bref et sous Charlemagne', in Erna Patzelt, *Die karolingische Renaissance* (Graz, 1965), 171–242.

—— 'La Réforme liturgique sous Charlemagne', in Braunfels, ii (1965), 217–32.

—— *Introduction aux sources de l'histoire du culte chrétien au moyen âge* (Biblioteca degli studi medievali, 1; Spoleto, 1966; 2nd edn., 1975; 3rd edn., 1981); trans. and rev. William G. Storey and Niels Krogh Rasmussen, *Medieval Liturgy: An Introduction to the Sources* (Washington, DC, 1986).

—— and Elze, Reinhard (eds.), *Le Pontifical romano-germanique du dixième siècle* (Studi e testi, 2267, 269; Rome, 1963, 1972).

VOGÜÉ, ADALBERT DE, *La Règle du Maître* (Sources chrétiennes, 105–7; Paris, 1964).

—— and NEUFVILLE, JEAN (eds.), *La Règle de Saint Benoît*, 7 vols. (Sources chrétiennes, 181–6; Paris, 1971–7).

VOLLAERTS, J. W. A., *Rhythmic Proportions in Early Medieval Ecclesiastical Chant* (Leiden, 1958, 2/1960).

VÖTTERLE Festschrift 1968 = Richard Baum and Wolfgang Rehm (eds.), *Musik und Verlag: Karl Vötterle zum 65. Geburtstag* (Kassel, 1968).

WADDELL, CHRYSOGONUS, 'Monastic Liturgy: Prologue to the Cistercian Antiphonary', in *The Works of Bernard of Clairvaux: Treatises—I* (Cisterican Fathers Series, 1; Spencer, Mass., 1970), 153–62.

—— 'The Origin and Early Evolution of the Cistercian Antiphonary: Reflections on Two Cistercian Chant Reforms', in *The Cistercian Spirit: A Symposium in Memory of Thomas Merton*, ed. M. Basil Pennington (Cistercian Studies Series, 3; Spencer, Mass., 1970), 190–223.

—— 'The Early Cistercian Experience of Liturgy', in *Rule and Life: An Interdisciplinary Symposium*, ed. M. Basil Pennington (Cistercian Studies Series, 12; Spencer, Mass., 1971), 77–116.

—— 'The Two Saint Malachy Offices from Clairvaux', in *Bernard of Clairvaux: Studies Presented to Dom Jean Leclercq* (Cistercian Studies Series, 23; Washington, DC, 1973), 123–59.

—— 'Peter Abelard's Letter 10 and Cistercian Liturgical Reform', in *Studies in Medieval*

Cistercian History, 2, ed. J. R. Sommerfeldt (Cistercian Studies Series, 24; Kalamazoo, Mich., 1976), 75–86.

—— 'St Bernard and the Cistercian Office at the Abbey of the Paraclete', in *The Chimaera of his Age: Studies on Bernard of Clairvaux*, ed. E. J. Elder and J. Sommerfeldt (*Studies in Medieval Cistercian History*, 5; Cistercian Studies Series, 63; Kalamazoo, Mich., 1980).

—— 'The Reform of the Liturgy from a Renaissance Perspective', in Benson, Constable, and Lanham 1982, 88–109.

—— *The Twelfth-Century Cistercian Hymnal* (Cistercian Liturgy Series, 1–2; Gethsemani Abbey, 1984).

—— 'The Pre-Cistercian Background of Cîteaux and the Cistercian Liturgy', in *Goad and Nail: Studies in Medieval Cistercian History*, 10, ed. E. Rozanne Elder (Kalamazoo, Mich., 1985), 119.

—— 'Epithalamica: An Easter Sequence by Peter Abelard', *MQ* 72 (1986), 239–71.

WAELTNER, ERNST LUDWIG, and BERNHARD, MICHAEL, *Wortindex zu den echten Schriften Guidos von Arezzo* (Bayerische Akademie der Wissenschaften, Veröffentlichungen der Musikhistorischen Kommission, 2; Munich, 1976).

WAGENER, H., *Die Begleitung des gregorianischen Chorals im neunzehnten Jahrhundert* (Regensburg, 1964).

WAGNER, PETER, *Einführung in die gregorianischen Melodien*, 3 vols.

Wagner I = *Ursprung und Entwicklung der liturgischen Gesangsformen* (Fribourg, 1895; 2nd edn., 1901; 3rd edn., 1911); trans. of 2nd edn. by Agnes Orme and E. G. P. Wyatt: *Introduction to the Gregorian Melodies, 1: Origin and Development of the Forms of Liturgical Chant up to the End of the Middle Ages* (London, 1907).

Wagner II = *Neumenkunde. Paläographie des liturgischen Gesanges* (Leipzig, 1905; 2nd edn., 1912).

Wagner III = *Gregorianische Formenlehre* (Leipzig, 1921).

—— 'Un piccolo trattato sul canto ecclesiastico in un manoscritto del secolo x–xi', *Rassegna gregoriana*, 3 (1904), 481–4.

—— 'Germanisches und Romanisches im frühmittelalterlichen Kirchengesang', in Congress Leipzig 1925, 21–34.

—— 'Aus der Frühzeit des Liniensystems', *AfMw* 8 (1926), 259–76.

—— 'Der mozarabische Kirchengesang und seine Überlieferung', *Gesammelte Aufsätze zur Kulturgeschichte Spaniens*, 1 (Spanische Forschungen der Görresgesellschaft, 1st ser.; Münster, 1928), 102–21.

—— 'Untersuchungen zu den Gesangstexten und zur responsorialen Psalmodie der altspanischen Liturgie', *Gesammelte Aufsätze zur Kulturgeschichte Spaniens*, 2 (Spanische Forschungen der Görresgesellschaft, 1st ser.; Münster, 1930), 67–113.

—— *Das Graduale der St. Thomaskirche zu Leipzig (14. Jahrhunderts)* (Publikationen älterer Musik, 5–6; 1930, 1932).

—— 'Zur mittelalterlichen Tonartlehre', in Adler Festschrift 1930, 29–32.

—— *Die Gesänge der Jakobsliturgie zu Santiago de Compostela aus dem sogennanten Codex Calixtinus* (Fribourg, 1931).

WAGNER Festschrift = K. Weinmann (ed.), *Festschrift für Peter Wagner* (Leipzig, 1926).

WALKER, G. S. M., *Sancti Columbani Opera* (Dublin, 1957).

WALLACE, WILFRID, *The Life of St. Edmund of Canterbury* (London, 1893).

WALLACE-HADRILL, J. M., *The Frankish Church* (Oxford, 1983).

WARREN, FREDERICK E., *The Liturgy and Ritual of the Celtic Church* (Oxford, 1881); rev. Jane Stevenson (Woodbridge, 1987).

—— *The Leofric Missal* (Oxford, 1883).

—— *The Antiphonary of Bangor* (Henry Bradshaw Society, 4 and 10 (London, 1893, 1895).

WASZINK, J., *Studien zum Timaioskommentar des Calcidius* (Leiden, 1964).

WEAKLAND, REMBERT, 'Hucbald as Musician and Theorist', *MQ* 42 (1956), 66–84.

—— 'The Beginnings of Troping', *MQ* 44 (1958), 477–88.

—— 'The Compositions of Hucbald', *EG* 3 (1959), 155–62.

—— 'Milanese Rite, Chants of', *NCE*.

WEINMANN, K., *Das Konzil von Trient und die Kirchenmusik* (Leipzig, 1919).

WEINRICH, LORENZ, 'Abelard', *NG*.

WEISBEIN, NICOLAS, 'Le "Laudes crucis attollamus" de Maître Hugues d'Orléans, dit le Primat', *Revue du moyen âge latin*, 3 (1947), 5–26.

WEISS, GÜNTHER, '"Tropierte Introitustropen" im Repertoire der südfranzösischen Handschriften', *Mf* 17 (1964), 266–9.

—— 'Zum Problem der Gruppierung südfranzösischer Tropare', *AfMw* 21 (1964), 163–71.

—— 'Zum "Ecce iam Christus"', *Mf* 18 (1965), 174–7.

—— 'Zur Rolle Italiens im frühen Tropenschaffen', in Stäblein Festschrift 1967, 287–92.

—— *see also* MMMA 3.

WELLESZ, EGON, *Eastern Elements in Western Chant* (Copenhagen, 1947).

—— *A History of Byzantine Music and Hymnography* (Oxford, 1949; 2nd edn., 1961).

—— 'Gregory the Great's Letter on the Alleluia', *Annales musicologiques*, 2 (1954), 7–26.

WELLESZ Festschrift 1966 = Jack Westrup (ed.), *Essays Presented to Egon Wellesz* (Oxford, 1966).

WESTRUP Festschrift 1975 = F. W. Sternfeld, Nigel Fortune, and Edward Olleson (eds.), *Essays on Opera and English Music in Honour of Sir Jack Westrup* (Oxford, 1975).

WEYNS, NORBERT I., 'Le Missel prémontré', *Analecta praemonstratensia*, 43 (1967), 203–25.

—— *Antiphonale missarum praemonstratense* (Bibliotheca analectorum praemonstratensium, 12; Averbode, 1973).

WHITE, ALISON, 'Boethius in the Medieval Quadrivium', in Gibson 1981, 162–205.

WHITEHILL, W. M., CARRO GARCÍA, J., and PRADO, G., *Liber Sancti Jacobi: Codex Calixtinus* (Santiago de Compostela, 1944).

WIESLI, WALTER, *Das Quilisma im Codex 359 der Stiftsbibliothek St. Gallen, erhellt durch das Zeugnis der Codices Einsiedeln 121, Bamberg lit. 6, Laon 239 und Chartres 47: Eine paläographisch-semiologische Studie* (Bethlehem Immensee, 1966).

WILKINSON, J., *Egeria's Travels* (London, 1971).

WILLIS, G. G., *Essays in Early Roman Liturgy* (Alcuin Club Collections, 46; London, 1964).

—— *Further Essays in Early Roman Liturgy* (Alcuin Club Collections, 50; London, 1968).

WILMART, ANDRÉ, 'Le Comes de Murbach', *RB* 30 (1913), 25–69.

—— 'Le Recueil des poèmes et des prières de Saint-Pierre Damien', *RB* 41 (1929), 342–57.

—— *Le Monachisme clunisien des origines au XVe siècle* (Paris, 1935; 2nd edn., 1970).

WILSON, HENRY AUSTIN (ed.), *The Pontifical of Magdalen College, with an Appendix of Extracts from Other English Mss. of the Twelfth Century* (Henry Bradshaw Society, 39; London, 1910).

—— (ed.), *The Gregorian Sacramentary under Charles the Great* (Henry Bradshaw Society, 49; London, 1915).

WINKLER, G., 'Über die Kathedralvesper in den verschiedenen Riten des Ostens und Westens', *Archiv für Liturgiewissenschaft*, 16 (1974), 53–102.

WINTERFELD, PAUL VON, 'Rythmen- und Sequenzstudien, I: Die lateinische Eulaliasequenz und ihre Sippe', *Zeitschrift für deutsches Altertum und deutsche Literatur*, 45 (1901), 133–47.

WOLF, JOHANNES, 'Ein anonymer Musiktraktat des elften bis zwölften Jahrhunderts', *Vierteljahrsschrift für Musikwissenschaft*, 9 (1893), 186–234.

WOODS, ISOBEL, '"Our awin Scottis use": Chant Usage in Medieval Scotland', *Journal of the Royal Musical Association*, 112 (1987), 21–37.

WORDSWORTH, CHRISTOPHER, *Ceremonies and Processions of the Cathedral Church of Salisbury* (Cambridge, 1901).

—— and LITTLEHALES, HENRY, *The Old Service Books of the English Church* (London, 1904).

WORMALD, FRANCIS, *English Kalendars before A.D. 1100*, i (Henry Bradshaw Society, 72; London, 1934).

—— *English Benedictine Kalendars after A.D. 1100*, i. Abbotsbury–Durham (Henry Bradshaw Society, 77; London, 1939).

—— *English Benedictine Kalendars after A.D. 1100*, ii. Ely–St. Neots (Henry Bradshaw Society, 81; London, 1946).

WRIGHT, CRAIG, *Music and Ceremony at Notre Dame of Paris, 500–1500* (Cambridge, 1989).

[YARDLEY], ANNE D. BAGNALL, 'Musical Practices in Medieval English Nunneries' (Diss., Columbia University, 1975; UMI 75–25648).

YARDLEY, ANNE BAGNALL, 'The Marriage of Heaven and Earth: A Late Medieval Source of the *Consecratio virginum*', in Sanders Festschrift 1990, 305–24.

YEARLEY, JANTHIA, 'A Bibliography of Planctus in Latin, Provençal, French, German, English, Italian, Catalan and Galician-Portuguese from the Time of Bede to the Early Fifteenth Century', *JPMMS* 4 (1981), 12–52.

—— 'The Medieval Latin Planctus as a Genre' (Diss., York, 1983; British Theses D48872/ 84).

YOUNG, KARL, *The Drama of the Medieval Church*, 2 vols. (Oxford, 1933).

I

Plainchant in the Liturgy

I.1. INTRODUCTION

Frere 1906; Fortescue 1917; Miller 1959; Miller, 'Liturgy', *NCE*; Lechner and Eisenhofer 1961; Gelineau 1962; Righetti 1964; Martimort 1965; *La Musique dans la liturgie*, 1971; Cheslyn Jones *et al.* 1978; Steiner and Levy, 'Liturgy and Liturgical Books', *NG*; Pfaff 1982; Martimort 1983; Ekenberg 1987; *Gottesdienst der Kirche*, iii, 1987.

Plainchant is liturgical music, music to be performed during the celebration of a divine service. The performance of the music is not, generally speaking, an end in itself but part of a religious ritual. Sometimes music assumes a prominent place within the performance of the ritual, sometimes a very minor one. Occasionally it has been cultivated with an exuberance and extravagance that seems to go beyond the needs of the liturgy; but this is not its normal role. Its function is to add solemnity to Christian worship. Liturgical texts which are sung, whether chanted on a monotone or to a highly melismatic melody, are more solemn, inspiring, and impressive, and a more worthy vehicle for human prayer and praise of God, than spoken words. (For a fuller discussion of its function see Gelineau 1962, *Gottesdienst der Kirche*, iii, 1987, and *La musique dans la liturgie*, 1971; for the views of early medieval writers see Ekenberg, 1987.)

Practically the whole of the plainchant repertory is music sung with a text. This is another reason why the music cannot always be discussed as a thing in itself: one has to see whether, and how, it articulates the texts being sung. Moreover, the texts were not usually chosen for their musical potential, in the sense of being particularly easy to sing, or having characteristics which showed to best advantage in musical performance. The nature of the music chosen for a particular text is determined by a sense of what was proper, that is appropriate, for the liturgical occasion. This is shown by the fact that some texts, particularly verses from the Book of Psalms, were used again and again in the liturgy; and they were sung in different ways according to what was proper at the particular point in the liturgy where they were being sung. The liturgical context determined the type of music that was to be sung, just as it had determined the texts.

For the most part, in this book I shall not discuss the theological and ritual bases of liturgical music. I describe the plainchant that has been sung, the state in which it has come down to us, and the way it was understood, not as ritual but as music. The book is written by a musician and not a liturgist, much less a theologian. But in order to understand the function of the music, why it was sung, and why it assumed the forms it did, the place of music in the liturgical services of the church must also be described. The first chapter of this book therefore gives a brief account of the liturgy of which plainchant is a part. In the sections which follow, I try to explain what a 'liturgy' is, why it is performed, and what it means; then comes an account of the yearly cycle within which are organized the services which go to make up the liturgy, followed by sections on the individual services themselves. This chapter is restricted to the Roman liturgy, that most widely used—in various forms—in Western Europe. Later in the book some mention is made of other Western liturgies and their chant, but, given the space available, discussion of these has inevitably had to be severely restricted.

Much of what follows describes a sort of 'norm' of liturgical practice. But every church in every age has had its own idiosyncrasies of practice, and it is difficult to steer a middle way between over-simplification of the variety that has always existed and the confusion which may result if too many of the variations are described. During this century it has become increasingly convenient to refer to the practice laid down in Vatican books as a standard. The *Liber usualis*, an amalgam of several official Roman service-books, which contains a large proportion of the plainchant needed for mass and the services of the office, is probably the compendium most used for the study of plainchant today. But in this and other books we see only one among many possible liturgies, a twentieth-century one, moreover, not a medieval one. From time to time some attempt will be made to indicate a few of the differences between uses, in liturgical as in musical matters.

I.2. LITURGY AND WORSHIP

Cheslyn Jones *et al.* 1978, 1–29.

Liturgy may be defined as the communal forms of worship of the Christian church. Christians have from the earliest days of the church ('church' in the sense of 'all Christian people') gathered together to praise and adore God, to pray to him, thank him, ask his assistance, and especially to relive in a symbolic way the events of Christ's life on earth, most importantly the Last Supper which Christ shared with his disciples. The liturgy is the more or less formal, organized way in which this is done. So the study of liturgy (sometimes called 'liturgiology') broadly includes the history of these formal acts of worship and their theological and social (ethical, anthropological) significance, the texts used, the music, the ceremonial actions, the special clothes (vestments) worn by those who officiate at the ceremonies, and the nature of the buildings in which worship take place (church architecture, internal decoration, and furniture).

Christians believe that God saved mankind from its natural state of sin through his son, Jesus Christ, who by his death paid the penalty for man's sins, and ensured eternal life for all mankind. In this, Christ was made a sacrifice for mankind. After Christ had risen to heaven, the Holy Spirit (the third Person, with God the Father and Christ the Son, of the Holy Trinity) came first to the Apostles, and then to the whole church, to continue the work of salvation ('saving'). By meeting together for worship Christians further the work of the Holy Spirit in a special way, since liturgical acts of worship bring them into particularly close contact with the events at the root of their religious beliefs. The acts of worship are an encounter with Christ which renews the Christian people, unites them with him, and makes them ready to receive God's grace (unmerited favour of being saved). As the Constitution on the Sacred Liturgy of the Second Vatican Council (1963) puts it:

> Thus not only when things are read 'which were written for our instruction' (Rom. 15: 4), but also when the Church prays or sings or acts, the faith of those taking part is nourished, and their minds are raised to God so that they may offer him their spiritual homage and receive his grace more abundantly. (Para. I. I. 33.)

The most important ritual act within the liturgy is the Eucharist, Holy Communion, or Lord's Supper. The word 'eucharist' derives from the Greek word *eukharistia*, meaning thanksgiving. It refers to the thanks Christ gave at the Last Supper, and also to the thanks of the church for God's work of redemption. At the last meal with his disciples

> Jesus took bread, and blessed it, and brake it, and gave it to the disciples, and said, Take, eat; this is my body. And he took the cup, and gave thanks, and gave it to them, saying, Drink ye all of it; For this is my blood of the new testament, which is shed for many for the remission of sins. (Matt. 26: 26–8.)

And in St Luke's account, Christ commands the disciples: 'This do in remembrance of me' (Luke 22: 19). It is through a re-enactment of this scene, a reliving of the Last Supper, that Christians come into closest union with God. In the eucharist the minister, or celebrant (that is the priest who is the chief actor in the ceremony, administering the bread and wine) is Christ's representative. But more than this:

> Christ is always present . . . in the Sacrifice of Mass not only in the person of his minister . . . but especially in the eucharistic species . . . he is present when the Church prays and sings, for he has promised 'where two or three are gathered together in my name there am I in the midst of them' (Matt. 18: 20). (*Constitution on the Sacred Liturgy*, I. I. 7.)

The word 'eucharist' is also used to mean the sacred elements, or species, that is, the bread and wine, which Christians eat and drink at the culmination of the service. The word 'service' is used here in its usual liturgical sense of 'one self-contained communal act of worship', so that the 'service of Holy Communion' would mean the actual taking of the eucharist by the Christian congregation, together with all the prayers, readings, and music surrounding that act. The service of Compline, to take another example, is likewise made up of a series of prayers, readings, and chants; and so for other services.

The word 'liturgy', interestingly, derives from the Greek *leitourgia*, originally meaning some gratuitous act of service to the public, such as the giving of money to pay for a military project, or public entertainment. In the Christian West the word has been used in several different ways. It can refer to the whole complex of forms of public worship, in its broadest sense. But it is also used to refer to specific parts of that whole, such as the part for a particular day (for example, 'the Good Friday liturgy'), or the forms used in a particular part of the universal church (for example, 'the Sarum liturgy', meaning the forms of worship used at Salisbury cathedral). In the East, 'liturgy' is usually restricted to meaning the Holy Communion service alone, and some books on 'the liturgy' restrict themselves according to the same convention (for example, Srawley 1947). Partly because of this, I shall use another common term for the Holy Communion service, the 'mass'. As the Latin form of the service developed in the fourth century onwards, it acquired the words of dismissal 'Ite missa est' ('Go, you are dismissed'), and from this the title for the whole service is taken.

As well as the consuming of the bread and wine—the 'body' and 'blood' of Christ—praying, singing, and reading sacred texts are important parts of the liturgy. Mass contains all these things. The other services in the liturgy—usually known as the canonical hours, office hours, or simply the office—which do not include communion, are composed almost entirely of praying, singing, and reading. The reading of sacred texts is important for bringing to mind the history of God's work. The writings now collected in that part of the Bible known as the Old Testament, compiled by Jewish teachers before Christ's ministry on earth, are used beside those of Christian writers (the Gospels of the four evangelists, the letters of St Paul, etc.), for Christians believe that the earlier history of Israel contains signs and prefigurations of later events. Eventually the writings of Christian figures from times much later than the apostolic age (the first century or so of Christianity) also came to be read at liturgical services. These were often patristic writings, the writings of the Church Fathers of the second to sixth centuries (that is, roughly to the time of Pope Gregory I 'the Great', d. 604).

The sense of divine history to which I have just alluded affects the character and content of the liturgy according to the time of year. Thus, although the act of communion does not vary in essentials from one day to the next, the prayers, readings, and also the texts of the chants vary in accordance with a cycle of commemorations of the events of the life of Christ and his most important followers. At the time of year when Christ was born, the texts, and some of the ceremonial actions, refer to Christ's birth; similarly for Christ's death, resurrection, and ascension into heaven, and the coming of the Holy Spirit. The whole year is full of special commemorations of this sort. For example, 10 August is traditionally the day in the year when St Laurence, a deacon of the church at Rome, was martyred by being roasted on a grid during the persecutions of Valerian in 258. (King Philip II of Spain, in thanks for a military victory won at Saint-Quentin on 10 August 1557, built his great palace, monastery, church, and college of El Escorial, supposedly in the shape of this (inauthentic) instrument of execution, a striking and macabre example of the strength of Christian historical symbolism.) In the Middle Ages passages relating his life and death would have been read during Matins, some of the prayers at mass would invoke his intercession, and some of the chants would also refer to him. Thus at mass, the alleluia would probably use the verse *Levita Laurentius*: 'The Levite Laurence has wrought a pious work, who by the sign of the cross enlightened the blind, and distributed to the poor the riches of the church.' This short text would also be used for some chants of the office (sung to different music, as befitting the particular liturgical function of the chant in question). Not all the texts would have been special to St Laurence, since he was but one of very many martyrs, though undoubtedly one of the more important in medieval eyes. The introit at mass uses a text suitable for almost any saint: *Confessio et pulchritudo* (from Ps. 96): 'Glory and worship are before him: power and honour are in his sanctuary', and indeed this introit was used for other saints as well.

The annual cycle of commemorations and symbolic re-enactments is of such importance that the next section of this book is devoted to it.

The brief indication of the liturgy's significance sketched above is only one way of looking at the matter, albeit one suggested by modern scholarship (Crichton in Cheslyn Jones *et al.* 1978, Bouyer 1956 and 1962, Vagaggini 1957) and the official teaching of the Church. The liturgy obviously played different parts in the lives of different Christians: priest and people, monk and clerk. It meant different things to different persons at different times and places. Communion, the actual consuming of the bread and wine, was strikingly rare in the Middle Ages, and later times as well—to the extent that it might take place only on Easter Day. On other occasions, although the form of the mass would have been followed, and all the texts and music heard, the sacred elements would merely have been displayed to those present, the only communicant being the priest. This sort of variation in practice obviously has consequences for the music needed during the ceremony (in this case, see Atkinson 1977). Just as striking as the differences in liturgical practice are the differences in interpretation of the liturgy. In the Middle Ages in particular, numerous allegorical explanations of the significance of liturgical actions were composed—often of what were in origin rather simple, functional actions. (See the account of allegorical interpretations of the mass, and a rapid sketch of the changes in the character of the ceremony, in Jungmann 1962, *Missarum*, Pt. I, chs. 9 and 11–14, and Pt. II, ch. 2.)

For example, the influential Amalarius of Metz (d. *c.*853) saw symbolic significance in every person, text, action, vestment, and property found in the liturgy. I quote here from Jungmann's presentation of Amalarius's summary of the contents of his *Expositio* or *Eclogae* (813–14) of the mass (for an extensive presentation of the much longer discussion of the mass by Amalarius in his *De ecclesiasticis officiis*, see Hardison 1965, Essay II; note, however, the reservations expressed by McKitterick 1977 148 ff.: Amalarius may have occupied a somewhat extreme position). The parts of the mass mentioned in this extract represent Christ's life 'from the first coming of the Lord to the time when he hastened to Jerusalem to suffer':

> The Introit represents the choir of the Prophets (who announced the advent of Christ just as the singers announce the advent of the bishop) . . . , the Kyrie eleison represents the Prophets at the time of Christ's coming, Zachary and his son John among them; the Gloria in excelsis Deo indicates the throng of angels who proclaimed to the shepherds the joyous tidings of our Lord's birth (and indeed in this manner, that first one spoke and the others joined in, just as in the Mass the bishop intones and the whole church joins in); the first collect represents what our Lord did in His twelfth year . . . ; the Epistle represents the preaching of John, the Responsorium the readiness of the Apostles when our Lord called them and they followed Him; the Alleluia their joy of heart when they heard his promises or saw the miracles He wrought . . . , the Gospel his preaching. (Jungmann 1962, *Missarum*, Pt. I, 118.)

One need not necessarily agree with Hardison's interpretation of Amalarius—he sees him as a striking example of the mentality which cultivated liturgical drama—to realize that the possible relevance of the ideas in Amalarius' writings to liturgical texts (tropes, and dramatic ceremonies) has to be considered. (For another example of the possible connection between theological/allegorical ideas and liturgical music, see Schlager 1983, 'Trinitas'.)

It is not only the theological or allegorical significance of the liturgy that has consequences for musical matters, but also a great deal of its practical detail. I have already alluded to the matter of frequency of communion, more precisely the number of communicants at mass. While this is of crucial importance in recent liturgical history, it has practical consequences for the singing of chants during communion: how many are required? To take another example, the introit chant, the first of mass, has at one time, it seems, consisted of the singing of most or all of a psalm, with an antiphon, then been reduced to a single psalm verse plus antiphon, then on high feasts been supplemented by numerous trope verses (making, in effect, a chant equivalent in length to one of the longer psalms). The introit is traditionally a chant sung during the entrance of the officiant (the priest who administers communion) and his assistants. Are we to imagine entrance routes of varying length, varying numbers of persons in the procession, as calling forth these varying forms of introit chant? Some answers to questions such as these will be found in the course of this book.

For many it is not easy today to gain a sense of the power of the liturgy, its slow rhythm from day to day, week to week, year to year, its seasons of grief and penitence, hope and joy. Those who attend church regularly are in a better position than most, for, albeit on a reduced scale, and interrupted by the numerous activities of the rest of the week, they may experience the sense of unity with the whole Christian church in communion, and follow in Bible readings, prayers, and the texts of musical compositions the annual cycle of Christ's birth, passion, and resurrection, supported by the witness of the Old Testament, and uplifted by the ancient lyric texts of the psalms. Yet, for those with imagination, even the mere perusal of liturgical books can awaken a sense of the immensity of the material involved, the majesty of the slow procession through the psalter, the steady flow of the great responsories, and again, the colourful juxtaposition of diverse forms and styles in the chants of mass. How much more strongly, then, must those whose lives were (and are) a continuous celebration of the liturgy have experienced these things! The reader is urged, therefore, to try to keep the musical material discussed in these pages in a liturgical perspective. Beyond that, a day spent attending the services in a monastic community teaches one more than many books. For while a chant may be discussed and dissected here as an object of study in itself, it must not be forgotten that it was composed in the creation of a complete way of life, the performance of the 'opus Dei', the work of God.

I.3. THE CHURCH YEAR

Frere 1930; McArthur 1953; Pascher 1963; Andrew Hughes 1982; Vogel 1986, 304–14.
   For special parts of the year, Hermann Schmidt 1956–7 and Jungmann Festschrift 1959 (Easter), Botte 1932 (Christmas and Epiphany), Willis 1964 (Ember Days).
   For information on the saints found in Western kalendars and service-books, there are numerous volumes, both great (the series *Acta sanctorum* and *Biblioteca sanctorum*) and small (Farmer 1978). The lists in Grotefend 1898 and Dalton and Doble 1941 are also very useful.

For different days of worship in the Christian year, different prayers and lessons are intoned and different chants are sung. At the beginning of a new year (usually reckoned to be the start of the Advent season, which looks forward to Christ's birth), the whole mighty cycle begins again, repeating the prayers, lessons, and chants of the previous year. In order to appreciate the rhythm and variety of the huge body of liturgical material which has come down to us, some sense of the way the church year is organized is essential.

   Although Christian worship is organized in a year's cycle, weekly units and cycles are also most important. Sunday, the day in the week when Christ rose from the dead, was of such significance that much of the liturgy revolves around Sunday services. To one seeking absolute regularity in adherence to the dates of a 365- (or 366-)day year, the Christian year is therefore an odd arrangement of fixed and (mostly) variable dates. It has its own compelling logic, however, attuned to the natural rhythm of the seasons, the waxing and waning of the moon, the summer and winter solstices, the week and the day.

   Many individual elements within the yearly cycle are organized on an annual basis, in the sense that the special liturgies of the days in question come round only once in the year. Even these, however, are not always fixed to an unvarying calendar date. Some of the days are given to the commemoration of events in the life and ministry of Christ and his followers, and also to remembering the witness of later Christians, particularly those martyred during the persecution of the early Church and those who carried Christianity into new lands. Holy days which commemorate saints are usually fixed to particular dates in the year, mostly that when the saint passed to his or her heavenly reward, or occasionally when the saint's relics were translated (solemnly removed) from one place to another. An important, and generally more ancient part of the yearly cycle, however, is governed by the date of Easter, when Christ rose from the dead. This date varies from year to year, so that those parts of the Christian year which prepare for, and follow on, from Easter are said to be 'movable'. In fact, the fixed-date part is governed by the solar cycle, the movable part by the lunar cycle: two different temporal cycles have to be reconciled.

Even when reference is being made to Christmas and Easter, however, the weekly unit may still be an organizing factor. Thus services may take place on 'the Sunday after', such and such a date or day. Since the calendar dates of Sundays are not the same from one year to the next, any day in the Christian year which is fixed on a particular date—such as Christmas Day, 25 December, or a saint's day—will fall sometimes on a Sunday and sometimes not, and some recognition of this will often be found in the liturgy. Another basic layer of liturgical material is formed by the weekly recitation of the office, where a seven-day supply of material is simply repeated week after week.

Hence many items in the liturgy are said to be 'proper', used only on one day, or certain days of the year, or 'ordinary', used on every or nearly every day. There is in fact a very large number of intermediate gradations between completely proper and totally ordinary, and it is impossible to give here more than a vague idea of the main outlines of the picture. (Andrew Hughes 1982 is a strikingly comprehensive and serious attempt to account for the patterns of repetitions throughout the year in later medieval books.) What should be borne in mind is the great variety of the possible arrangements. Formularies may be proper not simply to single feast-days, but also to seasons of the year, or to one day in the week (that is, not sung on the other six). Much office material is subject to complex variation of this sort: thus there is a different hymn for each day of the week for Lauds, but these may be displaced by a hymn proper to the season at certain times of the year. Particularly during Lent, the 'triduum' (the last three days before Easter Sunday), and the Easter season generally (Easter Sunday to Whit Sunday), many 'ordinary' items will disappear or change place. Music may also contribute to the 'properness' of a piece. While the text may remain the same for a certain period of the year, or even for the whole of it, the musical setting may vary from occasion to occasion. The best-known instances of this are the various melodies for some of the ordinary texts of mass, Kyrie, Gloria, Credo, Sanctus, and Agnus Dei. Some of these, however, were associated with particular occasions—Christmas, Easter, feasts of the Blessed Virgin Mary—and were to that extent proper. The usual hymn for Prime, *Iam lucis ortu sidere*, changes its melody according to season.

Rather few events in Christ's life are commemorated on unchanging dates: his birth (Christmas Day) has traditionally been celebrated on 25 December in the Roman church since the fourth century. (It was the pagan festival of the winter solstice when first adopted by the church.) Practically all other fixed days are those of saints, the Blessed Virgin Mary, and so on. The movable period dependent upon Easter contains such days as Ash Wednesday (the start of the Lenten fasting period), Palm Sunday (when Christ entered Jerusalem, the start of the week leading to his crucifixion and resurrection), Ascension Day (when he was taken up into heaven), and Whit Sunday (when the Holy Spirit came upon the Apostles). Since every Sunday in the church year had its own special prayers, lessons, and chants, the Sundays were usually reckoned in relation to the great feast-days, either those with a fixed date or those which were movable. Different places, periods, and liturgical books had their own ways of setting out the material required by this complex arrangement of holy days.

For a correct performance of the liturgy it was essential to know the correct calendar date, the days when Sunday occurred, and, most importantly, when Easter fell. It became customary in the Middle Ages to compile tables, sometimes known as 'compotus' or 'computus' tables, which set out this information in ready-reckoner form. These usually contain code letters and numbers for days and years which assist in calculating the repetition of calendric cycles over a span of decades or even centuries. Computus tables are sometimes found bound with liturgical books, and many liturgical books are provided with a kalendar which sets out the yearly cycle of saints' days and other feasts. (The spelling 'kalendar' is here employed for the actual document; 'calendar' is reserved for the general notion of a system of fixed dates.)

The calculation of the relationship between the seven-day week and 365- (or 366-)day year and the date of Easter necessitates a knowledge of the solar cycle (yearly alternation of summer, with longer days, and winter, with longer nights) and the lunar cycle (waxing and waning of the moon). The 365-day year, with 366 days every fourth year, is an arrangement promoted by Julius Caesar, hence the term 'Julian calendar'. The days of the month were reckoned in Roman use from the *Kalendae*, the first day (originally the first day of the new moon), backwards. The time of the full moon was the *Idus*, and nine days before that came the time of the quarter moon, the *Nonae*. To gain his 'leap year', Julius Caesar doubled the sixth before the Kalends of March, that is, the sixth day before the first day of March, 24 February. Such a year was also known, therefore, as 'bissextile'.

The lunar cycle coincides with the solar cycle every nineteen (solar) years: that is, nineteen years after a given new moon another new moon will appear, during which period 235 lunations will take place. Each of the nineteen years in the cycle (known as the 'Metonic' cycle) was given—in Greek antiquity and therefore in Roman and medieval times—a 'golden number' (Archer 1941, 4). To find the golden number of any year, the number of the year is divided by 19 and 1 is added to the remainder. Thus for 1990 the golden number is $(1990/19)_r + 1 = 14 + 1 = 15$.

The Christians also assigned to each day of the year an alphabetical letter from A to G, in order from 1 January (A) onward. Saint's days, always fixed to one date in the year, thus always corresponded to one particular letter. The letters were known as 'dominical' or 'Sunday' letters, for the relation between the occurrence of Sundays and the date of the day could be deduced therewith.

The predominant medieval system of finding the date of Easter was established at the Council of Nicaea in 325. The date depends first of all on the vernal equinox, then as now 21 March. Next one has to know which lunar cycle (28 or 29 days, beginning with a new moon) has its fourteenth day on or after 21 March. Easter falls on the first Sunday after that fourteenth day. Thus in 1990 the relevant new moon fell on Monday, 26 March, fourteen days after being Monday 9 April, Easter Sunday thus falling on 15 April. The earliest possible date for Easter Sunday is 22 March and the latest 25 April.

It would in theory have been possible for the Christians to have adopted a fixed date for Easter Day, that of the year of Christ's resurrection. Its occurrence on Sunday (the day after the Jewish Sabbath), however, and its relationship to the Jewish Passover (determined in a manner adapted by the Christians for Easter) made of it a movable feast. The correct calculation of Easter was one of the most important accomplishments of a medieval priest. It could symbolize the difference between true and deviant faith, as in the disputation between representatives of Roman and Celtic practice at the Synod of Whitby in 664 (Bede, trans. Sherley-Price, 186 ff.: the British used an 84-year cycle, and celebrated Easter on the fourteenth day if that was a Sunday). It was a vital part of the Carolingian ecclesiastical reforms, as witness Hrabanus Maurus' treatise *De computo*, of 820, revised from previous versions (possibly of Irish origin).

Not only Easter itself but also the previous and subsequent weeks were associated in the movable portion of the church year. Before Easter came the penitential season of Lent, notionally a fast of forty days, corresponding to periods of fasting by Moses, Elijah, and Christ himself. The actual number of days varied: in fourth-century Jerusalem there were eight weeks of five-day fasts (Saturday and Sunday were exempt); in the Eastern churches in the Middle Ages seven such weeks, plus Holy Saturday (that is thirty-six days); in the West six weeks, excluding only Sundays (thirty-six days). From the seventh century in the West, Ash Wednesday and the subsequent three days were added, making forty. The forty days were known also as Quadragesima, and that term was also given to the first Sunday of the period. By analogy, the previous Sundays became known as Quinquagesima (which is indeed fifty days before Easter), Sexagesima, and Septuagesima. The latter is a term first encountered in the Gelasian Sacramentary. Although not the start of the Lenten fast proper, it was marked in the Middle Ages by a change to purple vestments and the exclusion of the word 'alleluia' from all services. The terms Septuagesima, etc., were suppressed in Roman usage in 1969.

After Easter Sunday there are seven weeks, fifty days, until the Sunday of Pentecost. The important stage on the way is Ascension Day, forty days after Easter. Pentecost was, like other feasts in the Christian year, originally a Jewish holy day. The term 'Pentecost' (literally 'fifty days') at first referred to the whole period, but was soon restricted to the day itself.

The length of the time preceding Septuagesima, that is the period after Epiphany, must therefore be adaptable in order to allow for the varying date of Easter. If Easter is early, then there may be no more than one Sunday after Epiphany. If it is late, then up to six will be celebrated. Similarly, the fixed start of Advent will bring to an end the series of Sundays after Pentecost, whose number varies according to whether Easter is early or late.

From about the ninth century the first Sunday after Pentecost was often given over to a commemoration of the Holy Trinity. The long series of Sundays through the summer season of the year might therefore be reckoned as Sundays after Pentecost or after Trinity. (From the fourteenth century the Thursday after Trinity was usually celebrated as the Feast of Corpus Christi, officially instituted by Urban IV in 1264.) Interestingly, the enumeration of these Sundays is generally different according to whether a mass or an office book is involved. In missals and graduals it was customary to set out a cycle of Sundays after Pentecost (or Trinity), numbered 1 to 23 (24, 25, or however many were deemed to be necessary by the compiler of the book; the later Sundays might be omitted). Office books, however, usually tied the Sundays to specific dates of the calendar, and it was the early Sundays that were omitted if necessary.

Before the fixed date of Christmas comes the Advent season, which begins with the Sunday nearest St Andrew's Day, 30 November. At least four Sundays therefore precede Christmas. After Christmas, Epiphany on 6 January is celebrated as the day when Christ was 'manifested' to the Gentiles, that is, to the three Magi. It had additional significance as the day of Christ's baptism (that was its primary significance in the East, whence the feast originally came), and of the miracle Christ performed at the wedding feast at Cana.

At four times in the year special days of fasting and abstinence are observed, the so-called Ember Days (so called because of the ashes sometimes marked on the head at this time). Although at first their date was somewhat variable, and only three of them were observed, by the sixth or seventh century there were four, known often as the 'quatuor tempora'. They fell on the Wednesday, Friday, and Saturday of four weeks, after St Lucy (13 December; or, alternatively, during the third week of Advent), Quadragesima Sunday (that is, during the first week of Lent), Whit Sunday, and the Exaltation of the Holy Cross (14 September). The liturgy during these days is somewhat similar to those of Lent.

There are several other days of special observance, which require separate description in later chapters. Among these are the Rogation Days, the Monday, Tuesday, and Wednesday before Ascension Day (which is on Thursday). The preceding Sunday, actually the fifth after Easter, is thus known as Rogation Sunday. On the Rogation Days special prayers of intercession, including the 'greater Litany', are chanted (cf. Latin 'rogare': to ask), and processions are held. Processions will be discussed later, but it may be stated here that Palm Sunday is naturally distinguished by special processional rites, as befits the day commemorating Christ's triumphal entry into Jerusalem. Later in that week, Holy Week, the liturgies of Maundy Thursday, Good Friday, and Holy Saturday are unique, as well as very ancient in many respects. Some of the special ceremonies in these days concern the blessing and bringing into use of special symbolic objects: candles on the feast of the Purification of the Blessed Virgin Mary (2 February), penitential ashes on Ash Wednesday, palm branches on Palm Sunday, the New Fire, paschal candle, and font on Easter Eve.

When no special day occurred, the liturgy of the day reverted to a ferial form. (The Latin word 'feria' actually means 'feast'; in Latin Christian use it came to mean a day of worship; it was inevitably most often used when no other obvious designation could be given to a day, and therefore came to mean 'non-feast day'. Weekdays from Monday to Friday are 'feria ii' to 'feria vi'; Saturday is usually 'Sabbato' rather than 'feria vii', and Sunday, though actually 'feria i', is always called 'Dominica'.) In the celebration of mass, this meant repeating the formularies of the previous Sunday or feast-day, with some reduction of solemnity; in the office a regular weekly or ferial set of services, independent of any feast, was resumed.

Some especially solemn parts of the year, however, had proper formularies (prayers, lessons, chants, etc.) for each day of the week. These are the days of Lent, and the week after Easter. It should be said, however, that as time went on there was a tendency to provide more proper material (that is, prayers, chants, and particularly lessons special to a particular day); it is not uncommon to find lessons and prayers assigned to particular days in Advent (Wednesday, Friday, Saturday outside the Ember Week) and elsewhere in late medieval books.

Tables I.3.1. and 2 give lists of the two sections of the church year described briefly so far.

The two components of the church year sketched briefly above—the fixed part associated with Christmas and the movable part associated with Easter—were joined by a large number of commemorative days for saints and other holy persons. The decision as to which saints were to be remembered by special services was to a certain extent a local one, but all liturgical books have services for days of considerable antiquity, when the heroes of the early church are honoured. Then there will be remembered those who brought Christianity to the country, diocese, or area relevant

Table I.3.1 *The Church Year: the fixed part dependent upon Christmas*

1st Sunday in Advent—as near St Andrew (30 Nov.) as possible
2nd Sunday in Advent
Ember Days: Wednesday, Friday, Saturday—in the week after St Lucy (13 Dec.)
3rd Sunday in Advent
4th Sunday in Advent
Vigil of Christmas Day, or Christmas Eve (24 Dec.)
Nativity, or Christmas Day (25 Dec.)
Octave of the Nativity, or Feast of the Circumcision, or New Year's Day (1 Jan.)
Epiphany (6 Jan.)
Sundays after Epiphany (up to 6 as required)
Ember Days: Wednesday, Friday, Saturday—in the week after Exaltation of the Holy Cross
    (14 Sept.)

Table I.3.2. *The Church year: the movable part dependent on Easter*

Septuagesima Sunday (9th before Easter)
Sexagesima Sunday (8th before Easter)
Quinquagesima Sunday (7th before Easter)
Ash Wednesday
Ember Friday
Ember Saturday
Quadragesima Sunday (1st in Lent, 6th before Easter)
2nd Sunday in Lent (5th before Easter)
3rd Sunday in Lent (4th before Easter)
4th Sunday in Lent (3rd before Easter)
Passion Sunday (5th Sunday in Lent, 2nd before Easter)
Palm Sunday
Maundy Thursday
Good Friday
Holy Saturday
Easter Sunday
1st Sunday after Easter (Low Sunday)
2nd Sunday after Easter
3rd Sunday after Easter
4th Sunday after Easter
5th Sunday after Easter
Ascension Day (Thursday)
Sunday after Ascension
Pentecost Sunday, or Whit Sunday
Ember Days: Wednesday, Friday, Saturday
Trinity Sunday
Sundays after Pentecost or after Trinity (up to 25 as required)

to where the book is used. Finally, there may be commemorations peculiar to the particular church for which the book is compiled.

Frere (1930) is an account of how the cycle of sanctoral developed, and it is instructive to look through the kalendars edited by Wormald (1934, 1939, 1946), for example, to see the typical intermingling of universal and local commemorations. As an example I give in Table I.3.3. in an abbreviated form the entries of the month of October from the kalendar in an eleventh-century English manuscript, the so-called 'Portiforium of St Wulstan', Cambridge, Corpus Christi College 391. (Most of the text of this manuscript is edited in Anselm Hughes 1958–60, the kalendar in abbreviated form in Dewick and Frere 1921 and Wormald 1934. A facsimile of the October page is given by Dewick and Frere, pl. I.) To help the reader I have given the dates 1 to 31 in the first column. These do not appear in the original manuscript. In the second column (the original first column) appears a series of roman numerals which indicate when the full moon will appear in particular years of the nineteen-year cycle. The golden number for 1 October is 16: in the sixteenth year of the cycle the full moon will fall on that day. (This information is particularly important, of course, for the period when Easter may be celebrated.) The next two columns of letters in the original are not of liturgical significance and are omitted here. Then come the dominical letters for each date. After that is given the date not in modern but in Roman fashion, beginning with a large KL for the *Kalendae*, then the days before NON (*Nonae*), before ID (*Idus*), and before the *Kalendae* at the beginning of the next month.

Many of the saints' names inscribed for October on this kalendar might be found on almost any Latin Christian kalendar. Others are more local. I have separated these out in the lists given after the kalendar in Table I.3.3.

Some of these feasts are of especial importance. SS Simon and Jude have a Vigil marked: that is, the previous day will also be marked by liturgical material in their honour. (Some saints have the further distinction of an Octave, one week later, when their liturgy will be repeated once more.) On 31 October a Vigil is marked for the next day, 1 November, which is All Saints' Day. Some feasts are marked 'lc. xii' (twelve lessons) in red ink, to show their special solemnity: the Night Office will be celebrated with the full number of lessons and reponsories. Among the additions (preceded by an asterisk) the Translation of St Oswald of Worcester is entered in red ink (indicated by R in brackets in the table). (This way of indicating specially important feasts is the origin of the popular expression 'red-letter day'.)

Table I.3.3. *Kalendar for October: Cambridge, Corpus Christi College 391*
* denotes material added in a later hand

AEQUAT ET OCTOBER SEMENTIS TEMPORE LIBRAM

1	xvi	A	KL OCT	Sanctorum Remigii. Uedasti. Germani. * xii lc. * Obitus Læfgæuæ mater Godithe.
2	v	B	vi NON	Sancti Leodegarii episcopi.
3	xiii	C	v	
4	ii	D	iiii NON	
5		E	iii	
6	x	F	ii NON	* Sancte Fidis. virginis et martiris. xii lc. * Obitus Æadwi decani.
7		G	NON	Sancti Marci pape. Marcelli et Apulei.
8	xviii	A	viii ID	* TRANSLATIO SANCTI OSVVALDI ARCHIEPISCOPI. et Sancti Demetrii martiris. (R)
9	vii	B	vii	Sanctorum Dionisii. Rustici. et Eleutherii.
10		C	vi ID	Sancti Paulini episcopi.
11	xv	D	v	
12	iiii	E	iiii ID	Sancti Uuilfridi episcopi. * lc. xii
13		F	iii	

14	xii	G	ii ID	Sancti Calesti pape.
15	i	A	IDUS	* Commemoratio Sanctorum quorum reliquie hic habentur. (R)
16		B	xvii KL NOV	
17	ix	C	xvi	
18		D	xv KL	Sancti Luce euangeliste et Sancti Iusti martyris.
				SOL IN SCORPIONEM
19	xvii	E	xiiii	
20	vi	F	xiii KL	
21		G	xii	* Ordinatio sancti DVNSTANI archiepiscopi.
22	xiiii	A	xi KL	
23	iii	B	x	
24		C	ix KL	* Obitus Henrici episcopi
25	xi	D	viii	*Sanctorum Crispini et Crispiniani.
26		E	vii KL	
27	xix	F	vi	VIGILIA
28	viii	G	v KL	APOSTOLORUM SYMONIS. ET IUDE.
29		A	iiii	
30	xvi	B	iii KL	
31	v	C	ii	Sancti Quintini martyris. VIGILIA.
				NOX HORAS .XIII. DIES .X.

Notes. The feasts may be classified as follows:

(i) feasts of the Blessed Virgin Mary, Apostles, and other biblical persons: 18 Luke the Evangelist; 28 the Apostles Simon and Jude

(ii) early martyrs, bishops (usually of Rome, i.e. popes), and Church Fathers: 7 Pope Mark (336), the martyrs Marcellus and Apuleius; 14 Pope Calixtus I (d. 222)

(iii) saints local to North France or England: 1 Remi of Reims, Vedast of Arras, and Germanus of Auxerre; 2 Ledger of Autun; 9 Denis of Paris and his companions Rusticus and Eleutherius; 10 Paulinus of York (d. 644); 12 Wilfrid of York (c.633–709); 18 Justus of Beauvais; 31 Quentin of Saint-Quentin

(iv) Additions: 1 Obit of Læfgæva, mother of Lady Godiva, benefactress of Worcester and Bishop Wulfstan; 6 Faith, apparently a 3rd-c. martyr; obit of dean Æadwi; 8 Translation of St Oswald of Worcester, carried out by Wulfstan in 1089; also Demetrius, a 3rd-c. martyr; 15 Feast of the Relics at Worcester, instituted by Wulfstan on the Octave of St Oswald; 21 Dunstan of Canterbury (909–88); 24 Obit of Bishop Henry (d.1189); 25 the 3rd-c. Roman martyrs Crispin and Crispinian; one legend has it that they fled persecution to Faversham in Kent

All the indications of twelve lessons for certain feasts are in a later hand.

This kalendar was clearly consulted for some time after its compilation, for it has received numerous additional entries, all of more or less local significance. These also are listed after Table I.3.3. The obits (entries recording the deaths of prominent persons) are unlikely to be of liturgical moment. Rather the kalendar has been used as a convenient place to note these secular events.

There is a certain amount of astronomical information on the kalendar, the number of hours of day and night, the position of the sun in the zodiac. Not all kalendars include this, especially later medieval examples, but particularly from the thirteenth century onward kalendars were often much more specific about the degree of solemnity of the feast: not simply how many lessons were to be sung, but its grade according to liturgical dress (for example, whether or not copes were to be worn), the number of 'rulers' (leading singers) for the choir, and so on. The various possibilities were subsumed under a series of headings: feast of first class, second class, double, semi-double, etc., by which were understood all the possible distinctions of ritual at a particular institution. (Different churches had different systems of grading. For much useful information on the consultation of later kalendars, see Andrew Hughes 1982.)

Setting out the feasts to be celebrated in a form convenient for quick consultation, kalendars are often useful for determining the provenance of a manuscript, according to the local elements they may contain. The indication 'lc. xii' (twelve lessons) shows that Cambridge, Corpus Christi College 391 was used at a monastic institution, and its numerous Worcester peculiarities leave its provenance in no doubt. The rest of the manuscript comprises computus tables, and material for the Benedictine office: a psalter, hymnal, canticles, collectar, blessings, and a series of offices for saints, Sundays, and so on. (See III.6).

A list only of the more important saints' days, such as might be found in almost any medieval kalendar, is given in Table I.3.4. Some of these feasts (Margaret, Katherine, Nicholas, Thomas of Canterbury) became popular only later in the Middle Ages. (For some later feasts see Pfaff 1970.)

I.4. THE DAILY ROUND

Andrew Hughes 1982.

In the Roman Church a fairly constant daily pattern or succession of services was established by the eighth century, although certain days of the year differed from the majority in having special services of an individual character. This generalization holds good despite the inevitable modifications which occurred as time went on and the varying preferences of different regions and churches.

The most important service of the day was Mass, which usually took place in the morning. (The actual time varied from one season of the year to the other, and also, of course, from place to place. The same is true of all the other daily services.)

The start of the day, from a liturgical point of view, was anciently the evening before, when the Vespers service took place, notionally at sunset. (This convention was taken over by early Christians from the Jewish reckoning of time.) It was followed by Compline, a brief service before the community retired for the night. On Sundays and important feasts the liturgical day was reckoned to extend further and to include the next Vespers and Compline service, and this meant that for all the subsequent ferias Vespers and Compline were not the evening before, so to speak, but at the end of the day. (Compline had very little proper material, special to one day rather than another, so its assignment was not a sensitive matter.) If two feast-days fell on consecutive days, then usually Vespers of the second feast took precedence.

Table I.3.4. *Saints widely celebrated in the Roman Church*

JANUARY
14 Felix M
16 Marcellus P & M
20 Fabian & Sebastian MM
21 Agnes V & M
22 Vincent M
25 Conversion of Paul AP *
28 Octave (Second Feast) of Agnes

FEBRUARY
2 Purification of the BVM
5 Agatha V & M
22 Peter's Chair (Cathedra) *

MARCH
12 Gregory P
21 Benedict AB *
25 Annunciation of the BVM

APRIL
4 Ambrose EP & D *
23 George M

MAY
1 Philip & James APP
3 Invention of the Holy Cross *
6 John before the Latin Gate

JUNE
24 John the Baptist
26 John & Paul MM
29 Peter & Paul APP
30 Paul AP

JULY
10 Seven Brothers MM
11 Translation of Benedict AB *
20 Margaret *
22 Mary Magdalene *
25 James AP *

AUGUST
1 Peter's Chains (Vincula)
3 Invention of Stephen M and companions*
6 Transfiguration *
10 Laurence M
15 Assumption of the BVM
24 Bartholomew AP *
   Augustine EP & D *
29 Beheading of John the Baptist *

SEPTEMBER
8 Nativity of the BVM
14 Exaltation of the Holy Cross
21 Matthew AP & EV*
22 Maurice and his companions MM *
29 Michael the Archangel
30 Jerome D *

OCTOBER
9 Denis EP & M and his companions *
18 Luke EV *
28 Simon & Jude APP *

NOVEMBER
1 All Saints *
8 Four Holy Crowned Martyrs
11 Martin EP
22 Cecilia V & M
25 Katherine of Alexandria V & M *
30 Andrew AP

DECEMBER
6 Nicholas EP *
13 Lucy V & M
21 Thomas AP *
26 Stephen M
27 John the Evangelist
28 Holy Innocents
20 Thomas of Canterbury M *
31 Silvester P

---

* not in the 'Hadrianum', the Roman sacramentary sent to Charlemagne in the late 8th c.

AB	= Abbot	EV	= Evangelist
AP	= Apostle	M	= Martyr
BVM	= Blessed Virgin Mary	P	= Pope
D	= Doctor	V	= Virgin
EP	= Bishop		

During the night, the Night Office was sung, the longest of the services other than Mass. The service developed from the early Christian Vigils service, and was known by that name occasionally up until the eleventh century. Its most common name is Matins, but this name was also used in medieval times for the service at daybreak, Lauds, and is therefore avoided here. Yet again, the name Nocturns is occasionally found, reflecting the composition of the service as a number of 'nocturns'.

After Lauds, notionally at daybreak, there followed a succession of Day Hours, or Little Hours. These were Prime, Terce, Sext, and None (or Nones), which in theory took place at the first, third, sixth, and ninth hour of the day (beginning, in summer at least, at our 6 a.m.; the 'hours' were shorter, and began later, in winter).

The main mass of the day, often called the *magna missa* or high mass, usually came between Terce and Sext on Sundays and feasts, after Sext on ferial days. There were, however, often other masses of one kind or another (low mass), especially in religious communities. Medieval monasteries frequently celebrated a morning mass before the daily meeting in chapter. A daily mass for the dead was also often celebrated. Later in the Middle Ages, when many important churches had acquired altars and chantries specially endowed so that mass could be said (not usually sung) at them daily, the number of masses celebrated was staggeringly large. (Harrison 1963, 56 cites the Lincoln statutes of 1531, which indicate forty-four masses daily.) Almost all of these were of little musical significance, said by a single priest throughout. An important exception to this generalization was the Mass of the Blessed Virgin Mary, which from the thirteenth century became a weekly celebration, on Saturday, and was often embellished by the polyphonic setting of some of its chants. Such a mass, often taking place in a specially built Lady Chapel dedicated to the Virgin, is known as a votive mass. Sometimes a weekly cycle of votive masses was organized, for example: for the Holy Trinity on Sunday, the Holy Spirit Monday, Angels Tuesday, All Saints Wednesday, Corpus Christi Thursday, the Holy Cross Friday, and the Blessed Virgin Mary on Saturday. So ardent was veneration of the Blessed Virgin in the later Middle Ages that a complete cycle of Hours of the BVM, parallel with the main cycle, was often sung or said.

Some monastic observances at other times of the day were in many ways as formalized as a liturgical service. One such occasion was the daily meeting which monastic communities held in their chapter-houses. It usually included the reading of the martyrology, containing accounts of the saints to be commemorated on the day in question. Several versicles and responses were said, with Paternoster and Kyrie, and readings might include passages from the Rule of St Benedict, or from the gospels. After the business matters had been attended to, the meeting ended with a series of psalms for the dead, recited in the chapter-house or on the way to the church for mass. Little if anything in the meeting seems to have involved singing.

Among other monastic additions to the office (their description takes more than 100 pages in Tolhurst 1943) may be mentioned:

   (i) the Office of the Dead, that is Vespers, the Night Office, and Lauds, said in
       parallel with the main or canonical office; from the first word of its first

antiphon the Vespers service was known as 'Placebo', and the Night Office as 'Dirige';

(ii) a similarly formed Office of All Saints;

(iii) the fifteen Gradual Psalms (Pss. 119–33) said privately by each monk seated ready for the start of the Night Office;

(iv) the Trina Oratio, a threefold devotion performed three times daily, before the Night Office, before Prime in summer, and before Terce in winter, and after Compline, whose main constituent was the saying of three groups of penitential psalms, each group introduced by a prayer (oratio);

(v) the Psalmi Familiares, psalms said after each hour for the royal family.

The exact time of the services depended on their content (shorter days in winter might mean shorter day services; ferial services were shorter than festal ones), on whether a period of fasting was in force, and on the other obligations of the community. The extent to which a monk's waking hours would be occupied with religious devotions or services, in church or elsewhere, may be seen from modern editions of any medieval monastic 'horarium' (timetable), for example those of tenth- and eleventh-century England, in the *Regularis Concordia* and the statutes of Lanfranc, respectively, published by Knowles (1963, 714, 448 ff.; Knowles 1969, 217, based on Knowles 1951). Table I.4.1 shows the usual timetable at eleventh-century Cluny, the foremost Benedictine monastery of its time. For most days of the year, the succession of services followed this order. Several special days, such as the triduum, the three days preceding Easter Sunday, followed a somewhat different pattern. On many high feasts a solemn procession was made, most commonly before mass. An account of the most important of these special days is given later. First an outline of the more usual services is presented.

As with the constituent parts of the day, so the contents of each individual service varied from occasion to occasion, being especially dependent on the solemnity of the day. In the descriptions of each service in the sections below, as usual in this book, only a generalized account of 'normal' practice is therefore given. (Most of these services accumulated varied amounts of prefatory material in the form of prayers, versicles, and responses, but this is not described.)

Mass, the most important and complex of the services, is described first. Although it had some elements analogous to features of the office hours, its essential nature and history resulted in many unique forms. Whereas the office hours are static, contemplative services, Mass has elaborate opening ceremonies, and the elevated ritual of the eucharist itself. Its chants are exceptionally varied. The introit and communion resemble an antiphon with psalm verses simply intoned. The offertory is an elaborate chant with verses in highly ornate musical style. After the opening rituals, with the Kyrie litany chant, and the extended Gloria hymn, come lessons and melismatic chants, almost the only moments of repose in a seemingly dramatically unfolding performance. Two of its important chants, the Sanctus and Agnus Dei, are actually part of a long and elaborate series of prayers.

The most regular part of the office services, by contrast, was the singing of psalms.

Table I.4.1. *The liturgical horarium at Cluny in the late eleventh century (after Hunt 1967, 101–3)*

Winter (extra Lenten observances in brackets)	Summer
Trina Oratio last 32 psalms	Trina Oratio
Night office 4 *psalmi familiares* (2 psalms prostrate)	*Night office* 2 or 4 *psalmi familiares*
(Procession to Church of Our Lady) Lauds of All Saints, Lauds of the Dead, extra psalmody	
Lauds extra psalmody 4 *psalmi familiares* (2 psalms prostrate)	*Lauds* 2 or 4 *psalmi familiares* Trina Oratio
Prime (at dawn) 4 *psalmi familiares* (2 psalms prostrate, 7 penitential psalms) Litany 4 psalms	*Prime* (at dawn) 2 or 4 *psalmi familiares* Litany 7 psalms
	Missa matutinalis
Chapter 'private' masses	Chapter 'private' masses
Terce 4 *psalmi familiares* (2 psalms prostrate)	*Terce* 2 or 4 *psalmi familiares*
Missa matutinalis	*Missa maior*
Sext 4 *psalmi familiares* (2 psalms prostrate) Litany	*Sext* 2 or 4 *psalmi familiares*
Missa maior	
midday meal	midday meal
None 4 *psalmi familiares* (2 psalms prostrate)	*None*
Vespers 4 *psalmi familiares* (2 psalms prostrate)	*Vespers*
(Procession to Church of Our Lady) Vespers of All Saints, Vespers of the Dead	
evening meal	evening meal
(Vigils of the Dead)	
Compline Trina Oratio	*Compline* Trina Oratio

A weekly round of the Office would hear the singing of all 150 psalms, if uninterrupted by feasts of one kind or another, which would have their own proper psalms. Most of the first two-thirds of the psalms were sung during the Night Office, the last third during Vespers. Lauds had the last three psalms, which, since they begin with the word 'Laudate', are supposed to have given the service its name. Others were used at various hours because of their allusions to particular times of day, as Ps. 62 at Lauds: 'O God thou art my God: early will I seek thee'; or Ps. 4 at Compline, with its ninth verse: 'I will lay me down in peace, and take my rest . . .'. The rather complex deployment of the psalms across the week's services is outlined in both secular and monastic cursus by Andrew Hughes 1982, 52. A simplified outline is given in Table I.4.2. When it is stated below that a certain number of psalms are sung at such and such a service, it should be understood that 'psalm' might actually mean more than one psalm according to the numerical series, if a particularly brief psalm were involved. Correspondingly, longer psalms might be split into sections.

Table I.4.2. *Psalms allotted to office hours, ferial cursus*

Secular cursus:
Night Office: 1–3, 6–20, 26–41, 43–9, 51, 54–61, 63, 65, 67–88, 93–108
Vespers: 109–16, 119, 121–32, 134–41, 143–7
Prime, Terce, Sext and None: 21–5, 53, 117–18, 120
Lauds: 5, 42, 50, 52, 62, 64, 66, 89, 91–2, 99, 142, 148–50
Compline: 4, 30, 90, 133

Monastic cursus:
Night Office: 20–34, 36–41, 43–9, 51–5, 57–61, 65, 67–74, 76–86, 88, 92–108
Vespers: 109–16, 128–32, 134–41, 143–7
Prime, Terce, Sext and None: 1–2, 6–19, 118–27
Lauds: 5, 35, 42, 50, 56, 62–4, 66, 75, 87, 89, 91, 117, 142, 148–50
Compline: 4, 90, 133

On the numbering of the psalms in medieval usage, as opposed to the English Authorized Version of the Bible, see Andrew Hughes 1982, 51, 224–6.

The descriptions below begin with the Night Office, because it is the most substantial musically, distinguished from the other services by its lessons and associated great responsories. Next in size, and somewhat similar to each other, are Vespers and Lauds, while Compline and the so-called Little Hours of Prime, Terce, Sext, and None are comparatively simple and brief.

All except the four Little Hours contain not only psalms but one or more canticles, passages from the Old Testament or other sources in the nature of 'songs', and sung exactly like psalms. The best-known examples are the three New Testament canticles sung towards the close of Vespers, Compline, and Lauds, respectively: the Magnificat (Luke 2: 46–55, the Song of the Blessed Virgin Mary, when Gabriel announced to her that she was to bear Christ), the Nunc Dimittis (Luke 2: 29–32, the Song of Simeon, when he had seen Christ), and the Benedictus (Luke 1: 68–79, the Song of Zacharias, when his son John the Baptist was born). Fourteen others were sung on a weekly basis at Lauds, and one other (varying with the season) was sung in the Night Office in monastic uses on Sundays and feasts.

All the offices contain at least one responsory. These chants, consisting of respond and psalm verse, with alternating solo and choral sections, are of two kinds. Those of the Night Office, the 'great' responsories, are elaborate, melismatic chants, among the chief glories of the chant repertory. The other hours usually have a 'short' responsory, syllabic in style, whose simple character has occasioned the belief that they may contain vestiges of the most ancient recoverable psalmodic practice. At Vespers on high feasts it was sometimes the practice to use one of the great responsories. (They were commonly thus employed also in processions, at least in the later Middle Ages.)

The other musical element in the office services to which attention may be drawn in advance is the hymn, one of which was generally sung at each of the hours.

The musical genres just mentioned are those that are discussed in separate chapters later in this book. Naturally, less attention is paid to the numerous versicles and responses, brief dialogues between priest (or other person) and choir, which are intoned to simple formulas at many points in the services.

I.5. MASS

Jungmann 1962; Andrew Hughes 1982.

The actual start of mass was preceded by various ceremonies of preparation. On Sundays, these included the blessing and sprinkling of salt and water on the main altar of the church, with prayers, responses, and the antiphon with psalm verse *Asperges me* V. *Miserere mei* and *Gloria* (*Vidi aquam* V. *Confitemini* in Paschal time, from Easter Sunday to Whit Sunday). If several altars were to be sprinkled, the ceremony might involve a procession. After the priest and his assistants had vested themselves, to the accompaniment of more prayers and the recitation of Ps. 42, they moved to the altar.

The introit accompanied this solemn entry. Because it was part of the proper, a different introit being sung on each important day of the year, it was usually the first chant copied for a particular day in a gradual or missal.

The Kyrie and Gloria followed, their texts remaining basically the same throughout the year and therefore being reckoned part of the ordinary of mass. From a musical point of view, the distinction is not strictly appropriate, since a variety of melodies were available for both chants, and particular melodies became associated with particular feast-days, or classes of feasts (important, not so important, feasts of the Blessed Virgin Mary, etc.). Furthermore, tropes of one kind or another might make the Kyrie and Gloria proper to a particular feast.

The priest then intoned the collect proper to the day. This ended the opening section of the mass.

There followed the lessons and responsorial chants of mass. The lessons were the epistle and the gospel, between which were sung various combinations of gradual, alleluia, sequence, and tract, depending on the liturgical season. The gradual was a more or less constant item, except for the Saturday of Easter week and the days up to Whit Sunday, when an extra alleluia was sung. The alleluia was not sung from Septuagesima to the end of Lent, nor on penitential Ember Days outside Lent, though this was relaxed for the Pentecost Ember Days. The tract was sung only in penitential seasons: on Septuagesima and succeeding Sundays until Easter, and on Mondays, Wednesdays, and Fridays during Lent, beginning with Ash Wednesday. The sequence was sung only on the highest feasts, and not usually on those falling in Lent.

This part of mass was considerably extended on the Ember Saturdays. Four lessons were each followed by a gradual (an alleluia at Pentecost), then a reading from Daniel was followed by the canticle (sometimes labelled 'tractus' or 'hymnus') with refrain (*Benedictus es Domine* V. *Et laudabilis* in Advent, *Benedictus es in firmamento* V. *Hymnum dicite* in Lent, *Omnipotentem semper adorant* V. *Et benedicunt* in September. Modern practice has *Benedictus es* on all three occasions.) After that came the epistle, another tract, and the gospel. The Wednesday of the fourth week of Lent also had extra lessons: this was the so-called 'Day of the Great Scrutiny', when catechumens (see below) were examined before being admitted to baptism.

The gospel was intoned by the deacon. The Credo was then sung (from the eleventh century, and only on the most solemn days).

In the early centuries the mass up until the gospel had been regarded as a 'fore-mass' service, for all Christians to attend. What followed, however, was for baptized Christians only. By the central Middle Ages no trace remained of the dismissal of the catechumens (those undergoing preparation for baptism), which would previously have taken place here. The change in the character of the liturgy is noticeable, nevertheless, the entrance ceremonies having been succeeded by lessons and responsorial chants, which now gave way to the solemn prayers and chants of the eucharistic ritual.

The gifts were brought to the altar as the offertory chant was sung, and the music might last through the priest's preparation of the altar, reception of and prayers over the gifts, including the silent prayer known as the secret (proper to the day). There followed the preface, which began with short phrases sung in dialogue between priest and congregation and was continued by the priest alone, a long text with sections proper to the season. It led directly into the Sanctus chant, sung by the choir. The text of the preface always requires the Sanctus as its natural conclusion, for example:

(*Preface, ending*) . . . Et ideo cum angelis et archangelis, cum thronis et dominationibus, cumque omni militia caelestis exercitus, hymnum gloriae tuae canimus, sine fine dicentes: Sanctus, Sanctus, Sanctus Dominus Deus Sabaoth . . .

And therefore with angels and archangels, with thrones and dominations, and the whole host of the heavenly army, we sing the hymn of thy glory, evermore singing: Holy, Holy, Holy Lord God of Hosts . . .

Sanctus and Benedictus are musically linked, both concluding with 'Hosanna in excelsis', but from the later Middle Ages it was customary to split off the Benedictus and sing it during a later part of the prayers of this part of mass, for example after the consecration of the bread and wine. The prayers began with the canon, a long series of brief formulas which include Christ's words 'This is my body . . . This is the cup of my blood . . .' The Paternoster (Lord's Prayer) was sung shortly after, and the bread was broken. During the next group of prayers a small portion of the bread was mingled with the wine, and the Agnus Dei chant was sung by the choir. Further solemn prayers were sung while the bread and wine were consumed by the priest (they were rarely received by anyone else in the Middle Ages). The communion chant was sung during the clearing-up actions, after which the priest intoned the postcommunion prayer of thanksgiving (proper). During the dismissal of the congregation, the priest would intone another proper prayer, the Super Populum (literally 'over the people'), during Lent.

'Ite missa est' ('Go, you are dismissed'), the words which gave the mass its name (*missa*), was the usual form of dismissal sung by the priest, answered 'Deo gratias' ('Thanks be to God') by the choir. However, it was generally used only when the Gloria was sung, that is during Christmas, Epiphany, Easter, and the summer season. On other occasions the conclusion was 'Benedicamus Domino R. Deo gratias', and this formula was also used when another service followed immediately upon mass, as would be the case on several of the highest feasts with extended liturgies (Andrew Hughes 1982, 93).

Since early times it has been the custom, for various practical reasons, to preserve the bread (occasionally also the wine, also bread onto which a drop of wine has been placed) consecrated at mass but not consumed there. This is known as reservation of the sacrament. It may be kept in one of several places: the sacristy, outside the church; an aumbry in the wall of the church; a pyx suspended over the altar; or a tabernacle on the altar.

In the Middle Ages, and for centuries afterwards, communion by any person other than the priest was relatively unknown, outside one or two of the most important days of the church year. For theological reasons (and perhaps, for many, because of the lack of closer contact with the sacred elements) there developed from the thirteenth century on the practice of elevation. After it had been consecrated, the celebrant raised on high the host (that is, the bread—elevation of the wine was a later trend) for all to see. The singing of the Benedictus chant, musically a part of the Sanctus, was often delayed to coincide with this solemn moment.

Veneration for the sacred host eventually resulted in ritual forms, in particular Benediction. Here the host was exhibited outside mass, and a procession might be made with it, culminating in a general blessing with the host. Favourite accompanying chants were hymns such as *Adoro te devote* and *Tantum ergo sacramentum* (the fifth verse and following of *Pange lingua gloriosi*). As a result of their great popularity, many of these have multiple melodies. (The chief result of the increase in veneration for the host was the establishment of the feast of Corpus Christi, introduced in Liège in 1246, prescribed for the whole church by Urban IV in 1264.)

I.6. OFFICE

  (i) The Night Office (Matins, Vigils, or Nocturns) in Secular Use
 (ii) The Night Office in Monastic Use
(iii) Lauds in Secular Use
 (iv) Lauds in Monastic Use
  (v) Vespers in Secular Use
 (vi) Vespers in Monastic Use
(vii) Compline
(viii) Prime, Terce, Sext, and None

Tolhurst 1943; Andrew Hughes 1982.

(i) *The Night Office (Matins, Vigils, or Nocturns) in Secular Use*

Apart from the differences between festal and ferial forms of the Night Office, there are other important differences between secular and monastic forms. (By 'secular', 'Roman', or 'canonical' is understood the use of non-monastic churches such as most cathedrals, collegiate chapels, and parish churches.) As with most of the offices, the singing of psalms and antiphons forms an important part of the service, but the Night Office is distinguished by a group of great responsories, sung in association with the lessons. These numerous and lengthy responsorial chants are at once among the chief glories of the chant repertory and yet among its least known parts, since in modern times the singing of the Night Office in its medieval manner has all but ceased.

Before the Night Office began, Paternoster, *Ave Maria* and the Apostles' Creed (*Credo in Deum patrem omnipotentem*) were said silently.

The Night Office started with versicles and responses. The first was usually *Deus in adiutorium meum intende* R. *Domine ad adiuvandum me festina*, with *Gloria*, which began each of the office hours (it was often omitted during the triduum, that is the last three days before Easter Sunday, and during Easter Week). This was followed by *Domina labia mea aperies* R. *Et os meuim annuntiabit laudem tuam*.

The Invitatory psalm (Ps. 94, *Venite exultemus Dominum*) was sung, with its antiphon, the invitatory. A hymn followed (less commonly found in early medieval secular uses than in monastic ones). A further versicle and response led to the first Nocturn. This was a group of psalms and antiphons, lessons and responsories, with more brief versicles and responses. The number of nocturns depended on the solemnity of the day, less important days having only one.

Psalms and antiphons began each nocturn. There were different conventions about the number of psalms and antiphons. The singing of an antiphon preceded the whole group of psalms and concluded the series, but it might not be performed between individual psalms in the group. The same was true of the Gloria ('Glory be to the Father . . .'), added as a final verse to all or only to the last of the psalms. Some of the commoner schemes were the following (A = Antiphon, Ps = Psalm, Gl = Gloria patri):

On ferial days: A, two Ps, Gl, A—six times (total of 12 Ps, 6 A)
On Sundays: A, four Ps, Gl, A—three times (total of 12 Ps, 3 A)
On major feasts: A, Ps, Gl, A—three times (total of 3 Ps, 3 A)

A further versicle and response were sung, commonly to a tone which, like the opening *Deus in adiutorium*, recurred at other hours of the day.

The Paternoster was said silently until the final verse, which was said aloud. Then the lector asked for the blessing of the officiating priest. At the end of the lesson he would intone a further versicle.

After each lesson (with versicles), one of the great responsories was sung. In secular use three lessons and thus three responsories were sung in each nocturn. The usual form of the responsory was respond–verse–respond, but for the last responsory of the nocturn the Gloria would be used as a second verse, as it were, making the scheme R, V, R, Gl, R. (The repeats of the respond were usually progressively shortened.)

The lessons might not all be biblical. The first nocturn (on some days the only nocturn) had Old Testament readings (except during the Easter season), from different books according to season. The others commonly had sermons (commentaries on the appropriate liturgical theme by one of the Fathers of the Church, such as St Augustine of Hippo or St Gregory the Great) or homilies (commentaries, again by a Church Father, on a passage from the Gospels, which would therefore itself be intoned at the start of the lesson). The lessons on a saint's day might be drawn from his *vita* or *legenda* (account of his life, death and miracles—the terms mean literally 'life' and 'that which is to be read').

The number of nocturns to be sung according the above patterns depended upon the solemnity of the day. Sundays and feast-days usually had three. An important exception was Easter Sunday, which had only one (but some later medieval books give it three), and this was also sometimes the pattern for the rest of the Easter season up to and including Whit Sunday.

Although in general antiphons and responsories may be said to be proper, in the sense that each day usually had its own particular chants, there was a considerable amount of reuse within weeks or seasons. For example, for those parts of the year where nine responsories were sung on Sunday and three on each weekday, the Sunday nine might be re-employed during the week, in sets of three. (See Andrew Hughes 1982 for some of the complex schemes evolved.)

The Night Office ended with the singing of the Te Deum on Sundays and on most feasts except in Advent and Lent. In the later Middle Ages, the last responsory might be repeated if the Te Deum were not sung. If Lauds did not immediately follow the Night Office, then the closing formula (priest) *Benedicamus domino* V. (choir) *Deo gratias* would be sung. (The conventions about when it was sung are somewhat unclear: the preceding statement is derived from Andrew Hughes 1982, 66.)

(ii) *The Night Office in Monastic Use*

Monastic use (the Benedictine pattern is referred to here, but not all monastic uses were exactly the same) differed in many important details from the secular norm outlined above, though the basic pattern was similar. Only the chief dissimilarities are pointed out here.

The larger differences are that although the number of nocturns was generally three on Sundays and feasts, there were two on ferias (not one as in secular use). Furthermore, the nocturns were not of an identical pattern.

After the opening versicle and response, Ps. 3 was intoned, without any antiphon. The numbers of psalms and antiphons, lessons and responsories in each nocturn were substantially different from secular patterns. The antiphon–psalm pattern was the following:

Ferias, both nocturns:
either:	A, Ps, Gl, A—six times
or:	A, six Ps, Gl, A

Sundays and feasts:
first nocturn:	A, Ps, Gl, A—six times
second nocturn:	A, Ps, Gl, A—six times
third nocturn:	A, three canticles, Gl, A
	(total of 13 A)

On summer ferias, the rest of the first nocturn had but one lesson and responsory, and the second nocturn only a very brief lesson, no more than a verse of scripture, commonly called a *capitulum* (chapter). There followed a versicle and response, the short Kyrie litany, Paternoster, and a closing collect.

Winter ferias were similar except that in the first nocturn there were three lessons, each followed by a responsory.

On Sundays and feasts, the three nocturns each had four lessons with their responsories. Furthermore, after the Te Deum, and versicles and responses, a passage from the Gospel was intoned, making an entirely different climax to the service. On Christmas Day the gospel was the Genealogy of Christ from St Matthew's Gospel, on Epiphany the Genealogy from St Luke. The hymn *Te decet laus* was sung, then followed the blessing and final collect, with Benedicamus Domino when appropriate.

In some early antiphoners, for example the Hartker Antiphoner (St Gall, Stiftsbibliothek 390–391; PalMus II/1), the first two nocturns each have three lessons and responsories, the third nocturn also sometimes three of each, more often four. This is understood to be a survival of earlier arrangements wherein monastic use was not as strongly differentiated from secular as was later the case.

(iii) *Lauds in Secular Use*

Lauds and Vespers were two similar services, at sunrise and sunset, respectively. Of the two, it may be said that Vespers has the larger body of material proper to particular days of the week, or to various seasons of the year.

In secular use, Lauds began with versicles and responses, the first of which was, as usual with the office services, *Deus in adiutorium*. There followed five psalms, either each with its Gloria and antiphon, or else with only one antiphon and the Gloria said only after the last psalm.

In the weekly or ferial cycle, the psalms on Sunday were repeated during the week, except for the second, which was different on each of the other six days. Regarding psalms it should be understood that the third psalm was Pss. 62 and 66 together, the fourth was one of the lesser canticles, and the fifth was Pss. 148–50. Like the second psalm, the lesser canticle changed daily, and also according to season; there were fourteen in all (listed by Andrew Hughes 1982, 365). Feast-days would occasion the selection of other psalms and another canticle, but the pattern remained the same.

There followed a brief chapter, with choral response 'Deo gratias', a hymn, a versicle and response, and then the major canticle of Lauds, the Benedictus, with its antiphon (a relatively long and elaborate one). Final blessings and prayers followed and the Benedicamus Domino.

(iv) *Lauds in Monastic Use*

In monastic use Lauds had almost the same form as in secular use. The psalms were chosen, or arranged, differently; but the notional number of five (including the lesser canticle) was the same. Monastic Lauds, however, included a short responsory after the little chapter.

(v) *Vespers in Secular Use*

As mentioned above (I.4), Vespers on most Sundays and important feasts was celebrated twice, first in its ancient position as a vigil ceremony anticipating the succeeding holy day, secondly at the end of the same day. Most of the material of the two would be identical, though with a certain reduction in solemnity, by using common rather than proper items, omissions, and shortenings, at Second Vespers. (Andrew Hughes 1982, 69 ff. gives detailed information.)

Not only did Vespers have a structure similar to that of Lauds, but some of the items for any one day might be sung at both Lauds and Vespers: for example, the large antiphon for the major canticle, the psalm antiphons, chapter, and hymn.

Instead of the opening versicle *Deus in adiutorium*, omitted anyway around Easter, a ninefold Kyrie was sung during Easter week, the so-called 'Paschal Kyrie' (Vatican I, Melnicki melody 39).

The psalms for the weekly cycle form a continuous numerical sequence at Vespers, and there is none of the repetition during the week of Sunday items that characterizes ferial Lauds. The notional number of five psalms is the same, but a lesser canticle forms no part of the group. During Easter week, only three psalms were sung, with one Gloria and antiphon; they were followed by the Easter gradual *Hec dies*, with verse changing day by day, and an alleluia varied similarly through the week.

After the chapter on the more important days of the year, a responsory was sung (as in monastic Lauds, but here not on days of lesser importance). Occasionally this was one of the great responsories, rather than a short one.

Particularly notable among the antiphons for the Vespers canticle, the Magnificat, are those for ferial days before Christmas, the great 'O' antiphons, so called because their texts all begin with 'O', all of which use the same melody.

(vi) *Vespers in Monastic Use*

The essential difference between secular and monastic Vespers was the presence of only four psalms in monastic use. Only a smaller number of psalms could therefore be sung during the weekly cycle, the others being taken into the Little Hours (Table I.6.1.).

Table I.6.1. *Psalms at ferial Vespers*

	Secular	Monastic
Sunday	109–13	109–12
Monday	114–16, 119–20	113–16, 128 (115+116 as 1 psalm)
Tuesday	121–5	129–32
Wednesday	126–30	134–7
Thursday	131–2, 134–6	138–40 (138 as 2 psalms)
Friday	137–41	141, 143 as 2 psalms, 144 first half
Saturday	143–7	144 second half, 145–7
		(119–27 to the Little Hours)

At the end of both Lauds and Vespers it was sometimes customary to say or sing a series of antiphons, called 'suffrages', for the Blessed Virgin Mary and the patron saints of the church.

(vii) *Compline*

Compline, the last service of the evening, and the four short hours of the day, Prime, Terce, Sext, and None, were similar in length and content, although their histories were not identical.

Versicles and responses opened the service as usual, followed by four psalms, sung with one antiphon only and Gloria: A, 4 Ps, Gl, A. The hymn came next, then the chapter was intoned, and a short responsory was sung. After the versicle and response *Custodi nos* R. *Sub umbra* ('Keep me as the apple of thine eye . . .') came the canticle Nunc dimittis with its antiphon. On more important days the closing items would include the *preces*, beginning with Kyrie invocations (only intoned, as usual with prayers), Lord's Prayer and Apostles' Creed (both said silently until the last verse). There might also be confession and absolution. After the Benedicamus the service ended (at least from the later Middle Ages) with a votive antiphon to the Blessed Virgin Mary, and a collect for the same.

Rather little of Compline changed during the year, or even during the week. In Easter week, however, Ps. 90 fell out of the cursus, and the Nunc dimittis was moved to take its place. There followed the Easter gradual *Hec dies* V. *In resurrectione tua*, as at Vespers.

Monastic use differed in that there were but three psalms, sung without an antiphon (but usually each with its Gloria). More surprisingly, the Nunc Dimittis was not part of monastic Compline, except when sung during Easter week as a psalm. The service was thus even more like one of the Little Hours than was secular Compline.

(viii) *Prime, Terce, Sext, and None*

In each of these Little Hours there were generally three psalms with only one antiphon, but usually each with its Gloria: A, Ps + Gl, Ps + Gl, Ps + Gl, A. On Sunday and some feasts, however, Prime had nine psalms, except during the Easter season. Prime was also different in that after the psalms the Athanasian Creed *Quincunque vult* was usually sung, with an antiphon, except during the triduum and Easter.

Unlike other hours, these had the hymn after the opening formularies. After the psalms came the chapter with response, short responsory, but no canticle. There were the usual series of versicles and responses and other prayers at either end of the service, similar to the arrangement at Compline.

Monastic use had four psalms on Sunday at Prime (actually four sections of Ps. 118), and short responsories were not usually sung, but otherwise there was little difference from secular arrangements in the Little Hours.

I.7. PROCESSIONS

Bailey 1971.

Liturgical processions were a common feature of medieval worship on feast-days. Their use was particularly subject to local variation, since they were necessarily adapted to the design of the church buildings and the geography of the town where they took place, and the identity of the relics which often featured prominently in the proceedings. But although irregular and variable in frequency and form, processions were highly organized rituals, whose rubrics are set out in medieval books with as much care and exactitude as those of any other service. The main purpose of the procession was to visit some holy place within the main church (such as an altar) or without (another church in the town). Chants would be sung on the way there and on the way back. There a station was made, that is, the persons in the procession halted. Prayers would be said, chants sung, and usually some special ceremonial action performed, such as the sprinkling of the altar with holy water or the veneration of the relics of a saint preserved at the place of the station. Sometimes mass would be celebrated.

Thus, on the feast of the Purification of the Blessed Virgin Mary (2 February), it was the custom to make a procession with lighted candles, symbolizing Christ as the light of the world. The occasion in the gospels from which this ceremony takes its theme is the Virgin's visit to the temple in Jerusalem for the Jewish purification ritual. In Luke 2: 2–39 it is recounted how the child Jesus was presented in the temple (as ritual also demanded), and the aged Simeon recognized him as the Christ, declaring him to be 'a light to lighten the Gentiles'. After Terce, therefore, in medieval liturgies, an array of candles was blessed by the priest, with prayers, sprinkling them with holy water, and censing them. They were distributed to all those present, while the Nunc Dimittis canticle (Simeon's words themselves) was sung, with the antiphon *Lumen ad revelationem gentium*. The deacon turned to the people and sang the versicle 'Let us proceed in peace', and they answered 'In the name of Christ. Amen.'

The procession then set off, a typical order having the thurifer first, swinging the censer, then the subdeacon with the cross, with two acolytes with candles on either side, then all the clergy in order of rank, ending with the priest, and deacon on his left. During the procession, grand ceremonial antiphons would be sung, such as the *Adorna thalamum suum* (translated from a Byzantine original: see Wellesz 1947, 60 ff.—the processional ceremony apparently derives from Byzantine usage). The procession would perambulate the churchyard or cloister, and on returning the priest and clergy would change vestments for the celebration of mass, which followed immediately.

Very different was the atmosphere on Ash Wednesday, the beginning of Lent. As usual on a feria, Mass would follow Sext, but it was preceded by the ceremony of the laying of ashes on the heads of those present (originally of sinners who thenceforth wore sackcloth and were excluded from Holy Communion, until their period of penitence ended with a ceremony of reconciliation on Maundy Thursday). The service might begin with the singing of the Seven Penitential Psalms (Pss 6, 32, 38, 51, 102, 130, 143) with an antiphon. The priest would intone various prayers over a bowl of ashes, and versicles and responses would be sung. He would sprinkle holy water over the ashes; then he would first receive them on his own forehead in the form of a cross and then place them in the same fashion on the foreheads of the others present. The antiphon *Exaudi nos Deus*, with an intoned psalm verse *Salvum me fac*, often accompanied this action, and also ceremonial antiphons such as *Immutemur habitu in cinere et cilicio* ('Let us change our garments for sackcloth and ashes') and

Emendemus in melius ('Let us amend . . .') with verse *Adiuva nos Domine*. A procession would form, and move to the west door of the church, where the penitents would be expelled, one by one, kissing the hand of the official delegated for the task. The door would be shut and the clergy and choir would return to the east end of the church for mass. During the processional movement, there and back, ceremonial antiphons such as those mentioned would be sung.

Some processions were regular and oft-repeated ceremonies, for example the procession before the main mass held on Sundays and major feasts at many medieval churches. Others, being proper to a particular saint, would often contain special items of considerable musical interest, for example a responsory with a prosula, or even a chant set in polyphony. Much of the material sung or said in processions might be borrowed from other services of the day. A favourite practice in the later Middle Ages was to borrow one of the great responsories of the Night Office (which was why the Solesmes *Processionale monasticum* of 1893 has long been a useful source for the study of the responsories, poorly represented in modern editions). Several processional responsories or antiphons (the terms seem often to have been interchangeable) were of a different musical character, having possibly survived from Gallican liturgical uses of the time before the Carolingian move towards Roman practice. They were frequently long chants in ornate style, sometimes with a less lengthy verse (hence the designation 'responsorial').

Bailey (1971) lists several other points during the day, apart from the procession before mass, when processions were held in Sarum use: during Vespers in Easter week and on Saturdays through the summer; after Vespers on the eve of a saint with an altar in the church, and on the four saints' days after Christmas (SS Stephen, John the Evangelist, Holy Innocents, Thomas of Canterbury); during the Night Office in Easter week; after the Night Office on Easter Day; and after None on Wednesdays and Fridays in Lent, the Rogation Days, the triduum, and on the eve of Whit Sunday.

Some of the most ancient and solemn liturgies of the church year, such as those of the triduum, include processions, and these are mentioned in their place in the next chapters. The unique colour and splendour of many of the processions—with their special vestments, crosses, censers, and ranks of choir and clergy—seen against the steady regularity of the office hours, must have contributed greatly to the awe-inspiring solemnity of the medieval liturgy.

I.8. CEREMONIES OF HOLY WEEK

   (i) General
  (ii) Palm Sunday
 (iii) Maundy Thursday
 (iv) Good Friday
  (v) Holy Saturday or Easter Eve
 (vi) Easter Sunday

Tyrer 1932; Young 1933; Hermann Schmidt 1956–7; Hardison 1965; Andrew Hughes 1982.

(i) *General*

The ceremonies of Holy Week are among the most ancient and universal of the Church. Here the normal sequence of mass and office hours is frequently disturbed, and the content of the services is usually not like that of mass and the office hours on other days.

Lent, the period of the church year before Easter, is popularly understood to be a penitential season. The penance implied was in ancient times a very real observance, undertaken freely or by command of an ecclesiastical authority, and involved the exclusion of the person concerned from mass until a ceremony of reconciliation on Maundy Thursday. (The ceremony of exclusion has been mentioned above, I.7.) Lent was also the time of preparation for the catechumens, those being trained and instructed for baptism. After a series of 'scrutinies' on various Lenten weekdays (the most important, the Great Scrutiny, took place on the Wednesday of the fourth week in Lent), to ascertain who was suitable, baptism finally took place on Easter Eve. Numerous rituals of deep symbolism surrounded the final days before Easter, such as the extinguishing of candles for Maundy Thursday and then the lighting of the great Paschal Candle, the light of Christ, on Easter Eve. Coinciding as it did with the passage from winter to spring, this progress from penitential preparation and the death of Christ to his resurrection, and the rebirth in Christ of those newly admitted to communion, gave the Passiontide and Easter liturgies immense power and importance. When Christ had died on Good Friday, no consecration of bread and wine for communion, his body and blood, was possible until the Vigil Mass (or Paschal Mass) on Easter Eve. A previously consecrated eucharist had to be used, with corresponding alteration of the ceremony, which was therefore called the Mass of the Presanctified.

Some of the general differences in liturgical matters between Lent and the rest of the year were indicated briefly above, when the essential structure of each service was indicated (I.5–6). At a ceremonial level, a new note was struck on the fifth Sunday of Lent, known as Passion Sunday, when crucifixes, images, and pictures in the church were veiled in purple, and the Gloria patri was omitted from psalms, invitatories, and introits. On the next Sunday, Palm Sunday, the most important procession of the year was made, a symbolic re-enactment of Christ's entry into Jerusalem.

(ii) *Palm Sunday*

The procession on Palm Sunday took place between Terce and Mass, as usual on a Sunday. The main point of the action is that the priest, as Christ's representative, will make a ceremonial entry into his church from the outside, as if entering Jerusalem. Quite often the procession would cover a considerable distance, visiting other churches or special places, where a consecrated host might have been placed: this would then represent Christ on his entry. The amount of music normally provided in medieval service books—processional antiphons and hymns, and so on—is correspondingly generous. There is of course great variety between different books in the exact form of the procession and correspondingly in the choice of chants (see Edmund Bishop, 1918 for discussion of the practice at Rouen, Salisbury, and Hereford). What is sketched below is merely a typical order of ceremony.

The special objects used in the procession, palms, branches of olive trees, or other branches, must of course first be blessed, consecrated for the purpose. This was commonly done to the accompaniment of the short ceremonial antiphon *Hosanna filio David* (without verse), with prayers, sprinkling with holy water, and censing. The consecrated palms were then distributed, while verses from Psalms 23 and 46 were sung, with the antiphons *Pueri Hebreorum portantes* and *Pueri Hebreorum vestimenta* after every pair of verses. The gospel might then be read, before the procession formed up and left the church.

Some of the most notable of processional antiphons and hymns were sung during this ceremony, including the antiphons *Cum appropinquaret Dominus*, *Cum audisset populus*, *Collegerunt pontifices* V. *Unus autem ex ipsis*, and the refrain-hymns *En rex venit* and *Gloria laus et honor*. The antiphon *Ingrediente Domino* V. *Cum audisset* was commonly sung at the moment of re-entry into the church. A final station might be made before the principal cross of the church.

Mass would then begin. Two of its most important special features were the singing of the longest tract in the repertory, *Deus Deus meus*, with as many as fourteen verses (from Ps. 21), and the intoning of the Passion, that is the gospel account of Christ's betrayal and death, on this day from Matthew 26 and 27. The Passion was intoned from Mark's Gospel on Tuesday, from Luke on Wednesday, and from John at the Mass of the Presanctified on Good Friday, in each case to a tone different from that usual for gospel lessons.

(iii) *Maundy Thursday*

On this day Christ ate the Last Supper with his disciples, and instituted the Eucharist. He washed the feet of the disciples, and the words of the first antiphon sung during the medieval repetition of this ceremony, *Mandatum novum*, gave its name to the day itself.

The Night Office and Lauds on this and the next two days acquired the name of Tenebrae (literally 'shadows') because of the custom of extinguishing the lights during the two services. The Night Office would begin with only one set of fifteen candles before the altar. A candle was extinguished after each of the psalms, until the Benedictus of Lauds was reached, after which the one remaining candle was hidden behind the altar.

The Little Hours might be said and not sung, without lights.

On these days the Night Office had nocturns of three lessons even in monastic use, and on Holy Saturday only one nocturn (although later in the Middle Ages there was a tendency to make it a three-nocturn service.) The first three lessons of the Night Office (the only three on Saturday) were taken from the Lamentations of Jeremiah (they were not sung to a tone other than the usual one, however, although the blessings for the reader might be omitted).

Mass on Maundy Thursday, which took place after None (notionally 3 p.m.), was especially complicated by the ceremonies of the Reconciliation of Penitents and the Blessing of the Holy Oils. The former preceded mass, and involved the repeated genuflection and prostration of the penitents as they made their way from the west door of the church in procession. The Seven Pentitential Psalms were said; and the antiphon *Venite venite* was sung between each verse of Ps. 33: 'Come ye children, and hearken unto me: I will teach you the fear of the Lord.' The short Kyrie invocation was sung during the series of prayers, versicles, and responses.

Mass followed, and during the Gloria in excelsis the bells might be rung for the last time until Easter Eve. The Blessing of the Holy Oils had one special musical item, the refrain hymn *O redemptor sume carnem*, to accompany the bringing of the oils from the sacristy to the choir, and their return after the blessing. The point at which this took place was usually just before the consecration of the hosts, not only of the host for this mass but also that required for the next day. The oils were of a special type known as the chrism, hence the name for this mass of Missa Chrismalis. They were to be used for the anointing of the baptized in two days' time. The host which was to be 'reserved' for future use was carried to a special place either before the end of this part of the ceremony or after the Vespers section. This translation and reservation was carried out with considerable solemnity, with a properly formed procession, and, of course, chants, such as the hymn *Pange lingua*. The host might actually be buried in a sepulchre, thus anticipating the burial of Christ himself; a more solemn burial, or deposition, would take place on Good Friday.

As soon as the communion chant was over, Vespers began, without even the *Deus in adiutorium* versicle and response. Vespers consisted only of five psalms and the Magnificat canticle, all with antiphons. The postcommunion prayer for the end of mass was postponed until after this reduced Vespers, with Ite missa est to conclude.

Compline would follow later, but first the altars of the church would be stripped and washed with holy water (which would first be blessed), the whole procedure accompanied by some sequence of psalms with antiphons, responsories borrowed from the Night Office, or other ceremonial antiphons, and prayers.

Either now, as a separate ceremony, or occasionally at some point in the mass (such as after the gospel), the ceremony of the foot-washing, 'pedilavium' or 'Mandatum', was performed. It might be performed at any convenient spot, in church or perhaps in the chapter-house. Usually twelve chosen men (the number of disciples) had their feet washed by the officiating priest. The ceremony might be preceded by the relevant gospel reading, perhaps even a sermon, and it was followed by prayers and versicles. During the action of washing a series of ceremonial antiphons was sung, some with intoned psalm verses. The first in the series was *Mandatum novum do vobis*, but the choice thereafter varied widely. (An antiphon found in some uses, *Venit ad Petrum*, is occasionally provided with a long melisma on the word *caput*. In Sarum use the antiphon is last in the series, and the Sarum version of the melisma is famous as the cantus firmus of a polyphonic setting of the ordinary of mass once attributed to Dufay.)

After Compline a watch might be kept by the reserved sacrament.

(iv) *Good Friday*

The Night Office and Lauds had the same form as on Maundy Thursday, having the character of Tenebrae, and including further readings from the Lamentations.

Again as on Maundy Thursday, Mass took place after None, and contained even more special rituals, notably the Adoration of the Cross. As usual during the triduum, normal ceremonial was largely stripped away, and none of the usual opening items—introit, Kyrie, or Gloria—was sung. The service began straightaway with a series of lessons and responsorial chants. The order is: lesson, tract, prayer, lesson, tract, Passion (St John). (The tracts are often labelled 'responsory', and thus stand somewhat apart from the rest of the repertory, as do the 'cantica' of Holy Saturday.)

Nine solemn collects followed, prayers following a special formula. First the priest, beginning 'Oremus et pro . . .' ('Let us pray for . . .'), announced the purpose of the prayer. He then intoned the single word 'Oremus', and he or the deacon sang 'Flectamus genua' ('Let us bow the knee'). All prayed in silence until the command 'Levate' ('Rise').

The Adoration (or Veneration) of the Cross now took place. A cross with the image of the crucified Christ, veiled in purple, was brought with processional solemnity from the sacristy or wherever it had been kept. The priest uncovered it a little and the antiphon *Ecce lignum crucis* was sung three times, at successively higher pitches. After a few moments of silent adoration, the cross was completely unveiled, and the principal act of adoration took place. This involved genuflection before the cross and the kissing of the feet of the Saviour on the cross. The first two chants sung during this action were refrain chants of a special kind, known as the Improperia (Reproaches) after the character of the text. Each main verse had Christ, as it were, saying what he had done for his people (drawing on events in the Old Testament), and reproaching them with their cruel ingratitude. In both chants, some sort of alternatim scheme was used for the performance, either between soloist and choir or between two choirs.

The first had as refrain a group of alternating Greek and Latin invocations, 'Agios . . . Sanctus . . .', known as the Trisagion. As set out in modern Roman missals, the scheme is as follows:

two cantors in mid-choir: *Popule meus . . .* V. *Quia eduxi te de terra Egypti . . .*
choir 1: *Agios o Theos*
choir 2: *Sanctus Deus*
choir 1: *Agios ischyros*

choir 2: *Sanctus fortis*
choir 1: *Agios athanatos eleison imas*
choir 2: *Sanctus immortalis miserere nobis*
two soloists from choir 2: *Quia eduxi te per desertum* . . .
choirs 1 and 2: *Agios . . . Sanctus . . .* (etc. as before)
two soloists from choir 1: *Quid ultra debui facere tibi* . . .
choirs 1 and 2: *Agios . . . Sanctus . . .* (etc. as before)

The second of the Improperia had a simpler refrain, namely the respond *Popule meus* (without verse) of the first chant. This was sung in answer to each of a long series of verses beginning 'Ego . . .', chanted to a simple tone by cantors from either side of the choir in turn:

cantors of choir 1: V1
full: *Popule meus* . . .
cantors of choir 2: V2
full: *Popule meus* . . .
—and so on

The antiphon *Crucem tuam adoramus* was then sung, with simply intoned verse *Deus misereatur nostri*. The final chant of the group was the refrain hymn *Crux fidelis* V. *Pange lingua gloriosi*, where the refrain is divided into two, and each part in turn performed between the verses:

Crux fidelis (complete)
V1 *Pange lingua*
Crux fidelis (first part)
V2 *De parentis*
Crux fidelis (second part)
V3 *Hoc opus*
—and so on

Mass now began, sometimes with a procession to bring the presanctified host from where it had been kept, with accompanying antiphons such as *Salvator mundi salva nos*. The prayers for the eucharist were changed from those of other days to reflect the difference of the Blessed Sacrament, and the ritual was said rather than sung, except for the chanting of appropriate responsories or antiphons from the office of the day. The same might be used for a solemn procession to return the Blessed Sacrament to the sacristy or other place where it was kept, or even to 'bury' it again in a special tabernacle or 'sepulchre'. Many local versions of the ceremony of burying the host and/or a cross (the 'Depositiae hostiae/crucis') are known (described in Young 1933, chs. IV–V), together with their exhumation before the Night Office early on Easter morning.

All lights were extinguished. Vespers was omitted. There remained only Compline to sing.

(v) *Holy Saturday or Easter Eve*

At the Night Office, which followed the secular cursus even in monastic use, the Lamentations once more provided the lessons. Night Office and Lauds together again formed the Tenebrae services. The Little Hours and Vespers were sung in their turn, but the whole focus of the day was towards its end, and the momentous ceremonies of the Blessing of the New Fire, the sacrament of Baptism, and the Mass of the Paschal Vigil, usually timed so as to start at midnight.

The lighting of the Paschal Candle, symbolizing the light of Christ, was surrounded by complex and solemn ritual, involving the striking of fire from a flint, the placing of five grains in the candle, the lighting of the candle and, from it, all the others held by those present. Each action was accompanied by sprinklings and censing where necessary, and, as the Paschal Candle was lit and shown to the people, the solemn intonation 'Lumen Christi' was sung three times at successively higher pitches, and answered 'Deo gratias'. In some uses a procession would be made with the candle, calling for a chant such as the refrain hymn *Inventor rutili* found in Sarum books.

The chief glory of the ceremony was the Exultet chant, which now followed, so called from its first words, 'Exultet iam angelica turba celorum', and also known as the *Praeconium paschale* (Proclamation of Easter) or *Benedictio cerei* (Blessing of the Candle). Its text bears some resemblance to a preface at mass, while musically it is an immensely long and elaborate intonation, using several different formulas.

A short form of mass ensued, starting not with the opening chants and prayers but straight away with lessons (perhaps as many as twelve) and responsorial chants. In number they differed from one use to another, but the sequence: lesson—silent prayer—collect—lesson, and so on, with a tract (sometimes called here a canticle) replacing a lesson at intervals, seems generally to have been followed.

Attention then turned to the font, where baptism was to take place. The procession there and back was usually accompanied with the chanting of litanies, or a canticle, or a refrain hymn such as *Rex sanctorum angelorum*, found in Sarum use for the return procession. An elaborate ritual of blessing the water of baptism was carried out, including the dipping of the paschal candle three times into the water, symbolizing the descent of the Holy Spirit into it, and the mingling with it of baptismal oil. Although the prayer of blessing was textually like the Exultet, no special chant was sung.

The sacrament of baptism, with its solemn questions and promises, required no musical support. Those baptized, sometimes referred to henceforth as 'neophytes', were now ready to receive their first communion.

The mass which was then celebrated had no introit chant, for priest and clergy were already present. The Kyrie formed the natural conclusion to the processional litany, the Gloria was sung, and between the lessons the *Alleluia* V. *Confitemini Domino quoniam bonus* and the tract (or canticle) *Laudate Dominum*. There was no offertory chant or Agnus Dei or communion chant. As on the previous days, Vespers intervened before the final formularies of mass, having but a single psalm, Ps. 116 *Laudate Dominum*, with antiphon *Alleluia*. There was no chapter, hymn, or versicle, and after the Magnificat, the postcommunion prayer of mass and Ite missa est were sung.

A short Compline, lacking hymn, chapter, and short responsory, was then sung.

(vi) *Easter Sunday*

In the early Middle Ages the Night Office had but one nocturn, monastic use following (or reverting to) the secular form with three psalms and three lessons, with their respective antiphons and responsories. Its last responsory, commonly *Dum (Cum) transisset Sabbatum* V. *Et valde mane*, frequently introduced the enactment of the famous Visitatio sepulchri ceremony (discussed in II.25). There was sometimes a special ceremony to discover and display a cross and/or host buried on Good Friday (Young 1933, chs. IV–V), which might take place before or after the Night Office. As for all such ceremonies, chants specially selected or composed were performed, and occasionally they would amount to a regular Visitatio representation.

The other services of the day proceeded more or less according to their common form, although affected in character by the festive and joyful nature of the day.

I.9. THE 'FEAST OF FOOLS' AND RELATED CUSTOMS

Chevalier 1894; Chambers 1903; Villetard 1907; Wagner 1931; Whitehill *et al.* 1944; Arlt 1970; Hohler 1972; Arlt 1978.

At this point I interrupt the generalized account of liturgical usages in order to describe some customs which, in contrast to those so far treated, were by no means universally practised. Their outstanding musical importance necessitates this excursus. The customs are those associated with the so-called 'Feast of Fools'.

While the most important part of the year from the church's point of view is undoubtedly Easter, in the later Middle Ages the liturgy of the Christmas season was often enlivened by practices which are of special musical interest. Because of the licence involved, such a festal liturgy was often known as a 'Feast of Fools'. It was most often performed on New Year's Day, the Feast of Circumcision. While such liturgies are based on the normal services of the daily round, they are ornamented by numerous extra items, particularly the rhyming Latin songs known as conductus or versus. Five principal sources for a Feast of Fools liturgy, or something closely comparable, have survived: from Sens, Beauvais, Laon, Santiago, and Le Puy.

The official in a medieval cathedral who had charge of matters concerning the performance of the liturgy was usually the precentor (derived from *prae* + *cantor*, literally 'foremost singer'), whose responsibilities included the conduct and welfare of the choir and the management of chant-books. There was naturally variety from place to place as to the exact delegation of authority in these and other matters (see the survey in Kathleen Edwards, 1967); and in many institutions the practical rehearsal and execution of music was in the charge of a succentor (from *sub + cantor*). Within the choir itself, leading singers were detailed to sing solo parts of those chants which required them. The number of leaders (or *rectores*, 'rulers', as they were sometimes called) reflected the solemnity of the feast, more soloists (up to four) indicating the most important rank of feast (see Harrison 1963, 51 ff. and 104 ff. for useful information, again about English arrangements).

On certain days of the year, however, the normal hierarchy of dignitaries and officials in church might be reversed. On St Nicholas's Day (6 December) or on Holy Innocents' Day (28 December) some churches might elect a 'Boy Bishop' to preside over the liturgy of the day, and the boy choristers directed the services, detailing others to perform all the duties which boys were accustomed to carry out, such as intoning the first lesson of the Night Office, holding the book for the officiating priest, bearing crosses, censers, and other ceremonial equipment (Chambers 1903, ch. XV).

A similar transference of authority might take place on other days of the Christmas season, and in some cathedrals (mostly French) it was the custom for the deacons to order the services of St Stephen's Day (26 December—St Stephen was a deacon of the early church), priests on St John's Day (27 December), and the subdeacons on 1 January, the Octave of the Nativity or Feast of the Circumcision (when Christ was circumcised), known also, because of the extravagant licence with which the liturgy was performed, as the Feast of Fools, the Feast of the Ass, and so on. Another name was the *festum baculi*, after the *baculum*, the rod of office normally held by the precentor but on this day wielded by the subdeacon. Furthermore, one oft-quoted writer of the late twelfth-century, John Beleth (see Chambers 1903, i. 275; Arlt 1970, Darstellungsband, 40), speaks also of *tripudia* (literally 'dances') performed by the deacons, priests, choirboys, and subdeacons respectively, on the four above-mentioned days after Christmas. The subdeacons' feast might also be on Epiphany (6 January) or its octave.

Apart from the strictures issued by ecclesiastical authorities against it (see Chambers 1903, ch. XIII), we have little other evidence about dancing (and other horseplay) which marked the Feast of Fools, although a Sens precentor's book of the fourteenth century, Sens, Bibliothèque Municipale 6, notates a melisma in a processional responsory for the feast of the Invention of St Stephen (3 August— St Stephen was the patron saint of Sens cathedral) with the instruction that the precentor shall dance to it (Chailley 1949). There do survive, however, five principal sources of music for Feast of Fools liturgies, or a representative portion of it, as well as many minor sources. One is from Sens, already mentioned, and later Sens tradition ascribed its compilation to Pierre of Corbeil, archbishop of Sens. Paris records also attribute to Pierre the version of the Circumcision liturgy in use in Paris, though no copy of a special festal office has survived from Paris. It is known that the conduct of the feast at the cathedral of Paris (that is, of Notre-Dame) was reformed in order to control some of its abuses in the closing years of the twelfth century. It is therefore possible that what we see in the copy of the Sens office, is, as it were, a 'respectable' version of the Feast of Fools.

The Sens copy of a Circumcision office, Sens, Bibliothèque Municipale 46, dates from the early thirteenth century (ed. Villetard 1907). This is also the date of a similar office from Beauvais, with a somewhat richer provision of special festal material, London, British Library, Egerton 2615 (ed. Arlt 1970—the manuscript also contains the famous Play of Daniel, probably for performance on the same day, and has been bound with some polyphonic compositions; some of these were for performance during the Circumcision liturgy; some of them were composed in Paris). Between them, these two sources may indicate something of the (lost) Parisian Circumcision liturgy. A late twelfth-century source from Laon cathedral, Laon, Bibliothèque Municipale 263, contains much similar material rubricated for feasts of the Christmas period, concentrating especially on the feast of the Epiphany.

Outside North France the sources are sparser. This is not to say that manuscripts containing music typical of the Feast of Fools are lacking—such songs may be found in considerable quantities in, for example, the Norman-Sicilian tropers Madrid, Biblioteca Nacional 288, 289, and 19421; the Aquitanian tropers and song-books Paris, Bibliothèque Nationale, lat. 1139, 3549, and 3719; and the song-books Cambridge, University Library Ff. i. 17 (English?) and London, British Library, Add. 36881; all these are twelfth-century sources. But none of them sets out the material in systematic liturgical order or indicates the succession of items which will make up the complete festal liturgy. This occurs, outside North France, only in the well-known 'Jacobus' or 'Codex Calixtinus' of Santiago de Compostela (twelfth century; for St James Day, 25 July; ed. Wagner 1931, and Whitehill *et al.* 1944; see also Hohler 1972), and the Circumcision office of Le Puy (late medieval copies and supplements of an original going back, like the material in the other sources, to the twelfth century; text ed. Chevalier 1894; see also Arlt 1978).

All these sources contain chants and other material for the secular cursus of the office. The most important characteristic they share is the inclusion of conductus or versus, Latin songs, at various points in the services, either as replacements for a regular liturgical form or, more frequently, as extra festal items. Troped and farsed items are abundant, sequences replace hymns on several occasions. Among many services worthy of fuller description, the Epiphany *completorium infinitum* (endless Compline) of Laon should be mentioned, where, after a particularly lavish provision of conductus, the direction is given to conclude the office: 'tot Benedicamus quot novit quisque canamus' ('Let us sing all the Benedicamus [songs] we know') (see Arlt 1970, *Darstellungsband*, 220 ff.; also Steiner, 'Compline', *NG*). Polyphony is occasionally rubricked in the Beauvais office (the manuscript is now bound together with some polyphony), and appears in supplements to 'Jacobus' and the Le Puy office.

As an indication of the way in which special material was grafted on to the normal liturgy, Table I.9.1. sets out the course of the last service of the Beauvais office,

Second Vespers (Arlt 1970, Editionsband, 139 ff). On the left are placed the regular liturgical forms, on the right the songs, tropes, and so on.

Despite the licence which undoubtedly marked the celebration of the Feast of Fools, these offices do not, in their sober manifestation on medieval parchment or in modern edition, seem unduly disorderly. To those, however, who were accustomed to the regular unfolding of the liturgy, day by day and week by week, such efflorescences of markedly un-Gregorian chant (their style is discussed II.24) could not fail to have made a vivid, even disconcerting impression. Perhaps if we try to imagine correspondingly exotic irregularity in the ceremonial action and the visual presentation of the liturgy, we may be better able to appreciate the reasons for the episcopal displeasure evidently aroused from time to time. When the thirteenth-century Bayeux ordinal Bayeux, Bibliothèque Municipale 121 says that the four special Christmas feasts are to be celebrated 'quam sollennius possunt' (ed. Chevalier 1902, 65, cited by Arlt 1970, 48; 1978, 7) the term used, 'as solemnly as they [the clergy] are able', should not be interpreted in the modern sense as 'serious, sober'. The adjective *sollemnis* is conventional in ordinals as an indication of the character of the liturgies of feast-days, and should rather convey the idea of ritual splendour, richness, and variety; and, on the Feast of Fools, outbursts of extravagant and exotic music as well.

I.10. OTHER SERVICES: BAPTISM, CONFIRMATION, ORDINATION, CORONATION, MARRIAGE, BURIAL, DEDICATION

The account of the liturgy given in the preceding chapters has concerned itself with the regular cycle of services as performed day by day and year by year, with some added information about the special forms those services might take on certain days of unusual significance. The present chapter contains a few brief remarks on ceremonies of a different character, largely independent of the normal liturgical cycle, performed to sanctify a special occasion concerning a person, a church, or an ecclesiastical object of some sort. Many of them are episcopal functions: that is, the services are carried out by a bishop, and the prayers and chants sung are mostly found in the pontifical, the bishop's service-book.

Not all these services require the performance of music as an essential element. The role of plainchant varies according to how much ceremonial action is involved; where a procession of some sort takes place, as in the service for the dedication of a church, then it is not surprising to find a large provision of ceremonial antiphons. Some of the most important of all Christian rites, however, have little musical support.

Among those to be mentioned here are the services marking the sacraments. A sacrament was defined by Thomas Aquinas as 'the sign of a sacred thing in so far as it sanctifies men', and there are usually said to be seven: the eucharist, baptism, confirmation, penance, ordination, matrimony, and extreme unction (the anointing of

Table I.9.1. *Beauvais Circumcision office, First Vespers*

Arlt No.	Regular item	Replacement item	Addition
1			Verse *Lux hodie*
2			Conductus 'when the ass is brought along', *Orientis partibus*
3	Versicle *Deus in adiutorium*		
4	['Alleluia' acclamation after 'Amen']	Alleluia *Veni sancte spiritus*, 'or *Veni doctor previe* with organum'*	
5			verse sung behind the altar, *Hac est clara dies*
6			verse sung before the altar, *Salve festa dies*
7			prose sung by full choir, *Letemur gaudiis*
8			verse *Christus manens* 'in the pulpit with organum'*
9	A. *Ecce annuntio vobis* Ps. *Dixit dominus* ('all antiphons are to be started with "falseto"')		
10	A. *Hodie intacta virgo* Ps. *Confitebor*		
11	A. *Virgo verbo concepit* Ps. *Beatus vir*		
12	A. *Virgo hodie fidelis* Ps. *De profundis*		
13	A. *Nesciens mater* Ps. *Memento*		
14	Chapter *Populus gentium*		
15	R. *Confirmatum est cor virginis* (no V. given)		'in the pulpit with organum'*, with supplementary trope verses, one series set out as a prosula performed alternatim texted by leaders and melismatically by choir
16	[hymn]	sequence *Letabundus exultet fidelis chorus*	
17			verse *Ave virgo speciosa*
18	A. *Qui de terra* Magnificat		
19	[Benedicamus versicle]	verse *Corde patris genitus . . . Benedicamus domino, Super omnes alias . . . referamus gratias*	

*organum present in another part of manuscript.

the sick). The eucharist has of course been the subject of extended discussion above. The baptismal rite in its ancient place during the mass of Easter Eve has also been described briefly, and it was stated that processional music was sung during the progress to the font and on the return. A similar ceremony took place when the baptism was held on other days. It was one that had grown out of the early practice of adult baptism, whereas infant baptism was much more common in the Middle Ages, as now. The ritual surrounding this was considerably shorter. Confirmation was in many respects the same as baptism in its ritual requirements, and in general took place alongside baptism on Easter Eve. Unlike baptism, it remained a service which only a bishop could perform.

Public rites of penance consisted largely of formulas and prayers pronounced by the bishop, with accompanying actions. They centred upon the ceremony when the penance was laid upon those who had confessed their sins, on Ash Wednesday, and the complementary rite when the penitents were reconciled on Maundy Thursday. During Lent intercessions for the penitents were said, and it is possible that the tract *Domine non secundum* V. *Domine ne memineris* V. *Adiuva nos Deus*, sung first on Ash Wednesday and then on each Monday, Wednesday, and Friday during Lent, is a relic of such prayers.

The ordination rites of the church—whereby a candidate is admitted to one of the so-called 'major' orders of bishop, priest, deacon, or subdeacon, or one of the 'minor' orders of acolyte, exorcist, reader, or door-keeper, or to the monastic profession—are largely matters of prayers, exorcisms, and other intoned forms delivered by the bishop, and chants to cover an action are rarely necessary. The tonsuring of clergy would be accompanied by the singing of a psalm with its antiphon. The more important the rank being conferred, the more elaborate the ceremony, with perhaps a full preface prayer, a series of interrogations, blessings of the badges of office (bishop's ring and staff), and so on.

Since kings ruled by divine right, their coronation was a full-scale liturgical ceremony, with considerably more opportunity for ritual action, and hence music, than most special ceremonies of this type. Processional antiphons (*Firmetur manus tua*) and litanies would accompany the entrance of the monarch, moments such as the anointing with oil would have their ceremonial antiphon (*Unxerunt Salomonem Sadoch sacerdos*) and the singing of the Te Deum was a constant feature of a ceremony which naturally took several forms at diverse times and places.

In contrast to the marriage service, which in its normal form has no claims to attention as far as plainchant is concerned, the ceremonial surrounding a funeral is impressively elaborate. The Mass for the Dead, popularly known as the Requiem after the opening words of its introit, was something which might be sung, or at least said, at any time of the year as a votive commemoration. And in the later Middle Ages the performance of a complete Office for the Dead was also not uncommon, especially in monastic houses. The actual occasion of a person's death and burial might also call for the performance of these services. Before that, however, there would have been a procession to bring the corpse to the church for the final exequies, accompanied by the singing of penitential psalms with appropriate antiphons and the addition of the versicle *Requiem eternam*. On entering the church the responsory *Subvenite sancti Dei* V. *Suscipiat te* V. *Requiem eternam* was sung.

After mass (and any preceding services of the day) the absolution ceremony begins with the responsory *Libera me Domine*, with several verses. Versicles and prayers then lead to the procession to the grave, during which the ceremonial antiphon *In paradisum* is sung, with psalms as necessary. The grave itself may be blessed, and it and the body are sprinkled with holy water and censed. The Benedictus canticle, with antiphon *Ego sum resurrectio*, is the last important musical item, for the rest of the versicles and prayers, and psalms for the return to the church, are mostly said.

Finally, perhaps the most extended use of ceremonial antiphons is to be found in the ceremony for the dedication or consecration of a church, with its several perambulations and blessings, culminating, like several of these special services, in mass. Again, several forms were known, but the following are the usual features. The clergy assemble to the singing of psalms and versicles, and an antiphon (*Zachee festinans descende*) is sung at the entrance to the church. All then walk round the church three times, singing the major litany. Psalm 23 *Domini est terra* is also sung, with verse 7 as its antiphon *Tollite portas*, and this and the other final verses provide the text for the entrance ritual. The bishop knocks on the door of the church, and sings 'Lift up your heads O ye gates and be ye lift up ye everlasting doors: and the king of glory shall come in!' From inside comes the question: 'Who is the king of glory?' The bishop walks round the church again, and the same sequence is repeated. After a third perambulation bishop and clergy finally sing the answer: 'Even the Lord of hosts, he is the king of glory.' A litany may accompany the approach to the altar, and then the bishop traces with his staff the letters of the Greek and Latin alphabets across the corners of the church, to the singing of ceremonial antiphons (see Plate 10). The sanctifying of the building requires various substances—salt, ashes, water, hyssop—and their exorcizing, blessing, and sprinkling or other administration, in various parts of the church, which also calls for the singing of antiphons, or complete psalms with antiphon. The church will have holy relics of its patronal saint, and their installation in the high altar, or wherever they are to reside, requires further censing and other actions, prayers, and chants. When the ceremony is complete, the mass for the dedication, beginning with introit *Terribilis est locus iste*, is celebrated. This mass will be sung each year on the same day, together with a full office.

II

Chant Genres

II.1. INTRODUCTION

In this chapter I describe the chief genres, forms and styles, of chant in the Roman liturgy.

The main genres are arranged according to their musical type. I start with the simple recitation formulas for prayers and lessons (II.2), and then go on to psalm tones and the more elaborate recitations for invitatory and responsory verses (II.3). After that I discuss the much more complex chants which retain elements of recitation patterns by marking syntactical units with typical melodic gestures: responsories, graduals, and tracts (II.4–5). Short responsories are then considered (II.6). The next sections are about chants which generally use free-ranging melodies of varying complexity: first office antiphons (II.7), and then invitatory antiphons, processional antiphons, Marian antiphons, and other special types (II.8–10). After that I consider the 'antiphons' of mass: introits and communions (II.11–12) and offertories (II.13). Alleluias follow in II.14, then there is a switch to quite different genres, hymns (II.15) and chants for the ordinary of mass (II.16–21). The next sections treat various later medieval chants: sequences (II.22), the multitude of different types of trope (II.23), and liturgical Latin songs of various kinds, often with texts in verse (II.24). So-called liturgical dramas (II.25) and offices with verse texts (II.26) conclude the chapter, after which I have appended an excursus about some literary aspects of poetic texts (II.27), intended to be of assistance to the non-specialist.

Describing the musical make-up of medieval chants involves consideration of many features, not all of which can be given their due in the space available here. Some of the most discussed heretofore relate to form, in the sense of the layout of a melody: repeat structures, recitation patterns such as those of office psalmody which can sometimes be found in other chants, deployment of standard musical phrases for the delivery of varying texts. The function of chant as text declamation is fundamentally important, and has also often been treated, though the 'ground rules' for all genres of chant have not yet been defined. (Approaches of these types may be found in, for example, Johner 1953, Jammers 1965, 'Choral'.) Although the modality of chants has also been much discussed, some of their tonal aspects are less well understood: for example, the identity and role of structurally important and less important notes (which may not necessarily be directly related to the final). Nowacki (1977) has recently suggested ways of coping with questions of underlying structure and its embellishment. I have also found useful the distinctions employed by Hansen (1979) in what is to date the only comprehensive attempt at a tonal analysis of the complete repertory of proper of mass chants (as recorded in Montpellier H. 159). Hansen regards as 'secondary' those notes only approached and quitted by step, whereas 'main tones' are those approached and quitted by a leap. (Hansen employs supplementary criteria as well.) On this basis most of the chants can be divided into pentatonic or predominantly pentatonic melodies, and melodies where interlocking chains of thirds are structurally important. These distinctions override classifications according to liturgical function, mode, final, ambitus, and so on, so that chants of different genres and mode may have the same (pentatonic or tertian) tonal backbone. (Other promising computer-assisted analyses have been carried out, on very different repertories, by Halperin 1986 and Binford-Walsh 1990.)

It is nevertheless safe to say that the musical analysis of plainchant is in many respects still at a preliminary stage. In what follows different examples are therefore discussed in different ways, emphasizing now one possible approach, now another.

II.2. RECITATION FORMULAS FOR PRAYERS AND LESSONS

   (i) General
   (ii) Prayers
   (iii) Lessons

Léon Robert 1963; Huglo, 'Epistle', 'Exultet', 'Gospel', 'Litany', *NG*; Stäblein, 'Epistel', 'Evangelium', 'Exultet', 'Litanei', 'Passion', 'Pater noster', 'Präfation', *MGG*; Steiner, 'Lord's Prayer', *NG*.

(i) *General*

A great deal of the music performed during office and mass is extremely simple in character, being for the most part (sometimes entirely) a monotone, with melodic nuances at significant points in the text: usually at the starts and ends of phrases, and particularly at final cadences.

Even the simplest versicles and responses often illustrate some of the general principles by which recitations usually work. For the most part the versicle is sung to one repeated note, which may of course be sung as many times as required by the length of the text. Cadences may be marked by a fall in the voice, or a more elaborate gesture. Occasionally the voice rises from a lower pitch in order to attain the main reciting note, as it is usually called, or 'tenor'. This is what happens at the beginning of the ubiquitous office versicle *Dominus vobiscum*, quoted in its Sarum form in Ex. II.2.1. Here there is a double nuance at the cadence. It introduces a collect, sung completely monotone until the simplest of falling final cadences. Then *Dominus*

Ex. II.2.1. Versicles and responses (AS 4–5; Frere 1898–1901, pt.II, 210 and lxxvj)

vobiscum and its response are repeated, followed by *Benedicamus Domino* and its response: these take a different reciting note and have a lightly ornamented cadence.

In the melodic formulas used for singing prayers and lessons, a hierarchy of major and minor divisions in the text is often observed, corresponding roughly to our full stops, semicolons, and so on. Interrogative sentences may be signalled with their own special cadence. Some types of chant—one thinks immediately of the verses of the responsories of the Night Office (see II.3)—are sung to much more elaborate and musically sophisticated recitation formulas. Even these, however, retain one important feature of recitation technique: the length of the text to be sung is reflected in the number of times the reciting note is repeated.

In some of the examples discussed below, it will be noticed that an appreciation of the accentuation of particular words is important for the correct performance of the chant. There will be a departure from the reciting note at, say, the penultimate accented syllable, in order to mark a phrase-end. Sometimes accented syllables will be marked by a pitch higher than that of the reciting note—or better, there is a double reciting note, the higher note for accented syllables and the lower note for unaccented ones. There seems little doubt that these refinements of delivery were not universally practised, and that the cadential or other formula would often have been sung at the appropriate moment without respect for text accent. Similarly, most recitation formulas proceed without higher pitches for accented syllables. Thus the melodic inflexions for phrase-ends form a sort of musical punctuation. They mark syntactical and sense units, but do not necessarily reflect anything of the natural rise and fall of the human voice.

Although there is no doubt that the practice of reciting prayers and lessons in this way is an old one, medieval copies of recitation formulas are generally scarce, since they were known well enough for codification to have been unnecessary. Furthermore, most surviving music-books were made by and for trained musicians, and recitations of this type were not among their duties. On the other hand, more ornate and unusual lesson tones, such as those for the genealogies, are among the very earliest chants found with musical notation.

(ii) *Prayers*

The most important prayers to be recited in the liturgy are those of mass: the collect, secret, and postcommunion prayers, and the prayers of the eucharistic rite. The first three are relatively brief, and may be classed with many other occasional prayers for ritual actions, such as sprinkling holy water and blessings of various kinds. The *Liber usualis* gives a number of tones, plain or more varied to suit less or more important liturgical occasions. (There is as yet no convenient way of finding out how varied medieval practice was.)

The Preface prayer of mass (also sung, with a different type of conclusion, at special ceremonies of consecration of various kinds) is a much longer item, which begins with versicles and responses before launching into a long text varying with the liturgical occasion. More than one medieval tone for the prayer is known, but the one given in Ex. II.2.2 is typical. In it most clauses are sung to one formula, final clauses to another. As often in such formulas, the singer must know where the two final accented syllables fall. For the non-final clauses, he makes his inflections as follows: on the syllable after the penultimate accented syllable he quits the reciting note (*c*) for *b*; on the last accented syllable he goes even lower, so that a sub-tonal ending on *b* can be made (*ab b*). Final clauses have a slightly more ornate formula, depending only on the final accented syllable. Three syllables earlier the singer quits the reciting note to introduce a turning cadence which will settle on *a* (*aG, Ga b ab a*). Practically without exception, end syllables in these texts will be unaccented. If more than one syllable follows the accented one, then the *a* may simply be repeated (*ab a a*).

One prayer whose text has the character of a preface but whose music is much more elaborate is that for the consecration of the Paschal Candle on Easter Eve, the chant known from its opening word as the Exultet. It begins with a section not found in other prefaces, then come the versicles and responses, then the main part of the prayer. The first section is sung to a melody of four phrases repeated five times. The main part often uses the normal preface formulas, but introduces far more frequent cadences, and other more elaborate inflections. The degree of elaboration varies from one tradition to another (see Huglo, 'Exultet', *NG* for a summary of available

Ex. II.2.2. From the Blessing of the Font on Easter Eve (Oxford, Bodl. Lib. e Mus. 126, fo. 41ʳ)

Per omni-a secula seculorum.   A-men.

Dominus uobiscum.   Et cum spiritu tu-o.

Sursum corda.   Habemus ad dominum.

Gratias a-gamus domino De-o nostro.   Dignum et iustum est.

Vere dignum et iustum est equum et salutare

nos tibi semper et u-bique gratias a-gere

domine sancte pater omnipotens e-terne Deus

qui inuisibili potenci-a sacramentorum tu-orum mirabiliter o-peraris effectu

(two more phrases, then)   e-tiam ad nostras preces aures tu-e pi-etatis inclinas.

Deus ... (final clause) Ut omnis homo hoc sacramentum regeneracionis
                                    ingressus in uere innocentie

in nouam infanciam renascatur.

information and bibliography; also Stäblein, 'Exultet', *MGG* for a survey of melodies). The start of the chant is given in Ex. II.2.3 in two versions, from Paris and Salisbury respectively. Although both sources are from the thirteenth century, and not widely separated geographically, there are considerable differences between their versions. The Paris version is overall a fifth lower than the English one, but that is a relatively superficial distinction. More important is the simplicity of the Paris version compared to Salisbury. The contrast in elaboration is already apparent in the first part of the melody, the four-phrase unit which will be repeated several times. Then, when the recitation formula is employed in the main section of the prayer, the Salisbury version almost conceals the reciting note(s) under a wealth of melodic detail. The Paris version has *F* as its reciting note, with *E* for secondary cadences and *D* for main ones. For Salisbury, the possibility of a repeated *c* (or *b*) is hardly considered, and freely ranging phrases between *G* and *d* are often given. A simple fall *cb* becomes *cbbcb*; *ab* becomes *acbbc*. The assumption here is that the Salisbury version notates an ornate elaboration of some simple earlier pattern, an elaboration that has been developed in performance over the years. The solemnity of the occasion has resulted in a far higher degree of surface decoration than was normally practicable for such a prayer.

The prayers mentioned earlier were sung by the officiating priest, or celebrant, at mass. The Exultet differed from other prefaces in being sung usually by the deacon, hence the possibility of musical sophistication.

One other type of prayer, in which the whole assembly takes part, rather than having the priest speak for all, is the litany. Litanies were frequently sung in procession, for example during the ceremonies for consecrating a church. They were performed during the progress to and from the font for the solemn baptism ritual of Easter Eve (shortly after the Blessing of the Candle in which the Exultet is sung), and again, if that were the custom, in the echo of that service on Whitsun Eve. On the Rogation Days (the three weekdays before Ascension Day, which always falls on a Thursday) litanies were sung in procession before mass: this series was sometimes known as the Lesser litanies, while the Greater litanies were those sung in procession on St Mark's Day, 25 April. The normal form of such litanies is that of a long series of short verses, with a refrain. Invocation and refrain were sung first by soloists and then by the whole choir or assembly. Normally there will be introductory verses, often including the words 'Kyrie eleison, Christe eleison', then a series of saints (including the patron saints of the church) will be invoked (response 'Ora pro nobis': 'Pray for us'). Safety from ills and misfortunes will be requested ('Libera nos domine': 'Save us O Lord') and the saving events of Christ's life and ministry recalled; finally, Christ's aid for the church and people is requested ('Te rogamus audi nos': 'Hear us, we beseech thee'). The final versicles may include an Agnus Dei formulary and Kyries once again. Each section, with its refrain, is sung to a different musical formula. The amount of repetition might vary according to the custom of the church. The Easter Eve litanies were sometimes arranged in three series, called respectively by some such name as the Sevenfold (Septiformis), Fivefold (Quinquepartita) and

Ex. II.2.3. From the Exultet (left: Paris, Bibl. Nat. lat. 1112, fo. 96ʳ; right: *Graduale Sarisburiense*, 105)

Paris Bibliothèque Nationale lat.1112 f.96   Graduale Sarisburiense 105

E-xul-tet iam angelica turba celorum   E-xultet iam angelica turba ce-lo-rum

e-xultent diuina misteri-a   e-xultent diui-na miste-ri-a

et pro tanti regis uictori-a   et pro tan-ti re-gis uic-to-ri-a

tuba intonet salutaris.   tu-ba in-tonet sa-lu - ta - ris.

Vere qui-a dignum et iustum est   Ve - re qui-a dignum et   iustum est

inuisibilem Deum patrem omnipotentem.   inuisibilem Deum omnipotentem   patrem.

O ue-re be-a - ta nox   O   be - a-ta nox

que expoli-a-uit E-gypti-os   que expoli-a-uit E-gypti-os

di-ta-uit He-bre-os.   di-ta-uit Hebre-os.

Nox in qua terrenis celesti-a iunguntur. Nox in qua terrenis celesti-a   iun-gun - tur.

O-ramus er-go te do-mine ...   O-   -ramus te   do-mine ...

Threefold Litany (Tripartita Letania) according to the number of those leading the singing.

Ex. II.2.4 gives the opening of the Rogation Tuesday litany from a fifteenth-century York processional. As with most sources, the alternation and repetition scheme is not specified. Apart from the 'Ora pro nobis' phrase, which will presumably recur after each saint has been invoked, there are four refrain verses which, at first at least, are used in turn after each of the invocations. It is not clear whether or not this pattern is to continue throughout the subsequent series, which is not notated. (For a variety of other litany chants, many not using recitation formulas, see Stäblein, 'Litanei', *MGG*.)

Though the chanting of the saints' names or other invocations clearly involves a reciting note (*c* in Ex. II.2.4), the litanies may be relatively highly inflected. The number of notes other than the reciting note, combined with a relatively small

Ex. II.2.4. From a litany on Rogation Tuesday (Oxford, Bod. Lib. e Mus. 126, fo. 58ᵛ)

number of syllables, means that the reciting note is not repeated very much. The same is true of most melodies for the Lord's Prayer (see Léon Robert 1963, 127, and Stäblein, 'Pater noster', *MGG*, for various melodies). A widely known melody is transcribed from the fourteenth-century Cluniac breviary–missal of Lewes, Cambridge, Fitzwilliam Museum 369 in Ex. II.2.5. For the first part of the chant, the reciting note (the only one repeated) is *b*, but from 'sicut et nos' *a* becomes more prominent. The cadences are either on *G* (less important) or *a* (full stops).

Ex. II.2.5. *Pater noster* (Cambridge, Fitzwilliam Museum 369, fo. 241ᵛ/297ᵛ)

(iii) *Lessons*

The same principles of recitation as in those for prayers, and some of the same variety of musical elaboration, may be seen in the tones for chanting lessons. The tone for the short chapter of the office hours given in modern Roman books is a monotone without any initial rise, with flex, metrum, and full-stop figures. The longer lessons of the Night Office are sung in the same way, with the full stop coming at the lower fifth, for example:

　　reciting note *c*, full-stop figure *ć–G–á–F*, or
　　reciting note *a*, full-stop figure *á–F–Ǵ–D*

The epistle sung at mass, with reciting note *c*, has the following figures in modern Roman books (for a survey of various medieval formulas, see Stäblein, 'Epistel', *MGG*):

　　metrum: *a–ć–b–b́–c*
　　interrogation: reciting tone *b*, *a–b–bc*
　　full-stop: *ćd–b–á–b*

The end of the whole lesson is signalled by an accented *ac* inflection, then *a* fall to *b* for the reciting note and a final close *bc–c*. The gospel tone given as 'ancient' in modern Roman books (the other appears not to be older than the sixteenth century, being similar to the prescriptions of Guidetti's *Directorium chori* of 1582) is similar to this (see Stäblein, 'Evangelium', *MGG* for further medieval tones).

At various times and places, and even to a certain extent in modern service-books, more elaborate tones have been used for lessons on particularly solemn occasions in the year. The best-known instances are the tones used for the lessons of the first nocturn of the Night Office during the triduum, the three days before Easter Sunday. (See Prado 1934 for some remarkable Spanish examples.) Gospels at Christmas and Easter were occasionally also sung to a more elaborate tone (see Huglo, 'Gospel', *NG*), as for example the one from the Moosburg gradual, Munich, Universitätsbibliothek 2° 156 given in Ex. II.2.6. Here the hierarchy of main and secondary closes is not clear, and

Ex. II.2.6. From the Gospel on the feast of St John the Evangelist (Munich, Univ.-Bibl. 2° 156, fo. 226ʳ)

questions do not have a distinct inflexion. The reciting note is generally *G*, reached from an initial *E* or from a longer preliminary figure *G–a–E–F–G*, sometimes with pauses on the *E*. The endings are *E–(F–E)–DEG–G* and *FGa–GE–D–E*, until the conclusion of the lesson.

The intoning of the Passion—the Gospel on four days of Holy Week—was also performed in unusual ways. As far back as the ninth century copies of the texts were marked with the letters *c*, *t*, and *s* at the point where the evangelist, Christ, and other protagonists respectively were speaking. It appears that these letters are to be interpreted as in the series of notational aids known as significative letters (see below, IV.3), though in the later Middle Ages they were understood as labels for the actual persons. Other letters (see Stäblein, 'Passion', *MGG*) are also to be found occasionally:

> *c* = *celeriter* ('quickly'), later *cronista* ('evangelist') or *cantor*
>> *m* = *mediocriter* or *media* ('at medium pitch, medium voice') is also found
> *t* = *trahere* or *tenere* ('slow, sustained'), later written as a cross for Christ
>> *i* = *iusum*, *inferius* ('at a lower pitch') and *b* = *bassa* ('low voice') are also found
> *s* = *sursum* ('at a higher pitch'), later *synagoga* ('the people') or *succentor*
>> *a* = *alta* ('high voice') is found, and sometimes *ls* = *levare sursum* ('raise high') for the people is contrasted with *lm* = *levare mediocriter* ('raise moderately') for the disciples

The pitch-levels indicated by the letters are indeed to be found as three different reciting notes (usually *F*, *c*, and *f*) in copies in staff-notation, which are of course later than the oldest 'marked-up' copies of the text. In the late Middle Ages, if not earlier, different singers sang the different parts of the text, as appears from the fourteenth-century Sarum missal Parma, Biblioteca Palatina 98, where the 'parts' are marked *secunda vox*, *iii vox*, and so on (*MGG* 10, Tafel 61.2, after col. 960—this is the only passion in the source; furthermore, only the parts for the singers other than the evangelist are copied in full, mere incipits for the evangelist being given; the Sarum tone is edited in Andrew Hughes 1968, 184–6). An idea of the variety of tones used, within a general adherence to the principles of recitation technique and distinction between the persons of the story, can be seen from Stäblein's tables ('Passion', *MGG*). Particular attention was often directed towards Christ's words, for example the cry 'Eli, eli, lama sabacthani', and it is thus not surprising to see neumes entered specially above these words in the Jumièges evangeliary Rouen, Bibl. Municipale 310 (A. 293) (facsimile in MMS 2, pl. XLVII). For other extravagant passion tones, see Göllner 1975 (late Spanish sources).

Two other gospels of special ritual significance, and thus special musical character, were those which recounted the genealogy of Christ. That appearing at the start of St Matthew's Gospel was sung on Christmas Eve, and that according to St Luke on Epiphany. Monastic use, which had a gospel reading at the end of the Night Office, had the genealogies at that point in the liturgy. Gospels were not read there in secular uses as a general rule, but on these days an exception was made (they were intoned before rather than after the Te Deum). Usually the verses were grouped in pairs, with a different tone for each half of the unit. Occasionally more elaborate schemes are encountered in medieval sources (see Stäblein, 'Evangelium', *MGG*). Ex. II.2.7 gives a common tone, taken from a Reims missal of the thirteenth century. Ex. II.2.8 has a more complex melody for the St Luke genealogy, from a book of similar date but enigmatic provenance (it has elements of both Reims and Paris usage), Assisi, Biblioteca Comunale 695. Here two melodies are sung in alternation—they can hardly be called recitations any longer, though they are of course adaptable by note-repetition for names of differing numbers of syllables. Each melody has five distinct phrases; the shorter phrases of the St Luke's text lend themselves well to this treatment.

Ex. II.2.7. From the Genealogy according to St Matthew (Reims, Bibl. Mun. 224, fo. xi/16ᵛ)

Ex. II.2.8. From the Genealogy according to St Luke (Assisi, Bibl. Com. 695, fo. 37ʳ)

... Qui fuit Dauid, qui fuit Ihesse, qui fuit Obeth, qui fuit Bo-oz, qui fuit Salmon, ...

... Qui fuit Na-ason, qui fuit Aminadab, qui fuit Aram, qui fuit Esrom, qui fuit Pharens, ...

II.3. TONES FOR PSALMS AND OTHER CHANTS

   (i) Psalms
  (ii) Other Psalm Tones: The Parapteres, Tonus peregrinus
 (iii) Tones for the Canticles Magnificat and Benedictus
 (iv) Tones for the Psalm Verses of Introits and Communions
  (v) Tones for Responsory Verses
 (vi) Tones for the Invitatory Psalm
(vii) *Benedictus es Domine Deus patrum nostrum* in the Saturday Mass of the Ember
      Weeks
(viii) Te Deum laudamus

(i) *Psalms*

Bailey 1976, 'Accentual', 1979; Stäblein, 'Psalm, B', *MGG*; Connolly, 'Psalm, II, *NG*.

The cycle of psalms chanted daily during the office is performed to tones in a manner similar to the intonation of prayers and lessons. From an early date (the late eighth or early ninth century), perhaps under the influence of Byzantine practice (see below, III.14 and V.4), there were eight tones in the Gregorian system. The choice of tone for the performance of a particular psalm depended upon the mode of the antiphon with which the psalm was coupled in the liturgy of the day. In the course of a year the same psalm might be sung more than fifty times, nearly always with a different antiphon, and thus to a different tone. For most tones a variety of cadences (*differentiae*, *diffinitiones*) was available, in theory at least to cater for the different pitches with which the antiphon would start on its repetition after the psalm. Medieval sources regularly differ on the number of differentiae they provide, and their assignment of particular differentiae to particular antiphons. (Sometimes even the mode of the antiphon, hence the tone to be used for the singing of the psalm, was the subject of differing opinions: see V.4.) Thus the selection of differentiae given in Ex. II.3.1 below is not the same as that in modern Vatican books.

It is important to remember that the tone may be concluded on a variety of notes, and is in no way bound to finish on the same final note as the antiphon, that is, on what is normally reckoned to be the final of the mode. It is easy to confuse 'mode' and

Ex. II.3.1. The eight psalm tones (Piacenza, Bibl. Cap. 65, fos. 264ᵛ–267ᵛ)

1.

Glori-a pa-tri et fi-li-o   et spiritu-i sancto

Sicut e-rat in principi-o et nunc et semper   et in secula seculorum a - men.

2.

Glori-a pa-tri et fi-li-o   et spiritu-i sancto

Sicut e-rat in principi-o et nunc et semper   et in secula seculorum a-men.

3.

Glori-a pa-tri et fili-o   et spiri-tu-i sancto

Sicut e-rat in principi-o et nunc et sem-per   et in secula seculo-rum a-men.

4.

Glori-a pa-tri et fi-li-o   et spiri-tu-i sancto

Sicut e-rat in principi-o et nunc et semper   et in secula secu-lorum a-men.

(Ex. II.3.1. cont.)

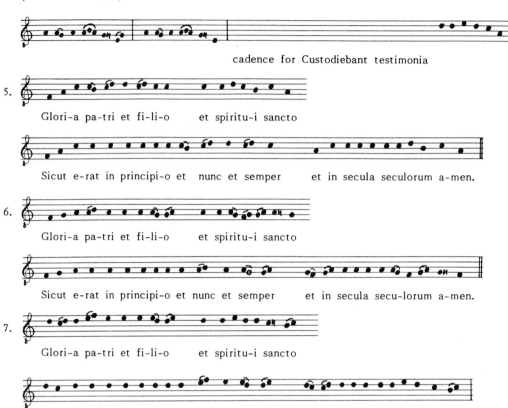

cadence for Custodiebant testimonia

5.

Glori-a pa-tri et fi-li-o   et spiritu-i sancto

Sicut e-rat in principi-o et nunc et semper   et in secula seculorum a-men.

6.

Glori-a pa-tri et fi-li-o   et spiritu-i sancto

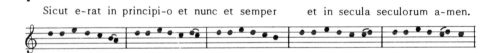

Sicut e-rat in principi-o et nunc et semper   et in secula secu-lorum a-men.

7.

Glori-a pa-tri et fi-li-o   et spiritu-i sancto

Sicut e-rat in principi-o et nunc et semper   et in secula seculorum a-men.

8.

Glori-a pa-tri et fi-li-o   et spiritu-i sancto

Sicut e-rat in principi-o et nunc et semper   et in secula seculorum a-men.

'tone', and it should therefore be borne in mind that while 'mode' is an abstract quality, having to do with the tonality of a chant, a tone is a sort of chant in itself, a melodic formula capable of supporting the performance of an almost infinite variety of psalm verses.

Ex. II.3.1 gives the eight psalm tones, and their cadences, as they appear in the tonary of a thirteenth-century compendium from Piacenza (Piacenza, Biblioteca Capitolare 65). (A tonary, which usually gives a list of chants in modal order, is often a more convenient source for tones than is a full antiphoner. Few antiphoners set out tones in full one after another. Indeed, to find out what tones and endings were used in a particular antiphoner it is necessary to check through each antiphon in the whole manuscript, as was done, for example, for the Italian antiphoner published in PalMus 9.) The cadences are given as six-syllable formulas, to fit the words *seculorum amen* (usually abbreviated 'e u o u a e'). At the end of the cadences for tone 4 is one for the antiphon *Custodiebant testimonia*, a chant which requires both *b* and *b♭* and which must therefore be notated with final on *a* instead of *E*.

If one were to turn to a medieval antiphoner in order to see how these tones were used, one might well be puzzled at first. At the appropriate point in the liturgy the antiphoner will probably give, not the text of the psalm to be sung, notated in accordance with the psalm tone, but simply a text incipit and the differentia. For example, in the antiphoner Karlsruhe, Badische Landesbibliothek, Aug. perg. 60, for Vespers of Christmas we find as first antiphon *Scitote quia prope* (*LU* 365, *AM* 237), a mode 8 antiphon (see Ex. II.3.2). It is followed by a text incipit for Ps. 112, *Laudate pueri*, and a differentia for the eighth tone. The full text of the psalm would be known to the performers, the monastic choir of Petershausen on the shore of Lake Constance, where this manuscript was used. The text could be found in a liturgical psalter of the church, as one of the psalms for Vespers on a Sunday. But even there it would probably not be marked up for singing, that is, the text would not contain any indication of when the singers should leave the reciting tone and make a cadence, for each verse of the psalm as required; they would be expected to know the practice by heart.

Ex. II.3.2. Antiphon *Scitote quia prope est*, with psalm-tone cadence (Karlsruhe, Badische Landesbibl. Aug. perg. 60, fo. 17ʳ)

Scito-te qui-a prope est regnum De-i   a - men dico uobis   qui-a non tardabit.

Ps. Laudate pueri.

Sometimes no psalm text incipit is given: the appropriate psalm would be ascertained from the liturgical psalter. Sometimes no cadence formula is given: the mode of the antiphon must then be decided upon, and the matching psalm tone used, with an appropriate cadence.

Modern books such as the *Liber usualis*, *Antiphonale Romanum* and *Antiphonale monasticum* give clear instructions on how the tone, the musical formula, is to be adapted for each verse of the psalms, whose texts are marked accordingly: the syllables where a change is to be made are given in bold or italic type. As far as the medieval practice goes, however, we are generally reliant on didactic texts, such as the *Commemoratio brevis de tonis et psalmis modulandis* of the ninth to tenth century. As Bailey (1976, 'Accentual') has pointed out, medieval practice cannot have been uniform. Some sources indicate that attention was paid to the accentuation of the text being sung, while others apply the cadence formulas mechanically. In Ex. II.3.3, taken from Bailey's edition of the *Commemoratio brevis* (1979, 52–3), the cadence has been adjusted so that the last accented syllable shall be sung to *G*, any further syllables to *F*:

Ex. II.3.3. Cadences from the *Commemoratio brevis* (ed. Bailey, 1979, 52–3)

..... fecit dominus ..... sanctum e-ius ..... iu-sti-ti-am su-am

If text accent were to be respected, then both the median cadence and the final cadence of the doxology would require adustments:

> . . . et fí-li-o   . . . spi-rí-tu-i sán-cto
> . . . nunc et sém-per   . . . se-cu-ló-rum. Á-men

We can look back at Ex. II.3.1 and see if there is a change at FI-lio (that would be an accentual cadence) or at fi-LI-o (a mechanically applied 'cursive' cadence). If the final change comes at spi-RI-tui, the cadence is accentual, if at spiri-TU-i it is cursive. In this case the test is met in the median cadences 2–5 and 8, but in the final cadences only of tones 5 and 7. Here Piacenza 65 always observes accentual cadences.

Tones for psalms, for the canticles, and for introit verses are given in parallel by Connolly ('Psalm, II', *NG*). As well as these, Stäblein gave responsory verse tones ('Psalm, B', *MGG*). Both authors draw upon both the Vatican edition and the *Commemoratio brevis*.

(ii) *Other Psalm Tones: the Parapteres, Tonus peregrinus*

Erbacher 1971; Bailey 1977–8; Atkinson 1982, '*Parapteres*'; Atkinson, 'Parapter', *HMT*.

The two halves of psalm verses are sung to the same reciting note in the eight common psalm tones, and most other tones, though both responsory verses and invitatory psalm tones have a change of reciting note. A rarely used psalm tone, the so-called 'tonus peregrinus' ('wandering tone'; the Latin term is first found in the writings of German theorists of the fourteenth century), also uses two different reciting notes. It is given from the late-tenth-century Aquitanian tonary Paris, Bibliothèque Nationale, lat. 1118 in Ex.II.3.4.

Ex. II.3.4. Tonus peregrinus (Paris, Bibl. Nat. lat. 1118, fo. 113ᵛ)

It is not the only tone of this sort. A number of them, called 'parapteres' (*paracteres*, *medii toni*, etc., possibly from Greek *para* + *aptō*, 'join alongside'), are mentioned in several early medieval treatises, from Aurelian of Réôme onward. They are usually cited in conjunction with a small group of antiphons of irregular modality, 'modulating antiphons', as they have been called. The most prominent groups of antiphons are those like *Nos qui vivimus*, and those designated by Gevaert (1895) as 'Theme 29', which, as the tenth-century treatise *De modis musicis* puts it, 'are not ended in the same way as they began'. (See II.7 for examples and discussion.) The psalm tone therefore reverses the modulation in the antiphon itself, so that there is no abrupt change of tonality.

Among the other tones, the *Commemoratio brevis* cites the one given in Ex. II.3.5 (the antiphon is completed from the Petershausen antiphoner). The higher second 'alleluia' of the antiphon is answered by the higher recitation on *a* in the psalm tone. Then the tone moves down to *F*, which is where the antiphon will begin when repeated after the psalm verses.

Ex. II.3.5. Antiphon with 'parapter' psalm tone (Karlsruhe, Badische Landesbibl. Aug. perg. 60, fo. 17ᵛ; *Commemoratio brevis*, ed. Bailey, 1979, 52–3)

As Bailey and Atkinson have both suggested, in these 'irregular' tones we seem to have vestiges of a more flexible psalmodic practice not limited by, and probably anterior to, the familiar eight tones. (The notion is reinforced by a comparison with Old Roman practice: see Dyer 1989, 'Singing', and below, VIII.3.) As to the tonus peregrinus, it has been remarked that antiphons requiring it (such as *Nos qui vivimus*) are often assigned as antiphons for the Benedicite at Lauds, which has led Steiner (1984, 'Antiphons') to ask whether the tonus peregrinus might possibly have been a special Benedicite tone. On the other hand, the tonus peregrinus was also regularly used during Vespers for Ps. 113, *In exitu Israel* (on Sunday in modern secular use, Monday in monastic use); it is even possible that the name of the tone was suggested by the psalm, which speaks of Israel's departure from the land of Egypt. The text *Nos qui vivimus* comes from this psalm.

(iii) *Tones for the Canticles Magnificat and Benedictus*

Where they are to be found in medieval sources, which is rarely, they are slightly more ornate than the common psalm tones. Most have a few more two-note groups than the psalm tones, but this is not so in all cases or in all manuscripts.

(iv) *Tones for the Psalm Verses of Introits and Communions*

Somewhat more ornate again are the tones for singing psalm verse(s) and doxology with the introit and communion at mass. The introit tones, in contrast to the usual practice for office psalm tones, have a new intonation after the median cadence. In this they resemble the tones for responsory verses.

At this point it becomes possible to illustrate without great difficulty the way in which text and tone were joined, for a number of medieval graduals copy out in full the psalm verse to be sung. Ex. II.3.6 gives some psalm verses from introits in the Chartres cathedral gradual Provins, Bibliothèque Municipale 12. The verses begin identically, and the first two are also alike after the median cadence, despite the fact that in the second 'iusticia' would normally be accented on the second syllable: the musical figure is independent of accentuation. In the first half of the verse, the rise to *d* is made on an accented syllable, and that means that in the second example only one *c* follows, because only three syllables remain altogether, instead of the four in the first

Ex. II.3.6. Psalm verses for introits (Provins, Bibl. Mun. 12, fos. 203ᵛ, 178ʳ, 186ʳ)

verses. The final cadence is applied mechanically, however, so that no change is made in applying the cadence formula: contrast the accentuation of 'nomini sancto eius' and 'et eripe me':

nó-mi-ni sán-cto éi-us

ét e - rí-pe mé

The third verse is very short and the two halves of the psalm tone are elided.

The same tones appear to have been used for the communion as for the introit. The communion psalm verse was in any case frequently borrowed from the introit.

(v) *Tones for Responsory Verses*

Far more elaborate are the tones used for the verses of office responsories. The same principles nevertheless hold true. Ex. II.3.7 gives the verses for the first three mode 3 responsories from the Petershausen antiphoner. Obviously, text accentuation determines where certain groups of notes will be placed. The figure *cdcc* is used for the first accented syllable, so that *Preóccupemus* begins with three unaccented *c*s, *A solis ortu* with only one, and *Tollite portas* with none at all. The rest of the word '(Tol)lite' has two unaccented syllables, so the figure *abcaaG* is split. The median cadence is the same in all three verses, but two different approaches are used: in

Ex. II.3.7. Verses of the responsories *Salvatorem expectamus*, *Audite verbum*, and *Ecce virgo concipiet* (Karlsruhe, Badische Landesbibl. Aug. perg. 60, fos.2ᵛ–3ʳ)

Preoccupemus and *Tollite portas* there is rather a lot of text to sing, mostly on *a*, with a couple of liquescent neumes in *Preoccupemus* and with accented syllables highlighted in *Tollite*; a more concise figure appears in *A solis ortu*. The second half of each verse recitation is on *c*; in *Preoccupemus* there is a short intonation figure. The final cadence is very ornate, stretching over five syllables (that is the whole text in *Ecce virgo!*). The music is applied here mechanically, for the three verses have different accent patterns, but the music is always the same, without any extra single notes or splitting of neumes.

It will be noticed that different reciting notes are used in the two halves of the verse, *a* and *c*. That is usually the case in the tones for responsory verses. (Copies from various sources may be found in Stäblein, 'Psalm B', *MGG*; Connolly, 'Psalm II', *NG*; Cutter, 'Responsory', *NG*; *Processionale monasticum*, and *AS*, the latter reprinted with a useful note in Hucke 1973.)

Most but not all tones—whether simple psalm tones or those of responsory verses—have a rising intonation and a falling cadence. Several rising cadences (the word 'cadence' is etymologically inappropriate) may be seen in Ex. II.3.1 above. A glance at the responsory verse tones in one of the editions just cited will reveal several falling openings.

As will be discussed later (V.4), the classification according to a system of eight modes is a relatively late development in Western chant. The possibility has just been mentioned that there were more than eight psalm tones, and the same possibility exists for the tones of responsory verses. A systematic enquiry has not yet been published, but it may be mentioned that Frere noticed a second verse tone for mode 8 responsories (*AS*, Intro., 60, facs. 171, 174, et.), and Hucke ('Responsorium', *MGG*, 320) reported that other verse tones were occasionally to be found, even amongst the old core repertory. Bearing in mind the fact that the notation of office chants does not seem to have been undertaken systematically before the later tenth century, one might well ask whether these other tones are the relics of a once greater number, or alternatively whether the restriction to eight was becoming relaxed by the end of the millennium.

(vi) *Tones for the Invitatory Psalm*

Frere in *AS*; Stäblein, 'Invitatorium', *MGG*; Steiner, 'Invitatory', *NG*.

Psalm verses usually consist of two hemistichs, and psalm tones therefore usually consist of two elements. At the start of the Night Office, however, Ps. 94, *Venite exultemus*, is sung in a different way, in units of five phrases. The text is not that of the so-called Gallican psalter, used for the common psalmody, but that of the so-called Roman psalter (see VIII.6).

There are both simple and ornate Venite tones. Sometimes the five elements are disposed as two times two phrases, with a recitation formula like a psalm tone, then a concluding phrase. In several of the tones, however, reciting notes are rarely to be heard. Some of the melodies, as we may call them, are quite ornate. The impression is not that of a 'tone' in the sense we have just been using, but rather of a long melody with a number of elastic points where expansion for a long verse, or contraction for a short one, may easily be effected.

Ex. II.3.8 gives the first three verses of a Venite from the Saint-Denis antiphoner Paris, Bibliothèque Nationale, lat. 17296. In the first phrase *c* can be heard as a reciting note (in verse 5 it is used more frequently), in the second phrase the note is *a*. Later verses have further repeated *c*s at the start of the third phrase, which is then supplanted by *d*. In the fourth phrase there are again two reciting notes, *c* at the beginning, then *a*, as in phrase 2, while *c* is the most important note in phrase 5. The mutual attraction of *a* and *c* is discernible throughout. The figure *a–G–a–c–b*, moving from *a* to *c* and leading back down again, contains the melodic essence of the piece (it is marked with a bracket in Ex. II.3.8). In the course of each verse it appears four times; practically everything else is recitation around *a* or *c*, before the final melisma elaborates the kernel figure for a last time. It is noticeable, however, that this figure appears sometimes at a break in the text ('salutari nostro'), sometimes elsewhere ('exultemus'), so it cannot be regarded as a cadential figure; neither does any other obvious cadential figure appear: the melodic material is disposed rather freely over the whole verse.

Ex. II.3.8. Invitatory tone (Paris, Bibl. Nat. lat. 17296, fo. 347^r)

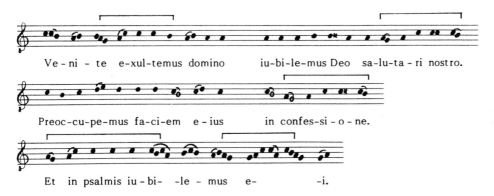

Few Venite tones are available in modern editions. Eight may be found in *Liber responsorialis*, others in PalMus 12. No comprehensive survey of the medieval repertory has been made, but it seems that invitatory antiphons were reckoned to belong only to modes 2–7, and consequently only six Venite tones were required. In practice some modes, especially mode 4, had several alternative tones. (For Cluniac practice, see Steiner 1987.)

(vii) Benedictus es Domine Deus patrum nostrum *in the Saturday Mass of the Ember Weeks*

Another canticle with its own tone—in this case one only—is the *Benedictus es* sung between the graduals and the tract at mass on Saturday in three of the four Ember Weeks of the year. The same melody is used for each verse. In modern service-books it is entitled 'Hymnus'. As a song of praise in a general sense it may qualify for this appellation, but the number of syallables in each verse differs, so that the melody, or tone, is stretched out by means of repeated notes, or shortened by elisions, as required. In its mobility the melody resembles some of the Venite tones. As in the biblical text itself, the second half of each verse forms a refrain.

(viii) *Te Deum laudamus*

Schlager, 'Te Deum', *MGG*; Steiner, 'Te Deum, 1–2', *NG*.

Like the Gloria of the mass, the Te Deum is a song of praise whose text is a compilation of heterogeneous parts: parallel verses at the start, a quotation from the Sanctus, a doxology (verses 11–13), then a section praising Christ, and finally verses drawn from the psalms.

Only one melody is known, in several variant versions. It is built upon psalmodic formulas, which change when a new section of text begins. This raises the possibility that the musical formulas are as old as their respective texts, at least in essence. An adequate study based on a comparison of manuscript sources is still needed.

Steiner has outlined the tonal problems of the melody, which is usually assigned to mode 4. From the conflicting or evasive versions in the available sources it looks as if the melody originally had both $F\sharp$ and $F\natural$ ($b\natural$ and bb at a higher transposition, $E\natural$ and Eb at a lower). $F\natural$ definitely seems right for the latter part of the chant, and earlier F can be avoided altogether if desired (cadencing G–G–E instead of G–$F\sharp$–E).

We may take the position with reciting note a, final E, for the purposes of discussion. The following information would have to be modified for individual sources. The reciting note in each phrase is circled.

The first part uses the psalmodic formula:

$$G\text{–}b\text{–}c\text{–}b\text{–}\textcircled{a}\text{–}b\text{–}a \parallel E\text{–}G\text{–}\textcircled{a}\text{–}G\text{–}ab\text{–}baG\text{–}G$$

After some less regular verses another formula takes over:

$$E\text{–}G\text{–}\textcircled{a}\text{–}G\text{–}a\text{–}b\text{–}a \parallel E\text{–}G\text{–}\textcircled{a}\text{–}G\text{–}a\text{–}b\text{–}G\text{–}E$$

At 'Aeterna fac' the whole tonal level sinks a fourth, which is where the decision about $F\natural$ or $F\sharp$ becomes important:

$$C\text{–}DE\text{–}\textcircled{E}\text{–}F\text{–}D\text{–}F\text{–}EDC \parallel C\text{–}E\text{–}F\text{–}\textcircled{G}\text{–}a\text{–}F\text{–}GFE\text{–}E$$

Sources like the Worcester Antiphoner (PalMus 12, 5) which start on D, with a reciting note G, now move up a tone to join the other versions.

At 'Per singulos dies' the previous level is regained. It should be emphasized that these formulas are handled with a good deal more flexibility than simple psalm tones.

II.4. THE GREAT RESPONSORIES OF THE NIGHT OFFICE

   (i) Introduction
  (ii) Repertory, Texts, and Form
 (iii) Music
 (iv) Centonization
  (v) Melismas

Frere in *AS*; Hucke, 'Responsorium', *MGG*; Hucke 1973; [Cutter], 'Responsory', *NG*.

(i) *Introduction*

The great responsories of the Night Office, like the graduals and tracts of mass considered in the next section, are long, ornate chants in which the same material can be found in several different pieces of the same tonality. This is comparable to what we have just seen in simpler chants. Underneath the surface detail of the ornate chants one may still recognize passages of recitation, and common ways of opening and closing the piece can be discerned, comparable to the intonations and cadences of simple psalmody. The melodies are, however, much more elaborate than the common psalm tones, and there are far more of them, in the case of responsories over a thousand in some medieval manuscripts. They may nevertheless be grouped in families, according to the musical material they use. And if we regard each melody-type as roughly equivalent to a simple psalm tone, then the disparity in numbers is not so startling.

Responsories, graduals, and tracts have fared well as far as musical analysis goes. Frere made a pioneering study of office responsories in the introduction to *AS*, identifying and labelling recurring musical figures to provide a taxonomy of a great part of the repertory. His shorthand method of characterizing pieces was extended by Apel (1958) to graduals and tracts. In the meantime Wagner (III) and Ferretti (1938) had also published analyses of the repertory. More recently Hucke has devoted several important studies to them ('Responsorium', *MGG*; 1955, 'Gregorianischer'; 1956; 1973; and further remarks, 1980, 'Towards'). Here it will be sufficient to illustrate the general principles at work.

(ii) *Repertory, Texts, and Form*

The responsories are not as well known as the graduals and tracts of mass, for they are far more numerous and less easily accessible in modern editions. There are the fascimiles *AS*, PalMus 9, and PalMus 12, and the texts of other antiphoners (and breviaries) have been published (notably the twelve in *CAO*). But because the Night

Office is no longer sung with the elaborate responsory melodies, Vatican versions are not available. The *Liber responsorialis* contains over 300, the *Processionale monasticum* over 100, and others are in *LU*.

Most medieval manuscripts contain two or three times this number. The Old Roman antiphoners contain over 600, and so does one of the earliest notated Frankish antiphoners, that of the monk Hartker of St Gall (St Gall, Stiftsbibliothek 390–391: PalMus II/1). Later medieval sources may contain over 1,200. As Hucke ('Responsorium', *MGG*) has pointed out, a basic core of the repertory is to be found in practically all medieval sources, but the order in which the responsories are assigned to the nocturns is very variable. (The same is true for office antiphons.) Comparing the order of responsories in various sources has therefore become a favourite way of detecting relationships between sources.

Responsory texts are mostly selected with respect to the readings of the Night Office which they follow. Thus there are blocks of verses excerpted from the Prophets in Advent, from the psalms (in numerical order) after Epiphany, from the Heptateuch in Lent, and so on. Groups of them were sometimes copied with titles denoting their source, as for example 'de Adam', 'de Noe' in Lent, or 'Historie' during the summer months (when Tobit, Judith, Esther, the Maccabees, and so on, were read).

Responsories consist of two main parts, here called the respond and the verse. The verses, usually sung to a rather elaborate tone, have just been discussed. The order of performance is basically respond–verse–respond, to which a Gloria and further repeat of the respond is sometimes added, giving the form R–V–R–G–R. When the respond is repeated it is usually shortened, in that only its second half is sung. For the repeat after the Gloria it may be further shortened, so that only the final phrase is sung.

The assignment of particular verses to particular responds is generally unstable. Early in the ninth century Amalarius of Metz reported that in Rome many responsories were sung with more than one verse, but in the earliest Frankish sources practically all have one only. Extra verses then become more frequent in the later Middle Ages. The variety between sources in choice of verses seems to reflect early practice, whereby the cantor selected his verses at will. This is an important point of difference with so-called antiphonal psalmody, where a complete psalm is sung.

Usually only the first part of the doxology was sung ('Gloria patri et filio et spiritui sancto'), though some sources stretch the same music over the complete text. The Rule of St Benedict calls for the doxology for the last responsory of each nocturn, the practice followed in most medieval books, though once again the practice varied from church to church.

Of the 634 responsories in Hartker's antiphoner (figures in Ferretti 1938, 246, and Cutter, 'Responsory', *NG*), about an eighth are in mode 4, about the same as mode 2, slightly more than in mode 3, and slightly less than in mode 1. Most (about a fifth each) are in modes 7 and 8, while the F modes have relatively few.

(iii) *Music*

We can gain an idea of the musical make-up of responsories by placing side by side four pieces from the Night Office for St Stephen (Ex. II.4.1). Remembering how responsory verses were performed, we can see here also standardized points of arrival and departure, and occasionally recitation notes (*F*); but there is greater freedom of movement than in the verse tones, and even room for excursions into foreign territory.

Frere used a system of labelling cadences which I have followed here, where possible (cf. *AS*, Introduction, 33–4). But in Frere's source (an antiphoner of Sarum use) the second phrase ends on *E*, whereas in the Klosterneuburg antiphoner transcribed here it ends on *F*. The same happens in three of the four responds in the third phrase.

The beginning and the end find the four responds in closest agreement, that is phrase Oa and phrase El.

In the second phrase, the first two responds are almost identical, the third respond has the same cadence, but the fourth respond merely has the same range and final note.

The third phrase is similar in all responds, though more loosely so than at the beginning of the responsory; the longer texts have extra recitation (*a ac a*, etc), while the shortest text, 'domine Iesu Christe', forgoes this altogether. It is interesting that all the four responds converge on the notes *aGFGaGaGF* (as for 'sanctorum' in the fourth of the responds), but two of them wander on beyond it: 'Christe' needs an extra four notes, 'ait' an extra five.

The fourth phrase sees the responds going along quite different paths. Three of them remain in the lower range, but the third respond rises up rather dramatically to *c* once again, perhaps in response to the words of Stephen himself: 'Behold, I see the heavens opened' (Acts 7: 56).

The second respond now has only one more phrase, the final phrase which is very similar in all four responds. In front of this, two of the responds have a preparatory phrase, cadencing on *G*. The third respond uses this phrase almost in passing, aiming instead for a melisma on 'stantem', which carried the music over into the final phrase.

These four are not the only responds using this musical material, as a glance at Frere's analyses will show. The final cadence is also used in mode 3 pieces, and so is that of the third phrase (which resembles the median cadence in the responds for this mode). Some phrases which were unique among our four responds may be found in other mode 4 responsories, such as 'Christi martyri' in the fourth respond or 'accipe spiritum' in the first. Other phrases remain unique: 'ecce video' and the 'stantem' melisma in the third respond. In view of the structural importance of the note *F*, one might expect some correspondence with F-mode responsories, for example, in the opening. But this is not the case; even though some F-mode responsories start with a phrase reciting on *F* and cadencing on *D*, they have a quite different approach.

Ex. II.4.1. Responds of four responsories (Klosterneuburg, Stiftsbibl. 1013, fos. 34^v–36^r)

The similarities and dissimilarities between responsories are easy enough to pinpoint, especially with Frere's analyses to hand. What does this tell us about the way these pieces were composed?

To answer this question, we should know among other things who sang the pieces. For the period immediately preceding the committing of these pieces to writing, that is, the eighth and ninth centuries, when the repertory was settling into the form we have in the earliest manuscripts, the situation is not altogether clear. The practice of the present day is that the leader(s) of the choir intone the first word (only) of the respond, and the choir completes the respond. The verse is sung by a soloist, or small group of singers. The repeat of the respond is sung by the choir. This practice is by no means a modern convenience, for it can be seen in the Parisian polyphonic settings of the twelfth to thirteenth century. If we allow the principle that soloists usually sing more difficult music than the choir, then this way of performing responsories seems somewhat contradictory. It would surely be easier to train a choir to sing the verses, adapting the well-known texts to the appropriate tone. We could imagine the cantor directing his singers where to make the appropriate inflexions, just as he might do for simpler psalmody. The responds, by contrast, are far less uniform, and one would expect them to have required more rehearsal.

As Helmut Hucke has pointed out (1977, 186; 1980, 'Towards', 452), in the Lucca antiphoner (PalMus 9) antiphons and responsories are usually marked with a cross at the point where the solo part ends and the choir takes over. (In responsories it is often, but not always, the same place as where the repeat after the verse starts.) The cross is equivalent to the asterisk found in modern chant-books, but it usually occurs far later than the modern asterisk. The choir apparently sang very little of the antiphon and only a small part of the responsory. In Ex. II.4.1 above I have placed crosses as they appear in the Lucca manuscript, although the Klosterneuburg manuscript has no such marks. (For the third respond the place of the cross is not clear: as far as I can see, one has been erased and another added). This is reassuring, for the crosses indicate that here the choir sings only the most frequently used phrases for mode 4 responsories, labelled G1 and E1. Admittedly, more extended rehearsal would still be needed for the third respond.

The division of the respond between soloist and choir may, however, be a relatively late practice. There are several indications that at an earlier time soloist and choir both sang the complete respond. Amalarius of Metz, writing *c*.830, reported both the Roman and the Frankish practice of his time (ed. Hanssens, iii. 55). The Romans, he says, sang responsories as follows:

Praecantor: Respond
Succentores: Respond
Praecantor: Verse
Succentores: Respond
Praecantor: Gloria
Succentores: Respond second part

Praecantor: Respond
Succentores: Respond

According to Amalarius, the Gloria had but recently been added in Rome (but in Ordo Romanus VI, of the eighth century, a Frankish monk said that all responsories were sung with the Gloria). The Franks, on the other hand, usually sang only the second part after the verse. (Complete repeats are nevertheless called for on some occasions in modern service-books.) At any rate, Amalarius does not speak of a solo start and choir continuation of the respond.

The choir's task is easier if the soloist has already sung the respond. Furthermore, in the Winchester polyphonic settings of the late 10th century the whole respond appears to be for the soloist.

The singing of the respond proceeded along clearly understood lines. The tonality (in this case mode 4) would be known, which brought with it numerous conventions guiding the performance: the notes used for reciting passages with longer text, how to start and end phrases, which intermediate goals to aim for (D at the end of the first phrase, E/F second and third), which phrases were opening phrases (Oa) and which closing phrases (G1–E1). Two decisions depend on a close reading of the text: how many phrases must be distinguished, and which syllables should carry musical figures reserved for accented syllables.

(iv) *Centonization*

Stäblein ('Graduale (Gesang)', *MGG*, 650–1) and Treitler (1975) have warned us against seeing conventional turns of phrase as fixed entities which a composer took 'off the peg', so to speak, and inserted at the appropriate moment. These are not (to adopt another modern analogy) 'identikit' compositions, put together from pre-existing segments. The music of each responsory is a new creation, which, by virtue of the fact that it is a responsory (and not, say, a gradual) and in mode 4 (not 3), follows conventions appropriate to those categories.

The term 'cento' has often been used to describe the musical make-up of these chants. The Latin word was commonly used in Roman times to mean a garment made of several pieces sewn together, a patchwork. But it also had a special literary meaning, being used as the title of a poem made up from verses of other poems. It seems to have come into modern circulation because John Hymmonides 'the Deacon' used it to describe the work of St Gregory himself: 'Antiphonarium centonem cantorum studiosissimus nimis utiliter compilavit' (PL 75, 90). If we understand the expression in its literary sense, 'cento antiphoner' describes very well the texts sung in the liturgy. Three of the four responsories just discussed have texts selected and adapted from Acts 6 and 7: I give here the Vulgate texts, printing the chant excerpts in capitals. The chant adds other words as well, and adapts others, for example through different conjugational endings, hence the half-capitalized words here.

Acts 7: 58–9: Et LAPIDABANT STEPHANUM INVOCANTEM ET DICENTEM: 'DOMINE JESU, SUSCIPE SPIRITUM MEUM.' Positis autem genibus, clamavit voce magna, dicens: 'Domine, NE STATUAS ILLIS HOC PECCATUM.'

Acts 6: 15: Et INTUEntes eum OMNES, QUI sedebant IN CONCILIO, VIDErunt faciem EIUS TANQUAM FACIEM ANGELI.

Acts 7: 55: Cum autem esset plenus Spiritu sancto, INTENDENS IN CAELUM, VIDIT GLORIAM DEI, et Jesum stantem a dextris Dei; ET AIT: 'ECCE, VIDEO CAELOS APERTOS ET FILIUM HOMINIS STANTEM A DEXTRIS DEI.'

At least some of these words are actually read as lessons during the office.

The use of conventional turns of phrase in singing these texts seems rather different, a procedure for which the term 'cento' is not really appropriate. These are neither pre-composed scraps to be sewn into place, nor quotations of an already existing composition.

Having come so far, we still do not know how the first decisions of all were taken, that is, how it was decided that this particular matrix of musical material, one of the mode 4 complexes, was the right one for these particular texts. It is conceivable that one piece led to another, for the close proximity of these four in the liturgy is surely not accidental. Other bunches of responsories linked both by musical similarity and by liturgical proximity may easily be picked out of Frere's tables, for example those using Frere's 'theme *a*' in mode 2, in Passiontide, the summer Histories, and for St Laurence (*AS*, Introduction, 7). Nevertheless, of the earliest stages of text selection and mode of performance we have no detailed knowledge.

There is perhaps a natural tendency to regard those responsories which use conventional turns of phrase as relatively old, those which are more independent as more recent in date. One associates a simple system with beginnings, deviations as later decadence. But our knowledge of the musical shape of the responsories dates only from the end of the tenth century, the date of the earliest notated antiphoners (the Mont-Renaud manuscript, PalMus 16, and the antiphoner copied by Hartker of St Gall, St Gall 390–391, PalMus II/1). Independent melodies occur even amongst the earliest attested pieces. To some extent comparison of the sources brings out different chronological layers. On this basis Frere did not hestitate to assign pieces to the ages of gold (ending soon after the time of St Gregory), silver (up to the tenth century), bronze, and clay. Later responsories occasionally have new music for their verses.

Frere's theme *a* in mode 2 (*AS*, Introduction, 5 ff.; Apel 1958, 332 ff.) and his theme *a* in mode 8 (*AS*, 52 ff.; Apel, 337 ff.) are two ways of singing responsories which deserve a designation other than 'matrix of musical material'. So many chants use them in such similar ways that they might well be called melodies. Even so, there are deviations from the commonest patterns. Apel actually divides the mode 8 group into four sub-groups. One must also remember that these 'melodies' have been identified as such not because of their melodic characteristics but because they have been adapted for several different texts. They stand, in fact, near one end of a seamless continuum between similarity and dissimilarity.

(v) *Melismas*

The great responsories of the Night Office are generally much less melismatic than the graduals and tracts of mass. In the earliest musical sources the graduals in particular already have long melismas on single syllables, and these are generally absent from responsories. Those which do appear are not present in all sources, or 'wander' from one responsory to another, or enter the written tradition at a relatively late stage. For this reason I have considered it best to discuss them as melodic additions to the earliest recoverable state of the responsories, and placed them with the sections on tropes (see II.23.iv).

II.5. GRADUALS AND TRACTS

   (i) Introduction
  (ii) Graduals in a: The 'Iustus ut palma' Group
 (iii) Graduals in F
 (iv) Other Graduals
  (v) Tracts
 (vi) Tracts in Mode 2
(vii) Tracts in Mode 8

On graduals: Stäblein, 'Graduale (Gesang)', *MGG*; Hucke 1955, 'Gregorianischer',
   1956, 'Gradual (i)', *NG*
On tracts: Hans Schmidt 1957, 1958; Hucke 1967, 'Tract', *NG*; Nowacki 1986;
   Kainzbauer 1991.

(i) *Introduction*

Graduals and tracts are more ornate chants than responsories, but much of the melodic embellishment is stylized, and they may more easily be classified in families sharing the same melody.

    They can be divided into distinct tonal groups. In Stäblein's list of the 115 graduals in the earliest sources ('Graduale (Gesang)', *MGG*) they break down as follows, according to final (omitting four cases where Stäblein registers doubt):

D (mode 1 only)   15
E (modes 3 and 4) 11
F (mode 5 only)   46
G (modes 7 and 8) 15
a ('mode 2')   24

Only the gradual has such a high proportion of pieces in the F mode. Tracts (listed in Hucke, 'Tract', *NG*) are even more selective:

D (mode 2 only)   6
G (mode 8 only)   15

Apart from their different musical materials, graduals and tracts are performed in a different way. Graduals were sung like the great office responsories, but with a complete repeat of the respond after the verse. Tracts were sung by soloists throughout (a practice sometimes referred to as 'direct psalmody'), without any section repeats.

Amalarius of Metz (*Ordinis missae expositio* I, 6; ed. Hanssens, iii. 302), writing about 830, reports that in the gradual there was a full performance of the respond by soloists, repeated in full by the choir. This seems to be supported by the copies in early cantatoria (books containing only the soloists' chants), where they are notated in full. By the beginning of the eleventh century, however, similar books (such as Oxford, Bodleian Library, Bodley 775) give only the first word of the respond, implying that the soloist simply intones the opening, the rest being sung by the choir. Later still the choir was allowed to sing the closing word or two of the verse, the practice set out in modern Solesmes/Vatican books. We know this, for example, through the polyphonic gradual settings from Paris in the twelfth and thirteenth centuries, where only the soloist's parts are set in polyphony.

(ii) *Graduals in* a: *The 'Iustus ut palma' Group*

The most closely unified group of graduals throws up some interesting questions. This is the family which includes *Iustus ut palma*, made famous by its presentation in over 200 facsimiles in PalMus 2 and tabulated there by Mocquereau, by Suñol (1935, fold-out), Ferretti (1938, 165), and Ribay (1988). Six of them are sung during the Ember Week of Advent. Both respond and verse of these graduals end on *a*. This pitch is usually reckoned as representing an upward transposition of a fifth from *D* (in order to avoid writing low *B♭* and both *F* and *F♯*), and the group is thus usually assigned to the second mode. The awkwardness of classification indicates that we have here a body of chant which was composed before (and therefore disregards) the eight-mode system.

Ex. II.5.1 is a transcription of *Tecum principium* V. *Dixit dominus* from the Aquitanian manuscript Paris, Bibliothèque Nationale, lat. 776. I have marked the individual phrases with the labels given by Apel (1958, 360), whose tables provide a convenient way of surveying the musical make-up of the gradual repertory. *Tecum principium* conveniently includes nearly all the musical material of the 'Iustus ut palma' family.

In the respond, the first and last phrases (A1 and A3) are those of practically all others in the family. (A3 was so well known that the scribe did not trouble to write it out in full: it is completed here from the first gradual of the year with this melody, *Tollite portas*.) Most graduals have only one phrase between these two, either A2 or F1; both appear here, with a linking phrase c1. *Haec dies*, however, has a different opening phrase (A4), then c1, F1, and A3.

Ex. II.5.1. Gradual *Tecum principium* (Paris, Bibl. Nat. lat. 776, fo. 12ʳ)

In the verse, all but a handful of graduals have the first phrase given here, D10. The others have a different D-phrase, so that the tonal direction of the melody was always the same. The second phrase, A10, is also practically always the same. For some verses, no more phrases were required, and the final invariable A12 concluded the piece. Most are four-phrase verses, however, and here one of the two F-phrases was employed: *Tecum principium* has F1, which has already been heard in the respond. F1 was generally employed when a fifth phrase, C10, was also present, as here; otherwise F10 was preferred. Just occasionally C10 is present without an F-phrase to follow. The double use of F1 goes hand in hand with the fact that A12 at the end of the verse is similar to A3 at the end of the respond, and c1 in the respond ends exactly like C10 in the verse. The overall form is therefore something like AB CB (repeat of respond AB again).

The ways in which the melody was used for this particular text, *Tecum principium*, are typical. The first phrase of text is exceptionally long, and thus is adapted in a way found only for two other graduals with similarly long opening phrases, *In omnem terram* and *Ne avertas*. (*Exsultabunt sancti*, by contrast, is exceptionally short and is therefore sung to yet another adaptation of the 'normal' form.) It is interesting that instead of simply repeating a reciting note to cope with the extra syllables, a different melodic excursion was included (extending roughly from '-pium' to 'virtu-'). Elsewhere repeated notes were indeed employed, for the number of syllables is not excessive: *a* and *c* in the respond, *d*, *c*, and lastly *a* in the verse. A3 and A12 begin with accented syllables, and here an unaccented preliminary *F* was sometimes sung.

It might be supposed that the melody is so constant because the phrases of text are all short (with the exception of the first), so that there was little need for variation, or because the melody is so ornate, necessitating careful regulation of detail. Yet on the one hand the text phrases in responsories are similarly brief, and on the other hand other graduals are similarly ornate but not so uniform in melody. For various reasons the 'Iustus ut palma' graduals have been regarded as very old: their uniformity, in which they display a similarity to the technique of singing tracts, which were also believed to be ancient; their use on days of long standing in the church year. On the other hand, Hucke has argued that the fashion for this melody might be relatively recent. A few of its characteristic turns of phrase can also be found in other graduals (compare C10 with C12 in the F-mode group, F10 with F17 in the F-mode group and F10 in the D-mode group). Hucke thinks this is because phrases from the newly popular melody were taken into the singing of older graduals. The argument cuts both ways, however, and depends on how exclusive the material for the different melodies or melodic complexes was reckoned to be. I myself incline to agree with Hucke, for it seems inherently more probable that a new melody might be used to meet a sudden demand for more graduals (perhaps in response to a liturgical and/or musical reform), than that an old and consistent practice should be superseded by something much less regular. But in the absence of firm historical data one cannot be dogmatic.

One thing is indisputable: the much more ornate surface detail of these and other graduals when compared with the office responsory. Most of this solemn splendour is concentrated on melismas for particular syllables, rather than being distributed evenly. End syllables are particularly favoured, whether accented or (more often) not. The musical matter is of a peculiar kind, reiterating particular notes or decorating them in a seemingly superfluous way. There is no firm direction to these musical ruminations: they are rather a means of making the performance more impressive and ceremonious. Some figures simply revolve around a central tone: 'princiPIum' and 'uteRO'. Others move further but return whence they came. Take for example the setting of 'die'. The melody moves from *d* to *a*, but introduces for ornamental effect a palindromic *dccaGaccd* before finally falling on to *a*. One may regard this either as an ornamentation of *d* or a double descent to *a*, but either way the redundancy is clear. The same happens on a much larger scale in the verse at 'meo' (twice down to *G* and back up to *e*) and 'meis' (twice down to *a* from *e*). 'Meis' is also interesting because of the oscillation between *c* and *a* in the middle. Such reiteration of *c*, and occasional touches down to *a*, are found not in graduals only, but also in tracts, introits, and offertories. The notes favoured for reiteration are *c*, *f*, and *g*, that is, those above the semitone step. The other notes touched upon, as it were, to provide extra impetus for the reiteration, are always a minor third lower: *a*, *d*, and *e* respectively.

(iii) *Graduals in* F

The largest group of graduals to be notated in the same mode is those with final on *F*. The melodic links between these graduals are much looser, however, than in the 'Iustus ut palma' group. One has only to scan Apel's table (346–7) to appreciate this, especially in the responds, where only the trio *Christus factus est*, *Exiit sermo*, and *Ecce sacerdos* are consistently similar. The verses are more easily assignable to families, but only one phrase, the final one (Apel's F10), is found almost throughout the repertory. Apel discerns eight groups in all, with several other graduals not assignable to any group. Between several groups there is almost no similarity at all.

Some general tendencies are nevertheless worth noting. Over half the verses start with a phrase ending on *a*. The next phrase is then either an *F*-, an *a*-, or a *c*-phrase. There are regular ascents to high *f*, even *g*, before the descent to phrase F10, whereas the respond stays much more consistently in the lower range *F–c*. In the lower range *b♭* appears regularly, whereas *b♮* is normal for most of the verse. Correspondence of musical material between respond and verse is restricted, when present at all, to the cadential melisma.

Although it is tempting to look for another basic melody, such as the 'Iustus ut palma' melody, at the root of the graduals in F, it cannot really be found. The variety of material used for performing these chants is much greater, the correspondences between the families of graduals consequently more informal. The responds give the impression of almost total lack of dependence upon standard turns of phrase. At the same time, one would not mistake the pieces for anything but graduals. For a start, F-mode pieces are relatively rare in other chant genres. Secondly, the ornamental character already noticed permeates these graduals as well. Somewhat more prominent than in the 'Iustus ut palma' group are recitations on or around one note, as in the closing phrases of *Anima nostra* (Ex. II.5.2). The final turn down to *F* at the end is almost perfunctory, made by using the only other common cadence figure in

Ex. II.5.2. From Gradual *Anima nostra* V. *Laqueus contritus est* (Paris, Bibl. Nat. lat. 776, fo. 17ᵛ)

these graduals (Apel's f11, found also at the end of the respond). (The manuscript has neither clefs nor coloured lines, and consequently does not distinguish the semitone step in the scale. I have assumed *b*bs by analogy with other manuscripts.)

(iv) *Other Graduals*

There are other smaller groups of graduals in D, E, and G. The D-mode graduals resemble the F-mode ones in that their verses resemble each other more closely than their responds, and also move in a generally higher range than the responds. The E- and G-mode graduals, on the other hand, are more loosely related to one another, and their verses are not much closer related than the responds. In all these graduals, then, one has the feeling that the singers were expected to range freely through the appropriate material, exercising their art with a degree of freedom not detectable in the 'Iustus ut palma' group or in the tracts shortly to be discussed. Once again many hallmarks of graduals are present, but the melodic gestures do not appear with such regularity. The striking melisma of Ex. II.5.3., where the melody seems to go into suspended animation, is found in only four graduals, its final phrase in only two (but the last six notes are a common way of cadencing on G in many chant genres).

Ex. II.5.3. From Gradual *Exaltabo te* V. *Domine Deus meus* (Paris, Bibl. Nat. lat. 776, fo. 54ᵛ)

sal – ua – sti me

The phrase given in Ex. II.5.3 includes some repeated figures (shown by brackets), but such closed forms are rare in graduals. They are even rarer if we set aside examples in this floating, reiterative manner. *Clamaverunt iusti* has two very mobile melismas in its verse, one with repeat structure AABBC, one unpatterned. *Viderunt omnes* has an example with AAB form in its verse. Such patterned melismas are common in alleluias, and are generally understood to be a relatively late development.

We have rather little evidence about the age of the music of the graduals, as is indeed the case with most chants for the proper of mass as it appears in the oldest music manuscripts. Hucke has suggested that the latest of the E-mode graduals may be *Iuravit dominus*, sung on a number of feasts introduced into the Roman kalendar in the second quarter of the seventh century: no gradual for a feast introduced later than this has an E-mode melody. The important F and a melodies would then be later than this. It would nevertheless be over-optimistic to believe that the melodies as we have them from the ninth century were sung in quite that way two centuries earlier. The known versions must be the result of a long process of stylization, adjustment to changing circumstances (not least the learning of Roman chant by the Franks), and, if Hucke is right, cross-fertilization from other families of melodies.

(v) *Tracts*

As already mentioned, tracts are sung by a soloist or small group of soloists, and include no return to a previous respond or other refrain section. They also have several verses. Those in mode 8 have up to five verses; the three mode 2 tracts designated as such in early sources are much longer: *Eripe me* (composed after the other two, according to Amalarius of Metz) has 11 verses, *Qui habitat* has 13, and *Deus Deus meus* 14. The technique of following a general pattern or mode (in the sense of 'manner', 'way') of singing a verse which we have already seen in the case of office responsories and graduals is also practised here, but now the correspondences may be observed not simply between different tracts but between different verses of the same tract. Since within the two tonal families they display considerable regularity of technique, they have been favourite subjects for discussion of centonization (a term now discredited) or, better, the development of ways of singing elaborate chants in the absence of written music (see especially Treitler 1974, 'Homer', where *Deus Deus meus* and other tracts are used for illustration). As usual, one may find one's bearings with the aid of Apel's tables.

The twenty-one tracts in the earliest sources comprise the following groups: eleven mode 8 tracts with psalm texts; four mode 8 cantica sung at the Paschal Vigil; three mode 2 tracts with psalm texts; three mode 2 tracts designated as graduals.

The first three cantica of the Paschal Vigil have texts which form a sequence with the lessons of this unusual mass ceremony (see above, I.8.v); they may even be regarded as 'sung lessons' (rather than simply intoned ones; of course their texts are much shorter than those of the other lessons). The fourth tract, *Sicut cervus*, accompanies a solemn procession to the font.

The three tracts called graduals in early manuscripts were presumably performed like graduals, that is, with the first verse repeated as a respond after each subsequent verse. This seems to have had an interesting effect on their use of cadence formulas: one particular formula (Apel's Dn) is usually reserved for the last cadence of all, but in these three pieces it is also used for earlier verses. This is presumably because the cadence of the first verse, of the respond, was now the final one (D15); there was no longer any need to reserve Dn for signalling the end of the performance.

(vi) *Tracts in Mode 2*

As may easily be seen from Apel's table, five of the mode 2 tracts follow the same general pattern in most of their verses. (*De necessitatibus* uses some turns of phrase characteristic of the group, but does not follow the pattern.) They usually have four sections, cadencing on *D*, *C*, *F*, and *D* respectively. Within these guide-lines there is nevertheless considerable variety of musical material.

All end the last verse in the same way; at the start of the first verse, there are two different opening phrases. But the opening phrases for other verses, *D*-phrases that is, number no less than ten; *Deus Deus meus* alone has eight different ones; to be fair, D10 usually functions as a preliminary phrase before D5.

The second section is less variable, though frequently only a cadence formula from the standard phrase is heard in its cadences; there are two alternative phrases. Most constant of all is the F-phrase, with no regular alternatives, though some small deviations. For the last phrase, ending on *D*, there are five main formulas, four of them used in *Deus Deus meus, Qui habitat*, and *Eripe me*, the other used in *Domine audivi* and *Domine exaudi*.

The variety of procedure, within the tonal guide-lines indicated, is quite remarkable, and it is clear that these tracts were no more bound in a strait-jacket than other chants. This is shown by the fact that no combination of the various *D*, *C*, *F*, and *D* phrases appears identically in more than one verse. (If we reduce the requirements simply to cadence formulas, a few identical sequences do emerge, but they are still remarkably rare.)

(vii) *Tracts in Mode 8*

The eleven tracts and four cantica in mode 8 do not display the regular succession of cadence points seen in the mode 2 tracts. Practically all opening and closing phrases have *G*-cadences, and usually there is an *F*-cadence somewhere between. For some verses two *F*-phrases suffice, three *G*-cadences are sometimes found, once four *G*-cadences, without any *F* cadences at all.

The number of different *G*- and *F*-phrases is restricted, more so than in the other tracts. Setting aside the invariable closing *G*-phrase for the very end of the tract, one may summarize Apel's table as follows:

5 *G*-phrases for the start of the tract, of which only one is used more than twice;
9 other *G*-phrases, of which only three are used more than thrice; one of them (Apel's G1) opens verses, another (G2) usually closes them, or appears as a penultimate phrase in the very last verse of the tract;
4 *F*-phrases;
1 *c*-phrase, a high-ranging melisma used at the start of two verses in *Commovisti*.

Within the group, the cantica for the Paschal Vigil form a particularly homogeneous set.

Ex. II.5.4 is the tract for Quinquagesima Sunday, *Iubilate domino*, a setting of verses from Ps. 99. When the psalm was sung during the Night Office the division of verses would usually have been somewhat different (the text would also have been that of the so-called Gallican psalter, rather than the older translation used here). Verse 3 would run from 'Scitote' to 'non ipsi nos'. But here there is a short verse 3 and a longer verse 4. The source here transcribed, Paris, Bibliothèque Nationale, lat. 776, indicates no break between verses 2 and 3, but in view of the usual function of phrase G2 as a verse-terminator, I have made the break indicated in other sources.

We can speculate about the way chosen for singing this tract. About the first and last phrases there can have been little pause for consideration, for these were the most

Ex. II.5.4. Tract *Iubilate Domino omnis terra* (Paris, Bibl. Nat. lat. 776, fo. 32ᵛ)

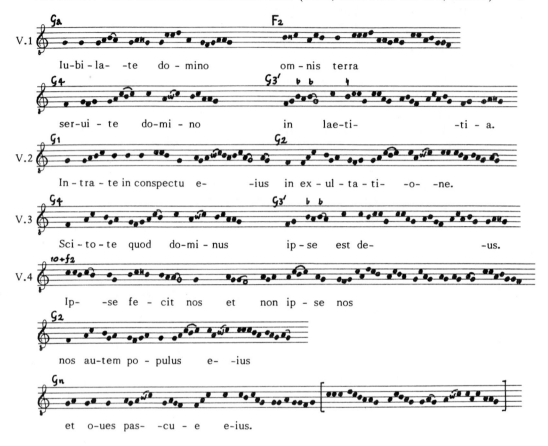

conventional parts of the chant. Since the first verse was divided into four phrases, it would be appropriate to move to an *F*-cadence next, and because the text is very short (only two words!) F2 is better than the longer F1. For a pair of short G-phrases, G4 + G3 is often used, G3 being a terminal phrase. They come round again for verse 3. Verse 2, on the other hand, has two longer phrases, so G1 + G2 is preferred. In fact G4 is similar to G2, the latter having a longer end-melisma in keeping with its terminal function.

The first phrase of verse 3, 'Scitote quod dominus', is shorter than 'Intrate in conspectu eius' in verse 2, but not yet as short as 'servite domino' in verse 1. To cope with 'Scitote' the opening of G2 is brought into play. The melisma at the end of the verse, in phrase G3, is usually an end-melisma, but is occasionally split to accommodate unaccented syllables, as here and in verse 1.

The last verse could have started with the popular pair of phrases G1 + F1, were the text not too short on this occasion. In any case, over half the tracts have unique music at this point, before using standard closing phrases. At the half-verse, for the melisma on the second 'nos', a conventional cadential formula is used, already heard in verse 1 in a similar position. The last melisma of all was so well known that it was not copied out in full by the scribe of the source used here: it is completed from the tract of this family for the preceding Sunday, *Commovisti*.

With rare exceptions, these tracts move in a very restricted range, hardly exceeding the sixth between *F* and *d*. Some could easily be classed in mode 7. They lack some of the splendour of the graduals, for internal melismas are rare and the pulsating reiterations of a structurally important tone are largely absent. Because of their procedural consistency several writers have seen similarities with simple psalm tones, pointing to intonations and recitations, flexes and mediants, and so on. Such an analogy should not be pressed too far, for then the temptation arises to strip the music down to some sort of basic tone, and see the tract as the result of historical development out of imagined simple beginnings. These are not 'variations upon a theme (in G or in D)' for there is no pre-existing theme, and the level of decoration, the degree of solemnity, might have been an essential part of the tract from the start. The conventions of articulation (starting phrases, terminal melismas) are a natural response to the need to mark off the major breaks in the text, found in very many chant genres and not necessarily deriving from simple psalmody. The tonality and range provided a musical frame of reference, and phrases of similar length and balance were sung in similar ways. The result is a mode of delivery flexible enough for the performance of multiple biblical texts, but musically characteristic enough of a particular liturgical moment to be recognizable for what it is and fulfil its proper liturgical function.

II.6. SHORT RESPONSORIES

Hucke, 'Responsorium', *MGG*; Claire 1962, 1975.

As well as the great responsories of the Night Office (the *responsoria prolixa*), much shorter responsories (*responsoria brevia*) were sung after the short lesson (capitulum, chapter) of the Little Hours (Prime, Terce, Sext, None, and Compline). In monastic use a short responsory was sung after the short lesson of Lauds and Vespers as well. The repertory is small, for single responsories did duty for whole seasons of the year, and rather few feasts had proper ones. In most churches, very few melodies seem to have been used, though some sources have more, florid versions of the usual simple melodies, newly composed ones, or melodies adapted from great responsories (see, for example, the Worcester antiphoner (PalMus 12), and the Nevers manuscript Paris, Bibliothèque Nationale, nouv. acq. lat. 1235 consulted by Wagner (III, 217–23). No survey of the repertory is available. The conventions for singing the doxology are obscure: some sources contain no cues at all; other specify Glorias of differing length (for example, up to 'sancto', or 'semper').

The musical style of most short responsories is very simple, chiefly syllabic, consisting of melodic formulas which are easily adaptable for a variety of texts. The form is that of all responsorial chants: solo respond, repeated by the choir, solo verse, choir respond (or the last part thereof), solo doxology, choir respond.

Hucke thought that the most popular melodies might be the result of a ninth-century Frankish recension. In one sense this is no doubt true, as it is of a large part of the chant repertory, but the simplicity of the melodies has encouraged speculation that a much older tradition lies behind them. A peculiarity of some of the melodies is that the formula used for singing the first part of the respond is the same as that for the whole of the verse. Musically, therefore, it is the second part of the respond which constitutes a refrain. This may be seen in Ex. II.6.1, a transcription of the first of the short responsories in the Lucca antiphoner (PalMus 9). The doxology is divided into two parts, the first rhyming musically with the first half of the respond (and the verse), the second with the second part of the respond.

Ex. II.6.1. Short responsory *Super te Ierusalem* (Lucca, Bibl. Cap. 601, p. 6)

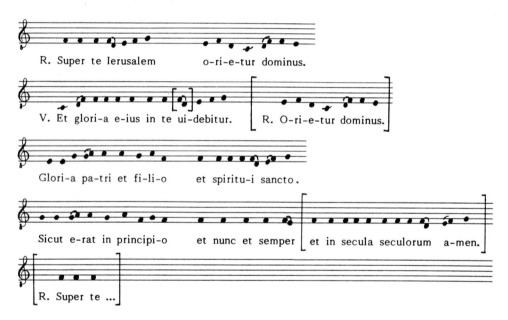

The melody of Ex. II.6.1 was the usual one for the half of the year from Advent onward. During the rest of the year a single F-mode melody in one of the three different variants sufficed. One of these is shown in Ex. II.6.2, transcribed from the Worcester antiphoner (PalMus 12). (The section of the verse 'alleluia, alleluia' is always written out complete, but presumably one did not then repeat the second half of the respond.) Once again the two phrases of the doxology end on the same notes as the respond, though here the shorter text is preferred.

In both the melodies quoted, the overall form is therefore:

Respond:	A	B
Verse, second part of Respond:	A	B
Doxology (cadence notes):	...a	...b
Respond:	A	B

Ex. II.6.2. Short responsory *Resurrexit Dominus* (Worcester, Chapter Lib. F 160, p. 142)

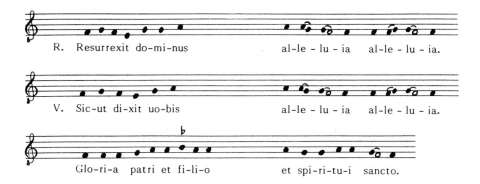

It would be interesting to know how widespread this pattern was, and whether, at some earlier time, the respond was simply the short second phrase.

The similarity of such simple melodies to some antiphons has often been remarked. Ferretti, for example (1938, 265) thought that the mode 6 antiphon melody for such pieces as *Ego sum vitis vera* and *Notum fecit dominus* (Gevaert theme 39) had given rise to the melody of Ex. II.6.2. Claire's work has tended to suggest the opposite: that the short responsories preserve relics of ancient melodic types, and families of antiphons were modelled on them. Some antiphons would derive from the second part (B above), some from the whole melody (AB). Claire draws parallels not only with Gregorian examples but also with Old Roman and Milanese ones. Particularly suggestive are the similarities with ferial antiphons, especially with the versions of these antiphons in manuscripts from Metz, Aachen, and Lyons. Two antiphons from Aachen, Dombibliothek, 20 (Ex. II.6.3) may be compared with Ex. II.6.1 above (compare the versions in Claire 1975, nos. 67bis and 44). Most other sources transform these into G-mode antiphons. Whether or not one agrees with Claire's

Ex. II.6.3. Ferial antiphons *Credidi* and *Portio mea* (Aachen, Dombibl. 20, fos 43ʳ, 47ʳ)

Credidi propter locutus sum.

Porti-o me-a domi-ne sit in terra uiuen-ti-um.

hypothesis, it is clear that these short responsories have no connection with any eight-mode or eight-tone system.

II.7. ANTIPHONS

   (i) Introduction
  (ii) Ferial or Psalter Antiphons
 (iii) Antiphons for the Psalms of Vespers, the Night Office, and Lauds
 (iv) Antiphons for the Magnificat and Benedictus
  (v) The Great O-Antiphons

Gevaert 1895; Frere in *AS*; Alfonzo 1935; Stäblein, 'Antiphon', *MGG*; Hucke 1951, 1953, 'Formen'; Turco 1972, 1979, 1987; Hourlier 1973; Claire 1975; Franca 1977; Udovich 1980; Huglo, 'Antiphon', *NG*; Crocker 1986; Dobszay 1990, 'Experiences'.

(i) *Introduction*

There are more pieces called antiphons than anything else in the chant repertory. The great majority belong to the singing of the office, where they are coupled to the daily, weekly, and yearly cycles of psalms and canticles. But some have no connection with psalms and are sung to accompany processions, or as free-standing votive anthems, most often in honour of the Blessed Virgin Mary. They are of a different musical character and are therefore discussed in a later section, as are the antiphons for the Venite of the Night Office, known as invitatories.

The many hundreds of pieces in the main body of office antiphons may be divided roughly into three groups.

   1. Generally short and simple in style are antiphons for the ferial office, that is, for the office hours on ordinary days (including Sunday) when no feast intervenes. Nearly 100 antiphons of this type, with texts drawn from the psalms they accompany, were generally required.

   2. The bulk of the repertory was sung on days with their own special liturgy, the dozens of feast-days of various types throughout the year. Over 1,000 antiphons in this category are usually to be found in medieval books, composed for Vespers, the Night Office, and Lauds (proper antiphons were not usually required for the Little Hours). Repertorially, there is a general distinction between secular and monastic books, because of the different numbers of pieces required: only at Lauds, where five antiphons were sung in both uses, is much agreement to be found, mostly in the Temporale.

   3. Antiphons for the two canticles of Vespers and Lauds, the Magnificat and Benedictus respectively (called gospel canticles because of their literary source), are generally longer and sometimes musically more elaborate than the others, at least on feast-days. Their texts are usually taken from the gospel at mass of the day, otherwise

they look to the Old Testament lessons of the Night Office, or, on saints' days, to the *vita* or life of the saint read in chapter and during the office itself.

Antiphons in these three categories will usually number about 1,500 in most medieval manuscripts. But there is considerable variety between sources, so that the total number used across Europe was enormous. The twelve sources whose texts were edited by Hesbert in *CAO* have well over 4,000 between them.

Antiphons are settings of prose texts; they follow no regular mode of delivery such as a psalm tone, but naturally, in view of the great numbers required, often display melodic identities, similarities, or standard responses to appropriate texts. Some melodies were very popular and were used with minimal variance for numerous different texts. Other melodies, or complexes of melodic material, were used with much greater flexibility; for example, Frere said of mode 7 antiphons: 'there is much similarity of material and method, which does not amount to a unity of theme'—there is of course room for disagreement as to what does constitute 'unity of theme' (see Nowacki 1977 for an analytical discussion).

The melodic style of most antiphons is relatively simple, with clear-cut phrases. There is no need of the melismas found in florid responsorial chants to mark crucial cadences or other structural features of the text. Within phrases one may find repeated notes, anacruses, and so on, in order to 'stretch' a melodic phrase over a longer text, but this is rarely so extended as to remind one of a psalmodic recitation. When a text has a relatively large number of phrases, however, extra phrases of music will be provided. (Something of this has already been seen in the responsorial chants.) A number of melodies seem to have been composed deliberately for two-phrase texts, three-phrase, four-phrase, and so on.

Since each antiphon preceded and succeeded the singing of a psalm or canticle, the choice of psalm tone was preferably tonally compatible with the antiphon melody. Lists of antiphons were drawn up in medieval tonaries (see III.14), where the antiphons are grouped according to mode and according to the psalm tone and differentia that they command. These groupings naturally bring together antiphons which are similar melodically. They do not, however, constitute a reliable thematic catalogue, for, if antiphons are to take the same tone and differentia, it is sufficient for their opening and their final to be similar: what happens in between is another matter.

Taking the tonary of Regino of Prüm (d. 915) as a starting point, the Belgian musicologist Gevaert published a melodic catalogue of over 1,000 antiphons under forty-seven 'themes' (Gevaert 1895). Except through passing remarks of Gevaert's, however, it is not easy to get an idea of how stable the 'themes' are in practice, since Gevaert usually cites only the opening of each antiphon. Apel's discussion of mode 7 antiphons demonstrates the value and drawbacks of Gevaert's presentation. So also does a comparison with Frere's analysis in *AS*: Frere identified fifty themes, established on the basis of the whole melody, not the incipits which Gevaert (like the medieval tonaries) cited, though Frere too was guided by the groupings of the Sarum tonary. Frere's themes are presented with comments and illustrations of their stability, or lack of it. In the index to the Sarum antiphoner which follows the introductory analyses, Frere marked about two-fifths of the 1,600 or so antiphons with the melodic labels assigned in his introduction.

The complex reality behind the 'themes' can be judged from, for example, some of the mode 8 antiphons. Gevaert's theme 12 antiphons are distributed across Frere's themes VIII*a*, *c*, and *j*. Antiphons of Gevaert's themes 13 and 15 are both in Frere's theme VIII*b*, but some go into VIII*l*. But if we actually check through the antiphons in Gevaert's theme 13, we find surprisingly few assigned to a theme by Frere at all. Gevaert has thirteen antiphons in his first epoch, that is, with texts from the psalter (including ferial antiphons listed in a footnote), and six of these are assigned by Frere to VIII*b*. But of Gevaert's twenty-nine second-epoch antiphons (with texts from other books of the Bible) and two third-epoch antiphons, not one is assigned in Frere. Some, it is true, were not sung in Sarum use, and others had variant forms of the melodies, but the lack of melodic similarity in Gevaert's group, beyond the opening, is still strikingly demonstrated.

For another example one may look at the transcriptions of twenty-two mode 8 antiphons in parallel given by Ferretti (1938, opposite p. 112). Of these, which are certainly closely related melodically, eight do not appear in the Sarum antiphoner. Three are not assigned to a theme by Frere, one is assigned to theme VIII*a*, six to VIII*e* and four to VIII*j*. of the twelve that appear in Gevaert's catalogue, eight are classified under theme 16 and four under theme 12 (not the same four as Frere's VIII*j*).

Later commentators have been more wary of seeing identity between antiphons. Huglo ('Antiphon', *NG*) mentions only seven 'prototype' melodies whose basic shape was largely unaffected by adaptation for different texts. Table II.7.1 compares his choice with Gevaert's and Frere's.

The most sophisticated morphology of the repertory so far has been achieved by Hucke, who distinguishes 'Lieder' (songs, the relatively unchanging melodies) from 'Strophen' (variable successions of standard phrases or groups of phrases) and from recitation types. Hucke's analysis has the merit of not being rooted in the concept of fixed tunes, and identifies features such as initia and other motifs which are independent of thematic families. The fact that only two-fifths of the antiphons were assigned to families by Frere, impressive though the number is, shows that this procedure alone cannot give a satisfactory account of the repertory. (Dobszay 1990, 'Experiences' has reported on a forthcoming new classification. Comparative studies with the Old Roman repertory, analysed by Nowacki 1980, and the Milanese repertory, analysed by Bailey and Merkley 1990, will undoubtedly shed further light on the Gregorian antiphons.)

The work of Claire (1975) and Turco (1972, 1979, 1987) has recently opened up important new avenues of inquiry, relating standard antiphon melodies to apparently ancient psalmodic practice (see also Jeanneteau 1985). There is space here only to mention once again the psalter antiphons.

Table II.7.1. *Antiphon themes or prototype melodies*

Mode			Number of themes		Prototype (Huglo)
			Gevaert	Frere	
D (Protus)	authentic	1	8	7	
	plagal	2	3	4	
	either		4		Gevaert 9, Frere II*e* (the melody of the 'O-antiphons')
E (Deuterus)	authentic	3	5	5	
	plagal	4	5	3	Gevaert 29, Frere IV*b*
F (Tritus)	authentic	5	3	2	Gevaert– , Frere V*a*
	plagal	6	3	5	Gevaert 39, Frere VI*b*
G (Tetrardus)	authentic	7	9	5	Gevaert 23, Frere VII*c*
	plagal	8	7	19	Gevaert 13/18, Frere VIII*b/l* Gevaert 12/16, Frere VIII*e/j*

(ii) *Ferial or Psalter Antiphons*

As mentioned in the previous section, Claire has suggested that some of the simple melodies for the chanting of antiphons during the ferial office may derive from the short responsories. Both would in fact spring ultimately from ancient responsorial psalmody, documented in, for example, the sixth-century Psalter of Saint-Germain-des-Prés (Paris, Bibliothèque Nationale, lat. 11947; see Huglo 1982, 'Répons-Graduel'). Another correspondence with old responsorial practice is that the texts of the ferial antiphons are taken from the psalm which the antiphon accompanies, often the first verse (this is also the case in over a quarter of the graduals of mass).

Many of the antiphons have a very limited ambitus, and their tonality was somewhat unstable, for one finds different versions in different modes in various medieval sources. This may also be an indication of antiquity, or at least of an origin before the advent of the eight-mode system. Some half-dozen melodies seem to have been particularly popular (though not always identical in all sources), those which appear in Gevaert (G) and Frere (F) as follows: G1, FI*c*; G2, FI*d* (these two are quite similar to one another); G14, FVIII*b*; G34, F III*e*; G40, FVI*c*; and G44, FVIII*n*. It is not uncharacteristic that the melody cited in the previous section as Ex. II.6.3, ending on *b* (in a sort of transposed E mode), should in the Sarum antiphoner on which Frere's analysis is based be a G-mode melody (VIII*n*).

Ex. II.7.1 gives three psalter antiphons from manuscript Aachen, Dombibliothek 20, all employing one of the mode 1 melodies (Gevaert 1, Frere I*c*). In this source they are assigned to various psalms of the Tuesday cycle. *Secundum magnam misericordiam* is an arrangement of the first verse of Ps. 50, *Miserere mei*. *Sitivit in te* is the second verse of Ps. 62, *Deus Deus meus*. *Adiutorium nostrum* is the last verse of

Ex. II.7.1. Psalter antiphons (Aachen, Dombibl. 20, fo. 44ʳ)

Ps. 123, *Nisi quia dominus*. Each is followed by the same differentia for the psalm tone and an incipit for the psalm itself (that of *Deus Deus meus* is in fact the mediant cadence, for the first half-verse consists of no more than the three-word incipit).

The three antiphons display in miniature several features of the genre as a whole. Unlike a psalm tone, where the reciting note or tenor is repeated as often as necessary for the number of syllables, the antiphon melody is more mobile, and repeated notes, extensions, and contractions may be employed wherever it is felt necessary. Not even the cadence, often the least variable part of a melody, is constant here. The basic shape of the melody may be summed up in terms of trichords: *F(G)a (aGF) EFG FED* (brackets enclose notes not always present).

The two-phrase structure is fairly universal among psalter antiphons, although in the briefest examples a division is hardly necessary.

(iii) *Antiphons for the Psalms of Vespers, the Night Office, and Lauds*

With all due regard for the variability of the melodic material itself and the sources in which it has come down to us, one may point to the melody Gevaert theme 29, Frere IV*b*, as relatively stable. It was a favourite for texts of four phrases. It is sometimes notated with final on *E*, but usually with final on *a*, and regarded as mode 4 transposed. The reason for this is that both *b* and *bb* are required, which would be *F♮* and *F♯* if the E final were preferred, and *F♯* was not part of the pitch-series used to notate chant. The first three phrases usually sound as if from a normal mode 3 melody (with *b♮*), whereas the final Phrygian cadence falls not on *E* but on *a*, approached through *bb*. Some sources nevertheless choose *c* instead of *b* or *bb*, or notate at the lower pitch with *F♮* throughout. And different antiphons may show different approaches to the tonality. Ex. II.7.2 demonstrates both the formal stability and the tonal difficulty of the melody. The melody is strongly represented in Advent and Passiontide, and the examples are taken from the latter season (in the Aachen manuscript they follow one another directly). (For other examples, presented synoptically, see Stäblein, 'Antiphon', *MGG*, 539, all on *E*.)

Ex. II.7.2. Antiphons with the melody Gevaert theme 29 (*AS* and Aachen, Dombibl. 20)

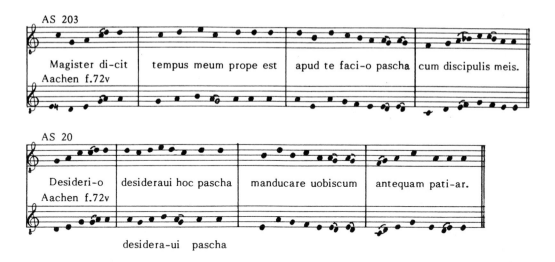

The most common melody, or melodic complex, in the regular E-mode is found among the antiphons of Gevaert's theme 36, Frere IIIc. Here the number of phrases is less regular, usually four or five. It has a relative among the psalter antiphons, and there is perhaps a temptation (which should be resisted) to see in these examples a steady expansion from simple beginnings (Ex. II.7.3). *Cunctis diebus*, the psalter antiphon, has but two phrases. Picking up the *G–a* opening of the second of these phrases, *Fidelis servus* brings in a central phrase, which cadences on *G*, in between the *b* and *E* cadences of the outside phrases. For a fourth phrase the *a–c*, *b–G* steps at the end of the first phrase may be developed: in *Herodes enim* this happens in the second phrase, in *Hic est discipulus ille* in the third. The last phrase may also generate an extension, as in *Nigra sum*.

One simple way of generating music for a longer text is to repeat the melody, which happens with some members of Gevaert 18, Frere VIIIb, and Gevaert 12, Frere VIIIj. Another group (Gevaert 39, Frere VIb), displays a simple ABAC form, where B and C are settings of 'alleluia', making an intermediate and final cadence respectively.

A large number of antiphons, however, draw upon a stock of phrases in a way somewhat akin to the procedures we have seen in responsorial chants. This has been demonstrated, for example, for a group of mode-1 antiphons by Ferretti (1938, 113–16; compare the antiphons having Ferretti's opening 1 with Gevaert theme 6 and Frere Ia). Some of the phrases were clearly associated with openings, some with antiphon endings; others may have a mediant function. Hucke pointed to the universality of this technique through many chant genres (1953, 'Formen'; see the example on pp. 16–19). A glimpse of it can be seen in Ex. II.7.3 above, where the outside phrases are constant but the inner ones vary.

Ex. II.7.3. Antiphons with the melody Gevaert theme 36 (Lucca, Bibl. Cap. 601)

Ex. II.7.4 shows something of the same technique persisting in a relatively late group of antiphons. They were notated in some sources, including the Lucca antiphoner used here, on *c*, in others on *F* with *b♭* throughout.

Gaudeamus omnes is from the liturgy of the Octave of the Nativity; exactly the same form of this melody was used for *Nesciens mater* and *Virgo hodie fidelis*, on the same feast. One may distinguish five phrases, labelled A to E. The third phrase C uses the same small decorative turn *cefed* as the second phrase B. The sense of the text carries us over the break between third and fourth phrase, though other antiphons in the family show a caesura here.

Pro eo quod non credidisti, for St John the Baptist, has a longer first phrase, which is given an ornamental opening O; the second phrase is as in *Gaudeamus*, but phrase C is absent and E is much shorter; D now echoes the opening; the last phrase has little room to expand up to the high *g*.

In *Modicum et non videbitis* (Easter week) the initial fall to *G* (O) is no longer ornamental. The second phrase (X) is now quite different, for the composer seems to want to reflect the parallelism of the text in his music. The D and E phrases are still

Ex. II.7.4. Antiphons (Lucca, Bibl. Cap. 601)

recognizably related to those in the other antiphons, E now consisting of two distinct units, the two 'alleluia' calls. But E also now resembles the opening phrase.

In *Que mulier* (post Pentecost) all the melodic material seen so far is used, with E again dividing into two units.

Finally, *O admirabile commercium* (Circumcision) needs an extended first phrase, makes a second out of it in fact. D is also more extended than before. E has the same form as in *Gaudeamus*.

Although details vary from one piece to another, there is obviously a clear sense of the character and direction of each phrase. A is a decorated recitation on *c*; B makes a decorative half-cadence which includes high *f*, then takes the high *f* as the starting point for a descent back to *c*. C recites on *c* then makes the same decorative cadence more conclusively on *d*. The function of D is to turn the recitation down to a low *G*. E reverses that movement, perhaps with enough energy to touch on high *g*. With quite simple ideas like this in mind, the singer could easily add more material of the same kind, omit or substitute an internal phrase.

(iv) *Antiphons for the Magnificat and Benedictus*

Much analysis remains to be done before we can see the full extent of the material shared by different antiphons. Perhaps a layering of the repertory, by season, or even chronologically, might then be discernible. The task demands not only analysis of the melodies themselves but also careful comparison of the repertories in sources of different provenance and date.

Considerable further difficulty arises because of the variety of musical readings in different sources. Even when sources have the same melody (disagreement even at the basic level is not uncommon) there is considerable difference of detail (see the parallel transcriptions by Udovich 1980).

At the moment it appears that for much of the yearly cycle antiphons for the Magnificat of Vespers and the Benedictus of Lauds shared melodic material and general style with the bulk of other office antiphons. Nevertheless, it is rare for these antiphons to have less than four phrases. Some antiphons, however, many probably of relatively late date, are far more extended and more florid, particularly those for the Sanctorale. An example will make this clear.

Ex. II.7.5 gives three Magnificat antiphons of this type, all in mode 1. All three relate part of the story of the martyrs in question, Andrew (30 November), Agnes (21 January), and Agatha (5 February). In each we are told where the saint is (Andrew comes to the place where the cross has been erected, Agnes stands amidst the flames, Agatha in prison) and then each saint speaks: Andrew addresses the cross itself, Agnes glorifies God, Agatha prays for strength and eventual reward in heaven. Despite the similarities in the text (particularly at the start of the Agnes and Agatha antiphons) the musical resemblances are not usually literal. As we should expect, however, the antiphons make use of phrases typical for this tonality, adapted to the text as necessary. Much of the movement is concerned with rising from *D* to *a* and

Ex. II.7.5. Magnificat antiphons (Lucca, Bibl. Cap. 601)

504 · DCFa · aD

Cum perue-nis-set be-a-tus An-dreas   ad lo-cum u-bi crux pa-ra-ta e-rat

aca · aFG · FC

excla-mauit et di-xit   O bo-na crux di-u de-si-de-ra-ta

CaD · aca · aFG

et iam concupiscen-ti a-nimo pre-pa-rata   se-curus et gaudens ueni-o ad te

(D)a · aFG · aC · CaD

i-ta ut et tu e-xultans   susci-pi-as me disci-pulum e-   -ius qui pepen-dit in te.

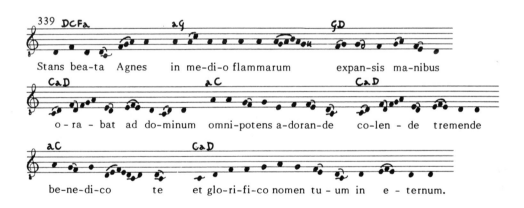

339 · DCFa · aG · GD

Stans bea-ta Agnes   in me-di-o flammarum   expan-sis ma-nibus

CaD · aC · CaD

o-ra-bat ad do-minum omni-potens a-doran-de   co-len-de tremende

aC · CaD

be-ne-di-co   te   et glo-ri-fi-co nomen tu-um in e-ternum.

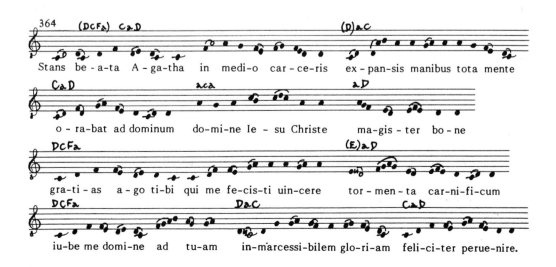

364 · (DCFa) CaD · (D)aC

Stans be-a-ta A-ga-tha   in medi-o car-ce-ris   ex-pan-sis manibus tota mente

CaD · aca · aD

o-ra-bat ad dominum   do-mi-ne Ie-su Christe   ma-gis-ter bo-ne

DCFa · (E)aD

gra-ti-as a-go ti-bi qui me fe-cis-ti uin-cere   tor-men-ta car-ni-fi-cum

DCFa · DaC · CaD

iu-be me domi-ne ad tu-am   in-marcessi-bilem glo-ri-am fe-li-ci-ter perue-nire.

falling back again. Sometimes the descent is to *C*, sometimes there are pauses around *a* or *G*. The following may be picked out here (the labels represent important pitches in each phrase and are also used in the transcriptions).

(i) *DFDCFa* + *a–D* and *c–a–D*. The penultimate phrase in all three descends from *a* to *C*, and the final phrase is an arch from *C* back to *a* and down to the final *D*. All three begin with a melodic shape seen at its simplest in the Agnes antiphon: *DFDCFa*. For Andrew there is then a long descending phrase back to *D*. Agatha only arrives at *a* at the beginning of the second phrase, but the overall progress has been the same as for Andrew. There is an exact parallel to the two opening phrases of the Andrew antiphon in Agatha's 'gratias . . . carnificum'. The opening phrase recurs yet again in Agatha's 'iube . . . tuam'. The rise from *C* to *a* and back to *D* is compressed into one unit for the final phrase; and it can be seen again in the Andrew antiphon for 'et iam . . . preparata', and twice in the Agnes antiphon; in the latter it forms an alliance with a descending *a–C* phrase, so that 'orabat . . . adorande' and 'colende . . . te' are the same; 'orabat ad dominum' is exactly as in the Agatha antiphon.

(ii) *a–C*. In Andrew 'o bona crux' forms part of a longer descent from *a* to *C* by 'desiderata'. This is also the purpose of 'discipulum eius'. We have already seen it twice in the Agnes ('omnipotens adorande', 'benedico te'), and noted its role as a penultimate phrase.

(iii) *a–c–a*. Andrew: 'exclamat et dixit', 'securus et gaudens'; Agatha: 'domine Iesu Christe'.

(iv) *a–G*. Andrew: 'o bona crux', 'venio ad te', 'suscipias me'; Agnes: 'in medio flammarum'.

Although the antiphons initially appear somewhat ornate, the basic melodic gestures are in fact quite simple. Similar phrases (indeed identical ones at the opening) may be seen in Ferretti's table of mode 1 formulas (Ferretti 1938, 113–14), though they are generally syllabic. If compared with their more numerous simpler sisters, these antiphons make more leisurely progress, and the whole effect is more solemn, as befits their role in the liturgy.

Magnificat and Benedictus antiphons were occasionally directed to be sung as processional chants. There is also a repertorial link with the *antiphonae ante evangelium*, possibly non-Roman survivals (see VI.5.v). All this may indicate that melodies of the Gallican rite have survived in the guise of Magnificat and Benedictus antiphons; but further study and analysis are required.

(v) *The Great O-Antiphons*

A special group of big, six-phrase Magnificat antiphons, all sung to the same unique mode 2 melody and beginning with the word 'O' (*O Sapientia, O Adonay, O Radix Iesse*, and so on) were sung on the seven days leading up to Christmas Eve. Others were later written in imitation of them, but the original seven appear to have been composed as a group, for they are all addressed to Christ, and may be linked by an acrostic: reading in reverse order the initial letters of each second word one finds the text 'ERO CRAS', which is interpreted as 'Tomorrow I shall be [with you].' The largest bell of the church was rung while they and the Magnificat were sung, and they were assigned in turn to the most prominent members of the ecclesiastical hierarchy: abbot, prior, cellarer, and so on.

II.8 INVITATORY ANTIPHONS

Frere in *AS*; Stäblein, 'Invitatorium', *MGG*; Steiner, 'Invitatory', *NG*.

The invitatory at the start of the Night Office comprised the singing of Ps. 94, *Venite exsultemus*, and an accompanying antiphon. Just as the Venite was sung to tones independent of the eight simple psalm tones, so the antiphons form a musical class of their own. In some respects they have more in common musically with the great responsories of the Night Office than with other office antiphons.

Ferretti (1938, 220–1) reckoned that twenty-nine invitatory antiphons belonged to the earliest layer of the repertory. A typical antiphoner will contain seventy or eighty. There is enormous variety in medieval sources as to the choice of antiphons and the Venite tone they command, and little systematic research on the repertory has yet been accomplished. This instability argues for a relatively late, expanding corpus; yet other features seem archaic. As reported above (II.3.vi), there are no antiphons (or very few) in modes 1 and 8. Furthermore, the antiphon seems to have been repeated after each verse of the Venite, the complete antiphon at the start and after verses 1, 3, 5, the second half of the antiphon after verses 2, 4, and the doxology. It is usually thought that in ancient practice all psalms were performed this way, or with complete repeats after each verse. Against this, it may be pointed out that certain processional hymns of recent and non-Roman origin were also sung thus in the Middle Ages (II.15.iv).

Ex. II.8.1. Invitatory antiphon *Quoniam Deus magnus* and part of responsory *Tolle arma tua* (Bamberg, Staatsbibl. lit. 25, fo. 47ʳ)

Most invitatory antiphons consist of some four phrases set in an ornate melodic style. As with antiphons in general, they sometimes use relatively stable typical melodies, sometimes draw upon material characteristic of the tonality but capable of more flexible use (Frere gives an account of the groupings in the Sarum antiphoner). An interesting feature of the mode 7 antiphons is their use of material also found in great responsories of the same mode. Ex. II.8.1 gives a Lenten invitatory antiphon and the start of the succeeding responsory (this fortuitous succession does not, of course, occur in all sources).

II.9. PROCESSIONAL ANTIPHONS

   (i) Introduction
 (ii) Rogation Antiphons
(iii) Palm Sunday Antiphons
(iv) Antiphons for Other Occasions

Bailey, 1971.

(i) *Introduction*

The antiphons for processions and the votive antiphons for the Blessed Virgin and other occasions, many composed in the later Middle Ages, require further study even more than do some of the other antiphon repertories just discussed. For processional antiphons a study by Bailey has fortunately prepared the ground. (In the absence of other editions, the facsimiles PalMus 12, 13, and 15 and Vecchi 1955 may be consulted.)

Medieval books are likely to have processional antiphons for the following occasions: (i) as part of the ritual accompanying the chanting of the litanies on 25 April (St Mark's Day) and on the rogation days on the Monday, Tuesday, and Wednesday before Ascension Day (Thursday); (ii) for the liturgy of Palm Sunday; (iii) in smaller groups as required for other days with special liturgies, such as those of Holy Week or the Blessing of the Candle on 2 February (Purification of the Blessed Virgin Mary); (iv) for the less elaborate processions instituted as a regular feature of many feast-days of the year, mostly before the high mass or after None. Most sources have about forty antiphons altogether.

Processional antiphons are generally ornate chants, some very long indeed, with lengthy melismas. Some have an equally ornate verse and were performed responsorially, that is, with a repeat of all or part of the first section, the respond, after the verse. Others appear with an incipit for the chanting of a psalm, but it is not clear how many verses would have been sung (presumably in alternation with the antiphon).

There is no doubt that the repertory contains chants of quite different origins. Over eighty can be found in Roman sources and may have originated in Rome, while

Frankish composition may be suspected for many others, and old Gallican relics may also be present. Yet the basic analyses of style which might help differentiate these three types, if they are really present, have not been carried out.

(ii) *Rogation Antiphons*

The earliest sources contain nearly 100 rogation antiphons, invoking God's aid in various times of trouble, or referring to the procession itself and the saints' images carried during it. Later sources usually content themselves with about twenty. Bailey's table of antiphons in twenty early sources, including those from Spain, Milan, and Rome, lists over 150 items (Bailey 1971, 122–7).

The antiphons with Roman counterparts usually display standard Gregorian features, the conventional cadences (for example *EGF FE* or *baGa aG*), occasional groups of repercussive notes (reminding one of introits or offertories) and the same gapped scale that provides a framework for many Gregorian chants. Occasionally there are hints of a different manner, for example in repeated musical phrases, but this is not a regular feature. (Compare the Gregorian version of *Non vos demergat*, Bailey, 57, which has a repeated ornate recitation formula, with the Roman one, MMMA 2, 573, without repetition. Other repetitions may be found in the 'alleluia' endings, but these are usually later additions).

Ego sum Deus, Ex. II.9.1, is an antiphon of this sort. One would expect a conventional *E*-cadence after 'eos', but it is made imperfect, and a twofold alleluia follows. Up until here the antiphon has the same general outline as the Roman version (MMMA 2, 545), but the 'alleluia' is quite different. In fact it is a 'wandering' addition, which also turns up at the end of several other antiphons in this mode.

Bailey has pointed out several examples of musical material found in more than one antiphon of the same mode, for example between *Ego sum deus* and *Populus Sion convertimini* (Bailey, 147). No network of interrelationships seems to be present, however, of the sort that has been determined for some groups of office antiphons and responsorial chants.

Ex. II.9.1. Processional antiphon *Ego sum Deus* (Provins, Bibl. Mun. 12, fo. 163^r)

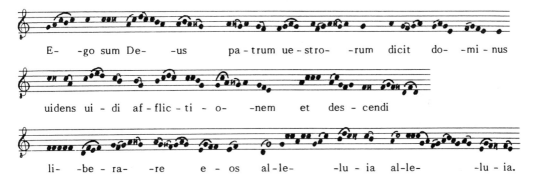

E- -go sum De- -us pa-trum ue-stro- -rum dicit do- -mi-nus

uidens ui - di af-flic-ti - o- -nem et des - cendi

li- -be - ra- -re e - os al-le- -lu - ia al-le- -lu - ia.

One rogation antiphon, *Deprecamur te*, has achieved special fame, since Bede reported that Augustine and his followers sang it as they first approached Canterbury, carrying a cross and an image of the Saviour, on their mission to England in 597 (Bede, trans. Sherley-Price, 70). The antiphon has been edited several times (for example, Stäblein, 'Antiphon', *MGG*, 542; MMMA 2, 565), and its various Frankish and Italian versions have been compared by Levy (1982).

(iii) *Palm Sunday Antiphons*

Of quite different dimensions and style are some of the grand antiphons for the Palm Sunday procession, such as *Collegerunt pontifices* V. *Unus autem, Cum appropinquaret Dominus Iesus*, and *Cum audisset populus*. (Most sources have some half-dozen chants.) *Collegerunt* (Ex. II.9.2) is occasionally found as an offertory for mass, possibly a reflection of earlier Gallican liturgical practice. It begins with a sweeping melisma, repeated almost at once, and other repeats are present (marked with letters). A richly ornamented recitation may be discerned, whereby the common cadence formula at 'Romani' and 'locum' is almost lost. Repetitions such as these may be a natural event in the setting of a long text, in melismatic style, when no vocabulary and syntax of the type developed for, say, graduals and tracts is available. We have already seen it in some later office antiphons, and the suspicion arises that it is a non-Gregorian characteristic.

Ex. II.9.2. From the processional antiphon *Collegerunt pontifices* (Provins, Bibl. Mun. 12, fo. 110ʳ)

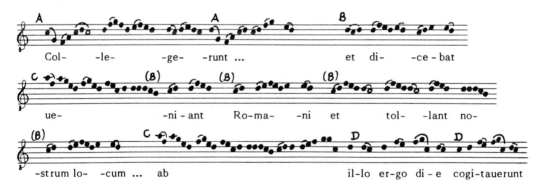

(iv) *Antiphons for Other Occasions*

Yet other stylistic features appear in the big ceremonial antiphons from other times of the year. Both *Ave gratia plena* and *Adorna thalamum suum* (Purification) were translated from Greek kontakia when the ceremony of blessing the candle was instituted in Rome by Sergius I (678–701). Byzantine melodies for these texts are not

Ex. II.9.3. Processional antiphon *Ego sum alpha et ω* (Paris, Bibl. Nat. lat. 903, fo. 74^r)

known, but the Latin antiphons are in any case not related to each other musically. *Adorna thalamum* has a very obvious repeat structure, since practically all lines follow the same highly ornate recitation formula (Stäblein, 'Antiphon', *MGG*, 542).

Different again are such antiphons as those usually assigned to Sundays after Epiphany or during Lent: *Ecce carissimi*, *Cum venerimus*, and *In die quando* (Bailey, 30–3). In these three big D-mode antiphons practically every cadence is made from the tone under the final or the dominant, C–D–D or G–a–a, a cadence often thought to be of Gallican origin, certainly not Gregorian, much used in sequences. In spite of an overall uniformity, caused by limited number of melodic goals, the only formal repetition occurs in one or two melismas. And not the least problem of deciding on the provenance of such pieces lies in the fact that the melismas may take on different forms in different sources, or be omitted altogether, and the cadences may take more normal Gregorian forms.

Easter processional antiphons include *Stetit (Sedit) angelus* V. *Crucifixum in carne* (two versions are discussed in Roederer, 1974) and *Christus resurgens* V. *Dicant nunc Iudei*. The two are nevertheless of different tonality, proportions, and musical character, *Stetit angelus* being in G, with a long respond and short verse and a very mobile, free-ranging melody, *Christus resurgens* being in D, with respond and verse equally long, cramped in range, and making use of the Gallican cadence.

One remarkable antiphon, *Ego sum alpha et ω* (Bailey, 42), is a meditation by Christ himself (Ex. II.9.3). Here yet another distinctive voice makes itself heard. Elaborate melismas are absent, with only a short alleluia with (in this source) brief repetition. The Gallican cadence is obvious, and the music of several lines is the same (7–10, 11–14, 15–18 form loosely three strophes; 19 and 21 start similarly); the alleluia has an AAB form. Comparison with other sources shows again, however, that not all these features are constant across the manuscript tradition (cf. Bailey, 42).

For many processions it was usual, at least in the later Middle Ages, to borrow one of the responsories from the Night Office of the day, or, since Vespers on feast-days might also have a responsory, that one instead. (The pieces in the *Liber responsorialis* are almost all such responsories.)

II.10. MARIAN ANTIPHONS

Harrison 1963, 81–8; Huglo, 'Antiphon', *NG*.

During the twelfth and thirteenth centuries the practice arose in many churches of singing an antiphon to the Blessed Virgin Mary as a devotional act in itself, independent of, though often attached to, one of the other services of the day. For example, in Roman and Franciscan usage since the thirteenth century, four antiphons have been sung at the end of Compline, one for each of four seasons in the church year: *Regina caeli*, *Alma redemptoris mater*, *Ave regina caelorum* and *Salve regina*. Harrison indicates the large number of different practices found in medieval England,

and this is no doubt typical of medieval Europe in general. Antiphons to the Saviour might also form part of the devotion. Collective names for such chants are votive, devotional, or commemorative antiphon. Similar antiphons might also be composed for other occasions, but those for the Virgin are the most numerous, because of the immense enthusiasm for her worship and the multiplication of services in her honour: a complete cycle of office hours, weekly or even daily mass, and so on.

In practice it is hard to draw a clear line between antiphons fulfilling these various functions. An antiphon in solemn (that is, ornate) style might serve as a Magnificat antiphon in one source, as the antiphon for a commemoration at the end of Vespers or Compline in another, or elsewhere be assigned to a separate ceremony.

Much work remains to be done in comparing the repertories of different sources, studying the rubrics in chant-books, ordinals, and customaries, and in analysing the style of the chants themselves. Few are available in print (some are published in *Variae preces*, *Cantus selecti*, and *Processionale monasticum*.)

Antiphons had of course been sung as part of the normal office hours on feasts of the Virgin since early times. Most early Marian antiphons are, as we would expect, simple, largely syllabic pieces indistinguishable musically from the rest. Such is, for example, *Sub tuum praesidium* (Ex. II.10.1), whose text can be traced back to the third century (Mercenier 1940) and which was sung in the Milanese as well as the Roman rite. A favourite source for such texts was the Song of Songs.

Ex. II.10.1. Marian antiphon *Sub tuum presidium* (Paris, Bibl. Nat. lat. 1139, fo. 126ᵛ)

Sub tu-um presi-di-um confu-gimus   de-i genitrix

nostras deprecati-o-nes ne despici-as   in ne-cessitatibus

set a periculis libera nos   sem-per   uirgo be - ne-di-cta.

Of more ample dimensions are antiphons which doubled as Magnificat or Benedictus antiphons. One such, *O virgo virginum*, was borrowed from the series of impressive O-antiphons. It was of course a simple matter to copy a psalm-tone cadence at the end of the antiphon if required, or omit it. The same antiphons might also be sung in procession, as for example, *Alma redemptoris mater* and *Ave regina caelorum* in Sarum use.

Older texts might be reset in a more ornate style to make them more appropriate for votive use. Ex. II.10.2 gives *Speciosa facta es* in two versions, the first as a simple office antiphon (in the Lucca antiphoner for the Night Office at Purification), the

Ex. II.10.2. Marian antiphon *Speciosa facta es* in two versions (Lucca, Bibl. Cap. 601, p. 347; *AS* 529)

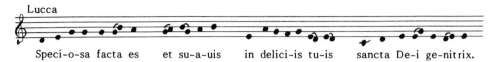

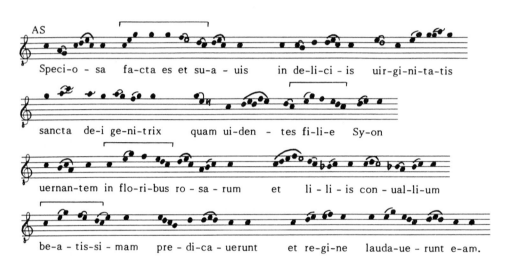

second reset and extended for processional use (in the Sarum antiphoner for Vespers of the Nativity of the Blessed Virgin). The great difference in musical style goes beyond the degree of decorative figuration. The simple Lucca version uses the popular 'theme 29' melody already seen in Ex. II.7.2 above. The Sarum melody gives the differentia for psalm tone 6 at the end, but because a bb is required (for the un-Gregorian bb–c–c cadences at 'liliis convallium') the chant is notated upon c. In almost every phrase the major triad c–e–g is outlined, usually followed by a stepwise descent g–f–e–d–c (bracketed in the example). All the cadences fall on c, e, or g.

Something of the same musical quality pervades *Alma redemptoris mater* (Ex. II.10.3), possibly the oldest of the four best-known Marian antiphons (Brunhölzl 1967), though not older than the ninth century. Its text, in hexameters, calls upon the Virgin's aid, in a manner not unlike that of many rogation antiphons. *Regina caeli*, with its joyful alleluias, is clearly an Easter antiphon. Its 'F-major' melody (F-mode but with bb throughout) features short melismas with repeat structure. *Ave regina caelorum* has a text in rhymed verse, in octosyllabic couplets, and is yet again in the 'major' mode, exactly like *Speciosa facta*, often pitched on c with an occasional bb; and as in *Speciosa facta* the triadic flavour of the melody is noticeable.

Salve regina is in the D-mode (Ex. II.10.3). The parallelism of the first two phrases, the rapid sweep through the whole octave d–D at 'misericordes oculos', and

Ex. II.10.3. From Marian antiphons *Alma redemptoris mater* and *Salve regina* (PalMus 12, 303 and 352)

the insistence on a restricted number of melodic goals (nearly all cadences fall on *D*) mark it out as a relatively late piece of music, of the same post-Gregorian vintage as the other three. In all of them, the scale is clearly divided triadically, with a strong insistence on the fifth and octave.

Another feature of the antiphons, and particularly of *Salve regina*, is the frequency of short attributive, exclamatory, or supplicatory phrases: 'Salve regina . . . spes nostra . . . O clemens, O pia, O dulcis virgo Maria'. This type of phrase lent itself well to the clearly oriented musical phrases. Ex. II.10.4 is another piece of this type. (Although we have just seen examples in this tonality notated on *F* and *c*, the sub-final here is always flat, so *G* has been chosen as final. A mode-6 differentia is nevertheless given once again. Transcription of the piece is not entirely without its problems.)

Text rhyme was an obvious way to bring such ejaculatory phrases into harmony. The history of the Marian antiphon here followed the same course as the office in

Ex. II.10.4. Marian antiphon *Aurei nominis Maria* (Paris, Bibl. Nat. lat. 1139, fo. 138ʳ)

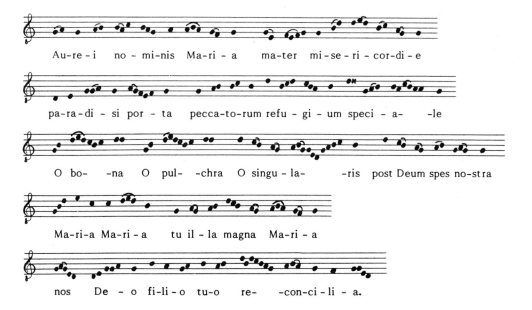

Ex. II.10.5. Marian antiphon *Trinitatis thalamum* (Paris, Bibl. Nat. lat. 1139, fo. 119ʳ)

general, and there are stylistic links with the Benedicamus songs, conductus, versus, and cantiones which became popular from the 11th century onward. Ex. II.10.5 gives an example of this type. (The D mode is indicated by the psalm-tone differentia for mode 1 at the end. Tonally it often behaves according to the pattern outlined in Ex. II.10.3 (see especially phrases 2–4 and 13–14), though a countering C–E–G is sometimes noticeable (lines 15–17). Another feature seems more archaic, however: none of the numerous short phrases is repeated.)

The most popular of the antiphons, not surprisingly, received attributions to such worthy musicians as Hermannus Contractus. Very few of these attributions have stood up to critical scrutiny (for information on individual pieces, see Szöverffy 1964–5 and 1983).

II.11. INTROITS

(i) Introduction
(ii) Introits in Mode 3
(iii) Comparison with Office Antiphons and Responsories

Cardine 1947; Froger 1948, 'Introït'; Stäblein, 'Introitus', *MGG*; Connolly 1972, 'Introits and Communions', 'Introits and Archetypes'; Steiner, 'Introit', *NG*; Hucke 1988, 'Fragen'.

(i) *Introduction*

The introit may be considered as a special festal type of antiphon, the more so because in earlier times it seems to have been sung with several psalm verses, perhaps even a complete psalm. In the earliest books with chant texts, of the eighth and ninth centuries, only one psalm verse usually remains. In some cases, however, another verse is to be found, the *Versus ad repetendum*, giving the overall form (I = Introit antiphon, Ps = Psalm verse, Gl = doxology, VadR = Versus ad repetendum):

$$\text{I–Ps–I–Gl–I [–VadR–I]}$$

These *Versus ad repetendum* appear in both the earliest Frankish sources and the Old Roman ones, also for the communion.

Practically all introit texts were taken from the Bible, two-thirds from the psalms. Those for the first seventeen Sundays after Pentecost have texts in the numerical order of the psalms.

Introit melodies are not easy to typify. They do not fall into melodic families, nor, apart from a few openings and cadences, are standard phrases to be found which link chants across the repertory. This means that they seem much more individual than most office antiphons. Of course, there are far fewer introits in most medieval chant-books, less than 150 in the earliest sources. Antiphoners and breviaries will contain ten times that number of antiphons, which makes multiple use of melodies or parts of melodies inevitable. With the introits it is almost as if there existed a conscious desire to make each chant recognizable in its own right, rather than simply part of a class or genre.

Connolly (1972, 'Introits and Archetypes') made the interesting observation that some Old Roman introit melodies are closely related to each other, whereas the Gregorian melodies for the same texts are not. Yet a similarity is discernible between each individual Old Roman melody and its Gregorian counterpart. This suggested to Connolly that introits once sung to the same basic melody (something like the Old Roman version) 'grew apart' on the way to achieving their Gregorian state. It could equally be argued that melodies once different (as the Gregorian ones are) grew more similar between the time of the Gregorian redaction (ninth century) and the date of the earliest Old Roman sources (eleventh century).

If standard phrases cannot be identified, how are we to come to grips with the substance of the melodies? What makes them appropriate for their liturgical function? The following should of course be regarded merely as preliminary remarks about only a small part of the repertory, which may suggest lines of approach for the rest. In describing melodies I have found it useful to concentrate on such features as the pitches most favoured (usually some sort of gapped scale is utilized, where certain notes, often E and b, are avoided or approached and quitted only by step), ornamental figures, the role of recitation, and the degree of mobility in individual phrases (their range and rapidity of movement).

Among the 140–50 introits in the earliest sources, there are about two dozen F-mode melodies, about thirty G-mode, about four dozen D-mode, and about four dozen E-mode melodies. The distribution throughout the year is not very even. For example, modes 1, 2, and 3 provide the majority of introits for saints' days. D-mode melodies are absent from Passion Sunday to Pentecost. Melodies in the same mode rarely occur in close proximity, an exception being the three mode 3 melodies on the Ember Days at Whitsuntide (characteristically, however, these are no more closely related melodically than others in the same mode).

(ii) *Introits in Mode 3*

In the oldest sources twenty-six melodies are assigned to mode 3. These may give some idea of the cohesion of the introit repertory as a whole. All quotations are taken from Hansen's transcription of the Dijon tonary, Montpellier, Faculté de Médecine H. 159. In this source the chants appear in tonal order, so that it is relatively easy to compare melodies for their likenesses and dissimilarities.

The openings are the most conventional feature of the mode-3 introits. Nearly all begin on G and rise to c in the first phrase, perhaps with an additional preliminary clause starting on E or F. Ex II.11.1 shows progressively more involved openings which follow this basic pattern:

    (*a*) simple rise G–c, the first syllable being acented;

    (*b*) rise through a; but *Benedicite* retains G (in this source: the treatment of such clichés occasionally differs from manuscript to manuscript);

    (*c*) a decorated form of (*b*), where the second syllable in a three-syllable word is accented;

    (*d*) preliminary E, ED, or EFD; the last two examples, *Ego autem* and *Ego clamavi*, use the decorative turn in (*c*);

    (*e*) the rise to c may be delayed within the lower tetrachord D–G;

    (*f*) shows a flourish around the final c; the substitution of adc for the simple ac is very common in other phrases also.

The choice of opening depends on the importance of the opening words in the text as a whole: in (*a*) the opening word is one of the most important in the phrase, whereas in (*e*) the key words come later. Then there are the usual adjustments for accentuation. Many of the ground-rules for introit openings have been outlined by Hucke (1988, 'Fragen').

Ex. II.11.1. Mode 3 introit openings (Montpellier, Faculté de Médecine H. 159)

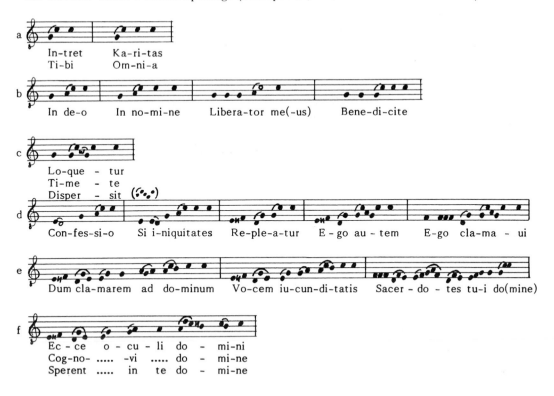

The cadences are not quite so conventional, except in the last few notes. Practically every cadence touches on the notes *E–G–F–E* in that order, naturally with repetitions or extra insertions; it seems important that the final *E* be anticipated or prepared. As we shall shortly see, many phrases tend to hang around either high *c* or middle *G*, and most cadences proceed from one or the other of these two. The cadences, illustrated in Ex. II.11.2, are grouped as follows.

(*a*) a rapid descent from *c*;

(*b*) more gradual descent from *c*; apart from the first example, 'mihi caro', these have the most common closing figure *EGFF FE*; some show an initial rise to high *c* before the run-in to the cadence;

(*c*) *bb* in the closing phrase;

(*d*) cadences which proceed from an orientation around *G* also favour the *EGFF FE* ending, otherwise *aGFGFE E*.

These examples do not, of course, exhaust all the openings and cadences of the mode 3 introits, only those which may easily be compared. What is noticeable, in fact, is that so many chants are dissimilar. A good deal of the likeness between responsorial chants was due to the presence of cadential melismas, and these are absent here. Of course, the phrases are moulded in accordance with the words being set to music, and

Ex. II.11.2. Mode 3 introit cadences (Montpellier, Faculté de Médecine H. 159)

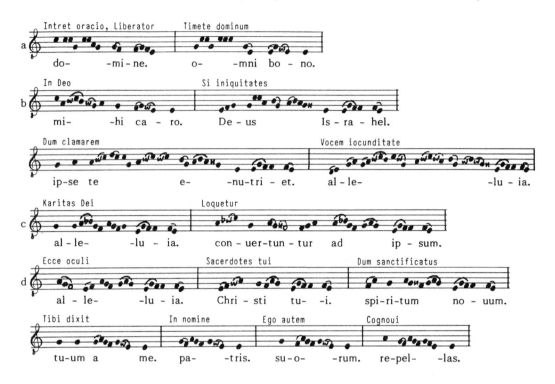

variety is only to be expected, but even so, surprisingly few cadences are actually identical apart from the last seven or eight notes.

Such similarity as exists concerns either small ornamental turns around one of the structurally important notes, or the general character of the phrase. By the latter I mean such features as the following: many phrases begin with a rise up to c, then dwell on c before falling down to a lower note, b, a, G, and E being the most common. Within this broad outline a great deal of variety is naturally possible. One cannot, however, characterize these phrases much more specifically. Among the mode 3 introits, phrases of this type outnumber all others, many introits having more than one such. A good proportion centre on G. Other phrases with a clear focal pitch are much rarer, and phrases which move directly from one pitch to another, or avoid a point of repose in some more convoluted way, are also relatively uncommon.

Ex. II.11.3 gives examples of phrases which dwell on c, almost as on a recitation tone, and fall to G. There seem to be several figures which may be used to decorate the c: ccc, $ccca$, cdc, $cdcc$, perhaps $acbc$ and $cabcbc$ as well. The ways of attaining c from a lower pitch have already been seen (Ex. II.11.1). There is little consistency in the descent to G, though nearly all progress through c–a–G, seen most simply in 'dominus', 'illis', and 'laude tua'. The last three examples are more highly ornamented than the rest.

Ex. II.11.3. Recitation around *c* in mode 3 introits (Montpellier, Faculté de Médecine H. 159)

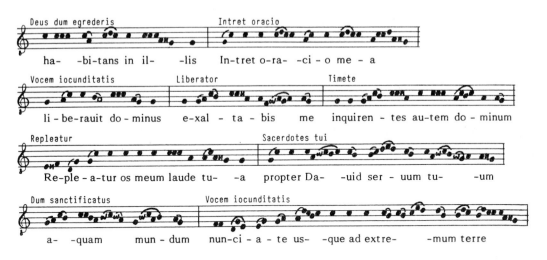

The recitation, if we may call it that, may not restrict itself to one note, but be a sort of oscillation between two poles. In Ex. II.11.4 the twin poles of *G* and *c* seem to have equal attractive force. It is no coincidence that *b* is rare in the last example, and that we have been discussing openings and cadences with *G*, *a*, and *c* as the most important pitches. This reflects the pentatonic orientation of very many chants, where *E* and *b* are often avoided or approached and quitted only by step. In effect, then, a large proportion of the phrases in mode 3 introits adumbrate the rise *G–a–c*, dwell on *c* as a recitation note, then fall to *b* or descend pentatonically to *a*, *G*, or *E*. The same sort of thing may be seen in other introits, except that the selection of structurally important tones will be different. For the plagal modes (mode 4 to an even greater extent than the others), *F* is important for recitation.

Ex. II.11.4. From introit *Ecce oculi* (Montpellier, Faculté de Médecine H. 159, p. 35)

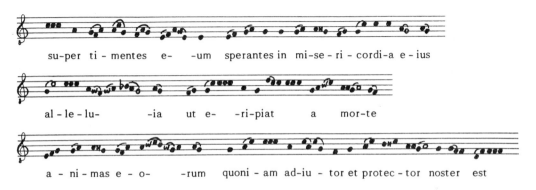

(iii) *Comparison with Office Antiphons and Responsories*

While the office antiphons share with introits many of the same structural tones and
melodic goals, their mainly syllabic style makes them less static. The introit can
achieve in a short melisma the melodic movement that would need several syllables in
an office antiphon. In the introit there is therefore more time to dwell on recitation
notes. Nor would the ornamentation of recitation notes be appropriate in an office
antiphon.

Ex. II.11.5 gives the mode 3 antiphon *Dominus legifer noster*, a member of a
melodic family already cited (Ex. II.7.3), and underneath it the introit *Timete
dominum*. It is not suggested here that the two melodies are directly related; but their
melodic outlines, at their very simplest, are at least comparable. The example is

Ex. II.11.5. Antiphon *Dominus legifer noster* (AS, k) and introit *Timete Dominum*
(Montpellier, Faculté de Médecine H. 159, p. 44)

intended to display the difference in musical manner between the two. The first phrase of each rises to *c*, dwells there, dips temporarily to *a*, then cadences from *c* down to *b*. The second phrases are similarly comparable, except that the introit cadences on *E* instead of *G*. The next introit phrase, a recitation on *G*, has no counterpart in the antiphon, but the last two phrases are comparable.

The ornamental figures in the introits occasionally recall those of responsories. Although it is generally true that Gregorian chant genres do not share melodic formulas, some introits have figures used elsewhere. For example, the more ornate cadences in Ex. II.11.2 and 3 may be found in office responsories. Compare:

Ex. II.11.2 (*d*) *Ecce oculi*, etc. with Frere in *AS*, Intro., 31 (E2);
Ex. II.11.2 (*c*) *Karitas dei* with Frere, 34 (E3);
Ex. II.11.3 *Sacerdotes tui* with Frere, 30 (g2).

This does not mean, however, that there is a generic link between introit and responsories (and even if there were, which would come first?). We are dealing here with the common coin of Gregorian chant.

Nevertheless, most of the ornate cadential figures which provide strong links between responsories are absent from introits, and the responsories are richer in short melismas of 7–11 notes. Furthermore, the groups of repeated *c*s or *F*s which characterize introits are rare in responsories. But these are in any case surface details. What of the melodic backbone? Scanning Frere's examples of mode 3 and mode 1 responsories (*AS*, Introduction, 29–32 and 17–28 respectively), one does indeed find similarities. Yet—and despite the obvious presence of recitation—most phrases in responsories are more mobile than those in introits, that is, they move more readily through the range of a fifth or more.

Much more space would be needed to characterize successfully the introits of other modes. Since it plays an important part, a few remarks on recitation may conclude this section. The role of either *F* or *c* as reciting notes in all modes deserves further study. Only these two notes carry groups of two, three, or more repeated notes as an

Table II.11.1. *Reciting notes in introit melodies*

Mode	Favoured reciting notes		Psalm verse
	lower	higher	
1	*F*	*a*	*a*
2	*F*	*F*	*F*
3	*G* or *a*	*c*	*c*
4	*F*	*G* (rarely *a*)	*a*
5	*a*	*c*	*c*
6	*F*	*G*, *a* or *c*	*F*, *a*
7	*c*	*d*	*d*
8	*G* (rarely *a*)	*c*	*c*

ornamental figure (the so-called *repercussi*). The favoured notes in each mode, usually divided between higher- and lower-lying phrases, are given in Table II.11.1 (together with the reciting note for the corresponding psalm tone). Since *F* is the final, and *a* and *c* are structurally important notes, mode 5 introits are particularly triadic. Mode 7 introits, on the other hand, display a tension between *c* and *d*, the lower phrases hanging around *c*, the higher ones not able to sustain any reciting note higher than *d*.

II.12. COMMUNIONS

(i) General
(ii) Groups of Communions with Psalm and Gospel Texts
(iii) Some F-mode Communions
(iv) Communions and Responsories

Stäblein, 'Communio', *MGG*; Murphy, 1977; Hucke and Huglo, 'Communion', *NG*.

(i) *General*

The text of one of the earliest communion chants was *Gustate et videte quoniam suavis est dominus*, from Ps. 33, documented in descriptions of many early rituals. (This verse survives for the mass of the eighth Sunday after Pentecost, though we have no means of knowing how its music relates to the chant of the early churches.) At some time a cycle of chants was developed so that each mass had its own proper communion. Like the introit, the communion chant appears with a psalm verse or verses in the earliest chant-books (list of sources in Huglo 1971 *Tonaires*, 401–2), but these fell out of use after the turn of the millenium.

Communions are among the most puzzling of chants. There are inconsistencies and irregularities in almost every aspect of the repertory.

Their texts are practically all biblical, but only a little over two-fifths are taken from the psalms. Most of the rest come from the Gospels, and often seem to sum up the theme of the mass in question; quite often they are therefore a verse from the gospel of the day, or at least from the same chapter of the New Testament. No other chant genre has this type of selection of texts.

Communions in the oldest sources are roughly as numerous as introits, but the proportions assigned to different modes are different (where the sources agree, for the modal tradition of many communions is unstable). D-mode melodies (over forty) are the most numerous, closely followed by G-mode (around forty). F-mode melodies (around three dozen) overtake E-mode pieces (less than thirty).

Early descriptions of the performance of the chant, and the fact that early sources have a psalm verse, permit one to call the communion chant an antiphon with psalm. In this it is directly comparable with the introit. Yet in musical character it resembles the introit only intermittently. Some communions are very short, others as long as the most ample introits; some are like simple office antiphons, others much more elaborate. Bomm (1929) discusses no less than forty-six communions over whose modality medieval manuscripts or theorists disagree (see also Fleming 1979). In some of these cases quite different melodies are in fact at issue. One group of Lenten communions with gospel texts appear to have had very simple melodies at first, which were frequently replaced by later editors in order to bring the chants musically into line with the somewhat more ornate style prevailing elsewhere. If this is what happened (it has been argued that the simple melodies may originate in the seventh century), it supports the idea that the communion repertory is composed of several stylistic layers (for the Lenten communions see VI.6.x below). Another peculiarity of the repertory is that a number of communions share a text with an office responsory, and sometimes the music of the two chants seems to be related.

Such is the variety among communion chants that one suspects they may have been culled from several different sources. Perhaps there was something haphazard about the compilation of the repertory, new antiphons being brought in from, say, the office, when a new mass formulary was introduced. Or perhaps older communions kept their honoured place when new pieces in more modern style were composed. The following remarks can pretend to do no more than highlight a few groups in this multifarious repertory.

(ii) *Groups of Communions with Psalm and Gospel texts*

There are several seasons of the year when communion chants with psalm texts predominate, others when a block of gospel texts appears. On two occasions the psalm texts are arranged in numerical order: for the weekdays of Lent and for the Sundays after Pentecost. These arrangements could be retrospective, that is, the original order could have been non-numerical (Murphy 1977, 144 ff. argues, however, that theme plays a role here as well).

In the case of the post-Pentecost Sundays, the numerical ordering affects the introits and offertories as well. In the communions the selection of texts is not quite consistent, for in both Lent and post-Pentecost series some gospel texts intervene.

The sources of the texts for most communions, as for other chants, can easily be checked in Wagner (see the table at the end of Wagner I). Besides the two long series of psalm texts in Lent and after Pentecost, there are other groups of psalm and gospel texts:

Psalms:
Sundays from Septuagesima to Quadragesima Sunday 4 (mode 1, 8, 1, 3, 5, 1, 4)
Gospels:
the three Sundays after Epiphany (mode 1, 6, 7)
Easter Sunday, and ferias, to the fourth Sunday after Easter (mode 6, 6, 7, 8, 7, 1, 2, 6, 2, 8, 8)
Sunday after Ascension to the end of Pentecost week (mode 4, 5, 7, 8, 5, 3)

From a musical point of view, however, there seems to be little stylistic unity within these groups. There are, of course, the common phrase openings and endings (though not the lengthy melismas which mark cadences in responsorial chants) which can be found throughout the chant repertory. But consistency of form is absent. As with introits, many phrases may be construed as intonation–recitation–cadence, but beyond this it is difficult to go, for the ways of elaborating this simple idea (and other basic shapes) are infinite.

(iii) *Some F-mode Communions*

The group of Eastertide communions with gospel texts includes three in mode 6 which may serve as examples of elaborated recitation (Ex. II.12.1). The only common cadence which the three use very often is *FGF F* ('veritatis', 'Petro' and the succeeding 'alleluia', 'alleluia' at the end of *Mitte manum*). Another cadence common in the F mode, *FGbbaGFG GF*, appears only once, at the end of *Pascha nostrum*, and then in modified form. There are three *C*-cadences ('epulemur', the first alleluia at the end of *Pascha nostrum*, and 'fidelis', the last two being the same), otherwise

Ex. II.12.1. Mode 6 communions (Montpellier, Faculté de Médecine H. 159)

practically all cadences are on F. It is also difficult to construe many of the phrases as anything other then decorated recitation around F. For example, 'itaque epulemur' has a momentary dip *DCD*, immediately balanced by touching on *aga*, so that the central tone *F* is not seriously challenged. In the first phrase of *Mitte manum*, G features as an alternative to *F*, mostly for accented syllables; *a* is secondary, using some of excess energy of the rise from *F* to *G* in ornamental turns. The basic shape of the phrase is bipartite, with both sections rising to *G*:

F F G G F G ‖ F F G F F F F F G

Mit-te ma-num tu-am ‖ et cog-nos-ce lo-ca cla-vo-rum

Had the use of standard phrases been the norm, the 'alleluia' phrases in these communions could easily have been identical. Those in *Pascha nostrum* are least striking. *Surrexit dominus* has already started out more adventurously. Its 'alleluia' then moves unexpectedly out of the *F–b♭* tetrachord down to the *C–F* range. There are repeated groups of notes in both these clusters: *ab aG–ab aG . . . FDE DC–FDE DC*. In the absence of a notational sign for *E♭*, the whole chant is transposed up a fifth in some sources. In *Mitte manum* (transposed for the same reason), the long *F–G* recitation just cited leads into an 'alleluia' which cadences on G by way of the trichord *E♭–G–b♭*. The final 'alleluia' phrases are different again.

Dicit dominus implete, for the second Sunday after Epiphany, has features in common with these, but develops an interesting sequel which seems to be a direct response to the text being set. The underlying notes in the first phrases seem to be:

Dicit dominus	*F–a . . . –G*
implete hydrias aqua	*F . . .*
et ferte architriclino	*F–G–a . . . –G–F*
dum gustasset architriclinus	*F–G–a–a–G–F–G–G–F*

Ex. II.12.2. Communion *Dicit Dominus implete hydrias* (Montpellier, Faculté de Médecine H. 159, p. 61)

The words of the guest, 'Thou hast kept the good wine until now', move right out of this range, which is why the communion is sometimes classed in mode 5 instead of the mode 6 suggested so far. A further surprise comes in the last line, a simple narrative statement: 'This beginning of miracles did Jesus before his disciples.' For the first time the chant is almost completely syllabic.

(iv) *Communions and Responsories*

Since both communions and responsories draw upon texts which have been read as lessons, instead of relying as heavily as other chant genres on the Book of Psalms, it is perhaps not surprising that they share a number of texts. In the Old Roman chant repertory some of the melodies are also shared (Murphy 1977, i. 481 ff. lists twenty-one cases of near identity). This is much less evident in the Gregorian repertory: perhaps the case could be argued for three or four melodies. One such is *Diffusa est gratia*, given in Ex. II.12.3 in its Sarum version both as communion and responsory (the verse of the responsory is omitted). It need not trouble us unduly that standard *F*-cadences appear in the communion, standard *E*-cadences in the responsory. The similarities of underlying phrase shape seem largely independent of this, except perhaps at 'propterea', where the communion rises a fifth *F–c* and cadences on *F*, the responsory rises a fifth *D–A* and cadences on *E*. The first phrase in both, 'Diffusa est gracia', dwells on *F* before descending to *D*. After 'propterea', already mentioned, the two melodies reunite during 'benedixit' and run similar courses to the end.

   Communions, therefore, are highly inconsistent in style, and point in many different directions, musically and liturgically. The impression remains of a fragmented repertory, the investigation of whose layers of material is one of the most intriguing tasks facing scholarship.

Ex. II.12.3. *Diffusa est gracia* as communion (Montpellier, Faculté de Médecine H. 159, p. 68) and responsory (*AS*, 663)

II.13. OFFERTORIES

(i) Introduction
(ii) Texts
(iii) The Melodies of the Offertory Respond
(iv) Verse Melodies

Ott 1935; Sidler 1939; Baroffio 1964; Steiner 1966; Hucke 1970; Dyer 1971; Kähmer 1971; Pitman 1973; Baroffio and Steiner, 'Offertory', *NG*; Dyer 1982; Levy 1984; *Offertoriale triplex*.

(i) *Introduction*

The splendid offertories of the medieval Gregorian repertory are still among the least known of chants. The long verses which were sung in the early Middle Ages, but which fell out of use in the twelfth to thirteenth centuries, do not form part of the modern Roman liturgy and were consequently not included in such books as the *Graduale Romanum* and the *Liber usualis*. Although edited by Ott (1935—his edition was recently reissued in *Offertoriale triplex*, with the addition of neumes from early sources), and despite Sidler's study of 1939, the verses are unfamiliar. But the first part of the chant (which I shall henceforth call the 'respond') is also relatively poorly understood, for, like the introit and communion, it does not rely on easily identified formulas. Discussion of the melodies has therefore mainly concerned their possible responsorial nature, and the repeat structures in the lengthy melismas which occur in both respond and verses.

For two reasons it has often been supposed that the offertory originally consisted of a psalm with antiphon, as many verses as were required to cover the liturgical action. The first part of the offertory, the respond, is sometimes labelled 'A(ntiphona)' in medieval sources. Furthermore, the function of the offertory somewhat resembles that of the introit and communion: a chant accompanying the solemn entrance (introit), the bringing of gifts to the altar (offertory), the clearing up after consumption of the sacred elements (communion). The melodies that are transmitted by the earliest sources are, however, nothing like office antiphons, or any other antiphons, but long, melismatic outpourings as impressive as anything in the Gregorian repertory. Some have no verse, but most have from one to three verses; four are occasionally found. After each verse all or part of the respond was repeated, as in the gradual or office responsory.

Because of their generally wide range and frequent change of register or even of tonality, not least between respond and verse, it is not always possible to distinguish between authentic and plagal modes. The Dijon tonary (Montpellier, Faculté de Médecine H. 196, PalMus 8, ed. Hansen 1974), for example, makes a simple fourfold division between chants in Protus mode (D; 28 melodies), Deuterus (E; 29), Tritus (F; 16) and Tetrardus (G; 31). (One or two of these are multiple textings of the same melody, but the number is about the average for early medieval sources.) The modern books do make the distinction, however, since they restrict themselves to the respond.

All examples in the present chapter are transcribed from Montpellier H. 196. Because of its notation of both *b* and *bb* it is a particularly valuable early witness, and it often appears to indicate chromaticisms and apparent modulations not found in other sources. Hansen's edition of the manuscript is the best way into the repertory. Ott's edition is based partly on Montpellier H. 159 and Trier, Stadtbibliothek 2254, the so-called Bohn Codex (see Steiner 1966 on the dangers of using Ott as a basis for melodic analysis). Facsimiles of sources with pitch notation are PalMus 13 (Paris, Bibliothèque Nationale, lat. 903), 15 (Benevento, Archivio Capitolare 34), and 19 (Graz, Universitätsbibliothek 807), and Thibaut, 1912 (St Petersburg, Saltykov-Shchedrin Public Library O. v. I. 6).

(ii) *Texts*

Most offertories have texts taken from the psalms (all but fourteen of the 107 in *AMS*, according to Hucke 1970). Hucke's study shows that more than a dozen offertories have for their verses the first verses of the psalm, the respond being taken from elsewhere in the psalm: this is a principle of antiphonal psalmody, where the whole psalm is sung, beginning of course with the first verse, whereas the antiphon is independent. In fifteen cases it is the respond which uses the first verse of the psalm, whereas in over twenty cases both respond and verse(s) select from later verses of the psalm: this method of selection resembles responsorial psalmody, where the cantor who sang the verses could select his texts as he pleased. Two dozen offertories have as first verse the first verse of the psalm, but for other verses select freely. Since the same types of text can be found among graduals and office responsories, this in itself does not tell us anything definite about the early history—antiphonal or responsorial—of the offertory.

The non-psalmodic texts have received special attention from Levy (1984), who argued for a musical connection between some members of the group and the old Spanish offertory, the *sacrificium*. Levy's hypothesis is that these are the descendents of the Gallican chant repertory (see further VIII.6). It is nevertheless difficult to see much musical difference between them and other offertories.

(iii) *The Melodies of the Offertory Respond*

The responds of the offertories vary considerably in length, but generally consist of at least four clauses, often eight or more. The melodic style is highly ornate, and melismas of considerable proportions sometimes appear (as they do in the verses). Even in the shortest offertory there will usually be at least a couple of syllables with melismas of ten or more notes. As in the introit, gradual, and communion, reiterations of a single pitch, *F* or *c*, are frequently present. Although standard phrases are absent from the offertory, some chants actually contain phrases borrowed from graduals: Baroffio and Steiner ('Offertory', *NG*) cite examples in the offertory *Super flumina*, which uses mode 1 gradual phrases (Apel 1958, 351, D1, somewhat extended, and c1). As far as openings, cadences, and underlying structures are concerned, however, the offertory is most easily compared with the introit, often seeming rather like a more ecstatic and wide-ranging expansion of the introit style.

As in introits and communions, many phrases are highly ornate recitations. Several mode 2 pieces are formed of little else but a constant oscillation between *D* and *F*, some mode 6 ones hardly make a single significant departure from *F*. Nearly a third of all the notes (over 190) in *Reges Tharsis* (mode 5) are *c*.

Ex. II.13.1. Offertory respond *Ad te Domine levavi* (Montpellier, Faculté de Médecine H. 159, p. 205)

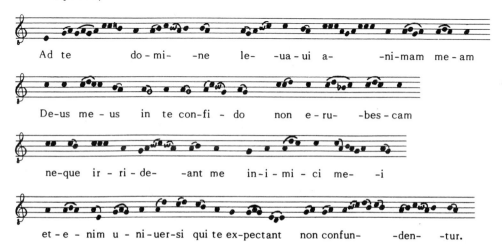

Ad te domine (Ex. II.13.1) is an offertory respond of this type. Since the melody requires both *bb* and *b♮*, and since the second verse will explore the lower region between *C* and *G*, the melody is notated with final on *a*. Most of the melodic movement is therefore between *a* and *c*. Although several syllables are centred on *c*, only one phrase actually cadences on *c*, 'non erubescam', transposing literally the *a*-cadence on 'animam meam' for the purpose. There are two cadences on *G*, and one on the next lowest note *E* (there is no *F*), the other four all being *a*-cadences.

The chromatic inflexion which appears briefly here is but a hint of the much bolder chromaticisms to be found elsewhere, a feature of offertories which evidently caused considerable problems for notators using pitch notation (see Sidler 1939 and Steiner 1966). Montpellier H. 159, the earliest source to distinguish between *b♮* and *bb*, preserves especially colourful versions of some melodies. In *In die sollempnitatis*, Ex. II.13.2, 'In die' at the start and 'alleluia' at the end have *b♮*, but the whole central section of the piece is notated with *bb*. The opening and ending have the same character as Ex. II.13.1, that is, D-mode transposed up a fifth, but elsewhere the music is that of the E-mode transposed up a fourth. The dominating pitches are

Ex. II.13.2. Offertory respond *In die sollempnitatis* (Montpellier, Faculté de Médecine H. 159, p. 215)

therefore *b♭* and *d*, a step higher than normal for transposed D-mode. (Some sources do notate the melody in D-mode, but without the *E♭s* which would result if the Montpellier version were transposed literally down a fifth: see Bomm 1929, 166 ff. and Jacobsthal 1897, 222 ff.)

As far as surface ornamentation goes, these offertories represent the more modest end of the repertory, with only a slightly higher overall degree of decoration than introits. Many offertories, however, go well beyond this. One finds phrases of text being repeated, usually with more elaborate music, a phenomenon practically unique in the chant repertory. The repeated notes can be spun out into passages as long as are to be found in graduals (compare Ex. II.5.3), and lengthy melismas may be introduced. All these may be illustrated by the respond of *Iubilate Deo universa terra*, a celebrated example which has been cited, at least in part, many times before, but which never fails to impress (Ex. II.13.3).

Tonally this composition is quite unproblematic. There is no modulation, and the verses (discussed below) do not move into a different range. *E* and *b* are mostly avoided or used with circumspection, so that the basic scale within which the music moves is pentatonic, *CDFGacd*; *b♭* appears only between two *as* (or, in the verses, at the peak of a motif such as *abbGF*).

After the standard opening, found in all sorts of D-mode chants, the first phrase is mostly concerned with *a* and *c*, including the notes which lead into *a* from below, *FGa*, and those that lead to *c*, *Gac*, or, just as often, *aGc*, which gives extra élan to the achievement of the top note. The text is then repeated with much more extended music: 'Deo universa' is the same as the first time, but 'terra' is more elaborate. The first word is set to a superb descending then ascending melisma; there is a rapid descent to *F* (note that the ornamental *GEF* is preferred to, say, *aGF*) then to *D* (ornamented *DCD*), followed by a gradual rise made up of little starts and pauses, rather like the ascending flight of a bird, hovering then climbing higher, *D–F–a–c–d* and just touching on *e* and *f* before tumbling back to 'Deo' as it was before. (For a

Ex. II.13.3. Offertory respond *Iubilate Deo universa terra* (Montpellier, Faculté de Médecine H. 159, p. 199)

similar case of elaborated repetition see *Iubilate Deo omnis terra*, cited by Baroffio and Steiner, 'Offertory', *NG*.)

The next phrase, 'psalmum dicite', is once again made up of *FGa* and *Gac* groups. The first rise to *c*, 'psalmum di-', and the second, 'nomi-', are actually the same, except for the extra ornament *bdc* the second time.

The section to be repeated after the verse now begins. The first phrase, 'venite . . . vobis', is not at all static in the way most of the music has been up until now. All syllables have more than one note, and the melodic line traces two broad curves. It is not really possible to speak of main and auxiliary notes here. The mobility persists to some extent until the end, for the repeated *c*s for 'omnes' (the music is the same as for 'no[mini]') are the last of their kind, and *F* takes over as the main centre of attraction.

(iv) *Verse Melodies*

Most of the features observed in offertory responds are present in verses, often somewhat exaggerated. There are repetitions of words and musical phrases, chromaticisms, and lengthy melismas, the latter being more frequent in verses than in responds.

A number of verses conclude with the same music as occurs in the respond just preceding the repeat section. The simplest form would thus be:

a   Respond, first part: soloist(s)
b   Respond, second part: choir
c   Verse: soloist(s), using cadence from *a*
b   Respond, second part: choir

Where there is more than one verse, perhaps only one of them will display this feature. In *Iubilate Deo universa terra* (Exx. II.13.3 and 6) both verses use the cadence from the respond.

There is frequently a hiatus between the ranges of verse and respond, perhaps even a change in tonality. Thus the D-mode *Deus Deus meus* (Hansen, no. 857) and its first verse *Sitivit in te* rise no higher than *a*, whereas the second verse, *In matutinis*, moves largely between *G* and *c*. *Benedicam dominum* is more clearly in mode 1 (authentic D-mode) throughout, as far as range goes; but the second verse, *Notas fecisti*, begins on *bb*, with a phrase which sounds as if it had been transposed up a fourth; and the whole of the second half of the verse moves in the range *a–e*, as if transposed up a fifth, before falling somewhat precipitately back to the final cadence on *D*. Sometimes different scribes in Montpellier H. 159 supplied different pitch notation, a clear sign of the difficulties caused by notating and singing melodies with unusual range. For example, in *Ascendit Deus in iubilatione* (Hansen, no. 866), the respond is unequivocally a mode 1 melody. But at the end of the first verse, *Omnes gentes*, the music starts to hang around *bb*, and ends with a cadence on *F*. The original letter notation had the subsequent two verses beginning on *a* and cadencing in identical fashion to the first verse on *F*. But a second set of pitch-letters was then added for most of the third verse and all of the fourth. The third verse starts off a fifth lower (on *D*), then about half-way through comes into unison with the first version, while at the end of the verse it rises above the first version, before cadencing like it on *G*. The third verse in its first version rose splendidly to high *g*. In its second version it is again notated a fifth lower at first, but the last phrase ('sub pedibus nostris') is a tone higher until the final cadence on *F*.

Tollite portas in many sources looks like a fairly typical mode 2 offertory, moderately elaborate, with only one large melisma at the end of the second verse. It is found in this form in, for example, Graz 807 (PalMus 19) and Benevento 34 (PalMus 15). Although verse 1 seems to end oddly—*FG Gb aG GF*—this is simply the lead into the repeated part of the respond, and reflects the corresponding passage in the respond itself. Verse 2, by contrast, occupies a higher register and would by itself be classified as mode 1; it ends on *D*.

Montpellier H. 159 classifies this offertory in mode 8. But at first the melody is notated with the same intervals as in the other sources, only a fifth higher. It therefore looks rather like Ex. II.13.1, and we should expect later *bb*s to be the reason for the choice of pitch a fifth higher. But the *bb*s appear in a quite unexpected manner. At the end of the respond, the melody suddenly dips down a tone, with *bb* replacing *c* as the usual 'reciting' pitch, and cadences on *G*, instead of the expected *a* (Ex. II.13.4). As Bomm (1929, 174 ff.) recognized, this is not likely to be a mistake, because (*a*) the same ending can be seen in St Petersburg O. v. I. 6, and (*b*) Frutolf of Michelsberg cites the D-mode ending in a way which suggests he is reproducing a correction for a well-known trouble-spot. (See also the discussion in Sidler 1939 43 ff.)

Ex. II.13.4. From offertory respond *Tollite portas* (Benevento, Archivio Cap. 34, fo. 13ʳ; Montpellier, Faculté de Médecine H. 159, p. 275)

Tol- -li-te por- -tas ..... rex glo - ri - e.

Verse 1 in Montpellier H. 159 is notated a fifth higher than in the other sources. But verse 2 brings further complications. Here Montpellier is only a fourth higher than the other sources, until the closing comments of the final melisma, when it moves up a step and cadences on *a*. This seems illogical, for if it had continued only a fourth higher than the rest, another *G*-close would have occurred, as in the respond. This cadence is, however, not the last of the piece, for the last part of the respond will now be repeated as usual.

Ex. II.13.5 gives the whole of verse 2 at the two contrasting pitches. Such examples make it abundantly clear that the adoption of pitch notation in the eleventh and twelfth centuries may often have resulted in the smoothing out of 'irregularities' in the tonal organization of many melodies.

The melisma in Ex. II.13.5 has the form AAB, extremely common in alleluias, moderately so in offertories. In the Montpellier source, which in this respect is representative, there are 103 offertories, with 216 verses in all. All but a handful of responds, and over ninety of the verses, have no lengthy melismas at all. Some verses have more than one. There is a rough balance (somewhat over eighty examples of each) between unstructured melismas and those which have a repeat form of some sort. There are about fifty melismas with AAB form, but few other forms are to be found regularly; there are, for example, only five instances of AABBC form. Among the more extended schemes may be cited AAAB–C–AAAB (*Super flumina*), AABBCCD (*Benedicam dominum*), AABBCCCD (*Deus enim firmavit*), AABBCDDE (*Benedictus es . . . in labiis*), and AABCCDCC (*Domine Deus meus*).

Ex. II.13.5. Second verse of offertory *Tollite portas* (sources as Ex. II.13.4)

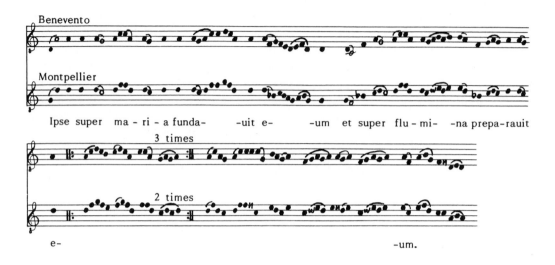

For examples of extended melismas, and of the carrying over of melodic material between respond and verses, we may return to *Iubilate Deo universa terra*, whose respond was given in Ex. II.13.3. The two verses appear as Ex. II.13.6. A blow-by-blow account of the verses is by now unnecessary. The reader will be able to locate without difficulty such features as the free recitation around particular notes (as at the start of verse 1). Sometimes the melodies become locked into repeated *c*s, with *G* and *a* in support (cf. the second 'tibi vota' in verse 1 and '[holocau]sta medullata' in verse 2). These ornate phrases lead into similar cadences: those for the second 'mea' and 'distinxerunt' in verse 1, and those for the second 'mea' and 'medullata' in verse 2 are all the same.

At the start of both verses the text is repeated: while the music in verse 1 is different the second time, in verse 2 the music is the same, except for the slightly more extended treatment of 'mea'. More surprisingly, this music is the same as the first line of the respond. Both verses have a lengthy melisma, and the two end identically. There is also some similarity at the start, where both melismas have brief repetition. Then verse 1 runs rather obviously from *F* to *c* and back again. The centre part of both verses hovers around repeated *c*s, much more extended in verse 2, with brief descents to *F* for relaxation of tension.

These are purely musical outpourings, where attention to the text is suspended and sheer joy in singing seems to take over, ecstatic and improvisatory (at least when compared to the repetitious schemes found, for example, in many alleluias). It remains unclear, nevertheless, what occasioned the composition of these glorious melodies. They are not assigned regularly to the high feasts of the church year. The contrast with that other great musical high point of mass, the sequence, could not be more pointed. Whereas sequences were sung only on the greater feasts, the

Ex. II.13.6. Verses of offertory *Iubilate Deo universa terra* (continuation of Ex. II.13.3)

assignments of the five offertories with especially extended melismas mentioned above are as follows:

Super flumina: twentieth Sunday after Pentecost*
Benedicam dominum: Monday of second week in Lent
Deus enim firmavit: second Mass of Christmas Day

Benedictus es . . . in labiis: Quinquagesima
Domine Deus meus (in very few sources): twenty-fourth Sunday after Pentecost*
(*the Sunday may vary between sources)

Iubilate Deo universa terra is universally assigned to the second Sunday after Epiphany.

II.14. ALLELUIAS

  (i) Introduction
 (ii) The Earlier and Later Styles
(iii) Rhymed Alleluias and Late Medieval Melodies

Stäblein, 'Alleluia', *MGG*; Schlager 1965; Treitler 1968; Jammers 1973; Schlager, 'Alleluia. I', *NG*; Bailey 1983, *Alleluias*; MMMA 7, 8.

(i) *Introduction*

The alleluia is a responsorial chant in that the first part of the chant ('Alleluia') forms a choral respond to be repeated after the verse, while singing the verse is a soloist's task (there may be more than one verse, and more than one soloist). It is conventional to divide the respond into two parts, the setting of the word 'alleluia' itself, and the vocalization, melisma, or *jubilus* on -*a*. The method of performance indicated in modern books is:

  Cantor: 'Alleluia'
  Choir: 'Alleluia' + jubilus
  Cantor: Verse (main part)
  Choir: end of Verse (which often includes a repeat of the jubilus)
  Cantor: 'Alleluia'
  Choir: jubilus

The early notated books known as cantatoria (because they contain only the music sung by the cantor, not that of the choir) contain complete alleluia melodies. This suggests the following manner of performance, where the 'Alleluia' call and the jubilus constitute an undivided respond:

  Cantor: 'Alleluia' + jubilus
  Choir: 'Alleluia' + jubilus
  Cantor: Verse
  Choir: 'Alleluia' + jubilus

Possibly a further two statements of the respond then followed, by the cantor and the choir respectively, as was originally the case with other responsorial chants (see II.4). The even more elaborate performance schemes of Old Roman and Milanese usage are also suggestive (see VIII.3–4).

    The outstanding work of Schlager in cataloguing and editing the medieval alleluia repertory (Schlager 1965 and MMA 7–8) means that the bases for study have been more firmly established than for almost any other chant genre. Nevertheless, a consensus on many of its aspects can hardly be said to have been achieved. The alleluia has attracted considerable musicological attention, partly because it has seemed a more purely musical genre than most others (with its wordless *jubilus*), partly because of its mysterious early history and the continuous expansion of the repertory through the Middle Ages, partly because of its connection with the sequence (an extended alleluia or not?). All these matters have been the subject of controversy.

In the earliest books with chant texts, those edited by Hesbert (1935), there are just over 100 alleluia texts. Not all have their own unique melody, however; Schlager reckons that around sixty melodies were used (see the list in Schlager, 'Alleluia', *NG*). Prominent among the melodies used for more than one text are those for *Dies sanctificatus* (third Mass on Christmas Day, nine other texts in the early repertory), *Dominus dixit ad me* (first Mass on Christmas Day, eleven other texts) and *Excita Domine* (third Sunday of Advent, six other texts). (For melodies with several texts, see the synoptic presentation by Madrignac 1981–6.) Over the two centuries up to about 1100 (the cut-off point for Schlager's catalogue) the repertory expanded dramatically. Schlager records 410 melodies, plus a considerable number of others in adiastematic neumes which could not be catalogued: a sevenfold increase. The number of texts had increased at the same time to over 600.

Just as the same melody might be used for several different texts, so might the same text be sung to several different melodies, a phenomenon far more widespread among alleluias than any other chant genre, with the exception of hymns and late medieval rhymed sequences. So the total number of different combinations of melody and text is far higher than that of the 600 texts alone. After the eleventh century composition continued. Schlager points to the striking efflorescence of creativity in South Germany and Bohemia in the fifteenth century and the frequent use of rhymed texts.

A corollary of the continuous compositional activity is the variety between manuscripts in their selection of alleluias, both melodies and texts (which means that the alleluia repertory is a prime resource for identifying the liturgical use of a manuscript and its nearest relatives). This instability extends right back to the earliest sources, so that one is tempted to hypothesize that the repertory in the eighth century, that is the period immediately before the copying of the extant sources, was rather small. Apel (1958, 379) listed the alleluias of the temporale whose assignment remained constant across the eighth- to ninth-century sources edited by Hesbert in *AMS*: only twenty-two in all (eighteen, discounting repeats). The number of melodies for these is only eleven. The agreement in the sanctorale is minimal. The supposition that not many alleluias were needed is strengthened by the state of the Old Roman and Milanese chant repertories, which used very few melodies right into the twelfth and thirteenth centuries.

(ii) *The Earlier and Later Styles*

Many of the melodies which have a constant assignment in the earliest sources share features of musical style, which suggests they belong to an early layer. The most important of these (Apel 1958, 391 discusses others) is a negative characteristic which becomes obvious when comparisons are made with the majority of later alleluias: the absence of repetition within the jubilus on *-ia*.

Ex. II.14.1. *Alleluia Dominus dixit ad me* (Paris, Bibl. Nat. lat. 776, fo. 12ʳ)

Ex. II.14.1 gives *All. Dominus dixit ad me* with its presumed original melody (melody 281 in Schlager's catalogue) from an Aquitanian manuscript. So well known is the melody that the scribe does not copy it in full, and one has to refer back to the first appearance of the melody, on the first Sunday of Advent for *All. Ostende nobis*, for the ending. There are in all three melismas of some length: the jubilus, on 'ho-[die]', and on 'te' at the end of the verse. The only one with a hint of repetition is the last, which might be construed as having paired phrases at the end, one ending 'imperfectly' on *F*, the other making the 'perfect' cadence on *G* (bracketed in Ex. II.14.1). In the respond, *c* is an obvious melodic goal. The melody keeps pushing up to it, then falling away, to *G*, to *F*, *b*, and finally *G* again. This is also what happens in the verse, with cadences on *b* ('me'), *G* ('tu'), *F* ('hodie'), and *G* again ('te'). In the final melisma *F* exerts a strong pull, so that the *G* cadence seems like a point of balance or repose, rather than a point of departure towards *c*. There are of course other ways of hearing the piece, but the similarity with other types of chant discussed so far, particularly the offertories, seems clear. The melisma on 'hodie' is particularly suggestive of the offertory, with its graded descent from *e* to *a*, then *d* to *G*, and finally *b* to *F*. It is not surprising, either, to find that the final groups of notes at the ends of the respond and verse are also found in G-mode graduals (cf. G10 and G1 in Apel 1958, 356).

Ex. II.14.2. Another melody for *Alleluia Dominus dixit ad me* (Paris, Bibl. Nat. lat. 776, fo. 12ʳ)

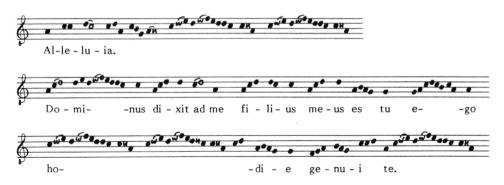

Al-le - lu - ia.

Do - mi- -nus di - xit ad me fi - li- us me -us es tu e- -go

ho- -di - e ge - nu - i te.

Perhaps because this melody seemed old-fashioned, another setting of the same text was copied immediately afterwards in the source used here, Paris, Bibliothèque Nationale, lat. 776. The melody is unique to this manuscript (Schlager 1965, No. 190), and it turns out to bear all the hallmarks of later composition (Ex. II.14.2). The music makes use of repetition to a high degree. Firstly, the jubilus has an obvious repeat (which should perhaps be made at the end of the verse as well). This music appears again for 'hodie', also repeated, and a good deal of the same phrase is used for 'Dominus' at the start of the verse. The music for 'Alleluia' is also used in the verse, from 'dixit' to 'meus', and one is even tempted to hear it behind the music for 'ego'. In other words, the music of the verse has been derived very largely from the respond. Since each phrase forms a melodic arch, no note establishes itself as a reciting pitch. (The manuscript has no clefs or coloured lines, and the choice of *a*-final, tantamount to D-mode with *b*♭s throughout, is a matter of opinion. Schlager transcribes the piece with *E*-final.)

Among the sixty or so alleluia melodies assigned to the early (ninth-century) repertory by Schlager there is a rough balance between those with and those without melismas having a repeat structure. Probably other alleluias could therefore be added to Apel's eleven 'earliest of all' (eighth century or earlier?) melodies, such as those for the common of saints or for Sundays, which by their nature would have no firm assignment to one particular day.

A melody just emerging into the new era, so to speak, is that for *All. Attendite*, for one of the summer Sundays (Schlager 1965, No. 224), Ex. II.14.3. (The semitone step is indicated often enough in Paris 903 to enable one to use B♭ as a key signature; Montpellier H. 159 notates the melody with a flat throughout.) There is a hint of repetition at the start of the jubilus, but a more obvious example is reserved for the middle of the verse, on 'meus'. Even without this repetition, one would be inclined to see a later composer at work than in *All. Dominus dixit ad me* (Ex. II.14.1), because of the easy way the melody moves in scale passages through the range *F* to *c*.

Ex. II.14.3. *Alleluia Attendite popule meus* (Paris, Bibl. Nat. lat 903, fo. 122ʳ)

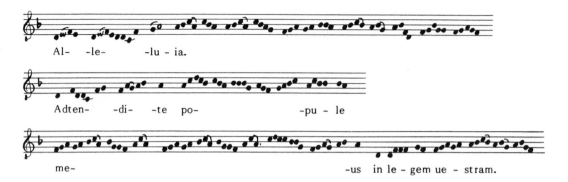

Al- -le- -lu - ia.

Adten- -di- -te po- -pu - le

me- -us in le - gem ue - stram.

One needs only to glance at the critical commentary to Schlager's edition to see how the shape of the longer melismas may vary from manuscript to manuscript. Simultaneously with the composition of new melodies, already existing ones were retouched, extended by repetition, made more regular. This affects not only the melismas but other phrases as well: for example, the start of the Alleluia for *All. Attendite* (Ex. II.14.3) in Paris 776 is like the start of the verse:

D FDDCF

Al-le-

Newer alleluias naturally exploited the full possibilities of repeat structures, and practically all of those which make up the rest of the repertory up to *c.*1100 include a repeated melisma, or reuse of respond material in the verse, or both. The scheme AAB is extremely popular, and forms like miniature sequences appear, for example AABBCCD, used for the following melodies (in some sources at least):

All. Beatus sanctus Martinus (Schlager 1965, No. 396)

All. Cum esset Stephanus (Schlager 1965, No. 102)

All. Surrexit Dominus et occurrens (Schlager 1965, No. 10)

Others display repetition with variation, as *All. Ego sum pastor bonus* (Schlager Melody No. 299, unique to Pistoia 120), which might be interpreted: AB A′B′ A″B A‴B″. The jubilus in *All. Nuptie facte sunt* (Schlager 1965, No. 111, found only in a few Aquitanian sources) displays a subtle alternation of like and unlike phrases (Ex. II.14.4). The small figure labelled 'x', or similar ones, appears four times in the jubilus. Four phrases in all end with an identical cadence, *CDD*, the so-called 'Gallican cadence' found in numerous sequences of this period. In the verse, the recurrent figure 'x' turns up once more, at the end of the 'Chana' melisma. Music from the 'alleluia' call is then reused at the end of the verse for 'mater eius', providing the lead into a repeat of the jubilus (indicated by a cue).

Thus, while the repetition in a short AAB scheme may be literal, the longer melismas tend to handle it in a more sophisticated way. In Ex. II.14.4 different

Ex. II.14.4. *Alleluia Nuptie facte sunt in Chana* (Paris, Bibl. Nat. lat. 903, fo. 18ʳ)

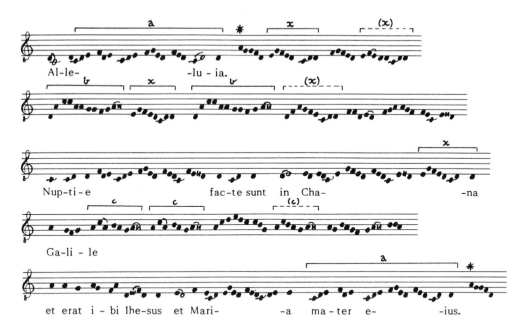

phrases were brought into relation by means of a common cadence, whereas Ex. II.14.5, *All. Veni sponsa Christi* (Schlager 1965, No. 35, likewise Aquitanian), shows similar beginnings leading to different endings, analogous to the first and second endings of later music. The jubilus consists of nine phrases, labelled 'b' to 'j'. The basic idea seems to be the alternation of phrases cadencing on *a* with those which move down to *D*. The *a* phrases are 'b' and 'e', which comes twice, the first time with a short, supplementary *a*-phrase 'f'. The *D*-phrases are 'c+d' (labelled separately because 'c' appears later alone) and 'g', while a longer ending includes a *G*-cadence ('h') before the final phrase ending on *D*. Within this pattern, however, there are others. The first notes of 'e' and 'g' are the same, and this causes 'e' and 'g' to be heard as a pair, with a half close (on *a*) and then a full close (on *D*). Three-note scale segments are common throughout the piece (a symptom of relatively late date): since there are pairs of them just before the cadence in both 'b' and 'd', one is inclined to hear these phrases as a pair as well, in an antecedent–consequent relationship.

The verse begins like the 'alleluia' call, but the next phrase, 'sponsa Christi', stands by itself. The four phrases of 'accipe coronam' make up an A-A-A′-A″ scheme: 'l' is actually a version of 'c+d', transposed up a fifth; the next phrase uses 'c' literally, up a fifth, with a new extension 'm'; together they might be regarded as an extended version of 'l'; the last phrase is made up of 'c+j', likewise transposed up a fifth. Is it stretching a point to call 'quam tibi' a variant of 'j'? and 'dominus' a variant of 'm', each with a new start (the scale segment *abcd*)?; 'j' then recurs at the end of 'preparavit'.

Ex. II.14.5. *Alleluia Veni sponsa Christi* (Paris, Bibl. Nat. lat. 903, fo. 104ᵛ)

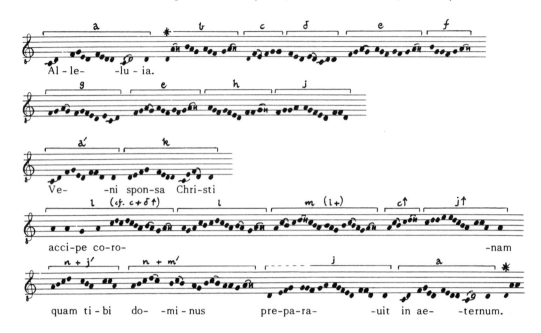

The last phrase, 'in aeternum', takes up the 'alleluia' call again, leading into a repeat of the jubilus, indicated by a brief cue.

For some, an analysis like this may be over-ingenious, while others may disagree over detail. The important point is that repetitions and echoes of the sort described are unthinkable in earlier chants and mark a radical departure in compositional technique. The whole composition is held together not simply by its overall responsorial form but by a network of internal references and patterning. It is important that the references are internal, that is, they do not relate the piece to others of its class but contribute to the individual identity of this one 'work of art'.

(iii) *Rhymed Alleluias and Late Medieval Melodies*

Very many alleluias composed after the ninth century remained local compositions, such as the Aquitanian ones just given as Ex. II.14.2, 4, and 5. The same remains true of most compositions of the later Middle Ages, from the twelfth century onwards. This is mostly because the basic liturgy throughout the year had its consignment of alleluias from an early date, and while new initiative might win local acceptance, it was hardly likely to cause wholesale revisions of the repertory right across Europe. Many later alleluias are compositions for local saints, whose cult was peculiar to a restricted area, even to a single church. One of the few classes of feast-day which was celebrated with increasing and universal enthusiasm was feasts of the Blessed Virgin

Mary, and it is among the vast corpus of Marian alleluias that a few widely known modern pieces may be identified. Many Marian alleluias were retexted for other liturgical occasions (particularly for virgin saints, of course), some more than twenty times. As with the earlier repertory, some texts are found with more than a dozen different melodies in different sources.

One of these is *All. Virga Iesse floruit* (for which Schlager lists nearly 200 sources in MMMA 8, 822 ff.). Its text is notable for the incipient rhyme, best called assonance, common to many new liturgical texts of the twelfth century. In what seems to be the earliest version, the last line is not assonant:

> Virga Iesse floruit
> virgo Deum et hominem genuit
> pacem Deus reddidit
> in se reconcilians yma summis

Other versions (MMMA 8, 564) replace the last line with:

> qui poli summa condidit
> orate (or 'Dei genitrix ora') pro nobis

Musically, this particular alleluia presents nothing radically new. Given the amount of repetition already present in earlier alleluias, however, it is not surprising that musical rhyme should occasionally be introduced in later pieces, either in support of, or in counterpoint with, the text rhyme. This occurs in, for example, *All. Ante thronum trinitatis*, another Mary alleluia, probably dating from the late thirteenth century. As a supplement to the alleluia with its usual text, transcribed by Schlager (35 ff.), I give the version adapted for St Barbara added in the sixteenth century to Paris, Bibliothèque Nationale, lat. 905 (from Rouen) in Ex. II.14.6. The musical repetitions affect not just the ends of each eight-syllable line, but also some of the half-verses. The text has the following rhyme scheme:

-is   -ata
-is   -ata
-atis   -iice
-atis   -iice

whereas the music proceeds as follows (the scheme is marked on the transcription):

a b   c
d e   c
a e   c
e b   (melisma with AAB structure, new ending)

The jubilus is by contrast relatively uninteresting (though it follows an AAB repeat pattern in some sources).

The melody is rather modest in scope. Phrases such as 'cunctis horis eiice' have an undoubtedly modern flavour, not so much because of the three-note groups as because of the rapid motion through the octave *f*–*F*. Every phrase begins or ends on *F*, *c*, or *f*, and most wing their way over the whole scale.

Ex. II.14.6. *Alleluia Ante thronum trinitatis* (Paris, Bibl. Nat. lat. 905, fol. 311ᵛ)

But there are plenty of more flamboyant melodies among late medieval alleluias. An example which one could call celebrated if it were better known is *All. Ora voce pia pro nobis virgo Maria* (MMMA 8, 393, in a handful of South German sources), with a staggering range of two and a half octaves from *G* to *d'*. Somewhat less spectacular, though still exotic by comparison with early medieval melodies, is *All. O Maria rubens rosa*. The piece is of manifold interest. Firstly, the numerous sources cited by Schlager (over 40: MMMA 8, 727 ff.) show that it was one of the most popular rhymed alleluias in south-east Europe; secondly, its E-mode melody is of a type characteristic for that area (Schlager thinks it might even have been the starting-point for the type); and thirdly, in about a quarter of the sources trope verses are inserted in the verse.

The turn around *E* which constitutes the most prominent fingerprint of this group of melodies stands out clearly in Ex. II.14.7. Other features of the melody are perhaps best described as typically late-medieval, only applied here to an E-mode melody, rather than the more common F- or G-modes (or transposed to *c*): the readiness to run through the whole octave, or beyond (here practically always descending), the octave leap *e* to *E*, the succession of leaps of a fourth and a fifth (line 8), yet again outlining the modal octave. There are again numerous melodic repetitions. The source transcribed here has no trope verses. Asterisks show where they appear in other manuscripts, and they are given in Ex. II.14.8. (cf. MMMA 8, 351 ff.).

The repertory of rhymed alleluias shares compositional features with other rhymed chants, such as the responsories for late medieval rhymed offices (see II.26). The parallel is not an exact one, however, because it was not the custom to compose whole new masses with rhymed texts: the alleluia and sequence (not necessarily as a pair) are usually the only such items at mass.

Ex. II.14.7. *Alleluia O Maria rubens rosa* (St Gall, Stiftsbibl. 546, fo. 327[r])

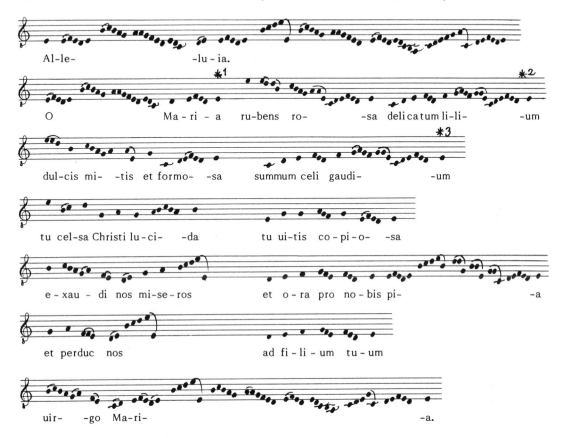

Ex. II.14.8. Trope verses for *Alleluia O Maria rubens rosa* (Munich, Univ.-Bibl. 2° 156, fo. 260[r])

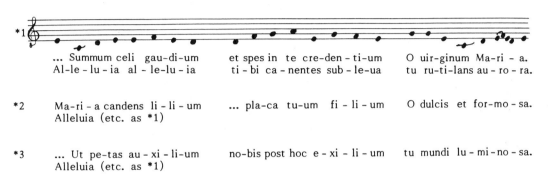

II.15. HYMNS

(i) Introduction
(ii) Texts
(iii) Music
(iv) Processional Hymns

Moberg 1947; MMMA 1; Szöverffy 1964–5; Gneuss 1968; Stäblein, 'Hymnus, B. Der lateinische Hymnus', *MGG*; Steiner, 'Hymn, II. Monophonic Latin', *NG*.

(i) *Introduction*

Despite the outstanding work of Stäblein—the edition and commentary in MMMA 1 and the article for *MGG*—a repertorial survey of a sufficient number of sources from over all Europe is still not available and much musical analysis remains to be done. Of Moberg's projected edition only the valuable comparative tables were completed. Stäblein edited the contents of a Milanese hymn collection and three French, one English, three German, and two Italian hymnaries, supplementing these melodies with others from additional French, English, German, and Italian sources. Over 550 melodies are edited. Lists of contents for all the sources Stäblein consulted were not published, however, and it is even difficult to compare the contents of his main sources. It can nevertheless be ascertained that very few hymns (less than thirty) were sung in all areas of Europe to the same melody; many popular texts were set time and again to different melodies, and many popular melodies were used for a dozen or more different texts. Some melodies became strongly associated with a particular season or liturgical position. Most widely known, and probably among the oldest, are the hymns of the weekly office cycle. But as soon as one moves beyond these and the main feasts of the year, medieval sources display great variety of choice. The majority of hymnaries contain between 80 and 100 pieces; those which make provision for local and lesser saints may have up to twice this number.

Hymns were reported by St Augustine to have been promoted in the church of Milan by St Ambrose (d. 397) for singing during the long night vigils. Augustine attributed *Aeterne rerum conditor*, *Deus creator omnium*, *Iam surgit hora tertia*, and *Veni redemptor gentium* to Ambrose. Only centuries later, alas, do the earliest Milanese musical sources appear, and there is no substantial agreement in the rest of Europe about melodies for these hymns. Indeed, *Iam surgit hora tertia* was rarely sung at all.

The development of the Milanese repertory has been outlined by Huglo *et al.* (1956, 85–103), and also its relationship to the Gallican repertory (Huglo, 'Gallican Rite, Music of the', *NG*).

The earliest notated hymnary appears to be the Kempten collection (edited by Stäblein), from around the turn of the millenium, with alphabetic notation. This means that for the crucial period of the Carolingian renaissance we are reliant on texts alone for our knowledge of the history of the repertory. Gneuss (1968) has provided valuable indications as to how it developed. He sees an early repertory dating back at least to the sixth century ('Old Hymnal' type I) drawn upon in Milan, by the rule of St Benedict and the respective rules for nuns and monks of Caesarius and Aurelian of Arles; this also survives in two early English sources. A new recension of the 'Old Hymnal' (type II) seems to belong to the Carolingian period. Part of it is found, for example, in the same source as the earliest tonary, probably from Saint-Riquier, Paris, Bibliothèque Nationale, lat. 13159, written in the closing years of the eighth century. But in the second half of the ninth century a new repertory (the 'New Hymnal') gained rapid and widespread acceptance. It forms the basis of all subsequent collections, including those in modern service-books. All notated hymnaries, therefore, are based on the 'New Hymnal'. We do not know if musical changes accompanied the changes in text selection. And, partly because hymns were usually copied in independent collections (often with the psalter), we cannot yet relate the history of the hymn to that of other office chants. Discussion of the music is also made difficult by the wide variance between sources in their transmission of many melodies. Hymns are of course not peculiar in this respect, but the lack of a strong central tradition seems to have resulted in unusually wide differences. Stäblein has noticed examples of the apparent 'modernization' of archaic melodies, the simplification of ornate ones, the regularization of note groups to reflect text rhythm, and differences of modal interpretation.

In what follows hymns are cited with their 'Old Hymnal' or 'New Hymnal' number, following Gneuss (24 ff., 60 ff.; if a distinction between the two types of Old Hymnal is possible, the appellation 'I' or 'II' is also given). Examples are mostly taken from an English thirteenth-century manuscript of Sarum use, Oxford, Bodleian Library, Laud lat. 95, whose repertory is naturally related to the French and English sources edited by Stäblein.

(ii) *Texts*

The most obvious textual and musical characteristic of hymns is their strophic form and metrical regularity. In view of this formal simplicity and the relatively small number of melodies required, it is hardly surprising that musical notation was not thought necessary in early sources. The last strophe of the hymn is usually a paraphrase of the doxology (the same doxology-strophe could be used for any number of hymns in the same metre). 'Amen' is usually sung at the end.

A large number of different poetic metres may be found, particularly among early hymns: asclepiads, distichs (usually in processional hymns), and so on (Stäblein surveys the commonest metres in *MGG*). But the majority of hymns are written in one of two metres.

Iambic dimeter x–∪–x–∪– four times) is the metre of the four hymns attributed to St Ambrose, and such poems were frequently known as 'Ambrosiani'. Strophes are of four lines, and there are typically four or eight strophes in a hymn. Stäblein reckoned that two-thirds of the repertory uses this metre. (Metres are explained below, II.27.)

Sapphic metre was the most popular of the antique metres used in medieval hymns. Each strophe has three full lines and a final half-line (as adonic):

–∪–––/∪∪–∪–∪ (3 times)

–∪∪––

Such poetic compositions were often sufficiently highly regarded as artistic creations for their authors' names to have come down (not all ascriptions are regarded as secure; AH 50 and 51 are collections of ascribed texts). Best known, besides those of St Ambrose, are:

Prudentius (d. *c*.410): *Inventor rutili, O crucifer bone, Pastis visceribus, Ales diei nuntius*, etc.

Venantius Fortunatus (d. *c*.600): *Agnoscat omne saeculum, Crux benedicta nitet, Pange lingua, Quem terra pontus aethera, Salve festa dies, Tibi laus perennis auctor, Vexilla regis*

Sedulius (first half of fifth century): *A solis ortus cardine, Hostis Herodes impie*

Not all these were originally designed as office hymns, some being extracts from longer poems, some written for special occasions. Thus *Hostis Herodes impie* is actually a later part of the same poem as *A solis ortus cardine*. The first three Prudentius hymns were sung during processions, not during the office.

(iii) *Music*

Many hymns are through-composed, without remarkable musical features. Others have simple repeat structures (aaba or abca, for example). In many cases—as a natural complement to some repeat structures—the highest notes of the strophe come in the third line, before the return to the tonic. In general hymns lack such musical features as repeated reciting notes, reminiscent of the delivery of prose texts. Quite often no strong tonal centre is apparent, and the lines may cadence on unexpected notes.

Christe qui lux es et dies (OH 30, NH 12) has been sung at Compline since the time of the 'Old Hymnal'. Its 'original' melody cannot be determined—if indeed it makes sense to think of one at all, for tunes are so readily interchangeable. It is given in Ex. II.15.1 with its commonest melody (No. 9 in Stäblein's edition). It is the only text for this melody—an unusual situation. The last line of melody repeats the first. The highest note, G, occurs in the third line, though the range is so limited that this makes only a subtle impression. Even so, this small gesture is necessary, is prepared for in fact, by the repetitive second line. It is not uncharacteristic that the melody starts on a note above the final, although the D final does not seem seriously in doubt.

More unpredictable tonally is *Somno refectis artubus* (NH 14), usually sung during

Ex. II.15.1. Hymn *Christe qui lux es* (Oxford, Bodl. Lib. Laud lat. 95, fo. 140ʳ)

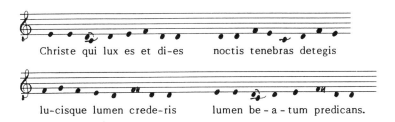

Ex. II.15.2. Hymn *Somno refectis artubus* (Oxford, Bodl. Lib. Laud lat. 95, fo. 137ᵛ)

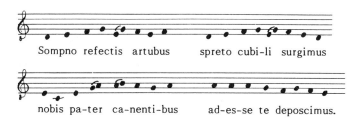

the Night Office on Monday, given in Ex. II.15.2 with its commonest melody (Stäblein No. 142; this is also the commonest melody for *Primo dierum omnium*, the equivalent Sunday hymn). The final E sounds very much like an 'imperfect' cadence, with perhaps D as the 'real' final. At any rate, that is the impression one has from the first two lines, which are the same but for their first-time/second-time cadence.

Both these hymns are 'Ambrosiani'. Whether their metre was reflected in performance in any way seems doubtful. In iambic dimeter long syllables could occasionally be substituted for short ones. In a performance with long and short notes (in effect, in triple time), either the regular rhythm would be disturbed or one would have to ignore the variance in the metre. Another consideration is the pattern of stressed and unstressed syllables (not the same as long and short). This varies so often from line to line that any attempt to reflect it in performance (again, perhaps through long and short notes) seems misguided. This is not to deny that the reflection of metric or accentual patterns in the music (through groups as opposed to single notes, or with tonic accents) can sometimes be found, particularly in late medieval melodies. But clearly it constituted no universal principle of word-setting. (See the discussion of text rhythm in II.27.)

The hymn *Ecce iam noctis* (NH 6), usually sung at Lauds on Sundays, was widely sung to melody 144 in Stäblein's edition. The metre is sapphic. As in the previous two examples, the simplicity of the melody may reflect high age, or it may be due to the lesser solemnity of the liturgical occasion. In the Sarum hymnary used for the examples in this section, two melodies are given for the hymn. Melody 144 is used for the second strophe, while for the first strophe melody 107 is given, as reproduced in Ex. II.15.3. The tonality of melody 144 is clearer than that of Ex. II.15.2, for *b* (*E* in

Ex. II.15.3. Two melodies for hymn *Ecce iam noctis* (Oxford, Bodl. Lib. Laud lat. 95, fo. 144ᵛ)

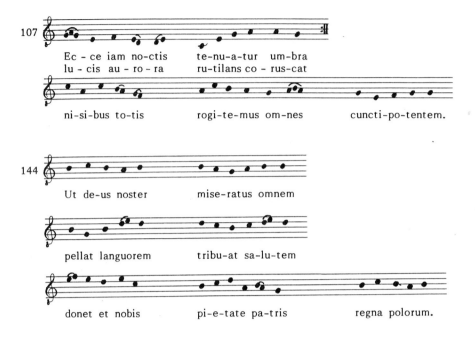

many sources which have the melody a fifth lower) is announced in the first line clearly enough for the repetition of the opening phrase at the end to be instantly recognizable.

Melody 107 was also fairly widely known, though not firmly attached to any particular text (*Iste confessor* in France, *Virginis proles* in Germany, and so on). It has a relatively modern air, partly because of the clear tonality, despite the way the melody moves easily through the whole octave, partly because of the 'Gallican' final cadence, the frequent steps of a third, and the repetition, not only of the first two lines but within the third also, where the tonal area above the final is explored for the only time. It is noticeable that groups of more than one note fall only on long syllables or at the ends of lines.

Because notation came so late to the hymn repertory it is impossible to know how old many melodies may be. For Fortunatus' well-known hymn *Pange lingua gloriosi proelium certaminis* (Ex. II.15.4) Stäblein melody 56 was sung all over Europe, though half a dozen other melodies gained local acceptance (see Stäblein 1950 for a possible predecessor of melody 56). This poem and the equally well-known *Vexilla regis prodeunt* appear to have been composed for Queen Radegunda at the convent of the Holy Cross in Poitiers, to whom the Emperor Justinian had given a fragment of the newly discovered True Cross. The poem has ten strophes. Like *Vexilla regis* it was taken up into the Passiontide liturgy, different parts often being sung on

Ex. II.15.4. Hymn *Pange lingua* (Oxford, Bodl. Lib. Laud lat. 95, fo. 141ᵛ)

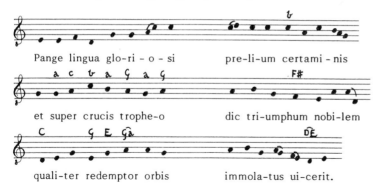

different days (the commonest divisions were made at strophe 6, *Lustra sex qui iam peracta*, and 8, *Crux fidelis inter omnes*). The popularity of the hymn entered a new phrase when St Thomas Aquinas (d. 1274) took the opening for his Corpus Christi hymn *Pange lingua gloriosi corporis mysterium*.

The text is in trochaic tetrameter with lines of 8+7 syllables ($-\cup-x-\cup-x\|-\cup-x-\cup-$). The melody contains traces of traditional Gregorian turns of phrase: the opening is a common mode 3 figure (compare Ex. II.11.1), *c* the usual reciting note for this mode; *a* then assumes its usual role as a goal for cadences above the final (with a link through *D* to the lower line 5 and an imperfect cadence for line 5 itself).

In Ex. II.15.4 the melodic variants for the next hymn in the Sarum manuscript, *Lustra sex*, are given over the stave. In this source the whole melody is given a fourth higher, to allow for both *b♮* and *bb* (*F♯* and *F♮* when transposed for the transcription).

Hymns of the later Middle Ages betray the same modern stylistic features as other chants of the period. In some churches a freedom to supplement the repertory was clearly felt, as with other genres. Most noticeable in the stylistically later hymns are the wide range of the melodies and the freedom of rapid movement within that range, the clear emphasis on tonic (final) and dominant (upper fifth), and the division of phrases between the scale-segments they command. These features are already foreshadowed in Ex. II.15.3, melody 107. They are more obviously present in Ex. II.15.5, one of five melodies provided in the Worcester compendium for *Sanctorum meritis* (NH 119).

Ex. II.15.5. Hymn *Sanctorum meritis* (Worcester, Cathedral Chapter Lib. F. 160; PalMus 12, 10*)

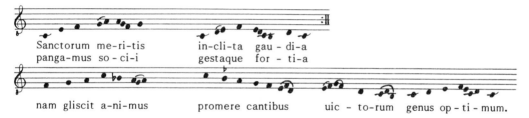

The universally known melody for this hymn is Stäblein's No. 159, which is mostly syllabic, tonally unfocused, though relatively mobile. It has the form ABC ABCD, with cadences on a, a, E, a, a, E, G (Worcester version, Stäblein, p. 198) or a, a, F, a, a, F, E (Nevers version, Stäblein, p. 102). In melody 420, given as Ex. II.15.5, there is no doubt about the primacy of C and G. After the repetition of the two opening lines an alternative tonal area is explored, F up to c with $b\flat$, rather than the upper segment of the C scale.

(iv) *Processional Hymns*

Hymns were often sung in processions. The repertory is not large (thirty-two in Stäblein's edition, of which not all are actually processional hymns). A large proportion of processional hymns have a refrain. Often the first strophe would be repeated after each subsequent strophe. *Pange lingua* was sung in this way (to a melody different from the one just given), using *Crux fidelis* as the refrain strophe. In some hymns each alternate repetition of the refrain was of its second half alone (R1+2 S1 R2 S2 R1+2 S3 R2, etc.). On the other hand, the refrains of some hymns are melodically and even metrically independent of the strophes.

Favoured metres are hexameters, distichs, and trochaic tetrameter. Fortunatus' *Salve festa dies* and Theodulf of Orléans's *Gloria laus et honor* are well-known examples in distichs.

The largest collection of processional hymns of this type that have come down to us is contained in St Gall, Stiftsbibliothek 381 (from St Gall, late tenth century), and smaller selections are found in other St Gall sources (e.g. St Gall 360). On pp. 23–50 and 142–66 of St Gall 381 there appear two series, of twelve and five items respectively, usually called *versus* (rather than *hymnus*), and frequently bearing a composer's name. Thus the first, *Sacrata libri dogmata*, is headed 'Versus Hartmanni, ante evangelium cum legatur canendi' ('A versus by Hartmann, to be sung before the gospel is read'; facsimile in Stäblein, 'Versus', *MGG*). The list of attributed pieces is as follows (see the remarkable 'footnote' surveying the repertory in Gautier 1886, 23–9; also, p. 159, an eleventh-century rhymed example from St Gall 382):

Hartmann (d. 925): *Sacrata libri dogmata*; *Salve lacteolo decoratum* (Innocents); *Cum natus esset Dominus* (Innocents); *Humili prece* (for feast-days generally, found in many adaptations with verses in honour of various local saints; melody in Wagner III, 481); and *Suscipe clementem plebs devotissima*, one of the final three versus of the collection which are for the reception of a king

Fortunatus: *Salve festa dies*

Ratpert (d. 884): *Ardua spes mundi* (see Stotz, 1982; used as *Humili prece*); *Laudes omnipotens* ('ad eucharistiam sumenda'—'as the eucharist is taken up'); *Aurea lux terra* (for the reception of a queen); *Annua sancte dei* (St Gall)

Notker (d. 912): *Ave beati germinis* ('about the Old Testament')

Waldramm (d. *c*.900): *Rex benedicte* (for the reception of a king)

Ratpert's *Ardua spes mundi* may serve as an example of the St Gall versus. (Rather few appear in later sources; all those that can be transcribed have been edited by Stäblein, MMMA 1, 478 ff.; texts in AH 50, unfortunately not always making the refrain arrangement clear; see also examples in Wagner III, 480–2). In the Rouen cathedral manuscript (Paris, Bibliothèque Nationale, lat. 904, thirteenth century) from which Ex. II.15.6 is transcribed, both *Ardua spes* and *Humili prece* are sung in the processions before mass on the Rogation Days leading up to Ascension Day (the weekdays following the fifth Sunday after Easter). In both chants verses in honour of Rouen saints are added to those already present; *Ardua spes* invokes Romanus and

Ex. II.15.6. From processional hymn *Ardua spes mundi* (Paris, Bibl. Nat. lat. 904, fo. 136ʳ)

Audoenus. The same melody is used for the refrain and the strophes (some other processional hymns have two different melodies).

II.16. CHANTS FOR THE ORDINARY OF MASS

In the study of chant, the term 'ordinary of mass' usually signifies those chants whose texts remain the same from one mass to the next, that is, Kyrie, Gloria, Credo, Sanctus, and Agnus Dei. There are a few other unvarying chants which are sung at festal masses: the antiphons *Asperges me* (outside Eastertide) and *Vidi aquam egredientem* (during Eastertide) with their psalm verses; these are not considered further here. The special case of the versicle *Ite missa est* and its relation, the *Benedicamus domino* versicle at the close of office hours, are discussed briefly at the end of this section. Looking at the liturgy of mass as a whole, however, the ordinary comprises all texts which remain the same for each mass: not only the chants but the prayers as well.

Although the texts of the five main ordinary chants remain the same, they were set to numerous different melodies. These were often associated with particular feasts or grades of liturgical celebration. Early sources are not often specific about the assignment of the chants, but there seems little doubt that each church would have its own established customs in this respect, even before rubricated collections became common in the thirteenth century. Large numbers of trope verses are known for all the ordinary chants, excepting only the Credo. These often refer specifically to particular feasts and thus make the chant 'proper' rather than 'ordinary'. (Tropes for ordinary chants are discussed below, II.23.)

Short sections are now devoted to each of the five chief ordinary chants. It should be noted in advance, however, that much research, transcribing, and analysing needs to be done before a rounded picture of these chants can be made up. For many years the best-known melodies have been those edited in the Solesmes/Vatican *Kyriale seu ordinarium missae* of 1905, which then passed into the Vatican *Graduale* and Solesmes *Liber usualis*. Eighteen sets were published, comprising Kyrie, Gloria, Sanctus, and Agnus, plus four Credo settings (later increased to seven) and other chants 'ad libitum'—eleven Kyries, three Glorias, three Sanctus, and two Agnus. These chants represent only a fraction of the medieval repertory, however, and the vast majority remain unavailable for study. It is fair to point out that many melodies had only a local circulation—many are known from but a single source—and the most popular are available in the Vatican selection. But until more are known it will be difficult to judge what is typical of the repertory as a whole, or of particular periods and areas, and what is eccentric.

The groundwork for future study has been established by the catalogues produced on the basis of the microfilm collection at Erlangen–Nuremberg University, by Landwehr-Melnicki, Bosse, Thannabaur, and Schildbach. The lack of British sources in the collection, and consequently in the catalogues, has been made good by Hiley

(1986, 'Ordinary'). Yet the earliest sources of all, those of the tenth century, were poorly represented in these catalogues, and several questions about the distribution of the oldest recorded melodies remain to answered. The gap has been more or less filled in the case of the Kyrie by Bjork, the Gloria by Rönnau, and the Agnus by Atkinson. A comparable catalogue exists for Credo melodies, by Miazga, but its scope and emphasis is different. Early sources for the Credo are sparse, for it does not seem to have been sung by the schola until relatively late, and even then by no means universally. Florid choir settings are rare before the fifteenth century. Miazga's lists are dominated by the great mass of eighteenth-century settings, whereas the other catalogues hardly go beyond the Middle Ages; and Miazga's chief concern was the tradition of Poland: later West European sources are less well represented.

At one period or another it has been the custom for the whole congregation to sing the ordinary chants. In the early centuries this was certainly the case, and the question as to whether any melodies might have survived from this time has fascinated scholars for many years. Some simple melodies, such as might have been sung by the people, have indeed come down to us. Yet the sources which contain them are invariably late, later than those for more elaborate chants. A degree of caution is therefore necessary when judging the claims to antiquity of these chants. On the other hand, it has to be remembered that our early sources are nearly all cantors' manuscripts, containing music for the trained schola and especially for soloists. Should we expect them to record congregational chants? We are not being asked to subscribe to the belief that elaborate chants grew out of simple ones: that would be a naïvely simplistic view of plainchant history. The question is one of function. Chants undergo changes, or new chants are composed, when the liturgical conditions change. When the schola takes over the singing of a chant from the congregation (it is still acting in lieu of the congregation), then we should expect different music to develop. But if the congregation still sang the chants, at least occasionally, there would have been no pressure to change. In these circumstances chants might well have survived through centuries of unwritten tradition. As Wagner said of Agnus Vatican XVIII, one of the simple chants in question: 'The indication in the Solesmes books that it originated in the twelfth century is certainly wrong. It may well be that it does not occur in any earlier notated chant-book; but that is not decisive as far as its age is concerned. Is our preface tone only as old as its earliest written occurrence?' (Wagner III, 448).

Stylistic analysis has seemed to offer insights into the problem. Thus Levy (1958–63) has made a bold attempt to link a Sanctus melody to other stylistically similar chants, Byzantine as well as Western, to reveal a 'modal area' from which the singing of the whole Byzantine ordinary is derived and the anaphora as well (that is, the prayers beginning with the preface, of which the Sanctus is a part, the memorial of the incarnation and words of institution). In the case of Kyrie and Agnus, litanies have been adduced as evidence.

In what follows some attention is therefore paid to possible early melodies. A few examples from the central repertory of the tenth to twelfth centuries are discussed (the biggest collections nearly all date from the twelfth century), and one or two quite different later chants (the fifteenth century marks a new high point in production) are also included.

Apart from those in the Vatican books, melodies have appeared rather haphazardly in facsimile or modern editions. The following may be consulted (mentioning facsimiles only of manuscripts in diastematic notation). Two Roman collections are available, the transcription from Rome, Biblioteca Apostolica Vaticana, Vat. lat. 5319 in MMMA 2, and the facsimile of Bodmer C. 74 edited by Lütolf (1987). The Beneventan manuscript Benevento, Biblioteca Capitolare VI. 34, is given in facsimile in PalMus 15. Transcriptions by Boe of the Beneventan repertory are in the course of publication. A facsimile of a Nonantola manuscript has been published by Vecchi (1955). French sources are poorly represented, only the Chartres melodies in PalMus 17 being available. On the other hand, two facsimiles of Sarum chants have been published, in *Graduale Sarisburiense* and MMS 4, and an edition has been made by Sandon (1984). For Eastern sources we have Marxer's transcriptions from St Gall, Stiftsbibliothek 546 (1908) and those from other manuscripts by Sigl (1911), and the facsimile of the *Graduale Pataviense* of 1511.

Few early sources indicate what music was used for singing the versicle *Ite missa est* and its response *Deo gratias*. It appears to have been popular to adapt a Kyrie melody for the purpose, or to share a melody with the office versicle *Benedicamus domino* (also sometimes sung at mass). The Benedicamus likewise relied heavily on borrowed music, for example, the melismas of office responsories. (See Harrison 1963, 74–6; Huglo 1982, 'Débuts'; Robertson 1988.)

II.17. KYRIE ELEISON

  (i) Kyrie eleison as a Litany
 (ii) Kyrie eleison after the Introit at Mass
(iii) Early Melodies
(iv) Italian Melodies
 (v) Melodic Types
(vi) Later Melodies

Stäblein, 'Kyrie', *MGG*; Landwehr-Melnicki 1955; Crocker, 'Kyrie eleison', *NG*; Bjork 1979–80; Boe 1989.

(i) *Kyrie eleison as a Litany*

'Kyrie eleison' is a litany formula and has been sung as a choral refrain in answer to petitions of one kind or another at least since the fourth century (as witnessed by Egeria in Jerusalem) and in the West at least since the fifth century. It can be found in numerous medieval litanies, as in these two examples:

Kyrie eleison. Christe eleison. Kyrie eleison. Pater de celis Deus miserere nobis. Fili redemptor mundi clemens miserere nobis. Spiritus sanctus Deus miserere nobis. Qui es trinus et unus Deus miserere nobis. Sancta virgo virginum ora pro nobis. [Then a series of saints' names, each followed by 'ora pro nobis'.]

Kyrie eleison. Christe eleison. Domine miserere. Christe miserere. Miserere nostri pie rex domine Iesu Christe. Christe audi nos. Sancta Maria ora pro nobis. [Then a series of saints' names, each followed by 'ora pro nobis'.]

Further 'Kyrie' invocations are sung at the end.

The invocation is a standard component of the preces sung in the Gallican liturgy, which survive in a few medieval sources (of which Paris, Bibliothèque Nationale, lat. 776 and 903 are the best known). Ex. II.17.1 gives the opening of one of these. Here 'Kyrie eleison' is repeated after each verse. All the verses have the same basic melody. (In Paris 903 the Kyrie invocation is sung to a melody more closely resembling the verses than is the case here.)

Ex. II.17.1. Litany (Cambrai, Bibl. Mun. 61, fo. 152ᵛ)

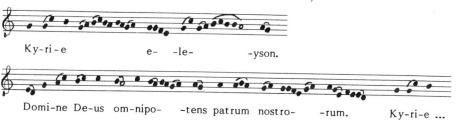

One composition with this form appears among the Kyries of early Roman sources. This is the Kyrie with Latin verses beginning *Devote canentes*, in the manuscript Rome, Biblioteca Apostolica Vaticana, Vat. lat. 5319. Only *Kyrie eleison* is sung as an invocation, not *Christe eleison*. (In other sources, however, *Christe* is sung—see Boe 1990, 'Italian'; the Kyrie is edited by Melnicki in MMMA 2, 587 and by Boe 1989, 74.) As Boe has pointed out, this Kyrie may, in some form or other, be very ancient, for its seventh Latin verse is a petition against the Arian heresy, which in Rome at least was defeated as early as the fourth century.

(ii) *Kyrie eleison after the Introit at Mass*

The usual place for a litany in the East was after the lessons at mass, and how it came to occupy its Roman position after the introit is unclear. It is found there in Ordo Romanus I (early eighth century), no longer, it seems, a song for the whole congregation, but one for the schola, who sang invocations until the pontiff gave the signal to stop. Only the simple Greek text is mentioned, but the possibility cannot be excluded that further verses were sometimes sung. That is certainly the import of remarks in the letter of Gregory I (d. 604) to Bishop John of Syracuse: Gregory writes that in Rome *Kyrie* is sung as often as *Christe*, and that on non-festal days only the Greek text was sung, not the longer verses sung on other occasions; the schola sings first and the congregation answers.

Ordo Romanus IV, a Frankish recension (surviving in a Saint-Amand source) of previous Roman ordines of the late eighth century, has the number of petitions fixed at nine. This is presumably the form with which we are familiar:

Kyrie eleison (Lord have mercy): three times
Christe eleison (Christ have mercy): three times
Kyrie eleison (Lord have mercy): three times

(The ending *eleison imas*—have mercy on us—for the final phrase is occasionally found in early sources.) The Ordo says nothing about Latin verses. Our earliest sources with music unfortunately date from much later (as is so often the case). Here we find not only melodies with the Greek text alone but also compositions with Latin verses. These have been traditionally regarded as tropes, later additions to the composition. In the sense that the Greek invocations form a nucleus around which Latin verses could be added, this is not incorrect. Yet from the musical point of view the matter is not quite so simple. As Crocker (1966; see also 'Kyrie eleison', *NG*) pointed out, for several melodies the first recorded appearance has them with Latin verses. We have no proof that the melody was first conceived for the Greek text alone. And in view of Gregory's statement and the existence of numerous types of litany involving both Greek and Latin verses, it seems prudent to admit the possibility that new Kyries might be composed from the start with Latin text.

Discussion of the various ways in which combinations of Greek and Latin verses were made is postponed until the section on tropes (see II.23.viii).

(iii) *Early Melodies*

From Bjork's survey of the earliest sources (1979–80) it is clear that very few melodies were known all over Europe at the earliest period for which we have definite information, the tenth century. The best known were the following (numbers from Landwehr-Melnicki's catalogue, followed by the number in modern Vatican books): 55 (Vat. ad lib. VI), 68 (Vat. XIV) and 155 (Vat. XV). Early Rhenish and Eastern sources also have: 39 (Vat. I), 144, and 151 (Vat. XVIII), while Western sources have: 47 (Vat. VI), 102 (Vat. ad lib. I), and 124. The number increased rapidly in subsequent centuries, as Melnicki's catalogue shows. (It lists 226 melodies, and as more sources are surveyed more melodies may be added.) Many were of local significance only. The modern Vatican books have a selection of melodies of widely differing age and provenance (Huglo 1958).

It is not clear if there survive among the oldest recorded melodies any compositions from earlier centuries. It would not be easy for any simple melodies such as the congregation might have sung to have survived when the schola assumed the singing for themselves, and during the further period through to the earliest sources (themselves designed to reflect the singing of the schola). Many of the early melodies give the impression of being sophisticated, carefully designed pieces, which speak rather for composition by Frankish musicians.

Kyrie 55 (Vat. ad lib. VI) displays features typical of early Kyrie melodies (Ex. II.17.2). The form of the piece is ABA CDC EFEx, where Ex signifies an extended version of E. The composer sets out very deliberately to explore different registers, so that B has a lower tessitura than A, D is lower than C, F than E. There is also a gradual rise through the three main sections. A rises to *c*, in C the melody touches *d* for the first time, while E moves up to the higher octave. The final extended invocation has itself an AAB form, a not uncommon way of building towards the climax of the composition.

Ex. II.17.2. Kyrie 55, Vatican ad lib. VI (Cambrai, Bibl. Mun. 61, fo.155ᵛ)

This sort of melodic design, balancing lower against higher phrases and gradually pushing towards higher melodic goals, is also found in many early sequences, and when the Kyrie melody is sung with a Latin text, one syllable per note, like the sequence, the resemblance is even stronger. (One might compare the last phrase of Ex. II.17.2 with the last phrase of the sequence melody known as 'Lyra', commonest text *Ecce pulchra*, Anselm Hughes 1934, 54; or with the seventh and the last phrase of 'Hodie Maria virgo', text *Aurea virga*, Hughes, 45.)

Ex. II.17.3 shows melody 68 in Landwehr-Melnicki's catalogue (Vat. XIV). The overall scheme is simpler: AAA BBB CCCx. The last dozen or so notes of A and B are the same, but B begins by moving confidently upward to *c*, to be followed by the even higher-lying C section. Characteristically, the brief expansion in Cx consists of a partial repetition of what has preceded, including the so-called 'Gallican' cadence, *cdd*.

The expanded final invocation is particularly liable to varying treatment in different manuscripts. Ex. II.17.4 shows melody 155 (Vat. XV), but not all sources have the

Ex. II.17.3. Kyrie 68, Vatican XIV (Cambrai, Bibl. Mun. 61, fo. 157ʳ)

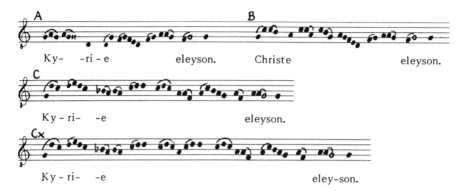

Ex. II.17.4. Kyrie 155, Vatican XV (Cambrai, Bibl. Mun. 61, fo. 159ᵛ)

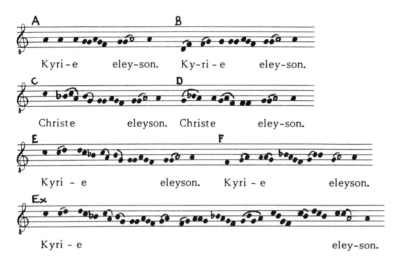

last verse in the form given here. Once again the form ABA CDC EFEx is used, but Ex actually combines E and F. A, B, C, D, and E all have the same ending. And yet again a consistent pattern of higher and lower phrases may be observed. In this case the progression from A to C to E is achieved simply by adding higher notes at the start of what is the same basic phrase throughout.

Composers evidently delighted in the possibilities offered by the thrice-three verse-scheme. All manner of repetition patterns and partial correspondence between verses may be discovered.

(iv) *Italian Melodies*

That these three melodies were so popular (and a number of others run them close when the greater number of sources from the eleventh and twelfth centuries is taken into account) seems to indicate that they answered well to the general sense of how a Kyrie should be sung. Their features are typical of a large number of early Kyrie melodies. But not all melodies have this character. Early Italian Kyries, for example, show little interest in repetition schemes and the upward surges so noticeable in Ex. II.17.2. The above-mentioned Kyrie with Latin verses *Devote canentes* etc., sung to melody 77 in Landwehr-Melnicki's catalogue, has the same melody for all invocations (and all Latin verses), and a static, repetitious quality. So does melody 52, associated with the Latin verses *Auctor celorum* etc. These two melodies are given in Ex. II.17.5.

Ex. II.17.5. Kyrie 77 and Kyrie 52 (Rome, Bibl. Vallic. C. 52, fos. 153ʳ, 150ʳ)

Since the earliest Italian sources date from the eleventh century (the earliest Roman one from as late as 1071), they are automatically excluded from the 'earliest' layer. This need not mean, however, that the melodies they contain are necessarily less ancient, only that they cannot be proven to be so. Stäblein ('Kyrie', *MGG*, Ex. 9) prints one melody (Landwehr-Melnicki 84) from a Roman source that is so simple that it might be centuries old; or it might be a relatively late composition deliberately made simple for ferial use, as Landwehr-Melnicki 7 seems to have been.

(v) *Melodic Types*

Not surprisingly, many melodies in this large repertory bear a family resemblance to each other, or make use of similar melodic gestures. Thus thirteen melodies open *Gab*, or some variant thereof. Several other melodies besides Landwehr-Melnicki 55 (Ex. II.17.2) fall to *E* after the *G*-opening (48, Vat. II, 102). Landwehr-Melnicki 39 (Vat. I), for example, sounds like a simplified version of 55; its Christe is the same as that of 55. Another family of short melodies has a first phrase ending on *b*, starting either on *G* or *b*, like Landwehr-Melnicki 155 (Ex. II.17.4): Landwehr-Melnicki 144 and 151, or, rather longer, Landwehr-Melnicki 124 and 142.

(vi) *Later Melodies*

From Landwehr-Melnicki's catalogue it can be seen that the composition of melodies continued unabated into the sixteenth century and beyond. Many of the newer items bear the marks of a later age: scale passages running through a fifth (usually from or towards the final), broken chord figures, and sudden moves to the upper octave. A number are couched in a somewhat sentimental 'F major' idiom, like the popular 'De angelis' melody (Landwehr-Melnicki 95, Vat. VIII, known mostly from French and

Italian sources). A melody of this type found in eastern sources (from Germany, Switzerland, Austria, Bohemia, etc.) is Landwehr-Melnicki 97, reproduced here in Ex. II.17.6.

Ex. II.17.6. Kyrie 97 (St Gall, Stiftsbibl. 546, fo. 38^v)

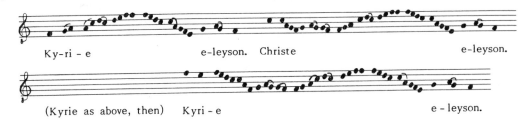

Local compositions include even more extreme examples in this vein, as for example Landwehr-Melnicki 139, one of ten unique melodies in St Gall, Stiftsbiblio-thek, 546, compiled at the famous abbey by Joachim Cuontz in 1507 (Ex. II.17.7).

Ex. II.17.7. Kyrie 139 (St Gall, Stiftsbibl. 546, fo. 38^r)

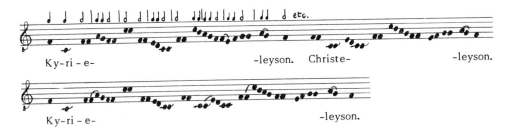

II.18. GLORIA IN EXCELSIS DEO

  (i) Introduction
  (ii) Recitation Types
  (iii) Through-Composed Melodies

Bosse 1955; Stäblein, 'Gloria in excelsis Deo', *MGG*; Crocker, 'Gloria in excelsis Deo', *NG*; Boe 1990, *Gloria*.

(i) *Introduction*

Some elements in the text of the Gloria in excelsis Deo are very ancient, dating back to early Christian times. Like the Te Deum, it is a non-biblical hymn of praise of irregular construction, an agglomeration of phrases of different date and origin. It seems to have been part of the office in the early centuries, but according to the *Liber pontificalis* it was introduced into mass in Rome on Sundays and saints' feast-days by Pope Symmachus (498–514). The earliest surviving version of the Gloria in Latin is no older than the so-called 'Antiphonary of Bangor', a late seventh-century Irish collection of texts mainly for the office. Variants in the text are still occasionally to be found in sources of the tenth century and later, that is, in the period from which we first have notated versions. Although originally a congregational chant, its execution was later reserved for the clergy, being intoned at the start by the pope, bishop, or priest as the occasion required. Among the surviving melodies some are easily singable by the whole congregation, as modern experience proves. But there is no way of knowing how old such melodies are. Some relatively ornate melodies seem more like music for soloists or the trained schola.

Bosse's catalogue lists fifty-two different melodies, to which a few may be added from sources not available to Bosse. Eighteen are to be found in the modern Vatican books. Bosse and Stäblein both survey the different types of melody so far known. Some are potentially quite old: that is, they may easily be imagined to date from the period of oral transmission. One type of melody may be described as a highly ornate recitation, with a constantly reiterated central pitch and cadential melismas; another type of melody consists of the constant repetition of a single phrase, adapted as required to verses of differing length. Other types may be more recent: some are through-composed, others are freely composed but with constant resort to particular motifs, which serve to bind the musical fabric together. It is not, of course, possible to assign all melodies absolutely to a particular category.

(ii) *Recitation Types*

One set of melodies is dominated by a single recitation note with neighbouring tones, delivered in a highly inflected, elevated manner which raises them above the level of, say, the introit or communion psalmody heard elsewhere in the mass. Ends of verses are often marked by cadential flourishes. We could imagine the melodies in this small group to be music for the choir, or even for soloists.

The most important member of the group is Gloria 39 in Bosse's catalogue, sometimes referred to as 'Gloria primus' or 'Gloria A', unfortunately not included in the Vatican selection. It is the grandest and most expansive of Glorias. Boe (1982) has found a version of it in the Old Roman Easter Vigil mass. Since it is usually found with trope verses—it was sung more frequently with tropes than any other melody—it is given elsewhere in full (see Ex. II.23.15). Suffice it here to point out its chief features. The melody has two recitation tones. In the first line and occasionally later the recitation is centred on *a* (typically in figures such as *Gab a*). Somewhat disconcertingly, the second line and most subsequent lines use *F* frequently (in such phrases as *aGF Ga a*), with *bb* as upper auxiliary. The most important cadential melismas are given in Ex. II.18.1a–d. The figure given as Ex. II.18.1e, which is also to be found at the start of the verse tone for mode 1 responsories, appears frequently from 'propter magnam' onwards.

Ex. II.18.1. Cadential figures (a–d) and intonation figure (e) from Gloria 39

Other examples of this type of melody may be seen in Bosse 13 and 22 (Vatican ad lib. II and III), centred around *D*, or the melody for the Greek Gloria *Doxa en ipsistis theo* (Huglo 1950, 'Gloria'). The latter begins with *b* as reciting note (but even more heavily ornamented than Gloria 39) with *G a c* as the common intonation. After several lines the cadences begin to fall on *E*, though most of the melodic movement still concerns the tetrachord *G–c*.

The simplest example of the type is Bosse 2, probably a south Italian melody (ed. Boe 1990, *Gloria*, 188) but given in Ex. II.18.2 from one of its few northern sources (presumably brought back to Normandy after the Norman conquest of south Italy and Sicily). It will be seen at once that practically every phrase opens with the intonation *F G a* and ends *a G*. A great deal of what happens in between stays within the same narrow range. The use of a higher opening for 'Domine Deus', 'Domine fili', and 'Cum sancto spiritu' may also be mentioned. Since it comes from a part of Italy where much ancient, non-Frankish material is preserved, one is tempted to see here one of the oldest surviving Gloria melodies. The sources are, however, too late in date to prove any hypothesis conclusively. In its usual northern version, which corresponds to Vatican melody XI, it was catalogued by Bosse as melody 51.

If some features recall the singing of office psalms, or responsory verses, the above melodies nevertheless remain essentially free of any simple, repeated formula. They are too elaborate (with the exception of melody 2) and too varied to subsume under psalmodic practice. There are, however, several other melodies which come closer to simple psalmody, consisting of hardly more than the constant repetition of one musical idea or formula.

The best-known example of this type of melody is Bosse 43 (Vatican XV). Each verse begins *E G a*, recites on *a* with inflections down to *G* or up to *b*, and cadences *a G a G E*.

More adventurous is Bosse 25 (Vatican V), with more than one musical idea. In the top half of the range comes the phrase *G b cd d b dcbaG ab aG G*. This is followed by a phrase which descends into the lower half of the range: *caG aG EFED*. The final is achieved through the phrase *D EFG GFa G*. Each phrase is of course treated quite flexibly, to accommodate the varying numbers of syllables and their varied accentuation patterns. These phrases, although deployed in a fashion resembling simple psalmody, have no musical resemblance to psalm tones. Nor has Bosse 38 (Vatican VIII), the melody which makes a musical pair with the Kyrie 'de angelis'. They are both much later, and much less widely known, than Bosse 43.

Ex. II.18.2. Gloria 2 (Paris, Bibl. Nat. lat. 10508, fo. 20ᵛ)

(iii) *Through-composed Melodies*

The majority of Gloria melodies are through-composed. This does not exclude the possibility of using recurrent motifs, which stamp the melody with a particular character without suggesting the technique of simple (or ornate) psalmody. Even when the motifs encompass whole phrases, their order is not predictable in the way that those of Bosse 25 are.

Few melodies actually lack recurrent motifs entirely. Bosse 24 (Vatican ad lib. I) is a well-known example. More typical is Bosse 56 (Vatican IV), the most popular of all Gloria melodies. The final, *E*, is heard at the end of practically every phrase, but it is not treated as a reciting note. A variety of musical ideas leads into it, from below (*F D C D E*) and above (*GF Ga GF E*), the latter sometimes preceded by a higher-lying motif (*G a c a*). The artistry of the composition lies in the judicious selection of one motif or another for the various verses of the text, balancing higher against lower, amplifying where necessary, as with the more expansive cadence heard four times: *DFGaGFG E*.

Another example of this type of melody is the little-known Bosse 5, found in Eastern European sources from around 1200 onwards (Ex. II.18.3). There is no need to attempt to isolate all the motifs which make up the basic melodic material; in any case, some instances of their use might no doubt be felt to be fleeting coincidences rather than deliberate employment. The motif *EDEG* for 'excelSIS', for example, occurs many times throughout, sometimes split to accommodate different syllables. Are these all conscious attempts to bind the piece together?

Ex. II.18.3. Gloria 5 (Munich, Univ.-Bibl. 2° 156, fo. 157ᵛ)

It is perhaps more worth while to follow the falling figure *bcbaG*, which is first heard twice in the second verse, 'Et in terra pax hominibus' (labelled 'x' in Ex. II.18.3). This has no rhetorical function: it may start or end a phrase, or stand in the middle. Among its many appearances are several where it is preceded by a fall and immediate rise of a fifth: at 'Adoramus te', 'Qui tollis' (the second one), and 'Tu solus dominus' (labelled 'y'). The leap of a fifth may bridge two different phrases, as at 'dexteram partris/miserere' and perhaps also 'peccata mundi/Suscipe'. The figure 'x' is frequently succeeded by a rise to *d*, as at 'terra pax' (labelled 'z'). And so on.

Glorias were usually sung with trope verses on high feasts from the tenth to the twelfth century (see II.23.xi), but because the trope verses were so often applied to different base melodies there seems to be no overwhelming reason for believing that tropes and melodies were composed at the same time. The number of known melodies is outnumbered two to one by the sets of trope verses. After Gloria tropes fell out of use in the thirteenth century, only one modest Marian example, *Spiritus et alme*, held its place. It and the Gloria melody it embellishes (Bosse 23) seem first to have become popular in the Paris region in the second half of the twelfth century.

Composition of new melodies was by no means as vigorous as for other ordinary of mass chants, though from Bosse's tables it can be seen that a flurry of activity took place in the fifteenth century. Several fine melodies still remain generally unknown.

II.19. SANCTUS

(i) The Oldest Melodies
(ii) Other Melodies

Levy 1958–63; Thannabaur 1962; Thannabaur, 'Sanctus', *MGG*; Crocker, 'Sanctus', *NG*.

(i) *The Oldest Melodies*

Much speculation has been occasioned by the problem of distinguishing a possible ancient melody, such as might have been sung from early times, when the Sanctus was a congregational chant, among the compositions which have come down to us in notated manuscripts from the tenth century onward. For many years it was confidently asserted that the simple melody Vatican XVIII, no. 41 in Thannabaur's catalogue, was a relic of this early period. Thannabaur then pointed out that the earliest source dates from the eleventh century, the melody becoming more widely known only in the thirteenth century. In fact, in Thannabaur's earliest source, Benevento 38, all the ordinary-of-mass chants are on added leaves of the twelfth century; and the melody appears there in a more elaborate version than was later to become usual (see *MGG*, 'Sanctus', Ex. 1; the ending should read *Gaf FG*). The next Italian source dates from the thirteenth century, and the rest are practically all fifteenth-century manuscripts. Lévy (1958–63, 27) cites another twelfth-century source for the incipit, at least, of the elaborate version. On the other hand, Boe (1982) has linked it with other ordinary chants for the Old Roman Easter Vigil mass.

This is not to deny that the Sanctus was probably once sung to a simple melody. The text is a continuation of the preface, which was sung by the officiant to a recitation formula. Thannabaur 41, Vatican XVIII, is hardly more than a continuation of one of the preface tones. And the participation of the congregation in the singing of the Sanctus is recorded frequently throughout the Middle Ages. At various times and places, however, the chant was performed by clerics. And trained singers must also have played a part, at least in the singing of trope verses, and perhaps also in the execution of some of the more elaborate melodies known from medieval sources.

It is therefore somewhat difficult to identify a possible early layer of melodies. One interesting candidate is the Greek Sanctus, or *Agios*, known from a group of tenth- and eleventh-century manuscripts, usually in conjunction with other ordinary-of-mass chants with Greek text (see the list in Atkinson, 1982, 'Missa Graeca'). Levy (1958–63) discovered a Byzantine relative of this and remarked that the same basic musical idea underlies the singing of the words 'sanctus, sanctus, sanctus' in the Te Deum. In fact, Levy pointed to the existence of a large family of melodies (or better, recitation formulas), both eastern and western, which may all have a common origin.

What unites them, to put it in most general terms, is their insistence on the scale segment *Gab*, with *c* and even *d* as upper options; *E* is sometimes used to start a phrase, but *F* is avoided. (The melodies are found in transposition with *FGa* or *CDE* as the basic notes.) The continuous oscillation within this narrow range is most familiar to us through the common tones for the preface and Lord's Prayer (see Stäblein, 'Präfation' and 'Pater noster', *MGG*). And as far as ordinary of mass chants are concerned, we have just seen examples of this type of melody in Italian Kyries (Ex. II.17.4) and Glorias (Ex. II.18.2). Ex. II.19.1 gives a version of the Greek Sanctus and another Italian melody in this idiom. (Compare also Boe 1982, Ex. 3.)

(ii) *Other Melodies*

The majority of melodies in Thannabaur's catalogue are not of this type. Over 230 are listed there, and others are to be found in sources not accessible to Thannabaur. Relatively few become known right across Europe:

Thannabaur	Vatican	earliest sources
32	XVII	11th c.
41	XVIII	12th c.
49	IV	11th c.
116	VIII	12th c.
177	XII	13th c.
202	XI	12th c.
203	II	12/13th c.
223	XV	10th c.

Ex. II.19.1. Greek Sanctus and Latin Sanctus 66 (Modena, Bibl. Cap. O. I. 7, fos. 108ʳ, 206ᵛ)

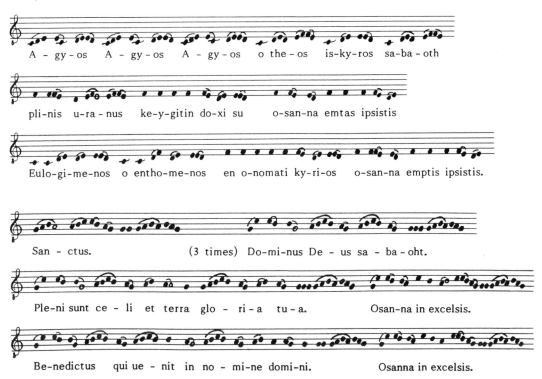

Melody 154 (Vatican I), widely known in France and Italy, is perhaps the only other melody which can really be called popular. It follows that the repertory as a whole is very diffuse, like that of other ordinary of mass chants, most melodies having purely local currency. Some of the oldest melodies are not represented in the above list. In three tenth-century sources, for example (these and several other of the oldest sources are not covered by Thannabaur), we find the following:

Cambridge, Corpus Christi College 473 (Winchester): 111, 154, 155, 216, 223

Paris, Bibliothèque Nationale, lat. 1118 (Aquitaine): 89, 111, 216, 223 (plus one not in Thannabaur)

St Gall, Stiftsbibliothek 381 (St Gall): 154, 216 (plus one not in Thannabaur, possibly 153)

Like the repertories of Kyries and Agnus, Sanctus collections were rather small before the eleventh century, when a deliberate effort seems to have been made to provide a set of chants for the festal liturgy. After the twelfth century production slackened, until a new peak was reached in the fifteenth century.

Most early melodies show some relationship between the various verses. The two settings of 'Hosanna in excelsis' are naturally often the same, and material from the first verse may reappear later in the piece. The three 'Sanctus' acclamations often stand apart, though their music too may be redeployed for the text that follows.

Sanctus 68, transcribed in Ex. II.19.2, is a fairly typical example. The music for the first 'Sanctus' recurs at 'Pleni sunt' and 'Benedictus'. The movement from *G* to *d* and back again at 'Sanctus dominus Deus' is heard again during 'Osanna' and 'qui venit'. Perhaps the short phrase for 'gloria tua' is an echo of 'Deus sabaoth'. The melody has a character quite different from that of the Greek Sanctus and the Italian melody given in Ex. II.19.1. The first 'Sanctus' stays within the segment *G–b*, the second touches *c*, and then the third rises to *d*, a gradual unfolding which is carried one step further later on in 'Domini'. There are also literal repeats, in 'sabaoth' and the melisma after 'gloria tua'.

Ex. II.19.2. Sanctus 68 (Paris, Bibl. Nat. lat. 10508, fo. 123ʳ)

(Osanna as above)

Melody 216 in Thannabaur's catalogue, found fairly widely among the earlier sources, has something of the same character, though it is cast in a more modest vein (Ex. II.19.3). The three acclamations rise gradually, then 'Deus sabaoth' echoes the opening, adding the so-called 'Gallican' cadence. That cadence recurs at the end of each verse. 'Pleni' and 'Benedictus' take up the music of the third 'Sanctus', but the 'Osanna' verse is independent.

Twelfth-century sources such as those used for the transcriptions so far have the largest of all collections of Sanctus melodies, excepting such retrospective anthologies as St Gall 546. Ex. II.19.4 is taken from this manuscript. Most of the new melodies of the fifteenth century and later are quite different in character from those seen so far, with arpeggiando figures and simple melodic repetitions. Thannabaur 98 is built almost entirely from the material of the second and third 'Sanctus' acclamation. The chant is notated mensurally in St Gall 546; each normal note is transcribed here as a crotchet, a double punctum becomes a minim, and the semibrevis a quaver.

Ex. II.19.3. Sanctus 216 (Paris, Bibl. Nat. lat. 10508, fo. 123ᵛ)

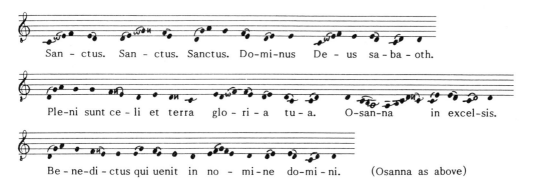

San - ctus. San - ctus. Sanctus. Do-mi-nus De - us sa - ba - oth.

Ple-ni sunt ce - li et terra glo - ri - a tu - a. O-san-na in excel-sis.

Be - ne-di - ctus qui uenit in no - mi-ne do-mi - ni. (Osanna as above)

Ex. II.19.4. Sanctus 98 (St Gall, Stiftsbibl. 546, fo. 68ʳ)

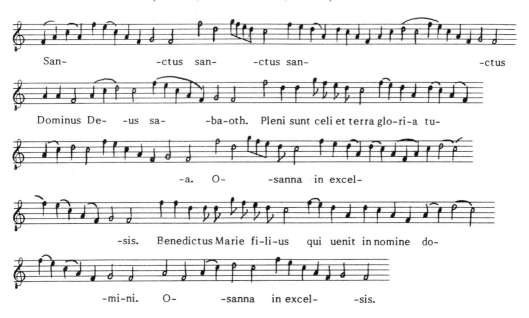

San- -ctus san- -ctus san- -ctus

Dominus De- -us sa- -ba-oth. Pleni sunt celi et terra glo-ri-a tu-

-a. O- -sanna in excel-

-sis. Benedictus Marie fi-li-us qui uenit in nomine do-

-mi-ni. O- -sanna in excel- -sis.

II.20. AGNUS DEI

Stäblein, 'Agnus Dei', *MGG*; Schildbach 1967; Atkinson 1977; Crocker, 'Agnus Dei', *NG*.

Investigation of the earliest melody or melodies for the Agnus Dei runs into the same problems as for the other ordinary-of-mass chants. According to the *Liber pontificalis*, Sergius I (687–701) introduced the chant into the Roman mass, as a piece sung by both clergy and people to accompany the breaking of the bread. Whether or not the attribution is secure, the approximate date can be supported indirectly. Ordo Romanus I (early eighth century) also understands it to be a fraction chant, but sung by the schola. Ordo Romanus III—the relevant part is Frankish and dates from the third quarter of the eighth century—indicates that the singing continues until the fraction ceremony is completed, rather in the way the early introit, Kyrie, and other chants were performed. Shortly after this, however, a number of Carolingian documents have the chant sung during the kiss of peace or communion itself. There arose concurrently the practice of using pieces of unleavened bread, and communion by the entire congregation was abandoned. The fraction ceremony diminished in significance. This seems to have resulted in a shifting of the performance of the Agnus towards communion and curtailment of its length, restricting it to the three petitions with which we are familiar. This is its form in the earliest extant document which is specific in such details, the Amiens sacramentary Paris, Bibliothèque Nationale, lat. 9432 (ninth century, second half). (Facsimile in Atkinson 1977, 10; Atkinson provides a detailed review of the evidence summarized here.)

The earliest musical sources are somewhat later (table in Atkinson, 13), and from these it is clear that only one melody had Europe-wide currency at an early date, no. 226 in Schildbach's catalogue, Vatican II. If we are looking for a hypothetical simple melody of the type which might have been sung by the congregation, repeated as often as necessary, according to the older practice, we might doubt whether this is it, or at least whether it appears in its former state. It has been suggested that two simpler melodies may be particularly ancient: Schildbach 101 (Vatican XVIII) has long been designated the oldest melody, partly because of its simplicity, partly because it follows easily from the preceding dialogue ('Pax domini . . . Et cum spiritu tuo'—see Wagner III, 449), partly because of musical identity (the text is different) with the 'Agnus Dei' section of the litany of the saints (Stäblein; *GR* 198, *LU* 838). Stäblein also directed attention to the Roman version of the melody (registered separately by Schildbach as no. 98), which appears in two Roman sources, Vatican Library, Vat. lat. 5319 and Archivio San Pietro B. 79. Whether the melody is much older than the earliest sources—which are much later than those for melody 226— remains open to question.

The oldest Roman source, Bodmer C. 74, has another relatively simple melody which belongs to the Italian type of chant discussed above in connection with the Kyrie, Gloria, and Sanctus. As Boe (1982) has shown, it belonged to the Old Roman Easter Vigil mass. Ex. II.20.1 gives this melody (not in Schildbach), and melody 226. The greater range and variety of the latter are obvious.

Many Agnus melodies may be compared with Kyries in that they suggest a tripartite form. For the above examples only one verse is copied out, to be sung three times. Other melodies may have a contrasting central verse, giving an ABA form, or even three different verses (for example, Vatican XI, Schildbach 220). Furthermore, there is often a similarity of range for 'Agnus Dei' and 'miserere nobis', whereas 'qui tollis peccata mundi' will rise to a higher level, giving an ABA shape to the verse. Thus Vatican XVII, Schildbach 34, has the overall form ABA A'BA ABA, the only

Ex. II.20.1. Old Roman Agnus (Cologny, Bibl. Bodmeriana C. 74, fo. 125ᵛ); Agnus 226 (Paris, Bibl. Nat. lat. 1119, fo. 249ʳ)

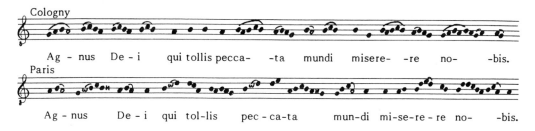

difference between the second verse and the others being the lower start. Such symmetry is a feature of the later rather than earlier repertory (the earliest sources for melody 34 are of the thirteenth century). Among the considerable variants often found between sources many affect the form of the piece. Thus while many sources have but a single verse of melody 226 (as in Ex. II.20.1), many have a contrasting second verse, most with the one published as Vatican II, others with that of Vatican XVI and XV.

The best known melodies were:

Schildbach	Vatican
34	XVII (less well known in France and England)
101	XVIII
114	IX
136	IV
164	XVI
209	XV
220	II

Of these, 34, 114, and 136 are relatively late F-mode melodies, with a particularly clear delineation of the *Fac* triad. Such Agnus melodies are not uncommon, and because of this preference the modal 'profile' of the repertory is different from that of Kyries and Sanctus. The rough percentages given here are derived from the tables of Landwehr-Melnicki, Thannabaur, and Schildbach. (The figures of course vary from area to area and century to century, as the commentary by those three authors shows. Some melodies are naturally found in different modes in different sources, and transpositions between G-mode, C-mode, and F-mode with B♭ are common. These are ignored here.)

	D	E	F	G
Kyrie	30	17	21	31
Sanctus	31	20	23	26
Agnus	36	19	27	19

As an extreme example in the F-mode vein, Ex. II.20.2 gives melody 144, found in a few fifteenth- and sixteenth-century sources. The mensural notation indicates a

Ex. II.20.2. Agnus 144 (St Gall, Stiftsbibl. 546, fo. 76ᵛ)

gentle, rocking rhythm. The first verse (repeated for the third) has an ABA form, the second begins differently, somewhat unsubtly taking up the music of 'Dei qui tollis', but then rejoins the music of the first.

For some melodies the earliest sources nearly all have trope verses. That is the case, for example, with melody 226. Following Crocker's prompting, one wonders if some melodies might not have been conceived from the first with trope verses. The case is different from that of the Kyrie, however; for the Agnus there is no previous record of singing extra verses, simply of repeating the same text over and over again. Atkinson has demonstrated that for melody 226, at least, the tropes are most probably later additions.

The text itself underwent a small change after about the turn of the millenium, when the third verse 'Agnus Dei . . . dona nobis pacem' begins to appear. Another common variant in the early sources, particularly in conjunction with melody 164, is the second verse 'Qui sedes ad dexteram patris miserere nobis', which seems to be a simple variant rather than an inserted trope verse.

II.21. CREDO

Huglo 1951; Stäblein, 'Credo', *MGG*; Miazga 1976; Crocker, 'Credo', *NG*.

Three different Credo texts were known in the Middle Ages. The Apostles' Creed ('Credo in Deum patrem omnipotentem creatorem caeli et terrae') was often said as part of the preparatory prayers before the services of the office. The only known musical settings are farsed ones from special festal liturgies (see Ex. II.23.19 below). Curiously, however, among the Greek ordinary-of-mass chants which are notated in a number of early sources (see VIII.2.v) is a Greek creed, and this is the Apostles' Creed. No source with diastematic notation is known. Likewise unknown in any notated source is the Athanasian Creed ('Quicumque vult salvus esse'), said at Prime.

The Nicene Creed ('Credo in unum Deum patrem omnipotentem factorem caeli et terrae') is so called because it is supposed to sum up the doctrinal beliefs established at the Council of Nicaea (325), though it was formulated somewhat later. It was originally part of the baptismal rite, the profession of faith of those about to be baptized. Eventually it was taken up in the mass, and is first found there in a Latin liturgy in Spain (Council of Toledo, 589). It achieved an official place (sung after the gospel) in the Frankish-Roman liturgy in a version by Paulinus of Aquileia (Council of Aachen, 798). It did not, however, become part of the liturgy in Rome itself until 1014, at the request of the German emperor Henry II, and then only for Sundays and those feast-days actually referred to in the text.

From the differing accounts and commentaries on the liturgy it seems that in the early Middle Ages it was sometimes recited by the congregation, sometimes sung by the clergy. Notated versions do not survive from before the eleventh century, and these are quite simple, either the syllabic chant known as Vatican I or variants of it. It was rarely copied with other ordinary-of-mass chants until the late Middle Ages. It was not usually troped, though farsed versions are known from the festal liturgies of Laon, Sens, and Beauvais. Only much later, particularly from the seventeenth century onward, were alternative musical settings composed in large numbers (many catalogued by Miazga).

Vatican I is built around two phrases, one lower in range, rising at the end *EGa*, the other higher in range, with $b\flat$ (*c* in eastern sources), ending *aGFG*. These are stretched or contracted freely as the verses of the text require. Often the first is preceded by a supplementary phrase with the basic shape *EFGFE*, and the second may be succeeded by another with the shape *EFGFaG*. *G* plays the part of reciting note, though recitation is of secondary importance in accommodating the lengthier verses, and the effect is different from that of simple psalmody. Recitation-type melodies may, however, be found in the Hispanic and Milanese chant repertories (Rojo and Prado 1929, 123, PalMus 6, 316).

Huglo (1951) has suggested a Byzantine link with melody Vatican I, though the relationship must have survived many centuries of unrecorded transmission. Vatican II, V, and VI use variants of the same typical phrases as Vatican I. And Stäblein has pointed out that a further melody in an Aquitanian source belongs to this group. This melody, reproduced here as Ex. II.21.1, has the opening word in its plural form 'Credimus' (as does Vatican VI in its original form in Paris, Bibliothèque Nationale, lat. 887). It is copied in the mass for Whit Sunday.

There are numerous variants in the text from the standard form. The music seems to combine formulas for openings and cadences fairly freely, with recitation on both *G* and *a*. Many verses simply begin on *G*, but there is one popular opening gesture (marked 'a' on the transcription) and another used only twice ('b'). The two chief cadences have the skeleton *aFaG* ('x') and *aFGE* ('z') respectively, and a third cadence figure *GaGFE* ('y') is also sometimes used. The similarity to Vatican I lies in the tonality, particularly in the dominant role of *E*, *G*, and *a*, in the syllabic word-setting, and in the similarity of some motifs. The resemblance is in fact often closer to Vatican II.

Ex. II.21.1. Credo (Paris, Bibl. Nat. lat. 776, fo. 92ᵛ)

Credimus in u-num De-um patrem om-nipo-tentem factorem caeli et terrae

uisibilium homnium et inuisi-bi-li-um conditorem et in unum dominum Ihesum Christum

filium De-i uni-genitum ex patre natum ante omni-a secula Deum uerum

de De-o uero genitum non factum homo-u-si-o patris per quem omni-a facta sunt

qui propter nos et propter nostram salutem descendit et incarna - tus est de spiritu sancto

Natus ex Mari-a uirgine homo factus passus est sub Ponti-o cruci-fi-xus mortu-us

et sepultus terci-a di-e surre-xit Ascendit ad caelos sedet ad dexteram patris

in-de uenturus in glo-ri-a iudica-re uiuos et mortuos cuius regni non erit finis

et in spiritum sanctum dominum et uiuificantem ex patre et fi-li-o procedentem

cum patre et fi-li-o a-dorandum et conglorificandum qui locutus est per prophetas

et unam sanctam catholicam atque apostolicam aecclesi-am confitemur unum baptisma

in remissionem peccatorum expectemus resurrectionem mortuorum uitam futuri seculi Amen.

The other Credos of the Vatican selection are a mixed bunch. Vatican VII is from the Sens Circumcision office, stripped of its farse verses. Vatican III is a triadic F-mode melody of the fifteenth-century, in a vein which matches the Kyrie 'de Angelis'. Vatican IV, known as the 'Credo cardinalis', is of similar age; it is frequently found with mensural notation (see Tack 1960, 50).

The Vatican selection gives no hint of the vast numbers of Credo melodies composed particularly from the seventeenth century onwards. Miazga's catalogue lists over 500 variants of the Vatican I melody, and over 700 other melodies. His figures for the different centuries record 57 for the fifteenth century (the first when new composition becomes strikingly evident), 44 for the sixteenth, 110 for the seventeenth, and no less than 424 for the eighteenth century! It is not yet possible to see this enormous production in a proper perspective, for the whole history of plainchant in these centuries is still poorly known. The catalogues of the other ordinary of mass chants concentrate almost exclusively on the medieval period, and the numbers of melodies they register are not comparable. Miazga's research has also concentrated on eastern European, particularly Polish, and Italian sources, and much remains to be done for other lands. For France, for example, the *Cinq messes en plainchant* (1669) of Henry Du Mont (1610–84) are relatively well known, being sung in France until this century. Less well known is a similar mass by Lully, still in use in the nineteenth century, of which an extract is given here from the gradual printed under the aegis of the ecclesiastical commission of Digne (Ex. II.21.2). The word-painting at 'descendit' and 'ascendit' will not pass unnoticed.

Ex. II.21.2. From Credo by Lully (*Graduel romain*, Marseilles, 1872, p. 116*)

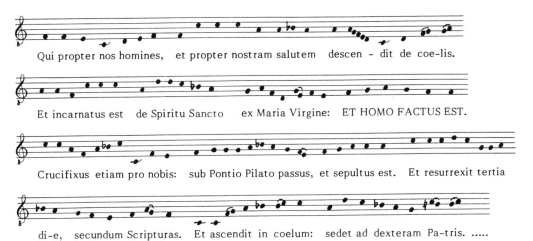

II.22. SEQUENCES

Stäblein, 'Sequenz', *MGG*; Crocker 1973; Crocker 1977; Crocker, 'Sequence', *NG*.
  Facsimiles: MMS 1, 3, 4; PalMus 15, 18; Vecchi, 1955.
  Editions: Misset and Aubry 1900; Drinkwelder 1914; Moberg 1927; Anselm Hughes 1934; de Goede 1965; Eggen 1968.

(i) *Introduction*

The origins of the sequence are so much disputed that I have chosen to present first an account of the sequences which appear in books of the ninth to eleventh centuries, the 'first epoch' of sequence-writing. The evidence about the early history of the sequence is reviewed in section vi below.

Sequences were sung at least from the ninth century onward after the alleluia at mass on feast-days (sometimes also elsewhere in the liturgy, for example as a substitute for the Vespers hymn). While an alleluia at a less important ceremony would be performed alleluia—verse—alleluia, on a high feast the pattern would be alleluia—verse—sequence. (See, for example, the rubrics in some Sarum books of the later Middle Ages: Dickinson 1861–83, cols. 9–10.,)

Most sequences were constructed in paired versicles, each line of music sung twice to different words. (In some sequences not all versicles are paired.) Some pieces, which may be called sequences because of their liturgical function (they follow the alleluia at mass on feast-days), are much shorter and are not constructed in parallel versicles (section iv).

The early sources of sequences are very disparate in both character and contents. The discussion of the early or 'first-epoch' sequence below is restricted to a few of the compositions which belong to the period up to about 1000. About 150 melodies were used in different parts of Europe up to this time. Many of them seem to have been local compositions which did not travel outside a particular area. The number of melodies known 'internationally', that is in both west and east Francia, is small, only about thirty. Since most melodies were sung with different texts in the different areas, the number of texts known internationally in the tenth century was practically nil, though the total number of texts was already quite considerable.

Little is known about the early stages of the Italian sequence tradition, but it seems as if the notion there of what constituted a sequence was rather different from that prevalent in the north. Italian compositions are treated briefly in a separate section below (v).

I have used the term 'sequence' to refer to the genre in general, with or without text. If I have wanted to be specific, I have referred to 'sequence melodies' and 'sequence texts'. Other writers (see Husmann 1954, 'Sequenz und Prosa') have preferred to reserve the word 'sequence' for the melody alone, calling the text 'prose'. Although this corresponds to the practice of some medieval manuscripts, it was not a universal custom. (I know of no witness earlier than the seventeenth century to the use of the word *sequela* for a sequence melody: see Gautier 1886, 14.)

(ii) *Early Sequences with Parallel-Verse Structure*

Some characteristics typical of many larger sequences of the early period are exemplified in Ex. II.22.1. This sequence is regular in structure, in that its verses are all paired. As in several other sequences, one of the verses has within itself an AAB structure: verse 6, *Virginum O regina*. Again like several other sequences, there is a distinctive change of register part way through the piece: verse 1 moves between *E* and *c*, verse 2 pushes a little higher to *d*, consolidated in verse 3. Then in verse 4 the cadence is made on *d* itself, which provides a base for another move, up to high *g* in verse 6. Verse 7 returns to the register of verse 5 (*a–f*), before the final string of verses in the highest register.

It is worth taking note of the neumatic notation of the melody. From the text alone it is clear that in the paired versicles two-syllable words are very often matched with two-syllable words in the parallel versicle, three-syllable words with three-syllable, and so on. The note-groupings of the melody when it is notated without its text (this is the usual practice in early sequence collections, but it died out in the twelfth century) also reflect the syllable groups. (On this oft-discussed phenomenon see Reichert 1949 and Schlager 1980.) Ex. II.22.2 shows verses 2 and 3 with each compound neume and its corresponding text.

Claris vocibus is the most widespread 'western' text for the melody of Ex. II.22.1; Ex. II.22.3 gives the start of *Eia recolamus*, the best-known 'eastern' one. Although *Eia recolamus* has no verse corresponding to verse 9 of *Claris vocibus*, the two are remarkably close in matters of note- and syllable-groupings, as can be seen from verses 2 and 3 once again. *Eia recolamus* has an 'extra' first verse (labelled 'A'), which texts the corresponding 'Alleluia' preceding *Claris vocibus*. This is a characteristic difference between western and eastern practice. The word 'alleluia' contains two instances of what is known as liquescence, where a special notational sign indicates a special manner of performance of the first *l* and the *i*. Exactly similar provision is made in the new text, for *i* in 'Eia' and *gn* in 'digna'.

While the correspondence between the western *Claris vocibus* and the eastern *Eia recolamus* in note- and syllable-groupings is close, this is not true for all the melodies known in both parts of northern Europe. Exx. II.22.4–5 give the western *Organicis canamus* and the eastern *Sancti baptiste* (the text is attributed to Notker Balbulus of St Gall), for the melody labelled 'Justus ut palma' in some early manuscripts. In order

Ex. II.22.1. Sequence *Claris vocibus* (Oxford, Bodl. Lib. Bodley 775, fos. 122^v, 141^r; London, Brit. Lib. Egerton 3759, fo. 73^v)

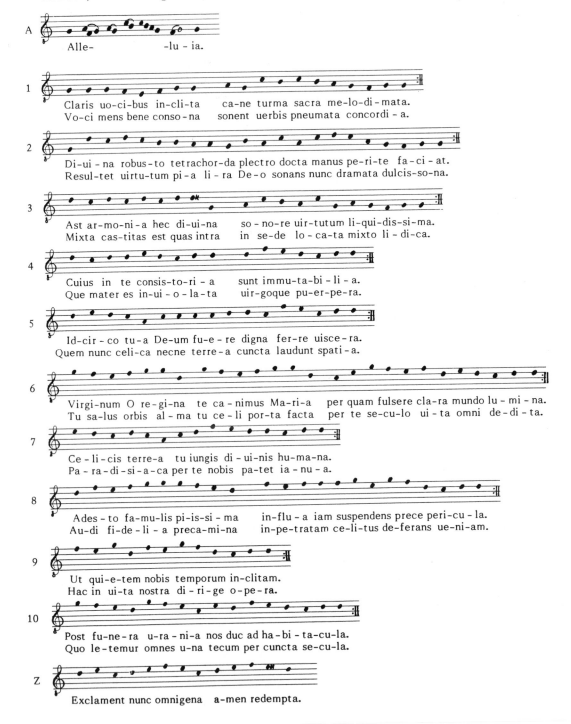

A　Alle- -lu - ia.

1　Claris uo-ci-bus in-cli-ta　ca-ne turma sacra me-lo-di-mata.
　　Vo-ci mens bene conso-na　sonent uerbis pneumata concordi - a.

2　Di-ui - na robus-to tetrachor-da plectro docta manus pe-ri-te fa-ci - at.
　　Resul-tet uirtu-tum pi-a li - ra De-o sonans nunc dramata dulcis-so-na.

3　Ast ar-mo-ni-a hec di-ui-na　so - no-re uir-tutum li-qui-dis-si-ma.
　　Mixta cas-titas est quas intra　in se-de lo-ca-ta mixto li - di-ca.

4　Cuius in te consis-to-ri - a　sunt immu-ta-bi - li - a.
　　Que mater es in-ui - o-la-ta　uir-goque pu-er-pe-ra.

5　Id-cir - co tu-a De-um fu-e - re digna fer-re uisce-ra.
　　Quem nunc celi-ca necne terre-a cuncta laudunt spati-a.

6　Virgi-num O re-gi-na te ca-nimus Ma-ri-a　per quam fulsere cla-ra mundo lu - mi - na.
　　Tu sa-lus orbis al - ma tu ce-li por-ta facta　per te se-cu-lo ui-ta omni de-di - ta.

7　Ce - li-cis terre-a tu iungis di - ui-nis hu-ma-na.
　　Pa - ra-di-si-a-ca per te nobis pa-tet ia - nu - a.

8　Ades - to fa-mu-lis pi-is-si - ma　in-flu - a iam suspendens prece peri-cu - la.
　　Au-di fi-de-li - a preca-mi-na　in-pe-tratam ce-li-tus de-ferans ue-ni-am.

9　Ut qui-e-tem nobis temporum in-clitam.
　　Hac in ui-ta nostra di - ri-ge o-pe-ra.

10　Post fu-ne-ra u-ra - ni-a nos duc ad ha-bi - ta-cu-la.
　　Quo le-temur omnes u-na tecum per cuncta se-cu-la.

Z　Exclament nunc omnigena a-men redempta.

Ex. II.22.2. From sequence *Claris vocibus* (Oxford, Bodl. Lib. Bodley 775, fos. 122ᵛ, 141ʳ)

2 Diuina   robusto   tetrachorda   plectro   docta   manus perite   faciat.
  Resultet uirtutum pia lira   Deo   sonans   nunc dramata dulcissona.

3 Ast armonia   hec diuina   sonore   uirtutum   liquidissima.
  Mixta castitas est quas intra in sede locata   mixto lidica.

Ex. II.22.3. Start of sequence *Eia recolamus* (St Gall, Stiftsbibl. 381, p. 336; Rome, Bibl. Cas. 1741, fo. 83ʳ)

A E - ia re-co - la-mus laudibus pi - is digna.

1 Hu-ius di - e - i car-mi-na   in qua nobis lux o - ritur gratis - si-ma.
   Noctis in-ter ne-bu-lo-sa   pe-re-unt nostri criminis umbra-cu-la.

2 Ho-di - e se-cu - lo maris stella est e - ni-xa no-ue sa-lu-tis gau-di-a.
   Quem tremunt baratra mors cruenta pa-uet ip-sa a quo peri-bit mortu - a.

3 Gemit cap-ta pestis an-ti-qua   co-lu-ber li - uidus perdit spo-li - a.
   Ho-mo lapsus o-uis abduc-ta   re-uo-ca-tur ad e-ter-na gau-di-a.

to facilitate discussion of the correspondences between the melodies I have numbered the lines in a special way. The eastern version, *Sancti baptiste* (Ex. II.22.5), is the more regularly parallel in structure, though there is a slight irregularity in verse 4, and 3X is like a modified repeat of one versicle only out of verse 3. The western version, *Organicis canamus* (Ex. II.22.4), has nothing to correspond with verse 1X of *Sancti baptiste*—or should one say that verse 1 of *Organicis* corresponds to verse 1X, not verse 1, of *Sancti baptiste*? Verse 4 of *Organicis* is complex. Its melody may be represented by the formula AAB, AAB, AAC, CD, where D is actually the same as the end of verse 3, transposed up a fourth. Compared to this one might say that *Sancti baptiste* ignores the first A, except for three notes in ascending order, which reappear at the end of the line. Stretching a point, one might thus represent the verse as A′ABA′, ABA′. Finally, in *Organicis* verse 5 is but a single versicle, and runs straight into verse Z.

Some of the differences within corresponding verses are also of interest. The rise into a higher register so obvious in *Organicis* does not happen in *Sancti baptiste*. Both

Ex. II.22.4. Sequence *Organicis canamus modulis* (Orléans, Bibl. Mun. 129, fo. 166ʳ)

1   Or - ga-ni-cis cana-mus modu-lis nunc Iohannes sollempni - a.
    Om-ni - ge-nis domi-no uo-cibus reddentes odas de-bi -tas.

2   Qui in sanctis su-is mi-ra - bi-lis ni-mis ... ...
    Nam et in ip-sis quasi quibusdam mu-...-si-cis

    multi-pli - ci uir-tutum flore e-os-dem deco-rat   ac mi - ri-fi-ce ad-ornat.
    instrumentis di-gi - to propri - o fi-des a - gi-tat   fi-des uirtu-tum so-norat.

3   Has numero-se percurrens singu-la
    Quam generat uir-tu-tem mater illa

    permiscens singulis di - a - tessa-ron   ... melli-flu-am me-lo-di-am.
    que a - li-is decenter compo-si - ta   reddet su-auem simpho-ni-am.

4   Qua si-ne cuncta...   fi-unt dis-so-na   necnon et fri-uo-la.
    Qua cum om-ni - a   fi-unt conso-na   necnon u-ti-li - a.

(4) Qua ius-ti be-ne   mora-ti ri - te   pe-tentes excel-sa ... po-li si-de - ra. .....
                                          A-lacres decantant no-ua canti-ca in cytha - ra

                                          Tre-i - ci - a.

5   Quorum a-gentes festa   consor-ci-um me-re - amur in ce-lesti pa - tri-a.

Z   A-   -men.

Ex. II.22.5. Sequence *Sancti baptiste* (St Gall, Stiftsbibl. 546, fo. 124ʳ)

A

Sancti bap-tis-te Christi preco-nis.

1

Solem-ni – a ce-le-brantes mori-bus ipsum se-quamur.
Ut ad ui-am quam predixit asseclas su – os per-ducat.

2

De-uo-ti te sanctissi-me ho-minum
Ap-parensque Zacha-ri-e Ga-bri-el

a – mice Ihesu Christi fla – gita-mus ut gaudi – a per-ci-pi – a-mus.
re-pro-misit qui tu-am ce – lebrarent ob-sequi – is na – ti-ui-ta-tem.

1X

Ut per hec festa e-terna gau-di – a ad – i-pis – camur.
Qua san-cti de-i sacris de-li – ci-is le – ticongaudent.

3

Tu qui prepa-ras fi-de-li – um corda
Te depos-cimus ut cri-mi-na nostra

Ne quid deui-um uel lu-bri-cus Deus in e-is in-ue-ni – at.
Et fa-ci – nora con-ti-nu-a pre-ce stude-as ab-sol-ue-re.

4

Pla-catus ut ip-se su-os semper in – ui-se-re fi-de-les.
Et mansi-o-nes in e-is fa-ce-re dig-netur.

3X

Et ag-ni uel-le-re quem tuo di – gi-to.
Mundi monstraueras tol-le-re cri-mi-na nos uelit in-du – e – re.

5

Ut ipsum mere – a-mur an-gelis as-so-ci – i.
In al – ba ueste se-qui per portam clarissi-mam.

Z

A-mi-ce Christi Io-hannes.

versions get as far as high *d*, but only *Organicis* presses on to *g*, hence the difference of a fourth at the final cadence. Almost equally noticeable is the way in which verses 1 and 2 in *Organicis* have a three-note cadence from the sub-final, *CDD*, which is absent in *Sancti baptiste*. The cadence is very common in sequences, somewhat more so in the west than the east, and is reputed to be of 'Gallican' origin. The text of *Organicis* has constant line-endings in 'a', which is usual in the western repertory, whereas *Sancti baptiste* does not, which is usual for Notker and some other eastern writers.

Just how the differences between the two versions arose is open to debate. There is close agreement about how verses such as 3 and 4 should proceed. Is the 'Gallican' cadence original or a later stylization? Many copies of *Organicis* (for example, the one presented by Crocker 1977, 288) have a double-versicle verse 5. Is that a later regularization of the melody? Crocker (1967, 'Sequences', and especially 1977) argues forcefully that regular parallelism was often imposed upon sequences at a later stage in their formation. Quite often it is the eastern versions which are less regularly parallel, and several of Crocker's presentations show how verses, single versicles, or segments within versicles, may have been added in the west, whereas the eastern versions preserve an earlier irregular state of the melody. (See especially Crocker 1977, chs. 3–4.)

The great variety of form displayed in the early sequence repertory makes the choice of what to mention here somewhat arbitrary. The obviously non-Gregorian character of the melodies has tempted some writers to suspect the influence of secular music, for example in the sequence with text *Plangant cigni* or *Clangant cigni* (and various later texts), entitled 'Planctus cigni' in some sources (see Stäblein 1962, 'Die Schwanenklage', also Stäblein 1975, 114; the melody is also in Anselm Hughes 1934, 63). This is one of many sequences which have a non-religious title; others are named after musical instruments (see Holschneider 1975). Sacred titles are usually alleluia incipits.

The repetition of segments of verses is very common in sequences. In some melodies, whole verses are repeated (e.g., *Laudes deo*, Crocker 1977, ch. 3; also the melody 'Hieronima' or 'Frigdola', Crocker 1977, ch. 6; and the melody of *Laudum carmina*, a sequence for St Benedict, given in Hughes 1934, 37). Special interest has focused on a few melodies where a group of verses is repeated as a block. An example of this is the well-known melody called variously 'Chorus', 'Concordia', etc., with texts such as the eastern *Hanc concordi famulatu* (attributed to Notker of St Gall) and the western *Epiphaniam domino* (the melody is discussed in Stäblein 1964 and Crocker 1977, ch. 5; see also Handschin 1954, 154). At its most repetitive (not all versions of the melody display this scheme), in *Gaude eia unica*, it has the form X, ABC, ABCB, A, Y (where X and Y are the non-repeated verses often found at the start and end of early sequences). A shorter example is the less well known *Pura deum* (Ex.II.22.6) (also known with texts *De sancto Iohanne* and *Pangat simul eia*, all restricted to the area of north France and England; Stäblein 1978 thought *Pange simul* was the original, perhaps composed by Hucbald). The most obvious feature of the melody is the large-scale repeat involving all the central section of the melody. Added to the fact that the last half of verse 1 reappears in verse 4, this means that

Ex. II.22.6. Sequence *Pura Deum* (Angers, Bibl. Mun. 97, fo. 104ʳ)

A

Al-le-lu- -ia. Pu-ra De-um laudet in-no-cen-ti-a.

1

Innocens Christus su-a quam sacrat infan-ti-a.
Par-uu-lorum ex morte martyrum pre-ti-o-sa.

2

Quos se-ui-ti-a ... ...He-ro-dis da-re ne-ci ius-serat.
Cum ex-tingue-re ni-ti-tur il-lum qui uitam da-re ue-nerat.

3 S

Natum namque regem audi-e-rat.
Statim mi-li-tum pergunt agmina.

T

Quem succe-de-re si-bi posse in-a-ni pa-uore for-mi-dabat
Ac tyran-ni-ca complent iussa mili-a mactantes in-nocentum

V

qui-a non ce-li regem no-ue-rat.
inter dul-ci-a matrum u-be-ra.

W

Mox in-ui-di-a ac-census et i-ra fer-ui-da infantes a bi-matu et infra.
O im-pi-e-tas O se-ua ni-mi-um fe-ri-tas O nunquam audi-ta cru-de-li-tas.

X

Oc-ci-di iubet in totam Da-uidis Bethleem pari-ter ex e-ius per cuncta confi-ni-a.
Inson-tes culpa non est que noxias fa-ci-at a-li-qua et mortis subeunt dis-crimi-na.

4

O Christi miran-da semper et in is-tis gra-ti-a.
Pro se passis e-ter-na quibus est largitus premi-a.

Z

Ip-si laus sit nunc et in se-cu-la. A-men.

nearly all the music is recalled at one point or another. What is more, the music of the recapitulated middle section contains internal repetitions, and also relates to the outside verses: 3S is the same as 3V and these start like the other phrases in verse 3; 3W and 3X are the same; the cadence of 3W and 3X is that of verse 2.

Pura deum, or rather its other text *Pange simul*, was hailed by Winterfeld (1901) as an example of the so-called 'sequence with double cursus'. Whether it should be so linked with the other members of this group is open to question, in view of the fact that, unlike the others, it seems to have been at home in the liturgy from the start and contains many melodic resemblances to other liturgical sequences. The double-cursus genre is discussed below (II.24.ii).

Several things, then, play a part in making a sequence recognizable by its musical style. The parallel-verse structure, and the text-setting on the principle of one syllable per note, are the most important features, and then perhaps recurrent figures at cadences and a few other points. Repeated notes are rather rare, and there is thus a constant impression of movement, occasionally locked in circles of repeated motifs but more often pushing purposefully towards a clear melodic goal: the frequent surges up a fifth into a new range are the most obvious manifestations of this. The melodic goals referred to are usually limited to the final, with occasionally a higher final which supersedes the first, and sometimes also the lower fourth or fifth. This tonal single-mindedness creates an impression of rather insistent enthusiasm; the melodies rarely sound reflective for long, especially not when the texts are sung, for the declamation of the words in syllabic fashion tends to emphasize individual notes at the expense of the melodic phrase as a whole.

(iii) *Notation; Performance; Partially Texted Melodies*

Sequence melodies were recorded in a number of different ways. In some early sources the melodies alone are copied, without texts, for example in Chartres, Bibliothèque Municipale 47 (PalMus 11) and St Gall, Stiftsbibliothek 484 (Crocker 1977, pl. 5; Stäblein, 'Sequenz', *MGG*, pl. 2). Another of the earliest major sources, by contrast, Paris, Bibliothèque Nationale, lat. 1240, has what became the commonest method: each syllable of text carries its own notational sign. In addition to the latter, most early West Frankish sources have a collection of melodies alone (see Crocker 1977, pls. I–II and Crocker, 'Sequence', *NG*, 142; also Frere 1894, *Winchester*; Holschneider 1978; Stäblein 1975, 115, 117).

East Frankish manuscripts usually place melody and text side by side in parallel columns, whereby each phrase of the melody (and the corresponding part of the text) is given a new line. The internal construction of the sequence is made particularly clear. (See Stäblein, 'Sequenz', *MGG*, pl. 1 and Stäblein 1975, 185, but especially Haug 1987).

A few sources, such as Rome, Biblioteca Angelica 123 (PalMus 18), have alternating verses of text and melody alone. A few north French sources (such as Laon, Bibliothèque Municipale 263 and Cambrai, Bibliothèque Municipale 78) give the complete melody immediately after the complete text.

Whether the notation of melody and text separately in these ways reflects some sort of alternatim performance (text—melisma—text—melisma, and so on) has been much debated (see e.g. Husmann, 1954, 'Sequenz und Prosa' and Smits van Waesberghe 1957 'Over het onstaan'). A systematic survey of rubrics in liturgical manuscripts might help clarify the matter, though such rubrics are rare before the thirteenth century. (The situation is thus less satisfactory than for the prosula, discussed by Kelly 1985, 'Melisma and Prosula'.)

The problem of the double notation, with and without text, is particularly critical in the case of a small number of sequences where in the midst of the melismatic version a few texted phrases are to be found. These phrases also appear embedded in the complete text for the whole sequence. If another text is provided for the sequence melody, then the same phrases are used again: they go with the melody. Their origin, function, and manner of performance are all puzzling and have inspired a good deal of speculation (see Blume in AH 49; von den Steinen 1946–7, 205–20; Husmann 1954, 'Sequenz und Prosa', 77–91; Stäblein 1961, 'Frühgeschichte', 8–33). The use of partial texts seems restricted to France (and north Spain) and England; the fashion for such pieces persisted longer in Aquitaine than elsewhere. (For an unusual Spanish example see Husmann, 1961, 'Ecce puerpera'.) It is not at all clear how an alternatim system of performance, or indeed any other system, would accommodate these special verses. (See, for example, Husmann 1954, 'Sequenz und Prosa', 86; on a peculiar method of notating the sequence *Terribilis rex alme* with partial text *Gloria victoria*, etc. in Beneventan sources, see Stäblein 1961, 'Frühgeschichte', 21.)

(iv) *Short Aparallel Sequences*

All early sequence collections are dominated by compositions where double versicles predominate, and where the number of verses is at least five and often ten or more: this means that the number of notes, even ignoring all repeats, often surpasses 200, and in the largest sequences, such as *Fulgens preclara* (Anselm Hughes 1934, no. 23), may pass 300. But most early collections also include a smaller number of quite different melodies lacking parallel-versicle structure and usually no more than about seventy notes long. Without the clear structural features of the larger sequence, they sound simply like an alleluia jubilus, and indeed, although some of them have a small amount of internal phrase repetition, this does not go beyond what one would expect of an alleluia jubilus.

Among the 150 or so sequence melodies known up to *c*.1000, only twenty fall into the short aparallel category; and only one was widely known (that is in Winchester, Aquitaine, and eastern sources): that known as 'Excita domine', with text *Qui regis sceptra*, in the west, and as 'Laudate Deum [omnes angeli]', with text *Angelorum ordo sacer*, in the east (the titles refer to the alleluias which begin like the sequence; both have the same melody). Most have been surveyed by Kohrs (1978).

Ex. II.22.7 shows three short aparallel sequences which take as their starting point *Alleluia Ostende*. The first sequence was widely known in the early period, and was usually sung with the text *Precamur nostras*. The other two are known only from Paris, Bibliothèque Nationale, lat. 1084 (Aquitaine, late tenth century), from which manuscript all three are edited, with the titles given there. Ex. II.22.7 is ordered as follows:

Ex. II.22.7. *Alleluia Ostende* and associated sequences (Paris, Bibl. Nat. lat. 776 (a–b) and 1084 (c–f))

(*a*) Alleluia, with jubilus

(*b*) Verse *Ostende nobis*

(*c*) Sequence (first)

(*d*) Text *Precamur nostras* for (*c*)

(*e*) Sequence (second)

(*f*) Sequence (third)

The pitch of the neumes of Paris 1084 is not always certain (cf. Kohrs 1978, 150). But the essential fact is clear enough, that the only melodic correspondence between alleluia and sequences lies in their opening. There are no internal repetitions in the first sequence (*c*), but sequence (*e*) begins with a repeated phrase, and sequence (*f*) has two repeated phrases. Both (*e*) and (*f*) end unexpectedly on an *a*, and it might be hypothesized that the melody should be completed by singing the jubilus of the alleluia. The effect of the text *Precamur nostras* is similar to that of an alleluia prosula.

(v) *Italian Sequences*

Brunner and Jonsson have recently emphasized that the regularity so conspicuous in (and occasionally imposed upon) sequences in northern lands was not as important a criterion for Italian composers or redactors. Several sequences of Italian origin display considerable irregularity of structure, and Italian versions of widely known sequences are often less regular than northern versions. Stäblein (1964, 379) was tempted to speak of a 'degeneration of the sequence-principle [i.e. parallel-verse structure] in the direction of the litany or psalmody'. In Italian sequences one finds frequent singles, as opposed to the regular doubles of most northern pieces, or even a disregard for paired versicles. The strictly syllabic word-setting of the northern compositions seems also to be less important; two notes are often coupled for a single syllable, and notes are sometimes repeated to accommodate extra syllables.

Even more striking is the appearance of a small number of sequences which display traces of what might be called a variation technique. Here the same melody is used, with variations, for most verses of the composition. Levy (1971) was the first to draw attention to this, in an analysis of the sequences *Lux de luce* and *Hodie dominus Iesus Christus*. A third sequence of this type, *Alma fulgens*, has been cited by Brunner (in an unpublished paper, Milan, 1984; for a bibliography of sequences in Italian sources see Brunner 1985).

Many northern compositions were sung in Italy, and many sequences in the more usual manner were composed in Italy. The examples mentioned here are rare. Yet the Italian type is historically significant. It is related to the Italian practice of composing Kyrie verses with Latin text, where the same melody is used for all verses; and a number of troped Agnus Dei chants are also constructed in the same way (see below II.23.viii and xi).

Ex. II.22.8 shows *Sancte crucis celebremus*, which betrays its Italian origin by the

Ex. II.22.8. Sequence *Sancte crucis celebremus* (Modena, Bibl. Cap. O.I.7, fo. 129ʳ)

frequent two- and three-note groups assigned to some syllables. The constant deviations from exact parallelism are also typically Italian. This is not an example of the 'variation-versus', although there is an echo of such pieces in the way all verses except 2 have similar cadential phrases.

When northern composers composed a new text for an old melody, they usually adopted the note-count and note-grouping of the melody exactly. This was not always the case in Italy. When the melody of *Sancte crucis celebremus* was used in the Modena manuscript for another sequence, *Sanctum diem celebremus*, phrases were stretched or contracted freely and new material was added.

The early history of the sequence in Italy remains obscure, for practically no Italian sources survive from before *c*.1000. The inevitable impression is that Italy was less productive than other lands. Nevertheless, some churches were certainly active in sequence composition (see the information on south Italian sources in Brunner 1981), and they were sung even in the Old Roman repertory by the eleventh century (see Lütolf 1987 and MMMA 2). It is at least possible that later Italian sources preserve compositions dating back as early as any in the north. On the other hand, the stylistic discrepancies between northern and Italian manners of sequence composition may suggest that Italian musicians had to assimilate the sequence as a foreign genre, and occasionally fell back on native habits of composition which assorted uneasily with the 'true' sequence.

(vi) *The Early History of the Sequence*

The earliest mention of the sequence by name comes in the late eighth-century manuscript Brussels, Bibliothèque Royale 10127–10144, the 'Blandiniensis', a portable booklet containing among other things the texts of chants to be sung at mass, probably written in north France (ed. Hesbert in *AMS*). Here six alleluias in a list of twenty-five for the post-Pentecost Sundays are rubricated 'CUM SEQUENTIA'. All six are found in later manuscripts either as alleluias with extensive melismas at the end of the verse (melismas different from the alleluia jubilus), and/or as short aparallel sequences. (See Stäblein 1961, 'Frühgeschichte', 4–7; Kohrs 1978, 78 ff.; Crocker 1977, 393 ff.) The six alleluias are all for 'ordinary' Sundays of the summer months, when no special commemoration or feast was involved. Sequences for these days are generally rare in later sources. Were longer sequences already being sung on the more important feast-days of the year? The Mont-Blandin manuscript makes no mention of them.

About 830, Amalarius of Metz, describing the alleluia of mass, refers to 'Haec jubilatio quam cantores sequentiam vocant' (ed. Hanssens, iii. 304; PL 105, 1123; Crocker 1977, 392). These melodies might be either short or long ones. *Sequentiae* are also mentioned in almost the same terms in the late ninth-century Ordo Romanus II (ed. Andrieu, ii 215) and by pseudo-Alcuin (PL 101, 1245).

Manuscript copies of sequences (melodies alone, or texts alone, or texts with notation) finally appear late in the ninth century, that is, at the same time as notated copies of most other chants, so that from this circumstance alone one cannot say that the sequences are a late development of the chant repertory. (See Stäblein, 'Sequenz', *MGG*, ex. 1; Stäblein 1961. 'Frühgeschichte', 7, with facs. facing 16; the melodies in Chartres 47, PalMus 11; and von den Steinen 1946–7, 252–63).

Again during the late ninth century the monk Notker Balbulus of St Gall (*c*.840–912) wrote a number of texts for sequences. (Documentary evidence exists for a few, but the rest are attributed only on stylistic grounds, the canon established by von den Steinen, 1948, having become generally accepted. See also Stäblein 1962–3; Crocker 1977 and Crocker, 'Sequence', *NG*.) The collection is usually reckoned to have been completed around 880, for it was dedicated to Bishop Liutward of Vercelli, counsellor and arch-chancellor of the emperor Charles the Fat, in 884.

Several manuscripts preface the collection with a remarkable prooemium, apparently written by Notker to explain and dedicate his work to Liutward (critical edition and German translation in von den Steinen 1948, English translations Crocker 1977, 1 and 'Sequence', *NG*; see also the discussion in Husmann 1954, 'St Galler'). The gist of it is that Notker as a boy had difficulty remembering 'melodiae longissimae'—which in the context must mean sequence melodies—and wondered what could be done to make them easier to learn. A monk, fleeing from the sack of Jumièges in what is now Normandy, came to St Gall with a chant-book in which Notker saw that texts had been composed to fit the sequence melodies ('versus ad sequentias erant modulati'). Notker thought he could compose better texts. He showed his first attempts to his master Iso for correction (Iso seems to have been familiar with the technique of texting melismas), then went on to produce a whole collection. His master Marcellus had them copied out individually on rolls to be shared out among the choirboys for learning.

In support of Notker's story, it has been demonstrated that texts found in the west Frankish repertory lie behind some of Notker's compositions (Crocker 1977). Sometimes he reworked ideas in previous texts, sometimes he developed his own original ideas.

If von den Steinen's list of Notker's texts is accurate, there are forty texts to thirty-three melodies; eight texts are for eight short, aparallel melodies; none contain special verses (partially texted melodies). This compares with twenty-nine melodies in Chartres 47 (ten short aparallel, no special verses), and with thirty texts for twenty-nine melodies in Paris, Bibliothèque Nationale, lat. 1240 (no short aparallel sequences, four with special phrases; listed in von den Steinen 1946–7, 126 and Crocker 1958, 'Repertory', 154–5). Thus there is no uniformity between these crucial early witnesses as to the type of sequence they favour.

Nor is there any uniformity in the assignment of sequence melodies to particular feast-days (see, for example, Table IV in Crocker 1977, 404–5). The impression is that it was often the text which made a sequence proper to a particular feast (several texts are nevertheless neutral in this respect, being generally laudatory, celebratory, or christological without implying a connection with any particular feast). The melody, by contrast, could be employed for any feast as the author of the text thought fit.

In addressing the vexed question of the connection between alleluia and sequence, it is often forgotten that the alleluia melodies themselves can have had little liturgical fixity before the ninth century. Alleluia and sequence may well have followed parallel paths, both starting from a small repertory, perhaps only a dozen melodies for each genre, the verses of the alleluia attaching it to specific occasions, texts doing the same for the sequence. The composition of different melodies, verses, and texts did not proceed uniformly across Europe. By the time Notker's hymn-book was completed (*c*.880) and the early collections of west Frankish texts were made (*c*.930 for Paris 1240), considerable divergence existed between the liturgical assignments of sequence melodies in different areas of Europe. Whatever connections may have existed between particular sequences and alleluias, these gradually dissolved, although they are still detectable in a few cases in the late ninth- to tenth-century repertory that has survived.

The state of the Old Roman alleluia repertory (earliest musical codification only in the eleventh century, however) and the Milanese repertory (musical record only from the twelfth century, however) lends some support to this hypothesis. The Roman repertory had only three widely used melodies, with some twelve others used less often, eight of them once only. At Milan, the same melody was sung on practically all major feasts, supplemented by extra *melodiae* for festal occasions (see VIII.4). In comparison with these, the chief distinguishing Frankish-Gregorian developments would be the rapid multiplication of alleluia and sequence melodies in the ninth century and the provision of texts for the sequences. (On supposed melodic relationships between alleluia and sequence, see Husmann 1955, 1956 'Alleluia, Vers und Sequenz', 1956, 'Mater-Gruppe', 1956, 'Iustus ut palma'.)

Crocker has argued on several occasions against this train of thought (see, for example, 1977, 400). For him, the sequence is essentially a new creation of the ninth century, arising out of the musical impulses and ambitions of Frankish musicians and only gradually assimilated to the alleluia of mass. There is thus no need to establish connections with hypothetical alleluia repertories, or to explain away conflicting liturgical assignments of melodies. Furthermore, while allowing for the possibility of contrafact texts, it brings the composition of melody and text into same period, in fact makes them part of the same compositional process. Thus Crocker believes only the short aparallel melodies were used in the mass before the mid-ninth century. The larger sequences belong to a radically different creative effort.

An indication of how things might have proceeded is suggested by the case of the melody 'Mater', already discussed by Husmann but revisited by Bower (1982). Bower was able to identify an alleluia melody (Schlager catalogue 274) which shares significant amounts of musical material with the sequence, but which cannot be shown to be any older than the sequence. Sources for both date only from the later ninth century.

It is nevertheless possible to imagine a compromise explanation, one which sees the alleluia as the genre from which the sequence developed but does not deny the originality of much of the new repertory. The rapid and concurrent expansion of both alleluia and sequence repertories in the ninth century which I have envisaged, into the dozens of melodies known by *c*.900, and the hundreds known by *c*.1000, allows both for a connection with the alleluia and energetic and inspired new composition. (None of these developments warrants the description of the sequence as a 'trope' of the alleluia.)

Crocker has argued that most sequences began life as the original creations of a poet/musician composing text and music together. Differences between one version of a melody and another would therefore be the result of deliberate revision. Yet the differences between eastern and western versions of many 'internationally' known melodies often suggests the end result of oral transmission of the melody alone (see

Exx. II.22.4–5 above). Here perhaps are some of the few, old melodies which circulated in different forms in different areas, before being texted. To provide a text would usually be to fix the shape of a melody in one of its different forms. Further texts composed in a particular area would then normally copy the familiar arrangement of verses and syllable-count in each verse.

Since sequence melodies often sound very un-Gregorian, explanations of their origin have often involved non-Gregorian music. Stäblein believed that some at least incorporated a strong secular element, particularly E-mode melodies (most are in G or D) like 'Planctus cigni'. Parallels have naturally been drawn between the sequences and secular forms employing the double-versicle scheme: the lai and the estampie (see Handschin 1929; Spanke 1938; and Stäblein 1962, 'Schwanenklage'), but lack of contemporary musical evidence about the secular forms remains a barrier. Byzantine influence is not now thought to have been a motivating factor, for although Byzantine pieces with occasional textual parallelism exist, their music is not syllabic but highly ornate. Stäblein ('Sequenz', *MGG*, 531) also pointed to examples of Irish sacred texts with textual parallelism, but no music for them survives.

We do not know what alleluias in the Roman chant repertory sounded like in the eighth century. If some already had new music (called *melodiae*) for the repeat of the alleluia after the verse, as they do in the earliest written sources of the eleventh century, this might have stimulated the Franks to compose similar melismas.

The Milanese alleluias constitute a comparable case, for by the twelfth century, when they were written down for the first time, *melodiae* of astonishing length were being sung, displaying intricate repeat structures (transcription and structural analysis by Bailey 1983, *Alleluias*). For some alleluias there are two sets of *melodiae*, that is, music for two extended repeats after the verse, and some even have *melodiae tertiae*. Here we cannot say if outside influences were at work in Milan, or exactly how old the *melodiae* are. But their history might well run parallel to the Frankish sequence.

The alleluias (*laudes*) of the Mozarabic rite also included lengthy melismas with repeat structures, sung with the repeat of the alleluia after the verse (reproduced and discussed by Brou 1951, 'Alléluia'). Their age is uncertain, for the earliest surviving sources date probably from the eleventh century.

Was anything sung in the Gallican liturgy (used in Gaul before the Frankish imposition of Roman use) which could have been taken up in the sequence? Since no Gallican liturgical book with music survives, the question cannot ultimately be resolved. The chief source of information about the liturgy is the so-called *Expositio antiquae liturgiae gallicanae* (ed. Ratcliff 1971), formerly attributed to Germanus of Paris, now thought to have been written in Burgundy in the early eighth century. It describes a threefold alleluia (called *laudes*): 'habet ipsa Alleluia primam et secundam et tertiam', which immediately suggests the form of the Milanese alleluias. It is not clear, however, whether these Gallican 'alleluias' are long melodies or merely brief acclamations. Among the Milanese *melodiae* is the chant called *francigenae* (ed. Bailey, 127). This has been taken as an example of the Gallican ('Frankish') alleluia.

But Bailey shows (*a*) that it is related melodically to another Milanese alleluia, and (*b*) it may have been composed no earlier than *c*.1000. The possibility nevertheless remains that in all four churches, Mozarabic, Roman, Ambrosian and Gallican, the alleluia might already in the eighth century have been sung with extended melismas for the repeats of the alleluia after the verse.

None of these hypotheses about origins should distract our attention from the sequence as we have them in their ninth-century forms, remarkable monuments to the inventiveness and inspiration of their composers.

(vii) *Rhymed Sequences*

At some time in the eleventh century a modest little Easter sequence was composed in eastern France or western Germany which was to enjoy longer popularity than any of the pieces discussed so far. *Victimae paschali laudes* is usually attributed to Wipo (d. *c*.1050), priest and chaplain to the emperors Conrad II and Henry III (see Szöverffy 1964–5, i. 372–4). It is only four verses long (3*a* is omitted in modern books because of its anti-Jewish sentiment). Its popularity may be due to the homely yet dramatic touch of including a question to Mary (usually interpreted as Mary Magdalene) and her answer in verse 2. Verses 2 and 3 are remarkable because of their internal- and end-rhyme, which links not the two half-verses but the phrases within the half-verse. It is one of the earliest sequences to exploit rhyme (Ex. II.22.9).

Again presumably because of verse 2, it was taken up into the liturgical drama, and achieved a place within the cycle of Easter sequences in most uses of late medieval

Ex. II.22.9. Sequence *Victime paschali laudes* (Cambrai, Bibl. Mun. 61, fo. 174ʳ)

Europe. It proved popular enough to withstand the post-Tridentine pruning exercise in the sixteenth century and is one of only five sequences found in modern Vatican books.

Regular rhythm is not a feature of *Victimae paschali*, but can be found in some sequences composed shortly afterwards. *Congaudentes exultemus* (Ex. II.22.10) is found in manuscripts from about 1100 onwards and may have been composed in response to the bringing of the relics of St Nicholas of Myra to Bari in 1087. Most of the twelve verses are in regular accentual verse, but verses 4 and 5 are not. In verse 6 the lines are trochaic with 8+8+7 syllables each; the rhyme scheme for each double verse is AABCCB, which became the most popular scheme of all in subsequent centuries. Like many earlier sequences, *Congaudentes* moves up to a higher register at verse 7, cadencing thenceforth always on *a* instead of *D*. The 'Gallican' cadence is likewise another feature common to many of both the older and the newer sequences.

Hodierne lux diei (Ex. II.22.11) is one of the earliest completely regular sequences, probably written in the first half of the twelfth century. Since all verses are of the same length and metrical/rhythmic pattern, the effect is that of a ten-strophe hymn with five different melodies. It too moves gradually higher from its plagal opening verse, to *a* in verse 2, *d* in verse 3. Verse 4 relaxes a little before the climactic verse 5, two of whose phrases start on high *d*. The first phrase of verse 5 is the same as the second of verse 3, and the cadence of that phrase is already present at the end of the first phrase of verse 3. The cadence of verse 1, phrase 2 reappears in verse 5, phrase 2. Such melodic cross-references and echoes are not, of course, absent from the older sequence. But now that verses were composed of regular short phrases, the possibilities for exploiting them were greater. All lines end FEDCD. Another new feature, in comparison with the old sequence, is the frequent occurrence of two or more notes per syllable.

A large number of rhyming sequences have changes of metre. This might be seen both as the counterpart of the constantly changing line-lengths of the older sequence, and as a display of technical accomplishment. There are indeed several dazzling examples of metrical variety. They can be found in several of the oldest rhyming sequences, such as *Stola iocunditatis* (which still has one non-accentual verse), *Mane prima sabbati* (for Mary Magdalene), and *Laudes crucis attollamus* (for the finding of the True Cross, perhaps written for the reception of a fragment of the True Cross at Notre-Dame, Paris, in 1109: see Husmann, 'Notre-Dame-Epoche', *MGG*, 1706; also Weisbein 1947). The melody of *Laudes crucis* became the most popular of all those for rhymed sequences, being used by Thomas Aquinas for the Corpus Christi sequence *Lauda Sion salvatorem*, also found in modern Vatican books. A vast number of contrafacta are known. Most verses proceed in the 8 + 8 + 7 scheme, but verse 3 is as follows:

```
3a   8 + 8 + 7 +  4 + 3  + 3  + 4 + 3 +  3  + 7
3b   8 + 8 + 7 +    7       +     7        + 7
```

Verses 9 and 10 are $8 + 8 + 8 + 7$, verse 11 is $8 + 8 + 8 + 8 + 7$. Not all contrafacta follow this plan exactly.

Paranymphus salutat virginem (Ex. II.22.12) also has a $4 + 3 + 3$ line, right at the beginning. (The scheme was very popular in the rhymed office.) Thereafter hardly two verses have the same scheme. The habit of ending a verse with a four-syllable word is best known from the Christmas sequence *Letabundus*. The inclusion

Ex. II.22.10. Sequence *Congaudentes exultemus* (Madrid, Bibl. Nac. 19421, fol. 81ᵛ)

(Ex. II.22.10 *cont.*)

Cum clamarent nec incassum   ec - ce quidam dicens adsum   ad uestra pre - si - di - a.
Statim au - ra da - tur grata   et tempestas fit se - da - ta   qui - e - uerunt ma - ri - a.

9

Ex ip - si - us tumba ma-nat   un-cti - o - nis co-pi - a.
Que in - firmos omnes sa-nat   per e - ius suffra - gi - a.

10

Nos qui sumus in hoc mundo   ui - ti - o - rum in profundo   iam passi nau-fra-gi - a.
Glo - ri - o - se Nycho-la - e   ad sa - lu - tis portum trahe   u - bi pax et glo - ri - a.

11

Ipsam nobis un-cti - o - nem   im-petres a do - mi - no prece pi - a.
Qua sa-nauit le - si - onem   mul-torum pecca-minum in ma - ri - a.

12

Cu - ius festum ce - le-brantes gaudeant per se - cu - la.
Et co - ro-net e - ius Christus post uite curri - cu - la   al - le - lu - ia.

Ex. II.22.11. Sequence *Hodierne lux diei* (Assisi, Bibl. Com. 695, fo. 51ᵛ)

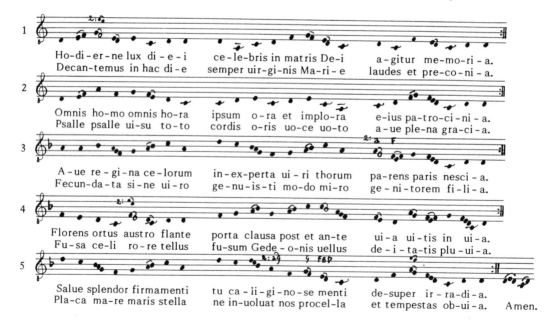

1

Ho - di - er - ne lux di - e - i   ce - le-bris in matris De - i   a - gitur me - mo - ri - a.
Decan - temus in hac di - e   semper uir - gi - nis Ma - ri - e   laudes et pre-co - ni - a.

2

Omnis ho - mo omnis ho - ra   ipsum o - ra et implo - ra   e - ius pa - tro - ci - ni - a.
Psalle psalle ui - su to - to   cordis o - ris uo-ce uo-to   a - ue ple - na gra - ci - a.

3

A - ue re - gi - na ce - lorum   in - ex-perta ui - ri thorum   pa - rens paris nesci - a.
Fecun - da - ta si - ne ui - ro   ge - nu - is - ti mo-do mi - ro   ge - ni - torem fi - li - a.

4

Florens ortus austro flante   porta clausa post et an-te   ui - a ui - tis in ui - a.
Fu - sa ce - li ro - re tellus   fu - sum Gede - o - nis uellus   de - i - ta-tis plu - ui - a.

5

Salue splendor firmamenti   tu ca - li - gi - no - se menti   de - super ir - ra - di - a.
Pla - ca ma - re maris stella   ne in - uoluat nos procel - la   et tempestas ob - ui - a.   Amen.

of the markedly shorter verse 2 may derive from the earlier unrhymed sequence. A gradual rise in pitch is again noticeable. The final verse is strongly reminiscent of *Hodierne lux diei*. Much work remains to be done in tracing such interrelationships across the repertory.

Ex. II.22.12. Sequence *Paranymphus salutat virginem* (Assisi, Bibl. Com. 695, fo. 75ʳ)

The advent of the rhymed sequence is contemporaneous with the composition of many other genres in rhymed, accentual verse: Latin Benedicamus songs and conductus (versus), for example, and religious dramas such as the 'Sponsus' play. Many different French centres will have played a part. One may speculate how influential were individual composers such as Peter Abelard (1072–1142: see Waddell 1986). The new type of sequence made particularly deep inroads into the liturgical cycle in Paris in the twelfth and thirteenth centuries: no other sequence repertories of the later Middle Ages have as high a porportion of rhymed sequences as the Parisian ones. A key part in the development seems to have been played by Adam of Saint-Victor. Two phases of activity have been distinguished by Husmann (1964) and Fassler (1984). A new repertory of rhymed sequences was created in Paris (probably on the basis of a few already existing pieces), perhaps as early as the first half of the twelfth century, though the earliest surviving sources are of the next century. Many of the texts contain echoes of the theological writings of Hugh and Richard of Saint-Victor, which argues for a contribution by a personage connected with Saint-Victor, the Parisian Abbey of Augustinian canons regular. That they were sung in all the churches of Paris must mean, however, that they were in the liturgy of the cathedral of Notre-Dame. The repertory then became known outside Paris and mingled with other local uses (see Husmann 1964, 191 ff.).

The liturgical assignment of these sequences is not always identical at Saint-Victor and Notre-Dame, and a number of texts are found only in Saint-Victor manuscripts, presumably composed there. Furthermore, there appears to have been a systematic revision of the melodies in use at Saint-Victor. That is, the ones we find in notated books of the abbey differ from those of Notre-Dame and the rest of the city. Sometimes they can be shown to rework the older melody (see Fassler 1984, 252–4). A large number of the ostensibly new melodies are based on material in *Laudes crucis attollamus*. This is the case not only in sequences which begin like *Laudes crucis* (see Apel 1958, 463), but in others where only internal phrases are reused (Fassler 1984, 258–60; incipits are therefore insufficient for checking interrelationships between repertories). Fassler believes that the resulting families of sequences were created in order to point up connections between ideas in the texts (Fassler, 1983). What makes this activity possible is of course the construction of the melodies in units of identical patterns, which, within limits, could be transferred from one text to another. (Misset and Aubry 1900 edit the earliest complete Saint-Victor source with music. Their melodic analyses do not distinguish between Victorine and Parisian melodies, but can be used with a list of specifically Victorine melodies to hand: see Husmann, 116, Fassler, 1984, 245).

As to the identity of Adam, Fassler connects him plausibly with the first period of sequence composition in Paris, and identifies him with the Adam who was precentor of Notre-Dame at least from 1107, who held a prebend at Saint-Victor from 1133. The problem remains of distinguishing sequences which originated in Paris (including Adam's) from others in similar style composed elsewhere. For another example of distinguished work in a similar vein by a contemporary of Adam one may point to the compositions of Nicolas of Clairvaux (see Benton 1962).

Rhymed sequences were composed in enormous numbers throughout the Middle

Ages. The vast majority are quite unknown. (Labhardt 1959–63 is a relatively isolated study.) Selecting typical examples is inevitably arbitrary, but one final sequence may demonstrate another of the multifarious formal schemes to be found. *Affluens deliciis* (Ex. II.22.13) appears to be a local composition. It has only three double-verses, but each is composed of an unusually large number of phrases, somewhat like

Ex. II.22.13. Sequence *Affluens deliciis* (Munich, Univ.-Bibl. 2° 156, fo. 194ʳ)

contemporary French lais (but without the first- and second-time endings of the lai; lais also have far more verses.) Melodically the piece betrays its origin by the constant use of the cadence figure *EFEDE*, which we have seen in an alleluia from this part of Europe (Ex. II.14.7) and which is also to be found in numerous ordinary-of-mass chants (see for example the group given by Stäblein, 'Agnus Dei', *MGG*, cols. 150 ff.), and other genres, as well as secular songs.

II.23. TROPES

Gautier, 1886; Paul Evans 1961; Crocker 1966; Paul Evans 1970, *Early*; Stäblein, 'Tropus', *MGG*; Steiner, 'Trope', *NG*.

   Facsimiles: Vecchi 1955; PalMus 15, 18
   Text editions: AH 47, 49; CT; Planchart 1977
   Music editions: Paul Evans 1970, *Early*; MMMA 3; Boe 1989 and 1990 (*Gloria*).

(i) *Introduction*

The great vitality and variety of trope composition in the Middle Ages creates considerable problems of definition and organization. I describe below three types of composition, all of which are essentially additions of one sort or another to pre-existing chant:

   (*a*) the addition of a musical phrase, a melisma, without additional text;
   (*b*) the addition of a text, a prosula, without additional music;
   (*c*) the addition of a new verse of chant, comprising both text and music.

(The Corpus Troporum research team in Stockholm has used the three terms (*a*) 'meloform', (*b*) 'melogene', and (*c*) 'logogene', respectively, to signify these three types of trope. See Marcusson in CT 2, 7 and Jonsson 1978, 102; also Huglo 1978, 7.)

Odelman (1975), in a survey of the rubrics in the sources themselves, showed how the term 'trope' (in its various Latin forms) and others such as *versus, laudes, prosa, prosula, verba*, were applied to various types of chant for the mass (not however, for office chants such as responsories). Medieval usage was not uniform.

Whatever the type of trope, they were practically all composed to embellish chants for the great feast-days of the church year. Many reasons for the medieval interest in these pieces have been discussed (see for example Gy 1983), but first and foremost would seem to have been the simple desire to make more splendid and solemn the performance of the liturgy (particularly mass) on the most important days of the year. Whatever the significance of tropes as didactic, theological, poetic material, and so forth, their prime effect on the liturgy is to make it longer (performance of a troped introit can last over ten minutes), to give richer opportunities for solo singers, often in alternation with the choir (most tropes are for soloists, and the genres troped are mostly choral chants), and generally heighten the rich complexity of the ceremonial (although we know rather little about the detail of the ritual performance of troped chants).

(ii) *Added Melismas in Introits*

Melodic extensions of introit phrases have been partially catalogued, edited, and discussed by Huglo (1978). They are restricted to two groups of manuscripts: on the one hand an Aquitanian group dating from the tenth to the twelfth century, and on the other a German-Swiss group of the tenth and eleventh centuries. (For modern editions of some of the Aquitanian ones, see MMMA 3, supplement, pp. 22, 27, and 35; Evans 1961, 127; Sevestre in CT 1, 287; Sevestre 1980, 34. See also Weakland 1958).

There seems to have been only one melisma or set of melismas for any one introit in the Aquitanian sources. Not only might the several phrases of the introit antiphon be extended thus, but so too might the psalm verse(s) and doxology (at the final 'Amen'), where they are usually rather modest in scope. Texts have not been found for these additional melismas. The case is quite otherwise, however, with the German–Swiss sources. Furthermore, for several introits these sources contain more than one set of melismas (none coincides with an Aquitanian one).

The provision of extra melismas, and texts for them, in St Gall, Stiftsbibliothek 381 for the Easter introit *Resurrexi* exemplifies the situation. St Gall has no less than ten different sets of melismas. The largest number of phrases in a set is ten: that is, in one set of melismas there are melodic extensions for 'Resurrexi', 'alleluia', 'me', 'alleluia', 'est', 'tua', 'alleluia', 'alleluia' (in the introit antiphon), 'meam' (end of the psalm verse) and 'Amen' (end of the doxology). While that particular set of melismas has no texts, several others have a text for every melisma in the set, where the music has been treated according to the well-known principle of one syllable per note, familiar from the sequence repertory and the standard method of procedure in prosulas. Practically all these melismas and their texts fell out of use before being recorded in pitch-notation. The set for the St John the Evangelist introit *In medio ecclesie* is a rare exception (Ex. II.23.1). Is it chance that the verse *milibus argenti* is a hexameter, that the right number of notes was available? or were text and music conceived together, as in the introductory verse *Dilectus iste domini*, which is in the form of an 'Ambrosian' hymn strophe? (For facsimiles of the German–Swiss melismas and texts see Gautier 1886; Weakland 1958; CT 1, pls. II–V; CT 3, pls. I–II; transcriptions in Handschin 1952, 166–7 and Stäblein, 'Tropus', *MGG*, Ex. 1 and Abb. 1 and 2; also Stäblein 1975, pl. 59 on 183.)

No other sources make so much of this type of introit trope as these early German–Swiss sources (the same is true of similar melismas for the Gloria, discussed next),

Ex. II.23.1. Trope verses *Dilectus iste Domini*, etc. (St Gall, Stiftsbibl. 484, p. 36; Pistoia, Bibl. Cap. C. 121, fo. 24ʳ) for introit *In medio ecclesie* (Einsiedeln, Stiftsbibl. 121, p. 39; Graz, Univ.-Bibl. 807, fo. 18ᵛ)

(Ex. II.23.1 cont.)

Dox.

GLORI-A PATRI ET FILIO ET SPIRITU-I SANCTO   SICUT ERAT IN PRINCIPIO

ET NUNC ET SEMPER   ET IN SECULA SECULORUM.   A-MEN.

which may mean that it originated at St Gall or a related centre. The textless melismas have been associated with the report of Ekkehard IV of St Gall (*Casus St Galli*, ch. 46: MGH, *Scriptores rerum Sangallensium*, 101) that Tuotilo (a monk at St Gall in the late ninth and early tenth century) composed trope melodies for the rotta (probably a type of psaltery).

Weiss noticed that in a very few cases additional trope verses were also extended by melismas; that is another way of 'troping a trope'. (See Weiss 1964, 'Tropierte'; also MMMA 3, no. 10, p. 13—the central melisma is not present in all sources—and the commentary to no. 153, p. 403.)

The principle of adding festal melodic extensions to traditional chants seems to recur throughout the history of plainchant. Weakland's and Huglo's belief that introit melismas were the oldest parts of the trope repertory cannot be substantiated, nor Stäblein's idea that the melody of Ex. II.23.1 above contains echoes of secular minstrelsy.

(iii) *Added Melismas in Glorias*

Other examples of such melismas may be found in the Gloria repertory, both in German–Swiss manuscripts and in French ones. In the German–Swiss sources, this type of Gloria trope predominates, indeed Rönnau (1967, *Tropen*, 197) stated that all the Gloria tropes he believed to have originated in St Gall were of this type; whereas for French ones added verses of text and music together are much more common. As in the case of the introit melismas, German–Swiss sources sometimes have texts for their Gloria melismas. The question always arises with this sort of double troping: is it certain that the melodic tropes preceded the provision of a text (or prosula, with one syllable per note)? could not both have been conceived together? Several of the St Gall sets of melismas have no texts, which suggests that texts were indeed additions.

The three French Gloria tropes of the type where both melisma and prosula exist have been discussed by Rönnau (1967, *Tropen*, 188–96, with transcriptions and facsimiles).

Another melisma connected with the Gloria trope repertory is of a different type. Within the trope verse *Regnum tuum solidum permanebit in eternum* several sources have a long melisma on the syllable 'per-'. Many manuscripts also have a text for this melisma, or more than one text. (For discussion, see Rönnau 1967, *Tropen*, 179–87 and Rönnau 1967, 'Regnum'.) These '*Regnum* prosulas', as they are often called in modern literature, although traceable back to the earliest sources (Paris, Bibliothèque Nationale, lat. 1240), may safely be said to post-date the melisma to which they are fitted, since the melisma is actually borrowed from a sequence melody, the one called 'Ostende (maior)' in early French and English sources (text *Salus eterna*, etc.) and 'Aurea' in German–Swiss ones (text *Clare sanctorum*, by Notker Balbulus). Rönnau adduced several reasons for supposing that the *Regnum* verse and its melisma were very old (ninth century): association with Gloria 'A', Bosse no. 39; association with the trope set *Laus tua*; the fact that both early 'western' and early 'eastern' sources know both these chants and the *Regnum* verse. (For examples, see below, section xi.)

The *Regnum* melisma is heard climactically in one of the last trope verses of a Gloria, a final flourish not dissimilar in its effect to the melismas which so often crown the performance of a festal responsory. Like responsory melismas, the *Regnum* and other climax melismas were frequently texted, and are thus mentioned below in the sections dealing with prosulas.

(iv) *Added Responsory Melismas*

Holman 1961, 1963; Steiner 1973; Kelly 1974, 1988.

At least one important melisma can be dated securely to the early ninth century, on the report of Amalarius of Metz, writing *c*.840. He stated that:

> In the last responsory [of matins of St John the Evangelist], that is *In medio ecclesiae*, in contrast to the practice for the other responsories, a neuma triplex is sung, and its verse and doxology are also prolonged by neumas, in contrast with usual custom . . . Therefore because the responsory is sung three times by the succentors [before and after the verse and after the doxology] the neuma is threefold . . . Furthermore, modern singers sing this neuma in the responsory *Descendit de caelis* [for Matins of Christmas Day] at the word 'fabricae mundi', with which word the neuma may very aptly harmonize.

(Ed. Hanssens, iii. 54–6; quoted in Hofmann-Brandt 1971, i. 12–13, Steiner 1979, 250; see also Handschin 1952, 142–5; Stäblein, 'Tropus', *MGG*, 811–12; Steiner 1970; Kelly 1988). In the sources that have come down to us, the melismas are usually found assigned to the responsory *Descendit de celis*, rather rarely to *In medio ecclesie*, and very occasionally to other responsories. The three melismas are usually arranged in an order of increasing length.

This is but the earliest traceable instance of a responsory melisma, or group of them, which (*a*) appears to be an addition to the original state of the responsory, for it is not found in all sources (though it could be argued that this is the result of suppression of an original practice because of its extravagance), and (*b*) is susceptible to transference from one responsory to another. In fact, such a triple set of melismas is very rare, and for parallels one has to look to the Ambrosian repertory, with its multiple responsory and alleluia melismas. By far the most common provision is a single melisma, usually on a syllable near the end of the respond, to be performed only during the final singing of the respond after the doxology. The responsory chosen is usually the last one of Matins, or the last responsory of an individual nocturn within Matins (the third and sixth in secular cursus, or the fourth and eighth in monastic use).

The extent of the repertory is difficult to assess, since this depends on extensive comparison of sources, in order to detect those melismas present in some sources but not others. (Holman 1961 and 1963 compares Worcester Cathedral, Chapter Library F. 160 with three other sources; see also David Hughes 1972.) That many of them might be considered as 'floating', assignable at will to any responsory of the appropriate mode, is suggested by the collection assembled by Jacobus of Liège as part of his tonary (ed. Bragard in CSM 3 vi. 256 ff.). Manuscripts vary widely in their choice of melismas and melodic details (Steiner 1973 and Kelly 1974 are excellent demonstrations of the instability of the repertory.)

(v) *Prosulas for Offertories and Alleluias*

Steiner 1969; Björkvall and Steiner 1982; Björkvall 1990; CT 2, 6.

Prosulas are the texts provided for melismas or predominantly melismatic music. The largest repertories that have come down to us are for alleluias, offertories, and responsories.

The texting of melismatic music, like the adding of melismas, seems to go back at least to the ninth century. One of the earliest of all examples of musical notation, Munich, Bayerische Staatsbibliothek, clm 9543, datable to the ninth century (perhaps even the first half of the century) transmits an alleluia prosula, *Psalle modulamina* for All. *Christus resurgens*. (See Smits van Waesberghe 1957, 'Over het onstaan' and 1959, and Stäblein 1963, 'Zwei Textierungen'; on the early history of texting melismas, see Stäblein 1961, 'Unterlegung'.)

Ex. II.23.2 uses three sources, Madrid, Biblioteca Nacional 288 (from Palermo, *c*.1100), with pitches derived from Montpellier, Faculté de Médecine H. 159 (Saint-Bénigne at Dijon, *c*.1025), and Paris, Bibliothèque Nationale, lat. 776 (Albi, second half of the eleventh century). The example shows one of the verses (the second verse in each of the three sources used) for the offertory *Benedixisti* (third Sunday of Advent). On the last syllable of the verse appears a melisma (one of very modest length, chosen for that reason: many are far larger), and in both Madrid 288 and Paris 776 a prosula is also given for this melisma. Although the two texts differ, they are clearly composed in the same way. Both start with the phrase 'da nobis' of the offertory verse, and the prosula of Madrid 288 ends with 'da nobis'. (It could be argued that 'tenebris' is assonant with it.) As many syllables as possible in both texts have been made assonant with '-is'. The syllable groupings reflect the note groups of the music—the note groups are easier to see in the French neumatic notation of Madrid 288, but even in the Aquitanian notation of Paris 776 they are not difficult to observe. Paris 776 reflects the note groups more faithfully at 'in caelis in terris'. Both

Ex. II.23.2. Verse *Ostende nobis* for offertory *Benedixisti*, with prosulas *Da nobis famulis* and *Da nobis potenti* (Madrid, Bibl. Nac. 288, fo. 121^r; Montpellier, Faculté de Médecine H. 159, p. 235; Paris, Bibl. Nat. lat. 776, fo. 8^r)

texts expand in a modest way the meaning of the parent offertory verse, taken from Psalm 84 (85):

Ps. 84: 2   Shew us, O Lord, thy mercy: and grant us thy salvation.

(Madrid 288)   Give unto us thy servants, I ask Christ, the reward in glory of the kingdom which thou hast promised thy saints, give it unto us.

(Paris 776)   Give unto us, thou who reignest in might in heaven and in earth, your strength, which once shone in our darkness.

Ex. II.23.3, an alleluia with prosulas, shows how a pre-existing text may be worked into the prosula, becoming completely assimilated into it. The prosula text is printed here with the pre-existing words (or parts of words) in capitals:

Ex. II.23.3. *Alleluia Concussum est mare* with prosula *Angelus Michael* (Rome, Bibl. Ang. 123, fos. 139ᵛ, 252ʳ; Modena, Bibl. Cap. O. I. 7, fo. 174ʳ)

AL -LE- -LU- -IA.

An-ge- -lus Mi-chael at-que Gabri-el si-mulque Raphael et omnes conciues

pol-lorum si-de-ris ag-mina nos con-cedentem in se-cu-la.

CONCUS- -SUM EST   Concussum ac percussum est

MA- -RE   Mare montes sa-xa et ar-ua

ET CONTRE- -MUIT   Et contremu-it montes et ex-pauit

TER- -RA   Draco-nem pes-ti-ferum serpes an-ti-quo

qui e-iectus est de ce-lo   di-mersus sub terra

U-BI ARCHANGELUS MI-CHA-EL DESCEN-DE-BAT DE CE- -LO

In mons Garga-ni-co uic-to-ri-am Christo

pugnantes cum Sathan duri-us   et ex-pu-lit e-um ex-in-de.

AN-GE-LUS MIcHAel atque Gabriel simulque Raphael et omnes concives polorum, sideris agmina, nos concedentem in saecula. CONCUSSUM ac percussum EST MARE, montes, saxa et arva, ET CONTREMUIT mundus et expavit draconem pestiferum, serpes antiquo qui eiectus est de caelo, dimersus sub TERRA, (UBI ARCHANGELUS MICHAEL DESCENDEBAT) in mons Garganico, victoriam Christo expugnavit cum Sathan durius et expulit eum exinde.

Alleluia. The sea was shattered and the earth was terror-shaken when the archangel Michael came down from heaven. *Prosula*. Angel Michael and Gabriel and Raphael as well and all the citizens of the heavens, the celestial host, together praise him who reigns for ever. The sea, the mountains, rocks, and land were shattered and battered, and the world was terror-shaken and grew afraid at the plague-bearing dragon, the ancient serpent who was cast out of heaven, engulfed below the earth. [When the archangel Michael came down from heaven] on Mount Gargano a victory for Christ he fought with Satan, more savagely, and drove him out from there. (Cf. Dronke 1985).

These examples show the two most important ways in which text was added to pre-existing chants. Most offertory prosulas were provided for melismas in the verses (see Björkvall and Steiner 1982 for a useful range of examples). In the absence of a comprehensive survey of the repertory, generalizations would be premature. Some indication of how matters stand may be deduced from the repertory of Paris, Bibliothèque Nationale, lat. 1118, the largest Aquitanian repository of 'tropic' material. Steiner (1969, 371) lists twenty-two offertories with prosulas in Paris 1118, and comments that most of them are for feasts not ornamented by tropes (of the text-plus-music type; nor by sequences). The number of prosulas is also far inferior to the number of tropes.

A few offertory prosulas became popular as compositions independent of their parent offertory. Schlager (1983, 'Tropen') has discussed the most celebrated example, *Letemur gaudiis*.

Alleluia prosulas are present in Paris 1118 in slightly fewer numbers (for twenty alleluias). Marcusson's edition of the texts (CT 2) has over 350 prosulas for eighty-six alleluias, from fifty-three sources. As with offertory prosulas, these numbers are far inferior to those of tropes.

The method of performance of offertory and alleluia prosulas is unclear.

Apart from isolated examples in secondary literature (see Schlager, 1967 for an interesting Beneventan alleluia prosula; also Steiner, 'Prosula', *NG*), offertory and alleluia prosulas have to be studied from manuscript facsimiles (PalMus 13, 15, 18, and Vecchi 1955).

(vi) *Responsory Prosulas*

Hofmann-Brandt 1971; Kelly 1977.

The texting of the threefold melismas for the Christmas responsories is apparently later than the provision of extra melismas. (The earliest surviving texts are in Paris

1084 and 1118, both Aquitanian sources of the late tenth century; for editions from various sources, see Stäblein, 'Tropus', MGG, Ex. 9, Steiner 1979, Kelly 1988.)

An interesting report places another example in the tenth century. Hofmann-Brandt (1971, i. 17) cites an anonymous eleventh-century chronicler of the bishops of Eichstätt in Bavaria, who says that a distinguished musician called Reginold composed an office for St Nicholas, and, after becoming bishop of Eichstätt (966–91), composed a further office for Eichstätt's first bishop, Willibald (d. 787). We are told that Reginold composed the office with long responsories, and placed melismas at the end of them ('in fine notulas apposuit'), and that he then put words beneath the melismas in the manner of a sequence: 'eisdemque notulis versiculos instar sequentiarum subiunxit'. (At least part of Reginold's very remarkable trilingual composition has survived: see Hofmann-Brandt 1971, i. 17–23.)

Apart from its early date, this report is interesting because it suggests that the melismatic 'cadenza' (the analogy is borrowed from Kelly 1974, 461) might have been thought distinct enough from the rest of the responsory to have been 'placed there' specially, like a jewel in the centre of a crown, one might say; and it suggests also that the provision of a text for the melisma might be more or less simultaneous.

Although Hofmann-Brandt was able to catalogue no fewer than 732 responsory prosulas, their dispersal among the surviving manuscripts is very thin, in the sense that few prosulas became widely known, and very few manuscripts have more than about twenty prosulas (the largest collection is that in Barcelona, Biblioteca de Catalunya M. 662, a Catalan antiphoner of the fourteenth to fifteenth century, which has forty; see Anglès 1935, 234–9 for several examples; the manuscript has now been studied by Bonastre 1982). Over 180 of Hofmann-Brandt's 496 sources actually contain but a single prosula. Very many are the special feature of an office composed for a local saint and not known elsewhere. Associated with these pious observances, the responsory prosula as a genre survived well beyond the Middle Ages.

The prosula technique was restricted, as far as responsories are concerned, to discrete melismas. Clear repeat structures became increasingly popular as time went on, and it even happened that a freshly composed prosula with repeat structure might supplant a prosula in less popular style. The best-known example of this occurred in the St Nicholas office (see Charles Jones 1963 and Hofmann-Brandt). Here a modest melisma occurs on the word 'Sospes' at the end of the responsory *Ex eius tumba*. The melisma is an integral part of the responsory as it is transmitted in scores of sources. Very rarely, it appears to have been texted (with the prosula *Sospes nunc efficitur*—but this prosula is found only in two sources of the eleventh to twelfth century from Saint-Maur-les-Fossés, near Paris, Paris, Bibliothèque Nationale, lat. 12044 and 12584). From the eleventh century onward, however, we have sources for a replacement melisma and text, which was usually performed in place of the modest former melody during the final repeat of the respond (after the doxology; at least, this is its usual place, and the usual place for a prosula). The new text, *Sospitati dedit egros*, is in regular fifteen-syllable trochaic lines, with a melodic repeat scheme AA BB CC DD (the original 'Sospes' melisma has no repeat structure).

Ex. II.23.4. Responsory *Beatus Nicholaus* with prosula *Oportet devota mente* (*AS*, 358)

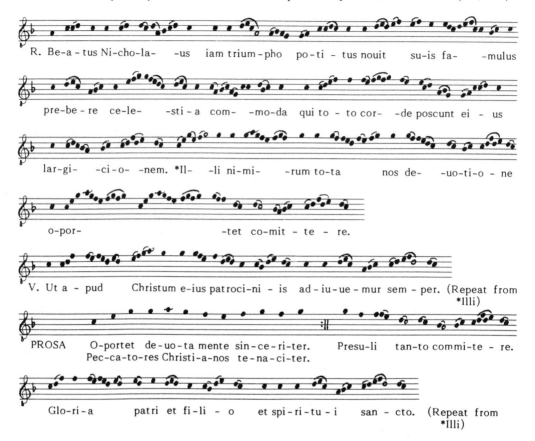

The new prosula—to all intents and purposes a newly inserted composition, text and music together—was extremely popular, and became the starting-point for numerous contrafacta (new texts to the same music). Since *Sospitati dedit egros* has been widely reproduced (for example, Stäblein, 'Tropus', *MGG*, Tafel 89; *AS* 360) I have chosen to illustrate these contrasts of style with other examples. (See also the sparkling demonstration in Kelly 1977.) Exx. II.23.4 and II.23.5 give two prosulas from Cambridge, University Library Mm. 2.9 (a thirteenth-century Sarum antiphoner; facs. *AS*). In Ex. II.23.4 the prosula is a simple texting of a modest melisma from a different St Nicholas responsory, *Beatus Nicholaus*. Although short it has the repeat structure AAB.

In Ex. II.23.5 the composer of the prosula has 'manufactured' a repeat structure out of a previously non-repeating melisma. Each phrase of the melisma is texted twice, and the repetitions are doubled again by the manner of performance indicated in the manuscript, the chorus being instructed to repeat each phrase of the prosula, vocalizing to the vowel *E*. The vowel itself is derived from the original melisma, 'suscip-E', and each half verse of the prosula also rhymes with *E*.

Ex. II.23.5. Responsory *O mater nostra* with prosula *Eterne virgo memorie* (AS, W)

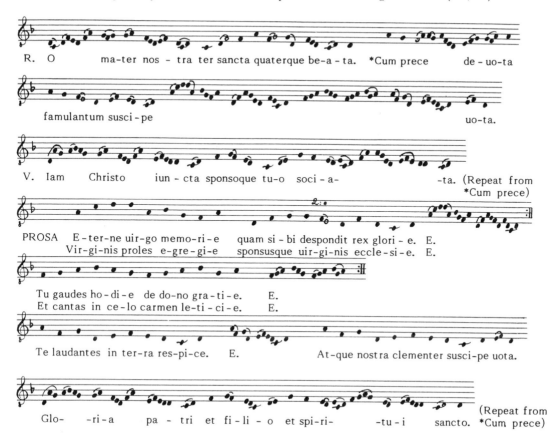

Many prosulas are, in effect, small sequences, self-contained musical items which could easily take on a life of their own. One extremely popular piece, *Inviolata integra*, which originated as a prosula for the Marian responsory *Gaude Maria*, was frequently used as a sequence at masses for the Blessed Virgin Mary. *Quem ethera et terra*, for the Christmas responsory *Verbum caro factum est* (ed. Irtenkauf 1956, 138), which was particularly popular among German and North Italian musicians, also achieved its independence.

By the later Middle Ages (and perhaps from the beginning) alternatim performance of responsory prosulas between singers performing the text and others singing the melisma was the norm. (See Kelly 1985, 'Melisma'; see Chailley 1949 for a rubric from a Sens manuscript instructing the cantor to dance during the prosula in a processional responsory.)

Ex. II.23.6. From Gloria 56 (Vatican IV) with trope verses *O gloria sanctorum*, etc., Regnum prosula *O rex glorie* (Madrid, Bibl. Nac. 19421, fo. 22ᵛ)

QUONI-AM TU SOLUS SANCTUS. TU SOLUS DOMINUS.

Ce-les-ti-um terres-tri-um et infer-no-rum rex.

TU SOLUS ALTISSIMUS.

Regnum tu-um so-   -lidum.

O rex glo-ri-e qui es splendor ... ... ac de-cus eccle-si-e.   E.
Quam decoras-ti tu-o quoque in cru-ce preci-o-so sangui-ne.   E.

Hanc rege semper pi-is-si-me.   E.   Per-ma-nebit in e-ternum.
Qui es fons mise-ri-cordi-e.   E.

IE-SU CHRI-STE.   Christe ce-lo-rum.

CUM SANCTO SPI - RI-TU IN GLORI-A DE-I PA - TRIS.   A-   -MEN.

Sal-ue uir-go uir-   -gi-num.

Ma-ri-a uir-go in-troce-de pro nobis ad do-minum. E.
Et pi-a pre-ce ro-ga De-um poten-ter per omni-a.   E.

Ut det nobis flo-ri-ge-ra se-de.   E..

Fru-i semper cum e-o.   E.   Et potens es in e-ternum.

(vii) *Other Prosulas*

Although the *Regnum* verse, its melisma on 'PER-manebit', and at least some of its prosulas were originally associated with the Gloria known as Gloria 'A' (Bosse 1955, no. 39), it was soon used with other Gloria melodies and with various sets of trope verses. Ex. II.23.6 gives the last section of Gloria 56 (in Bosse 1955; Vatican IV), with the trope set *O gloria sanctorum*, from Madrid, Biblioteca Nacional 19421 (Catania, Sicily, late twelfth century). This set of trope verses is adaptable to various saints' feasts, by the insertion of the required name: here 'Maria' is entered. The redactor of the manuscript has decided to use the *Regnum* verse and melisma with the popular prosula *O rex glorie*. He precedes it with another 'wandering' element, the trope verse *Celestium terrestrium*, which is likewise found in many other sets of Gloria trope verses (see below, section xi). At this point the melody dips for the first time to low G (G_1)—previously the range has been C–c—and the change of tessitura from this point on is quite noticeable (G_1–a). The main cadential note remains E, however, which is why no other pitch is possible for *Celestium* and *Regnum*. (In Gloria 39 there is no hiatus in the pitch when these trope verses are used.) After each phrase of prosula, the melody is repeated as a vocalization on 'E', derived from 'PERmanebit' and matched by the rhyme in the prosula.

In the manuscript, this Gloria with *Regnum* prosula is immediately followed by three more alternative prosulas with text in honour of, respectively, the Blessed Virgin Mary (obviously more apt for the Marian version of the *O gloria sanctorum* set of trope verses), Apostles, and the Blessed Virgin Mary again. Ex. II.23.6 shows only the first of these three alternatives. Interestingly, the principle of assonance has become rather confused. E is maintained for the melismas, though it no longer refers back to the first verse *Salve virgo virginum*, and the prosula has line-ends in u, a, e, and o.

The great majority of Sanctus prosulas are for the second 'Hosanna', again providing a climax for the composition. Like responsory prosulas they frequently assumed sequence form (see Thannabaur, 'Sanctus', *MGG*, Exx. 12–13; also Smits van Waesberghe 1962) and could even become independent liturgical items: *Trinitas unitas deitas* had an interesting career as a sequence, among other things (see Schlager 1983, 'Trinitas'), and *Voci vita* was used as a sequence at Hereford. (Two Hosanna prosulas form part of Ex. II.23.16 below.)

Very rare indeed are prosulas where the Sanctus melody itself is retexted: Thannabaur (1967) cites three examples. The type is limited to eastern sources of the fourteenth century and later.

Two chants rarely embellished by any kind of trope were the gradual and tract. Stäblein ('Graduale (Gesang)', *MGG*) gives examples of gradual prosulas, all from relatively late sources. Tract prosulas are found only in Italian manuscripts (see PalMus 15, fos. 75ᵛ, 90ʳ, 110ʳ, and Vecchi 1955, fo. 74 for examples in facsimile).

Ex. II.23.7. Kyrie *Te Christe supplices* (Laon, Bibl. Mun. 263, fo. 20ᵛ)

1. Te Christe suppli-ces e-xo-ramus cuncti-potens ut nostri dig-neris   e - ley - son.
3. O bone rex super astra qui sedes et do-mi-ne qui cuncta gubernas   e - ley - son.

1. Ky - ri - e-   -lei - son.
3. Ky - ri - e-   -lei - son.

2.   Te decet laus cum tripudi - o   pa-ter summe un-de te pe-timus e - leyson.

2.   Ky - ri - e-   -leison.

4. Tu-a de-uo-ta plebs implorat iu-gi-ter ut il-li dig-neris e-lei-son. ....
6. O the-os a-gy-e saluans ui-ui-fi-ce redemptor mundi e-... ... ... -ley-son.

4. Christe-   -lei-son. ....
6. Christe-   .... -lei - son.

5. Qui canunt ante te   tu pre-cibus an-nu-e   et no-bis semper e-ley - son.

5. Christe-   -lei-son.

7. Clamat incessanter nostra con-ci-o   dicens e - leyson.   Ky - ri - e-   -lei-son.

8. Mi-se-re-re fi-li De-i ui-ui nobis e-lei - son.   Kyri-e-   -lei - son.

9. In ex-celsis ... ... De-o ma-gna sit glo-ri-a e-ter-no patri   di-camus in-de-si-
    qui nos re-demit propri-o sangui-ne ut ui-ui-fi-caret a mor-te

-nenter u-na uo-ce e-ley - son.   Ky - ri - e-

-lei - son.

(viii) *Kyries with Latin Text, Kyrie Prosulas, and Kyrie Tropes*

Crocker 1966; Bjork 1979/80, 1980, 'Kyrie Trope', 1980, 'Early Settings', 1981.

Questions about what is added material and what constitutes the original form of the chant are particularly pressing in the case of many Kyries. Was each Kyrie with Latin text so composed in the beginning, or was the text added as a prosula to the ornate melody? If the earliest sources already have the Latin text, then we have no documentary evidence for supposing the untexted version existed earlier. Crocker and particularly Bjork have supported this reasoning with stylistic arguments.

Ex. II.23.7 is a transcription from Laon, Bibliothèque Municipale 263 (Laon, twelfth century) of a Kyrie with Latin text whose transmission goes back to both early west and east Frankish manuscripts. Although they are no certain test of prosula technique, the syllable- and note-groups do not suggest that this text has been fitted to a pre-existent melody. More important than that the syllable-groups do not follow the note-groups in this particular source (other sources show no better correspondence) is the fact that verses with the same melody (1=3, 4=6, 7+8=9) do not have the same syllable-groups. Against this could be adduced the *e* assonance in verses 5 and 6.

The Kyrie with text *Kyrie fons bonitatis* (Ex. II.23.8) is much more artful in these respects. The correspondence of note- and syllable-groups is quite striking, as also the constant *e* rhymes. This may simply reflect the taste of the eleventh century, as opposed to that of the ninth. But it possibly reflects also the fact that the text—we could justifiably call it a prosula—appears in the manuscript tradition later than the melody. The earliest sources, from Winchester and Arras, of the late tenth and early eleventh century, do not have the Latin text. (Previously available *Variae preces*, 165–7, *Cantus selecti*, 81*–82*, *Graduale Sarisburiense*, 2*v, together with several other Kyries of this type.)

Quite different in impact are the trope verses sometimes provided for these and other melodies, which were largely unknown until the examples in Bjork's articles appeared (principally Bjork 1980, 'Kyrie Trope', which contains practically all that can be reconstructed). Such trope verses contrast musically with the nine invocations of the main chant. Bjork (1980, 'Kyrie Trope', Table 2) lists some two dozen tropes, which are distributed thinly over early manuscripts from all areas of Europe. In early Swiss–German sources this type of Kyrie is preferred, while French ones contain a much greater proportion of Kyries with Latin text (or prosulas). Many fell out of use before being recorded in staff notation.

While many of these tropes consist of but a single introductory verse, others make up a set of eight verses deployed between the nine invocations of the main chant. A few sets contain three verses, distributed as in Ex. II.23.9. The set appears here with Kyrie 55 (in Landwehr-Melnicki 1955), also the base melody of Ex. II.23.7. Melodically it both complements it and contrasts with it, sometimes awkwardly: the verse *Iterum dicamus* rises to *d*, preparing us for the higher-lying *Christe* invocations, but then subsides on to the former low pitches. The last trope verse, *Et submissis vultibus*, again prepares for the move to higher pitches in the final Kyrie.

Ex. II.23.8. Kyrie *Fons bonitatis* (Laon, Bibl. Mun. 263, fo. 26ᵛ)

1. Ky-ri – e fons bo – ni-ta-tis pa-ter in-ge-ni-te   a quo bo-na cuncta pro-cedunt e-ley – son.
2. Ky-ri – e qui pa – ti natum mundi pro cri mi ne ip-sum ut sal-ua-ret mi-sis-ti   e-lei – son.
3. Ky-ri – e qui sep-ti-for-mi das dona pneumate a quo celum ter-ra re-pletur e-lei – son.

1. Ky-ri – e-   -lei – son. (also after 2. and 3.)

4. Christe   u – ni-ce De-i patris ge-ni-te quem de uirgi-ne nas-ci-turum mundo miri – fi – ce
5. Christe   a – gy-e   ce-li compos regi-e   melos glori-e   cu-i semper adstans pro numine
6. Christe   ce-li-tus adsis nostris precibus pronis mentibus quem in terris deuo-te co-li-mus

sancti pre-di-xerunt prophete e-ley – son.
an-ge-lorum decantat a-pex e-ley – son.
ad te pi-e Ihe-su clamantes e-lei – son.

4. Christe-   -lei – son. (also after 5. and 6.)

7. Ky-ri – e spi-ritus alme co-herens pa-tri natoque u-ni-us u-si-e   consis-tendo
8. Ki – ri – e qui bapti-zato in Iorda-nis unda Christo effulgens speci-e   colum-bi-na
9. Ky-ri – e ig-nis diui-ne pecto-ra nostra succende ut digni pa-ri-ter procla-ma-re

flans ab u-troque e-lei – son.
ap-pa-ru-is-ti   e-lei – son.
pos-simus semper e-lei – son.

7. Ki – ri – e-   -lei – son. (also after 8. and 9.)

A number of Italian Kyries with Latin verses are composed with the same melody throughout, that is, for all the acclamations and the alternating Latin verses. The result is not unlike the type of litany mentioned in section II.17.i above. (See Bjork 1980, 'Early Settings' and especially Boe 1989 for examples.) There is a connection here with other genres as cultivated by Italian musicians: sequences with the same melody for all verses have been mentioned above (II.22.v) and other troped ordinary

Ex. II.23.9. Kyrie 55 with trope verses *Christe redemptor*, etc. (Paris, Bibl. Nat. lat. 903, fo. 166ᵛ)

Christe redemptor mise-re-re no-bis kir-ri-e-lei-son

ei-a om-nes di-ci-te mi-se-re-re do-mi-ne kir-ri-e-lei-son.

Vo-ce corde proclamantes regem in-ui-si-bi-lem canen-tes il-li.

1.3. KIRRI-E   E-LEISON. 2. KIRRIE   E-LEISON.

I-te-rum dicamus omnes Christe elei-son et rogemus Christum deum una uoce dicentes.

4.6.CHRISTE   ELEISON. 5. CHRISTE   E-LEISON.

Et submissis uul-ti-bus depreca-mur tri-ni-ta-tem re-gem aeter-num canen-tes il-li.

7. KIR-RI-E   ELEISON. 8. KIRRIE   ELEI-SON.

f.163v

9. KIR-RI-E   E-LEISON.

of mass chants followed the same pattern. Ex. II.23.18 below is an Agnus Dei composed in this way.

(ix) *Benedicamus Chants with Extended Text, Prosulas, and Tropes*

Arlt 1970; Huglo 1982, 'Débuts'.

Many early Benedicamus chants have not simply the two phrases of text 'Benedicamus domino. Deo gracias', but longer texts. Of the seventeen examples which Huglo dates in the tenth or eleventh centuries, eleven have extended text. Are these to be regarded as tropes of a pre-existing chant, or prosulas for a melismatic chant, or are they an especially elaborate type of Benedicamus, composed expressly in that way, perhaps for festal liturgies rather than ferias? In many cases we shall probably never know, since they survive as *unica* (for example, many of those in Paris, Bibliothèque Nationale, lat. 887, discussed by Arlt 1970, *Darstellungsband*, 161–6).

Some of the compositions can be regarded as prosulas in that the melody is borrowed from elsewhere, and a new text fitted. Several Kyrie melodies were used as Benedicamus chants in this way, at least from the eleventh century onward; so was the 'Flos filius' melody, a melisma from the responsory *Stirps Iesse*. This is found widely, both with the simple Benedicamus text, with 'Benedicamus domino, alleluia, alleluia' (an Eastertide adaptation), and also with longer texts, for example: 'Benedicamus flori orto de stirpe Iesse die hodierna, qua processit virga virgo domino' (Arlt 1970, *Darstellungsband*, 169).

Prosula seems to describe less well the technique of *Eia nunc pueri* (Arlt, 165), which is not consistently syllabic. Although it may be the texting of a pre-existent melody, it seems less likely that a Benedicamus melody without extended text lies beneath this composition, for then one would expect the word 'Benedicamus' to start the piece (Ex. II.23.10).

Ex. II.23.10. Benedicamus chant *Eia nunc pueri* (Paris, Bibl. Nat. lat. 887, fo. 46ʳ)

With *Benedicamus . . . Verbi iungando* (Arlt, 163) we have what seems to be a self-sufficient Benedicamus phrase, to which a couplet of eight-syllable lines has been added. (The process is repeated for Deo gracias.) If the Benedicamus melody were to be located elsewhere without *Verbi iungando*, then one might be tempted to speak of a trope. And indeed it is found, with a different continuation, in the same source as for *Benedicamus . . . Verbi iungando*. The situation is complicated, however, by the fact that this time the new text, *Hodie surrexit*, is placed first, using the putative

Benedicamus melody, texting and extending it, and then 'Benedicamus domino' is spread over the whole new melody (Ex. II.23.11).

In fact, this explanation of how the compositions arose seems strained. They, and Ex. II.23.10, all come from the earliest substantial collection of such Benedicamus chants, Paris 887 (Aquitaine, eleventh century). I would prefer to believe that we are dealing with the result of a local campaign to extend the festal Benedicamus repertory, where several of the pieces have a family likeness not really attributable to a trope or prosula technique. They are newly composed from the start. It is only their prose texts, incorporating the words 'Benedicamus domino', and their predominantly traditional melodic style that tempt us to relate them to the trope repertory.

Ex. II.23.11. Benedicamus chants (Paris, Bibl. Nat. lat. 887, fos. 46ᵛ, 46ʳ)

(x) *Introit, Offertory, and Communion Tropes; Sequence Tropes*

Paul Evans 1970 'Early'; MMMA 3; CT 1, 3; Planchart 1977.

The earliest recorded composition of introit tropes, if the testimony is trustworthy, is by Tuotilo of St Gall (known between 895 and 912), who is said by the St Gall chronicler Ekkehard IV (*c*.990–1060) to have composed the introductory verses *Hodie cantandus est* in his youth ('plane iuvenis'—MGH, *Poetae Latini aevi Carolini*, iv. 1096; see also MGH, *Scriptores rerum Sangallensium*, 80—see Ex. II.23.14 below). Manuscript sources for them survive only from the tenth century onward (Paris, Bibliothèque Nationale, lat. 1240, Aquitaine, second quarter of the tenth century, is the earliest), but there seems little reason to doubt that they were well known in the ninth. It should nevertheless be noted that almost no introit tropes were known in both early 'western' and early 'eastern' sources, so that it is not possible to postulate a common basic layer on which later diverse collections were built.

Introit tropes were written in great numbers during the tenth and eleventh centuries, offertory and communion tropes in smaller quantities (many churches, including most Italian ones, appear to have ignored offertory and communion tropes almost entirely). The repertory is vast. Taking individual trope verses as units, there are 1,044 for the introit, 250 for the offertory, and 113 for the communion in CT I and III, that is, for the Christmas and Easter parts of the church year. There is a decline in sources from the twelfth century, and very few from the thirteenth century or later. This is critical in some areas, since very few north French and German trope collections with staff notation have survived.

Tropes for the introit, offertory, and communion (and the other chants discussed below in sections (xi) and (xii)) take the form of extra verses placed before or among the phrases of the main chant. The arrangement in Table II.23.1 is reasonably typical, for the Easter introit *Resurrexi*, in the tropers from Winchester, Cambridge, Corpus Christi College 473 and Oxford, Bodleian Library, Bodley 775 (late tenth and early eleventh century respectively) (for full texts see Frere 1894, *Winchester* and Planchart 1977). The trope incipits are given in lower case, main chant incipits in capitals (cf. St Gall, Stiftsbibliothek 376 in Gautier 1886, 139). Although in the Winchester sources the seventeen verses are laid out in this sytematic way, that is not the case in all sources. Some sources (for example, Paris 1118) have rather the character of an anthology. And the stylistic disparity between the trope verses so neatly arranged by the Winchester redactor shows that logic has been imposed upon what is an unstable repertory. Some of the verses are hexameters: the group for the second singing of the introit beginning *Ecce pater*, then the trope for the first psalm verse *Fregit inferni*, and then the group for the second singing of the introit beginning *Virgine progenitus*.

Trope verses in poetic metre of one kind or another are not at all uncommon. Hexameters dominate the field (in CT I, 40–2 it is stated that over 200 of the 766 verses in the volume are hexameters), and elegiac distichs, pentameters, and even verses like Ambrosian hymn strophes are also to be found. Stotz (1982) discusses examples in sapphics.

Ex. II.23.12 illustrates some of this variety. It is a transcription of the tropes for the introit *In medio ecclesiae*, for St John the Evangelist's Day (27 December). A troped version of this introit has already been given above (Ex. II.23.1), where St Gall and Pistoia manuscripts had an introductory metrical verse followed by several verses which seem to have originated as melismatic extensions of the introit melody. The Winchester collection of verses is quite different. There are almost as many verses as for the Easter introit above, except that the doxology and the second psalm verse have no trope. The first trope is a rhyming pair of trochaic fifteen-syllable lines; verse 5 is a

Table II.23.1. *Tropes for the Easter introit* Resurrexi *in Winchester sources*

	Trope verse	Main chant
(Introit)	Psallite regi magno	RESURREXI
	Dormivi pater	POSUISTI
	Ita pater	MIRABILIS
	Qui abscondisti	ALLELUIA
(Psalm verse)	En ego verus sol	DOMINE
(Introit)	Ecce pater	RESURREXI
	Victor ut ad caelos	POSUISTI
	Quo genus humanum	MIRABILIS
(Doxology)	Fregit inferni portas	GLORIA PATRI
(Introit)	Virgine progenitus	RESURREXI
	Quem non deservi	POSUISTI
	Ut per me	MIRABILIS
(Psalm verse)	Exsurge gloria mea	INTELLEXISTI
(Introit)	Postquam factus homo	RESURREXI
	In regno superno	POSUISTI
	Laudibus angelorum	MIRABILIS
	Cui canunt angeli	ALLELUIA

hexameter. Each trope or set of trope verses has its own history, part of which can be surmised from the sources known for these compositions (see CT I and Planchart 1977). The situation may be summarized as follows.

Trope verses 1–3 are found deployed in the same way in sources from all over Europe, some of the notable exceptions being the tropers from St Gall, St Emmeram at Regensburg, and Benevento.

Trope verse 4 is unique to England, appearing only in the two Winchester manuscripts and in London, British Library, Cotton Cal. A. xiv (early eleventh century). No transcription into modern notation is therefore possible.

Trope verses 5–7 appear in only a few French and German sources, and also in several Italian ones. The two Italian tropers from the Ravenna area (Padua, Biblioteca Comunale A. 47 and Modena, Biblioteca Capitolare 0.I.7) add two verses to the set, but can be used in conjunction with Paris, Bibliothèque Nationale, lat. 1084 (the only Aquitanian troper to be of help) to suggest pitches for the Winchester neumes.

Trope verses 8–10 are found in French sources only, apart from Ivrea, Biblioteca Capitolare 60 (which in any case has a high proportion of French material). Provins, Bibliothèque Municipale 12 (from Chartres, thirteenth century) and Paris, Bibliothèque Nationale, n.a.l. 1235 (from Nevers, twelfth century) are most convenient for deriving pitches.

Ex. II.23.12. Trope verses *Ecce iam Iohannes*, etc. (Oxford, Bodl. Lib. Bodley 775, fo. 35ᵛ) for introit *In medio ecclesie* (Worcester, Cathedral Chapter Lib. F. 160, fo. 299ʳ)

(Ex.II.23.12 cont.)

Dox.

GLORIA PATRI ET FILIO ET SPIRITUI SANCTO   SICUT ERAT IN PRINCIPIO

ET NUNC ET SEMPER   ET IN SECULA SECULORUM AMEN.

T8   Fons et o - ri - go sa - pi - en - ti - e   ad pro - pa - ganda su - e di - uini - tatis archana.

(I1 as above)

T9   Qui flu - en - ta e - uan - ge - li - i de ip - so sa - cro pec - to - re hausit.   (I2 as above)

T10   Vir - gi - ni - tatis quoque me - ri - to ma - tri uir - gi - ni fi - li - um con - ferens.

(I3 as above)

Ps2   IUSTUS UT PALMA FLOREBIT   SICUT CEDRUS QUE IN LIBANO EST MULTIPLICABITUR.

T11   A-   -mor an - ge - lo - rum et gau - di - um Christus Io - hannem di-   -ligens.   (I1 as above)

T12   Quo pan - de - retur om - ni - bus lux   gen - tibus uer - bi De - i.   (I2 as above)

T13   Et hunc ad ae - ter - num   ho - di - e uocans con - ui - ui - um.   (I3 as above)

Principal secondary sources for the trope verses: Provins Bibliothèque Municipale 12 (1–2, 8–10); Paris Bibliothèque Nationale lat.1084 (3, 5–7); Rome Biblioteca Casanatense 1741 (11–13)

Trope verses 11–13 survive outside England only in Nonantola manuscripts. It is fortunate that these have diastematic notation.

If one were to guess at the place of origin of these verses, one might suggest north France for verses 1–3, Winchester itself for 4, north France again for 5–7 and France for 8–10; for 11–13 it seems difficult to make even a guess. This is speculation. What is more important is that it indicates that the Winchester redactor probably had more than one exemplar to hand; these in their turn probably derived from a variety of exemplars; moreover, new compositions would continually enter the repertory, as they did at Winchester itself (verse 4). The result is a fascinating variety of 'case histories'. (Planchart 1977 is a magisterial discussion of the case histories of the Winchester tropes; Planchart 1981 considers the methods and problems of such research; see also Planchart 1977, ii. 104–10 on the version of Ex. II.23.12 in Husmann 1959, 138–9.)

Comparison of almost any two sources for the same trope reveals numerous small variants. These are not usually of major musical importance, except in so far as they suggest that the recorded versions rarely seem to be copied from any authoritative 'original' exemplar. Instead, the way in which the redactor of the source understands a trope, hears it with his inner ear, performs it in its due season, is more important at the moment the copy is made than exact adherence to an exemplar. Sometimes the deviation between sources is so wide as to suggest that the trope is being copied without any reference to an exemplar (see MMMA 3, 39, 123, or an extreme case, 294). And occasionally the music of two different sources will be so dissimilar that one may speak of two different melodies (see MMMA 3, 197, also the studies by Weiss 1964, 'Problem', and 1967). The grey areas between 'similar' and 'different' are sometimes a little hard to define (for a penetrating discussion see Treitler 1982 'Observations'). Occasionally similar music is used for divergent texts (MMMA 3, 62).

The ways in which trope verses introduce, amplify, explain the chants to which they are attached have been well surveyed by Husmann (1959), Stäblein (1963, 'Verständnis'), Paul Evans (1970, *Early Trope*) and Steiner ('Trope', *NG*), as well as in numerous articles on individual tropes. Husmann emphasized their role as invitations to begin the singing of a chant, commenting that this was something they shared with many Christian liturgical forms, eastern as well as western (those for the Gloria, Sanctus, and Agnus are discussed in the next subsection; even the Te Deum was given an introductory trope at St Gall: see Gautier 1886, 170). Muller (1924–5, 566–7) pointed out their close similarity in this respect to the (spoken) prefaces at the beginning of the mass of the Gallican rite. Some Byzantine counterparts are discussed by Strunk (1970). Musical relations between trope and base chant have also been discussed by Stäblein, Evans, and Sevestre (CT I and 1980). Some of the more obvious points may be illustrated with reference once again to Ex. II.23.12.

Introits customarily use verses drawn from the Book of Psalms, or sometimes another Old Testament book. Since these contain no direct references to New Testament events, all connections between an introit text and the day on which it is sung—Christmas, Easter, St Peter's Day, or whatever—are by special inference (it was central to Christian belief that the Old Testament contained numerous references to the coming of Christ). As far as the text is concerned, Ex. II.23.12 is typical of many tropes in that it establishes an unambiguous connection between introit text and feast-day. The introit is adapted from Eccles. 15: 5–6 in the Apocrypha, verses which in their original context had nothing to do with preaching the Gospel. Once placed at the start of mass for St John the Evangelist, they immediately take on a special significance, and for the believer they become one more among the myriad threads woven into the great design in which the coming of Christ was the culmination of all the previous history of mankind. But the tropist goes beyond this inferred meaning of the introit, and spells out in clear terms whose feast-day it is, referring also to John's role as evangelist and as the beloved disciple to whom Christ on the cross entrusted his mother Mary.

Although the trope verses are musically compatible with the introit, they cannot be said to mirror its style to any significant extent. Nowhere in them do we find the reiterated *F*s (repercussive notes) of the introit antiphon, nor do they echo the *FFFC* cadence of 'intellectus' nor the rise *ac* at 'eum', and in the psalm verses. The way in which both introit and tropes skip over *E* is simply a characteristic of chant melodies generally, not something which links these in any significant way. Stäblein (1963, 'Verständnis', 91) said that tropes were a sort of 'pseudo-' or 'neo-Gregorian' chant.

The texts of the trope verses in Ex. II.23.12 are typical of many which complement, but do not lead strongly into the introit text at any point. Other texts do lead on, in such a way that their texts are in no way self-sufficient. For example, the trope set *Deus pater filium suum* for the Christmas introit *Puer natus est* reads as follows:

Deus pater filium suum hodie misit in mundum, de quo gratulanter dicamus cum propheta: (CT I, 78)	Today God the Father sent his son into the world, wherefore we sing rejoicing with the prophet:
PUER NATUS EST NOBIS ET FILIUS DATUS EST NOBIS.	UNTO US A CHILD IS BORN, AND UNTO US A SON IS GIVEN.

This happens also to be one of many tropes in which a strong musical link seems to exist between trope and main chant. Several tropes for this introit begin with the same memorable rise of a fifth as the introit itself (cf. MMMA 3, 291–3, 299, 302–3; see also Stäblein 1963, 'Verständnis' for discussion of several other examples, and comments in Weiss 1965). Ex. II.23.13 illustrates this. And yet, as Treitler (1982, 'Observations', 92) has pointed out, such a melodic profile is conventional among mode 7 melodies (cf. introits *Aqua sapientiae*, *Oculi mei*, *Respice domine* for the same rise of a fifth—the whole textual and melodic relationship between *Puer natus est* and the *Deus pater* trope set is explored in depth by Arlt 1982).

Altogether exceptional seems to be the use of a sequence, the short aparallel *Ecce iam Christus* (melody 'Ostende (minor)', as an introit trope (see Strehl 1964 and Weiss 1965).

One set of introit tropes (if that is what one should all them) stands apart from the main chant. These are the *Versus ante officium*, relatively long compositions, consisting of several verses written for alternating sets of singers. In Cambrai, Bibliothèque Municipale 75 (from Saint-Vaast at Arras, early eleventh century) such pieces are labelled 'Ad processionem', implying that they were part of the pre-mass procession rather than the mass introit.

Ex. II.23.13. Trope verse *Deus pater filium suum* (Paris, Bibl. Nat. Lat. 1119, fo. 5ʳ) and start of introit *Puer natus est* (Paris, Bibl. Nat. lat. 1132, fo. 11ᵛ)

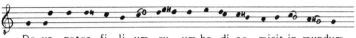

De-us   pater   fi - li - um   su - um ho - di-ae   misit in mundum

de quo gratu-lanter di - ca - mus cum prophe-ta

PU-ER NATUS EST NO-BIS   ET FI-LI-US DATUS EST NO-BIS.

Ex. II.23.14. Introductory verses *Hodie cantandus est*, etc. and start of introit *Puer natus est* (Graz, Univ.-Bibl. 807, fos. 167ʳ, 14ʳ)

f.167r

Ho-   -di - e   can-tandus est no-bis pu-er quem gignebat in-ef - fa - bi-li-ter

an-te   tempo-ra pater   et e - undem sub tempo - re ge-ne-ra-uit   in-clita mater.

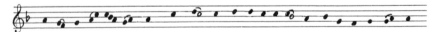

Quis est is-te   pu-er quem tam magnis preconi-is dignum uoci-fe-ra-mi-ni

di - ci-te no - bis ut col-lauda-to-res es-se pos - si-mus.

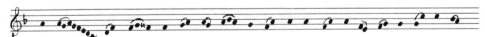

Hic e-   -nim est quem pre-sa-gus et e-lectus symnis-ta De-i ad terras

uen-tu-rum preuidens longe an-te   preno-ta-uit sic - que pre-di-xit.

f.14r

PUER NATUS EST NO-BIS   ET FI-LI-US DATUS EST NO-BIS

The most famous of them is one ascribed to Tuotilo of St Gall, *Hodie cantandus est*, which has three verses in the order 'Statement—Question—Answer'. This is a D-mode composition which contrasts strikingly with the G-mode *Puer natus est*. Some writers have thus been led to question its function as a direct introduction to *Puer natus* est (see the discussion of these versus by Planchart 1977, i. 234–6), while Stäblein (1966, 'Altrömische') went so far as to suggest that it was designed to introduce the Old Roman version of *Puer natus est*, with which it is tonally more compatible. (This implied that when Tuotilo composed the piece, Old Roman chant had not been fully replaced by Gregorian at St Gall, an idea that other scholars have been somewhat reluctant to accept.) Ex. II.23.14 is a transcription of *Hodie cantandus* from Graz, Universitätsbibliothek 807 (Klosterneuburg, twelfth century; PalMus 19).

(Two other famous sets of introductory verses, *Quem queritis in sepulchro* and *Gregorius praesul*, are discussed below, II.25.ii and VI.6.viii respectively.)

A rather small repertory of short introductory tropes for the sequence survives in the tenth- and eleventh-century Aquitanian tropers (Evans 1968).

(xi) *Gloria, Sanctus, and Agnus Tropes*

Rönnau 1967, *Tropen*; Falconer 1989; CT 4 and 7.

Like those discussed in section (x), tropes for the Gloria, Sanctus, and Agnus consist of complementary verses, some serving as introductions to the singing of the chant, some interposed between phrases of the chant. Most of their special features follow naturally from the different chants they complement. Rönnau catalogued over 100 Gloria tropes (or sets of trope verses), of which at least one probably dates back into the ninth century; and Hucbald of Saint-Amand (d. 930) is credited with the composition of another. Although the largest collection of Gloria tropes is to be found in a twelfth-century manuscript, Madrid, Biblioteca Nacional 19421 from Catania in Sicily, they seem to have passed out of use in most centres by the thirteenth century, with the exception of one or two relatively modern compositions (such as *Spiritus et alme orphanorum*, for the BVM). Sanctus and Agnus tropes may include some equally early compositions (on the earliest Agnus trope see Atkinson 1977) and groups of new compositions, especially for the BVM, continued to appear even late in the Middle Ages, many following the fashion for texts in regular rhyme and rhythm (some early Gloria, Sanctus, and Agnus texts are written in hexameters, distichs, etc.). From the catalogues of Thannabaur (1962) and Schildbach (1967) and other research, over 250 Sanctus tropes and over 140 Agnus tropes are known (one or two tropes are common to both categories).

While hardly any introit trope verses 'migrated', as it were, from one introit to another, many Gloria trope verses are found with more than one Gloria melody. And individual verses are associated now with one set of trope verses, now with another set. One such verse was used above in Ex. II.23.6, and it was also remarked there that a slight tonal disparity had resulted from the introduction of material originally designed for Gloria melody 39 into Gloria melody 56. In fact, the whole trope set *O gloria sanctorum*, in one arrangement or another, is found with Gloria 39, and its use with Gloria 56, as in the Catania manuscript used for Ex. II.23.6, is a peculiarity of a few English and Norman sources; in Nonantola it appears with Gloria 11 (Vatican XIV).

Rönnau (1967, *Tropen*, 84–6 and 246–9, with transcriptions) introduced the term 'Wandervers' ('wandering verse') to describe such a verse when it was used in Gloria 39 in the oldest sources: he believed not only that Gloria 39 was the oldest melody in the repertory, but also that trope verses associated with it in the oldest sources and found among several different sets of trope verses were a sort of 'original' layer of trope verses, which were used freely with Gloria 39, to be joined later by various sets of other verses. But it remains unclear whether this is correct, or whether these verses are not simply more mobile than most in a generally unstable repertory, because they are among the oldest, and therefore had longer opportunity to travel.

Some of these points may be exemplified in Ex. II.23.15, which is a complete transcription of the trope set *O laudabilis rex* as it appears in Paris, Bibliothèque Nationale, lat. 10508 (from Saint-Évroult in Normandy, twelfth century). The trope is found in sources right across France and in North Spain, Italy, and England as well. An inkling of the complexity of its transmission may be gained from Table II.23.2. This shows how the veses are arranged in Paris 10508, in Oxford, Bodleian Library, Bodley 775 (Winchester, early eleventh century), Paris 1240 (the oldest of all sources, which has two versions, from Limoges, tenth century), and Paris 1121 (Saint-Martial at Limoges, early eleventh century). As one might expect, given Rönnau's criteria, Paris 1240 has by far the most 'wandering verses'. The latest source, Paris 10508, has only one. The number of verses in Paris 1240 is unusually large. The table gives a good idea of the mobility of the verses, and the different verses which make up the set in each source. Simply saying that a source has the Gloria trope *O laudabilis rex* (or any other trope) therefore simplifies matters rather drastically. Rönnau called the four verses at the beginning which are common to all versions 'constitutive verses'.

O laudabilis rex is a Gloria trope which makes use of musical phrases identified by Rönnau (1967, 222–38) as formulas used in several other tropes as well (called by him 'Cento-Tropen'; Stäblein, 'Tropus', *MGG*, Ex. 14 is an example of a cento in the truer sense of the word, since its text is composed of distichs taken from Venantius Fortunatus' poem *Tempora florigero*). In Ex. II.23.15 the phrase marked 'b' is one of these. The others do not correspond exactly with Rönnau's demonstration, but show how a number of other melodic ideas recur:

'b' describes a double curve *abcbaGFGa*;
'x' moves from *a* to *F* and back again, corresponding to the latter half of 'b'; it is
    omnipresent in the Gloria melody itself;
in 'y' *bb* takes the place of *a* as reciting tone, falling to *F* and returning to *bb* or *a*;

usually it appears in a brief form *baGaGFa*; occasionally the reciting note is approached *FGb♭*;

'z' is a common mode 3 opening, here a whole tone lower

Much of the time, however, one has the impression of a highly ornate recitation around *a* or *b♭*, too fluid for convincing labelling. Significantly, the two trope verses which cannot be explained in this way are two unique to Paris 10508: *Misertus esto nostri* and *Nobis in terris* (T6 and T8).

Like several other genres for which trope verses were provided, Glorias have introductory verses which stand apart from the rest (see Kelly, 1984).

Among an interesting group of rather abnormal compositions noticed by Stephan (1956) is a Gloria whose trope verses are the Christmas song *Dies est leticie* (in Prague, University Library VI. C. 20, a fifteenth- to sixteenth-century song-book from Prague), but this is evidently a late sport. The same is true of the Sanctus with the song *Surrexit Christus hodie/Christ ist erstanden* as trope (in Erlangen, Universitätsbibliothek 464, a fifteenth-century south German miscellany). A pair of full-blown vernacular tropes were added on fos. 299ᵛ–300ᵛ of Limoges, Bibliothèque Municipale 2 (a gradual of the fourteenth century from Fontevrault): the Sanctus trope *Beaus peres* and the Agnus trope *Cist aigneaus*. (The manuscript's compiler evidently had a special interest in vernacular chants, for two French epistle farses appear elsewhere in the source—see section (xii) below.) Even more extraordinary than these is the use of a German dance (if Stephan is right) as a textless Sanctus trope, apparently notated mensurally in Erlangen 464.

Most Sanctus and Agnus tropes (edited by Iversen in CT 7 and 4 respectively) follow the pattern of introductory intercalatory verses. The repertory of prosulas for the Hosanna of the Sanctus has already been mentioned above (vii). Because of the way in which the reciting of the eucharistic prayer normally leads without a break into the singing of the Sanctus, introductory trope verses for the Sanctus are rather few (Thannabaur 1962, signals only four out of over 240 tropes; see Steiner, 'Trope', for examples of both Sanctus and Agnus introductions).

Ex. II.23.16 has supplementary trope verses for the three 'Sanctus' acclamations, and Hosanna prosulas in sequence form. The difference in musical character of the two types of trope is naturally very marked, though both seem to contain turns of phrase carried over from the main chant. The source used for the transcription, Madrid, Biblioteca Nacional 19421 (Catania, Sicily, second half of the twelfth century), records both prosulas one after the other, *Omnes tua gratia* referring to Christ and *Martyr Christi gratia*, based upon it, for the commemoration of a martyr.

Table II.23.2. *Gloria trope sets starting* O laudabilis rex

	Paris 1240, fo. 41v	Paris 1121, fo. 43v	Oxford 775, fo. 70r	Paris 10508, fo. 37r
GLORIA IN EXCELSIS DEO				
ET IN TERRA PAX	O laudabilis rex	O laudabilis rex	O laudabilis rex	O laudabilis rex
LAUDAMUS TE	Adonay benedicte	Adonay benedicte	Adonai benedicte	Adonay benedicte
BENEDICIMUS TE	O adoranda	O adoranda	O adoranda	O adoranda
ADORAMUS TE	Glorificande	Glorificande	Glorificande	Glorificande
GLORIFICAMUS TE	O bone rex (W)	Pax salus et vita (W)	Pax salus et vita (W)	Rex seculorum domine
GRATIAS AGIMUS	Plebs tua (W)	Sanctam maiestatem (W)		
PROPTER MAGNAM	Sanctam maiestatem (W)	Da pacem famulis (W)		
DOMINE DEUS	Pax salus et vita (W)	Aeternam cum sanctis		
DEUS PATER	Da pacem famulis (W)	Qui solus habes (W)		
DOMINE FILI	Pioque tuo amore			Misertus esto
IESU CHRISTE	Aeternam cum sanctis			

Gloria text			
DOMINE DEUS			
Magnus et fortis			
AGNUS DEI			
Rex pacificus (W)			
FILIUS PATRIS			
Redemptor universi (W)			
QUI TOLLIS	Rex seculorum domine		
Suscipe nunc (W)			
QUI TOLLIS		Rex seculorum domine	Pax salus et vita (W)
SUSCIPE DEPRECATIONEM	Caeli terraeque		
Qui super astra (W)			Nobis in terris
QUI SEDES			
O decus omnium (W)			O eterni sapientia
QUONIAM TU			
Eros poli (W)		Aeterni sapientia	Tu lux via
TU SOLUS DOMINUS			
Prolis O rutilis (W)		Tu lux via	O virtus honor
TU SOLUS ALTISSIMUS	Audi clemens		
Patri equalis		Rex regum (+ melisma)	Sceptrum tuum (+ prosula)
IESU CHRISTE			
Qui solus abes (W)			
CUM SANCTO SPIRITU			
Qui unus idemque (W)			
IN GLORIA DEI			

No Gloria melody is indicated in Paris 1240 or Paris 1121. Oxford 775 has Gloria 12, Paris 10508 has Gloria 39. (W) = wandering verse. The *Sceptrum tuum* prosula has the same melody as the *Regnum* melisma and its prosulas. The *Rex regum* melisma (in Oxford 775) is one of rather few melismas in the Gloria trope repertory which have a melody different from *Regnum*.

Ex. II.23.15. Gloria 39 with trope verses *O laudabilis rex*, etc. and Sceptrum prosula *Lumen eternum* (Paris, Bibl. Nat. lat. 10508, fos. 23ᵛ, 37ʳ)

AG- -NUS DEI.   FI -LI-US PATRIS.   QUI TOL- -LIS PECCA-TA MUN - DI

MI-SE-RE-RE   NO - BIS.   QUI TOL- -LIS PECCA-TA MUN - DI

T7   Pax sa - lus et ui -ta ho-minum ti - bi glo - ri - a.

SU-SCI- PE DE-PRE-CA-TI - O-NEM NO-   -STRAM.

T8   No-bis in ter-ris mi-se - re-re De-us al - me.

QUI SE-DES AD DEXTERAM PA- -TRIS MI-SE- RE-RE   NO-   -BIS.

T9   O   e-ter-ni sa-pi - en - ti - a pa-tris.

QUONI-AM TU SOLUS SANCTUS. T10 Tu lux ui - a   et spes no-stra.

TU SOLUS DO-MINUS.   T11 O   uirtus ho-nor De - i   pa-tris

et   glo-ri - a   om-ni-um ui-ta mo-ri uo-lu-is-ti pro cunctis O   bo-ne rex.

TU SO-LUS AL-TIS-SI-MUS.   T12 Sceptrum regni no-   -bi - le.

Lu-men... e-ternum qui splendor es sed de tu - o lu-mine.   E.
Sa-cre ec-cle-si - e so-ci-as-ti ad-mi-ra-bi-li do-te.   E.

(Ex.II.23.15 cont.)

Digna - re proles ab - sol - ue - re di - lec - te.   E.

Sponse tu - e di - uo pi - a - mi - ne.   E.

Perma - ne - bit in e - ternum.   IE- -SU   CHRISTE.

CUM SANCTO SPI - RI - TU IN GLO - RI - A DE - I   PA - TRIS.

A - MEN.

Ex. II.23.16. Sanctus 56 (Vatican III) with trope verses *Summe pater*, etc. and Osanna prosulas *Omnes tua gratia* and *Martyr Christi gratia* (Madrid, Bib. Nac. 19421, fo. 90ᵛ)

SAN-   -CTUS   Sum - me pater de quo mundi princi - pi - a   con - stant.

SANCTUS   Fi - li - us om - ni - po - tens per quem patris est pi - e   uel - le.

SAN-   -CTUS   Spi - ri - tus in quo par   uir - tus si - ne fi - ne re - fulget.

DO- -MINUS DE - US SA - BAOTH. PLENI SUNT CE - LI ET TERRA GLO - RI - A TU - A.

O- -SAN - NA IN EXCELSIS.   BE - NEDICTUS QUI UENIT IN NOMI - NE DO - MI - NI.

Omnes tu - a gra - ti - a   quos a morte re - demis - ti per - pe - tu - a.

(Ex. II.23.19 cont.)

Morte tu-a uis mortis cum principe proculcans ui-te nos re-pa-ras.
De-o pa-tri dans carum te pro nobis pre-ci-um et ui-uam hosti-am.

Tecum nos re-susci-ta.   Tecum in ce-lis col-lo-ca.   Et re-gni lar-gi-re consor-ti-a.
Te er-go de-posci-mus.   Ut cum iudex ad-ue-ne-ris.   Cunctorum dis-cerne-re me-ri-ta.

Nos cum ange-lis et sanctis so-ci-es.   O - SAN-NA IN EXCELSIS.
Cum quibus ti-bi ca-namus ... ...

Martyr Christi grati-a   cuius mortem sequens habes et gaudi-a.

Qui pro tu-is e-xoras-ti i-ni-micis ca-ri-ta-te plenus ge-mi-na.
De-i patris ... fi-li-um Ihesum pro te astantem ad patris dex-teram.

Ui-dis-ti magna gra-ti-a ... ... Cer-...-tus proti-nus dul-ci-a
Nunc er-...-go te poscimus martyr sancte ut de-uo-tis fa-mu-lis

tecum il-lo ha-bi-turum gau-di-a.
tu-a pos-tules ... ... consor-ti-a.

Simul ut an-ge-lis et sanctis so-ci-i.   O- -SAN-NA IN EXCELSIS.
cum...quibus De-o ca-namus ... ...

Some Sanctus tropes, and Agnus tropes as well, 'migrate' from one melody to another. With the Sanctus repertory this is perhaps not altogether surprising, since although the number of known Sanctus melodies is very large, they display little tonal variety, for the melodies will normally be compatible with the intoned prayers which surround the chant.

Prosulas for Agnus melodies are to my knowledge unknown. Ex. II.23.17 has both an introductory verse and verses before each 'Miserere nobis'. Its text is rhyming prose, where each first half-line of five to eight syllables rhymes with the next half, of six to fifteen syllables. This is not matched by musical rhyme, although certain melodic phrases are used several times. (It is admittedly often difficult to separate the 'common coin' of the melodic language from conscious melodic references.)

This source, as is very frequently the case with Agnus tropes in 'western' manuscripts, gives simply a 'miserere' incipit after the later (four) trope verses. It is not certain whether in these cases 'Agnus dei . . . mundi' should be supplied before each trope verse. This is in fact what usually appears in 'eastern' sources. But as it stands here, Ex. II.23.17 becomes a short litany-type composition, with leaders singing verses to which the choir responds with a refrain. The five 'miserere' are not uncommon in early French sources, although three is the usual number.

Ex. II.23.17. Agnus 15 with trope verses *Pro cunctis deductus*, etc. (Paris, Bibl. Nat. lat. 1134, fo. 106ᵛ)

Italian books prefer another arrangement, one which undoubtedly reflects a real difference in performance practice. Here the trope verse remains outside the Agnus verse. (Iversen in CT 4 makes clear these and other regional variations in practice.) Ex. II.23.18 displays both this and another fashion cultivated by Italian musicians, that of reusing the music of the main chant for the trope verses. When this happens more than once for the same melody, tropes with identical music result. The use of a pes (two-note ascending group) for 'secula' in the last trope verse of *Ad dextram patris* is typical of South Italian recitation passages. This making of trope melodies from the music of the main chant is also known from some Italian Kyries (see section (viii) above) and is related to other Italian composition techniques (on sequences see II.22.v).

Ex. II.23.18. Agnus 81a with trope verses *Ad dextram patris*, etc. and *Humanum genus*, etc. (Benevento, Arch. Cap. 34, fo. 181ᵛ, and 35, fo. 199ʳ)

(xii) *Farsed Lessons, Creeds, and Paternoster*

A different way of elaborating chants on special feasts was employed from the twelfth century onward for lessons and some other chants, especially in the highly individual liturgies of such feasts as Circumcision known from Beauvais, Sens, Laon, and other centres (see above, I.9). Phrases of pre-existent chants were inserted into the lesson. Such an insertion is often referred to as *farsa*, both in the Middle Ages and in modern writings, although 'farse' can refer to other types of troping as well. The epistle of mass was the chant most often farsed: over forty examples are known (nearly all listed, with sources, and with the epistle tone they accompany, in Stäblein, 'Epistel', *MGG*; see also Huglo, 'Epistle', *NG*). The Gospel, by contrast, was left unadorned (Stäblein, 'Evangelium', *MGG* cites three examples only, all found in German and Swiss sources). The Paternoster, Nicene creed, and (from Compline) the Apostles' Creed were also farsed.

The range of farsed chants sometimes present in the festal offices can be appreciated by listing those in the manuscript containing the Sens Circumcision office:

> for Circumcision: Paternoster and Apostles' Creed at Compline; Gloria, epistle, creed at mass
> for St Stephen: epistle
> for St John: epistle
> for Innocents: epistle

(Farsed chants present in the festal offices of Santiago, Sens, and Beauvais are edited in Wagner 1931, Villetard 1907, and Arlt 1970 respectively.)

Table II.23.3. *Farsed Apostles' Creed in Laon 263*

Farse verses	Source (all for Christmas season unless stated)
Solus qui tuetur	Sequence *Nato canunt omnia*
Sine quo nichil	?
Natum ante secula	Sequence *Natus ante secula*
Pro mundi remedio	Benedicamus song *Corde patris genitus*
[Natus] ineffabiliter	Hymn *Christe redemptor omnium*
Sol de stella	Sequence *Letabundus*
Ipse potestate	Hymn (St Peter) *Aurea luce*
Qui nulla perpetraret	Sequence (Sundays after Pentecost) *Stans a longe*
Gemit capta	Sequence *Eia recolamus*
Tirannum trudens	Hymn (Easter) *Ad cenam agni*
Unde descenderat	Sequence (Ascension) *Rex omnipotens*
Regna cuius iure	Sanctus trope *Perpetuo numine*
Reddens vicem	Hymn (Advent) *Verbum supernum*
Sine quo preces	Sequence (Pentecost) *Sancti spiritus*
Que construitur	Hymn (Dedication) *Urbs beata Ierusalem*
Angeli quorum	Antiphon (Innocents) *Angeli eorum*
Quibus deum	?
Inmortalitatem	?
Quem repromisit	Antiphon (Martyr) *Beatus vir*

The technique may be illustrated by Ex. II.23.19, the Apostle's Creed as farsed among the Epiphany chants of Laon, Bibliothèque Municipale 263 (Laon, late twelfth century). The piece is also present in the Beauvais and Sens offices. Villetard's list of the sources of the borrowed phrases is given in Table II.23.3).

Ex. II.23.19. Farsed Credo (Laon, Bibl. Mun. 263, fo. 139[r])

(Ex. II.23.19 *cont.*)

ASCENDIT AD CELOS.   Un-de descen-derat.

SEDET AD DEXTERAM DEI PATRIS OMNIPOTENTIS.   Reg-na cuius iu-re tenet pa-ri-li.

INDE UENTURUS IUDICARE UIUOS ET MORTUOS.   Reddens uicem per ab-di-tis

iustis-que regnum pro bonis.

CREDO IN SPIRITUM SANCTUM.

Sine quo preces omnes casse creduntur et indigne De-i auribus.

SANCTAM ECCLESIAM CATHOLICAM. Que construitur in celis uiuis ex la - pi - di - bus.

SANCTORUM COMMUNIONEM.   An-ge-li quorum semper uident faci-em patris.

REMISSIONEM PECCATORUM. Quibus Deum of-fen-dimus cor-de uerbis o-pe-ri-bus.

CARNIS RESURRECTIONEM.   In-mor-ta-li-tatem cum Christo.

UITAM ETERNAM.   Quem repro-misit De-us di-li-gen-tibus se.   A-   -men.

Probably the most widely known farsed epistle was the one sung all over Europe (see sources in AH 49, 171 and Stäblein 1975, 61, n. 601) on Christmas Eve, Christmas Day, or Circumcision, whose introductory verses are *Laudes Deo dicam per secula*, seemingly made for the purpose, in alcaic metre, and sung to a simple repeating musical phrase (Ex. II.23.20). (Some sources read *Laudem* and/or *dicant*.)

Ex. II.23.20. Introductory verse *Laudes Deo* for a farsed epistle (Limoges, Bibl. Mun. 2, fo. 26ʳ)

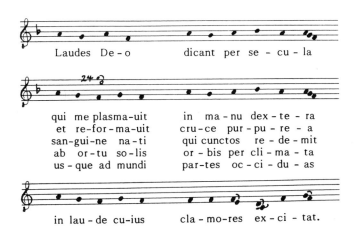

Laudes De - o   dicant per se - cu - la

qui me plasma-uit   in ma-nu dex-te-ra
et re-for-ma-uit   cru-ce pur-pu-re-a
san-gui-ne na-ti   qui cunctos re-de-mit
ab or-tu so-lis   or-bis per cli-ma-ta
us-que ad mundi   par-tes oc-ci-du-as

in lau-de cu-ius   cla-mo-res ex-ci-tat.

Ex. II.23.21. Start of farsed lesson with verses *Ce que Ysaies nos escrit*, etc. (Limoges, Bibl. Mun. 2, fo. 46ᵛ)

1. ... ... Ce que Y-sa-ies nos es-crit ...   de ... l'a-ue-nement Ihe-su Crist ...
2. ... ... Or nous leuons en contre lu-i   si comme orroit encore ... ...
3. Di-ex ap-parut ... c'est la pre-miesse   et l'estoille est l'autre lu-mie-re

1. bi-en nos do-it es-tre hu-i en re-meenbran-ce
2. hu-i doit es-tre chascuns ... ... es-cle-ri-ez
3. par cui i uindrent ... ... li troi roi ...

1. qui en di-eu auons no fi-en-ce ...
2. li seinz ... ... iors est re-pe-riez ...
3. la ti-er-ce lu-mie-re est la fo-iz.

1. car il ... en hor-te et semont   nos me-ismes et tout ... le mo-rit.
2. qui trois meunieres de clar-tez   nos a dou... ciel hu-i a-por-tez.

LECTI-O Y-SA-I-E PROPHE-TE.

Other farsing verses 'made to measure', so to speak, rather than lifted from elsewhere, are a small number of vernacular insertions found in a few French and Spanish sources of the fourteenth century onward (see Huglo, 'Farse', *NG*). Limoges 2 (Fontevrault, fourteenth century) is the most important of these, with five sets. The opening of the Epiphany farse, which continues as an alternation between more verses of the farse (all with the same melody) and lines of the epistle is transcribed in Ex. II.23.21.

Of several surviving Paternoster farses only *Fidem auge* in the Sens and Beauvais offices is relatively well known (edn. in Villetard 1907 and Arlt 1970). Stäblein (1977) has surveyed over twenty examples, of which one is polyphonic, for two voices, and two-thirds are copied without musical notation. Laon 263 has two, all other sources one only. Four are transcribed by Stäblein.

II.24. LATIN LITURGICAL SONGS

  (i) Introduction
 (ii) Versus in Early Aquitanian Manuscripts; Versus with 'Double Cursus'
(iii) Twelfth-Century Songs: Textual and Musical Style
(iv) Twelfth-Century Songs: Liturgical Function

Schmitz 1936; Stäblein, 'Saint-Martial', 'Versus', *MGG*; Emerson 1964; Harrison 1965; Treitler 1967; Arlt 1970; Crocker, 'Versus', *NG*; Planchart and Fuller, 'St Martial', *NG*.
   Editions: Villetard 1907; Treitler 1967; Arlt 1970.
   Facsimiles: Arlt and Stauffacher 1986; Gillingham 1987, 1989.

(i) *Introduction*

Benedicamus chants with extra text supplementing the basic 'Benedicamus domino: Deo gratias' have already been mentioned (II.23.ix). By the twelfth-century such compositions were often made with texts having rhyme and often regular rhythmic stress, features of new composition in many genres during that period. The texts are often strophic, and some have refrains. Very often the new style of the text (rhyme and regular stress-patterns) is matched by a new musical style, emphasizing steps of a third (tertian melodic style), presenting the fifth and octave above the final as melodic goals, with free and rapid movement through the chosen range, a tendency to compartmentalize movement of individual phrases into one or other segment of the octave, and the coupling of stressed syllables to those notes which form the tonal backbone of the music. Some pieces refer to liturgical formulas other than the Benedicamus text; a large number are free of liturgical references. Medieval names for these compositions include *versus, conductus, cantilena,* and *cantio.* Here they are called songs.

These songs share some characteristics with hymns and processional hymns (II.15). Like the songs, hymns have poetic texts in strophic form, and sometimes regular rhythm, with relatively simple music. Processional function and strophe-plus-refrain structure are further links. The new compositions did not, however, penetrate the traditional cycle of office hymns to any significant extent.

For many of the twelfth-century songs a function has to be surmised from circumstantial evidence. The same is also true of some other sacred songs in Aquitanian sources of the ninth to eleventh centuries, discussed briefly here before the twelfth-century repertory is addressed.

(ii) *Versus in Early Aquitanian Manuscripts; Versus with 'Double Cursus'*

Paris, Bibliothèque Nationale, lat. 1154 is a manuscript from Aquitaine probably of the second half of the ninth century, whose contents include (fos. 98–143) an important collection of Latin poems, some of which were notated (probably not before the tenth century). Most of the pieces are called *versus* in the manuscript, others *ritmus, carmen, planctus* (for two laments), *hymnus* (two of them are the processional hymns *Pange lingua* and *Tellus ac ethra*), and one *prosa* (a liturgical sequence for St Martial, *Concelebremus sacram*). Several of the pieces are abecedary hymns, that is, strophic compositions where each strophe begins with a different letter, progressing from A through the alphabet. A wide range of classical metres is represented in the collection, and the use of assonance is strikingly common.

Only four of these pieces have been found elsewhere as liturgical items, and the only reason for mentioning the rest is that their compositional techniques were presumably known to, and might have served as models for, the writers of liturgical pieces. Those found in liturgical manuscripts are the two processional hymns and the sequence just mentioned, which are copied together at the end of the collection, and another sequence, *Sancte Paule.* Many of the works are sacred at least in a general sense; some are topical, secular works (inventory and bibliography in Spanke 1931). Several of the poems are attributed to Gottschalk, Paulinus of Aquileia, and Boethius.

Scattered items of this sort are found in other sources. (See Fallows, 'Sources, MS, III, 2: Secular Monophony, Latin', *NG.*) In numbers they are more than outweighed by the fifty 'Goliard' songs in the 'Cambridge Songbook' (Cambridge, University Library Gg. v. 35, copied in the mid-eleventh century at St Augustine's abbey, Canterbury, from a Rhenish exemplar), certainly a secular repertory, but again one whose poetic techniques may not be without relevance for liturgical poetry. For the great majority of the pieces, no transcription is possible.

The importance of these works in the history of liturgical music has perhaps been exaggerated, since—unless more of the works can be shown to have fulfilled a role in the liturgy—to view them as any more than possible models for liturgical compositions seems over-enthusiastic. A considerable body of literature has nevertheless arisen around some of them, in particular *Sancte Paule*, and one other piece, *Dulce carmen*, labelled 'Versus de sco. Mauritio'. These are the two examples in Paris 1154 of what has been called (since Winterfeld, 1901) the 'sequence with double cursus' (Handschin's less happy term was 'archaic sequence'). These compositions are constructed at least partly in paired verses, like the sequence; but their later verses repeat the music of earlier ones, so that the text also must use the same structure: hence the term 'double cursus'. So far eight such pieces have been discovered. It should be pointed out that sets of four verses to the same melody, or even more, are almost as common as pairs (see the tabular summary of their structure in Phillips and Huglo 1982, with bibliography, to which can be added Stäblein 1978; on 'recapitulations' in early liturgical sequence melodies see above, II.22.ii).

The most famous of these double-cursus versus is *Rex caeli domine*, cited in the ninth-century treatise *Musica enchiriadis* and among the earliest known examples of polyphonic music (see Phillips and Huglo 1982 for corrections to the editions in Handschin 1929 and Stäblein, 'Sequenz', *MGG*). Given the lack of connection between *Rex caeli* (and most of the other double-cursus versus) and the liturgy, the assignation to such pieces of a critical role in the early development of the liturgical sequence by such writers as Dronke (1965) should be treated with caution. The same should be said of the early versus in general. There can be no doubt as to their artistic accomplishment: their texts display a considerable range of metre- and rhyme-schemes, and their music is frequently cast in short-breathed phrases to reflect the poetic designs. But in precisely these respects they are quite different in impact from those of the normal tenth- to eleventh-century liturgical sequence. Whereas the first-epoch sequence couples an expansive melodic exuberance with formal ruggedness, the versus prefers the deft interplay of small poetic and melodic units, without aiming for the monumentality of the sequence.

Ex. II.24.1, the opening of *Sancte Paule pastor bone*, shows how the versus in Paris 1154 (such as can be transcribed, that is) seem to proceed in short versicles which are interrelated musically. No line in the poem is longer than eleven syllables (in *Rex caeli* no more than thirteen syllables), and the average length is much shorter. The overall musical form of the piece, following de Goede's reconstruction, is:

AA   B1 B1   C C   B2 B2   D1 D2
AA   B1 B1   C C   B2 B2   D3 D4
B3

But this is only part of the story. As can be seen from the transcription (which is of AA B1B1 CC B2B2 in the above scheme) each verse is composed of smaller lines, indicated by small letters in the transcription. (The version of the 'D' section in Paris 1154 is corrupt according to Spanke, 1931, and thus the reconstruction presented by

Ex. II.24.1. Start of versus *Sancte Paule pastor bone* (Paris, Bibl. Nat. lat. 1154, fol. 129ᵛ; Benevento, Arch. Cap. 34, fo. 208ᵛ)

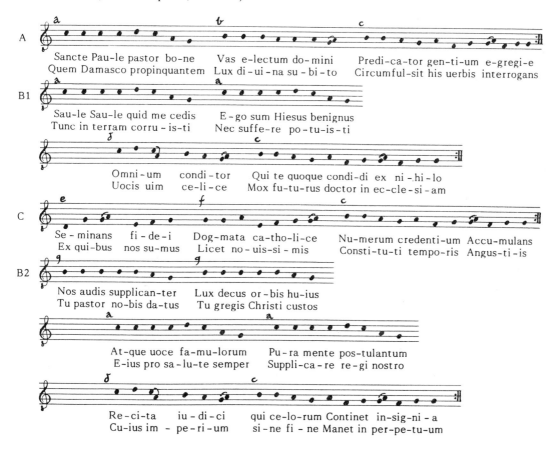

de Goede 1965, LXI is hypothetical. For further information on this and other such pieces see also Stäblein 1978.)

(iii) *Twelfth-Century Songs: Textual and Musical Style*

Some of the stylistic characteristics of the early secular versus (though not the double-cursus idea) are also found in the songs of the twelfth century. Whether it was that poetic and musical resources previously considered appropriate only for secular songs were now brought into liturgical use is not clear. (Such poets as Abelard—see Weinrich, 'Abelard', *NG*, and Huglo 1979, 'Abelard'—contributed to both liturgical and secular repertories.) Another question, which might seem to stand the previous idea on its head, is whether the liturgical songs of the twelfth century exercised any influence on the vernacular lyrics of the troubadours. The matter is really beyond the scope of this book, but readers may like to consider for themselves whether or not the typical features of the Latin liturgical songs described here are strongly reflected in the vernacular repertories. (No writer has subjected both textual and musical features of both repertories to an adequate analysis. Starting-points in the various areas are provided by Spanke 1930–1 and 1936, Treitler 1967, Arlt 1970.)

Ex. II.24.2 is a typical example of a twelfth-century liturgical song, one which includes the words 'Benedicamus domino' and 'Deo gratias', fitted neatly into the rhyme-scheme of the poem, in the extended third and sixth strophes of the piece. Melodic movement is almost completely tertian for the first four lines of the piece. At first the triad *a–c–e* is stressed, then after the rapid descent to low *D* and *C* there is some gentle playing off of the *D–F* interval against the *C–E–G* triad upon which the piece finally settles. The iambic rhythm of the text is so pervasive that slight irregularities (according to classical Latin usage) seem irrelevant. The quantities are less regular in alternating short and long. Yet it is noticeable that two-note groups always fall on what are felt to be stressed syllables in the second half of each strophe. Whether this reflects a performance in continuous up-beat 6/8 rhythm is of course open to question. What is beyond dispute is the regular rhythmic nature of the text, which the music highlights rather than disguises. (Regular note-grouping in the opposite sense, where it is the unstressed syllables which are given two-note groups, can also be found, for example in *Dei sapientia*, Arlt 1970, Editionsband, 148.)

Ex. II.24.2. Benedicamus song *Lux omni festa populo* (Madrid, Bibl. Nac. 289, fo. 132ᵛ)

```
a
1. Lux om-ni fes-ta po-pu-lo      re-cur-rit an-ni cir-cu-lo
b                                  1: F
   qua nun-ci-an-te an-ge-lo      ex-or-ta est re-dempti - o.
c
   Nostra-que li-be-ra-ci-o       ser-pentis ex a-cu-le-o.
```

2.   a   Dum omnia silencio   continerentur medio

      b   et nox inter altissimos   perageret curriculo.

      c   Sermo tuus O genitor   regali uenit solio

3.   a   Sponsus uti de thalamo   preceteris formosior

      b   ita de matris utero   processit orbis conditor.

      c   Pro seculi remedio   effectus est Deus homo.

      c   Quo circa nos in iubilo   BENEDICAMUS DOMINO.

4. a O matris alme uiscera   repleta Dei gracia

   b que genuerunt talia   tanta sacrata pignora.

   c Beata quoque ubera   que puer ille sugxerat.

5. a Cui tota celi curia   tremens in laude consonat

   b Cui talis est potencia   ut illi que sunt omnia

   c celestia terrestria   genuflectuntur subdita.

6. a Cuius misericordia   et admiranda bonitas.

   b A morte nos perpetua   aduentu primo liberet.

   c Secundo nos eripiat   ab infernali fouea.

   c Ut in polorum regia   dicamus DEO GRACIAS.

This simple and direct melodic idiom is the preferred style of what may be called the North French tradition in this international repertory, represented by the sources with festal offices from Laon, Sens, and Beauvais (I.9) and also the earlier tropers from Norman Sicily (Madrid, Biblioteca Nacional 288, 289, and 19421), all of which manuscripts date from the twelfth or early thirteenth century. Aquitanian twelfth-century sources occasionally record much more florid compositions. *Letabundi iubilemus*, Ex. II.24.3, is from Paris, Bibliothèque Nationale, lat. 1139, where it begins a series of pieces headed 'Ic incoant Benedicamus' ('Here begin the Benedicamus pieces'), although the text of this particular piece makes no reference to the ritual words. (In the Le Puy office it is rubricked as a 'conductus'.)

The text of *Letabundi iubilemus* is clearly of the newer type, with line-lengths far more varied than in Ex. II.24.2 (' = stressed or half-stressed at line-ends; = unstressed):

Letabundi iubilemus   '.'.'.

Accurate celebremus   '.'.'.

Christi natalitia   '.'.'

summa leticia   .'.'

cum gratia   .'

produxit   ' '

gratanter mentibus   .'.'

fidelibus   .'

inluxit.   ' '

The text consists of three regular strophes, set in an ABA (da capo) musical form. But the textual similarity of the three strophes is quite obscured by the very ornate setting for the central strophe (in this source, though not in the Le Puy version), as if the compressed energy of the syllabic lines suddenly explodes in a virtuoso vocal display. Chains of thirds are much less prominent, partly because of the ornate surface of the melody, though the strong *E–G* and *a–c–e* chains of the first three lines, and *a–c–e* in 'quod in nobis' and 'arcanum', are heard clearly enough. However, if the note-groupings within the melismas reflect phrasing in performance, then other thirds may be audible: in '(Eructa)vit' for example, *c–e*, *f–d*, *e–c*, *d–b*, etc. And such progressions as that at 'vis acerbum', partially concealed by the text, are also of

Ex. II.24.3. Latin song *Letabundi iubilemus* (Paris, Bibl. Nat. lat. 1139, fo. 58ʳ)

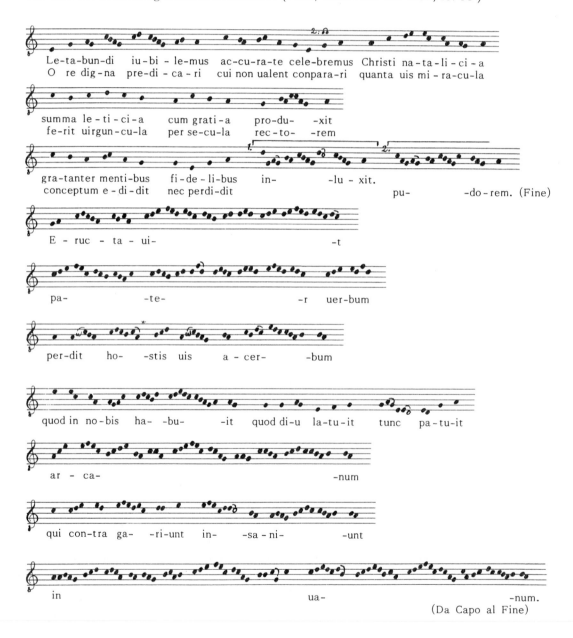

significance: *G–b*, *a–c*, *b–d*, *c–e*. An important element in the setting of the 'Eructavit' strophe is the roulade *cde–fed–edc–dc* in one permutation or another.

General speaking, the Benedicamus songs are the less adventurous in poetic metre and form, and few Benedicamus songs are in florid musical style, whereas songs with free texts often display rhythmic and melodic designs of considerable brilliance. The extravagance can be purely rhythmic, as in *Natus est, natus est* (Ex. II.24.4), where the repetitive melody highlights, by its very *naïveté*, the irresistible drive of the poem. The piece has two large strophes, with numerous internal repeats, which link it indirectly to compositions like Ex. II.24.1 above.

Ex. II.24.4. Latin song *Natus est, natus est* (Madrid, Bibl. Nac. 289, fo. 144ᵛ)

Poetic exuberance of a different kind is displayed in *Da laudis homo* (Arlt 1970, Darstellungsband, 212), where the repetition is textual, suggesting an almost rapt amazement at the wonder of the Incarnation.

Refrain songs are another characteristic feature of the repertory. Exx. II.24.5–7 show three examples. In Ex. II.24.5 the refrain consists simply of line endings.

Ex. II.24.5. Benedicamus song *Thesaurus nove gracie* (Madrid, Bibl. Nac. 289, fo. 134ᵛ)

1. The-sau-rus no - ue gra - ci - e
   est re-ue-la-tus ho-di - e

2. Tam cla-ris in na - ta - li - bus
   gau-den-dum con-stat om - ni - bus

3. Le - te-mur er - go pa - ri - ter
   psal-la-mus un - a - ni - mi - ter

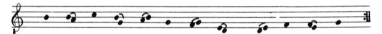

1. ce - les-tis flo-rem ger-mi - nis   E - ma-nu - el
   pro-du-xit al-uus uir-gi - nis   in Is - ra - el.

2. se-cun-da sa - lus om-ni - um   E - ma-nu - el
   cu-mu-ne fa - cit gau-di - um   in Is - ra - el.

3. et cor-de uo-ci con-so - no   E - ma-nu - el
   BE-NE-DI-CA-MUS DO-MI-NO   in Is - ra - el.

(most of strophes 4–6, for DEO GRACIAS, lost)

Ex. II.24.6. Latin song *Plebs Domini hac die* (Paris, Bibl. Nat. lat. 3549, fo. 167ᵛ)

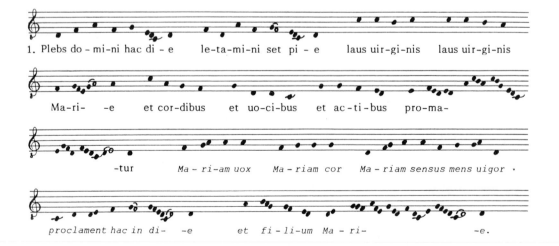

1. Plebs do - mi-ni hac di - e   le-ta-mi-ni set pi - e   laus uir-gi-nis   laus uir-gi-nis

Ma-ri- -e   et cor-dibus   et uo-ci-bus   et ac-ti-bus   pro-ma-

-tur   Ma - ri-am uox   Ma-riam cor   Ma - riam sensus mens uigor ·

proclament hac in di- -e   et fi-li-um Ma - ri- -e.

(full text AH 20 no.149 p.119)

8. Hoc sobrie de quanta
    stirps gracie gens sancta
    materie
    materie
    de tanta
    cum gaudio
    fit mentio
    iam lectio
    legatur.

Ex. II.24.6 has a refrain ('Mariam uox', etc.) equal in length to the strophes of the poem, and perhaps outweighing them musically by virtue of the melismas (it is characteristic of many songs to place melismas on the last stressed syllable of line; Ex. II.24.6 has both short and long ones). The poem ends with an invitation to begin the reading of the lesson, and is thus one of several such songs whose probable function was to accompany the procession of the subdeacon or deacon to the lectern to intone the Epistle or Gospel. (There is no indication for a repeat of the refrain after the final strophe.)

Ex. II.24.7. Latin song *Ave mater salvatoris* (London, Brit. Lib. Add 36881, fo. 32ᵛ)

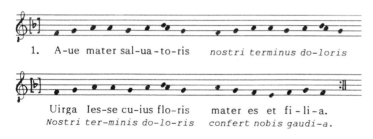

1. A-ue mater sal-ua-to-ris   *nostri terminus do-loris*

Uirga Ies-se cu-ius flo-ris   mater es et fi-li-a.
Nostri ter-minis do-lo-ris   confert nobis gaudi-a.

2. Moyses ardentem foris
    nostri terminus doloris
    uidit rubum sed ardoris
    non passus incendia.
    Nostri terminus doloris
    confert nobis gaudia.

3. Angelici uerbum oris
    nostri terminus doloris
    de supernis missum oris
    te repleuit gratia.
    Nostri terminus doloris
    confert nobis gaudia.

Ex. II.24.7 is an example of what has been called a 'Latin rondeau', by analogy with the French vernacular rondeau of Adam de la Halle and his contemporaries. It is easiest to represent its construction schematically. The music uses only two melodic units, in the order AAABAB, the text includes the first half of the refrain within the strophe, then the full refrain at the end:

strophe line 1
refrain line 1
strophe lines 2, 3
refrain lines 1, 2

In French vernacular rondeaux the practice was usually to sing the full refrain at the beginning also.

One other important form was used more rarely in the Latin liturgical song repertory than for secular Latin songs (the so-called 'Notre-Dame' conductus, for example): that of the sequence, or lai, as it has sometimes been called in this context. The paired-verse structure of the sequence is already visible in Ex. II.24.3. The best-known example is *Alto consilio*, found in the Beauvais festal office and several other sources (for editions see Arlt 1970, Editionsband, 157; Stäblein 1966, 'Musik'; Van Deusen 1982, 50.) What distinguishes such pieces from the more usual rhymed sequence is the rhythmic variety in their texts and the tendency to build strophes out of a succession of phrases of diverse length and rhythm.

(iv) *Twelfth-Century Songs: Liturgical Function*

Ex. II.24.8 is another song from Paris 1139, one of three in the manuscript in the vernacular, or mixing Latin and Provençal (inventories in Spanke 1930–2 and Sarah Fuller 1969). *Be deu hoi mais* neatly illustrates several of the problems surrounding many of the songs found in the twelfth-century sources. The composition has been called (since Gautier 1886, 169) a trope of the *Tu autem*, that is, of the versicle sung after the end of a lesson in the office. Yet only the start of the text 'Tu autem domine miserere nobis R. Deo gratias' is quoted, and there is no musical correspondence (see Tones for the Lessons in *LU*). It is therefore questionable whether 'trope' is an appropriate term to use. By analogy with other examples, it seems possible that *Be deu* was a substitute for the *Tu autem* versicle.

Paris 1139 is the first of three twelfth-century Aquitanian books containing considerable numbers of liturgical songs (the others are Paris 3549 and 3719), and it contains pieces referring to several liturgical forms. *De supernis affero nuntium* (fo. 32ʳ) invites the cantor to begin the singing of *Deus in adiutorium*, the opening versicle of all the office hours. *Nunc clericorum concio* (fo. 33ᵛ) bears an erased

Ex. II.24.8. Marcaronic song *Be deu hoi mais* (Paris, Bibl. Nat. lat. 1139, fo. 44ʳ)

Be deu hoi ma-is finir nostra ra-zos   un pauc soi las que trop fo aut lo sos

le-uen doi clerc que di-jen lo respos   tu autem deus qui est pai-re glo-ri-os.

Nos te pre-iam que t' re-mem-bre de nos   quant tri-a-ras los mals d'antre los bos.

'Benedicamus' rubric, and in fact announces that a lesson is about to be read. And this is the first of many twelfth-century examples of songs evidently intended to introduce the reading of a lesson. (One has already been given above as Ex. II.24.6.)

The ritual words with which the lector asks the officiant's blessing before reading a lesson in the office is the versicle *Jube domne benedicere* (*LU* 119–21). And the words 'Jube domne' do indeed appear at the end of one song, *Congaudentes iubilemus*, in the Norman-Sicilian troper Madrid 289 (called a *conductus* in the manuscript—it is the earliest source to carry this word as a rubric, which is equivalent to the word *versus* in Paris 1139). As with the 'Tu autem' song, the music of the original tone is not used. Three *Jube* songs also appear in the so-called 'Codex Calixtinus' (Santiago de Compostela), dating from the 1160s, some twenty years later than Madrid 289 (ed. Wagner 1931, 39–40, 94–6, all called 'Conductum' and ending 'Lector lege et de rege, qui regit omne, dic Iube domne').

Of all songs apparently designed to introduce a lesson, *Resonet intonet* was evidently the most popular. With its final strophe:

Munda sit	pura sit	hec ergo concio
Audiat	sentiat	quid dicat lectio

it seems equivalent to a *Jube* song, and appears in front of an epistle with farse *Laudes deo dicam per secula* in Madrid 289 and a Beneventan source of the same period in private hands (see Stäblein 1975, 144 n. 8). It is also found in the song-book Cambridge, University Library Ff. i. 17 (Norman or English, twelfth–thirteenth century). But in two other south Italian sources it is amalgamated with *Laudes deo* and used as a farse throughout the epistle (partial facs. and edn., Stäblein 1975, 144–5).

When a song is placed in front of the epistle in this way there can be no doubt that it functioned as a prelude to the reading of the lesson. But nearly all songs in these sources lack rubrics to make their purpose plain. Even in the 'Codex Calixtinus' the songs are copied separately in supplements to the main body of liturgical material. Fortunately, there is ample evidence in the Circumcision offices of Sens (Villetard 1907) and Beauvais (Arlt 1970) as to the use of this type of material. (The story continues in the Le Puy office, text ed. Chevalier 1894; see also Arlt 1978.) Thus in the Sens office, Benedicamus songs, as one would expect, conclude first Vespers, Compline, Lauds, Prime, Terce, Sext, None, and second Vespers. There are no songs for the lessons of Matins, but there is a 'Conductus ad subdiaconum' before the epistle at Mass (which has the *Laudes deo* farse), and a 'Conductus ad evangelium' before the gospel. Neither song includes ritual words or contains an invitation to begin the reading, but their function of accompanying the approach to the lectern of the subdeacon and deacon respectively seems clear. The Beauvais office makes even more colourful provision by having conductus for each lesson of Matins as well (table in Arlt 1970, Darstellungsband, 97), and one of these includes an invitation to the reader: *Nostri festi gaudium*, for the fifth lesson, ends 'Legatur in gaudio lectio' (Arlt

1970, Editionsband, 59). After each lesson, a 'Benedictio' is sung—the equivalent of, or substitute for, *Jube domne*—which actually lifts a section out of some appropriate chant, usually a sequence. Going one better than Sens, the Beauvais office splits the song for the gospel into two parts, so that the second part is sung while the deacon returns to the altar area after the gospel (Arlt 1970, Editionsband, 114–16). The capacity of the Beauvais redactor to farse his liturgy seems unlimited.

Prime sources for liturgical songs in later centuries are to be found in eastern manuscripts such as the 'Moosburg Gradual' (Spanke 1930; Stein 1956; Lipphardt 1957) the 'Seckau Cantionarium' (Irtenkauf 1956), and the Prague 'cantionales' (Orel 1922; Orel, Hornof, and Vosyka 1921; Nejedlý 1954–6; Bužga 'Kantional (tschechisch)', *MGG*). (See also Jammers, 'Cantio', *MGG* and *NG*; Stäblein 1975, 75 n. 768 lists some two dozen sources, with bibliography.) The terms *cantionale* and *cantio* used of these sources and their contents should not obscure the fact that the songs are generically and stylistically identical to their twelfth-century predecessors; indeed, the manuscripts preserve many of the same compositions, prolonging their useful life down to the end of the Middle Ages. As in the early days, Benedicamus songs are easily the most common of those that refer to a liturgical formula (see, for example, the fourteenth-century collection in Aosta, Seminario Maggiore 9. E. 19, discussed by Harrison 1965).

The Italian repertories of Latin laude (if that is the correct term) are also stylistically related, again including concordances with the northern repertories, although their function—as part of non-liturgical religious exercises—sets them apart. Most are written in ballata form, which is absent from the northern repertories, despite the predilection for refrain songs. (See Damilano 1963 on the most substantial collection, Turin, Biblioteca Nazionale Universitaria, F. I. 4, an antiphoner from Bobbio.)

The further the style of such pieces moves from that of the earliest parts of the 'Gregorian' repertory, the more it has inspired writers to surmise links with secular music and vernacular repertories. Given the scanty records of secular music (excepting the special courtly repertories of troubadour and trouvère songs), the difficulties of defining styles which might be typically secular are considerable. It can safely be asserted, however, that with the assimilation of the twelfth-century song into the festal liturgies of many prominent churches, the chant repertory achieved a colour and variety unimaginable in previous centuries and still a source of surprised joy to those fortunate enough to encounter it.

II.25. LITURGICAL DRAMAS

  (iv) Christmas and Epiphany Ceremonies; Rachel's and Mary's Laments
  (v) Rhymed Ceremonies; the 'Fleury Playbook'; the *Ludus Danielis*

E.A. Schuler 1951; Lipphardt, 'Liturgische Dramen', *MGG*; Stratman 1972; Stevens, 'Medieval Drama, II. Liturgical Drama', *NG*; Rankin 1981, 'Music'; Norton 1983; Norton 1987.

    Editions: Coussemaker 1861; Young 1933 (all ceremonies then known, texts only); Donovan 1958 (texts from Spain); Lipphardt 1975–81 (texts of all known Easter ceremonies); Rankin 1981, 'Music' (north French and English ceremonies); Campbell and Davidson 1985 (facsimile of Orléans, Bibliothèque Municipale 201).

(i) *Liturgy and Drama*

A great part of the liturgy is meditational; the sequence of office hours, and the first part of mass, the Ante-communion or Mass of the Catechumens. Prayer and praise are joined to readings from sacred texts to inspire devotion and recollection of God's work in the world. The second part of mass, the Mass of the Faithful, is of a different character, a commemoration of the Last Supper in which priest and congregation perform again the simple and essential actions first performed by Christ and his disciples. At a time when communion for most of those present was rare (as it was throughout the Middle Ages), one may ask how much the mass was thought of as a commemorative ritual act and how much as a spectacle performed to inspire the devotion of non-participants. Whatever the case may be, the principle 'Do this in remembrance of me' could be extended to justify other actions. Thus at the procession on Palm Sunday, branches were carried and the people called 'Hosanna', just as when Christ entered Jerusalem (see Amalarius' remarks: Hanssens, ii. 58–9, cited by Hardison 1965, 111). The washing of the feet at the Mandatum ceremony is another example. Here particular events are not merely the subject of meditation and interpretation, but re-enacted.

A number of symbolic acts run parallel to these literal representations. The Deposition ceremonies, where the host or a cross was buried on Good Friday, to be brought up again on Easter Day, are symbolic re-enactments of Christ's burial and resurrection (see Young 1933, ch. 4). Hardison gives ample demonstration of the propensity of Amalarius (a few other writers follow him) to interpret elements of the liturgy symbolically, though it is unlikely that all were originally intended to be symbolic.

Amalarius' writings are one manifestation of a desire detectable in many aspects of ninth- and tenth-century liturgical practice to make the liturgy more vividly present and topical. The addition of trope verses to introits and other chants is another example, where previously neutral texts are explicitly keyed to particular themes. It is perhaps against this background that the addition of a new representational act, centred on the *Quem queritis* dialogue, should be understood.

In view of the modern connotations of the words 'drama' and 'play', I have generally avoided them, for the roughly equivalent medieval title *ludus* is rare.

Medieval terms, where present, are usually *officium* or *ordo*. This may be translated as 'ceremony', a word which remains evocative of the liturgical context.

(ii) *The* Quem queritis *Dialogue*

Some time in the ninth century an adjunct to the Easter liturgy was composed, a simple dialogue between an angel and the Marys visiting Christ's tomb on Easter morning, the famous *Quem queritis* dialogue. About its origin and the way in which it was first performed we are ignorant. That it achieved wide distribution already in the tenth century suggests that it was composed in the ninth, but in this, as in its place of origin, scholars are in disagreement (from an extensive literature on the subject may be cited de Boor 1967; Rankin 1985, '*Quem queritis*'; Davril 1986). In the tenth-century sources it could clearly occupy one of several different places in the liturgy. Sometimes it apparently preceded the Introit *Resurrexi* of Mass on Easter Day. Sometimes it was part of the procession before mass, which could conveniently make a station by a 'sepulchre' where the dialogue would be sung. Since most early sources lack specific rubrics, it is usually impossible to distinguish between this usage and the previous. Another place was at the end of the Night Office on Easter morning, following the last responsory (usually *Dum transisset sabbatum*, which appropriately sets the scene), and preceding the Te Deum. (On the various liturgical assignments, see McGee 1976, and Bjork 1980, 'Dissemination'.)

The dialogue appears in many forms, where extra verses usually precede and/or succeed a relatively stable core. (What constitutes the 'original' core is also the subject of controversy.) The version in Ex. II.25.1 has one extra verse before (*Hora est*) and one after (*Alleluia*) the *Quem queritis* exchange. Other sources have *Quem queritis* and *Iesum Nazarenum* a fourth lower, in D-mode (as in the next example, Ex. II.25.2). The next verses, *Non est hic* and *Ite nuntiate . . . surrexit*, are then in the same D-mode, but sound relatively higher, like an exultant climax. The word 'dicente' is not always present, for it simply provides a link to the next verse. It is worth noting the musical similarities between *Quem queritis* and *Iesum Nazarenum*: both begin and end with sub-phrases cadencing *F–G* (*C–D* in transposition), while in between comes a sub-phrase which begins by outlining the triad *F–a–c*.

As a glance at the text edition in CT 3, 217–23 will show, the first verse *Hora est psallite* can be found not only in Italian but also in French, Catalonian, and German sources, though the distribution is thin. The distribution of *Alleluia resurrexit dominus* is equally wide, and a little less thin. Some sources have additional verses after it, some have the introit *Resurrexi*. In the Pistoia source used for the transcription (two Pistoia manuscripts are the only ones with this particular combination of verses) there is no unequivocal rubric. It seems that one group of singers, representing the angels, has taken up a position behind the altar (in the closely similar ceremony from Piacenza they are led by the cantor: see Young 1933, i. 216), while the schola, representing the Marys, sings in the choir. For the final verse all come together in the choir. Just when this ceremony would have been sung is not

Ex. II.25.1. *Quem queritis* (Pistoia, Bibl. Cap. 121, fo. 33ʳ)

Ho-ra est psal-li-te iu-bet domnus ca - ne - re   e- -ia di - ci - te.

Tunc dicunt .ii. aut .iii. stantes retro altare quasi interrogando.

Quem que-ri - tis in se - pul - chro chris-ti - co - le.

R scola

Ie - sum Na-za-re-num cru-ci - fi - xum   O cae-li - co - le.

et dicunt ipsi retro altare

Non est hic sur-re-xit sic-ut pre-di - xe-rat

i- -te nun-ti - a - te qui-a sur - re-xit di -cen-tes.

et dicunt ipsi uenientes in choro

Al - le - lu - ia al - le - lu - ia   resur-re-xit do- -mi-nus.

clear: in Piacenza it followed Terce, which means at the point when the procession before mass would have been made. It seems likely therefore that these chants were the last to be sung before mass itself, and would have been followed by the introit. (See the more explicit rubrics from other Italian churches given by Young, i. 215–16 and 228–9, McGee, 12–13.) But in this case they do not lead directly into *Resurrexi* as a trope.

It must be admitted that only one of the many verses which were used to conclude the little ceremony clearly needs a succeeding text: this is *Eia carissimi verba canite Christi* which in two sources leads into *Resurrexi*, in another into the Te Deum, and in a fourth manuscript has no indication of the next chant. Traditional liturgical antiphons (their text, that is, if not always their music) are sometimes included in the scene. A group of sources from the Rhine–Mosel area, and Winchester, have *Surrexit dominus de sepulchro* (*CAO* 5079), some Swiss and south German ones have *Surrexit enim sicut dixit* (5081), Winchester has *Cito euntes* (1813) and *Venite et videte* (5352). Another verse is clearly not sung by the Marys, for it sets the scene as well, being a literal setting of Mark 16: 3: 'Et dicebant ad invicem: Quis revolvet nobis lapidem ab ostio monumenti? Alleluia, alleluia.' It is in fact another office antiphon (*CAO* 2697) drafted into a new setting.

Much more explicitly representational than the Pistoia version and the other Italian ones just mentioned is the version known from late tenth-century England. The complete text with notation is preserved in the two tropers from Winchester, Cambridge, Corpus Christi College 473 and Oxford, Bodleian Library, Bodley 775, and a full description of the ceremony survives in two copies of the *Regularis Concordia*, guide-lines for English monastic life drawn up in the wake of a council held at Winchester *c*.973. This all-important description says specifically that the ceremony is performed 'in imitation of the angel seated on the tomb and of the women coming with perfumes to anoint the body of Jesus' (*Regularis Concordia*, ed. Symons 1953, 50). The monk representing the angel has a symbolic palm in his hand, the three brethren who are the Marys come 'as though searching for something'.

One should not over-emphasize the naturalism of the scene. The monks wear liturgical vestments and carry thuribles instead of real ointments. The ceremony is not a historical reconstruction of past events, but a vivid demonstration to the community that Christ is risen, there and then. When the scene is over Te Deum is sung, with the bells of the church ringing the while. What is intensely moving about this starkly simple scene has nothing to do with the character identification which is the essence of modern drama, but comes from the distillation of the Easter miracle into a few brief utterances. Realism is marginal, would even have been an impoverishment, for the more specific the representation, the more the symbolic overtones are lost. Nothing in the biblical narrative suggests that the angel holds a palm, but it brings to mind Jesus' triumphal entry into Jerusalem a week earlier, binds the scene to the ever-present sacred history. To dress the monks in women's garments would make them totally foreign to the religious community. When the angel sings 'Venite et videte locum' he is not merely enunciating the biblical text, but actually singing a liturgical antiphon with that text, making the ceremony part of the eternal liturgical cycle.

We are exceptionally fortunate in having so early a description of the ceremony, for earlier copies of the text and music (and many later ones) are quite unspecific in this respect. The question of whether the scene was acted more or less realistically from the beginning is, however, beside the point, for not even the most specific descriptions, such as the Winchester one, aim for realism.

Sources of the tenth century, and many later ones, do not go beyond the core dialogue, usually with a few accompanying verses. Already during this early period the verses were adapted for other seasons, Christmas (*Quem queritis in presepe*—

mainly French and North Italian sources) and Ascension (*Quem creditis super astra ascendisse*—mainly French sources).

(iii) *Easter Ceremonies from the Eleventh Century Onward*

From the later eleventh century onward the simple dialogue was often enlarged with extra material, while a few sources arrange new episodes from the Easter story. The latter tend to differ considerably from each other in text and music, which provides fascinating material for the study of interrelationships between the traditions and of differing concepts of representational ceremonies. Instead of attempting to mention each source of new material, I have chosen to describe briefly the reasonably typical contents of just one manuscript: Madrid, Biblioteca Nacional V. 20–4, one of the Norman-Sicilian manuscripts of the twelfth century.

Few sources have quite so much material altogether as this source. For the Night Office of Easter morning there is the traditional scene for the Marys and the angel, much expanded from the tenth-century versions (Ex. II.25.2). At the start each Mary sings a short lamenting verse, which leads into a setting of a more extensive biblical extract than was earlier the case. The core dialogue *Quem queritis* is not literally biblical. The text here takes the narrative in Mark 16: 1–7, which was also the Gospel at Mass on Easter Day, as its starting point. Some is set without alteration, some paraphrased:

Verse	*Mark 16*
Quis revolvet	3
Ecce lapis revolutus	cf. 4–5
Nolite timere	cf. 6
Quem queritis	—
Hiesum Nazarenum	—
Non est hic	6
Venite et videte	cf. 6
Ecce locus	6
Ite dicite	7
. . . precedet vos	7
Ibi eum videbitis	7

From then on the text is new composed, though the Marys repeat the words of the angel as they go to tell the good news.

The way in which the expansion of the older dialogue has been effected is sometimes all too clearly visible. The traditional antiphon *Venite et videte* already paraphrases the angel's invitation to see the place where Jesus had lain. *Ecce locus*, the biblical text unaltered, is almost superfluous, though we may imagine a pause between the two as the Marys approach the place more closely. (The lament at the start ends with an incipit for the *Quis revolvet* verse, as if the scribe was copying from two different exemplars and did not amalgamate them wholly successfully. I have omitted the incipit in the example.)

Ex. II.25.2. *Visitatio sepulchri* (Madrid, Bibl. Nac. V. 20–4, fo. 102ᵛ)

In die resurrectionis domini uersus mulierum

Priora

He – u   mi – se – re   cur conti-git ui – de – re mortem sal – ua – to – ris.

Secunda

He – u   redempti-o Is-rahel   ut quid tali-ter   a – ge-re uo-lu-it.

Tertia

He–u   con-so-la-ti-o nostra   ut quid mortem sustinu-it.

Insimul omnes   Uersus Marie mulierum

O   De-us   Quis re-uol-uet no-bis la-pi-dem ab hosti-o mo-ru-men-ti.

Ec-ce la-pis re-uo-lu-tus et in-uenis sto-la candi-da   co-o-per-tus.

Angelus

No-li-te ti-me-re uos di-ci-te   quem que-ri-tis ad se-pulchrum christi-co-le.

Mulieres

Hiesum Na-za-re-num cru-ci-fi-xum que-ri-mus O ce-li-co-le.

Angelus

Non est hic sur-re-xit si-cut predi-xe-rat   ue-ni-te et ui-de-te locum   u-bi

po-si-tus   e-rat do-minus.   Ec-ce lo-cus u-bi   po-su-e-runt e-um.

I-te di-ci-te   dis-ci-pu-lis eius quia sur-re-xit de se-pulchro   et ec-ce

prece-det uos in Ga-li-le-am   i-bi e-um ui-de-bi-tis sic--ut di-xit uo-bis.

Mulieres

E-a-mus nun-ti-a-re mirum quod ui-di-mus et gaudium quod ac-ce-pi-mus

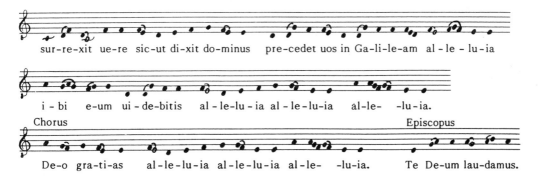

A few pages later in the manuscript further parts of the Easter story are represented. These are the encounter of two disciples with the risen Christ on the road to Emmaus, the 'Peregrinus' (pilgrim) episode, so called because the two do not recognize Christ at first, believing him to be a pilgrim (Luke 24: 13–32). Although Christ disappears at the end of the scene, he then returns in order to show the disciples his wounds (Luke 24: 36–9). The next episode in the manuscript is Mary Magdalene's encounter with Christ (John 20: 11–18). Finally, doubting Thomas is portrayed (John 20: 24–9).

The distribution in other manuscripts of settings of these episodes is somewhat patchy. Peregrinus plays are fairly widespread. Mary Magdalene episodes textually and musically related to this one are rarely found outside north French and English sources; similar scenes are present in sources from Germany and from Bohemia but were conceived quite independently. A scene with Thomas is given in only four other manuscripts: Tours, Bibliothèque Municipale 927 (from north France or England), Paris, Bibliothèque Nationale, n.a.l. 1064 (from Beauvais), Orléans, Bibliothèque Municipale 201 (the 'Fleury Playbook'), and Munich, Bayerische Staatsbibliothek, clm 4660a ('Carmina Burana'); in the Beauvais version the text is versified. Naturally there is a common recourse in these sources to dialogue drawn directly from the gospels. What tends to separate them is (*a*) the non-biblical material they include, (*b*) the musical setting, and (*c*) the casting of the text into verse.

There is no space here for a full demonstration of all these points. Fortunately, the Mary Magdalene episode has been lucidly analysed by Rankin (1981, 'Mary'). We have already seen that liturgical antiphons may be used in these scenes, understandable enough when so much of the biblical text was set in liturgical chants of this season (one needs only to browse through the Easter section of, say, the *Antiphonale monasticum* for this to become apparent; see Brockett 1977–8.) What is particularly interesting is that, as Rankin shows, some versions of the scene retain the traditional melodies for the antiphons thus drafted into use, while others replace them with newly composed music (Rankin 1981, 'Mary', Table 2 on p. 252).

The incorporation of pre-existing material raises questions about the musical consistency of the composition. The Thomas scene, as given in Madrid V. 20–4 (Ex. II.25.3), includes four liturgical antiphons:

	AM	CAO	mode
Vidimus dominum			D
Nisi videro			D
Data est michi	465	2099	G
Pax vobis! O Thoma			D
Misi digitum	480	3782	G
Quia vidisti me	479	4513	G
Surrexit dominus de sepulchro		5079	E

As mentioned above, the last item had previously been used in the simple dialogue scene. A conflict between it and the other traditional antiphons is understandable, but why were *Nisi videro* and *Pax vobis* not also made in G-mode?

Ex. II.25.3. Thomas scene (Madrid, Bibl. Nac. V. 20–4, fol. 107ᵛ)

Thomas autem non erat cum illis decem discipulis qui sunt in medio choro sed ueniens ex auerso ad illos decem discipulos et stabit. Qui surgentes dicant ei tribus uicibus:

Vi - di - mus do - mi - num.

Thomas respondit:

Ni - si ui-de-ro in manibus e-ius fi-xuram cla - uo-rum et mittam digitum meum

in lo-cum cla-uo-rum   et mittam manum meam in la-tus e - ius non credam al-lelu-ia.

Tunc ueniat Ihesus et appareat omnibus discipulis dicens:

Da-ta est mi-chi omnis po-testas in ce-lo et in ter-ra   al-le-lu-ia al-le-lu-ia.

Item dicat solummodo:

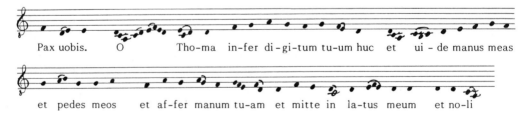

Pax uobis.   O   Tho-ma in-fer di-gi-tum tu-um huc et   ui - de manus meas

et pedes meos   et af-fer manum tu-am et mitte in la-tus meum   et no-li

es – se in-credu-lus sed fi – de-lis al – le – lu – ia.

Thomas uertat uultum suum ad populum dicat:

Mi-si di – gi-tum meum in fi-xu-ram cla-uorum et manum me-um in la-tus e-ius

et di-xi do – mi-nus me-us et De-us me-us al – le – lu – ia.

Tribus uicibus dicat *Dominus meus* adorans. Et hoc facto dicat Ihesus Thome:

Qui-a ui-disti me Thomas cre-di – di – sti be-a-ti qui non ui-derunt et cre-di-derunt

al – le – lu – ia. Tunc omnes discipuli uertant se ad populum. Insimul dicant
alta uoce:

Sur-re-xit do – mi – nus de se-pul-chro qui pro nobis pepen-dit in li-gno

al – le – lu-ia al – le – lu – ia al – le – lu – ia.

Such inconsistency, if we are to regard it as such, may still persist when the scene is
versified and set to new music. The Beauvais version has the following tonal structure:

	AM	CAO	mode
Pax vobis	(not 478)		E
Videte manus meus			E
Surrexit dominus de sepulchro		5079	E
Vere Thoma (rhymed)			D
Nisi fixuram (rhymed)			D
Pax vobis	478	4254	F
Thoma nunc vulnera (rhymed)			D
O Ihesu domine (rhymed)			D
Quia vidisti me	479	4513	G

Ex. II.25.4. Thomas scene (Paris, Bibl. Nat. n. a. l. 1064, fo. 10ᵛ)

Tunc ueniat Thomas qui defuerat. Et stanti in medio dicant ei duo pro aliis:

Ue-re Thoma   ui-di-mus do-minum
qui destru-xit mortis im-pe-ri-um.   Quibus Thomas:

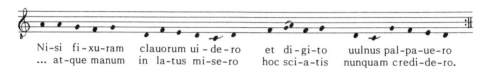

Ni-si fi-xu-ram   clauorum ui-de-ro   et di-gi-to   uulnus pal-pa-ue-ro
... at-que manum in la-tus mi-se-ro   hoc sci-a-tis nunquam credi-de-ro.

Tunc in medio ueniens dominus dicat omnibus:

Pax uobis   e-go sum   al-le-lu-ia   no-li-te ti-mere   al-le-lu-ia.

Deinde dicat Thome:

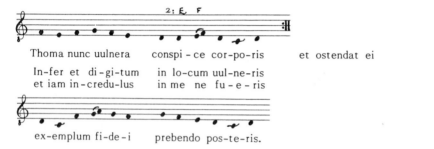

Thoma nunc uulnera   conspi-ce cor-po-ris   et ostendat ei

In-fer et di-gi-tum   in lo-cum uul-ne-ris
et iam in-credu-lus   in me ne fu-e-ris

ex-emplum fi-de-i   prebendo pos-te-ris.

The rhymed verses are all of a piece: syllabic D-mode items framed by liturgical antiphons in different modes. The first two items are musically similar, particularly in their use of the figure *FEDEFG* and the cadence *GaG–E* (found only in a handful of relatively late liturgical antiphons). Neither of these features is found in the antiphon *Surrexit dominus*. Ex. II.25.4 gives part of the scene, continuing from *Surrexit dominus de sepulchro*, already given in Ex. II.25.2. (The whole is edited by Coussemaker and Rankin.)

Rhymed verse is also present in the Madrid manuscript. At the start of the Peregrinus scene two openings are provided, one if the play is to be sung at Vespers on Easter Day, the other for Vespers on Easter Monday (the evening hour is appropriate because of the biblical setting). The first opening begins with two strophes, one for each disciple, of a rhyming hymn with eight syllables per line (text in Young 1933, i. 477). Another important feature of the play is the use of the sequence *Victimae paschali laudes* during the Mary Magdalene scene. The question to Mary which comes in the middle of the sequence (already given above, Ex. II.22.9) can be worked into a meeting between Mary and the disciples. The disciples sing the first verses of the sequence. Mary comes along and announces Christ's resurrection. The question-and-answer verses of the sequences follow, and the disciples sing the final pair. This is but one of many ways in which *Victimae paschali laudes* was worked into the Easter ceremonies. Sometimes only the question-and-answer verses were used; sometimes the sequence was made part of a scene with the three Marys.

In many eastern sources the basis for expansion away from the simple dialogue was rather different. The dialogue was recast textually and musically. Norton (1983, 1987) has identified the core of the new tradition as seven, instead of three verses, as follows:

Old tradition:	*New eastern tradition:*
	Quis revolvet nobis ab hostio lapidem
Quem queritis in sepulchro	Quem queritis O tremule mulieres
Iesum Nazarenum . . . O celicole	Iesum Nazarenum . . . querimus
Non est hic surrexit	Non est hic quem queritis
	Ad monumentum venimus
	Currebant duo simul (*CAO* 2081)
	Cernitis O socii

The new scene involves new action, for the angel's words *Non est hic*, telling the Marys to tell the good news to the disciples, are immediately acted upon. *Ad monumentum* is the Marys' announcement. *Currebant duo* tells how Peter and John run to the tomb. It is usually they who sing *Cernitis O socii*, an invitation to look at the grave-clothes and the empty tomb. Expansions of this core could include *Victimae paschali laudes*, increased dialogue for the disciples, and so on. The usual sequel is a chant of joy for the soloists or the schola, and often the Te Deum. Many later sources direct the vernacular hymn *Christ ist erstanden* to be sung.

The new music centres around *E* for the central verses, with *Quis revolvet* in A-mode and the last three verses in D. The transcription in Ex. II.25.5 is from one of several sources which transpose the central verses up a fourth to *a* (with *b*♭s). After the usual *Dum transisset*, the last responsory of the Night Office, a procession is to be made to a sepulchre, while the choir sings the antiphon *Maria Magdalena*. The seven core verses follow, then verses from *Victimae paschali* (*Dic nobis* is repeated as a choral refrain). *Christ ist erstanden* and the Te Deum conclude the ceremony. Although the liturgical antiphon *Currebant duo simul* is in indirect speech, it is allotted here to Peter and John.

Of the author of the new, German, form of the dialogue, we are as ignorant as in the case of the old *Quem queritis*. De Boor (1967) has summarized the main traditions, of which Norton (1983) sees Augsburg as perhaps the source. Among the earliest versions is one from Aquileia (Udine, Biblioteca Arcivescovile 234, a gradual of the late eleventh century), which corresponds very closely with later ones from Augsburg. This can only be explained by Aquileia's having received the ceremony from Augsburg in the first place, during the period when the influence of the German

Ex. II.25.5. *Visitatio sepulchri* (*Antiphonale Pataviense*, fo. 55ʳ)

Mulieres

Quis revolvet no‑bis ab os‑ti‑o la‑pi‑dem quem tegere sanctum cernimus sepulchrum.

Angelus

Quem queri‑tis O tremu‑le mu‑li‑e‑res: in hoc tu‑mu‑lo gementes.

Mulieres

Ie‑sus Na‑za‑re‑num cru‑ci‑fi‑xum que‑ri‑mus.

Angelus

Non est hic quem queri‑tis sed ci‑to e‑un‑tes nun‑ci‑a‑te dis‑ci‑pu‑lis e‑ius

et Petro: qui‑a sur‑re‑xit Ie‑sus.

Mulieres

Ad mo‑nu‑mentum ve‑ni‑mus gementes: an‑ge‑lum do‑mini se‑dentem vi‑di‑mus

et di‑centem: qui‑a sur‑re‑xit Ie‑sus.

Petrus et Ioannes

Cur‑rebant du‑o si‑mul: et il‑le a‑li‑us dis‑ci‑pu‑lus pre‑cu‑currit ci‑ti‑us

Pe‑tro et ve‑nit pri‑or ad mo‑numentum al‑le‑lu‑ia.

Petrus et Ioannes cum sudario

Cer‑ni‑tis O so‑ci‑i: ec‑ce lin‑the‑a‑mi‑na et su‑da‑ri‑um:

et cor‑pus non est in se‑pulchro in‑ventum.

Chorus

Dic no‑bis Ma‑ri‑a quid vi‑dis‑ti in vi‑a.

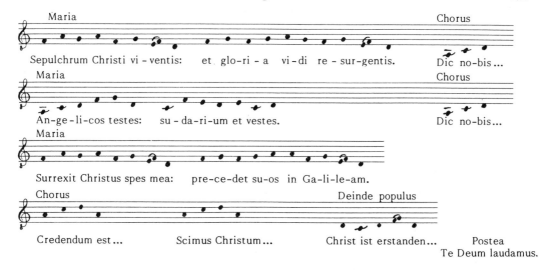

Maria

Sepulchrum Christi vi - ventis: et glo-ri - a   vi-di   re - sur-gentis.

Chorus

Dic no-bis...

Maria

An-ge-li-cos testes: su - da-ri-um et vestes.

Chorus

Dic no-bis...

Maria

Surrexit Christus spes mea: pre-ce-det su-os in Ga-li-le-am.

Chorus

Credendum est...   Scimus Christum...

Deinde populus

Christ ist erstanden...   Postea
Te Deum laudamus.

empire upon the Italian church was strong, probably in this case when the Augsburg canon Heinrich was patriarch of Aquileia (1077–84). Its later rapid diffusion can then be followed through the various ecclesiastical reform movements in the German church of the twelfth century.

(iv) *Christmas and Epiphany Ceremonies; Rachel's and Mary's Laments*

The adaptation of the *Quem queritis* dialogue for Christmas is found in relatively few sources, usually with extra verses and some indication of how the ceremony was to be performed. In manuscripts of the thirteenth and fourteenth centuries from Rouen a much fuller enactment of the Christmas story is to be found, luckily also preserved with music in the thirteenth-century Rouen gradual Paris, Bibliothèque Nationale, lat. 904. (Christmas texts are edited by Young, ii, ch. XVII; the music of the Rouen ceremony and the closely related one in the so-called Fleury Playbook are edited by Coussemaker 1861.)

Although some texts of the Rouen Christmas ceremony coincide with liturgical items, chants do not appear to have been quoted. Given the date of the manuscript, it is not surprising to find some rhymed items, and there are signs that these are later additions to an older version of the ceremony. An angel announces Christ's birth to the shepherds, and a choir of seven boys 'standing in a high place' sings 'Gloria in excelsis deo' (but not to its usual liturgical antiphon melody). The shepherds then sing the *versus Pax in terris nunciatur* (facsimile Young, ii, opposite p. 16), a rhymed song with two strophes and the refrain 'Eya! Eya! Transeamus, uideamus . . .'. This is immediately followed by a chant with the biblical text *Transeamus usque Bethleem, et videamus* . . . , in more traditional chant style. The text is superfluous after the song just sung; perhaps it was pushed into second place when the new song was introduced.

When the shepherds reach the manger two midwives ask them 'Quem queritis in presepe pastores dicite?', and when they have been shown the mother and child they sing a further short song, the two-strophe *Salve virgo singularis*. Curiously, although the whole of the rest of the ceremony is in G-mode, this little piece is copied in F, with B♭, perhaps another sign of late insertion.

Most interesting are the directions at the end of the ceremony and during the subsequent mass (which begins with a D-mode introit), which make it clear that the shepherds lead the choir, singing solo verses where proper. At the end another dialogue-chant is performed: the priest or bishop sings the verse *Quem vidistis pastores* to the·shepherds and they answer with the verse *Natum vidimus*. At the end of Lauds they also sing the Benedicamus song *Verbum patris hodie*.

The ceremonial nature of the shepherds' play, with its reminiscences of a liturgical procession and close integration with the subsequent mass, is also characteristic of the *officium stellae* ('ceremony of the star') in its Rouen version. It re-enacts the coming of the Magi to Herod and to Bethlehem, following the star. But this ceremony appears to have had quite different origins. Most of the early sources (which date from the first half of the eleventh century) are not liturgical books, and the ceremony contains relatively little of the biblical and liturgical material which one finds in Easter ceremonies, aiming rather to provide 'learned' new material in hexameters (or what might be called 'pseudo-hexameters', for they often offend against classical convention). The impression is of a ceremony conceived on an independent and original basis which different institutions inserted more or less loosely into the Epiphany liturgy. Thus some copies end with the cue Te Deum, some have no liturgical cue at all.

The various versions of the ceremony (there are over twenty) have been compared by Lipphardt (1963) and Drumbl (1981, 293–340). Most are either notated with adiastematic neumes or lack notation, but five copies with staff notation survive (for editions see Coussemaker 1861; Lipphardt 1963; Bernard 1965, 'Officium'). All the versions are closely related in content, though often they not only differ in small details of text and music but may also omit scenes from what appears to have been an original conceived on a grand scale. Thus the early source from Compiègne (Paris, Bibliothèque Nationale, lat. 16819) has parts not only for the three kings and Herod, but legates, a messenger and servant, scribes, and a knight, all in attendance on Herod, as well as the women at the manger and an angel. The action comprises numerous scenes: the Magi alone, their summons by messengers from Herod, their interview with Herod (which includes prophecy of Christ's kingship by the scribes), their arrival at the stable, the angel's warning, Herod's command to kill the children, and a final antiphon sung by the angel, *Sinite parvulos* ('Suffer the little children to come unto me . . .').

The Rouen redactor, on the other hand, evidently wished to integrate the ceremony as fully as possible into the liturgy. The processional element which is present to some extent in all versions (the three kings visiting various 'stations') is here strengthened by the inclusion of processional antiphons drawn from the liturgy. The occasion is the usual time for a procession between Terce and Mass, on the feast of the Epiphany. The first king appears behind the high altar, the others on either side. After the first chants, a procession is formed 'as on Sundays', in other words with the usual complement of choir, assistants, and so on. The cantor begins the processional antiphon *Magi veniunt* V. *Cum natus esset*. The procession makes its way out into the nave to the altar of the Holy Cross, where a curtain conceals an image of the Blessed Virgin. A crown is suspended before the altar 'in modo stelle'. When the midwives have drawn back the curtain the kings pray and feign sleep, whereupon the boy, representing the angel, sings the warning to them to return another way. This instruction is indeed carried out literally, for the procession now returns to the choir via the font and the left entrance into the choir, again according to usual custom of the cathedral. During it, another processional antiphon is sung, *Tria sunt munera* V. *Salutis nostre auctorem*, once again intoned by the cantor. Mass follows, with the kings leading the choir and singing solo verses where required.

Another Magi ceremony, from Limoges (Young, ii. 34), is placed after the offertory of mass, an obviously symbolic arrangement.

The Slaughter of the Innocents, traditionally commemorated on 28 December, is referred to in the Compiègne Epiphany ceremony and others, and at Laon forms a complete scene, but is one of many items apparently pruned for the special requirements of the Rouen ceremony. At Laon (unfortunately the manuscript has no notation) a moving lament for Rachel, symbolizing the mother of the children, is included, in dialogue with a 'Consolatrix'. In other sources another lament of Rachel has survived as an adjunct to the responsory *Sub altare dei*, and there are two full-blown Innocents' plays as well, both entitled 'Ordo Rachelis'. The simpler lament (in Paris, Bibliothèque Nationale, lat. 1139) is musically unadventurous; the part of the angel is in the metre which became popular in rhymed offices and was also used for some rhymed sequences and songs (Ex. II.25.6).

Although Bernard (1965, 'Officium') has argued for a certain amount of musical differentiation between the characters in the Epiphany ceremony, this cannot be regarded as a common feature of liturgical dramas. The melismatic exclamations of the weeping Rachel found in the Fleury Playbook version are, however, obvious attempts to impress the tragedy on the attention of the listener (Stevens, 'Medieval Drama', *NG*, 34; Coussemaker 1861, 170). This lament draws upon one of Notker's sequence texts, *Quid tu virgo mater*, and, though to a lesser extent, its melody as well.

A much bigger repertory of laments exists largely outside the liturgy and is generally subsumed under the Latin term *planctus*. As well as many secular planctus, Latin and vernacular (see the catalogue by Yearley 1981 and editions in Yearley 1983), there are a considerable number of laments of the Virgin Mary at the foot of the cross (German 'Marienklagen'). Although the circumstances of their performance are not always clear, it seems that such laments were often sung, at least in Germany and north Italy, beneath the crucifix on Good Friday. An appropriate liturgical place was after the reproaches during the Adoration of the Cross. Two early compositions of this sort,

Ex. II.25.6. *Lamentatio Rachelis* (Paris, Bibl. Nat. lat. 1139, fo. 32ᵛ)

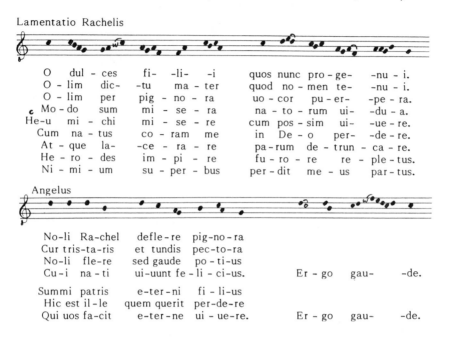

Lamentatio Rachelis

O dul - ces	fi- -li- -i	quos nunc pro - ge-	-nu - i.
O - lim dic-	-tu ma - ter	quod no - men te-	-nu - i.
O - lim per	pig - no - ra	uo - cor pu - er-	-pe - ra.
Mo - do sum	mi - se - ra	na - to - rum ui-	-du - a.
He-u mi - chi	mi - se - re	cum pos - sim ui-	-ue - re.
Cum na - tus	co - ram me	in De - o per-	-de - re.
At - que la-	-ce - ra - re	pa - rum de - trun - ca - re.
He - ro - des	im - pi - re	fu - ro - re re - ple - tus.
Ni - mi - um	su - per - bus	per - dit me - us par - tus.

Angelus

No-li Ra-chel	defle-re pig-no-ra	
Cur tris-ta-ris	et tundis pec-to-ra	
No-li fle-re	sed gaude po - ti-us	
Cu-i na-ti	ui-uunt fe-li-ci-us.	Er - go gau- -de.
Summi patris	e-ter-ni fi-li-us	
Hic est il-le	quem querit per-de-re	
Qui uos fa-cit	e-ter-ne ui-ue-re.	Er - go gau- -de.

which were often quoted in later laments, are *Planctus ante nescia* and *Flete fideles animae* (texts Young, i. 496 ff.; music Dobson and Harrison 1979, 84, 238; Gennrich 1932, 147; Stevens 1986, 131; Vecchi 1954, 56). Both are written in sequence form with paired strophes, but the length of the strophes and their varied length and rhythm is more reminiscent of the secular lai than the liturgical sequence. The Marienklage might be so extensive as to incorporate liturgical chants or vernacular songs.

Christ's passion was itself made the subject of dramatic representations, and it is not surprising to find laments being worked into the longer texts. For example, the Cividale Passion Play (Cividale, Museo Archeologico Nazionale CI, a fifteenth-century processional, music edited Coussemaker 1861, 285–97) borrows heavily from *Flete fideles animae* (the passages are identified by Young i. 512 n. 2). The manuscript is remarkable for its copious acting instructions, where almost every line is accompanied by a significant gesture.

(v) *Rhymed Ceremonies; the Fleury Playbook; the* Ludus Danielis

Rhymed songs have already been referred to several times. In dramatic presentations they first appear in considerable numbers in an 'Aquitanian manuscript of the early twelfth century, Paris, Bibliothèque Nationale, lat. 1139. Benedicamus songs and similar pieces from the manuscript have been cited in II.24. The dramas are the following:

fo. 32ᵛ	*Lamentatio Rachelis*	Latin strophes
fo. 53ʳ	*Hoc est de mulieribus*	*Quem queritis* dialogue
fo. 53ʳ	*Sponsus*	Latin and Occitan strophes
fo. 55ᵛ	*[Ordo prophetarum]*	Latin strophes

The *Lamentatio* (Ex. II.25.6), preceded by a responsory for the Holy Innocents, seems to have a liturgical place, and we may assume that the *Quem queritis* dialogue is sung on one of the usual occasions. The others remain unplaceable. The unique *Sponsus* play, where the story of the Wise and Foolish Virgins is enacted, is most remarkable. Its strophes are all of identical construction and are sung to one of four melodies (which do not, however, attach to specific characters). This, the presence of the vernacular, and the unusual subject matter, raise the possibility of non-liturgical origin. Perhaps it stands near the beginning of the tradition of the vernacular miracle play, played outside church. The music cannot, however, be distinguished from the repertory of Latin liturgical songs, of which this manuscript is one of the chief witnesses.

Whatever the relationship of *Sponsus* to the liturgy, rhymed verse and the vernacular can be found in several indubitably liturgical ceremonies. For example, the Easter ceremony in a fourteenth-century manuscript from the convent of Origny-Sainte-Benoîte (Saint-Quentin, Bibliothèque Municipale 86), is composed predominantly of regular rhymed strophes, French as well as Latin (Coussemaker 1861, 256–79). For the two episodes where French is used (the episodes of the Marys buying spices from merchants, and Mary Magdalene's dialogue with an angel) the music settles into constantly repeated strophes. Such repetition, as in the *Sponsus* play, seems therefore to be associated with the vernacular and non-biblical material. That is not invariably the case, however. The Easter play in the manuscript Tours, Bibliothèque Municipale 927 (from north France or England, thirteenth century) is composed almost entirely in Latin verse (see Coussemaker 1861, 21–48; Krieg 1956). But through composition of the music is rather the rule than the exception. The metre of the text is irregular, or when regular does not remain constant for long; the line-ends are often assonant rather than rhyming, which suggests a date of composition earlier rather than later in the twelfth century. Ex. II.25.7, the passage where disconsolate soldiers admit to Pilate that Christ is risen from the tomb, shows how different the music is from the plays in unchanging strophes. The music shapes the scene in that two of Pilate's utterances, 'Vos Romani' and 'Legem non habuistis', have the same melody; and the soldiers' 'Pro quo gentiles' and 'Nos veritatem' also form a pair. Pilate's final strophes are in iambic trimeter.

One source particularly rich in versified Latin dramas, the famous Carmina Burana manuscript (Munich, Bayerische Staatsbibliothek, clm 4660a), has only adiastematic notation (see the texts edited in Young, i. 432, 463, 514, 518, ii. 172, 463; Steer 1983

Ex. II.25.7. From the Tours Easter play (Tours, Bibl. Mun. 927, fo. 4ʳ)

Tunc milites surgant et redeant ad Pilatum tristi animo canendo

He- -u mi-se-ri quid fa-ci-mus quid di-ci-mus qui-a per-di-dimus

quem custo-di-mus. De ce-lo ue-nit an-gelus qui di-xit mu-li-e-ri-bus

qui-a sur-re-xit do-mi-nus.

Deinde dicat Pilatus ad milites

Uos Roma-ni mi-li-tes pre-cium ac-ci-pi-te et om-nibus di-ci-te

quod uobis sub-la-tum est.

Milites simul respondeant

Pro quo genti-les fu-i-mus se-pulchrum custo-di-ui-mus magnum sonum audi-ui-mus

et in terram ce-ci-di-mus.

Item dicat Pilatus

Legem non ha-bu-istis sed menti-ri po-tes-tis quod disci-pu-li ue-ne-runt

et e-um sustu-lerunt.

Milites simul respondent

Nos ue-ri-ta-tem di-cimus de ce-lo ue-nit an-ge-lus qui di-xit mu-li-e-ri-bus

qui-a su-re-xit do-mi-nus.

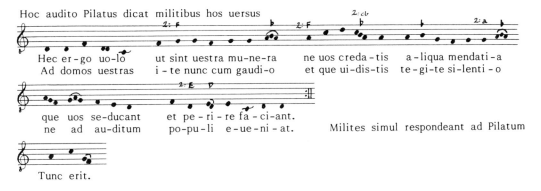

Hoc audito Pilatus dicat militibus hos uersus

Hec er-go uo-lo · ut sint uestra mu-ne-ra · ne uos creda-tis · a-liqua mendati-a
Ad domos uestras · i-te nunc cum gaudi-o · et que ui-dis-tis · te-gi-te si-lenti-o

que uos se-ducant · et pe-ri-re fa-ci-ant.
ne ad au-ditum · po-pu-li e-ue-ni-at.   Milites simul respondeant ad Pilatum

Tunc erit.

argues for provenance in South Tirol; facs. Bischoff, 1967). Even richer, however, is the equally famous Fleury Playbook (Orléans, Bibliothèque Municipale 201). The manuscript was kept for many years in the library of the celebrated Benedictine monastery of Saint-Benoît-sur-Loire, at Fleury, but there is no evidence that it originated there. The chief music scribe used an informal type of notation (similar to that, for example, in the Cambridge song-book Cambridge, University Library Ff. 1. 17 or Tours 927) which is practically impossible to localize, but which was doubtless used by many practising musicians of the time. (See the discussions by Corbin 1953 and Huglo 1985 'Analyse'; editions of all ten plays by Coussemaker 1861).

Some of the contents seem non-liturgical: four plays about miracles of St Nicholas, which make wide use of repeating musical strophes. At least for the *Tres clerici* play, where the same melody is used throughout the entire composition, one is tempted to suggest a connection with the method usually postulated for the performance of secular vernacular epics and romances (see Stevens 1986, 222 ff.). The same technique is used for a play about another miracle, the resurrection of Lazarus. By contrast, the Conversion of St Paul, though written throughout in lines of 4 + 3 + 3 syllables, is basically through-composed, with only infrequent repetitions.

The remaining plays mix traditional and apparently new material. There is an Epiphany play, which begins with the episode of the shepherds coming to the stable, an Innocents play, an Easter play, and a Peregrinus play. These are among the most extensive ceremonies for their respective occasions. Three end with the Te Deum, Peregrinus with *Salve festa dies*. Rhyme, where present, is not invariably associated with regular rhythm. One of the few regular pieces is a small processional song in Peregrinus, in sequence form with lines of 7 + 5 syllables, to be sung between the first appearance of the risen Jesus to his disciples and the scene with doubting Thomas (Ex. II.25.8). (The processional function of the song is even more clear in the Easter ceremony from Coutances: Young i. 408.)

Examples like this and the song which the shepherds sing at Rouen while processing to the stable fit well with our picture of the Latin liturgical songs of this period known as conductus (see above, II.24). They are relatively rare in liturgical dramas, however, and only one play makes systematic use of them. This is the famous

Ex. II.25.8. Song *Adam novus veterem* from Fleury Peregrinus play (Orléans, Bibl. Mun. 201, p. 228)

A-dam no-uus ue-te-rem   du-xit ad as-tra.
Cre-a - to - rem re-co-lit   iam cre-a-tu-ra.

Sed Ma-ri-a Ia-co-bi   cum Magda-le-na.
Et Ma-ri-a Sa-lo-me   ferunt un-guenta.

Qui-bus di-xit an-ge-lus   in ues-te al-ba.
Re-sur-re-xit do-mi-nus   mor-te cal-ca-ta.

Frac-ta linquens Tar-ta-ra   et spo-li-a-ta.
Re-fert secum spo-li-a   uic-tor ad as-tra.

Se de-monstrat pos-te-a   for-ma pre-cla-ra.
Di-lectis dis-ci-pu-lis   in Ga-li-le-a.

Comes factus in-cre-pat   la-tens in ui-a.
Et scriptu-ra re-se-rat   pi-us ar-cha-na.

Con-ui-uans ag-nos-ci-tur   pro-pri-a for-ma.
Pa-nis red-dit frac-ti-o   lu-mi-na cla-ra.

Si-bi laus et glo-ri-a.

Ludus Danielis of Beauvais, surviving in an early thirteenth-century manuscript (London, British Library, Egerton 2615), alongside a practically complete office for the Circumcision or 'Feast of Fools' on New Year's Day (see I.9). In the first half of the piece nearly every entrance or exit of characters is accompanied by vigorously rhythmic conductus (so called in the source). The melodies of some can be found elsewhere: thus *Iubilemus regi nostro*, sung as the sacred vessels from the temple are brought in to Belshazzar, turns up as *Iubilemus cordis voce* in the Laon Epiphany office; and the song which the nobles sing as they escort Daniel to the king, *Hic verus Dei famulus*, has the same melody as the Benedicamus song *Postquam celorum dominus* in Paris 1139, a key repository of both festal songs and liturgical dramas. *Hic verus Dei famulus* is notable not only for its macaronic text but also musically for its

Ex. II.25.9. From the *Ludus Danielis* (London, Brit. Lib. Egerton 2615, fo. 99ʳ)

tertian framework and the delaying of the tonic *c* until the very end of the song
(Ex. II.25.9).

Of the other elements mentioned up until now, there is little quotation of
traditional chant (two Lenten responsories, *Merito hec patimur* and *Emendemus in
melius*, when Darius' envious counsellors are about to be flung to the lions), and,
although there are strophic songs other than the conductus, there is nothing to
resemble the repetition technique of *Sponsus* and the miracle plays in the Fleury
Playbook. The text is mostly rhymed with regular rhythm, set in music of admirable
directness. Whereas much of the Tours and Fleury music moves at the leisurely pace
of traditional chant, Daniel often slips into syllabic, patterned phrases. Consider, for
example, the song of the envious counsellors as they inform Darius that Daniel is
worshipping his own god (Ex. II.25.10). In the first line there are three descending
melodic thirds (bracketed); and *F* occurs three times on stressed syllables; the same
happens in the third line, with three *G*s. In Darius' reply the thirds recur.

This clearly rhythmic, patterned music is complemented by a few items in a less
insistent style. When the counsellors have had their way, and Daniel has been
condemned to the lions' den, he laments as shown in Ex. II.25.11. The lines rhyme
but their metre is obscured by the 'Heu' exclamations and by deviations from syllabic
setting. Ignoring the exclamations, we have lines of verses of 5 + 9, 5 + 9, 7 + 8, and
again 5 + 8 syllables. The sudden change into the very highest register is rhetorically
inspired. So are the swoops down through the complete octave *cc* to *c* in lines 1 and
7–8 (running over the line-end), and perhaps the descent to *a* and *bb* for 'damnatio'.
The compartmentalization of phrases into either the lower or the upper part of the
octave (only *c* and *g* cadences are used) is typical of twelfth- and thirteenth-century
pieces.

Ex. II.25.10. From the *Ludus Danielis* (London, Brit. Lib. Egerton 2615, fo. 105^v)

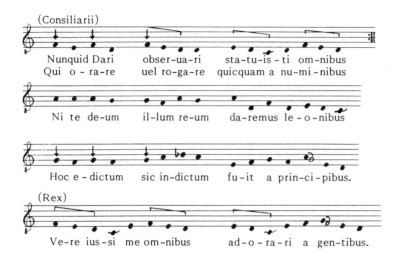

(Consiliarii)

Nunquid Dari   obser-ua-ri   sta-tu-is-ti om-nibus
Qui o - ra-re   uel ro-ga-re   quicquam a nu-mi-nibus

Ni te de-um   il-lum re-um   da-remus le - o-nibus

Hoc e - dictum   sic in-dictum   fu-it a prin-ci-pibus.

(Rex)

Ve-re ius-si me om-nibus   ad-o-ra-ri a gen-tibus.

Ex. II.25.11. From the *Ludus Danielis* (London, Brit. Lib. Egerton 2615, fo. 106^r)

1. He - u   he - u   he - u   quo ca-   -su sor - tis

2. ue-nit hoc dampna-ti - o   mortis.

3. He - u   he - u   he - u   sce-lus in - fan-dum

4. cur me da-bit ad la - ce - randum.

5. Hec fera tur-ba fe - ris   6. sic me rex perde-re queris.

7. He - u   qua morte mo-ri   8. me co-gis parce fu-ro-ri.

The *Ludus Danielis* is a unique and, to modern sensibilities, particularly happy mix of styles. No other work of the period moves with such directness and energy. Even in its time it must have seemed extravagant, not least musically. Little of the liturgical spirit informs it, although it ends with Te Deum. It stands at the opposite end of the scale from the Peregrinus ceremonies, with their copious use of biblical and liturgical material. It seems to have had no successors, and indeed, the composition of new dramatic liturgical ceremonies of any kind fell off rapidly after the thirteenth century.

II.26. OFFICES WITH VERSE TEXTS

Jammers 1929–30; Irtenkauf, 'Reimoffizium', *MGG*; Szöverffy 1964–5 (s.v. 'Reimof-fizien'); Andrew Hughes, 'Rhymed Office', *NG*, and 1985.

    Editions of texts: AH 5, 13, 17–18, 24–6, 28, 45

    Editions of music: Felder 1901; Auda 1923; Bayart 1926; Jammers 1934; Ottósson 1959; Falvy 1968; Hoppin 1968; Bernard 1977 and 1980; Epstein 1978; Deléglise 1983; Patier 1986; Edwards 1990.

The subject of the following remarks is the offices for saints with antiphons and responsories written in verse form. The topic is somewhat difficult to define and name. We are not dealing with a new liturgical genre, for the compositions in question form a normal part of the office; only their poetic technique distinguishes them from other antiphons and responsories. Nor are we dealing with a new type of music. While it is true that the chants of versified offices are usually arranged in modal order, offices with prose texts can also be found whose chants are in modal order. Other musical features of later medieval versified offices, discussed briefly below, are not exclusive to them, but can be found in other chants of the same period. Nor does the usual term 'rhymed office' seem adequate to contain the whole area, for rhyme is only one aspect of poetic technique which can be found in the offices. It is certainly true that the vast majority of versified offices were written from the twelfth century onward in accentual rhyming verse, but many earlier offices used rhyme more freely or not at all—one finds assonance, single rather than double rhyme, or rhymed prose; some texts are metrical rather than accentual. Thus the increasing and finally exclusive use of accentual rhyming texts is a matter of stylistic change, not of a new liturgical or musical form. It runs parallel to the use of accentual rhyming verse in conductus (which was a new genre), alleluias, sequences, and liturgical dramas.

    The term *historia* is used for the rhymed as for the non-rhymed office.

    The size of the repertory is very large, but rather few offices were widely known. The reasons are obvious. The most important saints, commemorated throughout Europe, already had offices with old prose texts. Saints of lesser, local importance might use chants from the Common of Saints, but many composers were inspired to compose a new office for the patron saint of the their church or diocese, using modern poetic techniques. The majority of offices with verse texts are therefore local demonstrations of piety. Only a few became widely known, such as those for St Thomas of Canterbury, St Francis, and St Dominic (the latter two naturally promulgated by their respective religious orders), which stimulated contrafacta for other saints, that is, the same music was utilized for new texts. Jammers (1929–30) surveyed ninety-nine offices in sources from the Rhine area. Of these seventeen were widely or fairly widely known outside that area, eight were promulgated by one of the religious orders, twelve were known in a restricted area, incuding at least part of the Rhineland, while the remaining sixty-two were sung in only a handful of churches or even just one. These figures seem typical of the general picture of distribution, as indeed they would be for many types of relatively new liturgical material.

One or two isolated metrical office antiphons and responsories, such as the antiphon *Solve iubente Deo* from the office for St Peter (text in hexameters, *AM* 993, *AS* 412; may be pre-Carolingian. Other metrical or rhyming or otherwise poetic antiphons and responsories are occasionally to be found among saints' offices of the ninth century. Hucbald of Saint-Amand (d. 930) is credited on indifferent authority with a number of texts (AH 13, 133 ff., 200 ff., 225 ff.; AH 26, 230 ff.). He appears to have a better claim to the antiphons of a St Peter office which has survived in transcribable notation (see Weakland 1959). This office is not versified. Another example, exceptional both because it was not composed in honour of a local saint and because it was adopted all over Europe, is the Trinity office composed by Bishop Stephen of Liège (901–20). (Parts of the Trinity office are in modern service-books, such as *AM* 535 ff.; facsimiles of a medieval one *AS* 286 ff.; a number of these early texts are discussed and edited by Jonsson 1968; there are facsimiles of eleventh-century metrical offices in Delaporte 1957 and 1959–60.)

It remains to be seen to what extent the music of these early offices differs from that of the rest of the repertory. Even at this early date the pieces may be arranged in modal order, as, for example, in the Trinity office:

	modes
Vespers antiphons	1–5
Night Office antiphons	1–8, 7
Night Office responsories	1–8, 1
Lauds antiphons (which all have verses)	1–5

Different arrangements were naturally required for Vespers and the Night Office (but not for Lauds) in the monastic form of the office, which required four Vespers antiphons and twelve Night Office antiphons and responsories. (See Andrew Hughes 1983 for a number of modal schemes.)

Another relatively well known example of a saint's office with chants in modal order but not in verse is that for St Nicholas, said to have been composed by Reginold of Eichstätt before he became bishop of the diocese (966–91) (see Charles Jones 1963; Hohler 1967; *AS* 354 ff.).

Chants in rhyming accentual verse are occasionally to be found in offices of the tenth century, becoming more frequent in the eleventh. An early example is *Gloriosa sanctissimi* for St Gregory, attributed to Leo IX (1048–54), the former Bishop Bruno of Toul (fac. Bernard 1977). Ex. II.26.1 gives the second responsory of this office, which is naturally in mode 2. Although the verse has a psalm verse for its text, it uses only the first half of the normal responsory verse tone. The respond has assonant rather than regularly rhyming lines, and their rhythm is varied.

Ex. II.26.1. Responsory *Videns Rome vir beatus* (Annecy, privately owned manuscript, fo. 139ʳ)

The music of the respond, by contrast with that of the verse, owes practically nothing to the melodic idioms of mode 2 responsories. Among other un-Gregorian features one could pick out the two 'Gallican' cadences ('pueros', 'angeli') and the final cadence. The figure at 'angeli' is convincing enough in its function as a cadence, but it can be found internally as well, in the first line. Two turns as at 'beatus' can be found in the traditional repertory, but not three as at 'illis'. Then there are the C–E–G triads, at 'Anglorum' and twice at 'iter salutis'. The fact that all but two lines cadence on *D* is also typical for this period, so that the phrases tend to become preoccupied with the tonal space above and below *D*: compare 'bene inquid' and 'vultu nitent ut', 'bene Angli' and 'angeli'. The tendency to repetition may extend over line-ends, so that 'vir beatus Anglorum forte pueros' is similar to 'illis monstrari iter salutis e-'.

Ex. II.26.2. Antiphon *Gregorius vigiliis* (Annecy, privately owned manuscript, fo. 140ʳ)

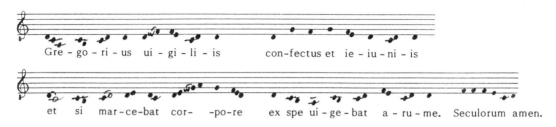

Equally untraditional is the antiphon *Gregorius vigiliis*, the first of the Lauds antiphons but also in mode 2 (Ex. II.26.2). There are two 'Gallican' cadences in the first line alone, and two later ones. Even 'ex spe vigebat' in the last line sounds likely to close this way (cf. 'Gregorius'). It is not surprising that phrases which explore the same restricted area should sound alike: line 1 is practically identical to line 4. (For a comparison with other offices attributed to Leo, see Bernard 1980, with further facsimiles.)

The restriction of melodic goals continues as a feature of the offices in completely regular accentual verse of the later twelfth century onward. Ex. II.26.3 gives the mode 1 antiphons from the office of St Dominic: the antiphon for first Vespers, the first and last antiphons for the Night Office and the first of Lauds. All cadences except one are on *D*, *a*, or *d*. What is more, the starting notes are almost as uniform, with two-thirds starting on *a*, and the rest on *C*, *D*, *F*, and *c*, with one *G*. Nearly all movement is by step rather than by leap. There is no repetition of notes, one of the biggest differences between these and the oldest Gregorian antiphons. All is activity, though on a small scale. There is no sense of longer periods, so that in the first antiphon, for instance, the music divides naturally after the first four syllables of each line, even though there is no text rhyme. The difference between phrases therefore tends to be a matter of how far in which direction, and how fast, the melody will go, between rather rigidly observed poles and in rather uniform motion. Surprisingly few lines are closely similar (compare 'dotata gloria' in *Gaude felix* and the last line of *Adest dies leticie*). Indeed, there seems to be a deliberate desire to avoid repetition (though the room for manœuvre is limited), when the constant recourse to the same few text patterns would have made repetition easy. Modern writers have nevertheless been tempted to speak of 'formulas' (Jammers) or 'variations' (Bernard). Compare, for example, the third antiphon of the Night Office for St Dominic with the same one for St Thomas of Canterbury, given in Ex. II.26.4.

Both antiphons make the usual opening progression through *D–G–a–c*, but whereas *Documentis artium* then takes another line to make a cadence on *b*, *Cultor agri* has already done so and comes swirling down (without leaps) to *E* again. That is what *Documentis artium* also does in line 3, while *Cultor agri* rises once again to *c*.

Deliberate musical rhyme may occasionally be found. Thus the St James office in the Codex Calixtinus includes some pieces in rhymed prose where musical rhyme

Ex. II.26.3. Antiphons from the office of St Dominic (Rome, Santa Sabina XIV lit. 1)

f.296r

Gau-de fe- -lix pa-rens Ys-pa-ni-a no - ue pro - le dans mundo gau-di-a

sed tu ma-gis plau-de Bo-no-ni-a tan - ti patris do - ta-ta glo - ri-a

lau- -da to-ta ma-ter ec-cle - si-a no - ue lau - dis a - gens sol-lem-ni-a.

f.296r

Pre - co no - uus et ce-li - cus mis - sus in fi - ne se - cu-li

pauper ful - sit Do - mi-ni-cus for-ma pre-ui-sus ca-tu - li.

f.297r

Li-ber car-nis uin-cu-lo ce-lum in-tro - i - uit

u - bi ple - no po-cu-lo gustat quod si - ti-uit.

f.297r

Ad-est di - es le - ti - ci - e quo be - a - tus Do - mi-ni - cus

au - lam ce-les - tis cu - ri - e ci-uis in - trat mag-ni - fi - cus.

Ex. II.26.4. Antiphons from the offices of St Dominic (Rome, Santa Sabina XIV lit. 1, fo. 296ʳ) and St Thomas of Canterbury (Edinburgh, Univ. Lib. 123, fo. 156ʳ)

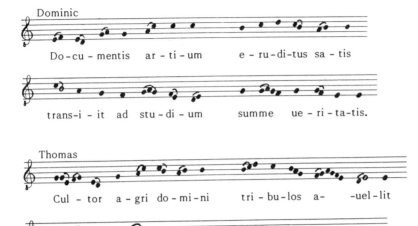

supports or counterpoints the text rhyme. For example, the responsory *Jacobe virginei frater* (ed. Wagner 1931, 73) has the following musical rhyme scheme (small letters indicate an identical cadence, capitals a whole identical line):

	text rhyme	musical rhyme
Jacobe virginei frater	a	a
preciose Johannis	b	B
qui pius Ermogenem	c	a
revocasti corde ferocem	c	B
ex mundi viciis	d	a
ad honorem cunctipotentis	d	B
V. Tu prece continua	e	C
pro nobis omnibus ora	e	D
(Qui pius, etc.)		
Gloria patri almo	f	C
natoque flamini sancto	f	D
(Ad honorem, etc.)		

The shortness of breath evident in the antiphons just discussed, or perhaps better the concentration on short, self-contained lines, extends also to responsories. Since they are more ornate, in the sense that single syllables often have melismas, it frequently transpires that a self-contained musical phrase may cover no more than a single word, or even a single syllable (see Stäblein 1975, 163). Ex. II.26.5 gives the fourth responsory of the monastic office of St Thomas. While one might argue for a single musical phrase encompassing both 'Post sex' and 'annos', others seem relatively self-contained: 'redit', 'vir', and 'stabilis', and obviously 'thesaurum'. These are each

Ex. II.26.5. Responsory *Post sex annos* (Edinburgh, Univ. Lib. 123, fo. 156ᵛ)

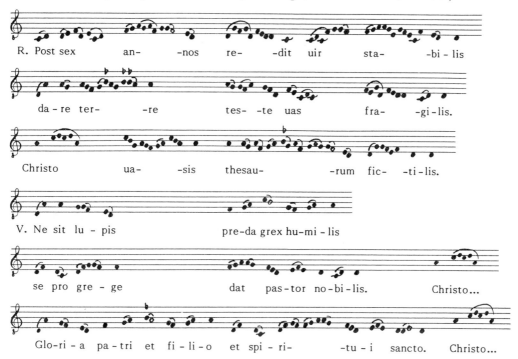

comparable with lines of antiphons, but are now largely melismatic rather than broken up by changes of syllable.

Although the rhythmic schemes employed in these texts are the same as in rhymed sequences and Latin songs, there is no opportunity here for stressed syllables to dominate the tonality, as they do in syllabic or near-syllabic pieces.

II.27. METRE, ACCENT, RHYTHM, AND RHYME IN LITURGICAL TEXTS

(i) Metre and Stress
(ii) Rhyme
(iii) Prose Rhythm (Cursus) and Prose Rhyme

Raby 1927; Gavel 1954; Beare 1957; Strecker 1957; Crocker 1958, 'Musica'; Norberg 1958; *Oxford Classical Dictionary*; Steiner, 'Cursus', *NG*; Fassler 1987.

ʹ = stressed syllable
. = unstressed syllable
− = long syllable
˘ = short syllable
/ = caesura

(i) *Metre and Stress*

Most liturgical texts are prose, taken from (or composed to be compatible with) biblical prose texts such as the Book of Psalms. Verse has, nevertheless, an important place in the liturgy, and any survey of plainchant must of necessity refer to technical aspects of Latin verse: metre, stress or accent, rhythm, rhyme, verse-forms, and so on. I offer here a very brief introduction to some of these matters, bearing in mind that specialists themselves are divided over many issues referred to, such as the nature of accent and ictus, and the nature and progress of the change from metrical to accentual verse.

A distinction has to be borne in mind between metre (which concerns syllable lengths) and stress. In classical Latin the two were independent. Although there is disagreement as to the nature of stress (or accent) in Latin at various periods of the classical age, it is at least certain that it played no part in the organization of classical poetry. Later, organization of poetry according to patterns of stressed syllables became more common, and poems are found in which stress and metre are co-ordinated, so that all stressed syllables might also all be long ones. Organization by stress was occasionally so dominant that syllable-lengths in their turn were ignored. In discussions about the texts of liturgical songs of the twelfth and thirteenth centuries there has been a certain amount of controversy about whether a stressed syllable should necessarily be counted a long one. Metre and stress are obviously of importance for the transcription and performance of these songs.

In classical Latin stressed syllables fall: (*a*) on the first syllable of two-syllable words; (*b*) in longer words on the penultimate syllable if it is long, otherwise on the antepenultimate.

Classical poetry was organized not according to stress but by syllable-length, where groups of long and short syllables were reckoned in 'feet'. The feet encountered most frequently in liturgical poetry are the following:

trochee	$-\smile$	anapaest	$\smile\smile-$	spondee	$--$
iamb	$\smile-$	dactyl	$-\smile\smile$	tribrach	$\smile\smile\smile$

It was common in classical times to accompany the recitation of poetic texts, in private or in teaching, with rhythmic movements, of the hand or a wand. Each foot was reckoned to have two parts: the *arsis* ('raising') and *thesis* ('setting down'), referring to upward and downward movement of the hand or stick. Marking of the feet in this way was referred to as marking the 'beat' (*ictus*, *percussio*, or *plausus*). Grammarians and authors of treatises on prosody do not agree as to how it was done, some calling each movement, arsis or thesis, an ictus, some assigning one 'beat' to the up+down combination, and some taking a pair of feet in one beat. There is also disagreement as to how the ictus was marked, if at all, in the declamation of the texts. And it is not known if the beats were spaced at regular intervals, so that, for example, a trochee took up the same time in performance as a spondee, perhaps in this way:

trochee = spondee

dotted crotchet + quaver = crotchet + crotchet

In classical Latin, ictus and stress were independent of each other, and neither seem to have been strongly heard in performance. (Beare 1957, 57–65 summarizes the problems and disagreement over ictus.)

The most common classical metres in liturgical poetry are the following (for others see Norberg 1958, ch. V):

(*a*) dactylic hexameters, where the first four feet are either dactyls or spondees, the fifth is a dactyl, and the last is a spondee (occasionally a trochee):

$$- \cup\cup - \cup\cup - \cup\cup - \cup\cup - \cup\cup - \, -$$

(or) $- - \ - - \ - - \ - - \ - \cup$

(*b*) the elegiac distich (distich = two verses, i.e. a couplet), where a hexameter is followed by a dactylic pentameter. The pentameter is in two halves: the first half has two feet, either dactyls or spondees, plus one long syllable; the second half has two dactyls and one further syllable:

$$- \cup\cup - \cup\cup - \ / \ - \cup\cup - \cup\cup -$$

(or) $- - \ - -$

(*c*) the sapphic stanza, whose last line is called an adonius:

1–3 trochee, spondee, dactyl, trochee, trochee or spondee

4 dactyl, trochee or spondee

(the best-known example is probably the hymn *Ut queant laxis*, attributed to Paul the Deacon).

(*d*) the asclepiad, in its catalectic (wanting a final syllable) or minor form:

spondee, dactyl, long syllable, dactyl, trochee or spondee, long syllable

(*e*) various trochaic and iambic trimeters (having three pairs of feet) or tetrameters (having four pairs of feet); a very popular one was the trochaic tetrameter catalectic, or trochaic septenarius, called the rhythm of the Roman legions' marching-songs:

$$- \cup - \ \cup \ - \cup - \ \cup \quad - \cup - \ \cup \ - \cup -$$

(or) $- - \quad - - \quad - -$

Hexameters have a long history in liturgical poetry. They are even to be found among the prayer texts in the so-called Mone Masses (Gallican, ed. Mone 1850 and Mohlberg, Eizenhöfer, and Siffrin 1958), though not in Roman formularies. Many trope texts of the ninth century onward are in hexameters, and so are some texts in traditional genres, for example the Marian introit *Salve sancta parens*.

Among lyric stanzas, one associated with the name of St Ambrose was far more popular than any classical model among composers of office hymns. This is in iambic dimeter (i.e. having two pairs of iambs; spondees are allowed for the first and third iambs). Ambrose's poetry is still metrical, rather than rhythmical: stress is not regular and does not coincide with long syllables:

$$- - \ \cup - - \quad - \ \cup - \qquad - - \ \cup - \ \cup \ - \ \cup -$$

Aeterne rerum conditor,   noctem diemque qui regis

$$. \quad . \quad \prime \quad . \quad \prime \quad . \quad \prime . \quad . \quad . \quad \prime \quad . \quad . \prime \quad . .$$
$$\_ \quad \_ \quad \cup \quad \_ \quad \_ \quad \_ \quad \cup \quad \_ \quad \cup \quad \_ \quad \cup \quad \_ \quad \_ \quad \_ \quad \cup \_$$

et temporum das tempora,   ut alleves fastidium.

The same lack of correspondence between long and stressed syllables can be seen in Venantius Fortunatus' *Pange lingue gloriosi* (trochaic tetrameter catalectic):

$$\prime \quad . \quad \prime \quad . \quad . \prime \quad . \quad \prime \quad . . \quad . \quad \prime \quad . .$$
$$\_ \quad \cup \quad \_ \quad \_ \quad \_\cup \quad \_ \quad \_ \quad \_\cup \quad \_ \quad \_ \quad \_ \quad \cup \quad \_$$

Pange lingua gloriosi proelium certaminis

$$. \quad \prime \quad . \quad \prime \quad . \quad \prime \quad . \quad . \quad \prime \quad . \quad \prime \quad . .$$
$$\_ \quad \cup \quad \_ \quad \_ \_ \quad . \quad \cup \quad \_ \quad \_ \_ \quad \cup \quad \_ \quad \_ \quad \_ \quad \cup \_$$

et super crucis tropaeo dic triumphum nobilem

$$\prime \quad . \quad . \quad . \quad \prime \quad . \quad \prime \quad . \quad . \quad . \quad \prime \quad . \quad \prime \quad . .$$
$$\_ \quad \cup \quad \_ \quad \cup \quad \_ \quad \cup \quad \_ \quad \cup \quad \_ \quad \cup \quad \_ \quad \_ \quad \_ \cup \_$$

qualiter redemptor orbis immolatus vicerit.

During the late imperial age, spoken Latin appears to have lost gradually its long- and short-syllable contrasts, and stressed syllables, previously only lightly felt, gained in intensity. A conflict between metre and stress pattern, previously of little or no significance, would presumably have become more noticeable. It also meant that, although poets continued to write metrical verses for centuries after, they would have had to acquire knowledge of correct quantities at second hand, from teachers and textbooks, rather than through natural experience of the language. And it became possible to write poetry which ignored quantities and was organized according to stress-patterns instead—possible, but for centuries not generally thought proper in educated circles. Norberg (1958, 92) cites the case of a ninth-century monk of Saint-Amand who asks that if his hexameters are not worthy to be called a 'carmen', they may at least be granted the name of 'rithmus', that is, a poem in accentual rather than metrical verse.

Rather than use the adjective 'rhythmic' to describe poetry organized according to stress-patterns, I have preferred the term 'accentual'. The word 'rhythm' is often used loosely nowadays with a wide variety of meanings; and *rhythmus* in antiquity and the early Middle Ages had a precise and special meaning not directly connected to stress or accent. As Crocker (1958, 'Musica') explains, *rhythmus* was the proportional relationship between the lengths of sounds (e.g. syllables, or feet, or groups of feet), whereas metre was concerned with the lengths of individual sounds (syllables).

The period roughly from the time of St Augustine (354–430) through to the ninth century saw increasing numbers of compositions where stress coincided with the metrical accent or *ictus*. And in other poems the metrical patterns are completely replaced by stress-patterns. The poetic feet are now:

trochee ′.   anapaest ..′   spondee ′.

iamb   .′   dactyl   ′..   tribrach ′..

There are examples of accentual texts in the sections on hymns and versus and conductus above (II.15 and 24; Norberg, ch. VI has very many more). One of these is the Christmas epistle farse which begins with the strophe *Laudes deo dicam per secula* (various sources have *Laudem* and/or *dicant*), which is in the accentual equivalent of alcaic metre (Ex. II.23.20), that is, it is the stress-pattern which reflects the alcaic metre, not the quantities.

Slight deviations from stress-patterns are common in accentual poetry. Norberg (1958) uses a descriptive code indicating simply how many syllables there are in a verse, and where the last stress falls. 'Oxytone' (on final syllable) stresses are unknown in Latin poetry, though some bear a moderately weak one. The stress is therefore 'paroxytone' (falling on the penultimate syllable) or 'proparoxytone' (on the antepenultimate). *Laudes deo* is therefore '5p + 6pp'.

One pattern popular from about the eleventh century is 4p + 6pp ('.'. '..'..), widely used in sequences, versus, liturgical dramas, and rhymed offices. Charles Jones (1963, 122–39) thought it German in origin, on the evidence of the Hildesheim source of the St Nicholas miracle play of the Three Daughters (London, British Library, Add. 22514: Young, ii. 311–14), which he believed to date from the early eleventh century. Hohler (1967, 47) objected that the manuscript is more likely to date from the early twelfth century, which means it is no earlier than Paris, Bibliothèque Nationale, lat. 1139, where other examples of the rhythm are to be found.

Another popular medieval pattern was the so-called 'Goliard' rhythm, 7pp + 6p (trochaic), which became widely practised in the twelfth century. It is used at the start of a versus in Paris 1139 (fo. 46ᵛ):

```
 ′    .    ′  . ′  .  .    ′   .  ′   . ′   .
Incomparabiliter   cum iocunditate
```

but was much more popular in secular verse, at least in the usual 'Goliard' strophes of two such verses with double rhyme. *Incomparabiliter* is more significant in that, like very many twelfth-century songs, it employs a variety of patterns within the same song, patterns which may be found elsewhere, to be sure, but not in this particular combination. The idea seems to be to have four main beats in each line: there is a clear pattern of alternating strong and secondary beats, with weak beats sometimes added as anacruses or subtracted at cadences.

Incomparabiliter	′.′.′.′
cum iocunditate	′.′.′ ′
gaudeamus pariter	′.′.′.′
in hoc sollempnitate	.′.′.′ ′
In festis beatorum	.′.′.′ ′
huius et aliorum.	.′.′.′ ′

(See Norberg 1958, ch. VI for a wide range of examples of other patterns.)

I referred above to a 'rithmus' as a poem in accentual rather than metrical verse. Actually, even this matter is disputed, for in a number of poems neither metrical feet nor recognizable stress-patterns seem to be present. It has been suggested (see Suchier 1950; Crocker 1958, 'Musica', 9) that the basic idea was to control the number of syllables in each line, regardless of quantity or stress, except perhaps at cadences (end of line or caesura). The theory of a 'syllable-counting' or 'numerical' poetry circumvents the need to explain apparent errors or irregularities in early medieval prosody.

It is a moot point whether or not the metrical accents (the 'ictus') or the stress-accents were reflected in musical performance or notation. Tropes written in hexameters and other metres usually have music too ornate to allow any appreciation of the quantities of the text. In syllabic text-settings stressed syllables could in theory be highlighted by stress in performance. And since stress-patterns replaced patterns of long and short syllables, one might argue that it would be illogical to posit a system of musical rhythm where stressed syllables took long notes and unstressed syllables took short ones. It is true that some late-medieval sources do notate hymns, sequences, and other poetic genres mensurally, with long and short notes (for example the Las Huelgas manuscript, ed. Anglès 1931 and Gordon Anderson 1982; and Michaelbeuern Cart. 1; see Lipphardt 1980). Yet these may be late exceptions which prove the rule.

(ii) *Rhyme*

Rhyme also underwent a substantial change in the early Middle Ages. It was used in classical poetry only for special effect, became common among hymn writers in the fifth and sixth centuries, held its own during the Carolingian renaissance (despite its 'unclassical' character), and reached a brilliant apogee in the songs for the festal liturgies of the twelfth and thirteenth centuries.

'Rhyme' here should be understood in the broad sense, including assonance (similarity of vowels) as well as identity of whole syllables. Not until the ninth century did pure monosyllabic rhyme become at all common, and disyllabic rhyme is hardly found before the eleventh (the sequence *Victimae paschali laudes* is an early example). Neither assonance nor rhyme is particularly common among the items in medieval hymn collections. Ambrose's hymns do not use it, while Venantius Fortunatus' do so occasionally (*Pange lingua*, cited above, has rather rare assonances, while *Vexilla regis prodeunt* is regularly assonant). *A solis ortus cardine* of Sedulius in the fifth century is worth quoting as an example of an abecedary hymn (each strophe begins with a different letter of the alphabet) with a rhyme which is consistently present though not in a consistent pattern:

A solis ortus cardine   **B**eatus auctor saeculi
  adusque terrae limitem   servile corpus induit,
  Christum canamus principem,   ut carne carnem liberans
  natum Maria virgine.   non perderet quod condidit. (etc.)

Some other uses of rhyme are less well known. It was possibly after Sedulius' example that rhyme in hexameters was used after the fifth century. Hexameters of the

Carolingian renaissance employ it rarely, but during the ninth century it became more common. In a hexameter, rhyme is effected between the end-syllable and the syllable at the caesura, that is, at the most prominent word division in the verse (usually after the first two and a half feet). Such a hexameter is called 'leonine' (after the twelfth-century Parisian poet/composer Leoninus, erroneously believed to have invented the technique). At Reichenau the following trope verse was sung with the introit of Holy Innocents' Day (CT 1, 146):

> Nos pueri· puero / resonemus carmina Christo (EX ORE . . .)

and the following distichs on the Octave (CT 1, 137, 146):

> Mentibus intentis / resonemus carmina laudis
> et cum psalmista / dicamus voce sonora (EX ORE . . .)
>
> Nunc fratres enesis / repetamus verbula dulcis
> quae psaltes regi / sic fert cantando perenni (EX ORE . . .)

(iii) *Prose Rhythm (Cursus) and Prose Rhyme*

Classical orators made use of metrical patterns in prose. A sentence-ending (clausula) employed a recognizable and approved succession of long and short syllables. When accentual rather than metrical verse became common, the metrical sentence-endings were transformed into accentual ones. In effect it was always the two final words (the separation is indicated by | below) that were affected. The commonest patterns appear to have been:

	Metrical	Accentual
'cursus planus'	$-\cup \mid --\cup$	$',\ '\ .'.$
'cursus tardus, durus' or 'ecclesiasticus'	$-\cup \mid --\cup-$	$',\ '\ .'..$
'cursus velox' (most popular)	$-\cup- \mid -\cup--$	$'..\ '\ ..'.$
'cursus trispondaicus'	$-- \mid \cup\cup--$	$',\ '\ ..'.$

(For rhythm in classical prose, see Norden 1923; in medieval prose Nicolau 1930 and Janson 1975; see also Beare 1957, 193–205.)

Another oratorical device was prose rhyme, where pairs of cola (phrases, separated by speech-pauses) had end-rhyme. (See Norden 1923, Anhang I; and especially Polheim 1925.) In the liturgy it is best known from the prayers of the Gallican and especially Mozarabic rites (see Porter 1958, 49).

The extent to which these are relevant to chant texts is doubtful. In a massive demonstration in PalMus 4 Mocquereau tried to show that the accentual endings of many chant texts were reflected in their musical settings. Although scholars seem agreed that cursus survived through the period from the late imperial age to the Carolingian renaissance, and was not revived from a moribund state by ninth-century writers, it is by no means clear that the texts of the older 'Gregorian' repertory deliberately cultivate cursus. Mocquereau's thesis was that accented syllables in sentence-endings displaying cursus were matched by relatively higher notes.

However, as Apel argues (1958, 297–301), the evidence is not convincing, especially not in the examples which use the cadences of psalm tones, introit verse tones, and responsory verse tones; these are formulas which are applied mechanically, and would seem by their nature incapable of reflecting different cursus rhythms.

Bewerunge (1910–11) and Ferretti (1913) believed that metrical, not accentual, cursus patterns were involved. As Apel again argues, Bewerunge's evidence is inconclusive and shows no systematic attempt by chant composers to match syllable-lengths with a greater or lesser number of notes. Ferretti looks for cursus patterns in the note-groupings themselves, independently of the text, but his demonstration is also unconvincing.

It would seem to be more profitable to investigate the occurrence of rhythmic prose and rhyme prose in the new liturgical prose texts of the ninth century and later, particularly in sequences and tropes. Their relevance to sequences was suggested by Polheim (1925, 350), a reference taken up by Crocker (1958, 'Repertory', 163). Except for the obvious insistence on line-endings in *a* in many sequences, end-rhyme appears to have played little or no part in their composition until the eleventh century. On the other hand, there is often a significant degree of internal assonance and alliteration. Crocker investigated instances of cursus in Notker's texts (Crocker 1977), where the rhythm of *cursus planus* (though not always its usual word-division) is often apparent (also occasionally *cursus velox*).

Examples of assonance, alliteration, and *cursus planus* may be found in, for example, Notker's *Johannes Jesu Christo*, where the five note cadence *F–G–a–a–G* occurs at the end of verses 4, 5, and 7. Five times out of six this goes with text in *cursus planus*:

    ′   ..   ′   ′   .   . ′   .   ′   .   .   ′   .

    filii dei   esse perenni   dedit custodem

       ′   .   ′   ′   .   ′   .   .   ′   .

    pater revelat   semper commenda

Assonance and alliteration in verse 5 are underlined:

    Te Christus in cruce triumphans matri suae dedit custodem,
    Ut virgo virginem servares atque curam suppeditares.

Crocker (1958, 'Repertory') made the suggestion that art-prose, with the repertory of ornaments (*tropi*) such as those just outlined, was a formative influence in the early sequence, a suggestion that had the advantage of treating the prose texts of sequences on their own terms rather than as derivatives of a musical form (prosulas). Crocker also pointed to the possibility of 'isocolon' (or 'antithesis') technique, that is, identical clause lengths, in the repertory of tropi at the writer's disposal; and he has repeatedly stressed that there are numerous irregularities in the double-versicle structure of early sequences. The effect of these observations is to narrow the gap between art-prose and early sequence texts. Despite this, the double-versicle arrangement seems too insistent for the sequence to be regarded as art-prose pure and simple. However, the resources of the latter were no doubt occasionally exploited in the creation of sequence texts.

III

Liturgical Books and Plainchant Sources

III.1. INTRODUCTION

Wordsworth and Littlehales 1904; Gamber 1968; Andrew Hughes 1982; Vogel 1986; Huglo 1988, *Livres*.

The books in which the plainchant has been recorded since Carolingian times are diverse in nature and complex in content. This is chiefly because different books contain material for different parts of the liturgical round and for the use of different personages involved in the celebration. The possible combinations are quite numerous, and at different times and places a multitude of different selections of material have been made. Some important distinctions which should be borne in mind are:

1. between books containing texts and music to be recited or sung, and books of instructions as to how the liturgy is to be celebrated;
2. between books for the mass, books for the office, and books for other ceremonies;
3. between books for the priest or other officiant, books for the cantor, and books for other persons.

The chief concern of this section of the book is naturally with sources containing music, which usually means the cantor's books. But chants might also be recorded elsewhere. For example, the gradual, precisely because it is a cantor's book, does not usually contain the music for the priest's recitation of the canon of mass. And the chants for ceremonies performed by a bishop are usually to be found in a pontifical, although most are choir chants. Then again, chant-books might be combined in various ways with books containing texts. The later medieval missal and breviary are combinations of this sort, though not the only ones:

combined in the missal:
   gradual (chants of mass)
   sacramentary (prayers of mass)
   epistolary and evangeliary (lessons of mass)

combined in the breviary:
  antiphoner (office chants)
  psalter
  hymnal
  collectar (office prayers)
  homiliary, lectionary, passionary (office lessons)

Some knowledge of liturgical books other than those with only music is desirable because they may contain valuable information about the liturgical context in which the chant was sung, or indicate by text incipit the state of the chant repertory. Older than any chant-book are a number of ordines, or books of instruction about how the liturgy was performed. These often describe the part played by the choir in the liturgy during a period (the eighth and ninth centuries) when first-hand musical evidence is lacking. Many later medieval ordinals, the descendants of the earlier ordines, specify exactly which chants are to be sung on which occasions. This may be invaluable if we have few or no other books from the church where the ordinal was used.

It goes without saying that some knowledge of the liturgy, and particularly of how the material is disposed over the church year, is essential when dealing with liturgical books. The reverse is also to some extent true. There is no better way to discover the intricacies of the medieval liturgy and enter into its modus operandi than by learning to find one's way around medieval liturgical manuscripts.

Vogel (1986) is an excellent introduction to the study of early medieval liturgical sources other than music-books and office-books (sizeable lacunae), with exhaustive bibliographies. Its strength lies in the coverage of early ordines, sacramentaries, lectionaries, and pontificals, but it ignores chant sources almost completely. Nor does it describe how the material in the books is put together, that is, which piece follows which, how the seasons of the liturgical year are disposed, in short, what the newcomer to medieval manuscripts might find when he or she first encounters the sources themselves. A heroic attempt at this was made, however, by Andrew Hughes (1982). Experience shows that Hughes's book is most useful as a reference tool and support for advanced studies. It deals chiefly with later medieval missals and breviaries (the latter are undoubtedly among the most complex of all service-books), and is a mine of information about the liturgy. It gives a number of brief tables of contents of selected sources (390–408). Gamber's *Codices liturgici* contains basic bibliographical information about over 1,500 early liturgical sources. Huglo's recent survey (1988, *Livres*) supplements his numerous articles on individual types of book in *NG*.

The best way to learn how to find one's way about a particular medieval liturgical manuscript and to read it fluently is to have a modern edition of a similar book to hand. Even those medieval books of roughly the same type, age, and provenance are not exactly the same, but they do not vary inordinately from one another. With the aid of a properly indexed modern edition of a missal or a breviary it is usually possible to locate oneself on the right feast-day and acquire a working knowledge of the heavily abbreviated or otherwise incomprehensible rubrics or titles of formularies or individual items. Breviaries and missals, since they also contain chants, can thus act as a guide for antiphoners and graduals. The most recommendable books for use as references in this way are (*a*) modern Roman/Solesmes service-books before the post-Vatican II reforms, particularly *LU*, *GR*, *AR*, and *AM*; (*b*) such editions as *Breviarium ad usum Sarum* (ed. Procter and Wordsworth 1879–86) and *The Sarum Missal* (ed. Legg 1916). Naturally, the usefulness of the modern edition will vary according to how closely related is the medieval use being compared with it. Since modern editions of medieval chant-books in their entirety are practically non-existent, either the modern Roman/Solesmes editions or well-indexed facsimiles have to be used when dealing with the music itself.

There is general agreement, but no universal standard, on the nomenclature of medieval liturgical sources. Since so many books combine heterogeneous material, it seems pointless to try to be over-specific in finding the right label. For example, books commonly referred to as tropers could contain such a diversity of material (certainly not just tropes) that the title is almost meaningless, and reference to a short list of the contents of each individual 'troper' is essential. Only then can one see what such books have in common with each other and where they differ. (For suggested typologies and nomenclatures, see Fiala and Irtenkauf, 1963, also the surveys in Ehrensberger 1887 and Cabrol 1930, and the list in Andrew Hughes 1982, 119–20.)

In the following sections most of the discussion centres on music-books and the earliest liturgical books. The last chapter is devoted to tonaries, which are not liturgical in the sense that they do not record a portion of the liturgical round as it would have been performed. As musical sources they nevertheless aided the performance of chant, and rather than place them in the chapter on theory I have brought them in here. Books of hours, being for private use rather than public worship, are not discussed.

III.2. ORDINES ROMANI

Andrieu 1931–61; Vogel 1986, 135–224.

In Andrieu's authoritative edition there are fifty Ordines Romani. They are descriptions of how various parts of the Roman liturgy should be performed. Their contents reach back to the mid-seventh century and forward into the ninth. Although based on the liturgical practice of the city of Rome itself, all surviving copies are Frankish, and some of the ordines have been adapted to accommodate northern usages. At the same time as the Franks were acquiring books of prayers (sacramentaries) and other material from Rome, they also required books of instruction on ceremonial. We have reports of Frankish pilgrims and churchmen seeking out liturgical books in Rome from as early as the seventh century, a practice which became royal policy in the mid-eighth century (Vogel, 147). Careful analysis has revealed which ordines are liturgically compatible with which sacramentaries. Some reflect papal usage (the

Lateran church), some the practice of other churches in or about Rome (such as St Peter's at the Vatican). The biggest groups describe, respectively: the celebration of mass (Ordines I–X), special rituals throughout the year such as processions, rogations, the ceremonies of Holy Week, etc. (Ordines XX–XXXIII), and ordinations (XXXIV–XL). The ordines vary greatly in length. One of the earliest, Ordo Romanus I (trans. Atchley 1905), gives a very full description of the papal eucharist in the early eighth century, which can be shown to have been known in Francia by about 750. Others are much shorter.

Although the Ordines Romani contain no music, and indeed do not often give more than text incipits for the few chants they may mention, they contain important information for the musicologist. They may reveal the various roles of soloists and choir in the performance of chants. They are especially valuable when put side by side with other witnesses to early Roman usage, for example, the writings of Amalarius of Metz, and can be compared with the much later Roman chant-books for liturgical peculiarities. This is the case, for example, with Ordo Romanus XII, a brief description of the office which mentions a number of chants by incipit. (Andrieu's edition points out the correspondences with Amalarius' description.) One important source of ordines, Brussels, Bibliothèque Royale 10127–10144, also has a copy of the chant texts of mass to be sung throughout the year, in other words a gradual without music. The contents of the manuscript, the famous 'Blandiniensis' (so-called because it belonged to St Peter's abbey on Mont-Blandin at Ghent) are as follows (see Andrieu, i. 91–6). the folio numbers give an idea of the length of the texts.

ff. 1^r–79^v	canonical texts, extracts of correspondence of popes
79^v	Ordo Romanus 13: list of the books of the bible to be read during the Night Office
80^r–82^r	end of a treatise on the ecclesiastical computus
82^v–84^r	Bede on the ordering of Easter ferias
84^r–85^r	Ordo Romanus 26: ordering of services from the fifth Sunday of Lent to Holy Saturday
85^r–86^r	Ordo Romanus 3: six short Roman and Frankish points of liturgical practice
86^r–88^v	Ordo Romanus 14: list of readings for the office at St Peter's, Rome
88^v–89^v	Ordo Romanus 30ʙ: arrangement of the office from Thursday of Holy Week to the Saturday after Easter
90^r–115^r	gradual
115^r–121^v	ordo for initiating a sick catechumen
121^v–124^r	blessings for holy objects and substances
125^r–135^r	proper prayers for eleven masses

Some of the material in the Ordines was eventually taken up in the rubrics of other types of book, and from the twelfth to thirteenth century ordinals appear with much fuller listing of items to be performed at services throughout the year. Much of the ceremonial described in the Ordines Romani was to be performed by the pope, or at least a bishop, and was therefore subsumed in the pontifical. Andrieu's Ordo Romanus L was actually copied side by side with the Romano-German pontifical of the 950s, which became standard in the Empire and in Rome itself. It is a somewhat heterogeneous miscellany of current and archaic Roman ceremonies, together with commentary, a collector's book rather than purely for liturgical use.

III.3. SACRAMENTARIES AND LECTIONARIES

(i) Sacramentaries
(ii) Lectionaries

(i) *Sacramentaries*

Gamber 1958, 1968; Hope in Cheslyn Jones *et al.* 1978, 224–8; Deshusses and Darragon 1982–3; Vogel 1986, 31–134.

The sacramentary or *liber sacramentorum* was the book containing the texts to be recited by the officiating priest (bishop, pope) at mass, for administering sacraments, for consecrations, ordinations, and other rites. It was, in effect, a book of prayers, although it takes its title from the rites represented in it. The chief prayers of mass during the central Middle Ages were the Canon, with its proper prefaces for different times during the year, and the collect, secret and postcommunion prayers, which were proper for each mass. Ordination formularies and blessings were also included in the sacramentary, but not lessons, choir chant texts or rubrics covering ceremonial.

During the early centuries prayers were probably mostly improvised and there is little evidence of any formal collections of texts. The first sign of these appears at the end of the fourth century and the beginning of the fifth. For example, Musaeus, a priest of Marseilles (d. *c*.460) is said to have compiled a sacramentary, together with a lectionary and a book of responsories (that is, their texts) (Gennadius, *Liber de viris illustribus*: PL 58, 1104; see Morin 1937; Gamber 1959; Vogel 1986, 302–3, 321; McKinnon 1987, *Music*, 170). Most early collections seem to have been no more than leaflets containing only a few masses. The so-called Leonine or Verona Sacramentary (in the manuscript Verona, Biblioteca Capitolare, LXXXV, of the early seventh century) appears to be a conglomeration of such leaflets, or *libelli missarum*, compiled perhaps for reasons of private piety rather than public worship; its material is Roman, from the fifth and sixth centuries (ed. Mohlberg *et al.* 1956).

The earliest full sacramentaries of which there is clear evidence date back to the seventh century. A few of the more important Roman and Roman-Frankish books may be mentioned here, because of both their intrinsic importance and the light they shed on the adoption of Roman use by the Franks in the eighth century. What actually survives of Roman use is a group of sacramentaries based on Roman books but written in Francia and containing certain amounts of Gallican material. The classes or types into which they have been divided are (following Vogel's characterizations) the following.

1. The 'Old Gelasian' sacramentary, based on a book reflecting the use of the titular churches of Rome of the period 628–715. It survives in Rome, Biblioteca Apostolica Vaticana, Reg. lat. 316 (plus Paris, Bibliothèque Nationale, lat. 7193), written at Chelles *c*.750 (ed. Mohlberg *et al.* 1960).

2. The 'Frankish' or 'eighth-century Gelasian' sacramentary. The several sources preserve various versions based on an archetype perhaps compiled at Flavigny in Burgundy, perhaps during the reign of Pippin III 'the Short' (751–68). The book combines Old Gelasian material, a papal sacramentary adapted for presbyteral use at the Vatican, and the usual local Gallican formularies. (The Sacramentary of Gellone, perhaps the most important of many Frankish-Gelasian sacramentaries, has been edited by Dumas and Deshusses, 1981.)

3. The 'Gregorian' sacramentary. This papal sacramentary was probably compiled under Honorius I (625–38) (rather than Gregory the Great, 590–604, whose name it bears), and was supplemented with new material as required during the subsequent century and a half.

A copy of the Gregorian sacramentary in its late-seventh-century state appears to have been known to Alcuin of Tours (d. 804), chief liturgical adviser to Charlemagne; perhaps it came with other books he had sent from his home city of York in 797. It then became known to St Benedict of Aniane (d. 821), the great reformer and adviser of Louis the Pious, and to Arno, bishop of Salzburg (785–821), former abbot of Saint-Amand and a friend of both Alcuin and Benedict; it was used in the compilation of the Sacramentary of Trent or Salzburg (ed. Rehle and Gamber 1970, 1973).

In its mid-eighth-century state the Gregorian sacramentary was copied and sent by Hadrian I (772–95) to Charlemagne between 784 and 791. This manuscript, which was kept at Aachen as an exemplar for further copies, has not survived, but a transcription made for Bishop Hildoard of Cambrai in 811–12 still exists (Cambrai Bibliothèque Municipale 164). (See the edition by Deshusses 1971–9.)

It is clear from this (i) that we owe our knowledge of these Roman sacramentaries to Frankish churchmen, (ii) a number of conflicting types were circulating simultaneously in Francia, to which we should add the material brought from Rome more or less privately in the form of *libelli*.

There also survive from eighth-century Francia four Gallican sacramentaries which contain small amounts of Roman material, the so-called 'Missale Gothicum', 'Missale Francorum', 'Missale Gallicanum vetus' and 'Missale Bobbiense'.

The sacramentary sent by Hadrian at Charlemagne's request was itself insufficient for Frankish purposes. Instead of the expected formularies for the cycle of Sundays after Christmas, Epiphany, Easter, and Pentecost, the 'Hadrianum' contained merely a pool of prayers to choose from as required. It also lacked prayers for consecrations and votive masses, which the Franks were accustomed to seeing in previous sacramentaries. It contained, in fact, only what the pope himself would have required while celebrating in the Lateran and the stational churches of the city of Rome. Benedict of Aniane therefore compiled an official supplement (with an engagingly frank preface: trans. Vogel, 87–8). The admixtures of Gelasian material in the subsequent generations of Gregorian sacramentaries bear witness to the less than total success with which Charlemagne promulagated 'Roman' use, at least as far as the sacramentary is concerned. Success was in any case hardly feasible, given the nature of the materials and the impossibility of publishing 'correct' books.

The relevance of all this to the history of plainchant may be stated as follows. Roman material of several different types came north both piecemeal and in more comprehensive codices. Is this how the chant repertory, at least for mass, was also transmitted? Since it seems unlikely that musical notation was used until later in the ninth century, we should presumably imagine books containing just chant texts, intermingling to a greater or lesser extent with previous practice. It may be that the date of the first recension of the gradual, containing chant texts for the year's cycle of masses, was compiled at about the same time as the first such sacramentary. We have no firm evidence for this, however. One can go no further than to suggest that the history of the two books may have run partly parallel.

The sacramentary as a separate book gradually went out of use after the late eleventh century, becoming part of the missal. Before and during that time, various combinations of sacramentary, lectionary, and/or gradual were made. These are discussed below where the missal is described. A number of early sacramentaries have chant-text incipits written in their margins, sometimes notated (two examples: Oxford, Bodleian Library, Bodley 579, the 'Leofric Missal', text ed. Warren, 1883; and Düsseldorf, Universitätsbibliothek D 1, facs. Jammers 1952; Stäblein 1975, 107).

No classification of post-Carolingian sacramentaries has been carried out—the task is of course immense—though comparisons of books from specific areas have been made (for example, by British scholars whose work appears in the volumes of the Henry Bradshaw Society).

(ii) *Lectionaries*

Frere 1934, 1935; Klauser 1935; Chavasse 1952; Gamber 1968; Vogel 1986, 291–355.

The lessons to be read at mass were taken from the Bible and were known as 'pericopes'. They were recorded in one of several ways. The Bible could be marked up with crosses or other signs in the margin to indicate the passages to be recited. Lists of pericopes were also drawn up, known as capitularies, where incipits and explicits were recorded. Later the lessons were copied out separately in lectionaries (a general term for books with lessons for mass), perhaps divided between the epistolary (lessons from the Epistles, sung by the subdeacon) and the evangeliary (lessons from the Gospels, sung by the deacon). Another early name for the lectionary, or list of readings derived from a pre-existing lectionary, is *comes* (*Liber comitis*, *Liber commicus*).

In the early centuries scriptural passages would have been chosen more or less on an *ad hoc* basis, and not before the late fourth century is there good evidence that yearly cycles of lessons were organized. The example of Musaeus of Marseilles has just been mentioned. Much of the other surviving evidence of this period also comes from Gaul: the information to be gleaned from the sermons of Bishop Caesarius of Arles (502–43) and the *Expositio antiquae liturgiae gallicanae*, and the earliest of all surviving Latin service-books: the palimpsest Gallican lectionary Wolfenbüttel, Herzog-August-Bibliothek, Weissenburg 76, dating from the fifth to sixth century (ed. Dold, 1936). The Wolfenbüttel source is doubly important because text incipits of chants to be sung after the lessons are also included (Dold, xciv ff.). Also non-Roman is the next earliest source, the epistolary of Capua (ed. Ranke 1868), written *c*.545 for Victor of Capua, taken to England probably by Abbot Hadrian of Nisida, who was one of the companions of Archbishop Theodore of Canterbury, and later used by St Boniface). Another famous codex is the Lindisfarne Gospels, written *c*.700 at Lindisfarne but based on a Neapolitan exemplar (ed. Skeat 1871–87).

The non-Roman Latin liturgies had three readings at mass (from the Old Testament, the New Testament, and the Gospels, respectively), and this may have been the original usage in Rome as well. The earliest Roman witness is the capitulary or *comes* of Würzburg (*c*.700 but based on a model of the time of Gregory the Great, d. 604; ed. Morin 1910 and 1911). For some masses this source has three lessons, but it seems possible that the extra lesson is a duplicate (see Martimort 1984). It was until recently widely held that the gradual and alleluia (or tract) of the medieval mass might formerly have been separated by another lesson (for otherwise there would be no point in having two chants here), but lately it has been argued that the gradual itself originated as a psalm reading (McKinnon 1987; see the discussion in VII.5 below). From at least the seventh century, at any rate, two lessons separated by chant seems to have been the rule in Rome.

It seems that the selection of epistles and the selection of gospels were made independently of each other and independently of the chants and prayers of mass. At least, no system combining them all according to a logical principle has been discovered. The history of the organization of the lessons, which can be traced back to an earlier date than that of prayers and chants, may throw light on the latter. For example, the lessons from Septuagesima to Easter appear to have been organized before the time of Gregory the Great (d. 604) and remain constant thereafter. Chavasse (1952) has demonstrated that the development of the lectionary resembles that of the gradual, rather than the sacramentary.

Of the many different arrangements, Roman and non-Roman, for which witnesses survive from the eighth century, a Frankish-Gregorian type first represented by the *Comes* of Murbach (ed. Wilmart 1913) emerged as the basis of later medieval usage. In this case, unlike that of the sacramentary, no document of papal usage was solicited from Rome, and the readings were arranged to fit the liturgical occasions required by the Frankish-Gelasian sacramentary.

Lectionaries were combined with sacramentaries and/or graduals in a number of ways, culminating in the missals of the later Middle Ages. These are discussed briefly below (III.9). As with the sacramentary, no classification of post-Carolingian manuscripts (apart from late sources of the older types) has been accomplished.

III.4. GRADUALS (MASS ANTIPHONERS) AND CANTATORIA

(i) Introduction
(ii) Graduals without Notation
(iii) Notated Graduals

Stäblein, 'Cantatorium', *MGG*; Melnicki and Stäblein, 'Graduale (Buch')', *MGG Graduel romain, II: Les Sources* (1957); *Graduel romain, IV: Le Texte neumatique* (1960), 1962); Steiner, 'Cantatorium', *NG*; Huglo, 'Gradual (ii)', *NG*; Emerson, 'Sources, MS, II', *NG*; Andrew Hughes 1982, 124–42, 157–9.

FACSIMILES OF GRADUALS AND NOTED MISSALS:
Bamberg, Staatsbibliothek, lit. 6: see Bamberg 6
Benevento, Archivio Capitolare 33: PalMus 20
Benevento, Archivio Capitolare 34: PalMus 15
Bratislava, etc., fragments: *Missale notatum Strigoniense*
Chartres, Bibliothèque Municipale 47: PalMus 11
Darmstadt, Hessische Landes- und Hochschulbibliothek 1946: *Echternacher Sakramentar und Antiphonar*
Einsiedeln, Stiftsbibliothek 121 (sequentiary omitted): PalMus 4
Gniezno, Archiwum Archidiecezjalne 149: *Missale planarium . . . Gnesnensis*
Graz, Universitätsbibliothek 807: PalMus 19
Laon, Bibliothèque Municipale 239: PalMus 10
Leipzig, Universitätsbibliothek 391 (sequentiary omitted): Wagner 1930–2
London, British Library, Add. 12194 (without sequentiary): *Graduale Sarisburiense*
Montpellier, Faculté de Médecine H. 159: PalMus 8
Paris, Bibliothèque Nationale 903 (troper and sequentiary omitted): PalMus 13
Paris, Bibliothèque Nationale 904: Loriquet *et al.* 1907
Passau, printed gradual of 1511: *Graduale Pataviense*
Rome, Biblioteca Angelica 123: PalMus 18
Rome, Biblioteca Apostolica Vaticana, Vat. lat. 10673: PalMus 14
St Gall, Stiftsbibliothek 339: PalMus 1
St Gall, Stiftsbibliothek 359: PalMus II/2
St Petersburg, Saltykov-Shchedrin Public Library O. v. I. 6: Thibaut 1912
Västerås, printed gradual of *c.*1513: *Graduale Arosiense*
Vienna, Nationalbibliothek, ser. nov. 2700 (with an office antiphoner): *Antiphonar von St. Peter*
private collection, the 'Mont-Renaud manuscript' (with an antiphoner): PalMus 16
private collection of Martin Bodmer C. 74 (Old Roman): Lütolf 1987

EDITIONS:
GR; *AMS* (texts of six early sources); MMMA 2 (Rome, Biblioteca Apostolica Vaticana, Vat. lat. 5319, Old Roman); Sandon 1984, 1986 (Sarum).

(i) *Introduction*

The term gradual has been used since the early Middle Ages to designate books containing the proper chants of mass. Its derivation is obscure. The earliest surviving books of this general type bear the title 'antefonarius' (see the sources edited by Hesbert in his appropriately named *Antiphonale missarum sextuplex*), as for instance in the title of Brussels 10127–10144 (late eighth century): 'In dei nomen incipit antefonarius ordinatus a sancto Gregorio per circulum anni'. Since the word 'antiphonarius' could also be used of a book containing office chants, gradual is usually preferred for the mass chant-book.

Hucke (1955, 'Graduale') pointed out that in several early occurrences of the word it appears as an adjective *gradalis*, referring to the day hours of the liturgy, as opposed to *nocturnalis*, the night hours. Perhaps there is a connection with the *canticum graduum*, St Jerome's translation of the mysterious Hebrew rubric for Psalms 120–34, which played an important part in the day hours of the Roman liturgy.

In five of the sources of the *Sextuplex* the gradual chant is designed 'Resp(onsorium) Grad(ale)', in source R (see the explanation of the siglum below) simply 'Grad(ale)'. This use of 'gradale' is usually explained as 'responsory sung on the altar-steps' (from *gradus* = step) or 'while the deacon ascends the steps of the ambo to recite the gospel'. An alternative explanation could, however, be 'day responsory', or perhaps here the word has no connection with the previous meaning.

In Ordo Romanus I (*c*.700) it is stated that the cantor 'cum cantatorio' goes up to sing the responsory (gradual) and alleluia or tract. The statement reappears in Ordines IV, V, and VI. From this a special class of gradual has been identified, containing only these chants. Such manuscripts do indeed survive from the ninth century onward, including the famous St Gall, Stiftsbibliothek 359, which notates graduals, alleluias, and tracts in full but gives only text incipits for the other chants.

Amalarius of Metz (*Prologus de ordine antiphonarii*, 18, ed. Hanssens, i. 363; PL 105, 1245), writing *c*.830, says that what the Romans call a *cantatorium*, the Franks call a *gradale*. The Romans collect responsories in a *responsoriale* and antiphons in an *antiphonarius*. If *responsoriale* and *antiphonarius* refer only to the office, *cantatorium* to the mass, then the antiphons of mass are left unaccounted for. In fact, however, all Amalarius' writings in this context refer only to office chants. Hucke argues therefore that the cantatorium/gradale is once again the book containing chants for the day hours. The Night Office requires far more chants, and the Romans divide them between two books, one for responsories and the other for antiphons. 'I have followed our usage', says Amalarius, 'and place both responsories and antiphons mixed together according to the order of the hours'.

At this early stage, therefore, the nomenclature of chant-books was somewhat fluid, and references to 'antiphonarius' and 'gradale' in early book-catalogues have to be treated with some caution (for example those edited by Gustav Becker 1885: see nos. 7, 8, 23, 24 for 'gradale/gradalis', and nos. 7, 18, 21, 33, 238 for 'antiphonarius').

(ii) *Graduals without Notation*

Although books containing chant texts for mass probably existed as early as any other mass books, they have survived in much smaller numbers. The earliest evidence for them that we have is possibly Gennadius' reference, already mentioned (III.3), to the books compiled by Musaeus of Marseilles (d. *c*.460) at the request of his bishop Venerius (d. 452), though he may mean office responsories, rather than the gradual responsories of mass. At the end of the same century, however, appears the first copy of mass chant texts to have survived, the incipits copied after various lessons in the palimpsest Gallican lectionary Wolfenbüttel, Herzog-August-Bibliothek, Weissenburg 76 (ed. Dold 1936). Dold was with great difficulty able to decipher sixteen of these incipits, some of which correspond with later gradual or offertory texts. They may refer the user across to another book containing the complete chant texts, unless we are to assume that these were known by heart.

The first evidence for the existence of chant-books in Rome itself is by report only and considerably later. At the same time as Bede was writing his history of the English church and people (completed 731), Egbert, a member of the Northumbrian royal family, was ordained deacon in Rome and was then appointed bishop of York (*c*.732). In a letter of 735 which still survives, Bede advised Egbert to apply for the pallium, which he obtained; he died in 766. His brother Eadberht had become king in 738, and with his support Egbert founded a cathedral school in York, where Alcuin was a pupil and eventually master. In two places in Egbert's *Dialogus ecclesiasticus institutionis* (PL 89, 377–451) he refers to a chant book ('liber antiphonarius') and sacramentary ('liber missalis') which he had presumably seen in Rome. He shared the popular belief that their author was Gregory the Great and that Augustine had brought copies to England over a century earlier: '. . . beatus Gregorius in suo antiphonari et missali libro per paedagogum nostrum beatum Augustinum transmisit ordinatum et rescriptum . . .' (PL 89, 440–2, cited Gevaert 1890, 80–1 and Ashworth 1958).

Contemporary evidence is to be found in the canons of the Council of Cloveshoe (747):

Ut uno eodemque modo dominicae dispensationis in carne sacrosanctae festivitates, in omnibus ad eas rite competentibus rebus, id est in baptismi officio, in missarum celebratione, in cantilenae modo celebrentur, iuxta exemplar videlicet quod scriptum de romana habemus ecclesia. Itemque ut per gyrum totius anni natalicia sanctorum uno eodem die, iuxta martyrologium eiusdem romanae ecclesiae, cum sua sibi conventienti psalmodia seu cantilena venerentur. (Haddan and Stubbs 1871, 137.)

The reference to a Roman exemplar covers each of the types of book mentioned, for the baptismal rites, for the mass (presumably a sacramentary), and for chant. The feasts of the Sanctorale, governed by the martyrology, are to be sung with the appropriate psalmody and chants. There is no definite implication that either this chant-book or those seen by Egbert contained notation.

The first graduals (or mass antiphoners) which have come down to us all lack notation and were written in north France. They were edited by Hesbert in *AMS* and their date and provenance have been summarized by Froger (1978, 'Critical'). I list them here with the customary sigla.

Brussels, Bibliothèque Royale 10127–10144 (B = Blandiniensis), abbey of St Peter on Mont-Blandin, Ghent; late eighth century; the manuscript also contains Ordines Romani (see above, III.2).

Zürich, Zentralbibliothek, Rheinau 30 (R = Rhenaugiensis), Nivelles; 790s; the parts of chants sung by soloists (gradual verses, etc.) are not given; the manuscript also contains a Frankish-Gelasian sacramentary.

Paris, Bibliothèque Nationale, lat. 12050 (K = Corbiensis), abbey of Corbie; 850s; the manuscript also contains a Gregorian sacramentary with Benedict of Aniane's supplement, copied for Rodradus of Corbie.

Monza, Basilica S. Giovanni CIX (M = Modoetiensis), NE France; second half of ninth century; the manuscript has only soloists' chants.

Paris, Bibliothèque Nationale, lat. 17436 (C = Compendiensis), abbey of Saint-Médard at Soissons (see Froger 1980), later at the abbey of Saint-Corneille at Compiègne; second half of ninth century; the manuscript also contains an office antiphoner, the earliest such to have survived; sometimes known as the 'Antiphoner of Charles the Bald', in the belief that it was presented by Charles (d. 877) to Saint-Corneille.

Paris, Bibliothèque Sainte-Geneviève 111 (S = Silvanectensis), written at the monastery of Saint-Denis for the cathedral of Senlis; 877–82; the manuscript also contains a sacramentary.

A number of fragmentary sources also belong to this early group, of which the most important are perhaps the Lucca fragments of the late eighth century, which have office chants as well (*AMS*, XXIV–XXVI; Froger 1979, 'Fragment').

The manuscripts do not all record chants in the same way. The Monza source has only graduals, alleluias, and tracts (and of these only the tracts are consistently written out in full) and can therefore be designated as a cantatorium. The Nivelles book (R) by contrast has only texts sung by the choir.

As already mentioned, these manuscripts have no notation, although the Corbie gradual has letters indicating the mode of chants (see below, III.14). They nevertheless tell us a great deal about the development of the repertory. Particularly interesting are those sections where they differ from one another or lack chants regularly present in later books. There is space here to mention only a few of these peculiarities (see the exhaustive study in *AMS*).

Most striking is the disunity in the assignment of alleluias to the various masses. In several places the Blandiniensis simply gives the rubric 'alleluia quale volueris', meaning that the singer must at this point dip into a pool of suitable alleluias gathered elsewhere (in this case at the end of the manuscript). This happens in Easter week, for the Sundays after Easter, Whitsun week, and the Sundays after Whitsuntide. There is the greatest disparity between later books in their assignment of alleluias to these occasions. The graduals of the summer Sundays also display peculiarities of ordering, which has given rise to extensive discussion (see Hesbert 1932–3; PalMus 14, 124–44; Chavasse 1952; Raymond Le Roux 1962; Chavasse 1984).

The Roman mass antiphoners known to the Franks had a number of features which were altered by them, but which can be found in some later Gregorian sources. Thus M has the Roman offertory for St Michael *In conspectu angelorum* instead of the Frankish *Stetit angelus*. Huglo (1954, 'Vieux-romain') drew up a list of Roman characteristics and identified their presence in several later sources, mostly from central Italy.

On the other hand, all Frankish books include a mass for St Gorgonius (9 September), a non-Roman custom initiated when his relics were brought to Metz by Bishop Chrodegang during the reign of Pippin the Short (751–68).

It may be asked how the Franks, seemingly ardently engaged in imitating Roman use as closely as possible, felt at liberty to deviate occasionally from what they found in Roman books. The answer might be along the lines of the apologia presented by Benedict of Aniane, prefacing his supplement to the Gregorian sacramentary: '. . . there are other liturgical materials which Holy Church finds itself obliged to use but which the aforesaid Father [Gregory] omitted because he knew they had already been produced by other people . . .'

(iii) *Notated Graduals*

The earliest notated graduals to have survived are from the end of the ninth century:

Chartres, Bibliothèque Municipale 47 (destroyed in 1944; facsimile PalMus 11), from Brittany.

St Gall, Stiftsbibliothek 359 (PalMus II/2), from St Gall; the manuscript is a cantatorium, with soloist's chants only.

Laon, Bibliothèque Municipale 239 (PalMus 10), from Laon.

Equally early fragments have also survived from these areas (that is, Brittany, Germany, and north France), but the earliest sources from other areas are somewhat later. England, as might be expected after the devastation wreaked by the Danish invasions of the eighth and ninth centuries, is represented first in the late tenth century. The earliest surviving graduals from Aquitaine are from the early eleventh century, but three tropers of the tenth century survive. Notated Italian sources are practically non-existent before the eleventh century.

By the time of the appearance of the first notated graduals, the assignment of particular chants to particular masses was fairly stable, the main differences between sources being the following:

1. the choice of alleluias for the Easter and Whitsuntide series;
2. the choice of psalm verses for introit and offertory;
3. the choice of chants for saints' days; these could be taken from the common pool, in which case different sources might make a different choice, or (rarely) newly composed.

In the sources of the *Sextuplex* there is no common pool, except for four masses for a 'pontifex' in K and S only. Nor is there one in many other early sources. There is simply a cross-reference from the one saint to another where the appropriate mass is to be found. As the Sanctorale became more heavily populated, however, the frequent cross-referring became inconvenient, and the chants were eventually transferred to a Commune Sanctorum section at the end of the gradual.

Another way in which graduals might differ is in their interweaving of the Temporale and Sanctorale. From later service-books we are used to the placing of all saints' feasts except those immediately following Christmas in a section of their own. The Roman pattern as found in, for example, Rome, Biblioteca Apostolica Vaticana, Vat. lat. 5319 (MMMA 2), is given in Table III.4.1.

Table III.4.1 *Interlocking Temporale and Sanctorale in Roman use*

Temporale	Sanctorale
Advent Sundays 1–2	
	St Lucy (13 Dec.)
Advent Sunday 3	
Ember Week	
Christmas	
	St Stephen (26 Dec.) to St Silvester (31 Dec.)
Sunday after Christmas	
Epiphany	
Sundays after Epiphany	
	St Felix (14 Jan.) to Annunciation of the BVM (25 Mar.)
Septuagesima to the 4th Sunday after Easter	
	SS Tiburtius and Valerianus (14 Apr.) to St Pudentiana (19 May)
Ascension	
Sunday after Ascension	
	St Urban (25 May)
Whit Sunday and Whitsun week	
	SS Marcellinus and Peter (2 June) to SS John and Paul (26 June)
Sundays 1–4 after Whitsun	
	SS Peter and Paul (29 June) to Octave of SS Peter and Paul (6 July)
Sundays 5–9 after Whitsun	
	The Seven Brothers (10 July) to SS Felix and Adauctus (30 Aug.)
Sundays 10–15 after Whitsun	
	St Hadrian (8 Sept.) to St Matthew (21 Sept.)
Ember Week	
	SS Cosmas and Damian (27 Sept.) to St Mennas (11 Nov.)
Sundays 16–24 after Whitsun	
	St Cecilia (22 Nov.) to St Andrew (30 Nov.)

As the sources of the *Sextuplex* and early notated graduals show, the Franks altered the disposition of the post-Whitsuntide period, as for instance in St Gall 339 (contents listed in Wagner I), displayed in Table III.4.2.

Thus the saints of the Epiphany period have been amalgamated with the Temporale, not a common procedure, but also not a confusing one, for the Sundays do not differ much in date from year to year. On the other hand, the complete Sanctorale from June to November appears as one block, followed by all the Sundays after Whitsuntide together.

It was a logical step to remove the winter saints from the first part of the gradual as well, so that only those at the end of December remained. The arrangement in the thirteenth-century manuscript London, British Library, Add. 12194 (*Graduale Sarisburiense*) is given in Table III.4.3.

Of the kyriale and sequentiary more will be said below. Graduals also frequently

Table III.4.2. *Disposition of Temporale and Sanctorale in Frankish use*

Temporale	Sanctorale
Advent Sundays 1–2	
	St Lucy (13 Dec.)
Advent Sunday 3 Ember Week Christmas	
	St Stephen (26 Dec.) to St Silvester (31 Dec.)
Sunday after Christmas Epiphany 1st Sunday after Epiphany	
	St Felix (14 Jan.)
2nd Sunday after Epiphany	
	St Marcellus (16 Jan.) to St Agnes (21 Jan.)
3rd Sunday after Epiphany	
	St Vincent (22 Jan.) to Annunciation of the BVM (25 Mar.)
Septuagesima to the 4th Sunday after Easter	
	SS Tiburtius and Valerianus (14 Apr.) to St Pudentiana or Potentiana (19 May)
Ascension Sunday after Ascension	
	St Urban (25 May)
Whit Sunday and Whitsun week	
	SS Marcellinus and Peter (2 June) to St Andrew (30 Nov.)
Trinity Sunday Sundays 1–24 after the Octave of Whit Sunday	

Table III.4.3. *Separate Temporale and Sanctorale in later medieval use*

Temporale	Sanctorale	Other material
Advent Sundays 1–3 Ember Week Advent Sunday 4 Christmas		
	St Stephen (26 Dec.) to St Silvester (31 Dec.)	
Sunday after Christmas to the Sunday before Advent Dedication of the Church		
	St Andrew (30 Nov.) to St Linus (26 Nov.)	
		Commune sanctorum Requiem mass Votive masses Kyriale and sequentiary

contain processional chants, which, though not part of the mass itself, were often the part of the liturgy which immediately preceded mass (see I.7). Not all books follow the Sarum pattern exactly, of course. Paris books, for example, put even the Christmas saints' feasts in the Sanctorale. Sometimes the year began with a saint other than St Andrew. Graduals which also include tropes, ordinary of mass melodies, and sequences have different ways of deploying the material: while most place them in separate sections (or have a separate book for them), several Italian books have them among the proper chants for particular masses. The brief descriptions by Emerson ('Sources, MS, II', *NG*) indicate many of these differences.

Apart from differences of organization or choice of chants, graduals also differ from one another in their musical readings, because of oral transmission and the freedom of local cantors to notate the melodies as they understood them, rather than slavishly following an exemplar. The end result is a host of differences between sources. Some are very obvious differences, such as a quite different melody; the vast majority are trivial as far as the basic shape and character of the melody is concerned.

A little over thirty years ago the monks of Solesmes began publication of material for a new critical edition of the gradual with Gregorian melodies, under the title *Le Graduel romain*. A list of nearly 750 sources appeared first, with brief descriptions. Then the results were published of several comparisons of the sources with one another, carried out by isolating places where the melodic readings tended to show disparity. By counting the number of agreements between sources at these 'lieux variants' it was possible to group the sources into families. (For further work of this kind, see Hiley 1980–1 and 1986, 'Thurstan'.) One of the most striking results to emerge from the Solesmes investigation was the remarkable unanimity between

German sources, compared with the relatively fragmented picture for France and Italy (see below, IX.3).

As can be seen even from the few manuscripts of the *Sextuplex*, early books of chant texts were often coupled to sacramentaries or lectionaries. Chant incipits were also given in many early sacramentaries. As sources of mass chants, graduals and cantatoria were later joined by notated missals, but persisted as independent types of book until the end of the Middle Ages.

III.5. ANTIPHONERS (OFFICE ANTIPHONERS)

  (i) The Earliest Antiphoners
 (ii) Types of Manuscript and Calendric Organization
(iii) ·Comparison of Sources

Stäblein, 'Antiphonar', *MGG*; Huglo, 'Antiphoner', *NG*; Emerson, 'Sources, MS, II', *NG*; Andrew Hughes 1982, 161–97, 238–9, 242–4.

FACSIMILES:
Berlin, Staatsbibliothek Preußischer Kulturbesitz, Mus. ms. 40047: Möller 1990
Cambridge University Library, Mm.2.9: *AS*
Durham, Cathedral Chapter Library B. III. 11: Frere 1923 (see also *CAO* 1)
Graz, Universitätsbibliothek 211: *Codex Albensis*
London, British Library, Add. 30850: *Antiphonale Silense* (see also *CAO* 2)
Lucca, Biblioteca Capitolare 601: PalMus 9
Passau, printed antiphoner of 1519: *Antiphonale Pataviense*
St Gall, Stiftsbibliothek 390–391: PalMus II/1
Vienna, Nationalbibliothek, ser. nova 2700 (with a gradual): *Antiphonar von St. Peter*
Worcester, Cathedral Chapter Library F. 160: PalMus 12 (only the antiphoner, processional, and hymnal)
private collection, the 'Mont-Renaud manuscript' (with a gradual): PalMus 16

EDITIONS: *AR, AM*
The text editions in *CAO* 1–2:
CAO 1: manuscripts following the secular or Roman cursus
  B = Bamberg, Staatsbibliothek, lit. 23
  C = Paris, Bibliothèque Nationale, lat. 17436
  G = Durham, Cathedral Chapter Library B. III. 11 (see also facs.)
  E = Ivrea, Biblioteca Capitolare CVI
  M = Monza, Basilica S. Giovanni C. 12/75
  V = Verona, Biblioteca Capitolare XCVIII
CAO 2: manuscripts following the monastic cursus
  H = St Gall, Stiftsbibliothek 390–391 (see also facs.)
  R = Zürich, Zentralbibliothek, Rheinau 28
  D = Paris, Bibliothèque Nationale, lat. 17296
  F = Paris, Bibliothèque Nationale, lat. 12584
  S = London, British Library, Add. 30850 (see also facs.)
  L = Benevento, Archivio Capitolare 21

CAO 3–4: edition of each individual chant text

CAO 5–6: repertorial comparison of 800 sources (see also Ottosen 1986)

(i) *The Earliest Antiphoners*

The word antiphoner is usually used to mean the book containing the chants of the office hours. This distinction is a relatively recent one, however, for up to the ninth century the word was regularly used to mean any kind of chant-book. If Amalarius be understood correctly (see the passage discussed above, III.4), the Romans may have gathered both office and mass antiphons together in an 'antiphonarius', and both office and mass responsories together in a 'responsoriale'. As Huglo points out ('Antiphoner', *NG*, 482) the Romans may thus have had mass and office chants amalgamated in one book, whereas Frankish practice was always to separate office from mass chants. For these reasons, the 'antiphonarii' which appear in library catalogues of the eighth and ninth centuries (cited by Huglo) cannot be identified definitely with office antiphoners. Whatever the name in the early centuries, 'antiphoner' is used here in its commonest modern sense, to mean a book with all the office chants.

Early books containing office chant texts have survived in very small numbers compared with those containing mass chant texts. The texts copied by Winithar in St Gall, Stiftsbibliothek 1399 in the late eighth century (ed. Dold 1940) and the Lucca fragment of the late eighth century (Lucca, Biblioteca Capitolare 490: ed. Froger 1979, 'Fragment') are the earliest surviving documents. They are followed more than half a century later by the so-called 'Antiphoner of Charles the Bald', which also contains a gradual (ed. Hesbert, *CAO* 1), and an Aquitanian manuscript of the same type, Albi, Bibliothèque Municipale 44. These sources were not notated.

The earliest antiphoners with notation are:

the 'Mont-Renaud manuscript' (in a private collection), abbey of Saint-Denis, text first half of tenth century, neumes added tenth–eleventh century (PalMus 16).

St Gall, Stiftsbibliothek 390–391, St Gall, copied by the monk Hartker, end of tenth century (text ed. Hesbert, *CAO* 2; facs. PalMus II/1).

Oxford, Bodleian Library, Auct. F. 4. 26, a fragment with Breton notation, tenth century (facs. Nicholson 1913, pl. IX).

The antiphoner which Amalarius of Metz compiled, described in his *Liber de ordine antiphonarii* (*c*.830), is lost. It seems to have been a private initiative without long-lasting consequences, for no surviving antiphoner reflects the ordering of pieces suggested by Amalarius (see Hesbert 1980, 'Amalar'). The actual Metz repertory can nevertheless be reconstructed in its entirety from the ninth-century tonary Metz, Bibliothèque Municipale 351 (ed. Lipphardt 1965), as well as from later Metz manuscripts.

Notated sources from other parts of Europe have survived only from a later period. Two lists of chants survive in tenth-century sources from Limoges (Paris, Bibliothèque Nationale, lat. 1085 and 1240); but notated Aquitanian sources are lacking before the eleventh century. Toledo, Archivo Capitular 44. 1 was written in south France in the eleventh to twelfth century and taken to Spain at a time when Roman use was replacing Mozarabic. Other early Spanish manuscripts are also witness to this change: London, British Museum, Add. 30848 and 30850 were both written at San Domingo de Silos in the late eleventh century (the former is a noted breviary; the text of the latter ed. Hesbert, *CAO* 2, and facs. *Antiphonale Silense*). The earliest surviving source from south Italy is likewise a noted breviary, Montecassino, Archivio della Badia 420. North Italian antiphoners of the eleventh century include Monza, Basilica S. Giovanni C. 12/75 (text ed. Hesbert, *CAO* 1), Monza C. 15/79, and Oxford, Bodleian Library, Misc. lit. 366.

(ii) *Types of Manuscript and Calendric Organization*

Although the amalgamation of office and mass chants in a gradual-antiphoner is very rare, the presence of both gradual and antiphoner between the same covers of a medieval manuscript is not unknown. The complementing of chants by office lessons and other texts produced the breviary, which is discussed below. At the same time some parts of the office chant repertory were copied separately: in psalters and hymnals, both also discussed below. Another category within the general antiphoner class is the diurnal, where only the chants for the day hours are recorded. The great bulk of the material to be recorded prompted several such subdivisions and subspecies of book. Since such books often included not only chants but also lessons and prayers, two examples are discussed below in the section on breviaries (III.10).

The arrangement of Temporale and Sanctorale in antiphoners is usually different from that of graduals. The responsories and antiphons for the Sundays after Whitsuntide are usually copied separately from each other, perhaps reflecting the fact that the chants had once been assembled in different books. Benedictus and Magnificat antiphons form another group sometimes copied separately. The tones for psalms and canticles, being so universally familiar, are rarely found, at any rate in earlier antiphoners, and are usually placed in a supplement at the end of the book. They are likely to appear in close proximity to a tonary, and are thus occasionally given in books which otherwise contain no office chants, such as the Saint-Martial tropers Paris, Bibliothèque Nationale, lat. 909 and 1121.

Two different arrangements of the chants may be seen in Table III.5.1. The two sources represented are: Hartker's antiphoner, St Gall, Stiftsbibliothek 390–391, of the closing years of the tenth century (facsimile, PalMus II/2; text edition, *CAO* 2), and Lucca, Biblioteca Capitolare 601, from the abbey of St Peter at Pozzuoli, a manuscript of the early twelfth century (facsimile, PalMus 9). Both books are monastic, though that is not of relevance to the way the year is organized. (Both editions have tables setting out the order in which the chants appear: PalMus II/2, pp. 61*–62*; PalMus 9, pp. 16*–20*, which also gives the number of chants for each occasion.) It will be noticed that neither manuscript contains a psalter or a hymnal. Not even incipits for hymns are given.

Table III.5.1. *Deployment of material in two antiphoners*

St Gall, Stiftsbibliothek 390–391	Lucca, Biblioteca Capitolare 601
Advent Sundays 1–2	Advent Sundays 1–3
St Lucy (13 Dec.)	
Advent Sunday 3	
Advent Sunday 4	Advent Sunday 4
Christmas Eve to the 5th Sunday after Epiphany	Christmas Eve to the Octave of the Epiphany
	1st Sunday after the Octave of the Epiphany and subsequent ferias: these chants are for Sundays and ferias throughout the year when no feast-day intervenes
responsories for the weekly office	Sundays after the Octave of the Epiphany
antiphons and short responsories for the weekly office	
Trinity	
St Sebastian (20 Jan.) to Annunciation of the BVM (25 Mar.)	
Septuagesima to Easter week	Septuagesima to the 5th Sunday after Easter
antiphons for Sundays after Easter	
SS Philip and James (1 May)	
responsories for Sundays after Easter	
Commune Sanctorum in Paschaltide	
Finding of the Cross (3 May)	
Exaltation of the Cross (14 Sept.)	
SS Alexander, Eventius, and Theodulus (3 May)	
Vigil of Ascension through Whitsun week	Vigil of Ascension through Whitsun week
St John Baptist (24 June) to St Andrew (30 Nov.)	
Commune Sanctorum	
Office of the Dead	
responsories and Magnificat antiphons for the summer Sundays, a group of responsories being followed by a group of Magnificat antiphons	responsories and Magnificat antiphons for the summer Sundays, a group of responsories being followed by a group of Magnificat antiphons
ferial antiphons for the Benedicite, Benedictus, and Magnificat	
antiphons for the summer Sundays	antiphons for the summer Sundays
	St Lucy (13 Dec.) to St Andrew (30 Nov.)
	Commune Sanctorum
	Dedication of the Church
	Office of the Dead

It was customary in antiphoners and breviaries to designate the summer Sundays after the biblical book which provided the lessons of the Night Office. In the Lucca manuscript these Old Testament books, or histories (*historia*), are assigned as follows:

the Book of Kings, for the first Sunday after Whitsun to first Sunday in August
the Book of Wisdom, for August
the Book of Job, for the first half of September
the Books of Tobit, Judith, and Esther, for the rest of September
the Book of Maccabees, for October
the Books of the Prophets, for November

In any antiphoner the number of proper chants naturally varies greatly from occasion to occasion. The provision for Ascensiontide in the Lucca codex is typical:

Vigil of Ascension
    Lauds: Benedictus antiphon (text incipit only: already available elsewhere)
Ascension Day
    first Vespers: antiphon for psalms; responsory (incipit only); versicle (incipit only); Magnificat antiphon
    Night Office: invitatory; six antiphons, versicle, and four responsories for the first Nocturn; the same for the second Nocturn (versicle incipit only); antiphon, versicle (incipit only), and four responsories for the third Nocturn
    Lauds: five antiphons; short responsory; versicle; Benedictus antiphon
    Little Hours: versicles (incipits only) for Terce, Sext, and None
    second Vespers: antiphon (incipit only); short responsory; versicle (incipit only); Magnificat antiphon
Friday after Ascension Day
    Night Office: Invitatory; one antiphon (incipit only) for each of the first and second Nocturns
    Lauds: antiphon for psalms; Benedictus antiphon (incipit only)
    Little Hours: antiphon for Prime; versicles (incipits only) for Terce, Sext, and None
Sunday after Ascension Day
    Lauds: Benedictus antiphon
    second Vespers: Magnificat antiphon

Understanding what has been omitted and has to be drawn from the common pool, and where to find it, poses considerable problems to the uninitiated.

(iii) *Comparison of Sources*

As in the case of the gradual, much time and effort has been devoted to studying the relationships between sources, by comparing their selection of chants and (to a lesser extent so far) the variant readings of their texts and melodies. As a glance at the twelve manuscripts whose texts were edited by Hesbert (*CAO* 1–2) will show, the selection of chants for the antiphoner is much less stable than for the gradual, and the same is true of their melodies. The variance in text selection prompted Hesbert to carry out a massive survey of 800 sources (including breviaries as well as antiphoners), comparing their responsories for the four Sundays of Advent (*CAO* 5 and 6; see also Ottosen 1986). Similar repertory comparisons have been made by Raymond Le Roux for the antiphons of the Night Office and Lauds at Christmas and 1 January (1961), responsories of Epiphanytide (1963), and the responsories of the Triduum Sacrum and Easter (1979). These surveys have proved strikingly capable of isolating families of related sources: those belonging to the same church or diocese, those linked by monastic reform movements, those of the religious orders, and so on (apart from the lengthy discussions in *CAO* 5 and 6, see also Hesbert 1980, 'Antiphonaire de la Curie', 1980, 'Antiphonaire d'Amalar', 1980, 'Sarum', 1982, 'Antiphonaires monastiques insulaires', 1982, 'Matines de Pâques', and Raymond Le Roux 1967). A classic study of a branch of the tradition, with facsimiles of surviving fragments, was made by Gjerløw (1979) for the church of Nidaros. (For extended studies of the contents of individual antiphoners, see Ossing 1966, Ledwon 1986, Möller 1990.)

Musical comparisons have taken place on a much smaller scale, understandably enough, considering the daunting magnitude of the repertory. Udovich (1980) provides parallel transcriptions of the Magnificat antiphons from the ferial office in nineteen sources. Underwood (1982) has compared variant readings in antiphons from a handful of English sources with a view to establishing the pattern of their interrelationships (see also Hiley 1986, 'Thurstan').

III.6. PSALTERS, HYMNALS, COLLECTARS, OFFICE LECTIONARIES

(i) Psalters
(ii) Hymnals
(iii) Collectars
(iv) Office Lectionaries

The other books beside the antiphoner needed for the celebration of the Divine Office are the psalter and hymnal (essentially for singing), the collectar (which contains prayers), and various books containing lessons: passionaries, homiliaries, and so on. Various different types and combinations of these books are known, assembled to suit the needs of a particular institution or person at a particular time, as convenience dictated. All might be gathered together in the breviary.

(i) *Psalters*

Steiner, 'Psalter', *NG*; Andrew Hughes 1982, 224–36; Dyer 1984.

The Latin text of the psalter or Book of Psalms, like the text of the Bible as a whole, was not unified for all times and places. In the early centuries several Latin translations of the Bible or parts of it were current, made from the Greek Septuagint. There was no special urgency to impose uniformity, for Latin did not replace Greek as the language of the liturgy until the late fourth century. The main rites of Western Europe, Milanese, Mozarabic, etc. had their own translations. As far as the psalter is concerned, the most important types are the following.

(i) The Roman Psalter. Once attributed to St Jerome, this version is still used in St Peter's, Rome, and was used both in other Roman churches and elsewhere in Italy until its replacement by the Gallican Psalter under Pius V (1566–72). Adopted by the Franks from Roman exemplars, it was also the basis of chants sung at mass, in Gregorian as in Roman chant.

(ii) The Gallican Psalter. Made by St Jerome *c*.392 from the Hexapla Old Testament of Origen, this became very popular in Gaul, perhaps partly under the influence of Gregory of Tours and later of Alcuin. From the Carolingian period this was the translation used for the singing of the office in most churches outside Spain and Italy.

(iii) The Hebrew Psalter. A new translation made by Jerome *c*.400 as part of his translation of all the books of the Bible from the Hebrew. Jerome's work forms the greater part of the Vulgate Bible ('editio vulgata' = 'common or popular edition'), but the Gallican Psalter was already too popular to displace and was therefore included in the Vulgate in all lands except Italy (the Roman Psalter) and Spain (the Hebrew).

In order to adapt the biblical psalter for liturgical use, various measures were adopted. The psalms might be left in their usual order, or they might copied in a rearranged order to fit the weekly office cycle. The rearrangement would take a different form according to whether the secular (or Roman) cursus was being followed or the monastic cursus. If the psalms were left in their usual order then notes in the margin might indicate on which day and at which hour they were to be sung. Many psalters contain not just the texts of the psalms but also those of the canticles, sung in a similar fashion at the same services. And the most complete type of liturgical psalter contains the antiphons which accompany the psalms, also invitatories, and perhaps hymn incipits (for which pieces a full hymnal would be needed) and the short lessons and versicles with responses. The whole material necessary to sing the weekly office would then be present.

Such material is not often given in antiphoners and breviaries. It was convenient to have it in a separate book, because the weekly office cycle was performed independently of the festal cycle, breaking off whenever a feast-day intervened and resuming when the feast-day was over. As Steiner remarks, 'the Proper of the Time and of the Saints . . . is essentially a more or less lengthy series of interruptions into [the Sunday and ferial] Office' ('Psalter', *NG*).

No facsimile of a liturgical psalter with notated chants has been published. Leroquais (1940–1) describes a large number. The sections covering the weekly office in the *Antiphonale Romanum* and the *Antiphonale monasticum* present the psalter in order as sung but omit the Night Office, while the *Liber usualis* first sets out the

Vesper psalms and Magnificat in all eight psalm tones but without antiphons, then proceeds to the hours on Sunday except for the Night Office, which are given in full, then to the Little Hours on weekdays. The new *Psalterium monasticum* of 1981 restores the Night Office (called Vigils).

(ii) *Hymnals*

Moberg 1947; MMMA 1; Gneuss 1968; Stäblein, 'Hymnar', *MGG*; Steiner, 'Hymn, II. Monophonic Latin', *NG*.

   Editions with music:

Ebel 1930 (with facsimile of Einsiedeln, Stiftsbibliothek 366: see MMMA 1)

Waddell 1984 (critical edition of the Cistercian hymnal from manuscript sources)

MMMA 1: Stäblein's edition of over 550 melodies, where the following hymnals are given complete:

   Milan, Biblioteca Trivulziana 347 (Milan, 14th c.)

   Heiligenkreuz, Stiftsbibliothek 20 (Cistercian, 12th–13th c.)

   Rome, Biblioteca Apostolica Vaticana, Rossi 205 (Moissac, 10th–11th c.)

   Paris, Bibliothèque Nationale, n.a.l. 1235 (Nevers, 12th c.)

   Worcester, Cathedral Chapter Library F. 160 (Worcester, 13th c.; facs. PalMus 12, 1*–12*)

   Klosterneuburg, Stiftsbibliothek 1000 (Klosterneuburg, 1336)

   Zürich, Zentralbibliothek, Rheinau 83 (Kempten, early 11th c.)

   Einsiedeln, Stiftsbibliothek 366 (Einsiedeln, first half of the 12th c., mid-13th c.; facs. Ebel 1930)

   Verona, Biblioteca Capitolare CIX (102) (Verona, 11th c.)

   Rome, Biblioteca Casanatense 1574 (Gaeta, 12th c.)

The hymnal was the other chant-book for the office hours more often copied separately from the antiphoner than included in it. Occasionally a hymnal forms part of quite a different book: Paris 1235, for example, is a gradual with a large supplementary section giving tropes, sequences, and ordinary-of-mass chants for the main feasts; in between these come a processional, a collection of short responsories, a tonary, and a hymnal.

The arrangement of hymnals usually resembles that of antiphoners. The hymns for the weekly office usually appear at Epiphanytide. The much-repeated Compline hymns will be given at the beginning of the collection. As an example of the arrangement of a thirteenth-century hymnal, the Sarum manuscript Oxford, Bodleian Library, Laud lat. 95 may be described. There are 100 hymns, distributed as follows (only for the first few are the assignments given in detail):

1	first Sunday of Advent: Vespers
2	Compline except on double feasts and during Lent
3	first Sunday of Advent: Night Office
4	first Sunday of Advent: Lauds
5	Christmas Eve: Vespers

6	Compline on double feasts except on the Annunciation of the BVM and from Easter to Whitsuntide
7	Christmas: Night Office
8	Christmas: Lauds and Vespers
9	St Stephen: Vespers and Night Office
10	Epiphany: Vespers and Night Office
11	Epiphany: Lauds
12–19	first Sunday after Epiphany: first Vespers, Night Office, Lauds, Prime Terce, Sext, None, second Vespers
20–37	Monday to Saturday: Night Office, Lauds, Vespers (no Vespers on Saturday)
38–65	Quadragesima Sunday to the first Sunday after the Octave of Pentecost
66–88	Sanctorale from St Vincent (22 Jan.) to St John the Evangelist (27 Dec.) with a supplement
89–98	Common of Saints
99–100	Dedication of the Church

Stäblein's edition of the contents of several hymnals, with other miscellaneous items, in MMMA 1 is the single most important reference tool for the subject, and contains a list of over 360 sources. Many of these are antiphoners, breviaries, or psalters with attached hymnals. The earliest extant hymnals, which have no musical notation, date back to the eighth century; the earliest notated collections are from the eleventh century. While Gneuss (1968) has analyzed a number of early collections and established their interrelationships, Moberg (1947) has carried out almost the only comparable work on later books. The earliest collections, referred to as the Old Hymnal type I, unite in a broad way the monastic uses both of Gaul (St Caesarius of Arles) and Italy (St Benedict) and can be linked to the Ambrosian repertory. Two recastings of the Old Hymnal are then traceable: the Old Hymnal type II is found in Frankish sources of the eighth and ninth centuries; the New Hymnal, which may be connected with the reforms of St Benedict of Aniane (d. 821) is found in Frankish sources from the ninth century onwards and constitutes the basis of most later collections. An exception is the hymnal of the Cistercians, which was a conscious attempt to return to the supposed hymns of St Ambrose: Milanese sources were used for the purpose, though they had to be supplemented by others better known to the French monks (see the discussion in Waddell 1984).

(iii) *Collectars*

Gy 1960 (with list of sources to *c*.1400).

The collectar (or orational) is a book containing the collects and often also the short lessons (capitula or chapters) used in the office hours. The same person usually recited both. One each of these prayers and lessons was said at each of the day hours, that is, at all except the Night Office. The lesson followed the psalms, the collect came at the end. Collectars commonly include other material as well, such as the versicle and response and incipits for antiphons, responsories, and hymns. If the book provides the chants in full it has become, in effect, a sort of diurnal.

The earliest surviving collectar is not for the Roman but for the Old Spanish (Mozarabic) rite: the so-called Verona Orational from the early eighth century (Verona LXXXIX; ed. Vives and Claveras 1946).

Two fine collectars from eleventh-century England show what the collectar might encompass.

(i) London, British Library, Harley 2961 ('Leofric Collectar'), Exeter, third quarter of eleventh century (text ed. Dewick and Frere 1914–21):

> collectar, including antiphons and responsories, often in full hymnal
> eight sequences, with the start of a ninth before the manuscript breaks off incomplete

(ii) Cambridge, Corpus Christi College 391 ('Portiforium of St Wulstan or 'Collectar of St Oswald), Worcester, second half of eleventh century (most of text ed. Anselm Hughes 1958–60):

> kalendar
> psalter, canticles, litany
> hymnal
> monastic canticles (for the third Nocturn of the Night Office)
> collectar: Temporale and Sanctorale (from St Stephen, 26 Dec., to St Andrew, 30 Nov.), including notated chants
> supplementary collects for a much fuller Sanctorale (St Silvester, 31 Dec., to St Thomas, 21 Dec.)
> collectar: Commune Sanctorum and Dedication of the Church (without notation)
> blessings and ordeals

At this point a new hand has copied private prayers, Latin and Anglo-Saxon. Then a new section begins in yet another hand (or perhaps the first hand in a later phase), giving parts of the office as in a breviary, including the Night Office, with full lessons and notated chants:

> Commune Sanctorum
> Sunday Vespers and Night Office
> summer Sundays
> Trinity Sunday
> commemorative offices of the Holy Cross and the Blessed Virgin Mary on Saturday; Office of the Dead

At the end of the book a later hand has copied O-antiphons.

The thinking behind this particular assemblage of material is not always clear, and the manuscript in its present state may unite books originally prepared apart. At any rate, it allows some insight into the flexibility with which the office texts could be codified.

(iv) *Office Lectionaries*

While the day hours had only the short chapters for lessons, the Night Office had nine (secular or Roman cursus) or twelve (monastic cursus) lengthy lessons. These were taken from various sources, and books were made to contain each type of lesson. The commonest pattern was: first Nocturn, the Bible; second Nocturn, sermon by one of the Church Fathers; third Nocturn, homily (a short disquisition upon a passage of scripture), usually by one of the four great Latin Church Fathers (Ambrose, Augustine, Jerome, Gregory). For many feasts of the Sanctorale, passages were read from the *Vita* of the saint in question, which might be gathered together in a legendar, passional, or martyrology.

Basic work on homiliaries has been done by Grégoire (1966, 1980). Saints' lives are published in the massive series *Acta sanctorum* (a convenient way to find texts is at the entry for the saint in question in *Bibliotheca sanctorum*). Perhaps the two most important martyrologies are the Hieronymian Martyrology of the fifth century (*Acta sanctorum* Nov. II, i. 1894; new edition *Acta Sanctorum* Nov. II, ii. 1931) and Usuard's Martyrology of the ninth century (PL 123, 453–992; PL 124, 9–860; Dubois 1965). (On martyrologies, see Quentin 1908.)

III.7. SEQUENTIARIES, TROPERS, AND KYRIALES

RISM B/V/1; Emerson, 'Sources, MS, II', *NG*.

FACSIMILES OF MANUSCRIPTS CONTAINING SEQUENCES, TROPES, AND ORDINARY-OF-MASS CHANTS WITH TROPES:
Aachen, Bischöfliche Diözesanbibliothek 13: MMS 2 (omits gradual)
Bamberg, Staatsbibliothek lit. 6: see Bamberg 6
Bari, Biblioteca Capitolare 1: MMS 1 (omits ordinary-of-mass chants)
Benevento, Archivio Capitolare 34: PalMus 15
Cambridge, University Library, Add. 710: MMS 4
Chartres, Bibliothèque Municipale 47: PalMus 11
Rome, Biblioteca Angelica 123: PalMus 18
Rome, Biblioteca Casanatense 1741: Vecchi 1955
private collection of Martin Bodmer, C. 74: Lütolf 1987

EDITIONS WITH MUSIC:
Paris, Bibliothèque Nationale lat. 1121: Paul Evans 1970
Rome, Biblioteca Apostolica Vaticana, Vat. lat. 5319: MMMA 2

OTHER COMPREHENSIVE EDITIONS OR STUDIES:
sources from Winchester: Frere 1894, Planchart 1977
Apt, Cathédrale Sainte-Anne 17 and 18: CT 5
Wolfenbüttel, Herzog-August-Bibliothek, Guelf. 79 Gud. lat.: CT 6

There are few written traces of sequences or tropes in the ninth century, though they were certainly sung then. The earliest trope and sequence collections of moderate size to have survived are in manuscripts of the beginning of the tenth century onwards. Although this is not so different from the situation of other chants, the textual tradition of the latter goes much further back. The collection of sequence melodies in Chartres, Bibliothèque Municipale 47 (Brittany, late ninth century), the sequences and tropes in Paris, Bibliothèque Nationale, lat. 1240 (Limoges, 930s) and the tropes in Vienna, Nationalbibliothek 1609 (Freising? early tenth century) stand near the beginning of the written tradition.

Sequences, tropes, and ordinary-of-mass chants had their own individual history and their own special function within the liturgy, and hence were often copied in collections separately from other chants. Sometimes they were grouped alongside each other, so that combined volumes of 'special' material were created. These might include other chants as well, such as offertory verses, alleluias, processional chants, liturgical dramas, and so on. The usual name for such a book is 'troper', although tropes in the strict sense are not the only components. Books containing only sequences, or sequentiaries, are less common; and books containing only ordinary-of-mass chants, known at least since the sixteenth century as kyriales, are even rarer. On the other hand, after the classical tropes for the proper chants of mass had fallen out of use in the twelfth century, it is common to find combined sequentiaries and kyriales. The use of the term 'troper' for these collections is less appropriate, for usually the only tropes present are a few for the ordinary-of-mass chants.

Most of the special material is for soloists to sing. By analogy with the older book containing the soloists' parts of the proper-of-mass chants (gradual and alleluia verses, tracts), the troper is sometimes referred to as a cantatorium; it is better to avoid this usage.

Like sequences and ordinary chants, most tropes are for the mass. Just as frequent as the collection of the special chants in a book apart is their inclusion in the gradual (or later in the missal). Since tropes for the office are restricted to prosulas for the melismas of responsories, the antiphoner was easily able to take these into its fold, and collections of prosulas in tropers are not common. When sequences, tropes, and ordinary-of-mass chants were included in an expanded gradual, they were usually copied in sections of their own. Some manuscripts, however—mainly Italian—give tropes and ordinary chants at the appropriate liturgical place among the proper chants. Sequences can occasionally be found among the proper chants outside Italy as well.

The general trend was away from separate collections of special material and towards its integration into the gradual or missal. After the thirteenth century tropers are rare, partly also because troping itself fell out of fashion.

The manuscripts described by Husmann in RISM B/V/1, *Tropen- und Sequenzen-handschriften*, do not include graduals (with one or two exceptions) or antiphoners. No representative listing of sequence sources is available, though a provisional one could be compiled by combining Husmann's information with that in *Le Graduel romain, II: Les Sources*, which covers graduals and missals. For the proper tropes of mass, there are comprehensive lists of sources in Corpus Troporum 1 and 3 (ninety-two in the latter). Ordinary-of-mass chants are equally divided between tropers, graduals, and missals, again usually collected in sections apart, but occasionally amalgamated with the proper chants, or at least cued to their appropriate liturgical place by an incipit. Between them, the catalogues of Landwehr-Melnicki (1955), Bosse (1955), Rönnau (1967), Thannabaur (1962), and Schildbach (1967), supplemented by Hiley (1986, 'Ordinary'), list over 500 sources of ordinary chants. 496 sources of responsory prosulas are listed by Hofmann-Brandt (1971).

The chief contents of few more or less typical sources may be listed briefly, in order to give an idea of their nature and variety.

(a) Tropers

1. Oxford, Bodleian Library, Bodley 775 (Old Minster, Winchester, mid-eleventh century):

graduals and proper tropes for the whole year, sometimes with alleluias. For each
   mass the usual order is: trope verses for the introit, gradual verse, alleluia (often
   only incipit), trope verses for the offertory, trope verses for the communion

ordinary-of-mass chants, troped then untroped

alleluias for the whole year

tracts

sequence melodies

sequence texts, notated

At the beginning, after the sequence melodies, and at the end are further sequences and ordinary-of-mass chants, most with tropes, added in the twelfth century.

2. Munich, Bayerische Staatsbibliothek, clm 14322 (abbey of St Emmeram, Regensburg, 1030s):

litany hymns, Exultet

notated sequence texts with melodies in margin

graduals and tracts

alleluias

proper tropes

ordinary-of-mass chants (Kyries, troped then untroped, Glorias, troped then
   untroped, etc.)

offertory verses (only the incipit of the respond is given)

Ite missa est chants

3. Madrid, Biblioteca Nacional 288 (chapel of the Norman rulers of south Italy and Sicily, *c.*1100):

tonary

processional

Kyries, troped then untroped

Glorias, troped then untroped

alleluias

sequences

offertory verses, some with prosulas

Sanctus, troped then untroped

Agnus, troped then untroped

responsory prosulas

Benedicamus chants, some with tropes, Benedicamus songs

liturgical dramas

offices of SS Julian, Egidius, and Mary Magdalene

4. Paris, Bibliothèque Nationale, lat. 909 (abbey of Saint-Martial at Limoges, *c*.1025–30):

proper tropes

chants for the ordinary of mass, troped then untroped (Kyries and start of Glorias lost)

sequences

tracts

communion psalm verses

processional

alleluias

offertory verses

tonary

office antiphons for summer Sundays, Trinity office

Between the proper tropes and ordinary chants material was added which includes the offices of SS Martial, Valeria, and Austriclinianus of Limoges.

5. Paris, Bibliothèque Nationale, lat. 1118 (Auch?, late 10th c.):

combined proper tropes and ordinary-of-mass chants, distributed in liturgical order feast by feast

tonary

prosulas (alleluia and offertory prosulas, also the *Fabrice mundi* responsory prosulas, all in liturgical order)

sequence melodies

notated sequence texts

6. Paris, Bibliothèque Nationale, lat. 10508 (abbey of Saint-Évroult, Normandy, early 12th c.):

incipits for introit, offertory, and communion throughout the year

Kyries, troped then untroped

Glorias, troped then untroped

sequences, often with gradual and alleluia for the mass in question

Sanctus, troped then untroped

Agnus, troped then untroped

The rest of the manuscript contains six theory treatises, including Guido of Arezzo's *Micrologus*.

7. Rome, Biblioteca Casanatense 1741 (abbey of Nonantola, late 11th c.):

ordinary-of-mass chants (Kyries, troped then untroped, Glorias, troped then untroped, etc.)

fraction antiphons

expanded cantatorium. For each mass the order is usually: trope verses for the introit, gradual, tract (in Lent, etc.), alleluia prosula, sequence, *antiphona ante evangelium*, trope verses for the offertory, trope verses for the communion.

(*b*) *Graduals*

8. Modena, Biblioteca Capitolare O. I. 7 (Forlimpopoli, near Ravenna, 11th–12th c.):

gradual, with tropes, sequences, etc. integrated into each mass in liturgical order. As an example of the most elaborate festal mass, that on Easter Day includes the following: troped introit, troped Kyrie, troped Gloria, gradual, alleluia with prosulas, sequence, troped offertory, troped Sanctus, fraction antiphon, troped Agnus, troped communion.

supplement of festal pieces (ordinary-of-mass chants with tropes, sequences, proper tropes, etc.)

9. Paris, Bibliothèque Nationale, lat. 903 (abbey of Saint-Yrieix, near Limoges, second half of 11th c.)

gradual, with prosulas for alleluias and offertories

processional antiphons, *preces*

proper tropes

ordinary-of-mass chants (Kyries, troped then untroped, Glorias, troped then untroped, etc.)

sequences

In 2 and 4 each genre is separate; in 1 only the graduals and proper tropes are integrated. Proper and ordinary tropes are combined in 5. Sequences are to some extent combined with graduals and alleluias in 6. More comprehensive is the integration in 7, before 8 amalgamates the material completely. In 3 and 6 it may be seen that the various chants have been placed in their correct liturgical order (Kyrie, Gloria, sequence, Sanctus, etc.), though the collections are not amalgamated with one another. There are collections of processional chants in 3, 4, and 9; tonaries in 3, 4, and 5 (and theory treatises in 6); and saints' offices in 3 and 4.

III.8. PROCESSIONALS

Gy 1960; Bailey 1971; Huglo, 'Processional', *NG*.

FACSIMILES:

Worcester, Cathedral Chapter Library F. 160 (PalMus 12, 232–410); *Processionale Sarum*

TEXT EDITIONS:

W. G. Henderson 1875 (York); W. G. Henderson 1882 (Sarum); Legg 1899 (Chester); Wordsworth 1901 (Sarum)

INVENTORIES:

Allworth 1970 (Donaueschingen 882—Dominican); Floyd 1990 (English monastic processionals)

The processional contains processional chants: antiphons (some with verses) and hymns (usually with refrains). The processions where they were sung took place (i) on certain days of special observance, (ii) before mass on feast-days (or at some other time during the day). The repertory was not large, and most early sources copy the chants required into the gradual at the appropriate place. Another place for them was in tropers, as has been seen in the previous chapter. From about the thirteenth century separate books for processional chants become more common, partly because the rubrics for performing the services become much more explicit about this time. The earliest surviving separate processionals date from the twelfth century.

Many of the processions for feast-days borrowed a responsory from Vespers or the Night Office of the day in question; these are not usually copied into the processional but would have been taken from the antiphoner. Conversely, the processional

Table III.8.1. *Contents of the Castle Acre processional*

Special ceremonies	Feast-days
	chants for processions after Vespers and the Night Office from the 1st Sunday of Advent to Epiphany
	A. *O crux benedicta* for procession to the rood after Vespers on Sundays from Octave of the Epiphany to Passion Sunday and for the summer Sundays
	chants for procession before mass from Advent to Septuagesima Sunday
Ash Wednesday (Blessing of the Ashes and Dismissal of the Penitents)	
Palm Sunday (Blessing of the Palms)	
Maundy Thursday (Mandatum)	
Good Friday (Reproaches, Adoration of the Cross)	
	Easter Eve
	1st Sunday after Easter
Rogation Days	
	Ascension Day to Corpus Christi
Purification of the BVM (2 Feb.) (Blessing of the Candles)	
	St Benedict (21 Mar.) to St Nicholas (6 Dec.)
	further chants for procession in Advent, Lent, etc.

attracted to itself chants that formed part of ceremonies other than mass and the office hours, for example the Maundy antiphons, chants for the Veneration of the Cross on Good Friday and for the Easter Vigil (the Exultet, the hymn *Inventor rutili*). And because liturgical dramas also contain a processional element these too were sometimes recorded in processionals. As an example of what a well-stocked book might include, the contents of the fifteenth-century processional of the Cluniac priory at Castle Acre, Norfolk (Norwich Castle Museum 158.926.4e) may be summarized. Table III.8.1 lists the occasions in manuscript order in two columns one for the special ceremonies, one for the feast-days for which a processional chant is recorded. On feast-days the processions took place after Vespers, after the Night Office or after Mass.

No comprehensive survey of the sources has yet been published (but Huglo is preparing one for RISM). Meanwhile Huglo's article 'Processional' (*NG*) cites several dozen sources, and Bailey (1971) lists the principal chants of many sources in tabular form.

III.9. MISSALS

Leroquais 1924; *Le Graduel romain*, II: Les Sources; Huglo, 'Missal', *NG*; Emerson 'Sources, MS, II', *NG*; Andrew Hughes 1982, 143–59.

FACSIMILES:

Benevento, Biblioteca Capitolare VI. 33: PalMus 20
Bratislava, City Archives, EC. Lad. 3, EL. 18 and other fragments: *Missale notatum Strigoniense*
Missale Aboense (Dominican): Parvio 1971

TEXT EDITIONS (*signifies noted missal):
printed missal of Hereford of 1502: W. G. Henderson 1874, *Herfordensis*
London, Westminster Abbey, 'Lytlington missal': Legg 1891–7
* Le Havre, Bibliothèque Municipale 330: Turner 1962
* Palermo, Archivio Storico e Diocesano 8: Terrizzi 1970
* Paris, Bibliothèque Nationale, lat. 1105: Anselm Hughes 1963
* Sarum: Legg 1916 (from three early manuscripts)
Sarum: Dickinson 1861–83 (from printed editions)
York: W. G. Henderson 1874, *Eboracensis* (from printed editions)

SACRAMENTARIES WITH CHANT INCIPITS IN MARGIN:
* Oxford, Bodleian Library, Bodley 579: Warren 1883
Rome, Biblioteca Apostolica Vaticana, Ottob. lat. 313: Wilson 1915

JUXTAPOSED SACRAMENTARY AND GRADUAL:
Paris, Bibliothèque Nationale, lat. 2291: chants listed Netzer 1910
Paris, Bibliothèque Nationale, lat. 12050: *AMS*
Zürich, Zentralbibliothek, Rheinau 30: *AMS*

The term 'missal' is best reserved for the book containing all three of the main components of the mass: chants, prayers, and readings. 'Noted missal' can then indicate a book where the chants are notated. Noted missals can be found from as early as the tenth century, but they are common only from the twelfth. The oldest is Baltimore, Walters Art Galley M. 6, which contains only festal and votive masses, written probably at St Michael's, Monte Gargano in the tenth century. Other early examples are also Italian, including Benevento, Biblioteca Capitolare 33 (PalMus 20).

Other countries did not take up the idea with any urgency. Many early northern books combine sacramentary, lectionary, and gradual in less integrated fashion. The juxtaposition of the gradual and other books, side by side but not amalgamated, is known as early as the ninth century. Huglo ('Missal', *NG*) has cited many examples. Numerous early sacramentaries from north France have chant text incipits copied in the margin, sometimes notated. A very few books survive where for each mass first the chants are given and then the prayers. Even rarer are books where the chants and prayers are intermingled in the correct liturgical order. Huglo is able to cite only three examples of the juxtaposition of gradual and lectionary. (See Plate 2 for the remains of one.) A few books are also known where gradual, sacramentary, and lectionary are juxtaposed.

There was at first no great practical benefit to be had from amalgamating gradual, sacramentary, and lectionary in one volume (though the situation whereby sacramentaries began at Christmas, graduals at Advent, could be rationalized, usually in favour of the gradual). But at the end of the eleventh century it became obligatory in Rome, and then customary elsewhere, for the celebrant to recite for himself the texts of all chants (Ludwig Fischer 1916, 80–1). And from the thirteenth century several leading churches, the most important being the papal chapel itself (others include Paris, Salisbury, the Dominicans) revised their liturgies and codified them in comprehensive rubricated manuscripts, which served as models for other uses.

Many missals were not designed to serve the cantor and remained without notation. A large number have notation only for the Canon of the Mass, sung by the celebrant. Another type of missal contains only masses for the highest feasts (sometimes called a 'missale festivum'). Such missals were usually for the use of a bishop or abbot, who might be expected to be present at his 'home' church on such high festivals.

Many missals with notation survive from churches whence no gradual is known, and are crucial witnesses for our knowledge of regional chant traditions (see the descriptions in *Le Graduel romain* II and Emerson). Unnotated missals are also important for the study of chant repertories.

III.10. BREVIARIES

Bäumer 1905; Batiffol 1911; Leroquais 1934; Salmon 1959; Salmon 1967; Huglo, 'Breviary', *NG*; Andrew Hughes 1982, 197–224, 238–44.

TEXT EDITIONS:
Cambridge, Magdalene College F. 4. 11: Collins 1969
printed breviary of Hereford, 1505: Frere and Brown 1904–15

Oxford, Bodleian Library, Rawl. Lit. e. 1* and Gough Lit. 8: Tolhurst 1932–43
printed breviary of Salisbury, 1531: Procter and Wordsworth 1879–86
printed breviary of York, 1476: Lawley 1880–3

The breviary unites all the chants, prayers, and lessons of the office hours in one volume. The reason for assembling such a cumbersome volume—cumbersome both from its sheer size and also because of the frequent cross-referring always necessary in the office—seems to have been the urge to provide standard exemplars of the liturgical use of various churches. The obligation of travelling priests and monks to recite the office for themselves may also have led to the production of such books—the Rule of Chrodegang of Metz (d. 766) already directed that when a cleric could not attend one of the office hours he must say it privately. Most surviving early breviaries are nevertheless for use in monasteries.

A few early sources are juxtapositions of the component parts of the breviary, rather than amalgamations with all the material in correct liturgical order (see Huglo).

Many breviaries contain shortened lessons, a practice which gave the book its name. This practice is known from the papal court from the thirteenth century, where the office was in any case only recited privately. The breviary drawn up under Innocent III (1198–1216) is of this abbreviated type (see van Dijk and Walker 1960).

Noted breviaries are as valuable as antiphoners for the investigation of regional musical traditions. Breviaries without notation are also vital for our understanding of the development of local repertories, as Hesbert (*CAO* 5–6) has demonstrated.

III.11. COMPENDIA

The division of office from mass was not practised in the Mozarabic or Milanese rites, and their liturgical books amalgamate what in Gregorian usage would have been the antiphoner and gradual. This is rare among Gregorian sources, though a number of early sources juxtapose an unnotated gradual and an antiphoner: Albi, Bibliothèque Municipale 44 and Paris, Bibliothèque Nationale, lat. 17436, both of the ninth century, are two such. The 'Mont-Renaud manuscript' (PalMus 16) is a book of similar sort of the mid-tenth century (probably from Corbie), to which notation was then added about half a century later. Paris, Bibliothèque Mazarine 384 (Saint-Denis, early eleventh century) contains both a notated gradual and a list of chants of the antiphoner. Paris, Bibliothèque Nationale, lat. 12584 is a combined antiphoner and gradual (the antiphoner text edited by Hesbert, *CAO* 2) of the twelfth century from Saint-Maur-les-Fossés. Again, the two sections are not amalgamated. The book also includes a tonary. It follows Cluniac use, significant in view of the apparent interest of the Cluniacs in creating compendia to include the entire liturgy.

Such compendia seem to have been attempts to codify the complete use of a particular church for reference purposes. They would have been of particular value to the precentor of the church, who normally had charge of the performance of the liturgy.

One of the earliest combined noted breviary–missals is Cluniac: Rome, Biblioteca Casanatense 1907, from San Salvatore on Monte Amiate, of the early eleventh century. Another Cluniac example is the combined noted breviary–missal of Lewes, from the thirteenth century, Cambridge, Fitzwilliam Museum 369. Although a thick book with 517 folios, it uses a minute script, perhaps so that it would be portable, for use as a *correctorium* when the prior of Lewes visited other English Cluniac houses (see Leroquais, 1935; Holder, 1985). Another Cluniac combined breviary–missal, without notation, has survived from Pontefract or Wenlock: London, British Library, Add. 49363. (For a further noted breviary–missal, not Cluniac, see Salmon 1964.)

The newer religious orders also established master exemplars of their liturgy. Part of the Cistercian exemplar survives in Dijon, Bibliothèque Municipale 114 (see Choisselet and Vernet 1989, pl. 1 after p. 52 for the original list of contents). The master exemplar for the whole Dominican order still survives in Rome, Santa Sabina XIV. lit. 1, the 'Codex of Blessed Humbert of Romans'. This immense book (997 folios measuring 48 × 32 cm.) was compiled *c*.1260 to codify the liturgy revised by Humbert of Romans, master-general of the order, 1264–77. London, British Library, Add. 23935 contains the same material (minus the ordinal, breviary, and private missal) in small format, presumably for use as a *correctorium* on visitations in the English province.

The fourteen sections of the Santa Sabina manuscript are titled as follows:

Ordinarium
Martyrologium
Collectarium
Processionarium
Psalterium
Breviarium (for private recitation of the office, with only short lessons, psalm incipits, and so on)
Lectionarium (for the office)
Antiphonarium
Graduale
Pulpitarium (for one, two, or four friars, depending on the liturgical grading of the services, in the pulpit in mid-choir; it contains invitatories, responsory verses, gradual verses, tracts, and the Litany of the Saints)
Missale conventuale (for High Mass)
Epistolarium
Evangelistarium
Missale minorum altarium (for private masses)

From the same century three other English monastic compendia have survived. Cambridge, University Library Ii. 4. 20 is a combined breviary–missal from Ely (Benedictine). The two others are fully or almost fully noted. The best known is

Worcester, Cathedral Chapter Library F. 160, substantial portions of which were published in facsimile in PalMus 12. The sections of the manuscript are:

antiphoner (Temporale)
Venites
processional
antiphoner (Sanctorale, Commune Sanctorum, Office of the Dead)
Magnificat and Benedictus tones
(an added section of the fourteenth century with the office of the Visitation of the
    BVM and Corpus Christi)
kalendar
psalter with canticles, litany, and collects
hymnal
collectar
Kyries and Glorias
gradual
sequentiary (mostly lost)
Sanctus and Agnus
Laudes regiae
mass ordinal

For the performance of the liturgy only the prayers and lessons for mass and the lessons of the office would be required in addition to this.

The principal contents of London, British Library, Add. 35285, from the Augustinian priory at Guisborough, north Yorkshire, are as follows:

tonary
missal (without notation)
kalendar
psalter (beginning lost)
four ordines from the rituale
Venites
antiphoner (Sanctorale and Commune Sanctorum)
office lectionary
tonary
processional

Possibly several originally separate codices were here bound together.

The stimulus to make a complete copy of a church's chant repertory might come from a liturgical or institutional reform, such as the rebuilding of the church in question. This was the case at Piacenza, for example, where after the cathedral had been rebuilt, the compendium Piacenza, Biblioteca Capitolare 65 was copied from 1142 onward. This contains a tonary and theoretical material, sets of tones for office and mass psalmody, a psalter, hymnal, antiphoner, gradual, and collections of tropes and sequences (see Grégoire 1968, 557; RISM B/III/2, 79; Huglo 1971, *Tonaires*, 174, etc.).

III.12. PONTIFICALS AND RITUALS

(i) Pontificals and Benedictionals
(ii) Rituals, Manuals, or Agenda

(i) *Pontificals and Benedictionals*

[Hiley]: 'Pontifical', *NG*; Vogel 1986, 225–71.

Pontificals are books containing the texts for services performed by a bishop outside mass and the office: dedication of the church, confirmation, ordination and consecrations, blessings of sacred objects, the sacring and crowning of monarchs, and so on. Chants are sung at these ceremonies—the series of antiphons sung during the lengthy ceremony for the dedication of a church are notable—and these are frequently notated in pontificals. Several of the earliest examples of neumatic notation survive in pontificals, for they were often splendidly decorated and carefully preserved from early times.

The early development of the pontifical has been illuminated by Rasmussen (1978) and its history described by Vogel, with emphasis on the development of the Roman-German type first compiled at St Alban's abbey, Mainz, in the 950s and adopted in Rome within a few years under imperial influence. (See the editions by Vogel and Elze, 1963, and Andrieu 1938–41. For descriptions of pontificals in French libraries see Leroquais 1937; for English ones see Frere 1901; the texts of numerous pontificals have been edited, particularly in the series of the Henry Bradshaw Society.)

One special action of a bishop was to pronounce a blessing at Mass after the Lord's Prayer and before *Pax domini semper vobiscum*. Collections of these benedictions have been preserved from the Middle Ages, in books known as benedictionals. Some benedictionals for the use of abbots have also survived. (See Baudot, 'Bénédictionnaire', *DACL*; editions by Moeller 1971–9, and several volumes in the series of the Henry Bradshaw Society.)

(ii) *Rituals, Manuals, or Agenda*

Gy 1960; Vogel 1986, 257–64 (with list of editions).

The counterpart of the pontifical for priests is the ritual or manual, which contains the material for non-episcopal functions outside mass and the office. Prominent among the formularies are benedictions of all sorts and the services for baptism, marriage, and burial. Early rituals are often found in combination with collectars or sacramentaries, which gave the texts to be recited by the priest during the office or at mass respectively. Some rituals reflecting monastic use also survive.

III.13. ORDINALS AND CUSTOMARIES

Neither medieval nor modern nomenclature is consistent on this point, but a general distinction may be made between (i) ordinals, which regulate the performance of the liturgy, specify which items are to be performed and sometimes the manner of their performance (singers involved, position, etc.), and (ii) customaries, which regulate the administrative organization and way of life of a religious community, secular or monastic. The ordinal may give invaluable information about the repertory of a particular church, and may even contain notated chant incipits, as well as giving details affecting performance. Special items such as liturgical dramas, or the use of polyphony, may be described. The customary may explain the duties and training of the cantor and singers.

Most surviving ordinals date from the thirteenth century or later. This is also the period when service-books were first regularly provided with copious rubrics governing the performance of the liturgy, and it is no doubt symptomatic of a general desire to codify aspects of the liturgy previously left to the practical knowledge of those involved.

No distinction is made between customaries and ordinals in the handlist of sources for over 120 institutions listed in *Le Graduel romain* II, 189–96. Others are given by Hänggi (1957, xxv–xxxvi) and Jacob (1970). Among modern editions those in the series Corpus consuetudinum monasticarum (CCM) and Henry Bradshaw Society are the most important. See also Angerer (1977) and Fassler (1985).

III.14. TONARIES

(i) Definition and Function
(ii) Type-Melodies for the Eight Modes
(iii) How Psalm Tones were Specified

Huglo 1971, *Tonaires*; Bailey 1974; Huglo, 'Tonary', *NG*.
MODERN EDITIONS OF MEDIEVAL TONARIES:
(i) manuscripts:
Leipzig, Universitätsbibliothek 1492 (15th-c. copy of a Reichenau exemplar of *c*.1075): Sowa 1935
Metz, Bibliothèque Municipale 351 (Metz, 9th c.): Lipphardt 1965
Naples, Biblioteca Nazionale VIII. D. 14 (Auvergne, 12th c.): DMA A.I
Paris, Bibliothèque Nationale, lat. 1118 (Auch?, late 10th c.) and 1121 (Limoges, early 11th c.): C. T. Russell 1966
Paris, Bibliothèque Nationale, lat. 12050 (Corbie, 9th c.): reconstructed in *AMS*, cxxiii–cxxvi; cf. the reconstruction co-ordinated with the Mont-Renaud manuscript and Laon, Bibliothèque Municipale 118 in Huglo 1971, *Tonaires*, 94–101
Paris, Bibliothèque Nationale, lat. 13159 (for Saint-Riquier, late 8th or 9th c.): Huglo 1952, 'Tonaire', Huglo 1971, *Tonaires*, 26–8, Planer 1970

Rome, Biblioteca Apostolica Vaticana, Pal. lat. 1346 (German, 12th–13th c.): Donato 1978

St Gall, Stiftsbibliothek, 390–391 (St Gall, late 10th c.): facs. PalMus II/1

various Sarum sources: Frere 1898–1901

Tonaries compiled from the contents of medieval antiphoners were published in PalMus 9 and 12 and Udovich 1985

(ii) theorists:

pseudo-St Bernard of Clairvaux: *Tonale sancti Bernardi*: GS ii. 265–77; PL 182, 1154 ff.

Berno of Reichenau: GS ii. 79–91

Frutolf of Michelsberg: Vivell 1919

Jerome of Moravia: CS i. 1–94; Cserba 1935

John ('Cotton' or 'of Afflighem' or of Passau): CSM 1; trans. Babb 1978

Odo of Arezzo: GS i. 248–8 (introduction); CS ii. 81–109

Odorannus of Sens: Bautier *et al.* 1972

Petrus de Cruce of Amiens: CS i. 262–8; CSM 29

pseudo-Odo in Saint-Dié, Bibliothèque Municipale 42 (Franciscan, 14th c.): CS ii. 117–42

Regino of Prüm: CS ii. 3–73

Rudolf of Saint-Trond or Franco of Liège: *Quaestiones in musica*: Steglich 1911

Walter Odington: CS i. 182–250; CSM 14

(i) *Definition and Function*

A tonary (*tonarius* and, sometimes later, *tonale*, are the medieval terms employed) is a list or series of chants in tonal order. The earliest of them, from the late eighth and ninth centuries, are as old as any liturgical books of chant texts. Being an abstraction from everyday musical practice, though frequently serving a practical need, they were often compiled by theorists who are known also for other theoretical writings.

In tonaries chants in the same mode were listed in series, disturbing the order in which they would have been sung in the church year. It was particularly common to make lists of pieces sung with psalm verses: introits and communions of mass, antiphons of the office. A further subdivision of the series was usually effected, so that those pieces were grouped together which required the same cadence (differentia) for their psalm verse(s). Ex. III.14.1, an extract from the tonary in Paris, Bibliothèque Nationale, lat. 776, shows how the series were usually presented. The manuscript has listed chants in the first two modes, and has now begun a series for mode 3. A tone for singing the Gloria patri is copied (given at the start of Ex. III.14.1), followed by the incipits of fifty-nine office antiphons. Ex. III.14.1 gives the last eight of this series, starting with *Tenemus ecce arma*. There follow the first eight examples for the second 'varietas', and all the incipits for the other three sets.

Such lists are found in a great variety of sources. The chief contents of Paris 776, from which Ex. III.14.1 is transcribed, are a gradual; tonaries were often copied in other such liturgical music books, as for example in:

Ex. III. 14.1. Extract from a tonary, part of the third tone (Paris, Bibl. Nat. lat. 776, fo. 152ʳ)

INCIPIT TERCIUS TONUS .....

Glo-ri-a patri et fi-li-o   et spiri-tu-i sancto

Sic-ut e-rat in princi-pi-o et nunc et semper   et in secula seculorum   A-men.

..... Tenemus ecce ar - ma.   Sancta legi-o Agau(nensis).   Fidelis seruus et.

Iustus germinabit. Tollite portas prin(cipes). Uideo uirum. Leto animo. Sancti qui sperant.

VARIETAS .II.

Glo-ri-a   seculorum A-men.   In e-o-dem namque. Conuocatis er-go.

Quando natus est. Qui de terra est.   Surrexerunt autem.   Hierusalem Iherusalem.

Ualerius igitur.   Beatus Uincentius. .....

VARIETAS .III.

Seculorum A-men.   Ecce concipies. Hec est quem nesciui. Ascendit   Simon Petrus.

.IIII.

Glori-a   seculorum A-men.   Sapientia magnificat. Cum ergo.

.V.

Seculorum A-men.   Fac benigne. Homo quidam. Du-o homines. Cum repente.

Quasi u-nus de. Mari-a er-go. Domum istam. Pastor bonus. Ego sum pastor. Simon Petrus.

Monza, Basilica S. Giovanni C. 12/75 (gradual and antiphoner, Monza, early 11th c.)

London, British Library, Add. 30850 (monastic antiphoner, Santo Domingo de Silos, early 12th c.; facsimile *Antiphonale Silense*)

Eichstätt, Bischöfliches Ordinariatsarchiv, pontifical of Bishop Gondekar II (1057–75) (the inclusion of a tonary in a pontifical is altogether unique)

Many tonaries appear in tropers, indicating their place among material for the musician with special skills and interests:

Paris, Bibliothèque Nationale, lat. 1118 (Auch?, late 10th c.)

Bamberg, Statsbibliothek, lit. 5 (Reichenau, 1001)

Madrid, Biblioteca Nacional 288 (Palermo, *c*.1100)

Several authors of theoretical treatises also compiled tonaries. Regino, abbot of Prüm and later of St Martin in Trier, wrote a treatise and tonary *c*.901 for Archbishop Radbod of Trier, in the form of a letter to Radbod. Regino's is the first tonary by a known author. The treatise of Berno, abbot of Reichenau 1008–48, was written for Archbishop Pilgrim of Cologne (1021–36).

It may be pointed out, in view of the confusion present in the older literature, that Odo, abbot of a monastery in Arezzo in the late tenth century, does not come into this category. Odo appears to have been the author of the most widely known of Italian tonaries (CS ii. 78–109, edited from Paris, Bibliothèque Nationale, lat. 10508, from Saint-Évroult, twelfth century, unfortunately a version far removed from the archetype). But he did not compose the well-known treatise *Dialogus de musica* (written by an anonymous monk in north Italy, early eleventh century; GS i. 252–64). The tortuous history of these texts was elucidated by Huglo (1971, *Tonaires*, 182–224, and 1969). The misattributions so rife in this case result from the common practice of including tonaries in compilations of theoretical writings. Such a compilation is Munich, Bayerische Staatsbibliothek, clm 14272, containing Boethius, *Musica enchiriadis*, *Commemoratio brevis*, *Alia musica*, and a treatise on the monochord, as well as a tonary (St Emmeram at Regensburg, mid eleventh century).

Theorists who discuss the eight-mode system frequently cite examples, which may amount almost to a small tonary in themselves, though rarely giving the actual recitation tones. The celebrated anonymous treatise *Musica enchiriadis* (north-east France, late ninth century), for example, gives the incipit of just one antiphon per mode. The tenth-century Einsiedeln source of the *Enchiriadis*, Einsiedeln, Stiftsbibliothek 79, then takes the step of attaching a proper tonary to the treatise (see Huglo 1971, *Tonaires*, 62)—or one might say that the treatise has been brought in as a preface for the tonary, since many tonaries carry an introductory text explaining the modes. (For another example, where passages of the treatise *Dialogus de musica* are drafted into the tonary of the fifteenth-century Utrecht antiphoner Utrecht, Bibliotheek der Rijksuniversiteit 408, see Huglo 1971, *Tonaires*, 183. The twelfth-century tonary in the Utrecht antiphoner Utrecht 406 borrows from both the *Dialogus* and Berno of Reichenau: see Huglo 1971, *Tonaires*, 203.) Even closer to the tonary in spirit is the *Commemoratio brevis* (ed. Bailey 1979), 'a veritable treatise on psalmody', as Huglo puts it (1971, *Tonaires*, 64).

One commonly adduced reason for the compilation of these lists of chants in tonal order is that they were a convenient codification, for the purposes of reference and learning, of the tone and differentia to be used when singing a psalm with each antiphon listed. Given the huge number of antiphons used during the liturgical year (3,000 or more), and the restricted number of tones and endings (around forty), it is not surprising that musicians felt the need to set down in some convenient form the decisions about intonations which they were accustomed to taking. But this simple explanation is not adequate for all types of tonary.

The order in which the compiler of a tonary encountered the chants would have been a liturgical one. Most tonaries, once they have broken down the original order into groups by tone and differentia, disturb the liturgical order no further. Thus within each group there remains a liturgical succession. For example, the ninth-century Metz tonary Metz, Bibliothèque Municipale 351, lists the following twenty-two antiphons under the third differentia (or *diffinitio*, as the manuscript calls it) of tone 3 (taken from edn. by Lipphardt, 1965, 34–5; * denotes addition). I have appended the liturgical occasion for each chant:

Tu Bethleem	4th Sunday in Advent
Ascendente Ihesu in navim	4th Sunday after Epiphany
Fac benigne in bona	3rd Sunday of Quadragesima
Si in digito Dei	3rd Sunday of Quadragesima
Lignum vite	Inventio S. Crucis (3 May)
Tuam crucem adoramus	Inventio S. Crucis (3 May)
Claudius quidam	St Peter (29 June)
Petrus autem servabatur	St Peter (29 June)
Et respicientes viderunt	Easter Sunday
Et intravit cum illis	Easter Monday
Multa quidem et	1st Sunday after Easter
Mercennarius est	2nd Sunday after Easter
Ego sum pastor bonus	2nd Sunday after Easter
Si quis diligit me	Pentecost Sunday
Si oportuerit me mori	Palm Sunday
Beatus vir qui inventus	St George (23 Apr.)
Domine spes sanctorum et turris	Vigil of All Saints (31 Oct.)
*Serve bone et fidelis**	Confessors
Iste sanctus pro lege Dei sui	a Martyr
Omni tempore	*de libro Judith*
Quidam homo fecit	2nd Sunday after Pentecost
*Sancti qui sperant**	Martyrs

In fact there are three series here. The first follows the year from Advent to midsummer, the second goes back to Easter and continues to Pentecost, and the third doubles back again, to Palm Sunday. (The two added antiphons are slightly out of place: *Serve bone*, for Confessors, should follow antiphons for Martyrs, and the last antiphon should follow *Iste sanctus*.) The three groups appear to be distinguished by their openings:

G ab b, *G a Gb b*, etc.
G G Gc a G, etc.
a mixture of the two previous?

There is a certain small disadvantage in retaining a liturgical order within the various groups of chants, for it is not always easy to locate any one piece, at least within very big series like those for tone 1 and tone 8. An alphabetic order would be more useful if a little-known chant were being sought. However, medieval musicians seem not to have felt a need for this, for only in Reichenau sources was it adopted: Bamberg, Staatsbibliothek, lit. 5 (written 1001); the tonary of Berno of Reichenau (second quarter of eleventh century); and the tonary of *c*.1075, notated in Hermannus Contractus' interval notation, now surviving only in a fifteenth-century copy (ed. Sowa 1935; see Huglo 1971, *Tonaires*, 254). (One version of Berno, Vienna, Nationalbibliothek 1836, from Austria, eleventh to twelfth century, actually restores the antiphons to their liturgical order—see Huglo 1971, *Tonaires*, 272.)

Perhaps, then, the tonary is not so much a reference book for locating the odd troublesome item as a check-list to aid the rehearsal of pieces with similar musical characteristics. Few tonaries attempt the classification of all the antiphons of the year, and most contain only short lists, citing only a few dozen chants from the bigger groups and only a handful of the less common varieties. Thus the majority of tonaries, of the short type, serve neither to specify tones for the whole antiphon repertory, nor to classify melodies into melodic families; their more modest aim seems to be simply to remind musicians of the melodic idioms of the main families. The function of specifying tones was in any case carried out effectively by other means (see (iii) below).

The majority of tonaries contain lists of antiphons. A significant number have introits and communions, and responsories are sometimes included—for they too have a set of elaborate verse tones. But it is clear from even the earliest tonary that more was involved in their compilation than the assignation of psalm tones, for several contain chants that do not have psalm verses. Thus Paris, Bibliothèque Nationale, lat. 13159, from Saint-Riquier in the late eighth (Huglo 1952, 'Tonaire', and 1971, *Tonaires*, 25–9) or early ninth century (Planer 1970), has a few examples in each mode of the following categories: introit, gradual, alleluia, offertory, and communion. This source seems to be intended as a demonstration of the eight modes, rather than as a catalogue of tones. Paris, Bibliothèque Nationale, lat. 780 (Narbonne, late twelfth century) has alleluias, offertories, invitatories, and processional antiphons beside the more usual chants. The extensive listings in Montecassino, Archivio della Badia 318 (Montecassino, twelfth century) even include sequences (studied by Brunner 1981).

In such sources one may discern a didactic or scholarly intention beyond the specification of psalm tones. An altogether exceptional compilation which shares this character is the famous 'Dijon tonary', possibly in part the work of William of Dijon (d. 1031), Montpellier, Faculté de Médecine H. 159 (facs. PalMus 8; edn. Hansen 1974). In this source, after a short tonary for antiphons and responsories, a complete repertory of chants for the proper of mass is copied in full. The pieces appear in the following order:

> introits and communions: modes 1 to 8 (that is, introits mode 1, communions mode 1, introits mode 2, communions mode 2, etc.)
> alleluias: modes 1 to 8
> tracts: mode 8
> graduals: modes D to G (no separation of authentic from plagal modes)
> offertories: as for graduals
> tracts: mode 2

For two groups of pieces, the range is specified in the margin, and there is also an indication as to whether the piece uses B or Bb—these are unique to this source. Within each modal group, the chants are ordered according to two more criteria: (i) according to their starting note, from lowest upwards; (ii) then, according to their highest note. This remarkable manuscript may be regarded as a master copy of the repertory, of a special kind, organized not according to liturgical principles but abstract musical ones.

The name tonary is also used for a simple set of psalm tones or other tones, without indication of any of the chants which control their use.

(ii) *Type-Melodies for the Eight Modes*

The aim of bringing to mind some of the characteristic melodic contours of melodies in a particular mode seems also to lie behind three peculiar sets of chants known almost exclusively from tonaries and theoretical sources. They are placed at the head of each group of antiphons or other chants in a particular mode.

1. The first, particularly intriguing, set of phrases (only about a dozen notes in length) has peculiar pseudo-Greek texts: 'Nonanoeane', 'Noeagis', etc.

2. A second set of longer melodies resembles single-phrase antiphons in melodic style, and they have Latin texts which allude to the number of their tonal group. Thus for tone 1, the chant usually given is *Primum quaerite regnum Dei* ('First seek ye the kingdom of God', cf. Matt. 6: 36). A different set of such chants is found in St Gall, Stiftsbibliothek 390–391 and related sources, where the first text is *Primum mandatum amor Dei est* ('The first commandment is the love of God', cf. Deut. 6: 5). Another set found in later German sources is given in Huglo 1971, *Tonaires*, 421.

3. Finally, both the Latin and the pseudo-Greek chants may be suffixed by a melisma or neuma, covering similar melodic ground. (On the various meanings of 'neuma' see below, IV.I.iii.)

Ex. III.14.2. Type-melodies from a tonary (Paris, Bibl. Nat. lat. 1121, fos. 201ᵛ–205ᵛ)

Ex. III.14.2 is a transcription of all three sets of melodies as they appear in the early tenth-century Limoges troper Paris, Bibliothèque Nationale, lat. 1121. The neuma is attached to both the pseudo-Greek and the Latin pieces. After each batch of type-melodies comes a set of antiphon, responsory, introit, and communion incipits.

Bailey, who has studied and made a critical edition of all the three sets of type-melodies from a comprehensive range of sources, dates the introduction of the pseudo-Greek phrases to the early ninth century (see Bailey, 1974, 5–8), and believes (with other scholars) that they are of Byzantine origin. Aurelian of Réôme, who is the first to mention them, in the mid ninth century, speaks of them as Byzantine; and melodies of remarkably similar type are known from Byzantine sources (albeit of the twelfth century and later), the *echemata* (see Raasted 1966, and the parallel copies in Huglo 1971, *Tonaires*, 384 and Bailey 1974, 12). Bailey points out that the Byzantine *echemata* might be sung by a cantor to bring to mind the characteristic melodic idioms of a mode, before he embarked upon a liturgical item; a similar function may be intended for these Western *echemata*.

The neumae are probably slightly later in date, while the Latin compositions are not known from earlier than the mid tenth century.

Of all the type-melodies, only the neumae were used liturgically, as special prolongations of the antiphons for the Benedictus, Magnificat, and Quicunque vult, and for the final antiphons of Vespers, Lauds, and the nocturns of Matins (see Frere 1898, ii. 209, for the instruction in Sarum ordinals). In time they seem to have acquired an almost mystical significance, as representations of the 'breath' of the Holy Spirit (see Huglo 1971, *Tonaires*, 389). Not surprisingly, perhaps, in the context of extravagant liturgies of the Île de France of the twelfth and thirteenth centuries, they acquired prosula texts and polyphonic upper voices (motets in the Montpellier and Bamberg manuscripts). New melodies of similar type might even be composed. In Laon, Bibliothèque Municipale 263, containing a festal Epiphany liturgy, the kyriale is preceded by a set of tones for the Gloria of the introit. To each tone a melisma is attached, and then there follows a prosula for the melisma. Ex. III.14.3 gives those for tones 5 and 6 (1–5 are short, 6–8 long).

(iii) *How Psalm Tones were Specified*

The tonary is only one of several ways in which tones and differentiae were codified. Very many antiphoners actually copy after each antiphon the start or ending to be used, over some abbreviation of the 'Gloria patri' such as 'e u o u a e' ('sEcUlOrUm AmEn'), or simply 'Amen'. Many old sources use code numbers (I to VIII) or letters to indicate the same thing. For example, three graduals of the early tenth century, none originally intended to have notation, Paris, Bibliothèque Nationale lat. 12050 (Corbie), Mont-Renaud (Corbie, notated later in tenth century), and Laon, Bibliothèque Municipale 118 (St Denis), use the following system to indicate the psalm tone to be used for introit and communion (complete integrated lists in Huglo 1971, *Tonaires*, 94–101):

AP or APR	= Authentus Protus
PP, PLP, or PLPR	= Plagalis Proti
AD	= Authentus Deuterus
PD or PLD	= Plagalis Deuteri
ATr	= Authentus Tritus
PTr	= Plagalis Triti
ATD	= Authentus Tetrardus
PTD or PlTD	= Plagalis Tetrardi

An important group of Swiss sources (studied by Omlin, 1934, with additions by Huglo 1971, *Tonaires*, 232–51) uses vowels to indicate the tones: a, e, i, o, u, H, y; ω, and consonants for the differentiae: b, c, d, g, h, k, p, q. This system is first seen

Ex. III.14.3. From the Gloria tones (Laon, Bibl. Mun. 263, fo. 21ᵛ)

in the early eleventh century, in the tonary attached to Hartker's antiphoner (St Gall, Stiftsbibliothek 390–391 copied, fortunately before Hartker's work lost many pages, in St Gall 388), then in many later St Gall sources, and books from Rheinau, Engelberg, Einsiedeln, Disentis, and also from churches in south Germany, Austria, and north Italy. The system survived into the seventeenth century and was actually revived in the *Antiphonale monasticum secundum traditionem Helveticae Congregationis Benedictinae* of 1943.

The Breton gradual Chartres, Bibliothèque Municipale 47 (*c*.900) uses Roman numerals to specify introit tones, as do many medieval sources. But it also uses a shorthand method of indicating the melodic peculiarities of some melodies. Huglo (1971, *Tonaires*, 105–8) explains the most common entries as follows:

the letters *i*, *m*, and *ul* refer to the start (initium'), internal cadence ('medium'), and final syllable ('ultima') of the chant;

the mode is then stated, with the words *i* ('in') or *em i* ('eminet in' = 'stands out as').

Thus 'i em i IIII' means that the chant starts as a mode 4 piece; 'm i VII' means that it has an internal cadence suggesting mode 7.

There are many medieval antiphon lists whose chief function, besides specifying the selection of chants for the liturgical year, is to codify the psalm tones and endings to be used. These are easily confused with tonary lists, but their true nature is clear when it is realized that the chants are listed in liturgical, not tonal order. (See Huglo 1971, *Tonaires*, 22–3.)

III.15. IDENTIFYING AND DESCRIBING CHANT-BOOKS

Each different type of liturgical book has its own content and characteristics. Without claiming to cover all important points, I attempt here a brief summary of the features of chant-books which are customarily the object of investigation, and on which one hopes to find information in catalogues, inventories, and handlists.

The three most important pieces of information about the book are: the type of book; its provenance; its date. Manuscript sources almost never carry any title or explicit to make these clear.

The type of book is of course determined by its contents, which after the preceding chapters hardly needs further comment. Experience shows, however, that missals and breviaries should always be described as 'noted' or 'without notation', and the extent of the notation made clear.

The date is partly a matter of palaeography, the accurate chronological identification of the text and music scripts. Partly it is a matter of estimating the significance of the saints represented in the Sanctorale or kalendar, or the date of the latest items in the source.

The biggest problems are often posed by the provenance of the source. A threefold distinction may be observed here, between:

1. the derivation of the source, that is, the liturgical tradition to which it belongs;
2. the place where it was prepared, which may be different from
3. the place where it was used.

Thus Rouen, Bibliothèque Municipale 277, a noted missal of the mid-thirteenth century, can be shown to have been copied in Paris (Branner 1977), but follows the use of Rouen cathedral and was presumably prepared for use there.

Rubricated books may contain information about the personages involved in the performance of, for example, the ceremonial of Holy Week. For example, if a bishop and 'clerici' are involved, the book would have been used in a cathedral rather than a monastery. For graduals and missals the distinction between secular and monastic use is often difficult to make. Kalendar and litanies (if present) and the Sanctorale should be inspected to see if particular attention is paid to monastic saints (especially St Benedict and his sister St Scholastica). Office books will naturally follow either secular or Roman cursus or the monastic cursus. Other clues may be gleaned from processional ceremonies, for example those of Palm Sunday, which visit different stations or churches; these may be specified in the source. The date of the service for the dedication of the church may also be marked in the kalendar or be deducible from its position in the Sanctorale, and in rare instances this can lead to an identification of the provenance of the source.

The saints to whom attention is paid are important evidence for the derivation and destination of a source, not so much those who were universally venerated, such as the apostles or early martyrs, as saints of purely local importance. They may be found in three places: (i) in the kalendar; psalters, missals, and breviaries often have a kalendar, whereas graduals, antiphoners, tropers do not; (ii) the litanies of the saints in Holy Week and Ascension week; (iii) the Sanctorale; not only graduals and antiphoners (as well as missals and breviaries) have a Sanctorale, but also collections of sequences and tropes. Although a psalter has no Sanctorale, it may be joined to a hymnal which does.

Repertorial comparisons between sources are a powerful tool for the investigation of derivation and destination. Although it would be possible to list the complete contents of a source and compare it with others, in practice it is necessary to be selective and concentrate on those parts of the repertory which most often differ from source to source. As far as chants are concerned, the following may be picked out.

For mass:

1. Alleluias, including their melodies, which may be identified with the aid of Schlager's catalogue (1965). Text incipits of the series for Easter week and the Sundays thereafter, Whitsun week and the Sundays thereafter should always be noted. Lists for comparison with other sources have been accumulated by various scholars but are unfortunately not available in print in large numbers (for examples see Husmann 1962 'Studien', and Hiley 1980–81).

2. Sequences, including, for rhymed sequences, their melodies (the melodies of older sequences do not vary).

3. Ordinary-of-mass melodies, and tropes, which should be identified with the aid of the catalogues of Landwehr-Melnicki (1955), Bosse (1955), Thannabaur (1962), and Schildbach (1967), supplemented by Hiley (1986, 'Ordinary').

4. Tropes and any other festal material.

For the office:

1. The choice of office chants varies greatly from source to source. In practice it is advisable to restrict one's attention to parts of the repertory already investigated, such as the Advent responsories and their verses listed by Hesbert (*CAO* 5–6).

2. Hymns, whose melodies should be identified with those published by Stäblein (MMMA 1).

The way in which repertory comparisons can illuminate the derivation of a source may be demonstrated with the aid of the twelve sources edited in *CAO*. Let us assume for the moment that the provenance of the gradual and antiphoner published in facsimile in PalMus 16, the 'Mont-Renaud manuscript', is unknown. In order to establish the affiliations of the manuscript we should ideally compare it with as many manuscripts as possible, which is what Dom Hesbert did in volumes 5 and 6 of *CAO* (manuscript 326/728). For present purposes the twelve sources edited in *CAO* 1 and 2 will suffice.

The responsories of the Night Office for St Andrew (30 November) provide a suitable point of comparison. The twelve sources of *CAO* 1–2 have between them sixteen different responsories (formulary 120, 117). I have given them in Table III.15.1 in the order in which they appear in the oldest source, the Compiègne gradual–antiphoner Paris, Bibliothèque Nationale, lat. 17436. Figures indicate the order in which they appear in the other manuscripts. The Mont-Renaud manuscript (fo. 111^r–^{v}) appears in the last column.

Table III.15.1 *Responsories in the Night Office of St Andrew:* CAO *and the Mont-Renaud manuscript*

	C	B	E	M	V	H	R	D	F	S	L	MR
Dum de/per-ambulant	1	1	1	1	1	1	1	1	1	1	1	1
Venite post me	2			2	7		2	2			2	2
Mox ut vocem	3	2	2	3	2	2	3	3	2	2	3	3
Oravit sanctus Andreas	4		10		11		9	8	7	3	10	8
Homo dei	5	3	4	6	4	3	4	4	5	5	4	4
O beata/bona crux	6	7	5	7	5	7	8	7	6	6	9	7
Doctor bonus	7	4	3	4/5	3	4	5	5	4	4	7	5
Expandi manus	8	5	6	8	6	5	10	10	8	7		10
Salve crux	9	6	9			6	7	6	10	10	8	6
Dilexit Andream	10	8	8	9	8	9	12	11	12	12	11	11
Vir iste	11	9	7	11	9	10	11	12	11	8		12
Cum videret/vidisset	12					8	6		3	11	5	
Dum penderet	13											
Videns/Vidit crucem	14		11	10	10						6	
Beatus Andreas de cruce								9	9	9		9
Cives apostolorum											12	

In most sources the responsories are grouped for the three nocturns in three, fours, or larger groups, sometimes reflecting the secular or Roman cursus, sometimes the monastic, but sometimes with several 'extra' responsories, which could be selected as the precentor saw fit. This is most obviously the case in the Compiègne gradual–antiphoner. It has almost all the responsories found elsewhere. Three of the sources simply copy all responsories together after the antiphons. (The shelf-marks of the sources are given at the start of III.5; only their provenance is given here; 'G', Durham B. III. 11, has a lacuna at this point.)

C	= Compiègne	4 + 3 + 7
B	= Bamberg	3 + 3 + 3
E	= Ivrea	3 + 3 + 5
M	= Monza	4 + 4 + 3
V	= Verona	3 + 4 + 4
H	= St Gall (Hartker)	3 + 3 + 4
R	= Rheinau	4 + 4 + 4
D	= Saint-Denis	12
F	= Saint-Maur-des-Fossés	4 + 4 + 4
S	= Silos	4 + 4 + 4
L	= Benevento	12
MR	= the Mont-Renaud manuscript	12

It may be seen that only one other manuscript has the order of responsories in the Mont-Renaud book, and that is the Saint-Denis antiphoner. The Mont-Renaud source may also come from Saint-Denis, or, slightly more likely, from Corbie, whose liturgy and chant were in most respects identical with that of Saint-Denis. So the little test carried out here accords with other findings about the manuscript. (It may be said that not every formulary finds these two sources in agreement.)

It looks as if the manuscripts all select responsories from a common pool of favourite compositions, practically all of which were already available early in the ninth century (at least in time for the Compiègne source to include them). So the comparison sheds a little light on the historical development of the repertory, which one could follow further by consulting the Old Roman sources.

The importance of comparison of musical variants cannot be overestimated. Not only do they act as a litmus test for the relationships between sources, they constitute valuable evidence as to the mode of transmission of chant. The only extensive work so far carried out is that of the monks of Solesmes, reported in *Le Graduel romain* IV, for proper chants of mass. (For examples of work on other mass chants, office antiphons, and sequences, see Underwood 1982 and Hiley 1980–1 and 1986, 'Thurstan'.) This work is laborious, for its effectiveness depends both on the accumulation of a large number of points of musical variance and their tracking through a large number of sources.

Much of the information elicited by the above methods of investigation has to be reported in articles rather than brief catalogue descriptions of sources, though the latter can summarize the most important points. Catalogue descriptions should, however, include a codicological and palaeographical description of the source and describe the nature and extent of the musical notation. (The latter point is often overlooked: for example, has a missal notation only for the canon of the mass, or for all chants?) A photograph of the notation often says more than a paragraph of description.

There are by now a great number of descriptions of chant sources, particularly in the catalogues of individual libraries. Their quality and the space they devote to each source naturally varies greatly. As examples of widely differing concepts of what should be reported, the reader may consult Frere (1894–32), Anglès and Subirá (1946), Husmann in RISM B/V/1 (1964), Bernard (1965–74), Arnese (1967), Stenzl (1972), Plocek (1973, Fernández de la Cuesta (1980), and Gottwald's catalogues of music manuscripts in Augsburg, Freiburg im Breisgau, Munich, Nuremberg, and Stuttgart. Fine catalogues of liturgical manuscripts (that is, not just music-books) are now available for several areas of Europe, for example Radó (1973) for Hungary, Grégoire (1968–73) for Italy, and the volumes of Iter Helveticum. Other essential research tools include catalogues of medieval libraries, either original documents (G. H. Becker 1885; the *Mittelalterliche Bibliotheks-Kataloge* published for Germany, Switzerland, and Austria; Nortier 1957–8 for Norman monasteries) or lists of surviving sources from medieval centres (Ker 1964, 1987 and Krämer 1989).

IV

Notation

IV.1. INTRODUCTION

(i) Preliminary
(ii) The Signs in Montpellier H. 159
(iii) Neume

Wagner II; PalMus 13; Suñol 1935; Jammers 1965, *Tafeln*; Stäblein 1975; Corbin 1977; Corbin, 'Neumatic Notations, I–IV', *NG*; Hiley, 'Notation, III, 1', *NG*; Huglo 1990.

(i) *Preliminary*

In this chapter first the notational signs used in the earliest chant sources are discussed, the signs generally known as neumes. It is not easy to come to grips with the different styles of notation, their beginnings and historical development, without some basic knowledge of the system or systems involved. First of all, therefore, a brief description of the signs in one particular manuscript is given, a manuscript chosen (*a*) because it is available for further study in facsimile in a modern transcription, and (*b*) because it has a letter notation, specifying pitches, as well as the signs commonly known as neumes, which specify mode of delivery. This is the well-known Dijon tonary, Montpellier, Faculté de Médecine H. 159 (PalMus 8; Hansen 1974). It should be clear from the outset, however, that other sources have signs for which there is no counterpart in Montpellier H. 159, and that it has some which they may lack, and that no two manuscripts correspond exactly in the way they employ the signs.

Other styles of music-writing are then illustrated, after which the beginnings of notation are discussed, a topic which has caused much controversy, not least in the last few years. Then the rhythmic nuances indicated in the refined notation of some important early sources are discussed. After that the development of pitch-notation from the eleventh century onward is treated. Some systems found only in theoretical treatises are mentioned, and then the notations of the later Middle Ages and subsequent centuries. Finally, a few thoughts are hazarded about performance and transcription.

Although it has become customary to call the early notational signs 'neumes', the word was more often used in a different sense in the early Middle Ages, to mean a melodic phrase. A neuma could be as short as one note or as long as 101 notes. The meaning of the term is discussed below. Like Cardine (1968, etc.) I have usually spoken simply of 'signs'.

The primary function of musical notation is usually regarded as the indication of pitch, but this was evidently not the case with the earliest notations. It is essential to realize that the music represented was already known by heart as far as its tonality and pitch content were concerned. The notation reminded the singer of details of phrasing, rhythm, and dynamic, together with some refinements of performance associated with the use of the signs known as the oriscus and quilisma, whose significance is not now fully clear. Even after notation began to be used, most performance of chant continued from memory. A notated book was for refreshing the memory, for *recordatio*, as it was known, in the song-school or before a particular service. We should not, therefore, expect early manuscripts to tell us the same as modern editions such as the *Graduale* or *Antiphonale Romanum*, nor regard their notation as some sort of primitive, groping, and flawed start on a long road to ultimate perfection.

(ii) *The Signs in Montpellier H. 159*

Table IV.1.1 sets out the signs in Montpellier H. 159, together with corresponding signs in the modern Vatican/Solesmes editions and in the transcriptions given in this book. At the beginning of each line the melodic gesture involved is stated, and the name of the sign in the later medieval sign-tables. (These tables, which appear from the twelfth century onward, have been edited and discussed by Huglo 1954, 'Noms'. Most of the tables are of German origin; for two Italian ones see Ferretti 1929. Only the better-known names are given here.) Montpellier H. 159, but not many other manuscripts, uses the slanting virga (1B) for a note which comes at a low point in a phrase, but is not low in the overall range of the piece. And very few other sources use the descending quilisma. The interpretation of the special signs 11–15 is discussed in IV.3.

Ex. IV.1.1 shows how some of the signs are used. Each sign is labelled with the number given in the Table. The signs usually give no indication as to the size of the intervals involved. The liquescent sign 15B is used indiscriminately here for a second, a third, and a fourth. The scandicus (4) can just as well be used for *Gbd* as for *Gab*. Only the signs for single notes indicate something of relative height, so that 'hic fert' has virga–punctum. The punctum is usually reserved for the lowest of a group of single notes, so that there are more virgae than puncta: thus the 'qui' at 'qui manet' has a virga, not a punctum, likewise 'et' and '-go' at 'et ego'. By analogy one would then expect the scandicus and climacus to consist of but one punctum and two virgae, but the reverse is the case. The slanting virga (1B) is used three times, for *a*, *b*, and *b* respectively; but the punctum can also be found for these notes as well, or even for *c*

Table IV.1.1. *Notational signs in Montpellier, Faculté de Médecine H. 159*

Melodic gesture (H = higher note, M = middle note, L = lower note)	Name in medieval sign-tables	Montpellier	Vatican (* = no special sign)	Transcription
1. single notes				
A. H	virga		*	
B. L (relatively high in range)	——		*	
C. L	punctum			
2. L–H	pes, podatus			
3. H–L	clivis, flexa			
4. L–M–H, etc.	scandicus			
5. H–M–L, etc.	climacus			
6. L–H–M–L	pes subbipunctis			
7. L–H–L	torculus			
8. H–L–H	porrectus			
9. H–L–H–L	porrectus flexus			
10. L–H–L–H	torculus resupinus			

11.	repeated notes			
	A.	bivirga, trivirga		
	B.	bistropha, tristropha		
	C. L–H and H–L + repeated note	——		
12.	H–H–L	trigon		
13.	see below, IV.3	oriscus (not found alone)		
	A. followed by a punctum			
	B. joined to the punctum			
	C. followed by a virga			
14.	see below, IV.3	quilisma (always between two notes a 3rd apart)		
15.	liquescent neumes (selection)			
	A. punctum + ascending liquescence = liquescent pes (L–H)	epiphonus		
	B. virga + descending liquescence = liquescent clivis (H–L)	cephalicus		
	C. clivis + descending liquescence = liquescent climacus (H–M–L)	ancus		

Ex. IV.1.1. Communion *Ego sum vitis vera* (Montpellier, Faculté de Médecine H. 159, p. 82)

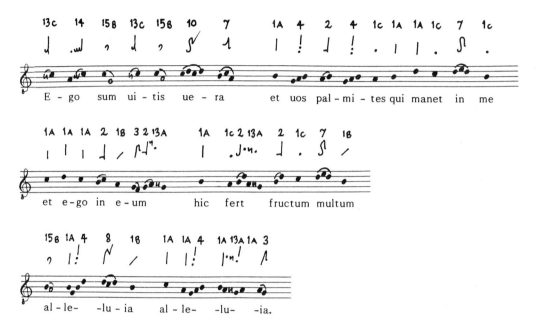

('maNET'). Clearly there is flexibility in the use of the signs and much depends on melodic context.

As already mentioned, the extant tables of notational signs date only from the twelfth century onward. The names given to the individual signs are of uncertain derivation, and most are undoubtedly late Latin or pseudo-Greek neologisms. They are probably better known now than they were in the Middle Ages. The literal meanings of those given on Table IV.I.1 are as follows. It will be seen that most are descriptive either of the actual shape of the sign or of the melodic gesture involved. (For further discussion, see Huglo 1954, 'Noms'.)

virga	Lat.: 'staff'
punctum	Lat.: 'point'
pes	Lat.: 'foot'
podatus	from Gk. *pous, podus*: 'foot'
clivis	from Lat. *clivus*: 'slope'
flexa	possibly an abbreviation of Lat. accent 'circumflexa'
scandicus	from Lat. *scandere*: 'ascend'
climacus	from Gk. *klimax*: 'ladder'
pes subbipunctis	a pes with two lower puncta
torculus	Lat. 'screw of a wine-press'
porrectus	Lat. 'stretched out'
porrectus flexus	a porrectus turned back down

torculus resupinus	a torculus turned back up
bivirga, trivirga	two, three virgae
bistropha, tristropha	two, three apostrophes
trigon	from Gk. *trigonon*: 'triangle'
oriscus	from Gk. *horos*: 'limit'?
quilisma	from Gk. *kylisma*: 'rolling action'
epiphonus	from Gk. *epi + phone*: 'added sound'
cephalicus	from Gk. *kephalis*: 'little head'
ancus	from Gk. *agkon*: 'curve'

(iii) *Neume*

There are two Latin spellings of the word, with meanings really distinct but often confused in the Middle Ages: the Latin word *neuma* (from the same word in Greek) means a 'gesture'; *pneuma* (also from the same word in Greek) means 'breath'. The latter form was commonly used to signify the Holy Spirit, and this meaning became applied by transference to neuma, thus linking music to the divine creative force itself.

Although neuma did eventually come to mean a notational sign, that was not its primary musical meaning. More often it was used to refer to a sounding melody, or phrase, in particular one which has no words. Amalarius of Metz, writing about 830, speaks in a well-known passage about a *neuma triplex*, a set of three melismas, textless melodies, added to certain responsories of the Night Office (ed. Hanssens iii. 54–5). The textless melody of the sequence was also often referred to as the neuma:

> Post alleluya, quaedam melodia neumatum cantatur, quod sequentiam quidam appellant (Udalricus of Cluny: PL 149. 655)

> Pneuma sequentie quod post alleluia cantatur, laudem eterne glorie significat (Johannes of Avranches: PL 147. 34)

(See Bautier-Regnier 1964, 5; Hiley, 'Neuma', *NG*). Neuma is also the 'model' melody for each mode found in tonaries or added to antiphons as a festal extension (see III.14.ii).

The use of neuma to mean the written sign seems not to be as old as these. In the controversial eleventh-century treatise *Quid est cantus?* (which will be referred to again in the discussion of the origins of neumatic notation) it is stated: 'De accentibus toni oritur nota que dicitur neuma' (Wagner 1904, 481–4).

The usual word for written musical signs in the Middle Ages was *nota*. As a term for alphabetic or comparable signs, this is found as early as the ninth century, in the anonymous treatise *Musica enchiriadis* and in the treatise of Hucbald of Saint-Amand. The author of *Musica enchiriadis* has been speaking of the special signs of 'dasian' notation, and ends: 'Sunt et alia plura plurium sonorum signa, inventa antiquitus, quibus evitatis, has faciliores hic inserere curae fuit notas.' ('And there are very many other signs for sounds, found in antiquity, which have been avoided, and our care was to insert here these easier notes.') Eventually *nota* signifies the non-alphabetic signs as well—as in the treatise *Quid est cantus?* mentioned above.

By the eleventh to twelfth century, there had arisen a manuscript tradition of listing neumes in didactic tables. There are two types of table, one short and the other long. One thirteenth-century source of the short table (Huglo 1954, 'Noms', 55, source R) entitles its list 'Nomina notarum', while the usual title for the long table is 'Nomina neumarum'. The last verse of the short table runs: 'Neumarum signis erras qui plura refingis.' ('You err with signs of the neumes when you fashion more.) Thus here 'neume' still means the melody itself, not the sign.

As a final witness to the meaning of neume as a melody, rather than a written sign, Guido of Arezzo's *Micrologus*, written about 1030, may be cited. In ch. 15 Guido draws an interesting analogy between the construction of metrical verse and that of a melody. The terms he uses are the following. The individual letters of verse (*litterae*) are comparable to the individual sounds of music (*phthongi* or *soni*). As syllables (*syllabae*) are composed of letters, so musical syllables (*syllabae*) are composed of one or more sounds. Syllables are made up into parts or feet (*partes* or *pedes*); one or more musical syllables make up neumes or parts (*neumae* or *partes*). The bringing together of feet makes a verse (*versus*), which is equivalent to a section (*distinctio*) of a musical composition, composed of several neumes.

The term neume, meaning a notational sign, has by now become so embedded in the musical literature that it would be pointless to try and restrict its meaning to the other medieval sense of the word. I have nevertheless avoided it when speaking of notation.

IV.2. REGIONAL STYLES

   (i) French and German Notation (see Plates 4, 6–8)
   (ii) Palaeofrankish, Laon, Breton, and Aquitanian Notations (Plates 1–3)
   (iii) Types Related to French–German Notation (Plate 5)
   (iv) Other Italian Notations (Plates 11–13)
   (v) Examples

PalMus 2–3; Bannister 1913; Suñol 1935.

This chapter gives brief descriptions of some of the more important types of chant notation. It will be recognized, of course, that sources habitually classified together do not always use absolutely identical shapes. The signs given in the tables below are copied from one more or less typical manuscript, rather than being regularized on the basis of the generality of sources. Nor is an attempt made to give all the variant shapes within each source: only the most common are selected.

(i) *French and German Notation (see Plates 4, 6–8)*

    Facsimiles:
Bamberg, Staatsbibliothek lit. 6 (German/St Gall): see Bamberg 6
Einsiedeln, Stiftsbibliothek 121 (German/St Gall): PalMus 4
Montpellier, Faculté de Médecine H. 159 (French): PalMus 8
Paris, Bibliothèque Mazarine 384 (French): MMS 5
St Gall, Stiftsbibliothek 339 (German/St Gall): PalMus 1
St Gall, Stiftsbibliothek 359 (German/St Gall): PalMus II/2
St Gall, Stiftsbibliothek 390–391 (German/St Gall): PalMus II/1
private collection, the Mont-Renaud manuscript (French): PalMus 16

Although commonly regarded as distinct, French and German signs are similar in many basic ways. The habit of classifying them apart has come about because whereas up-strokes of the pen in French notation are more or less vertical, those of German sources are usually slanted. Furthermore, the deservedly high degree of attention paid to the most interesting sources of German notation, the St Gall manuscripts, with a wealth of notational detail not found in French sources, has created the impression of a system which is different from that of French notation. If, however, St Gall notation is regarded as exceptional, and a comparison is made between simpler German and French examples, the similarities are obvious.

Table IV.2.1 sets out the usual forms for French and German notation, but places between them the forms found in an intermediate area, as it were. The sources on which the table is based are the Mont-Renaud manuscript (from Corbie, late tenth century), Leipzig, Universitätsbibliothek, Rep. I. 93 (from the Mosel area, tenth century), and London, British Library, Add. 19768 (from Mainz, tenth century). This table may be compared with Table IV.1.1 in the previous section, for Montpellier H. 159 belongs to this family of notations.

The varieties of French notation found in Cluniac sources have been surveyed by Hourlier (1951, 'Remarques'). The notation of the St Gall sources has long been the subject of intensive study. Because of its especially sophisticated character some of its features are discussed below, IV.3 and IV.5.

(ii) *Palaeofrankish, Laon, Breton, and Aquitanian Notations (see Plates 1–3)*

    Facsimiles:
Düsseldorf, Universitätsbibliothek D. 1 (Palaeofrankish): Jammers 1952
Laon, Bibliothèque Municipale 239 (Laon): PalMus 10
Chartres, Bibliothèque Municipale 47 (Breton): PalMus 1
Paris, Bibliothèque Nationale lat. 903 (Aquitanian): PalMus 13

Unlike the widely established French–German type, Palaeofrankish signs were known in only a small area of north-east France, and survive in but a handful of sources, none of which is a full gradual or antiphoner (for a list of sources, see Hourlier and Huglo 1957). Handschin (1950) gave it the name 'Palaeofrankish',

Table IV.2.1. *French and German notational signs*

	Mont-Renaud	Leipzig Rep. I. 93	London 19768
1. virga	•	/	/	/
2. punctum	•	-	. -	-
3. clivis				
4. pes			!	
5. porrectus				
6. torculus				
7. climacus				
8. scandicus				
9. quilisma				
10. oriscus				
11. liquescents				

believing it to be the oldest type of notation developed in the Frankish empire. Instead of writing pes and clivis with two distinct movements of the pen, this notation uses only one slanting or slightly curved stroke. The porrectus and torculus also tend to form a single arc or angle, rather than a tripartite figure. Quilisma and oriscus do not appear to be present in the earliest examples. Since so many of the examples are brief, often entered in manuscripts not originally designed to contain notation, generalizations about the signs are somewhat hazardous. Those entered in Table IV.2.2 are copied from Düsseldorf, Universitätsbibliothek D. 1 (from ?Korvey, tenth century).

The area where Palaeofrankish notation was used is also that where so-called 'Messine' signs were used. The name 'Messine' refers to the supposed origin of the type in Metz, famous in the early ninth century for its chant. No ninth-century notated sources from Metz have survived, and indeed, those from later centuries are also relatively scarce. Corbin (1977, and 'Neumatic Notations', *NG*) argued that

'Lorraine' was a better name for the type, since the area where it was used corresponded to the ancient territory of Lotharingia (from which descends the modern name 'Lorraine'). 'Lorraine' is, however, no more appropriate than the name 'Messine'. Map IV.2.1 shows the centres from which sources with this notation survive. Neither the ninth-century kingdom nor the tenth-century duchy of Lotharingia included the more westerly centres, such as Lille, Noyon, Laon, or Reims; but they did include cities such as Trier and Aachen, and other territory as far as the Rhine where the notation was unknown. The area corresponds better, though still not exactly, with the archdiocese of Reims. Metz is on its periphery. In this book I have called it 'Laon' notation, after its most famous representative, manuscript Laon, Bibliothèque Municipale 239, and because Laon is reasonably central to the area. (For a survey of sources see Hourlier 1951, 'Messine'.)

The most characteristic feature of Laon notation is the small hook often used for the punctum (sometimes called an uncinus). The virga is usually a long flat **S** shape,

Table IV.2.2. *Palaeofrankish, Laon, Breton, and Aquitanian notational signs*

		Düsseldorf D.1	Laon 239	Chartres 47	Paris 1084
1.	virga	•	[⟋]	⟋	[⌐]
2.	punctum	•	•	• ⌐	• •	• ⌐
3.	clivis	• ⟍	⟍	⌐ ⌐	⌐ ∩	⫶
4.	pes	⌐•	⟋	⌐ ⟋	⟋ ⟋	⌐
5. porrectus	• •	Ⲩ	Ⲩ	Ⲩ	⫶⌐
6.	torculus	• •	∩	⋀	⋀	⋀
7.	climacus	• •	⟍	⫶ ⫶ (etc.)	ⲇ ⲇ (etc.)	⫶ ⫶ (etc.)
8.	scandicus	• •	⫶	⟋ ⟋	⫶⟋	⫶ ⌐
9.	quilisma	•	⟋	⟋	⌐ η	⌐
10.	oriscus	ц		⌀ N		m
11.	liquescents	• • • •	∩	ꝺ ⟍	⌐ ⌣	ꝺ ⟍

Map IV.2.1. Centres where Laon notation was used

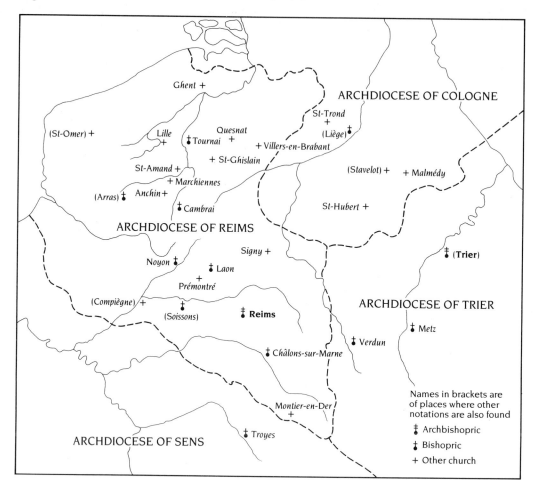

rather than a straight line; it is hardly ever seen except in combination with the punctum, to form a pes or other sign.

The notation of the manuscript Laon 239 is of a complexity and sophistication comparable only with that of the St Gall manuscripts, and more is said about it below (IV.3 and IV.5).

French–German notation starts its clivis with an upward stroke, but Laon and Breton notation begin instead with a short, almost horizontal movement of the pen, sometimes a shallow arc which is curved in towards the following down-stroke. Breton notation takes its name from the area of north-west Frence, Brittany, whence practically all its sources originate. (For a survey of sources, see Huglo 1963, 'Domaine'.) In many respects it bears a resemblance to Laon notation, though much more compact in overall appearance, and lacking the characteristic Laon hook. It does not have a quilisma. Two signs are used in places where, from other manuscripts, one would expect to see a quilisma or oriscus, but the conventions governing their use in Breton notation are slightly different. Table IV.2.2 shows the Breton signs of Chartres, Bibliothèque Municipale 47 (late ninth century; destroyed in 1944: facsimile in PalMus 11).

Although many notations use combinations of disjunct puncta and virgae as a special variant instead of their more usual conjunct forms for clivis, flexa, etc., few go as far as to make the disjunct forms the rule. This is so, however, with Aquitanian signs (see the survey in PalMus 13, by Ferretti). Again the virga is usually found only 'in composition' with other elements. The signs in Table IV.2.2 are drawn from Paris, Bibliothèque Nationale, lat. 1084 (?Aurillac, then Limoges, late tenth century).

It will be noticed that Laon, Breton, and Aquitanian notations all rise diagonally and fall vertically (compare the climacus and scandicus forms).

(iii) *Types Related to French–German Notation (see Plate 5)*

Facsimiles:
London, British Library, Add. 30850 (north Spanish): *Antiphonale Silense*
Rome, Biblioteca Angelica 123 (Bologna): PalMus 18
Studies:
Barezzani 1981 (Brescia).

Signs of the French–German variety were also used outside their 'home' area. Upright French shapes were copied in England in the late tenth century (summary of sources by Rankin 1987); and also in north Italy, at centres such as Novalesa and Asti, and, with slightly more individual character, at Brescia and Mantua. By contrast, Monza and Bobbio used the slanting German shapes.

Besides this, signs bearing a more general resemblance to the main French–German type are found in both Spain and Italy. Of the two main Spanish types, that used in north Spain had upright shapes, that of Toledo almost horizontal neumes. Both, however, are recognizably related to French signs, in their use of a virga, and an upward/forward movement of the pen at the start of clivis and porrectus. Further research is needed before the identification and use of 'ornamental' signs in the Spanish notations—quilisma, oriscus, and liquescents—is properly understood.

The long, bold strokes used in a well-known source probably from Bologna, Rome, Biblioteca Angelica 123 (eleventh century), are also close relatives of the French–German type. The notation is also found in a source possibly from Modena (flyleaves of Modena, Biblioteca Capitolare O. I. 13), and another seemingly from far to the south, possibly a monastic house in the Abruzzi, Rome, Biblioteca Apostolica Vaticana, Vat. lat. 4770 (closely related by repertory to Benevantan sources).

Sources from the Catalan area of north-east Spain have signs which bear surprisingly little resemblance to those of the Aquitanian area. For example, three vertically aligned puncta signify the ascending scandicus, not a descending climacus as in Aquitanian notation; and there is a preference for conjunct forms. Catalan signs are unlike other types in ascending vertically but descending on a very shallow incline.

Table IV.2.3. *Signs related to French–German notation*

	León 8	Toledo 35. 7	Angelica 123	Girona 20
1. virga				
2. punctum				
3. clivis				
4. pes				
5. porrectus				
6. torculus				
7. climacus				
8. scandicus				
9. quilisma				
10. oriscus				
11. liquescents				

Catalan notation has been studied by Mas (1988). The signs in Table IV.2.3 are drawn from Girona, Sant Feliu 20 (eleventh to twelfth century).

(iv) *Other Italian Notations (see Plates 11–13)*

    Facsimiles:
Benevento, Archivio Capitolare 33 (Beneventan): PalMus 20
Benevento, Archivio Capitolare 34 (Beneventan): PalMus 15
London, British Library, Add. 34209 (Milanese): PalMus 5
Lucca, Biblioteca Capitolare 601 (central Italian): PalMus 9
Rome, Biblioteca Casanatense 1741 (Nonantolan): Vecchi 1955
Rome, Biblioteca Apostolica Vaticana, Vat. lat. 10673 (Beneventan): PalMus 14
private collection of Martin Bodmer, C. 74 (Roman): Lütolf 1987
    Studies:
Huglo *et al.*, 1956 (Milanese)
Moderini 1970 (Nonantola)

Besides French or German shapes, other notations were also adopted in Italy. At Como, Laon signs influenced the notation (see Sesini 1932), Breton ones at Pavia (Huglo 1963, 'Domaine'). Disjunct forms (not at first sight similar to Aquitanian signs, however) were preferred in the very individual style practised at Nonantola. This is but one of many centres whose notation does not reflect the influences discernible in its repertory, for the monastery took over much of the German/Swiss trope and sequence repertory. The notation is related to that found at Torcello, and also appears occasionally in Verona manuscripts. The signs in Table IV.2.4 are drawn from Rome, Biblioteca Casanatense 1741 (twelfth century). A peculiar feature of Nonantolan notation is the way in which letters in the text were provided with long extensions to a height appropriate for the note being sung (which could only be done, of course, for the first note of each syllable).

Sources from many Italian centres have hardly survived at all before the time when staff-notation came into use (eleventh century or later). Generally speaking, the use of a staff often seems to have affected the notational signs, causing the appearance of

Table IV.2.4. *Other Italian notations*

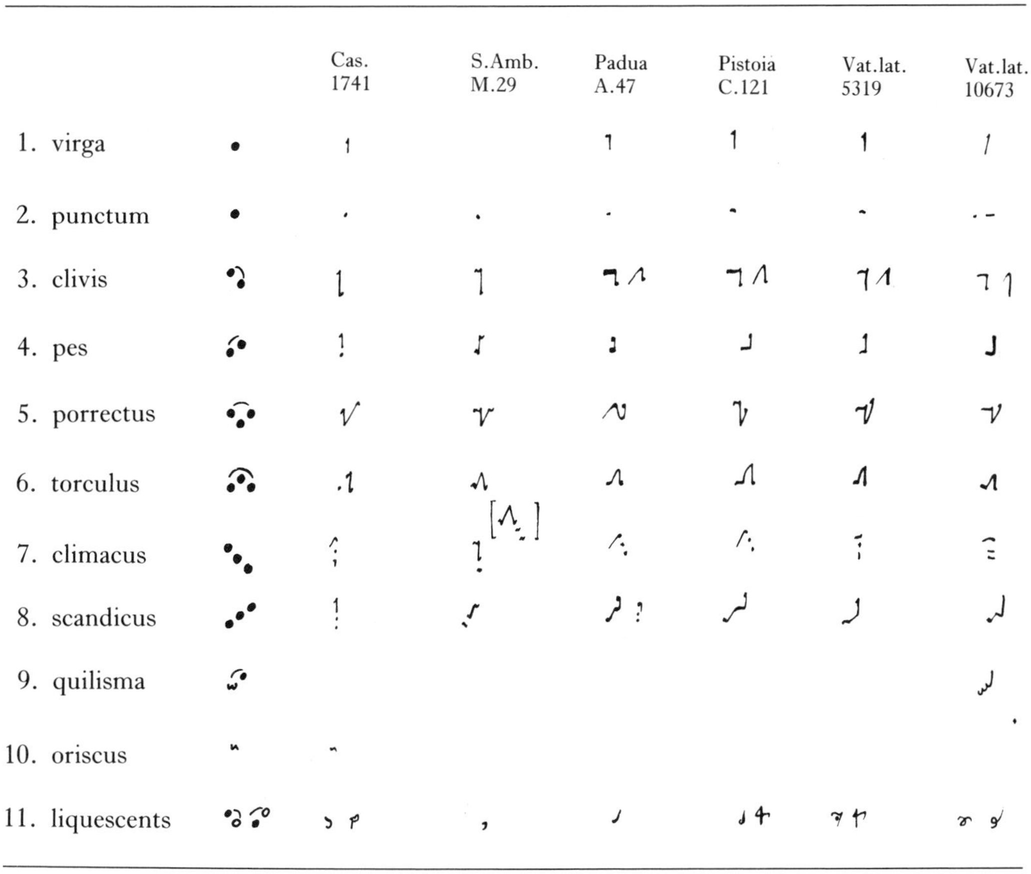
Cas. 1741
S.Amb. M.29
Padua A.47
Pistoia C.121
Vat.lat. 5319
Vat.lat. 10673
1. virga
2. punctum
3. clivis
4. pes
5. porrectus
6. torculus
7. climacus
8. scandicus
9. quilisma
10. oriscus
11. liquescents

more obvious heads and other precisely placed pen-marks. Sources from Milan show this, their signs often being little groups of tiny waved strokes, joined by fine lines. (See examples in Table IV.2.4, from the twelfth-century manuscript Milan, Sant'Ambrogio M. 29.)

Many Italian sources are distinguished by a conjunct scandicus, and the horizontal bar at the start of the clivis is another diagnostic feature. In most respects there is little to choose, as far as the basic shape of the signs is concerned, between the majority of sources written south of a line drawn between Lucca in the west and Ravenna in the east, excepting those types already mentioned. For want of a more accurate term, I have referred to this family as central Italian notation. Of course, there are differences, such as the great variety of liquescent shapes in manuscripts of the Montecassino and Benevento area. Variation in such matters as the thickness of the pen often has a strong effect on the visual aspect of the signs. One would not readily confuse the bold strokes of later Beneventan sources for any others.

Alongside the Nonantola and Milan examples already mentioned, Table IV.2.4 gives some typical shapes from sources from Ravenna (Padua, Biblioteca Capitolare A. 47, twelfth century), Pistoia, Biblioteca Capitolare C. 121 (Pistoia, twelfth century), Rome, Biblioteca Apostolica Vaticana, Vat. lat. 5319 (Old Roman, twelfth century), and the region of Bari (Rome, Biblioteca Apostolica Vaticana, Vat. lat. 10673, eleventh century). The variation between sources in the angle of ascent and descent is important (particularly in the climacus; the Milanese source usually descends vertically, but after a torculus the points continue diagonally).

(v) *Examples*

Ex. IV.2.1 gives another specimen of French notation, like that in the previous chapter, but from a different source: the Mont-Renaud manuscript (Corbie, tenth century: see above, Table IV.2.1). Two other versions of the same antiphon, *Beatus ille servus*, in staff notation are copied in parallel with it: Worcester, Cathedral Chapter Library F. 160 (Worcester, thirteenth century) and Paris, Bibliothèque Nationale, lat. 17296 (Saint-Denis, late twelfth century). The notation of modern Solesmes/Vatican books was developed from the signs in manuscripts such as these.

The neumes of the Corbie manuscript include both punctum (dot) and tractulus (dash), always found either for relatively low isolated notes, or as the lower element in a pes. In some sources it seems that the tractulus signifies a longer note than the punctum, or a stressed note (see IV.4). It will be noticed that there are proportionately more virgae than puncta: evidently the scribe preferred to reserve the punctum/tractulus for only the lowest note(s) of a section, rather than reserving the virga for only the highest note(s). The punctum+virga form of the pes is used when the pes begins at a lower pitch than the preceding note. The form with two virgae is used when there is a rise to the start of the pes.

Between the two later versions in staff-notation there are interesting discrepancies. (These are common enough for office chants: see the much more disparate version in

Ex. IV.2.1. Antiphon *Beatus ille servus* (Mont-Renaud manuscript, fo. 119ᵛ; Worcester, Cathedral Chapter Library F. 160, p. 427; Paris, Bibl. Nat. lat. 17296, fo. 282ᵛ)

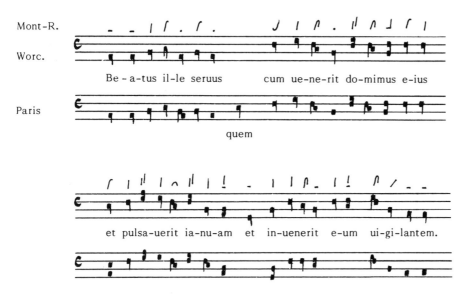

AM 671, based on the St Gall version.) The Worcester source is generally closer to what is indicated by the signs of the Corbie manuscript (liquescent form + punctum for *ille*, etc.; see also the differences in wording). But it does not usually employ the punctum (square) as opposed to the virga (square with tail). The Saint-Denis source is not so close textually and melodically, but makes use of the punctum for the low notes.

Not all notations use the virga as an isolated note. Ex. IV.2.2 gives the communion *Quis dabit ex Sion* from five different manuscripts. At the foot are the pitches given in letter notation by the eleventh-century manuscript from Saint-Bénigne at Dijon, Montpellier, Faculté de Médecine H. 159; above it are the neumes of the same manuscript (occasionally differing slightly from the letters). Above these are the neumes of four other manuscripts of the late ninth to the eleventh centuries. There is no isolated virga in either Paris, Bibliothèque Nationale, lat. 776 (Aquitanian) or Laon, Bibliothèque Municipale 239 (Laon). The apostropha (see 3 and 40) is to be found only in Einsiedeln, Stiftsbibliothek 121 (German) and Montpellier H. 159 (French). There is no quilisma (see 41) in Chartres, Bibliothèque Municipale 47 (Breton), and only Montpellier H. 159 uses the oriscus (end of 36, middle of 44).

Ex. IV.2.2. Communion *Quis dabit ex Sion* (P = Paris, Bibl. Nat. lat. 776; C = Chartres, Bibl. Mun. 47; L = Laon, Bibl. Mun. 239; E = Einsiedeln, Stiftsbibl. 121; M = Montpellier, Faculté de Médecine H. 159)

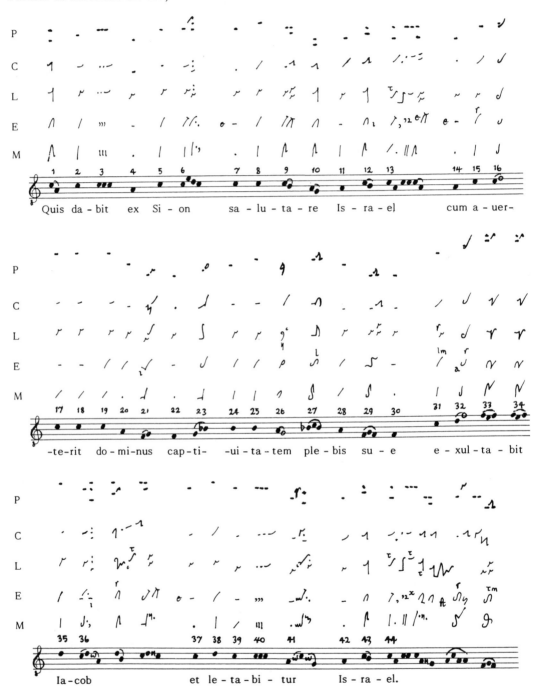

IV.3. LIQUESCENCE, ORISCUS, QUILISMA; OTHER SPECIAL SIGNS

(i) Signs for Liquescence
(ii) Quilisma
(iii) Oriscus
(iv) Virga strata, Pressus, Pes stratus, Pes quassus, Salicus

(i) *Signs for Liquescence*

Mocquereau in PalMus 2; Freistedt 1929; Göschl 1980; Hiley 1984.

Looking once again at Ex. IV.2.1, one notices that there are no liquescent signs in the Saint-Denis version, and only one in the Worcester copy, at 'ille'. The Corbie manuscript has that one, and others at 'ser(vus)', 'cum', 'ei(us)', and 'et'. Signs for liquescence occur most often where in the text there is a conjunction of a liquid consonant followed by another consonant. The two *l*'s of 'ille', 'rv' in 'servus', and 'mv' in 'cum venerit' are typical examples. Diphthongs usually call forth a liquescent sign, as at 'eius'. Quite often only one consonant may be present. It is not uncommon also to find liquescents with the syllable 'et', and there are other instances where neither a liquid consonant nor a diphthong seems to be involved. This may, however, indicate that the consonants in question were actually liquescent when the manuscript was being written—such differences of pronunciation are still common today from area to area, so that, for example, in Berlin *g* is pronounced like an English *y*: 'Wenn die Jemsen springen über Berjesjipfel . . .'

The manner of performance of liquescent signs is not entirely clear. A passing-note of some sort seems to be involved, since many manuscripts will record an ordinary neume (say, of two notes) where others have a liquescent (one note + 'passing-note'). The liquescent note itself may in some cases have been sung 'through' the consonant, as it were, vocalized to *l*, *m*, *n*, or *r* (the most common situations). As Guido of Arezzo said: 'At many points notes liquesce like the liquid letters, so that the interval from one note to another is begun with a smooth glide and does not appear to have a stopping place en route' (trans. Babb 1978, 72).

Comparisons between sources indicate that the use of liquescent signs was inconsistent, as in Ex. IV.2.1. This too is stated by Guido, who continues: 'If, however, you wish to perform [an example] more fully, not liquescing, no harm is done, indeed, it is often more pleasing.' So the signs are there as reminders, though no two scribes would put them in exactly the same places, rather in the way that the rhythmic nuances discussed below (IV.5) are not recorded identically in different manuscripts. Montpellier, Faculté de Médecine H. 159 seems to be a moderately generous source, but is easily surpassed by, for example, the early St Gall sources, and especially by those of southern Italy, which have an unrivalled variety of liquescent shapes.

The written form of the note in square notation of the thirteenth century and later came to be called the *plica* ('fold').

(ii) *Quilisma*

Wiesli 1966.

The sign-tables depict the quilisma as a wavy shape ending in a longer ascending stroke. This is translated as two notes in the Montpellier letter notation, the first being the special wavy one, the second higher. The name seems to refer to the wavy note in particular.

In most manuscripts the quilisma is nearly always used for the middle note in an ascending-third formation, sometimes only for a minor third. The name seems to derive from the Greek *kylisma* ('rolling'), *kylio* ('to roll'). Medieval writers on music do not elucidate its nature very much. For example, in the middle of the ninth century, when notation itself was relatively rare, Aurelian of Réôme uses the word *tremula* to describe a passage usually later notated with a quilisma. Speaking of the gradual *Exultabunt sancti* V. *Cantate domino*, at the passage beginning 'can[ticum novum]' he says (Ch. XIII): '. . . flexibilis est modulatio, duplicata, quae inflexione tremula emittitur vox non gravis sonoritas . . .' ('. . . the melody is flexible, repeated, where a sound, not low, is given forth with a tremulous inflexion . . .'). Ex. IV.3.1 gives the signs in Montpellier H. 159, Laon, Bibliothèque Municipale 239, and St Gall, Stiftsbibliothek 359.

It has been suggested that the quilisma might involve a portamento delivery, or one with semitone steps filling in the larger intervals, or a gruppetto ornament. In those manuscripts where its use is restricted to the minor-third progression, it constitutes a sign for the semitone step, and may therefore have been used specifically to locate the melody tonally. The idea that the quilisma was primarily of rhythmic significance (Wiesli, Cardine 1968, etc.) does not seem sufficient to explain the special shape of the sign.

Ex. IV.3.1. From gradual *Exultabunt sancti* V. *Cantate Domino*

(iii) *Oriscus*

The derivation of the word is obscure: pseudo-Greek *oriscos* ('a little hill') has been suggested, or *horos* ('a limit'). The oriscus does not usually stand alone, but, like the wavy component of the quilisma, comes in formation with other elements, typically in a figure such as *a–a–G*, where the middle note is an oriscus (see three instances in Ex. IV.1.1).

The oriscus is used in the Montpellier codex only as a repeat of the previous pitch: at least, so it seems when one compares other sources. In fact, in the letter notation underneath the musical signs in this manuscript, there is no letter for the oriscus, only a wavy line to show that 'something' is sung. It should also be borne in mind that repeated notes can be shown by other means, for example by placing a virga or punctum close to the preceding sign. It has therefore been suggested that some special vocal delivery was involved. If not, why use the special sign?

Wagner believed that a non-diatonic pitch was involved. This was because of figures like those in Ex. IV.3.2, where the oriscus is now on *b*, now on *c*, as if the 'true' note were actually in between. Once again the semitone step is involved.

Cardine believed its significance was rhythmic, directing rhythmic weight to the succeeding note, which is very often lower. Whether this rhythmic nuance is sufficient justification for the use of a special sign is debatable (see IV.5).

Ex. IV.3.2. Examples of oriscus

from Gradual <u>Eripe me</u> V. <u>Liberator meus</u>

St.Gall 359 p.85

Montpellier p.161

exalta–bis

from Canticum <u>Attende caelum</u>

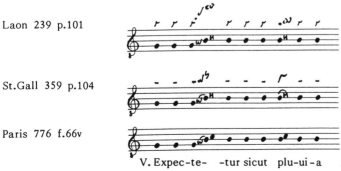

Laon 239 p.101

St.Gall 359 p.104

Paris 776 f.66v

V. Expec-te- -tur sicut plu-ui-a

In many sources the oriscus can be found in places where the text suggests liquescence, perhaps therefore a sort of liquescence at the unison.

(iv) *Virga strata, Pressus, Pes stratus, Pes quassus, Salicus*

Cardine 1968.

A number of special signs peculiar to the St Gall manuscripts and those closely related to them appear to involve the oriscus. A virga followed by an oriscus was usually run together into one sign, the virga strata. This sign has also been called the franculus, but Huglo (1954, 'Noms', 65) questions the correctness of this identification. The name for this sign on the longer sign-table is gutturalis ('throaty'), which may imply some special vocal effect.

On the shorter sign-table, the oriscus followed by a punctum is called a pressus minor, the virga strata followed by a punctum a pressus maior.

A pes followed by an oriscus is run together into one sign, called a pes stratus, just as happens to the virga alone.

The pes quassus appears to be a pes of which the first note is an oriscus. Sometimes, however, three notes are found in other sources, so that the pes quassus would be a pes with an extra oriscus between the two normal notes. It is often found at the beginning of phrases in mode 3 introits, where other manuscripts have an *E–E–F* opening. So the semitone step is again involved.

The salicus is a three-note ascending figure where the middle note is an oriscus. Its nature is puzzling, because it can be found in situations where other sources have: (*a*) two unison notes followed by a higher one; (*b*) a normal scandicus, that is, three notes in ascending order; (*c*) a quilisma in three-note ascending formation. Cardine (1968) believed its chief significance was rhythmic, and that the oriscus, as usual in Cardine's interpretation, directed rhythmic weight on to the succeeding note. To distinguish it from the quilisma, which Cardine interpreted similarly, he pointed to its tonal position: the oriscus element in the salicus falls chiefly on *f* and *c*, the wavy element of the quilisma by contrast falls chiefly on *E* and *b*; furthermore, the note preceding the wavy element in the quilisma is usually long or stressed. (See the comparisons published by Wiesli 1966, ch. 6; studies of the salicus in Laon 239 and St Gall 359 have been published by Picone 1977 and Ponchelet 1973, respectively.)

Without attempting to adjudicate among the different possible interpretations here, the most common characteristics associated with all the neumes involving the oriscus, and also the quilisma, are (i) frequent occurrence at the semitone step, (ii) special vocal delivery, (iii) Cardine's idea of rhythmic drive to the succeeding note. The signs are given in Table IV.3.1, which shows the St Gall form, the usual equivalents in Laon 239, and the form of transcription used in this book.

Wagner called all the liquescent neumes and those involving apostrophe, oriscus, quilisma, with the trigon as well, 'Hakenneumen' ('hook-neumes'), because in the St Gall sources their shapes nearly always include at least one small semicircle or hook

Table IV.3.1. *Special signs in St Gall and Laon notation*

	St Gall	Laon	Transcription
virga strata/gutturalis			
pressus minor			
pressus maior			
pes stratus			
pes quassus			
salicus			

(Wagner II, ch. 8). Thus the apostrophe (used principally St Gall, less frequently in other French–German notations) is used principally for repeated notes on *c* and other pitches at the semitone, in such chants as graduals. The trigon (three dots set out in triangular formation) was used for a figure such as *b–c–a* or *c–c–a*. These signs all derived, Wagner believed, from the apostrophe sign ('Haken' in German). For Wagner their frequent occurrence at the semitone step was crucial, and he believed that chromatic progressions, perhaps even involving quarter-tones, were indicated. While one must certainly reckon with the possibility that the matching of the plainchant repertory to the diatonic scale may have been imperfect, so that many melodies included non-diatonic progressions, the evidence of the plainchant sources themselves has so far been inconclusive, and the statements of medieval writers on music too insubstantial to support Wagner's theory.

Although the neumes discussed in this section have been called 'special' (sometimes 'ornamental'), they are fully integrated into the notation of the earliest sources. It is only from the modern point of view—they had generally fallen out of use by the end of the Middle Ages—that they are 'abnormal', because it is hard to match them to the unequivocal but restricted vocabulary of later notation.

IV.4. THE ORIGINS OF CHANT NOTATION

(i) *Introduction*

The origins of musical notation are obscure. Four main types of evidence may be considered. The first is the evidence of actual examples of musical notation. How old are they? and does anything about them suggest a previous stage of development? The second type of evidence is the information given about notation by medieval writers. The third lies in the existence of parallel or comparable systems of signs to guide vocal performance, which might possibly have suggested or led to the neumatic notation of Western chant: punctuation signs, oratorical accents, Byzantine ekphonetic notation, Byzantine chant notation. The fourth type of evidence is circumstantial. The earliest notation is a record of plainchant melodies. These melodies had previously been performed, learned, and transmitted without the aid of any written record (and thus they continued, to a considerable extent). In these circumstances, one would expect different singers to perform the melodies in different ways. And conversely, when a written transmission became normal, one would expect the differences to have diminished, or even disappeared. If we compare our medieval manuscripts, do we see evidence of the unity attributable to written transmission, or the variety produced by oral tradition?

(ii) *Early Examples*

The earliest examples of chant notation are the earliest musical notation of any kind in Western Europe. They are generally dated in the ninth century. Only isolated chants or small groups of pieces are notated. The reason for copying them is probably that they were unusual in some way; either (*a*) they were an unusual addition to the normal repertory, and thus required a special record to be made, or (*b*) their music was exceptional in comparison with what was normally expected of the singer. At first sight, therefore, it does not look as if it were a general custom to notate the chant repertory as a whole in the ninth century, although Levy (1987, 'Origin', 1987, 'Archetype') has argued that a notated exemplar for mass chants was made as early as around 800. The three earliest surviving books containing a year's cycle of chants date from around 900: Chartres, Bibliothèque Municipale 47 (from Brittany), Laon, Bibliothèque Municipale 239 (from Laon), and St Gall, Stiftsbibliothek 359 (from St Gall). Mirroring the diversity of the earlier examples, these three manuscripts also use different types of neumes.

The dating of early examples to the ninth century usually depends upon palaeographical evidence, and it is not surprising that different scholars have given different estimates. Various lists are available, none containing the same series of sources (see, for example, the twenty-one in Hiley 'Notation, III, 1', *NG* 346, or the eleven in Corbin 1977, 30 ff.). Further palaeographical study seems essential. The earliest datable specimen—at least according to one opinion—is a copy of the alleluia prosula *Psalle modulamina* in Munich, Bayerische Staatsbibliothek clm 9543. The book in question is a copy of St Ambrose's writings, which ends half-way down the final page of the manuscript. At the foot appears a scribal explicit, which tells that the book was copied by one Engyldeo, a cleric at the monastery of St Emmeram in Regensburg from 817 to 834. The blank space in between the main text and the explicit was used to copy *Psalle modulamina*. (For reproductions see MGG 9, 1625, Smits van Waesberghe 1957/1958, Bischoff 1981.) Bischoff (1974–80, i. 203–4) believed the same hand to have written both main text, prosula text, and neumes, though not all scholars have accepted this (Corbin 1977 omits it from her list of eleven; Stäblein 1975 is notably silent; but see the recent discussion in Möller 1990, 'Prosula'). The date of this one example is not so important, but the difficulties it presents are symptomatic. The general picture remains the same with or without it: sporadic activity through the ninth century, preceding the notation of complete books somewhere towards the end of the century.

The types of neumes represented among the early examples are those known as French, German, and Palaeofrankish. (It is also possible that Spanish examples may be dated to the ninth century: see Huglo 1985, 'Notation Wisigothique'.) Extending the time-limit to the middle of the tenth century, we can add Breton neumes (in Chartres 47), Laon (in Laon 239, and the slightly older flyleaves of Laon 107), and Aquitanian (in Paris, Bibliothèque Nationale, lat. 1154 and 1240, which have both been dated to the early tenth century. It is also noteworthy that Laon and St Gall 359 are not only among the earliest manuscripts with a comprehensive repertory, but also use the most complex and sophisticated systems known. It seems common sense to suggest that they were preceded by less sophisticated codifications of the repertory. Yet the effort does not seem to have been centralized, otherwise the types of neumes would not be so different. Different centres must have had sufficient independence to develop their own styles.

Much has been made of supposed difference of principle underlying the different types of signs. For example, accent-neumes have been held to constitute a different species from point-neumes. Accent-neumes indicate several notes with a single stroke of the pen, as for example the French–German torculus or porrectus. French–German and related types are supposed to belong to this family. Point-neumes tend to indicate each separate note by a separate mark; Aquitanian notation is the obvious example of this type, and Breton and Laon notation have been grouped with it. Yet the distinction is by no means watertight. A glance at Table IV.2.2, with signs of the so-called point-neume notations, shows plenty of signs of the accent-type, for example for the torculus and porrectus again. Palaeofrankish notation, in which a single stroke can represent two notes, has been felt to be qualitatively different to such an extent that Levy, for example (1987, 'Origin'), has seen it as the notation for a first written archetype of chant in the Carolingian period, later superseded by one in another notation. The evidence for this distant period nevertheless remains tantalizingly scanty, and attempts to construct hiererchies and 'family trees' of notations seem likely to remain hypothetical.

(iii) *Early References to Notation*

We have many accounts of the pains taken by the Frankish rulers of the eighth and ninth centuries to make Roman liturgy and its chant normative in their rapidly broadening empire. To this end it would obviously have been useful if the Franks could have obtained from Rome written copies of the music which had to be sung in their churches. But were they able to acquire such copies? It has been suggested that they were available as early as 747, for in the records of the Synod of Cloveshoe (somewhere in Britain) reference is made to various aspects of worship, including chant, which are to be performed according to a written Roman exemplar (see above, III.4). From the context it is not, however, certain that the reference is to notated books.

There is a similar difficulty in interpreting the instruction in Charlemagne's *Admonitio generalis* of 789 that boys should be taught, among other things, *nota*— 'signs', but of what sort? (MGH *Capitularia regum francorum,* i. 60). The word is found qualified by a gloss (121) which connects it with the *notarius*, the secretary, rather than the writer of music, which implies simply that the boys should be taught how to write. Nor do we find clear references to musical notation in the *Institutiones cánonicorum* resulting from the Councils of Aachen of 816 and 817, and these do include sections on the duties of the cantor (MGH *Concilia aevi Karolini*, i, Pars 2, 414).

From the ninth century come our first Frankish treatises on music, and for the first time there are distinct references to musical notation. Indeed, by the end of the century, when Hucbald of Saint-Amand wrote his *De harmonica institutione*, different types of neumes were clearly in use (GS i. 117, Babb, 36–7). Back in the middle of the century, Aurelian of Réôme's treatise *Musica disciplina* clearly envisages the presence of written musical signs: 'Plagis proti melodia . . . habet notarum formas . . .' (CSM 21, xix. 34, trans. Ponte 1968, 48) and 'haec consistit figura notarum . . .' (xix. 42, Ponte 1968, 49).

It has been suggested that some of Aurelian's references to acute and circumflex accents also refer to written signs, but this is not explicitly stated: the passages read more easily if we assume that the vocal inflexion, but not a written representation of it, is meant (xix. 10–11, Ponte 1968, 47).

The evidence of documentary references, therefore, seems to indicate beginnings in the first half of the ninth century, which is the same period as suggested by the surviving neumes themselves.

(iv) *Parallel Systems: Oratorical Accents, Punctuation, Ekphonetic Notation*

Three parallel systems of signs which indicate features of vocal delivery have been compared with neumatic notation and put forward as its predecessor. These are the prosodic accents of classical Greek literature, the punctuation signs used in copying texts of the early Middle Ages, and the ekphonetic signs used to indicate the delivery of liturgical lessons.

(a) Prosodic Accents

Thompson 1912, 61–4; Laum 1928; Schwyzer 1939.

'De accentibus toni oritur nota quae dicitur neuma' ('It is from the prosodic accents that there comes a sign called a neume') states the anonymous author of the treatise *Quid est cantus?* in the eleventh-century Vatican manuscript Pal. lat. 235. The accent-theory of the origins of Western chant notation was put forward by Coussemaker (1852, 154); it gained wide currency through the explanation in Pothier's *Les Mélodies grégoriennes* (1880), and then in Suñol's handbook of chant notation (1935). Not all modern authors have accepted it, however. The main problem has been that it is difficult to detect any continuous use of prosodic accents in Western literary manuscripts up to and including the time when music was first notated in the ninth century.

Prosodic accent-signs were used by the Greeks from early times, but more especially from the third century BC, when Greek gradually became an international language of learning throughout the Near East, and non-natives needed guides on the correct pronunciation of written texts. Aristophanes of Byzantium and his pupil Aristarchos developed the system in Alexandria. In the second century AD Herodian led a resurgence of interest in grammatical studies, including accents, and they are common in second-century papyri, though not always uniform in type. Accents of the type still used today are especially prominent in minuscule hands of the eighth century onward.

The acute accent (*oxeia*) signified that a syllable should be pronounced with a higher pitch. The grave accent (*bareia*) signified a lower pitching of the syllable; it is much less common, being generally reserved for syllables which would normally be acute, but for some reason (for example, not ending a clause) should be lower, or at 'normal' pitch. The circumflex accent (*perispōmenē*) signified a higher pitched syllable falling lower.

The signs were divided into groups: the tonoi (pitch-signs), chronoi (duration), pneumata (breathing), and pathe (mutations). Starting a word with a 'rough breathing', that is an H-sound, was signified by the *daseia*; absence thereof, 'soft breathing', by the *psile*. Long and short were the *makra* and *bracheia*, respectively. The *hyphen* linked words, denoting elision. The *diastole* separated them. The *apostrophos* indicated a short break. The prosodic accents are shown in Table IV.4.1, together with their Latin names.

Such prosodic signs are very rare in Latin sources. Almost the only one found is an acute over an exclamatory monosyllable such as *o*. On the other hand, many Latin grammarians (in this as in many other ways heavily indebted to Greek models) include brief accounts of prosodic signs. They were certainly known in Carolingian

Table IV.4.1. *Prosodic accents*

Greek		Accent	Latin	
tonoi			*toni*	
prosodeia	oxeia	⊂	accentus	acutus
"	bareia	⌒	"	gravis
"	perispōmenē	⸍	"	circumflexus
	(oxybareia)			(later ∼)
chronoi			*tempora*	
"	makra	⊢	"	longus
"	bracheia	⊣	"	brevis
pneumata			*spiritus*	
"	daseia	⌐	"	asper (later ㄴ c)
"	psilē	—	"	lenis (later ⌐ ⊃)
pathe			*variationes*	
apostrophos		∪ ∨	apostrophus	
hyphen		⟍	conjunctio	
diastolē (hypodiastolē)		/	separatio	

times, for there exists an eighth-century treatise *De accentibus* falsely attributed to Priscian (Keil, lii. 517–28); and among writings attributed to Alcuin is a grammar in dialogue form, which, drawing on Priscian, mentions the three *toni* and two *spiritus*. The question arises as to whether they reflected or influenced the actual pronunciation practised by Latin speakers at various periods. At least they might have contributed to an emerging practice of notating music.

(b) Punctuation

Early medieval punctuation systems have also been cited as possible catalysts for the new musical notation. Either the signs could have been taken over literally, with modifications of shape or significance, or the general principle of marking syntactic units with a sign might also have been of importance. Chant singers made wide use of well-known melodies which could be adapted to many different texts. Punctuation might therefore guide the application of standard melodic phrases to units of text. (See Bischoff 1986, 224–5.)

Grammarians of antiquity up to Isidore of Seville used a point, placed low, in the middle, or high, to indicate pauses of different length:

comma (short pause)
colon (medium pause)
periodus (end of a sentence)

In Carolingian times two systems were prevalent:

	usual Carolingian	later Carolingian ('Isidorian')
short pause		
long pause		

A variety of signs for the question mark were known (see Bischoff 1986, 197), some of which bear a distinct resemblance to one of the several quilisma signs:

Treitler (1982, 'Early'; 1984) has tried hard to see notation from this angle. He has started from the fact that the notation in early chant-books does not record pitches, and must therefore have a different function, one of guiding the articulation and delivery of the text. The signs used to build up a music notation should therefore not be sought among prosodic accents, which were concerned with pitch, but among punctuation signs, which concern the articulation of the texts into its syntactic units. And among other things he has drawn attention to the similarity of the forms of question mark found in late eighth-century manuscripts and forms of the quilisma in the next century. Both indicate a rising vocal gesture.

(c) Ekphonetic Notation

A special type of punctuation was used to guide the intonation of liturgical texts, lessons in particular. The Christian world knew many different systems. The Syrian church had a system employing dots even before the schism of the sixth century. The Byzantine system can be traced back at least to the ninth century, and it then gave rise to the Slavonic, Georgian, and Armenian traditions. Hebrew Bibles also use a system traditionally supposed to have been invented in the ninth century.

That this is something different in degree from simple punctuation may be seen in Byzantine examples (Wellesz 1961, 249 ff., pls. 1–2; Thibaut 1913, 39, 41); see Table IV.4.2. Although the exact meaning of the individual signs is not always clear, the way they function is not in doubt: in general they give an indication of the way each phrase should be delivered, usually by placing one sign at the beginning of the phrase and another at the end. They were usually positioned beneath the appropriate syllables, and might be written in red ink, thus being distinct in both ways from prosodic accents. The signs do not merely mark off the text in syntactic units, they carry a range of pitch signification well beyond that of the punctuation signs. The degree to which they were developed from prosodic accents is debatable, and it is probable that the prosodic accents were themselves known only academically at the time the ekphonetic system was developed, that is, they no longer in practice corresponded to the normal manner of declaiming Greek texts. This would not, however, have prevented their utilization in a newly developing system of lection notation.

Table IV.4.2. *Byzantine ekphonetic signs* (after Wellesz, 1961, 252)

Oxeia (diplai)	/ //	the voice rises to a higher pitch and stays there until the second Oxeia
Kremastē	✓	rise of the voice with a slight emphasis
Bareia (diplai)	\ \\	fall of the voice with a certain emphasis
Apostrophos	ʼ	lower pitch without emphasis
Apesō exō	ʼ .... ✓	lower . . . higher
Kentēmata	∴	rise
Synemba	◡	joining two words smoothly
Hypokrisis	.... ʼʼ ʼʼʼ	separation, shorter or longer
Teleia	.... +	pause
Syrmatikē	~	undulating movement
Paraklitikē	γ	in a beseeching manner
Kathistē	ᴧ	normal starting tone

There seems little doubt that Byzantine ekphonetic and Byzantine neumatic notation are related to each other, for some of the signs are identical. The oldest stages of Byzantine neumatic notation that can be recovered, however, the so-called Palaeobyzantine types from the tenth century onward, are to be distinguished from ekphonetic notation in that each sign (or group) refers to one syllable, whereas the basis of ekphonetic notation is the phrase. Yet this does not mean that in neumatic notation every syllable has its melodic sign: that stage was not reached until the eleventh century, and in older sources many syllables are left without a sign. Another distinction is that neumatic notation was entered over the text.

It does not seem possible, therefore, to speak of a steady development from accents to ekphonetic notation to neumatic notation, despite the signs held in common. It seems safe to speak of three different systems which shared some ideas and graphic shapes (see Table IV.4.3). The principal reason why they are not more closely related is that they served different purposes. Ekphonetic aids were needed to specify a range of intonation formulas not discernible simply from the normal punctuation of the text. Neumatic notation was needed to suggest much more varied melodic outlines, where each syllable might be delivered with an individual melodic gesture.

The role of ekphonetic notations in the West is still somewhat obscure. Ekphonetic signs of many different types, that is, signs which seem to affect the delivery of complete phrases, rather than separate syllables, may indeed be found in Western books, but they seem to be no earlier than neumes, nor to adhere to any unified tradition or traditions. What is more, no one manuscript seems to use more than eight or nine signs, so that in scope it is something like an extended punctuation system. Indeed, for the experienced singer of a clearly punctuated manuscript, such extra

Table IV.4.3. *Accents, ekphonetic signs, and neumes (Byzantine)*

Prosodic accents		Ekphonetic signs		Palaeobyzantine neumes	
Oxeia	/	Oxeia	⌐	Oxeia	/
		Kremaste	✓	Petaste	✓
		Kentemata	∴	Kentemata*	••
Bareia	\	Bareia	\	Bareia	\
		Apostrophos	ɔ	Apostrophos	ɔ
Oxybareia	∧ ∩				
		Apeso exo	ɔ ..... ✓	Apeso exo*	ɔ✓
Makra	—				
Bracheia	◡				
		Kathiste	∿		
		Syrmatike	~	Syrma*	∿
		Paraklitike	Ⅴ	Paraklitike*	∿ ε
Daseia	⊢ ⌐ ⊂				
Psile	⊣ ⌐ ɔ				
Apostrophos	'	Hypokrisis	₃ ³	Katabasma*	} }
		Teleia	+	Stavros*	+
Hyphen	◡	Synemba	◡	Klasma*	◡
Diastole	)				

*not recognized by all experts

signs were superfluous. There seems little evidence, therefore, to suggest that Western neumatic notation might have used ekphonetic signs as a springboard for the development of its system. (For examples see Suñol 1935, 181; Jammers 1965, Tafeln, 105; Ekberg, 'Ekphonetic Notation', *NG* 102; Rankin in Fenlon 1982, 90–4.)

(d) Byzantine notation

From the previous section it will have become clear that a Western adoption or imitation of Byzantine chant notation is very unlikely to have occurred. The idea has nevertheless had its supporters. The Greek names of many neumes, the similarity of some shapes, even if they signified something different, have exercised a considerable fascination. Thibaut, whose studies of Byzantine ekphonetic notation have just been referred to, believed that Western chant notation derived from the Byzantine ekphonetic signs. More recently, Floros (1970) has attempted to reconstruct the earliest Byzantine chant notation and then show that in this early form it was adopted in the West. The evidence on which the theory rests appears shaky, however (see Haas 1975).

(v) *The 'Cheironomic' Theory*

Before the final body of evidence which might illuminate the early history of chant notation is considered, mention should be made of another theory: that early neumes reflect the manual gestures of the cantor as he directs his choir in performance; they are cheironomic. The idea appears to have originated with Mocquereau (PalMus 11, 96), while discussion of the practicalities of chant-conducting in the Middle Ages began with Kienle (1885; see also Gindele 1951). Nearly all writers on chant notation have at least paid lip-service to the idea (Wagner II, 17; Jammers 1965, Tafeln, 23; Stäblein 1975, 28), and it forms part of, for example, the theory of neumatic separation elaborated by Cardine (discussed in the next section). A variety of evidence of cheironomic systems was reviewed by Huglo (1963, 'Chironomie'), and most recently Hucke (1979) has subjected the whole matter to stringent criticism.

Cheironomic systems as commonly understood actually specify the melodic intervals to be sung, and, as Hucke points out, there is no evidence that Western chant notation was linked to any method of this sort. Even literary references to cantors conducting with hand movements are extremely sparse. The theory that chant notation—or at least the St Gall or Laon neumes usually invoked in this context— depicts cantorial hand gestures has therefore no concrete support either from what is known of cheironomy in other contexts (the music of other cultures, modern Western systems such as the Kodaly method) or from contemporary documentation. This does not mean to say that it is erroneous, only that it cannot be proven. Practical experiment suggests that it is possible to reproduce, say, St Gall neumes as hand gestures in conducting a choir; so the reverse is theoretically possible: that conducting gestures were reproduced as written signs. As a historical explanation it lacks all foundation, however, and scholarly caution demands that it be treated, at best, as a picturesque analogy.

(vi) *The Early Transmission of Chant*

So far I have suggested that (*a*) the earliest specimens of chant notation date back to the early ninth century; (*b*) the earliest unequivocal references to notation come from Aurelian of Réôme, writing around 850 (the complicated composition and transmission of Aurelian's treatise make caution advisable); (*c*) some of the materials which might have helped in the development of a basically newfangled notation in the West were available in the Carolingian period: the accents of classical prosody and punctuation signs, with a faint possibility that Byzantine ekphonetic signs or even the beginnings of Byzantine musical notation might have existed to spur the imagination of the Franks.

Before notation was employed to codify chant melodies, and provide an aid to learning and preservation, the music was performed from memory. This state of affairs can actually be deduced from the style of the music itself. A large part of the chant repertory consists of melodies which have been adapted to fit different texts, or which use stock ways of beginning and cadencing, and share common ways of delivering texts of a particular type. It might have been expected from this that singers would have had some room for manœuvre in the performance of such chants. So that for the performance of the tract *Jubilate deo* (Ex. II.5.4), for example, different manuscripts—representing the decisions of different cantors about how this text should be performed—would employ different formulas, at least occasionally, so that there would be a certain amount of interchange between G1, G2, and so on, as the standard phrases of Ex. II.5.4 are labelled. But this is not so. All sources draw upon the same traditional phrases at the same time, follow the same procedure when the length of the text is different, and at first sight, seem much more like copies of the same exemplar than different rememberings and reproductions.

And yet differences, small but persistent, are to be found. They concern liquescent notes, as we might expect, and the use of such signs as the quilisma and oriscus, how to deal with notes at the semitone step (for example: a pes *bc* or two virgae *cc*?), and so on. In other words, they concern surface detail, not basic structure.

On the other hand, some other types of chant show bigger differences between sources, for example, many antiphons (see those transcribed by Udovich 1980 and Fickett 1983).

This reinforces the impression that we have to do not with a repertory transmitted in writing, but one remembered and later codified differently in different places. For some chants the learning process had been so thorough and exact, the procedures for remembering so diligently rehearsed, that hardly any difference can be seen between sources. For other chants the remembering was not so uniform. (For much valuable comparative material, with discussion, see Van der Werf 1983.)

(vii) *Some Conclusions*

When trying to decide when the chant repertory was first systematically codified with musical notation, one should bear in mind that early manuscripts were not 'performing scores' in the modern sense, but works for study and reference. There must always have been a greater or lesser discrepancy between the music sung in church and the master exemplar in the song-school, at least in soloists' music.

If our earliest manuscripts are independent of each other—containing different readings, as well as using a different notation—then they do not suggest a date for the beginnings of notation any earlier than their actual date of copying. For their particular churches they might well be the first such books ever made. Whatever the previous uses of notation, it does not seem to have been used for a whole gradual before the late ninth century (and not until a century later for office chants). We might envisage a period when the various notation-types were becoming established, in the ninth century, and were then finally employed for a major codifying project, the whole mass repertory.

If, as Levy believes, a standard exemplar existed earlier, perhaps compiled at Charlemagne's instigation, then later cantors and scribes must have felt free to copy it (*a*) using their own locally developed notations and (*b*) incorporating their own 'interpretation' of details such as those mentioned above. In other words, they would be working partly from memory, partly from an exemplar. This way of making chant-books might well have been one of the commonest throughout the Middle Ages, for an exactly analogous situation arose in the eleventh and twelfth centuries when churches all over Europe wanted to make new records of their repertory in staff-notation.

The essence of early chant notation must lie, as Treitler has indicated, in its ability to link the appropriate vocal gesture to the text being delivered. This is easiest to envisage in the case of highly standardized melodic units. Treitler uses responsory verses to illustrate the technique (Treitler 1984, 172, reproduced from Rankin 1984). The examples of psalm verses from introits provided by Hucke (1985) can be read as a parallel demonstration. In Ex. IV.4.1 are the openings of some mode 3 introits already cited above (II.11: compare Ex. II.11.1). I transcribe the versions in the *Graduale triplex*.

Ex. IV.4.1. Opening of mode 3 introits (St Gall notation)

We could imagine a scenario something like the following. The cantor knows that a mode 3 melody is traditionally sung for the following texts: *Repleatur os meum* (*GT* 246), *Ego autem sicut oliva* (*GT* 424), *Vocem iocunditatis* (*GT* 229), and *Benedicite dominum* (*GT* 607). He will be aiming for *c* as the reciting note in the first phrase, usually achieved directly from *G*, or with an ornamental gesture around *G* and *a*, and, if the text warrants it, some preliminary movement around *D* and *E*. That is, he will know a number of stereotyped openings appropriate for introits in this mode. The neumatic notation reminds him which opening is best for each particular text. *Benedicite dominum* is the simplest, for only one upward movement is indicated. *Repleatur os meum* has three ascending gestures, which must be for *EFE*, *DG*, and *Gc* respectively. The ornamental flourish on 'au[tem]' in *Ego eutem* is instantly recognizable, and the more leisurely progress of *Vocem iucunditatis* is also, in its way, quite clear, for it is bound to start with *EF*, *DG*, the arrival at *c* is unmistakable, and in between *E*, *G* and *a* are almost certainly the only notes available. The notation therefore guides the controlled delivery of the text towards the first structural crux. (The length of its stay there is also clear in this notation from the apostrophes, associated in this mode principally with *c*.) It does not specify pitches: the matrix from which they will be drawn is already known. It specifies what words belong to the opening gesture, and how the syllables are disposed within it, its function therefore resembling that of punctuation more than that of modern pitch notation. But it is very highly developed type of punctuation system, one which indicates not simply the main syntactic divisions in the text but the shaping of much detail within those divisions. And, if some signs have been suggested by punctuation signs, others may well have come from oratorical accents, or simply from a sense of the contour of a melody (Duchez 1979, 'Représentation').

Once the decision had been taken to record each and every movement of the melody, the latter aspect of chant notation became predominant. The fact that no single point of departure leads to a complete repertory of signs for chant notation, and that different centres employed different styles of writing, suggests a period of vigorous but controlled, new initiative. In this context it seems unnecessary to look outside the Carolingian realm for a pre-existing system.

We might see the early history of notation in three principal stages. Adiastematic notation is best adapted to represent the standard idioms and formulas of the earliest 'Gregorian' layer of music. Its use became most urgent precisely when the old musical types were increasingly being challenged by newer, non-formulaic, compositions. For these, pitch-notation was eventually indispensable. The third stage, the specification of rhythm, was not needed until the music for two or more voices both moving in regular rhythm had to be coordinated; this stage concerns not plainchant but the Parisian polyphony of the twelfth to thirteenth century.

IV.5. THE NOTATION OF RHYTHM

    (i) Rhythmic Elements in Early Notations
    (ii) Rhythm in Simple Antiphons
    (iii) Cardine's 'Gregorian Semiology'
    (iv) The Evidence of Theorists
    (v) Conclusions

The notation of several early chant manuscripts includes indications of rhythmic differentiation between individual notes or groups of notes. Richest in such indications are the early sources from Switzerland and south Germany (such as St Gall, Stiftsbibliothek 359 and 339, from St Gall, Einsiedeln, Stiftsbibliothek 121, from Einsiedeln, and Bamberg, Staatsbibliothek, lit. 6 from Regensburg) and Laon, Bibliothèque Municipale 239, but very many other manuscripts indicate at least some degree of rhythmic differentiation. There are several ways in which this was done: (*a*) the normal shape of a sign was altered in such a way as to suggest a rhythmic alteration, (*b*) in the Eastern manuscripts, extra elements were added to the normal signs, usually the short bar known as the episema, (*c*) the way in which several notes were comprehended in one sign was altered, separating what might have been joined and joining what might have been separate, (*d*) so-called 'significative letters' were placed adjacent to the sign.

The medieval tables of musical signs are later than the early sources just mentioned, and it is not surprising that they provide no explanation of durational significance. Only the Italian tables in Montecassino, Archivio della Badia 318 (Coussemaker 1852, pl. XXXVII; Ferretti 1929, 193, pl. II) and Florence, Biblioteca Magliabecchiana F. 3. 565 (Ferretti, 195) indicate a differentiation: thus in Montecassino 318 we find *percussionalis brevis* (a dot or punctum) and *percussionalis longa* (a dash or tractulus).

(i) *Rhythmic Elements in Early Notations*

The meaning of the significative letters, at least as they were understood at St Gall, is given in a letter attributed to Notker (see PalMus 4, 10 and Pl. 5. B-D; also Suñol 1935, 134; *NG* 13, 132; critical edition by Froger 1962). St Gall tradition ascribed their invention to the Roman cantor 'Romanus', who was supposed to have brought chant-books from Rome to St Gall in the eighth century. Hence the name 'Romanian letters', coined by Schubiger (1858). But we have no contemporary explanation of the other features mentioned above. The only way in which we can approximately understand the notation is by careful comparison of the sources among themselves.

Significative letters can be found in many manuscripts up to the early eleventh century from different areas of Europe, in Breton and Aquitanian sources and the early Winchester manuscripts, for example. But they are commonest in the abovementioned Eastern and Laon manuscripts. Smits van Waesberghe's special study of them (1938–42) reported that Einsiedeln 121 contained the staggering total of 32,378 significative letters. Notker's letter ascribed a meaning to almost every letter of the alphabet, but several cannot be traced in extant manuscripts, or are very rare. Preferences naturally differ from one area to another, and we have no Notker to explain the system in, say, Laon 239. Not all the letters refer to rhythmic properties of performance, some concerning dynamic or melodic features. The commonest in the Eastern sources are the following:

dynamic: *f* = *cum fragore seu frendore*: with harsh attack
           k = *klenche* (Gk.?) or *clange*: with ringing tone
melodic: *a* = *altius*: higher in pitch
           e or *eq* = *equaliter*: at the same pitch
           i, io, or *iu* = *inferius, iosum,* or *iusum*: lower in pitch
           iv = *inferius valde*: much lower

> $l = levate$: rise to a higher pitch
> $s = sursum$: ascend to a higher pitch
rhythmic: $c = cito$ or *celeriter*: quickly
> $t = trahete$ or *tenete*: drag, hold
> $x = expectate$: wait

The preferences of Laon 239 are for the following:

melodic: $eq = equaliter$: at the same pitch
> $h = humiliter$: at a low pitch
> $s = sursum$ = ascend to a higher pitch
rhythmic: $a = augete$: lengthen
> $c = cito$ or *celeriter*: quickly
> n or nl or $nt = naturaliter$: normal
> $t = trahete$ or *tenete*: drag, hold

In both traditions, m or $md = mediocriter$: moderately, can qualify either a melodic or a rhythmic indication, or even appear on its own, when its meaning has to be deduced from the context.

(More complete lists of letters may be found in studies of the individual notations, and in Suñol 1935, 134–8, 141, 143–4, etc.)

Among the letters indicating melodic features, e is particularly important, for it specifies the relationship between two disjunct note-groups. The others usually (though by no means exclusively) serve to warn of intervals larger than a tone (or semitone). Ex. IV.5.1. shows a typical use of the letters e, i, m, and s (medieval s looks like an elongated r).

Ex. IV.5.1. From offertory *Benedictus es Domine* (GT 277, notation of Einsiedeln 121)

in   la - bi - is   me - is   pronun-ti - a-   -ui

Rhythmic differentiation depends almost entirely on c and t. They are the key to understanding the notational signs that have been altered from their usual form or given an episema, for a fairly consistent correspondence (though it is by no means complete) can be seen between the use of, on the one hand, the letter c and the normal shapes for signs, and, on the other hand, the letter t and the altered or supplemented shapes. Typical normal and altered shapes in St Gall notation are given in Ex. IV.5.2, together with two passages from *Alleluia Pascha nostrum*, with the signs in St Gall 359 and 339 and in Einsiedeln 121.

The letter c is always used in conjunction with the normal signs, whereas the letter t is usually allied to an episema (in the same or one of the other sources). Striking confirmation of the sense of the episemas comes from the notation of Laon 239. This

Ex. IV.5.2. Normal and modified signs in St Gall notation (selection)

	normal	modified
clivis	∧	
pes	◡	
porrectus	𝒩	
torculus	∿	
climacus	/..	
scandicus	..✓	

from Alleluia Pascha nostrum (notation of St.Gall 359, 339 and Einsiedeln 121)

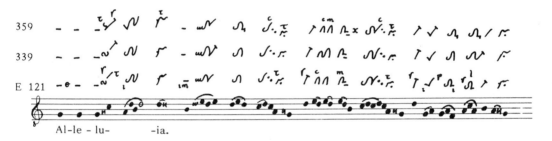

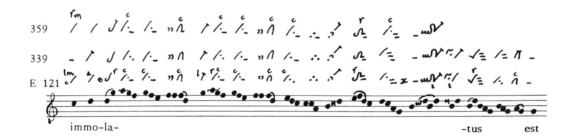

source does not have episemas, but makes much more frequent use of the letters *a* and *t*, both meaning a lengthening of some sort. Ex. IV.5.3 gives first the 'normal' and the 'longer' signs in Laon 239, then the same extracts from *Alleluia Pascha nostrum*.

One important feature of the second part of Ex. IV.5.2 in particular is that the notes are grouped in exactly the same way—single notes, twos, threes, and so on—in all the sources. The phrasing of the melody was obviously important enough to have been transmitted with a fair degree of uniformity. In a strikingly large number of

Ex. IV.5.3. Normal and 'longer' signs in Laon notation (selection)

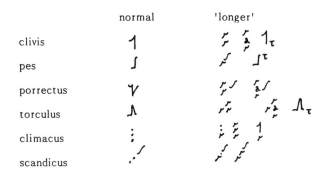

from <u>Alleluia Pascha nostrum</u> (notation of Laon 239)

instances, the group ends with a longer note. The rhythmic significance of the point of separation between the groups has been studied in depth by Cardine (his term for it was 'coupure neumatique').

It is enlightening to observe the consistency with which one of the standard phrases of the repertory is reproduced in these early sources. Ex. IV.5.4 gives a standard phrase from the graduals of mode 5 (phrase F1 in Apel's analysis) with the neumes of Laon 239 and St Gall 359. The notation is given complete for the phrase in *Iustus non conturbabitur*, whereas for five other graduals only deviations are recorded. (Three of them are not present in Laon 239.) The lack of disagreement is obvious.

The question immediately arises, how slow is slow (*tenete* or *augete*) and how fast is fast (*cito* or *celeriter*)? There is unfortunately no clear answer. And with how many gradations of duration do we have to reckon (however approximately)?—two (since the signs on their own, without letters, can be roughly divided only into 'normal' and 'longer'), or perhaps more? Do the significative letters simply warn about something

Ex. IV.5.4. A standard phrase from mode 5 graduals (Apel Fl) with Laon and St Gall notation

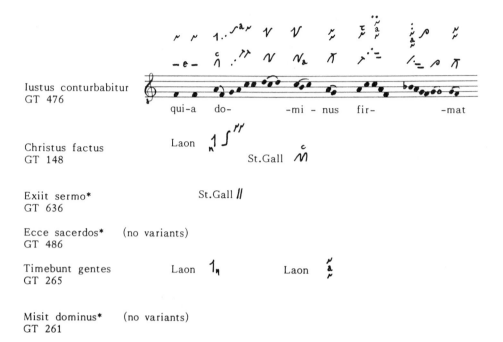

Iustus conturbabitur
GT 476

qui-a   do-   -mi - nus   fir-   -mat

Christus factus   Laon
GT 148   St.Gall

Exiit sermo*   St.Gall //
GT 636

Ecce sacerdos*   (no variants)
GT 486

Timebunt gentes   Laon   Laon
GT 265

Misit dominus*   (no variants)
GT 261

in the notation anyway, or do they mark an extra degree of slowness or fastness? What shades of difference divide the signs given in Ex. IV.5.5?

It is noticeable that in the Eastern sources *c* is used far more with the clivis than with the pes, whereas in Laon *c* appears only with the pes.

Ex. IV.5.5. Possible gradations of duration in pes and clivis (Laon and St Gall notations)

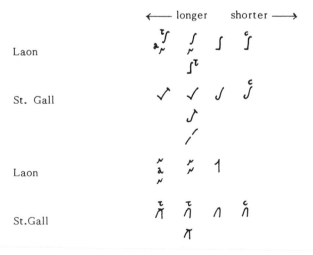

← longer   shorter →

Laon

St. Gall

Laon

St.Gall

Some writers have argued that the shorter and longer notes stand in a strict mensural relationship to one another, so that Murray (1963) transcribed the above passage from *All. Pascha nostrum* as shown in Ex. IV.5.6.

Ex. IV.5.6. Mensural transcription of passages from *Alleluia Pascha nostrum* (after Murray 1963)

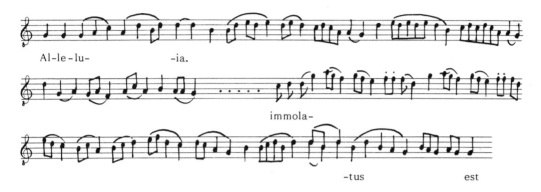

The Vatican gradual of 1908, produced under Pothier's leadership, did not reflect any of the rhythmic detail of the above-mentioned manuscripts, for Pothier believed it was only of local significance. The editions produced by Solesmes under Mocquereau's direction, however, included dots and horizontal bars, indicating lengthening, which derive from the early sources. In Solesmes practice, these are regarded as rhythmic nuances, not susceptible to rigid measurement. (The Solesmes interpretation under Mocquereau and Gajard, famous from many gramophone recordings, is explained in Mocquereau 1908–27 and Gajard 1951.) At another extreme, it is possible to regard the shorter notes as a type of ornamental figuration, to be performed like the ornaments in Baroque music, or, better, the roulades in some non-European musics. A precursor in some wise of this type of modern performance was Dechevrens, whose fanciful transcription of *All. Pascha nostrum* is shown in Ex. IV.5.7 (taken from Decheverens 1898, iii/2, 465).

There can be no doubt that rhythmic differentiation was an essential element in the practice of those choirs for whom the St Gall, Laon and other sources were written. The fact that the Laon source is widely separate geographically from the others suggests that this way of singing chant was quite widespread. How long it persisted is unclear.

Several comparative tables of signs from early sources have been published, displaying the range of characters used for each melodic progression. The most detailed tables for St Gall notation have been published by Cardine (1968 etc.), and for Laon 239 in the English translation of Cardine. In PalMus 11 (55–8) and Suñol (1935, 147–52) signs from St Gall, Laon 239, and Chartres, Bibliothèque Municipale 47 are presented.

Ex. IV.5.7. Mensural transcription of passages from *Alleluia Pascha nostrum* (after Dechevrens 1898, iii/2, 465)

Al-lelu - ia.

..... immo-la-

-tus   est .....

(ii) *Rhythm in Simple Antiphons*

Interesting approaches to the rhythm of office antiphons have been developed by Jammers (1937) and Lipphardt (1950). The most valuable part of their work has consisted in comparing antiphons of the same melodic family in order to see if they had a common rhythmic tradition, using the evidence of the notation in Hartker's antiphoner, St Gall 390–391 (PalMus II/1). This has helped establish conventions for the relationships between the signs for single notes and those for more than one note, at least in predominantly syllabic passages. Jammers provided transcriptions of (among other chants) antiphons from Gevaert's Themes 1–11, Lipphardt from Theme 29. Briefer demonstrations of the same principle can be found in Cardine's work (for example in *Sémiologie grégorienne*, ch. 3). Consider Ex. IV.5.8, which develops one of Lipphardt's examples.

   If we assume that the melody proceeds in basically equal notes, we can see that the angular pes at the beginning of phrase two equals two syllables with single notes. The virga with episema for '[sa]cra[ta]' likewise seems equivalent to two simple virgae. So also the clivis with episema at the end of the third phrase, which in turn suggests that the clivis with *c* in phrase two should be twice as fast. At the beginning of phrase three, it is particularly interesting to see how the same notes are adapted to two syllables (a 'slow' pes and 'slow' clivis), three syllables ('slow' pes and two virgae), four syllables (four simple notes) or even five syllables ('glori-' with a *c*, that is quickly). It is not difficult to make a transcription in 2/4 or 4/4 time of these antiphons (Lipphardt opts quixotically for 6/4), arranging stressed syllables to coincide with the beginnings of bars. One wonders if the melody had an inherent rhythm, so to speak, for which

Ex. IV.5.8. Antiphons with the same melody (notation of St Gall 390–391, PalMus II/1; pitches derived from Bamberg, Staatsbibl. lit. 25) these texts were deliberately composed. Or were there traditional ways of 'rhythmiciz-ing' texts to bring them into line with this and other melodic schemes? It has to be said that this group of antiphons displays a more regular structure than many others, and attempts to find similar rhythmic schemes elsewhere often appear improbable. Furthermore, different performance conventions may have existed for more ornate chants.

Be that as it may, Cardine developed a theory of syllabic equivalence, relating the length of notes to the normal delivery of a syllable with a single note. This time-unit is the 'syllabic beat' ('temps syllabique'). Such a note is normally notated in St Gall sources as an ordinary virga or tractulus. It can be lengthened with an episema or shortened by the letter 'c'. The angular pes and the clivis with episema are then equivalent to two syllabic beats, whereas the normal pes and clivis will represent shorter durations.

(iii) *Cardine's 'Gregorian Semiology'*

It will by now have become clear how much can be learned from the careful comparison of manuscripts with one another, and of similar musical passages in the same manuscript. Particularly comprehensive and painstaking work of this kind has been accomplished by Cardine and students at the Pontificio Istituto di Musica Sacra in Rome. (There are summaries of these studies by Albarosa 1974, 1977, 1983, and lists in the Cardine Festschrift 1980 and Albarosa 1983. Many have been published in the journals *Études grégoriennes* and *Beiträge zur Gregorianik*. Cardine summed up his own work in the book *Semiologia gregoriana*, 1968, which subsequently appeared in French and English versions, 1970 and 1982 respectively).

Cardine dubbed these studies 'semiology', in the general sense of 'study of signs'. (The term is not usually used of the mechanics of writing-systems but of modes of human communication, particularly in the area of linguistics, and also recently in the analysis of music.) Most of the studies have taken a single sign and explored its use throughout one or more of the early manuscripts referred to above. The main emphasis of the work has been on the rhythmic weight of the notes represented by the signs—light or heavy, part of a fluid progression or marking a pause in melodic movement—and their articulation, with special reference to the grouping and separation of notes. It has shown that the early scribes displayed considerable finesse in adapting signs to reflect nuances of performance.

To summarize all the findings to date of this type of research is naturally impossible here, but some idea of the insights gained may be seen through one example, Ponchelet's work (1973) on the use of the salicus in St Gall 359 (the book is a cantatorium; since the salicus is not used in tracts, this means in effect that the investigation concerns graduals and alleluias).

If followed by other notes in a large group, the salicus has an episema on the final virga 104 out of 120 times; but when there is a change of syllable after the salicus the episema is added only six out of 239 times. (All these instances are displayed in tables, arranged according the melodic context.) This seems to suggest that the last note is always prolonged, for a change of syllable will naturally cause at least a slight delay on the note. For the salicus followed by further notes in a large group, further analysis distinguishes between cases where it is followed by a higher note, a note at the same pitch, or a lower note. Cases are traced where the same melodic phrase is found over one syllable or split between two or more syllables (for example in the cadence formula for the verses of the 'Iustus ut palma' group of mode 2 graduals). Standard phrases of this sort naturally provide prime material for investigation.

The results do not, of course, necessarily bear on other manuscripts. As Ponchelet points out, St Gall 339 is far more sparing of the episema on the last note of the salicus. Does this indicate a change of performance practice at St Gall? How widespread, then, was the type of performance indicated in St Gall 359? St Gall 359 stands near the beginning of the written tradition at St Gall, perhaps near the beginning of the singing tradition; one might surmise that later scribes felt less need to include episemas, the performance tradition being by then more firmly established. The episema in the later manuscripts would by this reasoning carry greater weight. Comparisons with other traditions, those of Laon 239 and Chartres 47, are illuminating. In cases where the salicus in St Gall 359 is followed by other notes, these sources have a salicus on only about one in eight occasions; where the salicus precedes a change of syllable, the proportion is about half. When not notating a salicus, Laon 239 usually has a scandicus with *t* by the final note, while Chartres 47 has a simple scandicus.

Ponchelet can therefore claim that in the salicus in these early sources rhythmic weight is always directed on to the final note. For Chartres 47 one has to assume that in many cases the weight is felt simply because it is the final note of the three, even in the absence of any graphic emphasis (the oriscus element, or a significative letter). The nature of the sign is investigated purely in rhythmic terms, as is also the case with the quilisma in the studies by Cardine (1968) and Wiesli (1966). This means inevitably that the distinction between all the signs which comprise three notes in ascending order, scandicus, salicus, and quilisma, is a very fine one. If the oriscus element in the salicus (or the 'trembling' element in the quilisma) has any other meaning—say, purely hypothetically, the non-diatonic pitch suggested by Wagner— then this type of study will not reveal it, for its principal tools of investigation are purely rhythmic: episema, significative letters, articulation into note-groups. It has to be said that if the oriscus element really does represent some extraordinary element of performance, its relatively rare occurrence in Laon 239 and Chartres 47, in the contexts investigated, is more difficult to explain.

The early editions published at Solesmes of music-books for the modern liturgy— the *Liber usualis* is still the best known—included numerous dots, bars, and other signs, inspired by the notation of the early sources. The work of the 'semiologists' has resulted in a new sensitivity to these matters, and it is common for modern performers to use the *Graduale triplex*, where the notation of Laon 239 and one of the early Eastern sources is copied into the Solesmes edition of the *Graduale Romanum* of

1974. Manuals of the new type of chant performance have been produced by Agustoni (1963) and by Agustoni and Göschl (1987).

(iv) *The Evidence of Theorists*

Early medieval writings on chant help but little in understanding how note-lengths were differentiated. One group of related writings, which was discussed at length by Vollaerts (1960), among others, has little to do with the matter, for it refers to the final notes of chants, of their phrases and sub-phrases. As Bower has explained (1989, 'Model'), this is done by analogy with the grammatical structure of the text being sung: the musical delivery should reflect the grammatical structure, by making the hierarchy of text-units clear. This group of writings includes the *Scolica enchiriadis* (ed. Hans Schmid 1981, pars 1, 86–7), Guido of Arezzo's *Micrologus* (ch. 15), and the commentaries on the latter by Aribo (CSM 2. 48–50, 65–70) and in the *Commentarius in Micrologum* (ed. Smits van Waesberghe 1957). Several of Guido's comments refer also to proportional relationships between 'neumes', that is, between phrases of chants; some neumes will be of equal length, while the lengths of others will stand in a proportion of one to two, one to three, two to three, or three to four. This does not refer to the durations of individual notes.

Two other passages have also been adduced in support of a metrical interpretation of the different note-lengths. The *Commemoratio brevis* of the late ninth century contains a number of references to the different speeds of singing appropriate to different occasions or types of chant, to the maintainance of a steady tempo (except for controlled variations such as a 'rallentando'), and also to long and short notes. Towards the close of the passage it is said that these are in the proportion 2:1:

> Breves must not be slower than is fitting for breves; nor may longs be distorted in erratic haste and made faster than is appropriate for longs. But just as all breves are short so must all longs be uniformly long, except at the divisions, which must be sung with similar care. All notes which are long must correspond rhythmically with those which are not long through their proper inherent durations, and any chant must be performed entirely, from one end to the other, according to this same rhythmic scheme. In chant which is sung quickly this proportion is maintained even though the melody is slowed towards the end, or occasionally near the beginning (as in chant which is sung slowly and concluded in a quicker manner). For the longer values consist of the shorter, and the shorter subsist in the longer, and in such a fashion that one has always twice the duration of the other, neither more or less. While singing, one choir is always answered by the other in the same tempo, and neither may sing faster or slower. (Ed. Bailey 1979, 103.)

Since the *Commemoratio brevis* belongs to the same manuscript tradition as the *Scolica enchiriadis*, one naturally asks if the long notes referred to here are those at the ends of grammatical units, the 'syllables', 'neumes', and 'periods' of chant.

Another relevant text, unrelated to those mentioned so far, is the treatise *Quid est cantus?* in the Vatican manuscript Pal. lat. 235, discussed by Wagner (1904, also

Wagner II, 355–8) and Baralli (1905). This short treatise, which appears to date from the eleventh century, has already been cited because of its well-known statement that 'the sign known as a neume comes from accents' (see above), and much else in the text relies on grammatical terminology. Just as there are acute, grave, and circumflex accents in the melody (*tonus*), says the anonymous author, so there are long and short syllables in the feet of the text. He goes on to describe some of the basic shapes of the musical signs, and gives instances from the chant repertory where they are to be found: long and short, the various 'accents', the 'tremula' (quilisma), 'triangulata' (trigon) and so on. On the basis of the analogy with metrical feet, a 2:1 relationship between long and short could be said to exist here.

(v) *Conclusions*

Whatever the truth behind the matter, some reservations are in order. It seems most unlikely that practice was uniform everywhere. It is true that sources are in general surprisingly uniform in their grouping of notes and even in the placing of such special features as quilismas. And the correspondences between Laon 239 and Eastern sources (and other early sources to varying degress) in matters of rhythmic detail cannot be overlooked. But the agreement is general, not exact. Any claim to have identified an 'authentic' performance tradition should be treated with caution.

Another point has been raised by Hucke (1958, 'Rhythmus'): can we necessarily assume that all types of chant were sung in the same way? We could at least imagine a difference between, say, hymns, antiphons, and responsories, to mention only the office chants. And the author of the *Commemoratio brevis* repeatedly mentions variety even within the antiphons, for example:

> The repetitions of the antiphons which occur between the verses (of psalms) should be at the same speed as the psalms, but when the psalm is finished the antiphon is to be slowed by exactly half to its proper tempo. There is an exception in the case of the Gospel Canticles, which are sung so slowly that their antiphon should follow at the same tempo, and not be further protracted. (Ed. Bailey, 107.)

Finally, our experience of the rhythmic characteristics of music outside the tradition of Western art-music has opened our ears to the possibility of much more flexible patterns than can be recorded easily with conventional Western notation. One has only to look at transcriptions of, say, the chant of the Coptic church (*NG* 4. 731, col. 1) to become suspicious of simple 'equalist' or 'mensuralist' interpretations. Might not the singing of the ninth century be equally difficult to capture in modern written form? Meanwhile the minute investigation and recording of the ways in which signs were employed provides essential evidence for a better understanding not only of rhythmic matters but also of other aspects of chant notation.

IV.6. PITCH-NOTATION

(i) From the Ninth Century to William of Dijon
(ii) Guido of Arezzo
(iii) Staff-Notation in Different Lands

Smits van Waesberghe 1957 'Origines'; Crocker 1979; Browne 1981.

(i) *From the Ninth Century to William of Dijon*

In Ch. V, where writings on music theory which concern plainchant are discussed, the crucial importance is emphasized of the way in which the plainchant repertory was matched with the pitch-system borrowed through Boethius from classical teaching, the Greek Greater Perfect System. The link seems to have been made in the ninth century: at least, its clearest early formulation is found in the writings of Hucbald of Saint-Amand (d. 930); and the group of treatises using dasian notation—*Musica enchiriadis*, *Scolica enchiriadis*, and *Commemoratio brevis*—possibly earlier than Hucbald, also use a pitch-system. Dasian notation is a pitch-notation, and Hucbald too proposed the use of a pitch-specific notation. His idea was to use a series of letters (borrowed from the Alypian 'vocal' and 'instrumental' letters, again through Boethius), which would be placed by the side of the traditional notation. Hucbald was quite explicit about the reason for wanting both notations: the traditional signs indicated phrasing, longer or shorter notes, trembling notes, and so on, while the letters would make clear the pitch of the notes.

Although the dasian pitch-symbols enjoyed a certain restricted currency in some branches of the theoretical literature, they were never used in regular chant-books. (They are discussed in the next section with other notations used only in the theoretical literature.) Hucbald's set of symbols does not seem to have been taken up at all. But from the latter years of the tenth century there survive a number of sources supplied with pitch-symbols. Instead of the dasian symbols, or Hucbald's even more arcane ones, simple letters of the alphabet were used.

One of the earliest documents of this nature is the famous Winchester manuscript containing organal voices, Cambridge, Corpus Christi College 473. In the fascicle where sequence texts with their music are copied, several compositions are provided with alphabetic letters side by side with the usual signs. Letters are provided only for the first of the double versicles, for the music repeats itself in the second versicle. Conventional cadences are often left without letters. (See Holschneider 1968 and 1978.) The series of letters coincides not with our modern *A–G* but with a scale starting on *C*. Thus the Winchester *A–B–C* is modern *C–D–E*, and so on. This arrangement of letters was already described by Boethius, and Hucbald too mentions hydraulic organs whose lowest note would be our *C*, and whose scale corresponds to our modern major scale. Hucbald does not actually assign letters to this scale, but Holschneider and others have connected the Winchester letter-notation with the famous Winchester organ described by Wulfstan, who was also probably the copyist of the relevant portion of Cambridge 473. Coincidentally, this type of letter-notation is again used for a sequence in St Gall, Stiftsbibliothek 380 (p. 176, *Carmen suo dilecto*). Whether the notation was primarily to aid vocal polyphony, or instrumental performance, remains unclear.

The earliest evidence that some chant-books may have been provided with alphabetic notation corresponding to modern nomenclature comes from Italy. In the first chapter of the anonymous *Dialogus de musica* (once attributed to Odo of Cluny), written in Italy at the end of the tenth century, the author recommends that the appropriate pitch-letters deduced from the monochord be copied against any antiphon that has to be learned. The teacher will pick out the melody on the monochord, write the pitch-letters into the book, the pupil will then learn by playing the melody on the monochord, reading from the pitch-letters. This replaces the old method of learning by listening to the teacher singing, 'and after a few months' training, they are able to discard the string and sing by sight alone, without hesitation, music they have never heard' (trans. Strunk 1950, 105).

Most surviving books with letter notation are not Italian, however, but French. The most famous is the Dijon tonary, Montpellier, Faculté de Médecine H. 159, written in the first half of the eleventh century, probably at the monastery of Saint-Bénigne at Dijon. Its notation has already been used to exemplify French notation at the beginning of this chapter. Its genesis is bound up with the activities of the great abbot William of Saint-Bénigne, an Italian from Volpiano in Lombardy. After studying and taking the monk's habit in north Italy he entered the monastery of Cluny, and in 990 was made abbot of Saint-Bénigne, with the commission of reforming its observances. In 1001 he was invited by Duke Richard II of Normandy to revive monastic life in the duchy. The results of this reform can be traced clearly in the chant-books of Dijon and Normandy (see below, IX.5), and the alphabetic notation that is found in the Montpellier manuscript can be seen in a number of Norman manuscripts (see Corbin 1955, and, for the fullest list of sources so far, Browne 1981). The alphabet runs from *a* to *p*, corresponding to modern convention as follows:

Ex. IV.6.1. Alphabetic notation *a–p*

It is possible that William learnt the idea of using alphabetic notation in Italy. The *a–p* system is also mentioned by the theorist Odorannus of Sens (985–1046) in a letter of 1032 or 1033 (ed. Bautier *et al.* 213–14 and cited by Browne 1981, 12).

Practically all the sources using this notation contain relatively little music, a saint's office, a few antiphons or hymns, and so on, many added in blank space in a manuscript prepared for other purposes. The flyleaves of the Hereford noted breviary Hereford, Cathedral Chapter Library P. 9. vii, however, appear to be the remains of a complete antiphoner notated in this way (see Frere 1894–1932, *Bibliotheca* i, pl. 2). The famous Montpellier manuscript is by far the most comprehensive source in the group, containing the complete proper chants of mass, arranged in tonal order. It also surpasses the other sources in the sophistication of its alphabetic notation, containing special signs for liquescence, the oriscus, and the quilisma, and five other signs, which have gained some notoriety. They occur where one would expect to find the letters at semitone steps in the scale (see Table IV.6.1). Soon after the discovery of the manuscript Vincent (1854) claimed that they represented quarter-tones, a claim which received comprehensive support from Gmelch (1911). Froger, however, has showed the evidence to be insubstantiable (1978, 'Quarts'). An alternative explanation, that they are represent longer notes, carries rather little conviction. The signs remain mysterious.

Table IV.6.1. *Special signs at the semitone step in Montpellier, Faculté de Médecine H. 159*

at *B*	⊢
at *E*	⊣
at *a* (below b♭)	Γ
at *b*	⌐
at *e*	⌐

   Although it does not seem to be related to the latter signs, another notational feature draws attention to the semitone steps in chant. In a number of eleventh- and twelfth-century French and English sources the punctum occurring on the lower note of the semitone step was given a special shape, usually rather like the uncinus of Laon notation. It is found in the staffless notation of some north French (mostly Norman) sources of the eleventh century, and in books with staff-notation (where it was, strictly speaking, redundant) in Norman and English books of the twelfth century (see Plate 10). It is also to be found in eleventh- to twelfth-century Aquitanian manuscripts (see Colette 1990), and passed thence into Spain and Portugal, where it survived until the end of the Middle Ages (Corbin 1952, *Musique*).

(ii) *Guido of Arezzo*

The pressure to adopt a pitch-specific notation for chant-books was so great, it seems, that no sooner was the ink dry on the first alphabetically notated books, so to speak, than another type of notation was invented, Guido of Arezzo's staff-notation. As is well known, this became the basis of the musical notation used in Western Europe for centuries to come. Guido tells how he came to invent the notation, and how he was subsequently called to Rome by John XIX (1024–33) to demonstrate it, in a letter to a former fellow monk, Michael, in the abbey of Pomposa (north of Ravenna) (*GS* ii. 43, trans. Strunk 1950, 121 ff.). The notation itself is described in what appears to have been conceived as the foreword to the antiphoner (*GS* ii. 34; DMA A.III; trans. Strunk 1950, 117 ff.). Although no manuscripts survive which can be linked directly with Guido, his intentions are fairly clear. Stave-lines were scratched onto the parchment with a dry stylus, the *F*-line was coloured red and the *c*-line yellow. Guido's choice of *F* and *c* as the principle pitches to be identified is understandable, for these mark the upper note of a semitone step. As an alternative way of designating the pitch of the lines, or in addition to colouring the lines, the letters *F* and *c* could be set at the start of the stave as clefs. Guido does not specify the number of stave-lines: most often four were employed, for this number accommodates melodies with the range of up to a ninth, sufficient for most purposes, while extra notes can be dealt with by changing the clef or coloured line.

Although Guido does not state it outright, it is possible (and happened in practice) for a coloured line to be drawn through a space: for a chant with high tessitura, if the line for middle C is coloured yellow, then the red *f* above it will fall in a space. Many early sources with staff-notation use other colours—green is common—and it eventually became general practice to draw four lines in black (sometimes red) ink and use clefs alone to designate pitch. Practice with clefs also varied. Some sources diligently place a letter against every line. *F* and *c* are by far the most common, and sometimes high *g* is found. Much rarer is a gamma-clef for the lowest *G*. English manuscripts use *D* and *b♭* clefs more frequently than their continental neighbours, and even *b♮* is used as a clef.

Smits van Waesberghe (1951, 'Notation'), following the early history of Guidonian notation in the strict sense through a range of both Italian and foreign sources, could list almost none of the eleventh century. One of the earliest is the Old Roman gradual of 1071, Bodmer C. 74 (facsimile ed. Lütolf 1987). Staff-notation even in a general sense became the norm only in the twelfth century, at least in France, England, and Italy.

In at least one curious instance, the so-called 'Wolffheim Antiphonal' (now University of California at Berkeley 748), the staff was used for a neumatic notation which was not placed accurately upon it; in other words, the staves might as well never have been drawn in (Wagner 1926; Emerson 1958–63).

(iii) *Staff-Notation in Different Lands*

The shape of the signs deployed on the stave continued to vary from area to area, for the traditional signs could simply be set out in the new way on the page. Only in German lands does the adoption of staff-notation seem to have entailed a radical change. Many German scribes remained faithful to the older staffless notation centuries after it had been adopted in most other centres. The fourteenth-century graduals of Passau, St Florian, and Seckau, for example, are still written without staves (see the convenient list in *Le Graduel romain* II, 158–71). German notation was usually strongly sloping in character, and relatively few manuscripts show a vertical ductus. But for staff-notation the vertical alignment became almost universal.

A striking example of the change in appearance of German notations can be seen within a single source, the antiphoner from Petershausen near Konstanz, Karlsruhe, Badische Landesbibliothek, LX Aug., probably dependent on the monastic tradition of Reichenau. The manuscript was from the beginning notated in staff-notation, originally fine little sloping signs of the twelfth century. At the end of the manuscript the Office of the Dead was notated in a different hand, still fine but with practically vertical ductus. Some of this survives, but in the fourteenth century much was erased and replaced by upright 'Gothic' forms. Together with several additions, these scripts provide a fascinating cross-section of hands (studied by Hain 1925).

There appears too to have been a distinct rapprochement between the newer German notation and the later stages of Laon notation, so that in at least some areas it is possible to speak of hybrid forms. The signs shown in Table IV.6.2 can all be found in varying mixtures, a 'Laon' flexa side by side with a 'German' virga, for example.

The 'German' virga, looking somewhat like a horseshoe nail, has given its name to this phase of German notation as a whole: 'Hufnagelschrift'. In later centuries it was commonly written with a broad-nibbed pen (as were other late medieval notations), and in this form is also known as 'Gothic' notation.

Table IV.6.2. *Late forms derived from Laon and German notation*

	Late Laon	Late German	Bohemian	Hungarian
single notes	↗ ↗	↑ •		
pes	⌐	⊣	↗	
clivis	⌐	⋂ ⋂		
porrectus	⋎	⋂⋃		
scandicus	⋰ ⋰	⋰	⋰	
climacus	⋰	↑⋰		⋰ ⋰

Manuscripts showing predominantly or even completely Laon forms were written in many German centres: Leipzig (PalMus 3, Pl. 174) and Hildesheim (Pl. 175), for example. For reasons which are not yet clear, Laon notation was used in a group of sources usually assigned to Klosterneuburg (for example, Graz, Universtätsbibliothek 807: PalMus 19, which nevertheless has German strophici). Further east and north other forms related to Laon and German notations were used. Typical of Bohemian sources is a zigzag form of the pes and scandicus, while Hungarian sources often have a sort of double punctum at the head of a climacus (presumably descended from a right-facing virga). Signs similar to some of these can actually be seen in south and central Italian sources, and the possibility of influence from that quarter should perhaps be considered. (See the remarks on East European notation in Szendrei 1985.)

Much work remains to be done on the regional varieties of Eastern notations. In particular, the situation in Germany, Austria, and Switzerland remains to be clarified. On the other hand, Hungarian sources have been magisterially surveyed and classified by Szendrei (1983 and 1988, with over 130 plates) and the main lines of investigation for the East as a whole sketched out (Szendrei 1986). Centres of influence need to be identified, for example among the religious orders, which might explain the adoption of different signs in different churches and the reasons for such phenomena as the 'invasion' of 'German' territory by 'Laon' forms.

In the eleventh century in south France and south Italy many manuscripts were written with exactly heighted notation yet without adopting the Guidonian staff. Aquitanian notation, consisting very largely of separate points, was comparatively easy to set out in such a way that the intervals between notes were accurately indicated. It was common for scribes to rule pages as for a text manuscript, then use each alternate line as a guide for the musical notation. Clefs were not used, so that it was necessary to have an idea of the start of the piece in order to get going. Beneventan notation of the eleventh century followed a similar course; but it developed a bold, angular style, using a broad-nibbed pen, which gives it a quite distinctive surface appearance, however similar to other Italian styles it may be at root (see Pl. 12, Beneventan, and Pl. 13, Roman).

There was in most cases no great difficulty in adapting traditional 'unheighted' signs to the stave. Even before staff-notation was introduced in north France, for example, the virga and pes were commonly written with a head at the end of the up-stroke, and the flexa often had a little foot at the end of the down-stroke (see, for example, Pl. 9). Gradually these heads and feet became larger and more prominent, at the expense of the strokes in between, reduced to mere hair-lines. In Parisian manuscripts of the thirteenth century, nearly all the signs are composed of squares joined by these thin lines—forms known as ligatures in the literature on Parisian polyphony of the same period, which naturally took the local notational shapes as its starting-point. The exceptions are the porrectus, with a descending diagonal as its middle element, and the climacus, with rhomboid puncta. Square or quadratic notation, as it is commonly called, naturally existed in many regional varieties (see Plates 14–15, and Suñol 1935, ch. 16). That it became so widespread is due to the cultural pre-eminence enjoyed by Paris in the thirteenth century. Books were produced there for churches in a wide area around Paris (see Branner 1977). The master exemplars of the Dominican liturgical books were produced in Paris and set a pattern for others far and wide, even in Eastern Europe.

Everywhere in Europe in the later Middle Ages there was a move towards larger and thicker notational forms, a general 'Gothicization', it has been remarked. Another general trend was the resolution of all shapes into vertical, horizontal, or diagonal, curves being eschewed. By the end of the period, nearly all centres were using either a form of square notation or one of the varieties of Eastern notation. One of the last centres to retain its local forms was Milan, whose notation had undergone its own evolution from dainty early medieval signs to impressive squares and lozenges (Pl. 16). Here as elsewhere, the adoption of a large format was necessary when it became the practice for choirs to sing from the written page instead of from memory. (The growth of this practice has yet to be comprehensively documented.)

Notations have commonly been divided into adiastematic (or non-diastematic) and diastematic (from the Greek *diastema* = interval according to whether or not they indicate clearly the intervals between notes. Similar terms are 'unheighted' and 'heighted' (of signs). 'Intervallic' notation is another. Where some intervals remain doubtful the diastematy is said to be 'incomplete', or even (with pejorative overtones best avoided) 'imperfect'. In theory nothing of the flexibility of the older notational phase need have been lost when the pitch-specific mode was adopted. This is clearly borne out by the staff-notation of relatively late German sources, where such features as the quilisma and strophici can be seen (PalMus 3, 132–7). The gradual adoption of staff-notation occurred, however, at a time when many niceties of notation (and presumably of performance) were declining. The salicus is rare outside the earliest manuscripts, the strophici are hardly distinguished notationally outside Eastern sources, the quilisma hardly survived the eleventh century in Western sources.

IV.7. THEORISTS' NOTATIONS

Crocker 1979.

A brief survey is given here of notational systems found only in writings on music theory of the early Middle Ages. They are of as much relevance to chant theory as to the notation of plainchant for service-books. Theorists' notations and notations for practical purposes were naturally related; but theorists felt the need to illustrate theoretical concepts with unambiguous pitch symbols earlier than singing-masters felt the need to give their pupils pitch-specific chant-books to study.

The earliest systems, found in the ninth century, have already been mentioned. Hucbald's notation, which seems to have had no other users, uses the signs shown in Ex. IV.7.1. They were borrowed in the first instance from Boethius (iv. 3–4), who was reproducing the notation of the Greek writer Alypius (3rd–4th c. AD; see Warren Anderson, 'Alypius', *NG*; Jan 1895, 369; Isobel Henderson 1957, 358; Potiron 1957; Babb 1978, 9, 38). Hucbald applied these signs to the notes of the Greek Greater Perfect System. For the most part, he, and many writers subsequently, referred to the notes of the Greek system by their Greek names, also given in Ex. IV.7.1. In prose discussions this remained the preferred method for well over a century. The real advantage of pitch-symbols was in diagrams and musical examples, where there was often little room for the cumbersome Greek names. It is important to realize that

Ex. IV.7.1. Hucbald's letter-notation

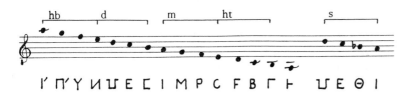

tetrachords:		Greek pitch names:		
hb	hyperboleon		aa	nete hyperboleon
d	diezeugmenon		g	paranete hyperboleon
m	meson		f	trite hyperboleon
ht	hypaton		e	nete diezeugmenon
s	synemmenon		d	paranete diezeugmenon
				or nete synemmenon
			c	trite diezeugmenon
				or paranete synemmenon
			b	paramese
			b♭	trite synemmenon
			a	mese
			G	lichanos meson
			F	parhypate meson
			E	hypate meson
			D	lichanos hypaton
			C	parhypate hypaton
			B	hypate hypaton
			A	proslambanomenos

Hucbald did not use the A–G nomenclature. In fact, the modern equivalent pitches shown in Ex. IV.7.1 could have started with any note, as long as the sequence of tones and semitones was preserved.

The same is true for the dasian notation employed in the 'Enchiriadis' group of treatises: *Musica enchiriadis*, *Scolica enchiriadis*, *Commemoratio brevis de tonis et psalmis modulandis*, and a few other brief texts (all edited by Hans Schmid 1981). The author of *Musica enchiriadis* (usually dated to the late ninth century) first describes a series of tetrachords with reference to their intervallic structure, then designates the individual pitches with the dasia signs. If we take low *G* as the modern equivalent of the lowest note, the scale shown in Ex. IV.7.2 emerges. Nearly all the signs are derived from the Greek aspirant sign, the *daseia* (modern 'h' sound), which is given extra curls and strokes (like *s* and *c*), and turned upside-down and back to front.

Ex. IV.7.2. Dasia signs in the *Enchiriadis* group of treatises

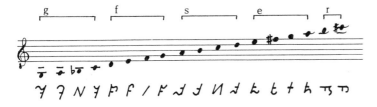

tetrachords:	
g	graves
f	finales
s	superiores
e	excellentes
r	residui

The presence of low *B*♭, high *f*♯, and *c*♯′, and the absence of middle *b*♭, remain something of a puzzle. Since few of the musical examples touch on the extremes of the range, the outer chromatic notes are not usually a problem in the treatises themselves. Not every note here has a corresponding octave, however, so that when the author discusses octave doublings he simply omits the dasian signs if necessary (Schmid, 27, 32, etc.). Phillips (1984) has argued that the notes were intended to create a more flexible system than the Greek diatonic series for the recording of chants with chromatic notes.

The various demonstrations and diagrams in Boethius' treatise employ a number of alphabetic series. For example, at one point Boethius gives a diagram including all the notes of the Greater and Lesser Perfect System, not only for the diatonic genus but including the chromatic and enharmonic genera as well. To denote all these pitches, a series running from A to Z, AA to LL was required. If the notes of the diatonic scale are extracted from the complete series, the following scale is represented:

Ex. IV.7.3. Boethius' pitch-letters

This series is actually used in the late ninth century treatise *Alia musica*.

The A–G alphabetic series with which we are familiar can also be found in Boethius (iv. 14), actually extended beyond G to make a second octave H–O (omitting J). On another occasion Boethius posits a complete two-octave series A–P (iv. 17), but the argument he propounds does not require the specification of individual pitches and intervals. The earliest medieval adaptation of this appears to be that in Part ii of *Scolica enchiriadis*. (The *Γ–d′* series in *Musica enchiriadis*, Schmid, 30; GS i. 162, is peculiar to Paris, Bibliothèque Nationale, lat. 7202, of the tenth to eleventh century.) Here there is an unexpected and fleeting reference to a two-octave range A–H–P (Schmid, 90–1; GS i. 184–5). The notes in between are not specified here, but the example for which they are used demands that A, H, and P all correspond to modern *E*, *e*, and *e′*. Then in Part iii of the same treatise an A–P scale is set out complete (Schmid, 147; GS i. 209), but once again it does not correspond to the modern *A–G*, *a–g*, *a′*, but to modern *C–b*, *c–b′*, *c′*; it is in fact the series known from the Winchester notation for some sequences, described above, and connected in some way with organs and other instruments (see IV.6).

The series A–P applied to what we would call *A–G*, *a–g*, *a′* finally appears in an anonymous treatise (GS i. 338–342, Anonymous II), apparently written in south Germany about the end of the millenium. Whether there is any connection with the practical use of this series by William of Dijon (see again IV.6) is unknown. Another variant of the series is to be found in Gerbert's Anonymous I (GS i. 330–8).

The repeating series *A–G*, *a–g*, and so on is first found in the anonymous *Dialogus de musica*, written in north Italy in the late tenth century. It achieved rapid acceptance, being used also by Guido of Arezzo. The lowest G was given the Greek letter gamma (*Γ*). For the third octave double small letters were used, up to ā̊ in the *Dialogus*, up to e̊ in Guido's writings (Ex. IV.7.4).

Ex. IV.7.4. Pitch-letters in the *Dialogus de musica* and Guido of Arezzo

Γ A B C D E F G a b   c d e f g ā̊ (b̊ c̊ d̊ e̊)

Although Guido's writings had become well known in south Germany by his time, Hermannus Contractus (1013–54), monk of Reichenau, invented one further notational system, which did not, however, achieve any wider use. This was also a letter-notation, but specified intervals rather than pitches. A dot underneath the letter indicated downward motion. Ex. IV.7.5 lists the letters and transcribes the part of Hermannus' treatise where they are explained. (For a facsimile, see Stäblein 1975, 223, where a variant of the system in another south German theory treatise is also reproduced.)

Byzantine notation is also an interval notation, but there is no connection between Hermannus and Byzantine music.

IV.8. PRINTED CHANT-BOOKS

Riemann 1896; Molitor 1901–2 (Italian books), 1904 (German); Tack 1960.

Chant first appeared in printed books in several forms. The creation of musical type (or any other means of printing music) was very time-consuming. Movable type could be prepared either with stave-lines for each character, or printed in a second run on to staves already prepared, both cumbersome procedures. Not surprisingly, many early printed liturgical books simply left space for staves and music to be written in later by hand. Others had printed staves but no notes. Such partially handwritten books can be found as late as the eighteenth century.

The earliest use of movable musical type appears to have been for a gradual printed probably in Constance in 1473. The dating depends upon the fact that the same type was used to print the text of a breviary from Constance, and one copy of this exists with rubrics entered in 1473. (See A. H. King, 1968, with two plates from the only complete exemplar, in the British Library; also Molitor 1904, pl. II, from fragments in Tübingen, Universitätsbibliothek.) The earliest datable book with printed music is the *Missale* of 1476 printed in Rome by Ulrich Han of Ingolstadt (Molitor 1904, pl. I;

Ex. IV.7.5. Hermannus Contractus' interval notation (Munich, Bayerische Staatsbibl. clm 14965b, fo. 22ʳ)

A. H. King, pl. III). Whereas the 'Constance' gradual has German notation, the Rome missal uses square forms. Initials in both were entered by hand.

In Han's book, as in many others of the fifteenth century, relatively little music was required, for only the priest's chants are notated. For such short items wood blocks or even metal ones might have been used, but in liturgical books movable type was preferred. The earliest liturgical music-book printed from woodblock appears to be the *Obsequiale* printed in 1487 for the Augsburg diocese by Erhard Ratdolt (Molitor 1904, pl. VII).

The two books just mentioned are the oldest containing printed music of any sort. By the end of the century over 250 liturgical books with music had appeared. (Molitor gives short accounts of many printers in Italy and Germany.)

The notational forms were for the most part the same as in contemporary manuscript books of the corresponding locality. While there is no trace of the quilisma in German prints, strophici can still be found. Both had disappeared from square notation before the fifteenth century. German printers cleverly used the strophicus for liquescent notes as well, simply adding a strophicus after the principal sign. Liquescent notes do not appear in printed square notation; they had in any case practically disappeared from most manuscript books using square notation. On the other hand, numerous mensural rhythmic signs were employed as the occasion demanded: this development served reformed chant, which utilized proportional rhythmic notes.

The basic distinction between music-type using square forms and that using Gothic or German forms has survived until the present century, but the tendency to follow the example of Rome, where square notation was the rule, has led to the general disappearance of Gothic notation. Both forms have been susceptible to wide local variety. For example, several early printers of square notation maintained the distinction between virga, punctum, and rhomb, and some created the diagonal shape necessary for the porrectus. But the virga is comparatively rare in the Medicean edition of the early seventeenth century and its imitators, and the diagonal is absent. The virga was then entirely eliminated in many books (principally French ones) from the later seventeenth century onwards. Signs containing more than one note were then made up simply by placing the required number of square and/or lozenge shapes next to one another. (See Tack 1960, 49–57 for several facsimiles.)

As part of the restoration of the repertory to something like a medieval state, Pothier and the firm of Desclée in Tournai created a new font which was capable of reproducing the basic shapes of fourteenth- to fifteenth-century square notation, as it was known from north French chant books. To these were added a new sign for the quilisma. The first publication with this new music-type was Pothier's book *Les mélodies grégoriennes* of 1880, the first chant-book Pothier's *Liber gradualis* of 1883; the earliest Solesmes and revised Vatican books then used the same font. It was expanded for the *Antiphonale monasticum* of 1934 by the addition of signs for the oriscus, apostropha (distropha, tristropha), and liquescent punctum (that is, for liquescence by augmentation of the punctum, rather than diminution of the pes or clivis). The new *Antiphonale Romanum* has introduced a new batch of signs, designed to reflect even more closely the detail of early notations (principally that of Hartker's antiphoner, St Gall, Stiftsbibliothek 390–391, PalMus II/1; see the table on p. xii of the *Liber hymnarius* of 1983).

Ex. IV.8.1 gives the same passage from a fourteenth-century manuscript and three modern editions: the Ratisbon *Graduale Romanum* of 1898, the already partially restored *Graduel romain* printed in 1872 in Marseilles under the aegis of the Digne Commission, and the restored Vatican *Graduale Romanum* of 1908. The differences between the notes are of course striking, but also notable is the greater variety of signs in the Desclée font.

Ex. IV.8.1. Communion *Pascha nostrum* from *Graduale* (Rome, 1898), *Graduel romain* (Marseilles, 1872), *Graduale* (Rome 1908), and Rouen, Bibl. Mun. 250 (Jumièges, 14th c.)

IV.9. MODERN TRANSCRIPTION

    (i) The Liturgical Context

    (ii) Transcription

    (iii) Transcription from Staffless Notations

From the foregoing chapters it will have become clear that chant notation was from the start quite different in character from modern notation. The absence of pitch in the earliest sources presents special problems of transcription, for to arrive at any pitch-specific interpretation one must have recourse to another manuscript (or manuscripts), often considerably later in date than one's main source. Yet one must preserve as much as possible of the original, possibly to the extent of making a literal copy of its notational signs.

    Even the later forms of chant notation contain little of the information about performance practice that we normally expect from a musical score, such as indications of tempo and dynamic. They share this condition with practically all medieval music and a good deal of later music as well. Music sung all day and every day could hardly include extravagant performance practices, for according to its very nature chant was sung in a traditional way, which needed no specification. Customaries and ordinals therefore contain plenty of information about who shall sing what chants, on what occasion, and where in the church. But niceties of performing style were not committed to writing. If such aspects of performance as rhythm, and even pitch, are left out of account in the notation, it is only to be expected that other features should also be lacking.

It has to be admitted, therefore, that we know very little about the performing styles of medieval chant. In fact it is only since the advent of sound-recording techniques—first used for plainchant at the 1904 Gregorian Congress in Rome (Discant Recordings DIS 1–2)—that we can trace the history of chant performance in some of its most essential aspects: rhythm, dynamic, voice production, and so on. This is not the case, of course, with the new types of chant which arose in the seventeenth century and later. On French *chant figuré*, for example, we have both practical manuals and the eyewitness accounts of outsiders (see below, X.6). But there is no unbroken line of tradition back to the Middle Ages.

The method of chant performance is clearly of great importance in the church today. Plainchant has been singled out by several statements from the Holy See in terms such as 'the chief and special sacred chant of the Roman Church' (*Sacred Music and Liturgy*, §16), and church musicians have felt a special duty to sing the 'right' notes in the 'right' way. The musicologist has the luxury of not having to decide on what is 'right', indeed cannot even see the problem in such terms. This book stresses rather the variety which is to be found in medieval (and later) chant-books. If one seeks to recreate a particular melodic or rhythmic tradition (say that of St Gall) for modern use, that is to select but one tradition out of a multitude.

This book is not a manual of modern plainchant performance, for which one must look to such books as those by Agustoni and Göschl (1987), or at least take note of the directions prefacing the Solesmes/Vatican editions. Even to sketch a historical survey of what is known about chant performance in past centuries would need a book of its own. (For some recent contributions, see Berry 1965–6; Dyer 1978; Brunner 1982.) Instead, a brief consideration is given to three points of particular relevance to performance practice and transcription.

(i) *The Liturgical Context*

Any chant recorded in any source is an element in the worship of a particular church. Thus the introit *Puer natus est nobis* in the manuscript Paris, Bibliothèque Nationale, lat. 904 has to be understood as part of the liturgy for High Mass on Christmas Day in Rouen cathedral in the thirteenth century. We can, if we like, simply regard the version in that source as typical for the Middle Ages as a whole, but it cannot be assumed a priori that all other medieval manuscripts transmit the melody like Paris 904. We might also assume that the introit was performed in more or less the same way in most churches throughout the year. But experience shows that one should expect special ordinances to apply at the most important mass of Christmas Day, and these may not be the same at Rouen as at other churches. Since Paris 904 has a quantity of rubrics, we are better informed for thirteenth-century Rouen than we are for, say, tenth-century St Gall. For further information we should consult any ordinals that may have survived from Rouen (they may not be thirteenth-century ones, however; in fact there are at least nine, ranging in date from the late thirteenth to the fifteenth centuries: see *Le Graduel romain II*, 195). These considerations are of course especially relevant if one wishes to reconstruct the whole liturgy of a medieval church. Even if only the chant is important for the investigation in hand, it should still be borne in mind that a medieval source records only what was understood to be 'right' at a particular place and time.

(ii) *Transcription*

Any modern chant transcription should represent the original as faithfully as is compatible with the very notion of transcription. As already stated, performance from memory was for centuries the rule rather than the exception. But where a transcription is needed, either for performance or for study, how should it best be made?

The answer has seemed to many to lie in the Solesmes/Vatican books of the present century. Especially in the latest volumes of the *Antiphonale Romanum* there is a variety of signs which can cope with almost all features of early medieval originals: that is, most signs of the medieval source have a counterpart, so that with the modern edition in front of one, one could imagine how the original looked. The biggest advantage of transcription into a sort of square notation is that the grouping of notes in medieval sources can be preserved. This is, however, equally possible with transcription in modern round notes on a five-line stave. The Solesmes/Vatican square notation has by now the aura of tradition and ecclesiastical approbation, and is indissolubly linked with chant in an almost spiritual way. Despite its aptness, it should nevertheless not be forgotten that it is a late nineteenth-century revival, with modifications, of a notation from the high Middle Ages. Its resuscitation is comparable to the revival of Gothic architectural styles (with modifications) as the model for nineteenth-century churches.

Since transcriptions have to be made from a wide variety of medieval notations, a system which is flexible enough to reflect that variety is desirable. In this book I have therefore used round note-heads, with special signs for liquescence, the oriscus, the quilisma, and the strophici. Especially in view of the research of Cardine and his students it seems vital to indicate note-groupings as accurately as possible, and make clear exactly which notes are joined to each other in the same sign. Slurs are used here for that purpose. This book contains plenty of examples with both medieval signs and transcription in parallel, which demonstrate what I believe to be a reasonable method. Note-grouping can be reflected simply by placing notes close together, rather than barring them like modern quavers or placing them all under a general phrase mark. The practice of covering all notes sung to a single syllable with a single phrase mark is regrettable, especially in the case of long melismas.

(iii) *Transcription from Staffless Notations*

For any transcription from a staffless notation on to the modern stave, a source has to be selected to suggest possible pitches. The later source (or sources) should be acknowledged, even when it is a modern Solesmes/Vatican book. Ideally one would find a source in staff-notation whose melodic tradition is as close as possible to the earlier manuscript. The monks of Solesmes working on *Le Graduel romain* have had to wrestle with this problem. Whereas the earlier *Graduale* and *Antiphonale* cite no manuscript sources to justify the given pitches, the new project has first identified the most important early manuscripts with staffless notation, and then matched each one with a source with staff-notation (see Froger 1978, Edition; Berry 1979).

The copious detail of some of the earliest sources often demands special solutions. The best, but also the most cumbersome, is to copy the signs of the earlier manuscript over the stave for the transcription. As early as 1876 Michael Hermesdorff issued a *Graduale* with neumes printed over the stave. The invaluable *Graduale triplex* is an example of this method, with two sets of handwritten neumes over the typeset notation. Otherwise some means has to be devised of representing in the transcription such features as significative letters, episemas, and the distinction between punctum and tractulus (punctum and uncinus). It seems unwise to ignore any detail of the document being transcribed, especially when one is unsure of its precise significance.

PLATES

FACSIMILES OF CHANT BOOKS

INTRODUCTION

The illustrations on the next pages are intended to display both different types of liturgical books containing music and different types of notation. In eighteen photographs it is obviously impossible to give more than an arbitrary selection, but these facsimiles provide at least partial support for Ch. III, on liturgical books and plainchant sources, and Ch. IV, on notation.

I have used pages only from sources in British libraries, these being on the whole less well known than those of some other European countries (but see Briggs 1890 and Nicholson 1913: five sources used by Briggs are found here).

The manuscripts in order of presentation:

Plate 1.   London, British Library, Harley 1117
Plate 2.   Oxford, Bodleian Library, Lat. liturg. d. 3
Plate 3.   London, British Library, Egerton 857
Plate 4.   London, British Library, Harley 110
Plate 5.   London, British Library, Add. 30850
Plate 6.   London, British Library, Add. 19768
Plate 7.   Oxford, Bodleian Library, Canonici liturg. 350
Plate 8.   London, British Library, Arundel 156
Plate 9.   London, British Library, Royal 8, C. xiii
Plate 10.   Cambridge, University Library, L. 2. 10
Plate 11.   London, British Library, Add. 10335
Plate 12.   London, British Library, Egerton 3511
Plate 13.   London, British Library, Add. 29988
Plate 14.   London, British Library, Add. 17302
Plate 15.   Oxford, University College 148
Plate 16.   Oxford, Bodleian Library, Lat. liturg. a. 4
Plate 17.   Edinburgh, University Library 33
Plate 18.   London, British Library, printed book IB. 8668

Ꝝ luxuit dns̄. Ṽ Dixit dns̄. Ꝝ Magnificauit illum ⁊ seruauit illi testamentũ.
Ꝝ O . . . . bectum pr̄e sulem cuthberh . tum qui gratia diui
ni muneris ui gens multipli . cium morborum ualitu dines
de . . . . pulit prophetiae spi ritu pol lens plurimo . tum ac
su um ip si uso . . . . . bicum ⁊ pre
sciuit ⁊ prenuntiauit. Ṽ Admirandus cunctis operibus ⁊ uerbis di
uina sapientia uitam composu it . prophetiae. IN MATVTINIS LAVD
Ꝑ Cristi foras hic athleta ac uerus anachorita mundi tempsit haec in
fima quo caeli caperet summa. Ꝑ Qui de rupe prompsit aquam post
muuium uertens eam hoc utrumque donum suo caro contulit
cuthberhto. Ꝑ In episcopatu suo iam exacto biennio ut soli uacaret
deo dilecto se reddit antro. Ꝑ Hinc tinguntur artus sacri corporis
morbo loccali sciens uero se resolui conforcat ouile xp̄i. Ꝑ Mox
pater suos affatur quisque uestrum tempnat mundum Ameo
cristum colat bonum sic supernum scandet regnum. IN EVANG.
Ꝑ Languor adcrescens indies artus uexabat fragiles ipse sacras fundens preces
caelicas pregustat laudes dum terrenus reddit finis aeternum cuthberhto
diem salutares sumit dapes sic supernas scandit sedes. Ꝑ O magne presul
cuthberhte cui xp̄s fuit uiuere cui mori lucrum per enne dum post mor
tem uiuis uere signis diuinis inclite languidos sanans alabe hoc rogamus
pia prece tu pro nobis intercede.

London, British Library, Harley 1117, fo. 44ʳ. Lives of St Cuthbert, chants for saints' offices, from Canterbury, Christ Church, second half of 10th c. Breton neumes. The plate shows chants for the office of St Cuthbert.

PLATE 1   407

The manuscript contains Bede's prose and verse lives of St Cuthbert, and chants for the Night Office and Lauds of St Cuthbert (fos. 43ʳ–44ʳ), St Benedict (fos. 63ʳ–65ʳ), and St Guthlac (fos. 65ʳ–66ᵛ). (The start of the Cuthbert office may be seen in PalMus 2, pl. 81; fo. 63ʳ is in Briggs 1890, no. 6 and part of it in Suñol 1935, pl. 82.) It therefore contained material for both reading and singing, and served as a supplement to the other office-books of the church for which it was written.

At the start of the page are the final responsories of the Night Office. Two are given simply as text cues (*Iuravit dominus*, *Magnificavit illum*) because they would have been available complete in the section for the common of saints in an antiphoner. The final responsory begins on line 2, and is copied complete: R. *O beatum presulem Cuthberhtum* V. *Admirandus cunctis operibus*.

After the rubric IN MATUTINIS LAUDIBUS there follow five antiphons for the psalms to be sung at Lauds: *Cristi fortis hic athleta*, *Qui de rupe prompsit aquam*, *In episcopatu suo*, *Hinc tanguntur artis sacri*, and *Mox pater suos affatur*. The rubric IN EVANGELIA announces the antiphon for the Benedictus canticle, *Languor ad crescens in dies*. The last antiphon, *O magne presul Cuthberhte*, is for the Magnificat at Vespers or for subsequent commemorations.

The texts of the chants are rhymed (end-stressed), as for example the first antiphon:

> Cristi fortis hic athleta
> ac verus anachorita
> mundi tempsit haec infima
> quo caeli caperet summa.

These are classic Breton neumes, fully comparable with those from Brittany itself. The following may be noticed:

(*a*) both conjunct and disjunct pes and clivis:

(*b*) the last note of the climacus often has a long, slender, vertical tail:

(*c*) liquescent neumes:

(*d*) the oriscus, either alone, or joined to a subsequent lower note:

(*e*) the letter *l* ('levate') for the rise of a fifth in line 2 at 'qui':

The same office chants appear in the Worcester compendium (PalMus 12, 292 ff.), facilitating the following parallel transcription of the start of the responsory.

PLATE 2   409

Oxford, Bodleian Library, Lat. liturg. d. 3, fo. 3ʳ. Fragment of a combined lectionary and gradual, from Aquitaine, 10th c. Aquitanian neumes. The plate shows parts of the masses for Monday and Tuesday in Holy Week.

This fragment was originally one page, with text and music set out in two columns. It was then cut down to a suitable size for making guard leaves for a later manuscript. It was folded down the middle, but part of the initial N can be seen on the left of the fold. Text and music are missing up above what is now preserved, so altogether only the lower right-hand quarter (approximately) of a bifolium is preserved.

How does one identify the material on this fragment? The rubrics COMMUNIO and FERIA III on the right indicate the communion chant at the end of one mass, and the day on which the next mass is celebrated. The chants, the communion *Erubescant* and the introit *Nos autem gloriari*, can then be located in a modern chant-book, or any other properly indexed edition. The manuscript contains texts as well, so that a missal or the *Liber usualis* is necessary for identifying everything here visible. It can then be ascertained that on the left is part of the Gospel for the Monday after Psalm Sunday. The last few lines of the lesson and most of the Offertory chant are missing with the top half of the page. Only the closing words of the offertory *Eripe me* V. *Exaudi me* are visible: '(intr)es in iudicio cum servo tuo domine', with the cue 'Doce me' for the partial repeat of the respond. After the communion *Erubescant* the mass for Tuesday of Holy Week begins: the introit *Nos autem gloriari* and the start of the first lesson, from Jeremiah.

None of the proper prayers is given (they would come after the offertory, communion, and introit) so it looks as if the book combined lessons and chants only. The fragment is listed in Gamber 1968, no. 1381 (under the incorrect shelf-mark Lat. liturg. a. 6), alongside two other manuscripts of this type, one also from Aquitaine and one from Saint-Omer.

In order to achieve a transcription, the Aquitanian neumes may be read off against those in Paris, Bibliothèque Nationale, lat. 903, reproduced in PalMus 13. Paris 776 was used here to guide the transcription of the second half of the communion.

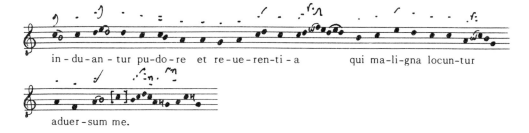

in - du - an - tur pu - do - re et re - ue - ren - ti - a   qui ma - li - gna locun - tur

aduer - sum me.

R Uenite filii V Accedite Alle lu ia

Te decet ymnus de us in syon
et tibi reddetur uo tum in hierusale

m OS Sicut in holocaustum arietum et tauro
rum et sicut in milibus agnorum pingui um sic fi at sacrifici um
no strum inconspectu tuo hodie ut pla ceat tibi quia non est confu
sio confitentibus inte domi ne TO Inclina aurem tuam accelera
ut erua s nos DOM· VIII· CAP· C· Lxxxiii·

R· cap· xxxi·   R· cap· lxvi·

SVSCEPIMVS deus PS Magnus dns R Esto michi in dm

V D eus inte sperauit Alle lu ia

Atten dite po pule me us

in legem me a m OT Populum humit CO Gustate et uidere qua
niam suauis e st dominus beatus uir qui sperat in eo· S·XIIII·

ECCE de us adiuuat me et DOM· VIIII· CAP· C· Lxxv
dominus susceptor est anime mee auerte mala inimicis me is in
ueritate tua disperde illo s protector me us s domine PS Deus
innomine tuo R Domine dominus noster quam admirabi le
est nome n tuum inuniuersa terra ES Quoniam eleuata
e st magnificentia tua super ce lo
Alle lu ia Domine deu s
salutis mee indie clama ui et nocte coram te

PLATE 3   411

London, British Library, Egerton 857, fo. 52ᵛ. Gradual, from Noyon, 11th–12th c. Laon neumes. The plate shows chants for the proper of mass.

There are no prayers or lessons on this page, for the book is a gradual. Chants are given from three masses of the post-Pentecost season. In the manuscript the masses are numbered, so that when a chant is to be repeated from some previous occasion, a reference to the earlier occurrence is given. The chants on this page are:

(end of mass for the seventh Sunday after Pentecost): gradual *Venite fili* V. *Accedite* (incipit only; reference to 'Capitulum lxviii', mass 68); *Alleluia Te decet hymnus*; offertory *Sicut in holocaustum*; communion *Inclina aurem tuam*.

'Dominica VIII'—eighth Sunday after Pentecost—mass 143: introit *Suscepimus deus* Ps. *Magnus deus* (incipit only; reference to mass 31); Gradual *Esto michi* V. *Deus in te speravi* (incipit only; reference to mass 66); *Alleluia Attendite popule meus*; offertory *Populum humilem* (incipit only; reference to mass 70); communion *Gustate et videte*.

Ninth Sunday after Pentecost—mass 144 (last letters not visible on photograph): introit *Ecce deus adiuvat me* Ps. *Deus in nomine tuo*: gradual *Domine dominus noster* V. *Quoniam elevata est*; *Alleluia Domine Deus salutis mee*.

Facsimiles of other pages will be found in *NG* xvii. 622 (masses 174–80); Briggs 1890, no. 8 (masses 135–6); PalMus 3, pl. 158 (masses 13–15).

The neumes are not in all instances exactly of the Laon type, hardly surprising when the chief point of reference, manuscript Laon 239, is older by two centuries. The favourite single note is more like a short, sloping, German virga with a head to the left, and the true Laon uncinus is largely absent, except for low notes at the start of a group, where there is little space for a long stem. The Laon virga, like an elongated S, is not used, but the pes looks rather like this. (The pes is somewhat rare in Laon notation, uncinus + virga being preferred.) The flexa is likewise preferred to a pair of vertically aligned uncini; it resembles more the French than the Laon form. The circular quilisma and oriscus like a double *o* can both be seen in following transcription of the communion *Gustate et videte*. To arrive at a plausible version, one should if possible compare the neumes with a later manuscript in staff notation from Noyon. Luckily one such has survived: Abbeville, Bibliothèque municipale 7, a notated missal of the 13th–14th century, which therefore guided the choice of pitches in the transcription. (Compare the version in the *Liber usualis* or *Graduale Romanum* at 'GusTAte' and 'QUOniam'.)

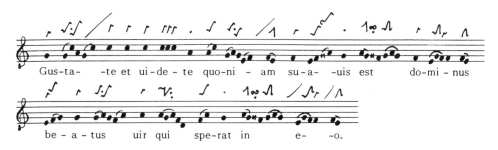

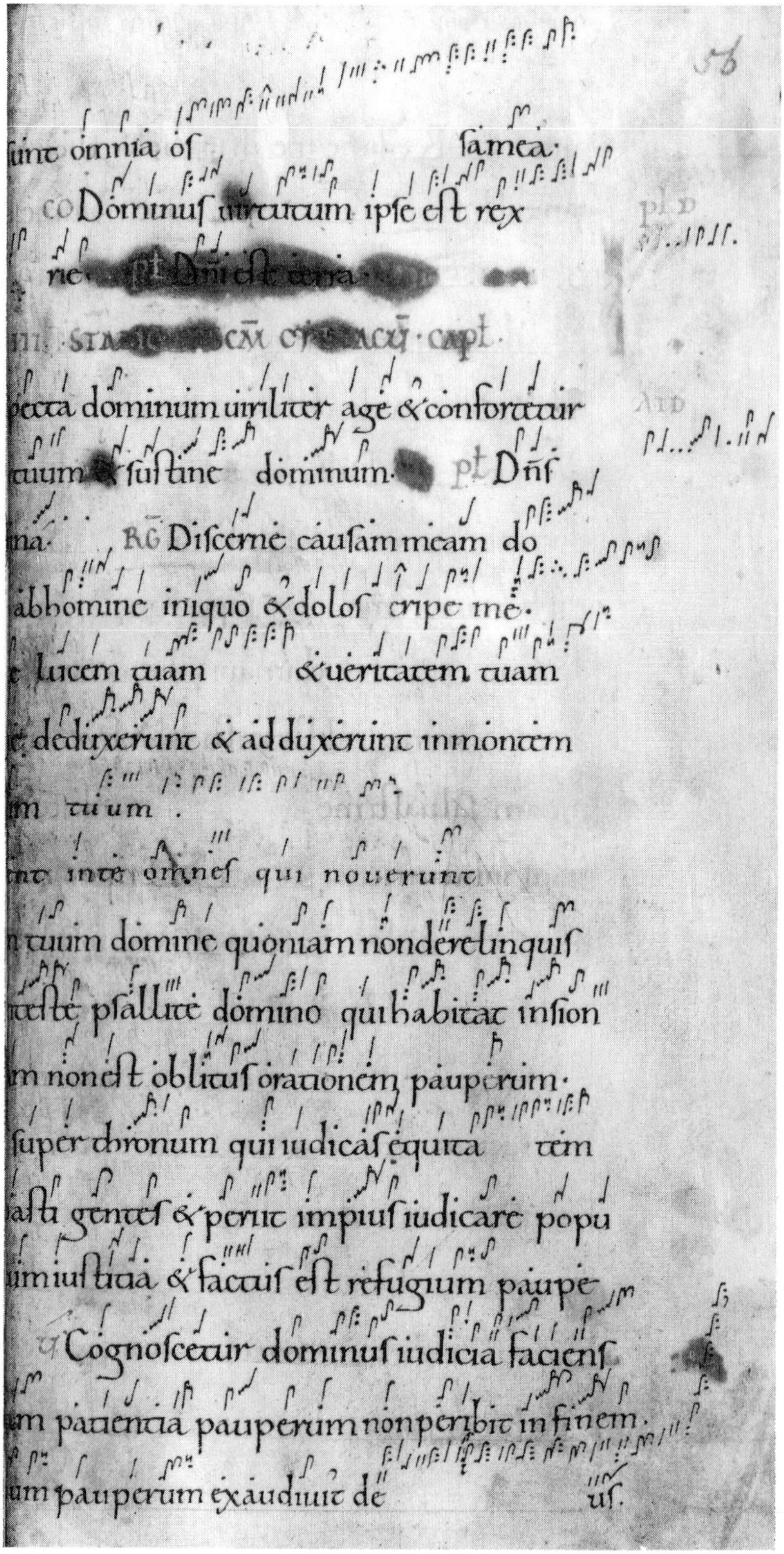

sunt omnia os sa mea.

CO Dominus uircucum ipse est rex

ne pt Dni est terra

III. STA mea cam cr senacu capt.

pecta dominum uiuiliter age & confortetur

tuum & sustine dominum pt Dns

ma RG Discerne causam meam do

abhomine iniquo & dolos eripe me.

lucem tuam & ueritatem tuam

dedixerunt & adduxerunt inmontem

m tu um .

no inte omnes qui nouerunt

n tuum domine quoniam nonderelinquis

teste psallite domino qui habitat insion

m nonest oblitus orationem pauperum.

super thronum qui iudicas equita tem

asti gentes & periit impius iudicare popu

um iustitia & factus est refugium paupe

y Cognoscetur dominus iudicia faciens

m patientia pauperum nonperibit infinem.

um pauperum exaudiuit de us.

PLATE 4 413

London, British Library, Harley 110, fo. 56ʳ. Fragment of a gradual, from Winchester, mid 11th c. English neumes. The plate shows chants for the proper of mass.

As with Plate 2, the use of this leaf for binding a later book has resulted in the loss of valuable notation, in this case on the left-hand side. The surviving labels 'CO' (*Communio*) and 'RG' (*Responsorium gradale*) show that the fragment is from a gradual, and consultation of a modern edition brings one to the masses for Monday and Tuesday after Passion Sunday. The chants are:

(Monday of Passion Week): end of offertory *Domine convertere* V. 1. *Domine ne in ira tua* V. 2. *Miserere mihi domine*; communion *Dominus virtutum* Ps. *Domini est terra*.

[FERIA] III STATIO AD SANCTUM CYRIACUM. CAPITULUM: introit *Expecta dominum viriliter age* Ps. *Dominus illuminatio*; Gradual *Discerne causam meam* V. *Emitte lucem taum*; offertory *Sperent in te omnes* V. 1. *Sedes super thronum* V. 2. *Cognoscetur dominus*.

The rubric 'Station at St Cyriacus' does not, of course, mean that the mass was sung at a non-existent church of St Cyriacus in Winchester. The mass was celebrated at such a church in Rome, and the rubric was simply carried over, even though no longer relevant, when Roman service-books were copied in Francia and elsewhere.

The neumes are exactly like those in one of the famous Winchester tropers, Oxford, Bodleian Library, Bodley 775, perhaps even written by the same hand. If the book had survived complete it would have formed an invaluable complement to the troper, containing in full the proper chants which are either absent or incomplete in the other book. No complete gradual from the Old Minster, Winchester, has survived.

There are numerous features of special interest. The communion has a psalm verse, and both it and the introit are supplied in the margin with the cadence for the psalm tone and with a label indicating the mode: plagalis deuterus (mode 4) and authentus deuterus (a mistake for authentus tetrardus: mode 7), respectively. Such labels are rather uncommon, but are found also in sources from Corbie and Saint-Denis (see Huglo 1971, *Tonaires*, 90–102). The following transcription of the communion was made with the aid of the Worcester compendium Worcester, Cathedral Chapter Library F. 160 (compare the version in *GR* or *LU*).

dicens quinque sicut multierū inquerentur lumen uttrepiā ueniet et uidete alleluia

alle luia. V Ecce precedit uos in galileam ibi eum uidebitis sicut dixit

uobis. uenite IhS Dum transisset sabbatum maria magdalene et maria

iacobi et salome emerūt aromata ut uenientes ungerent ihm alleluia

alle luia V Et ualde mane una sabbatorū ueniunt ad monu

mentum orto iam sole. alleluia Gloria patri et filio et spiritui sancto alle

+ Te dm laudamus. nō dicantur et in saecllō post uitā tuū incipua

menteur huc in modulata uoce. Surrexit dns de sepulcro. RS omi

expletis qro dicant hunc alleluia Et sic incipiatur one ber D Sin dns laudauit

e tū pans et vna Alleluia mox incipiatur A Angelus autem dni descendit

de celo et accedens reuoluit lapidē et sedebat super eum alleluia alleluia

T D nr regnauit. Gloria seculoy amen A Ecce terre motus factus est magnus

angelus autem dni descendit de celo alleluia P Iubilato gloria seculoy amen

Erat autem aspectus eius sicut fulgur uestimenta eius sicut nix alleluia

alleluia. D ds miseratur et miserabitur nobis ad hunc et tu gloria seculoy amen

Quem queritis i sepulcro hoc o christicole R dehpt e iuentu I hesu nazareno crucifixum

o celicole Iam resurrexit non est hic iuxta sicut locutus est I et nunciate quia

surgens fallelia A Surrexit +

PLATE 5 415

London, British Library, Add. 30850, fol. 106ᵛ. Antiphoner, from Santo Domingo de Silos, 11th c. Spanish neumes (northern type). The plate shows office chants for Easter Day, with the *Quem queritis* dialogue added at the foot.

The entire text of the manuscript has been edited by Hesbert (*CAO* 2, manuscript S), and published in facsimile (*Antiphonale Silense*).

The chants are Gregorian, not Mozarabic, but the old Spanish notation used for Mozarabic chants has been retained, in preference to the Aquitanian neumes found in some comparable Spanish sources (and in the additions to the first pages of this manuscript).

After the final responsory of the Night Office (*Dum transisset sabbatum*, line 3) there is a text incipit for the Te Deum (line 7). A cross in the margin directs the singer to three lines added at the foot of the page, the *Quem queritis* dialogue. The rest of the main text is taken up with Lauds antiphons.

Verse and Gloria for the responsory *Dum transisset* use a version of the standard tone for mode 4. The first half is easily transcribed, but for the rest one would have to locate an antiphoner of the same tradition with diastematic notation.

A provisional transcription of the Easter dialogue might be attempted with the aid of contemporary Aquitanian manuscripts (Paul Evans 1970, *Early*, 155, MMMA 3. 246); but some text variants (such as 'loquutus est' instead of 'praedixerat') are rare, and after 'o celicole' the melody follows an individual line.

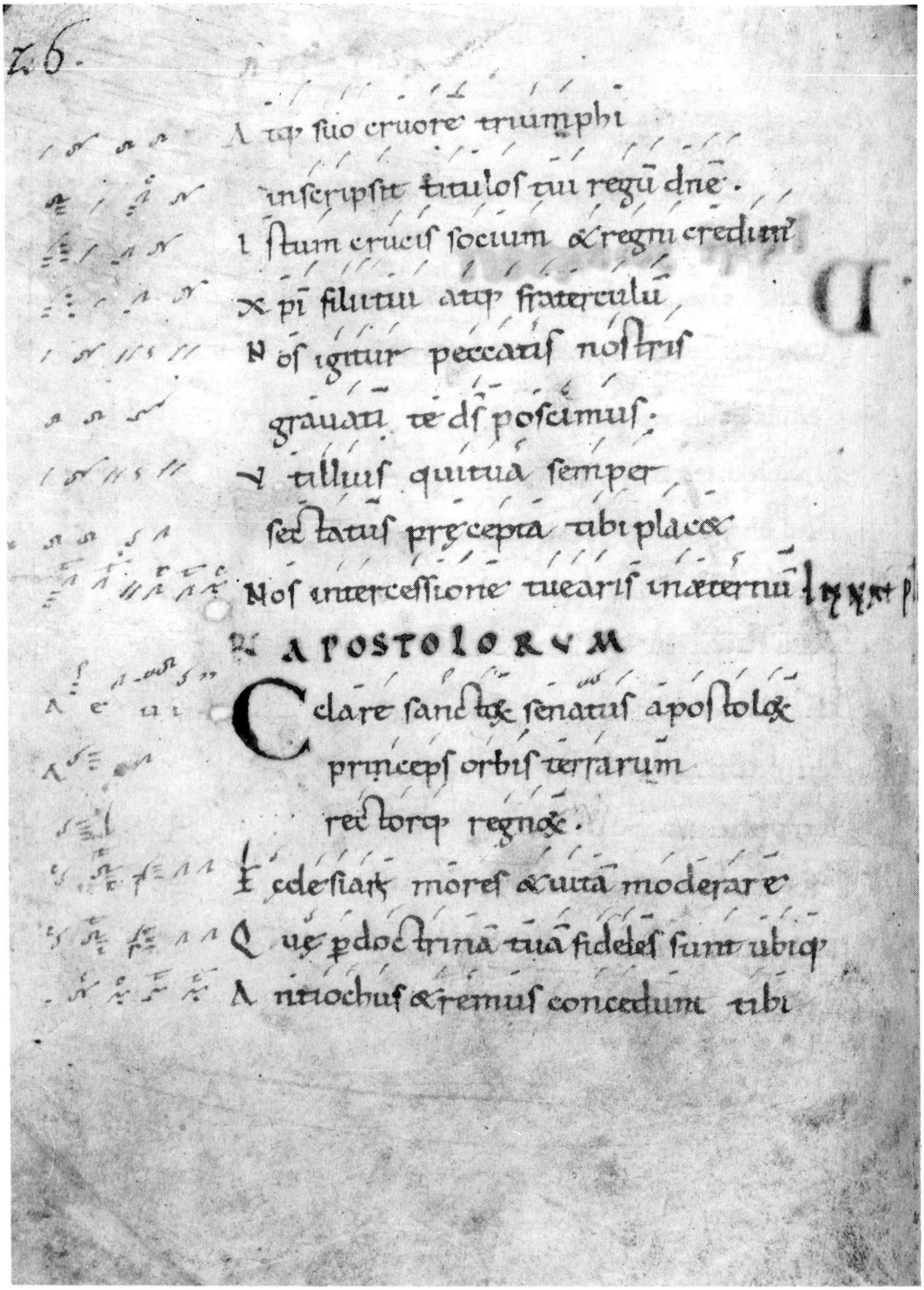

A tuo suo cruore triumphi

inscripsit titulos tui regu dne.

Istum crucis socium & regni credidit

xpi silutiu atqp fraterculu

Nos igitur peccatis nostris

grauati te ds poscimus:

ut illius quitua semper

sectatus precepta tibi placecc

Nos intercessione tuearis inaeternu

APOSTOLORVM

Clare sanctae senatus apostolae

princeps orbis terrarum

rectorqp regnae.

Ecclesiastap mores & uita moderare

Quae pdoctrina tua fideles sunt ubiqp

Antiochus & remus concedunt tibi

PLATE 6 417

London, British Library, Add. 19768, fol. 16ᵛ. Tropes and sequences, from the monastery of St Alban at Mainz, mid 10th c. German neumes. The plate shows sequences.

The page shows the end of the sequence for St Andrew, *Deus in tua virtute*, and the start of the sequence for Apostles (IN NATALE PLURIMORUM APOSTOLORUM), *Clare sanctorum senatus apostolorum*, whose text is attributed to Notker of St Gall. In addition to the neumes entered over each syllable of text, the melody for each phrase is written in the margin. The notation in the margin includes occasional significative letters, for example in line 9. The next line of melody, the first for *Clare sanctorum*, is underlaid with the vowels of the word 'Alleluia'.

The earliest German copies of sequences in staff-notation are nearly all of the 13th century or later, and one is therefore tempted to look to earlier sources for a guide to transcription. In the example (left-hand side) the start of the melody for *Clare sanctorum* is transcribed with pitches derived from Rome, Biblioteca Casanatense 1741, fo. 118ʳ (facsimile in Vecchi 1955). The significative letters used in lines 13–15 are: *e* ('equaliter'): lines 13 and 14 must start with G

t ('trahere'): more deliberate delivery

c ('cito'): quicker delivery

i ('iosum'): interval greater than a second at descent

s ('sursum'): interval greater than a second at ascent

Plate 18, which shows the same sequence in the printed Mainz gradual of 1500, gives other pitches, which do not correspond as well with the neumes of London 19768 (right-hand side in the example here). In this case the use of an earlier source, as opposed to a later one from the same region, is justified. London 19768 is nevertheless a monastic source (from the Benedictine abbey of St Alban) and one's first recourse should be to a later manuscript from the same monastery or one in close relation to it.

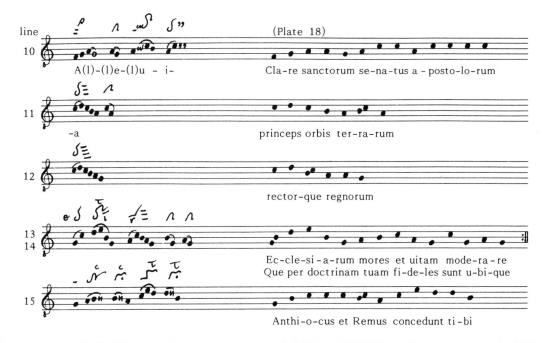

taris humana reppulit uetustate. Sedt. P̄ cōm

Sacrasti dn̄e familiā tuā muneribꝰ sacrisci. q̄s semp

intuencione nos resoue. cuius sollēnia celebramus P̄

HODIE cantandus est nob puer quem ignebat Admissa MAIOREM

ineffabiliter ante tempora pater eeundem subtempora generauit inclitamai.

℣ Quis est iste puer quem tam magnis preconus dignum uociferatis dicere

nobis ut con laudatores esse possimus. ℣ Hic enim est quem presagus ӕ electus

sim nīsta dei adterras uenturum preuides logne ante prenotauit sique p̄dixit.

UGRNATVS GST NOBIS

ӕ filius datus est nobis cuius imperium sup humerum

eius ӕ uocabitur nomen eius magni consili angelus.

P̄ Cantate domino cam nouū q̄ mirabilia fecit. EVOVAE

Kẏ rie kleyson. Kyrie leyson ẏrieleẏsō.

X peristele yson X peristele yson. X peristele yson.

kẏr rie leyson kẏ rie leyson. Kyrie leẏson.

Gloria in excelsis deo Et interra pax hominibus bone uoluntatis

laudamus te benedicimus te grās agimus tibi p̄pter magnā

gloriam tuam domine deus agnus dei filius patris qui tollis peccata

mundi misere nobis. qui tollis peccata mundi sostipe deprecacio

nem nostram. qui sedes ad dexteram patris miserere nobis

PLATE 7   419

Oxford, Bodleian Library, Canonici Liturg. 350, fo. 4ʳ. Notated missal, Aquileia?, early
12th c. German neumes. The plate shows proper chants from the third Mass of
Christmas Day, with introit trope, Kyrie, and Gloria.

The illuminator never entered the large initial in the middle of the page, but its shape
is unmistakable. At the top of the page is the last line of the postcommunion prayer
Huius nos domine quesumus sacramenti, then (line 2) another postcommunion,
Saciasti domine familiam tuam. These prayers conclude the second Mass on
Christmas Day. The rubric for the third Mass follows: 'Ad missam maiorem'. First
comes the dialogue trope attributed to Tuotilo of St Gall, *Hodie cantandus est nobis*,
for the introit *[P]uer natus est nobis* with Ps. *Cantate domino canticum novum*. Then
the Kyrie (Melnicki 1955, no. 74) and Gloria (Bosse 1955 no. 21) are notated.

The fullness of the material is surprising. Notated missals are not common at this
date, and the inclusion of tropes and ordinary-of-mass chants in place beside the
proper chants in order as they were sung is also unusual. It is, however, sometimes
encountered in Italy, a clue to the provenance of the book. The neumes are German,
but these were also used in a few North Italian centres: Bobbio, Monza, Moggio. On
the basis of the local saints honoured in the Sanctorale, the manuscript has been
assigned both to Brescia and Vercelli. If the identification of Kyrie 74 is correct,
however, another link emerges, for this rare melody is listed by Melnicki only for
sources from Moggio and Aquileia. A transcription may be made with the help of
these later sources.

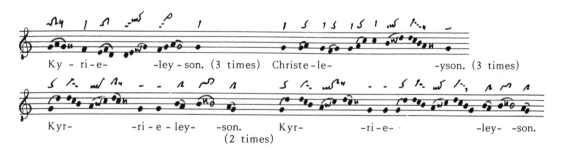

Sca Scolastica intercede pnobis. O m̅ nes sancti intercedite

pnobis Exaudi deus Voces nr̄as Exaudi xp̄e O rationem nr̄am

Exaudi deus Miserere nobis. Accendite

Kyrie leison Ae v ia & Confitemini

d̅o mino quoniam bonus quo niam insec̅ulu̅ misericor

dia ei uf Tract Lauda te dominum Ordo sepulch̅.

Quis reuoluet nobis abhostio lapidem quem tegere sanctum cer

nimus sepulchrum Angel̅ Quem queritis insepulchro o

xpicole Mulier̅. Ihm nazarenu̅ crucifixu̅ o celicole Angl̅s

Non est hic surrexit sicut predixerat ite nunciate quia surrexit

de sepulchro Angl̅s Non est hic quem queritis sed cito euntes nun

ciate discipulis eius & petro quia surrexit ih̅c a Venite & uidete

a Cito euntes a Ad monumentu̅ uenimus gementes angelum

domini sedentem uidimus & dicentem quia surrexit ih̅c Chor̅

Currebant duo simul Angl̅s z Mulier̅. Dicant nunc iude

i quomo do milites custodientes sepulchrum p̅diderunt re

gem aut lapidem positioni̅s quare non seruabant pe tram

iusticie aut sepultu̅m reddant aut resurgentem adorent no

biscu̅ dicentes aeuia a e v ia Seniores duo.

Cernitis o socij ecce lintheamina & sudarium & corpus non est i̅

sepulchro inuentu̅ a Surrexit dn̅s. Te deum laudam̅

PLATE 8   421

London, British Library, Arundel 156, fo. 35ʳ. Gradual, sacramentary, and lectionary, Würzburg diocese, early 13th c. German neumes. The plate shows Mass on Easter Eve and the 'Ordo sepulchri' liturgical drama in the Night Office.

Although the book basically contains material for the celebration of mass, an 'Ordo sepulchri' ceremony has been entered (rubric end of line 6), which ends with the Te Deum, showing that it was performed at the end of the Night Office. At the beginning of the page are the concluding chants for the special short form of mass at midnight between Holy Saturday and Easter Sunday: the end of a litany, Kyrie (only the incipit; Melnicki 1955, no. 39), *All. Confitemini domino quoniam bonus* and tract *Laudate domium omnes gentes* (incipit only).

The Visitatio is not fully notated, for reasons that are unclear. The text of this ceremony has been edited by Lipphardt (1975–81, no. 371).

The German neumes are thicker than those in the previous plate, and in some cases already have the form known as Gothic or 'Hufnagelschrift' (see Plate 18). The pes is still round, but the scandicus (e.g. line 9, 'crucifiXUM') consists of two lozenges joined lightly to a horseshoe-nail, as in many examples of Gothic notation.

The example transcribes the start of the Alleluia, with pitches derived from Graz, Universitätsbibliothek 807 (PalMus 19). The apostrophe and quilisma are still to be found, but the oriscus has disappeared and looks like a virga (horseshoe-nail), as in the penultimate note of the example.

A(ll)e-   -(l)u-   -ia.   V. Confi - temi - ni   do-   -mi-no

London, British Library, Royal 8. C. xiii, fo. 6ᵛ. Ordinary-of-mass chants and sequences, from Normandy, 11th–12th c. French neumes. The plate shows a Gloria with trope verses.

PLATE 9   423

The notation in Plates 9, 10, and 15 can be seen in a chronological progression, with allowance for the differences of provenance. Many of the neumes on Plate 9 have obvious heads or tails (virga, pes, clivis), which then makes clear their position on the staff in Plate 10.

In lines 11–13 there are several examples of the pes stratus, or pes + oriscus. They occur here in the melismas with associated prosulas which occur towards the end of this set of Gloria trope verses. The page begins in the middle of a Gloria (no. 39 in the catalogue of Bosse 1955). On the previous side there have been copied the opening verses of the trope set *Laus tua Deus*. At the end of line 4 comes the trope verse *Ut* (or *Et*) *hominem celo redimeres*, as far as I know unique to this manuscript. Trope and Gloria verses then alternate until line 10. Here, after the trope verse *Conditor generis*, another trope verse begins, *Sapientia dei patris*. It is syllabically composed, and each of its phrases is repeated melismatically to the syllable 'E'. The first melisma actually bears the syllable 'Per-', the start of a word completed at the beginning of line 15: '[P]ermanebit in eternum'. The reason for all this is that the 'Permanebit' melisma derives from another context (see above, II.23.iii and vii) and was only subsequently given the 'Sapientia' text.

Although London 8. C. xiii has many close relatives among Norman and Norman-Sicilian tropers, the latter do not contain *Sapientia*. A version very close to London 8. C. xiii in staff-notation does not seem to exist. The transcription is therefore a compromise version based on both Aquitanian and Italian sources. (Compare the 'O rex gloriae' prosula, Ex. II.23.6.)

iserere nobis pie rex dñe ihu xpe. Xpe audi nos.

ac surgat eps aboratione. ñ dicat dñs uobiscu s;
tantum ñ choc & dicat oremus. & diaconus fe
ctamus genua. leuate. O R A T I O.

Magnificare dñe ds ñr ñ scis tuis. & hoc in
templo humane edificationis psentia spi
ali benignus appare. ut qui oñia infi
nis adoptionis oparis. ipse semp in tua heredita
te lauderis. p. Deinde scribat pontifex sup pauim
ti cum cambuta sua. abecedariu grecum. ab angu
lo orientali usq; in dextrum angulu occidentale.
dicens hanc antiph'.

Fundamentum aliud nemo potest po
nere pter illud denique. quod positum est a xpo doñino. fund.

lfa. beta. gaina. delta. e. fima. zeta. eta. teta. iota. cappa. lapda. m
A B Γ Δ E S T H Θ I K Λ ⲁ ⲉ

in. psi. o. pi. ro. si. tau. gui. fi. xi. cofo. longu. chi.
ⲍ ⲟⲛ ⲡ ⲣ ⲧ Ⲩ φ ⲭ ⲱ ⲱ Λ

chile.
ⲁ Et ad extro angulo orientali scribat similit
usq; in sinistru occidentem dicat hanc antiph.

Hec aula acceptat a deo gram benedictionis & misediam a xpo ihu. Mag.d.
a b c d e f g h i k l m n o p q r s t
u x y z &. Post ueniat in medium ecclie &
dicat orationem hanc. Dñs uobiscum.

S qui scm moysen precunctis inilibz ista

PLATE 10   425

Cambridge, University Library, Ll. 2. 10, fo. 19ʳ. Pontifical, from Ely, mid 12th c. Early Anglo-Norman square notation. The plate shows chants for the Dedication of a Church.

This pontifical is one of a group of closely related manuscripts written in the century or so after the Norman Conquest of England. The text of one of them was edited by Wilson (1910), and the page shown here, part of the ceremony for the Dedication of a Church, corresponds largely to p. 105 of his edition. The end of a litany is visible at the top of the page. The next three lines (written in red ink, though this is hardly detectable in the photograph) are a rubric saying that the bishop shall rise from praying for the next part of the ceremony, and instructing him on the opening words. The prayer ('Oratio') *Magnificare domine* follows. On line 9 another rubric begins, for the bishop must now trace the Greek alphabet with his crozier (*cambuta*) from the left (i.e. north) eastern corner to the right western corner of the church, singing the antiphon *Fundamentum aliud*. The Greek alphabet is spelt out for him in the middle of the page: not only are the shapes depicted that he must trace, but a guide to pronunciation is also given. The bishop then traces the Latin alphabet from the right eastern to the left western corner of the church, singing the antiphon *Hec aula accipiat*.

Although many problems of transcription disappear with the introduction of the staff, manuscripts do not lose all their individuality of script and arrangement. The use of a wavy punctum for a note below the semitone step, the mi-neume (see IV.6.1), is characteristic of many Norman and English sources of the twelfth century: there is one in the first antiphon at 'NEmo'. The use of a b-flat clef in both antiphons is typically English. The opening of the first antiphon is given in transcription.

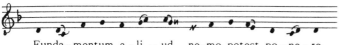

Funda-mentum a-li-ud   ne-mo potest po-ne-re

Sancte iohannes meritorum tuorum copias nequeo digne canere

Quod itaq̃: de hac ratione factu̅ e̅. ex decim̃ib; posse fieri
mille diximus̃ e̅. Sed ne grauis tibi imponat necessitas. qd ad
hunc modu̅ ꝰ usq; cuilibet simphonie minus quinq; accidant
uoces. & ipsas quinq; transgredi sepe ad uoces inuenietur. ut
tibi paulo liberius liceat euagari. alius n̅e uersu̅ subtinge
uocaliu̅. sed ita sit diuersu̅. ut a tercia loco prioris inci
piat hoc modo. FABCDEFGa hcdefga

a e i o u a e i o u a e i o u a
o u a e i o u a e i o u a e i o

Vbi cu̅ duob; ubiq;: sub sonis in quib; quinq; habeant uocales.
cu̅ uidelicet cuiq; sono & una subter & altera. satis e̅ liberior
facultas accedit. & p̃ductiori & contractiori plurimaru̅ moru̅
incedere. Vnde & hoc nunc uideamus. quale̅ simphoniam
huic rithmo sue uocales attulerint.

FGabcdef

e i o u a e i o
u a e i o u a e

G G G a G a a a a h G a c h c d
Lingua̅ refrenans temperet ne lixis orror insonet
e d e d c a c G a c h G a f o o
uisum ouendo contegat ne uanitatis auriat
insola enim ultima parte hoc argumentu̅ relinquimus. ut
meliꝰ suo retardo conueniens redderem. cu̅ itaq; suis

PLATE 11   427

London, British Library, Add. 10335, fo. 9ʳ. Theoretical writings and tonary, north Italy, early 12th c. Italian notation on dry-point lines, and alphabetic notation. The plate shows part of Guido of Arezzo's *Micrologus*.

The rather wide vertical spacing of the notes at the top of the page is caused by the use of the lines scratched for text as stave-lines.

Guido of Arezzo's famous treatise *Micrologus* has been edited by Gerbert (GS ii. 2–24) and Smits van Waesberghe (CSM 4), and translated into English by Babb (1978, 57–83).

The passage explains Guido's peculiar method of composing chants by deriving notes from the vowels of the text to be sung. The text *Sancte Iohannes* has just been set in this manner, whereby the five vowels *a*, *e*, *i*, *o*, and *u* correspond to the notes C–G. But Guido intimates that such a small number of notes is rather hampering, and he then sets out a table allowing greater freedom of melodic movement. The whole scale from Γ to $\overset{a}{\overline{a}}$ is set out, and the five vowels assigned twice over, the second time a third higher than the first. An example, *Linguam refrenens*, follows, in which Guido has stayed within the compass *F–f* suitable for mode 7. Guido says that he has allowed a deviation from the system at the end ('In sola ultima parte hoc argumentum relinquimus') in order to bring the melody back to the tetrardus mode ('ut melum suo tetrardo conveniens redderemus'); but it is not clear to which note he is referring. (On Guido's 'automatic' method of composition, see Smits van Waesberghe 1951, 'Improvisation'.)

& ubiqz gtaf acf agere;

Domine sancte pater omni-

potens eterne ds p xpm

dominum nostrum. Per

quem maiestatem tuam

laudant angeli adorant

dominationes tremunt po-

testates; Celi celorumqz

uirtutes ac beate Seraphin

socia exultatione concel-

ebrant; Cum quibz &

nostras uoces ut admitti

iubeas deprecamur sup-

pliter confessione dicentes

Et ideo cum angelis

& archangelis cum thro-

nis & dominationibus;

Cumque omni militie cele-

stis exercitus ymnum

glorie tue canimus sine

fine dicentes; In sollem-

nibz. manibz.

per omnia

secula

seculorum. amen.

Dns uobiscum. & cu

spu tuo. Sursum corda

PLATE 12 429

London, British Library, Egerton 3511 (*olim* Benevento, Biblioteca Capitolare, VI. 29), fo. 174ᵛ. Sacramentary with priest's chants notated, from Benevento, 12th c. Beneventan notation on dry-point lines. The plate shows prefaces.

The chief contents of a sacramentary are the proper prayers of mass: collect, secret, and postcommunion. It may also contain the prefaces proper to various types of feast and seasons of the year, and occasionally these solemn chants are notated. The page shown here contains mostly the preface 'In festivitatibus', and towards the end of the right-hand column begins the preface 'In sollemnitatibus'.

Both text and music are written in the characteristic Beneventan scripts. The music is not elaborate and the full available range of neumes is therefore not called for. Several of the characteristic Beneventan liquescent neumes are present:

Pes made liquescent; the end of the tail indicates the pitch of the second note.

Virga plus descending liquescence; the final short turn downward indicates the direction of melodic movement.

Punctum plus descending liquescence; the end of the tail indicates the pitch of the second note.

Double punctum (?) plus liquescence.

The start of the preface 'In sollemnitatibus' may be transcribed as given below. I have assumed that the cadences fall on *a* or *G*, as is usually the case for these chants. But this also involves assuming that the first custos is a tone too low.

Per om – ni – a se – cu – la   se – cu – lo – rum. A – men. Domi – nus uo – bis – cum.

Et cum spi – ri – tu tu – o.   Sur– –sum   cor – da.

rita uel Jnevaa. ito euntes picite discipulis qui

A surrexit dominus alleluia alleluia P Magnificat 0r

Concede qs omips ds. ut qui resurrectionis dnice
sollempnia colimus. Innouatione tuj spc amorte
anime resurgamus. P eunde

Isi die resurrectionis me e dicit dominus al le lu ia

congregabo gen tes et colligam reg na et ef fundam

super uos aquam mundam alle lu ia pfine Allelu

ia aeua aeua al le luia alleluia p Laudate dn Allelu

ia Orynos kebasileu sen e u prepia e

ne dica to enedicxi ton kyrios di na min kerie pte zo

ca ton Kegaresteos seit inicume

ni j tis uskleuthise te Jnevaa. Uenite et ui

dete locum ubi positus erat dominus alleluia Evouae 0j

Pta qs omips ds. ut qui resurrectionis dnice sollep
nia colimus. ereptionis nre suscipe letitiam mere
amur. P eundem dnm

PLATE 13   431

London, British Library, Add. 29988, fo. 74ᵛ. Antiphoner, Rome, possibly St Peter's, mid 12th c. Central Italian notation on dry-point lines with red F-line. The plate shows part of papal Vespers on Easter Eve.

The Vespers service celebrated by the pope on Easter Eve, with its magnificent series of Greek and Latin alleluias, constituted one of the chief glories of medieval Roman chant up until the 13th century. The form of the whole ceremony has been set out by Stäblein (MMMA 2, 111*–118*), and such of the music as appears in another Roman source, Rome lat. 5319, is edited in the same volume. The service consists of several parts, performed at different stations, hence the multiple singing of the Magnificat. The side shown here corresponds to pp. 114*–116* of Stäblein's table (for the previous side of the manuscript see *NG* i. 488), near the end of the first station. The chants are:

line of music
    1: Magnificat antiphon *Cito euntes*
(Second Station)
    3: processional antiphon *In die resurrectionis mee*
    5: antiphon *Alleluia alleluia*
    6: *Alleluia* V. 1. *O kyrios* V. 2. *Ke gar estereosen*
  10: Magnificat antiphon *Venite et videte locum*

The transcription is of the processional antiphon *In die resurrectionis mee* (cf. MMMA 2. 526). One of the few problems which arise in this, as in other manuscripts with central Italian and Beneventan neumes, is to differentiate between the different types of liquescence. In the transcription, each liquescent neume is given above the stave and numbered: From the size of the original sign there seem to be at least two 'degrees' of liquescence. The tiny commas used in 2, 4, 5, 6, and 9 seem to demand less emphasis (melodic or dynamic) than the others. It might be thought that they indicate a light liquescence at the unison, but after 4 comes the form of clivis implying a previous lower note, as at 5, so a lower liquescent note is implied.

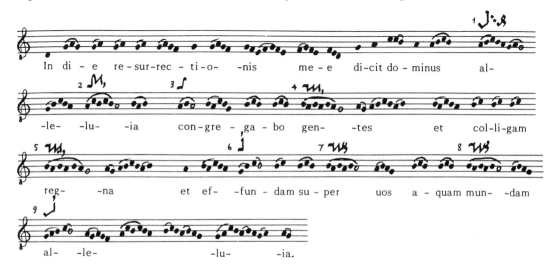

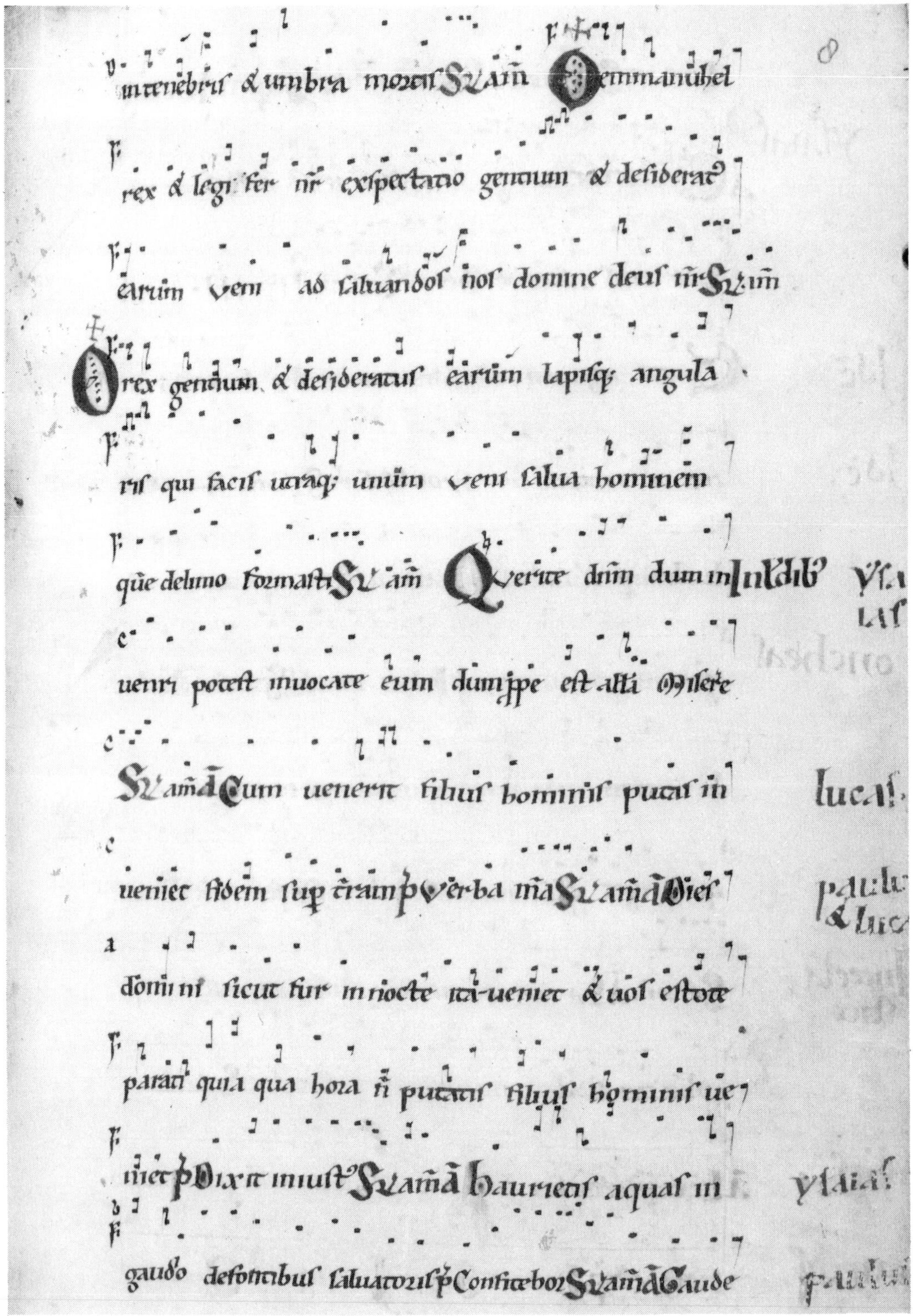

in tenebris & umbra mortis. Euouae. O emmanuhel

rex & legifer nr exspectatio gentium & desiderat

earum veni ad saluandos nos domine deus nr. Euouae.

O rex gentium & desideratus earum lapisq; angula

ris qui facis utraq; unum veni salua hominem

que de limo formasti. Euouae. Querite dnm dum in

ueniri potest inuocate eum dum ipse est alti. Miserere

Euouae. Cum uenerit filius hominis putas in

ueniet fidem sup terram p verba mea. Euouae. Dies

domini sicut fur in nocte ita ueniet & uos estote

parati quia qua hora n putatis filius hominis ue

niet p Dixit iniust. Euouae. Haurietis aquas in

gaudio de fontibus saluatoris p Confitebor Euouae. Gaude

ysaias
luc

lucas

paulus
& luc

ysaias

paulus

PLATE 14   433

London, British Library, Add. 17302, fo. 8ʳ. Diurnal, monastic, Carthusian, 12th c.
Square notation on dry-point lines. The plate shows the end of the series of O-antiphons.

The variety of square (or quadratic) notation known best nowadays is that of books written in Paris in the second half of the 13th and the early 14th century, used not only for plainchant but also. for polyphonic music and the songs of the troubadours and trouvères, and then adapted for the Solesmes/Vatican books of the last hundred years. Other varieties also existed, including that shown here. It is in fact more consistently square than the Parisian type, for even the climacus is written with three square note-heads. Another variety of square notation found in a number of Carthusian books can be seen in London, British Library, Add. 31384 (Briggs 1890, pl. 11), Naples, Biblioteca Nazionale VI. E. 11 (PalMus 2, pl. 47), and Toledo, Archivo Capitular 33. 24 (PalMus 3, pl.206). English notators, on the other hand, tended to favour the use of lozenges (see Plate 15). Compare the following types:

	Paris	London 17302	other Carthusian	England
virga				
punctum				
pes				
clivis				
climacus				
porrectus				

D, F, a, and c clefs may be seen on this page, but none of the dry-point lines has been inked in.

The chants shown here are:

(end of the O-antiphons)
  A. *O oriens splendor* (end only)
  A. *O Emmanuhel*
  A. *O rex gentium*
'In laudibus' (Lauds of Monday in Ember Week of Advent)
  A. *Querite dominum*
  A. *Cum venerit filius hominis*
  A. *Dies domini*
  A. *Haurietis aquas*
  A. *Gaude* . . . (first word only)

The texts of the O-antiphons are of course non-scriptural, but the sources of the other antiphons are indicated in the margin: Isaiah, St Luke, etc.

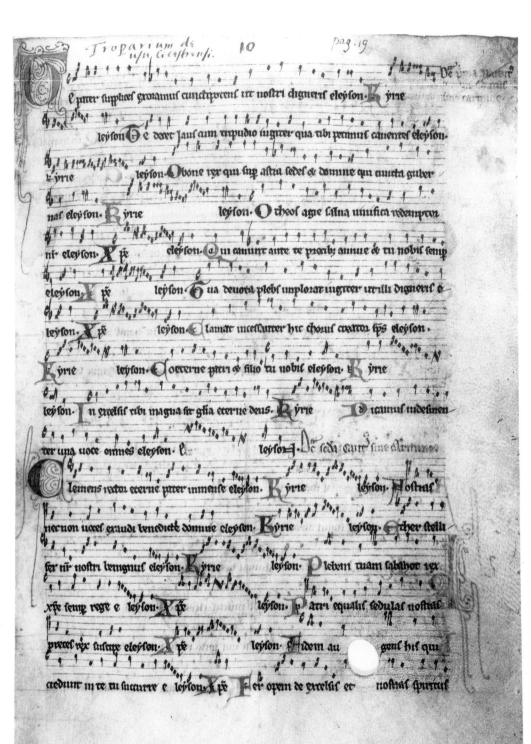

De una persona

Et pater supplices exoramus cunctipotens ut nostri digneris eleyson. Kyrie

leyson. Te decet laus cum tripudio iugiter qua tibi premimus canentes eleyson.

kyrie leyson. O bone rex qui sup astra sedes et domine qui cuncta guber

nas eleyson. Kyrie leyson. O theos agye salua uiuifica redemptor

ni eleyson. Xpe eleyson. Qui canunt ante te precibus annue et tu nobis semp

eleyson. Xpe leyson. Qua deuota plebs implorat iugiter ut illi digneris e

leyson. Xpe leyson. Clamat incessanter hic chorus creator tps eleyson.

Kyrie leyson. O eterne patri et filio tu nobis eleyson. Kyrie

leyson. In excelsis tibi magna sit gloria eterne deus. Kyrie Dicamus indesinen

ter una uoce omnes eleyson. De secunda cum sine eleyson.

Clemens rector eterne pater inmense eleyson. Kyrie leyson. Nostras

necnon uoces exaudi benedicte domine eleyson. Kyrie leyson. Ether stelli

fer nr nostri benignus eleyson. Kyrie leyson. Plebem tuam sabahot rex

xpe semp rege e leyson. Xpe leyson. Patri equalis sedulas nostras

preces rex suscipe eleyson. Xpe leyson. Fidem au gens his qui

credunt in te tu succurre e leyson. Xpe Fer opem de excelsis et nostras spiritus

PLATE 15   435

Oxford, University College 148 (kept in Bodleian Library), fo. 10ʳ. Ordinary-of-mass chants and sequences, from Chichester, 13th c. English square notation. This plate shows the start of the Kyries.

As already mentioned (see Plate 14), some English notators of the 13th century favoured the diamond or lozenge shape as a basic notational element. Even the head of the virga tends to slant. The liquescent forms can be seen most easily at 'eLEYson':

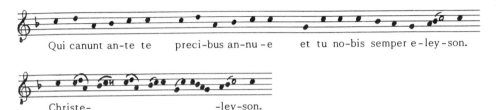

The oriscus as a repetition of the previous note (as in the pes stratus; cf. Plate 9) can still be detected:

The transcription gives the fifth verse of the first Kyrie on the page (line 5):

Qui canunt an-te te   preci-bus an-nu – e   et tu no-bis semper e-ley-son.

Christe-   -ley-son.

This is the first page of a collection of Kyries, with and without Latin verses (listed Hiley 1986, 'Ordinary', 39). The rubrics are still faintly discernible at the end of lines 1 and 10:

first Kyrie, *Te pater supplices*: 'Dominica prima adventus domini cantus sine carmine'
second Kyrie, *Clemens rector*: 'Dominica secunda cantus sine carmine'

Te pater supplices was frequently sung on Christmas Day, which is no doubt why it appears first here. The rubric indicates, however, that the melody alone ('cantus'), stripped of the Latin verses, was also to be sung on the first Sunday of Advent; similarly with *Clemens rector* on the second Sunday.

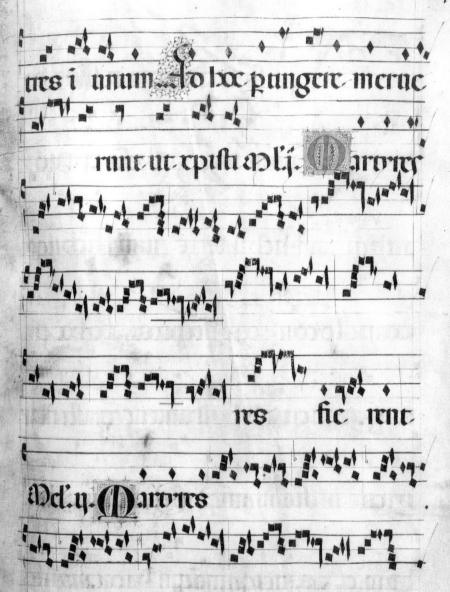

PLATE 16   437

Oxford, Bodleian Library, Lat. Liturg. a. 4, fo. 130ʳ. Chants of the Milanese rite, summer part, 1399. Milanese notation. The plate shows part of the responsory *Hi sunt qui secuti* for SS Gervase and Prothase.

The scribe of this manuscript, the priest Fazio de Castoldis, wrote at the end that he had finished work on 31 January 1399, at the age of 40.

Hi sunt secuti is a responsory 'cum infantibus', one of those sung after the hymn at Vespers on the most important saints' days (discussed by Moneta-Caglio 1957). The performance of the responsory involved much repetition and the inclusion of two long melismas, the *melodiae primae* and *melodiae secundae*, during the successive repeats of the respond. The overall form seems to have been:

Solo	Respond	*Hi sunt qui secuti*
		(second part) *Ad hoc pertingere meruerunt ut Christi martyres fierent*
Schola	Respond	*hi sunt qui secuti* (as sung by soloist)
Solo	Verse	*Ecce quam bonum*
Schola	Respond	(second part) *Ad hoc pertingere* with *melodiae primae* at 'martyres
Solo	Respond	*Hi sunt qui secuti* (first part)
Schola	Respond	(second part) *Ad hoc pertingere* with *melodiae secundae* at 'martyres'

Or possibly during the final repeats, the respond was sung complete by the soloist and complete again by the schola, with *melodiae*.

The schola was constituted by the *magister pueri* and the boys themselves, hence the nickname of such responsories. Sometimes a boy might have been entrusted also with the soloist's part.

The notation in this massive book (55 × 39 cm.) is large enough for singers to perform from, and presents no difficulties of transcription. The small virga-like note which follows a larger one of the same pitch, as in line 1 at 'per-TIN-GE-re' is a repeated note, not a sign of liquescence. A transcription of the *Melodiae primae* is given. Lines divide the melody into phrases, which I have labelled *a–d*; it transpires that the *melodiae* repeat themselves.

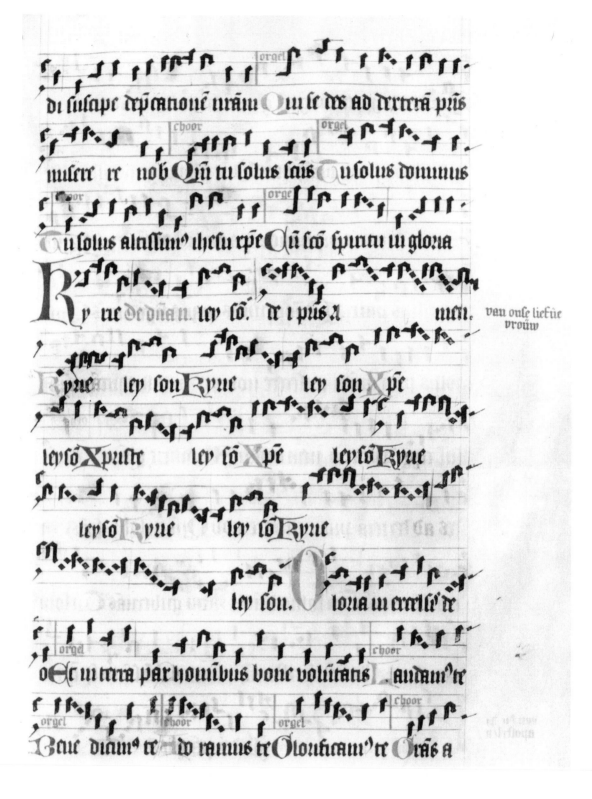

di suscipe deprecatioem nram Qui se des ad dexteram pris

misere te nob Qm tu solus sancis Tu solus dominus

Tu solus altissim' ihesu xpe Cm sco spiritu in gloria

Ky rie de dña ii. ley son. de i pms A men.

van onse liefue
vroüw

Kyrie ley son Kyrie ley son Xpe

leysô Xpuste ley sô Xpe leysô Kyrie

leysô Kyrie ley sô Kyrie

ley son. loria in excelsis de

o Et in terra pax homíbus bone voluntatis Laudam' te

Bene dicim' te Ado ramus te Glouficam' te Oras A

PLATE 17   439

Edinburgh, University Library 33, fo. 138ʳ. Gradual, from Holland, 15th c. Gothic notation. The plate shows chants for the ordinary of mass.

Many graduals contain not only the proper chants of mass but also everything else that cantor and schola should sing: tropes (in the earlier Middle Ages), ordinary-of-mass chants, and sequences. The present manuscript concludes with a collection of ordinary-of-mass chants, some of which are provided with indications for dividing the performance between choir and organ (the latter presumably using the chant as a cantus firmus for improvisation). The two Glorias on this page are examples of this. The melodies are well known as nos. I and IX in the Vatican/Solesmes service-books. In between comes a Kyrie, with the Latin rubric 'De domina ii' (a second Kyrie, or Kyrie–Gloria pair, for Our Lady), and, in the margin, the Dutch rubric 'van onse liefue vrouw'. This Kyrie–Gloria pair were frequently sung at Marian masses, and are assigned thus in the Vatican/Solesmes service-books.

The notation is characterized above all by the virga shaped like a horseshoe-nail ('Hufnagel'). A transcription of the opening of the Kyrie is given:

Ky-   –ri – e-   –ley – son.

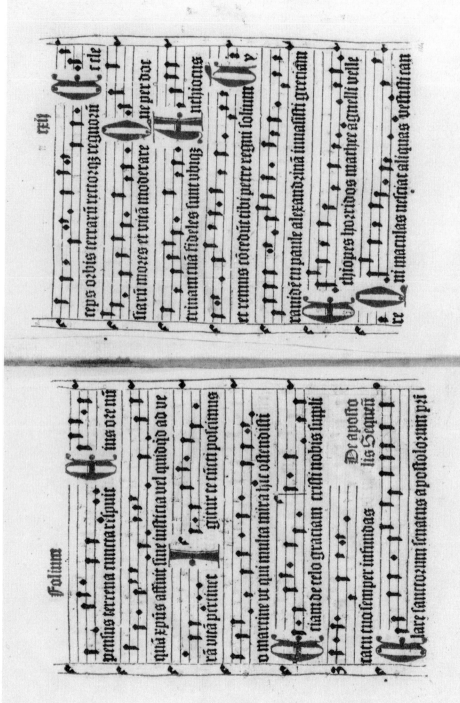

PLATE 18   441

London, British Library, printed book IB. 8668, fos. 113^v–114^r. Printed gradual of Mainz, 1500. Gothic notation. The plate shows sequences.

Early printed service-books with music are no different in arrangement from their manuscript predecessors. The present volume contains the proper chants of mass, then a collection of seventeen sequences (an extremely conservative collection, in which *Victime paschali laudes*, of the 11th century, is the most recent item), and finally Kyrie–Gloria pairs, Sanctus, and Agnus Dei chants.

The end of the sequence for St Martin, *Sacerdotem Christi*, occupies the first six lines of the left-hand page. On the last line begins the sequence for apostles, Notker's *Clare sanctorum senatus apostolorum*, which we have seen in a Mainz version of some 540 years earlier on Plate 6.

V

Plainchant and Early Music Theory

V.1. INTRODUCTION

Vivell 1912; Riemann 1920; Smits van Waesberghe 1969; Gushee 1973; Palisca, 'Theory, Theorists', *NG*; Schueller 1988; Dyer 1990.

In a general sense a great part of the theoretical writing of the Middle Ages is of relevance to the plainchant repertory. For example, much could be made of the place of 'music' in the history of ideas in the Middle Ages, in the medieval cosmology. But this is not the place for such a demonstration, and no attempt can be made here to summarize all the possible themes. Instead, the following sections will comment on a restricted number of topics which touch directly upon chant composition and practice. The two most important of these are: (i) the way in which the chant repertory was aligned with the pitch system of Greek antiquity, and the concomitant coupling of the modal system with particular pitches; and (ii) the identification of what constitutes the modal character of particular chants. Some of the material treated here borders on the speculative part of music theory (music as the embodiment of celestial harmony), some of it on the pedagogical (how to notate music, how to read it), but no full account of these is attempted.

The growth of a music-theoretical literature from the ninth century onward was stimulated by two factors. The first was the interest in and the desire to emulate the writings of late antiquity that had survived to the time of the Carolingian renaissance. The second was the need to abstract from the practice of music in the liturgy some rules to regulate that practice and make it more uniform from place to place.

The concept of 'musica' inherited by the Middle Ages from classical antiquity was that of musical science rather than practice. The Greeks themselves did not write about their own practical music. So it is not to be expected that 'musica', music theory, should have been of direct relevance to Christian chant. The chant repertory had of course grown up quite independently of any theoretical system, and it is interesting to see how medieval writers wrestled with the heritage of antiquity (mainly as transmitted by Boethius), drew from it what was of practical use in the daily performance of the liturgy, and used some of its simpler doctrines as the foundation for speculative theory, all in a completely new intellectual and musical environment.

The other fascinating process at work through the ninth to twelfth centuries is the bringing into line of the chant repertory with what had at first been theoretical concepts, which then exerted a retroactive influence on chant itself. It was a short step, though a momentous one, from the recognition that the majority of chants in such and such a mode displayed this or that tonal behaviour, to the deliberate alteration of chants which, as it were, stepped out of line with the majority.

The next section gives a brief account of the musical theory of antiquity which was known to the early Middle Ages. Subsequent sections describe the way in which the pitches sung in chant were assigned places in the Greek Greater and Lesser Perfect System, together with the growth of a literature about the modal quality of chants.

V.2. THE LEGACY OF ANTIQUITY

The principal writings on music of late antiquity known to the Middle Ages were, in chronological order, those of Censorinus (3rd c.), Calcidius (4th c.), Augustine (354–430), Macrobius (early 5th c.), Martianus Capella (early 5th c.), Boethius (*c*.480–*c*.530), Cassiodorus (*c*.485–*c*.580), and Isidore of Seville (*c*.599–636). Of these the last four were the most influential, Boethius providing by far the most substantial body of material. Censorinus, in three chapters of his work *De die natali*, mostly borrowing from Varro (1st c.), was concerned with speculative music theory: music and the cosmos, and its ethical influence on man (Censorinus, ed. Rocca-Serra 1980, ed. Sallmann; see also Sallmann 1983). Material of this sort is common to most of the above writers and was repeated by many medieval authors. Also repeated in the Middle Ages, but hardly expanded, were explanations of the harmony of the spheres, such as the one in Macrobius' commentary on the *Somnium Scipionis* of Cicero (ed. Willis 1970, trans. Stahl 1952). Less commonly encountered is the type of material in Augustine's *De musica* (PL 32. 1081–1194, trans. Taliaferro 1947, Knight 1949). Music is seen here, as elsewhere, as an embodiment of proportion. The concept was generally illustrated with reference to the harmonic intervals between notes—the octave, fifth, fourth, and so on. But in Augustine's treatise the proportions are those displayed in the metres and rhythms of metrical verse (see Crocker 1958, '*Musica*'). Music's power over living beings—not only humans but also animals and even plants—is again a principal preoccupation in the musical part of Martianus Capella's *De nuptiis Philologiae et Mercurii*, which then proceeds to borrow material by Aristides Quintilianus (3rd–4th c.) on the classification of musical knowledge, including its pitches, scale systems, species, and intervals (ed. Willis, trans. Stahl; see also Stahl 1971). The music-theoretical texts of Cassiodorus (ed. Mynors 1937) and Isidore (ed. Lindsay 1911) are likewise mainly concerned with the classification of musical knowledge in groups of topics: of the topics themselves there is relatively little discussion (On Cassiodorus see also Holtz 1984, Strunk 1950; on Isidore Strunk 1950.)

The classic exposition of the relationship between music, proportion, and the universe is in Plato's *Timaeus*, known to the Middle Ages in the reworking by

Calcidius (ed. Waszink 1962; Waszink 1964; for Plato see Arnoux 1960, Cornford 1937). Here it is explained how the demiurge (easily transmuted by later writers into the Christian God) created the world-soul from a substance which combined that which is indivisible and unchanging (eternal verities) and that which is divisible, material, and subject to change. This substance was divided into sections whose lengths were in strict proportional relationship with each other, 1 : 2, 2 : 3, 3 : 4 and so on. Since the proportions were those of Pythagorean harmonic theory, music could be understood as a reflection of the creation of the world and its divine order.

The divine order was reflected not only in the ordering of the Earth and heavenly bodies, the macrocosm, but also in man, the microcosm, in the balance between his spiritual and physical being. Hence the influence of music upon the state of his soul and his behaviour.

Although Boethius' *De institutione musica* (ed. Friedlein 1867, trans. Bower 1989) includes chapters about music's ethical power, the harmony of the spheres, Pythagoras' discovery of musical proportions (the story of the hammers of different sizes in the smithy) and other matter, its chief substance is a comprehensive exposition of proportional theory, scales, and species. (On Boethius see also Bower 1967, 1978, 1988, Caldwell 1981, Gibson 1981, White 1981, Chadwick 1981.) Much of Boethius' exposition is set out with reference to the monochord, a method taken up in the Middle Ages, at least by those writers who go so far as to demonstrate the physical basis of the simple proportions. Boethius appears to have wished to attempt a summary of Greek musical learning in the Latin language, at a time when the subjects of the quadrivium (arithmetic, geometry, music, astronomy) were neglected in favour of the rhetorical arts, the trivium (grammar, rhetoric, and dialectic), partly out of fear that Greek learning was in danger of being lost to the world of the late Roman Empire. His plan (which remained uncompleted) appears to have been a reworking of Nicomachus' handbook of harmonics, a lost introduction to music by Nicomachus, and Ptolemy's harmonics (Bower 1978).

Boethius' demonstration of the correct intervals between the notes which make up the Greek Greater and Lesser Perfect System was of great importance. Since he demonstrated not only the intervals of the diatonic scale (the equivalent of, say, our modern *A–a'*, without accidentals), but also those of the chromatic and enharmonic genera, Boethius provided far more information than was useful to medieval musicians: at least, there was no attempt in the Middle Ages to relate the intervals of those two genera to the chant repertory. Boethius' division of the monochord was accomplished in two stages, the first establishing the first, second, and fourth notes of each tetrachord, the second stage filling in the notes which differed from one genus to another. Ex. V.2.1 shows how the first stage proceeded, and then gives the second-stage division of the highest tetrachord. For convenience, the modern equivalent pitches *A–a'* are shown beneath the line representing the string of the monochord. The complete diatonic scale, its component tetrachords, and the Greek names of each note are given in Table V.2.1, again with hypothetical modern equivalent pitches.

Equally important was Boethius' demonstration of the different species of octave,

Ex. V.2.1. Boethius' division of the monochord

First division of the monochord

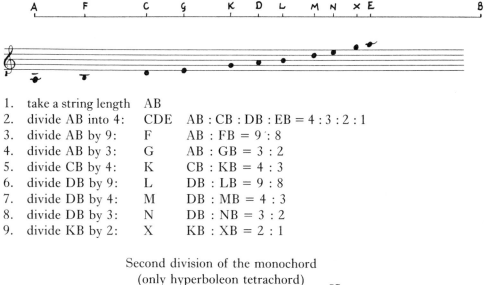

1. take a string length   AB
2. divide AB into 4:   CDE   AB : CB : DB : EB = 4 : 3 : 2 : 1
3. divide AB by 9:   F   AB : FB = 9 : 8
4. divide AB by 3:   G   AB : GB = 3 : 2
5. divide CB by 4:   K   CB : KB = 4 : 3
6. divide DB by 9:   L   DB : LB = 9 : 8
7. divide DB by 4:   M   DB : MB = 4 : 3
8. divide DB by 3:   N   DB : NB = 3 : 2
9. divide KB by 2:   X   KB : XB = 2 : 1

Second division of the monochord
(only hyperboleon tetrachord)

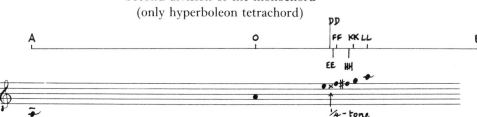

1. divide AB into 4: let the centre be O, three-quarters be LL
2. divide LL to B by 8; measure the same length (i.e. one-eighth of LL–B) in front of LL: KK.   KK–B : LL–B = 9 : 8
3. divide KK to B by 8; measure the same length in front of KK: FF.
      FF–B : KK–B = 9 : 8
4. divide LL to B by 3; measure the same length in front of LL: DD
      DD–B : LL–D = 4 : 3
5. divide KK to LL by 2; measure the same length in front of KK: HH
      HH–B : KK–B = 19 : 18
6. divide DD to FF by 2: EE
      DD–B : EE–B : FF–B = 512 : 499 : 486

fifth and fourth which can be found within the system (set out in a different order from that found in extant Greek writings). Each species is distinguished by the arrangement of intervals within it. The interval of the fourth, for example, occurs between five pairs of notes, modern equivalents *A–D, B–E, C–F, D–G, E–A.* Since, however, the semitone falls at the same point in the tetrachord for *A–D* and *D–G*—it

TABLE V.2.1. Greater and Lesser Perfect System, Diatonic Genus

Tetrachord	Pitch-name	Medieval equivalent
Greater Perfect System		
hyperboleon	⌈nete hyperboleon	a̿
	paranete hyperboleon	g
	trite hyperboleon	f
diezeugmenon	⌐nete diezeugmenon	e
	paranete diezeugmenon	d
	trite diezeugmenon	c
	⌊paramese	b
meson	⌈mese	a
	lichanos meson	G
	parhypate meson	F
hypaton	⌐hypate meson	E
	lichanos hypaton	D
	parhypate hypaton	C
	⌊hypate hypaton	B
	proslambanomenos	A
Lesser Perfect System		
synemmenon	⌈nete synemmenon	d
	paranete synemmenon	c
	trite synemmenon	b♭
meson	⌐mese	a
	lichanos meson	G
	parhypate meson	F
hypaton	⌐hypate meson	E
	lichanos hypaton	D
	parhypate hypaton	C
	⌊hypate hypaton	B
	proslambanomenos	A

forms the middle step of the three—and at the same point in the tetrachord for *B–E* and *E–A*, there are only three different species of fourth. In Ex. V.2.2 the species of fourth, fifth, and octave are given with hypothetical equivalent modern pitches, and the semitones are marked with the number of the step at which they occur. To this example have also been added the names of the eight octave species. (Boethius actually sets these out in a different way, imagining them all to occur between the same two pitches when the complete scale is itself transposed to different pitch levels: see *NG* xii. 379.) The difference between this nomenclature and the common medieval one is discussed below.

Ex. V.2.2. Species of intervals

species of 4th in diatonic genus:

species of 5th in diatonic genus:

species of octave in diatonic genus:

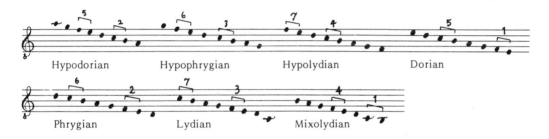

Hypodorian   Hypophrygian   Hypolydian   Dorian

Phrygian   Lydian   Mixolydian

V.3. A PITCH-SYSTEM FOR PLAINCHANT

(i) Introduction
(ii) Hucbald of Saint-Amand
(iii) The *Enchiriadis* Group of Treatises

(i) *Introduction*

Ecclesiastical chant plays no part in any of the writings referred to in the last section, and the first evidence that music theory and the practice of chant were on a converging course, so to speak, comes with the Carolingian renaissance of the late eighth and ninth centuries. It is known, for example, that Alcuin, Charlemagne's chief adviser on educational matters, conceived of a liberal arts programme at the imperial court and probably introduced it at his abbey of St Martin's, Tours; there is evidence of it at Orléans, under Theodulf, and especially at Fulda, under Rhabanus Maurus. The revival of classical learning was exactly contemporary with the establishment of the Roman liturgy and its chant in the empire.

Nevertheless, original new writing about music theory is not found until the next generations of scholars, beginning with Aurelian of Réôme, whose treatise *Musica disciplina* is usually dated *c*.850. Another group of scholars was connected in various ways with the court of Charles the Bald (823–77, king from 840, emperor from 875), and the schools at Saint-Amand, Laon, Reims, and Auxerre. These are Johannes Scotus Eriugena (*c*.810–*c*.877), Remigius of Auxerre (*c*.841–*c*.900), and Hucbald of Saint-Amand (*c*.840–930). Regino of Prüm, writing at the turn of the century, the anonymous treatise known as *Alia musica*, and the *Enchiriadis* group of treatises (*Musica enchiriadis*, *Scolica enchiriadis*, and the *Commemoratio brevis de tonis et psalmis modulandis*) complete what is altogether a very important, even astonishing achievement in creating a body of new music theory, borrowing occasionally from the authorities of the past but often startlingly independent, all datable to the second half of the ninth century.

Johannes Scotus Eriugena (or John Scot) and Remigius of Auxerre left no music treatise in the normal sense. Johannes was born in Ireland about 810 and came to Francia about 845. He is thought to have taught grammar and dialectic at the court of Charles the Bald (though the existence of a formal palace-school is not secure: see McKitterick 1983, 213–14, where Compiègne is proposed as Charles's favourite palace). He was perhaps the only scholar of his age who had a mastery of Greek, and in the early 860s he translated several Greek works for the king. Music is treated in John's work only parenthetically in his masterwork on transcendental philosophy, *Periphyseon* or *De divisione naturae*, and otherwise only in his commentary on Martianus Capella. Martianus' work seems first to have become known again during Charles's reign, after a long period of oblivion (see Huglo 1975, 142 ff.). Remigius also composed a commentary on it. (On John see Lutz 1939, Bower 1971, Uhlfelder 1976, Contreni 1978; on Remigius see Lutz 1956, 1957, 1962–5.)

(ii) *Hucbald of Saint-Amand*

Editions: GS i; Babb 1978; Traub 1989.
Literature: Müller 1884; Potiron 1957; Weakland 1956, 1959; Chartier 1973, 1987.

Hucbald's treatise *De harmonica institutione* (the title is actually Gerbert's) occupies a key position among these writings, not because it was the earliest or the best known (relatively few copies are known), but because it sets out most clearly the way in which classical theory could be explained with the aid of the chant repertory, and could in turn supply an intellectual foundation for a better understanding of chant. Hucbald had a monastic education, at Saint-Amand, Nevers, and Saint-Germain at Auxerre, where he was a fellow student of Remigius of Auxerre. He returned to Saint-Amand as master of the school, founded another school at Saint-Bertin, and in 883 was called with Remigius to revive the cathedral school at Reims. His remaining years were spent mainly at Saint-Amand.

It may well have been among the circle of Remigius and Hucbald that examples of plainchant were first used to illustrate theoretical concepts. Remigius, in his commentary on Martianus Capella, glossing Martianus' definition of the *symphoniae* (consonances), quotes three introits to illustrate the intervals of the fourth and fifth (*Tibi dixit cor meum* and *Deus in adiutorium meum* both begin with a rising fourth, *Oculi mei semper* with a rising fifth). Hucbald goes far beyond this, however, citing over sixty chant passages to illustrate intervals and scale-segments. In this the monochord plays no part. The intervals, for example, are not established according to their proportions in terms of length on the string of the monochord; Hucbald simply cites eighteen examples which illustrate nine different intervals in ascending and descending motion. He does not at first name the intervals in question, wishing only to establish the simple notion that there are intervals of differing size. He then shows how the larger intervals consist of combinations of the tone and semitone. The nine intervals which occur in the chant repertory are: semitone, tone, minor and major third, perfect fourth, tritone, perfect fifth, minor and major sixth.

The rudiments which Hucbald is next at pains to demonstrate are: the consonant intervals of the fourth, fifth, octave, octave plus fourth, octave plus fifth, and double octave; and the difference between a tone and a semitone. In the latter connection he cites two examples of chant passages which run through a hexachord, with the semitone in the middle: in modern notation *C–D–E–F–G–a* and then the reverse, *a–G–F–E–D–C*. This corresponds, says Hucbald, to the arrangement of strings on the six-stringed cithara, and he suggests that one should practise writing chants on six lines, representing these strings, with the semitone in the right place (see Ex. V.3.1).

Ex. V.3.1. Antiphon *Ecce vere Israelita* notated by Hucbald on six 'strings'

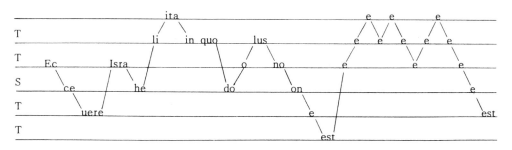

So far nothing has been said about pitch-names, neither letters nor the Greek names found in Boethius (whose work Hucbald knew and occasionally referred to). This continues in the next part of Hucbald's treatise, where he describes the arrangement of tones and semitones in a two-octave scale. The arrangement actually corresponds to that of the diatonic Greater Perfect System, which can be represented by modern *A–a'*, but Hucbald does not immediately state this. His next remark is that water organs and other instruments follow a different arrangement, one which corresponds to our modern major scale (Hucbald again simply gives the order of tones and semitones). Next he explains the classical arrangement of tetrachords within the double-octave scale. This is how Boethius arranges them, he says, and he goes on to cite chant segments where the notes of the tetrachord may be heard, once again in descending and ascending order.

   Hucbald's next step is crucially important. He suggests we now imagine the two-octave scale from the bottom up as a series of tetrachords of a different species, the species with semitone in the middle: modern *A–B–C–D*, *D–E–F–G*, and so on, and he gives an example from the chant repertory where this scale-segment can be found. At this point Hucbald establishes the place of the synemmenon tetrachord. For Boethius this would have corresponded to modern *d–c–db–a*, but for Hucbald, since he has introduced the different species of tetrachord, it corresponds to modern *G–a–bb–c* (Ex. V.3.2; compare Table V.2.1 above). In giving plainchant examples to

Ex. V.3.2. Hucbald's arrangement of ascending tetrachords

illustrate the distinction between diezeugmenon and synemmenon, the disjunct and conjunct tetrachords, he appears to imply that modern bb and b♮ can occur practically side by side. The version of the introit *Statuit ei dominus* described by Hucbald would seem to be something like Ex. V.3.3. Only at this point does Hucbald give the Greek names of all the notes in two-octave scale, for which he necessarily has to revert to the Greek tetrachords.

Ex. V.3.3. Introit *Statuit ei* as described by Hucbald

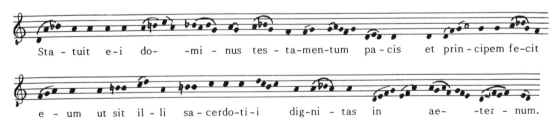

   It is now that Hucbald introduces the idea of pitch-symbols for the individual notes of the scale. As has already been explained (IV.5), the symbols he suggests were borrowed from Alypius via Boethius, but not subsequently taken up by any other writer. Once again, chant examples are given, among them the pseudo-Greek type melody for mode 1 (Ex. V.3.4).

Ex. V.3.4. Hucbald's letter-notation

As Hucbald himself says, the point has now been reached to which everything looked forward from the beginning, and we are about to see 'the fruit which springs from the seed just sown'. Every aspect of music theory discussed so far, intervals, tetrachords, the complete scale, and the pitch-symbols, has been illustrated by chant examples, and this is tantamount to saying that plainchant 'fits' the scale system just described. This is also the justification for Hucbald's introduction of a different tetrachord species, for the second tetrachord in his arrangement, modern *D–E–F–G*, corresponds to the four finals of the chant repertory. Hucbald gives the names of the four 'modes or tropes, which we call tones': protus, deuterus, tritus, and tetrardus; and he explains that each of the four notes 'reigns' over an authentic and a plagal trope.

Hucbald's last task is then to make clear the relationship between the starting-note of chants and the finals. A last batch of examples, notated with his pitch-symbols, illustrates a series of starting pitches in chants (nearly all antiphons) which end on each of the four finals. For the tetrardus mode, for example, *Ecce sacerdos*, starting on modern *d*, is cited, followed by *Beatus venter* and *Quomodo fiet*, both starting on *c*, *Dixit dominus domino meo*, starting on *b*, *Erumpant montes* and *Beati quos elegistis*, both starting on *a*, and so on.

It is clear that Hucbald had envisaged at least a large part of the chant repertory as being compatible with the diatonic two-octave scale plus middle b♭. If some chants were not capable of being fitted into that framework, and, by implication, of being notated with his pitch-letters, Hucbald does not say so. One wonders, however, whether the Greek system was taken over simply because there was no alternative. The main restriction was that the alternative chromatic step was available only at one point, a few steps up from the finales: *D–E–F–G . . . bb–b♮–c*. If a chant needed the alternative chromatic step at another point relative to the final, it had to be envisaged as occupying a different place in the system. This is the case, for instance, with a number of chants which need the alternative just above the final: they have to be notated with final *a*: *a–bb–b♮–c*. Nevertheless, once the idea gained a hold that chants stood in this relationship to a clearly definable pitch-system, and once teachers began to pick out chant melodies on monochords where the notes of the Greek system were marked off, pressure must have been felt to eliminate notes which did not fit that system.

Hucbald was probably not the first to make this connection between chant and theory. At any rate, it is also implicit in the anonymous *Alia musica* and the *Enchiriadis* group of treatises (which has its own pitch notation, the dasian symbols).

But nowhere else is the connection so painstakingly and clearly explained. Aurelian of Réôme, writing about thirty years earlier, has no notion of a pitch-system, and Hucbald's treatise has the aura of introducing something radically new. It is one of the crucial documents of Western music.

Hucbald was undoubtedly one of the leading churchmen and teachers of his time. Luckily a number of compositions by him appear to have survived. His office for St Peter is to be found in many medieval sources (*AS* 439), and his Gloria trope-set *Quem vere pia laus* occurs in several early manuscripts (Rönnau 1967, Tropen 240, to the melody Bosse 1955, no. 39). (See Weakland 1959.)

(iii) *The* Enchiriadis *Group of Treatises*

GS i; Le Holladay 1977; Bailey 1979; Hans Schmid 1981; Phillips 1984.

In the manuscript Paris, Bibliothèque Nationale, lat. 7202 the treatise *Musica enchiriadis* is attributed to Hucbald, and in Gerbert's edition not only that work but its *Scolica* and the treatise known as *Alia musica* were also credited to Hucbald. Müller (1884) and Mühlmann (1914) were able to disprove the attributions. Meanwhile, no other author has appeared to claim *Musica enchiriadis* (or the others), and even its date and provenance are unclear (see Phillips 1984). Until decisive evidence turns up, it is probably safe to regard it as a work written in north-east France or west Germany, perhaps even within the orbit of Hucbald's activity, in the second half of the ninth century.

Musica enchiriadis and the *Scolica*, which subjects some of the topics in *Musica enchiriadis* to more extended discussion in dialogue form, are most famous for their demonstration of organum of various types. Hucbald also referred to organum by way of illustrating the concept of consonance, and it has been argued that the main point of the discussion of organum in *Musica enchiriadis* is to illustrate the perfect consonances (Fuller 1981). In the first part of the treatise, on the other hand, the author presents eight chant examples, which are each typical of one of the eight ecclesiastical modes. This is done by means of the dasian notation set out in the opening chapters of the book (see above, IV.7). Each pair of examples is preceded by a brief formula which runs through the notes of one of the four species of fifth (modern equivalents *a–d*, *b–E*, *c–F*, *d–G*: it should be stressed that, as in Hucbald's work, the alphabetic series with which we are familiar is not connected with the dasian symbols). Like Hucbald's, the examples are set out on lines, though here the pitch of each line is made clear by a dasia sign. The examples for the protus mode are shown in Ex. V.3.5.

The *Enchiriadis* author, therefore, has an approach diametrically opposed to Hucbald's subtle and gradual method. The pitch-symbols appear at the very start of the work so that the author can immediately illustrate whatever concepts he choose without recourse to memorized chants.

While chant is not the only interest of the *Enchiriadis* author, the *Commemoratio brevis* is a veritable plainchant primer. Notated in dasian notation, and thus fixed in

Ex. V.3.5. Examples in protus mode from *Musica enchiriadis*

Modulatio ad principalem protum modum et subiugalem eius

Al‍le‍lu /F/ ia‍. Lau / da‍te‍ Do‍ mi / num‍ de /‍ cae‍ lis‍.

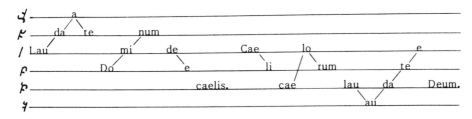

Cae / li‍ cae‍lo/ rum‍ lau‍da‍te‍! De‍um‍.

Al‑le‑lu‑ia. Lau‑da‑te Do‑minum de ce‑lis. Caeli cae‑lorum lau‑date Deum.

pitch notation over a century before pitch notation was used in chant-books themselves, are nearly seventy musical examples from the standard chant repertory, including the psalm-tone formulas with their multiple cadences, and also the didactic melodies associated with the different modes: Noeanne, Noeais, and so on. The plan is roughly as follows:

(*a*) pseudo-Greek (Noeanne) melodies and melismas (neumae) and psalm tones (Gloria patri), plus one or two antiphon incipits, for the gospel canticles in each of the eight modes ('used where the performance of a chant is slower, as for instance for the gospel canticles, when time allows');

(*b*) eight psalm tones 'for the faster chants';

(*c*) a series of tones which fall outside the regular eight;

(*d*) different final or median cadences for the psalm tones.

In this work the practical intent is plain. The *Commemoratio* author even introduces a sign for the quilisma into the dasian notation and a dot to separate note groups, adapting it more closely to the task of notating chant examples (see Bailey 1979, 19). If the given examples are learned, the student will have a thorough grounding in office psalmody and the psalm-tone formulas governed by various typical antiphons. This is not music theory but material for the teaching of chant. Even the additional commentary confines itself to practical matters: the speed of singing, the co-ordination of antiphonal choirs, and so on.

V.4. THE MODES

Powers, 'Mode', *NG*.

(i) *General*

The definition of 'mode', in the familiar sense of the eight plainchant modes, is not a simple matter. The Latin term *modus* itself was not universal in the Middle Ages, for it competed with *tonus* and *tropus*. There is no space here for a detailed examination of medieval terminology and practice. For the sake of clarity, I have regarded the eight modes as categories for the classification of melodies according to their tonality (implying range and prominent notes, including reciting notes and final notes) and melodic type. It is arguable that the latter feature was originally the deciding factor. It appears that it was the Byzantine system of echoi which inspired Western musicians in the Carolingian period to classify their own chant repertory, and also provided some technical terms for the classification. The echoi are primarily melodic types: 'In general, a hymn composed in an echos will contain a set of melodic turns (motifs, formulae or melody-types) peculiar to that echos; and these structural devices will be found in other hymns composed in the same echos' (Velimirović, 'Echos', *NG*). There is plenty of evidence to suggest that the eight-mode system was imposed in the West on a chant repertory not originally so conceived. Western chant did not develop within an eight-mode framework, with melodic material assignable to eight families or types. The eightfold division must therefore have relied primarily on the more abstract criteria mentioned above: range and prominent notes.

That this was so is suggested by the contrast between two medieval works. The tonary attributed to Regino of Prüm (*c*.900) classes a number of antiphons in a particular mode because of their opening melodic formula, but with the remark that they close in a different mode. About a century later, the author of the *Dialogus de musica* states categorically that the final note determines the modality: 'Tonus vel modus est regula, quae de omni cantu in fine diiudicat.' ('A tone or mode is a rule which classifies every melody by its final.') In the next section of the *Dialogus*, the author prescribes ranges for melodies in the eight modes. In so far as these two writers are typical, they suggest that theoretical abstractions, rules of thumb—final and range—replaced melodic characteristics as criteria for determining modality.

While such rules of thumb were no doubt useful at an elementary level, musicians continued to look for more comprehensive definitions. It was necessary, for example, to identify the final, which could be done with reference to the intervals on either side of it. Thus *D* has a tone then a semitone below it, and a tone then a semitone above it, whereas *E* has two tones below it and a semitone then a tone above, and so on. This focused attention on the scale-segments within which melodies of the various modes moved. Furthermore, more than one note might have the same pattern of intervals around it: *D* resembles *a*, *E* resembles *b*, *F* resembles *c* in this respect. There developed, therefore, a discussion in the theoretical literature about the tonal affinity between notes and hence between the modes they governed.

What was happening, when such a doctrine was developed, was that an abstract scale-segment (the hexachord was the most widely used), with which particular melodies could be compared, replaced a knowledge of the individual melodies or melodic families. We could postulate three (among many) stages: (i) The singer says: 'This melody (like its sisters in the family) belong to this mode . . .'. (ii) The singer says: 'This chant resembles the melody 'Noeagis', therefore belongs to this mode . . .'. (iii) The singer says: 'This chant ends on *re* of the hexachord, therefore belongs to the protus mode . . .'. In the first stage the singer has to know the modal assignation of all melodies, or at least of groups of similar melodies. In the second stage he has to know the Noeanne melodies and be able to select the right one for comparison. In the third stage he has to match his chant with but a single scale-segment.

The definition of modality was not a purely theoretical exercise in the Middle Ages. For chants with psalm verses sung to simple recitation formulas (introits and communions at mass, antiphons during the office) the selection of psalm tone and its cadence depended on the tonal character of the introit/communion/antiphon. It was convenient to be able to classify each chant in one of the eight modes, so that the selection of the corresponding psalm tone could follow as a matter of course. Watertight rules for determining modality therefore facilitated the execution of a large proportion of the chant repertory. In these circumstances we can easily understand how Regino of Prüm, at the end of the ninth century, was inspired to put pen to paper by what he describes as widespread confusion in singing caused by the improper selection of psalm tones ('propter dissonantiam toni').

None of these concerns was directly connected with the heritage of classical music theory. But as soon as chant could be thought of in terms of the pitches of the Greek two-octave system, concepts such as species of octave, fifth, and fourth could be employed in definitions of modality. The two developments went hand in hand, as late ninth-century writings bear witness.

(ii) *Aurelian of Réôme*

GS i; Gushee 1963 and CSM 21; Ponte 1961, 1968.

The earliest of the ninth-century writers, Aurelian of Réôme, shows very clearly how gross a division existed between the heritage of classical music theory and the needs of practical musicians. His *Musica disciplina* appears to have been written in the decade before 850 for Bernard, abbot of the monastery of Saint-Jean-de-Réôme (not far from Langres in eastern France). From the complicated make-up and transmission it is not clear how much Aurelian himself may have contributed to the treatise in its fullest form. (For example, the start in ch. 8, 'De octo tonis', is found separately elswhere, being attributed in a thirteenth-century source, and thereafter by Gerbert, to Alcuin. While that attribution is false, Aurelian may not have written the passage. Other material is borrowed, sometimes without much adaptation, from Boethius, Cassiodorus, and Bede.) The best-known passages in the treatise are those where Aurelian explains that Charlemagne added four modes to the usual eight because the Greeks (Byzantines) boasted that they had perfected the eight modes; whereupon the Greeks added four more themselves; and how Aurelian asked a Greek the meaning of the Noeanne syllables. (This actually constitutes some of our earliest evidence that Byzantine chant had a modal organization.) More important for the present discussion are the succeeding chapters, where he discusses melodies belonging to the (eight) different modes in purely practical terms.

Aurelian's particular concern is the link between main chant and verse in introits, offertories, and communions at mass and responsories and antiphons of the office. The closest thematic parallel is with the *Commemoratio brevis*, but the manner of presentation is entirely different, for Aurelian disposes of no pitch-notation. He begins, for example, by describing the three cadences for the psalm tone used with mode 1 introits:

> The first authentic mode has several *varietates*. For the introits it contains three *varietates*, the first of which is this: Ant. *Gaudete in domino semper*. The end of the verse flows directly into the beginning (of the introit), and equals it, and is neither pushed upwards nor pressed down.

This corresponds to the modern Solesmes/Vatican reading, where the cadence for the psalm verse falls on *D*, the starting-note for the introit. The other two joins are described thus:

> The second is this: Ant. *Iustus es domine*, whose psalm verse ending is raised up on high to coincide with the beginning (of the introit). The third is this: Ant. *Suscepimus Deus*, where the last syllable of the psalm verse dwells a little longer than is the case with the first and second (*varietates*), and is then raised up on high to fit well with the beginning of the introit.

The third cadence is evidently that which we find in the Solesmes/Vatican books, although the delay on *F* is just as long for *Gaudete in domino*. For the second introit, however, the Solesmes/Vatican books have the same cadence as for *Gaudete*; Aurelian has a different one in mind (rarely found in the modern books: see *Salus autem*, or the list of introit psalm-verse tones among the 'Common Tones of Mass' in *LU*). The three cadences and beginnings of the introits are given in Ex. V.4.1.

Ex. V.4.1. Psalm-verse cadences and openings of mode 1 introits cited by Aurelian of Réôme

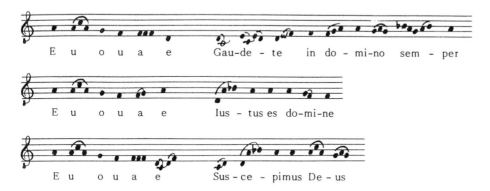

It is not always clear to us now why this or that cadence should have been selected, and a good deal of variety exists between medieval sources in this respect. Evidently Aurelian had also considered the matter, though his explanation is far from convincing:

> Because of this the mind of the eager reader is troubled why one single tone should have so many sorts of varieties (*varietatum modos*) . . . He who wishes to know everything else about this science, him we send to 'musica' (that is the science of music), and if he wishes to become versed in the latter, let him turn his eyes to the consonance of proportions and the contemplation of intervals, and also to the certitude of numbers.

Despite this pronouncement, Aurelian then proceeds to offer a more down-to-earth explanation. This is done with reference to the *tonus* of these introits, which seems to correspond to the chief reciting note, *a*. In *Gaudete*, he says, the tonus is reached on the third syllable of the introit, '-te'; in *Iustus* it occurs on the first syllable, in *Suscepimus* on the second.

It has puzzled modern readers that Aurelian should include offertories in his discussion, for the verses of these chants, at least since the time they were first recorded in notation (late ninth century) have been sung to elaborate melodies. Yet Aurelian says: 'Quod versus offertoriarum per tonos in ipsis intromittantur, cantor nemo (est) qui dubitet.' ('No cantor doubts that the verses of the offertories are allowed to pass into the offertories through the the tones.') This has been interpreted to mean that offertory verses are sung to psalm tones. But Aurelian never describes cadence formulas for the offertory verses, as he does for the introits and communions. His remarks resemble those about the office responsories, and concern the point at which the verse returns to the response: does it fit musically and make sense textually?

By *tonus* in the above passage, therefore, Aurelian may simply have meant the reciting note, or other important note, which had to be borne in mind when effecting the return to the response.

At the end of his treatise Aurelian includes some edifying remarks about the ethical power of music, and relates, among other stories, one about a monk from Saint-Victor, near Le Mans, who made a pilgrimage to the basilica of St Michael on Monte Gargano on the Adriatic coast of Italy. There he was blessed by hearing angels sing the responsory *Cives apostolorum* V. *Emitte domine spiritum*. The monk was able to sing this from memory to the 'clerks of the Roman church' ('Romanae Ecclesiae clericis'), who changed the verse to *In omnem terram*. The way in which Aurelian can move from music's power over birds and dolphins to a legend about the origins of a particular chant parallels the way in which the rest of his treatise juxtaposes borrowings from the musical knowledge of late antiquity with quite original observations on the practice of plainchant. This is a clear indication of the driving force behind much medieval writing on music: the necessity of defining codes of practice, with or without reference to 'musica' in the classical sense.

(iii) *Regino of Prüm*

GS i; Hüschen 1962; M. P. Le Roux 1965; Chartier 1965; Bower 1971; Bernhard 1979, *Studien*.

Something of the same dichotomy is present in the treatise of Regino of Prüm. Since the second half of the eighth century, the monastery of Prüm in the Eifel (western Germany) had been something of a royal foundation. A new monastery church is said to have been dedicated in 799 by Pope Leo III in Charlemagne's presence. Emperor Lothar I died there in 855, having assumed the monk's habit six days previously. As many as 300 monks may have lived there at the peak of its ecclesiastical and cultural achievement. But it suffered grievously at the hands of the Northmen in 882 and 892. Its recovery was due to the leadership of Regino, abbot from 882 to 899. Its wealth made it the object of political intrigue, however, and in 899 Counts Gerhard and Matfred of Hainault forced Regino to leave and had their brother Richarius, a mere boy, made abbot in his place. Regino removed to Trier and placed himself under the protection of Archbishop Rathbod. He accompanied the archbishop on visitations, restored the abbey of St Martin, Trier, and died in the abbey of St Maximin in 915.

Regino's writings fall into the Trier period. The *Epistola de harmonica institutione* is couched on the form of a letter to Rathbod. It is linked to a tonary with the title 'Octo toni de musicae artis cum suis differentiis'. In about 906 he wrote a book of canon law, *De ecclesiasticis disciplinis*, for use during visitations, dedicated to the primate of the east Frankish church and imperial chancellor Archbishop Hatto of Mainz. In about 908 Regino completed a chronicle of the world, dedicated to Bishop Adalbero of Augsburg, tutor to the royal children. Regino was one of the foremost churchmen of his time, and therefore a quite different figure from the obscure Aurelian.

An assessment of his *Epistola de harmonica institutione* and its tonary is again complicated by the transmission. Both have come down in a longer and a shorter form. The two principal sources both date from the tenth century, but neither stems directly from Regino. Brussels, Bibliothèque Royale 2750–65 (connected with the Belgian monastery of Stavelot) appears to transmit Regino's tonary in its original state, but has a short version of the *Epistola*, one which restricts itself to practical musical matters. Leipzig, Universitätsbibliothek, Rep. I. 93 (probably from Prüm) has a revised version of the tonary, and includes in the *Epistola* several chapters of speculative music theory, which, as Bower (1971) has shown, depend on the ideas of John Scotus.

Regino's tonary is one of the largest to have survived. This painstaking compilation has already been mentioned because of the numerous chants which Regino reckoned to begin in one mode and end in another. The work is of the highest value for our understanding of the modal assignment, but, despite the *Epistola*, we have to study the tonary itself to discover the principles involved. Regino, like Aurelian before him, does not define what constitutes modal identity. To see how this was attempted in the Middle Ages we should first inspect the terminology developed from the ninth century onward.

(iv) *Nomenclature*

Hucbald and the *Enchiriadis* treatises, and Aurelian of Réôme before them, were already familiar with a nomenclature for the eight modes. Following the spelling of the *Commemoratio brevis*, the terms are:

protus	{ authentus / plagis }	(modern D-mode)
deuterus	{ authentus / plagis }	(modern E-mode)
tritus	{ authentus / plagis }	(modern F-mode)
tetrardus	{ authentus / plagis }	(modern G-mode)

The first witness to this terminology in the West is the (incomplete) Saint-Riquier tonary, Paris, Bibliothèque Nationale, lat. 13159, of the late eighth or ninth century. The terms are derived from Byzantine practice, with the difference that the Greeks did not use the word 'authentic': there were four principle modes (echos protos, echos deuteros, etc.) designated *a*, *b*, *c*, *d*, and four plagal ones, designated plagios *a*, plagios *b*, plagios *c*, and plagios *d*. By good fortune evidence of the Byzantine nomenclature survives from about a century before its adoption in the West. There is manuscript witness to its use from possibly as early as the seventh century (see Huglo 1975, 140). The proper chants of the office of Saturday evening and Sunday in the Byzantine rite were organized in an eight-week modal cycle, each week having chants in a different mode. Such chants were copied in the book known as the oktoechos.

Cody (1982, 103) has reported a Melkite Syrian witness of the seventh or eighth century to the modal ordering of Sunday chants. Husmann (1971) has disproved the earlier belief that the hymns of Severus of Antioch, d. 538, were arranged in an oktoechos. Medieval tradition attributed the composition (more likely, organization) of the oktoechos to St John of Damascus (d. *c*.749). Strunk (1960) accepts the authorship of St Theodore Studites and a date of 794–7 for one particular early cycle of antiphons of the oktoechos, which also argues for the establishment of the system in the eighth century, if not earlier.

Huglo (1978, 138–42) has suggested that the coining of the term *authentus* was caused by the difficulty of finding an apt translation for *echos*, since *modus*, *tropus*, and *tonus* were already invested with slightly different meanings. The nearest equivalent to *echos* would have been *tonus*, which in the ninth century is generally used of the eight modes of plainchant: but there were only four echoi. A solution can be found in Hucbald, who speaks of the four finales (our *D*, *E*, *F*, and *G*) which make the appropriate endings for 'quatuor modis vel tropis, quos nunc tonos dicunt', and each in their turn reign over 'geminos . . . tropos'. So four 'tones' command eight 'tropes'. But Hucbald's usage is not that of the majority of early treatises and tonaries, which prefer to speak of eight tones: four principal or 'authentic' ones and four subsidiary or 'plagal'.

Another solution is that found in ch. 8 of *Musica enchiriadis*, where chant incipits in each of the eight modes are presented under the title: 'Modulatio ad principalem protum modum et subiugalem eius' ('Melodic phrases for the principal protus mode and its subordinate'). That is how 'mode' is commonly used nowadays (and in this book). But this is not the only terminology found in *Musica enchiriadis*, which, as Atkinson (1989, '*Modi*') has recently stressed, illustrates vividly the difficulties of marshalling terms borrowed from classical theory. We still have the same sort of contradictions: 'tone' can mean the interval of a tone, any single note, a psalm tone, or even (a meaning with no ancient etymology) quality of sound (harsh, soft, etc.).

Boethius usually used the word *tonus* to mean the interval of a whole tone. He was aware, however, that both *tonus* and *tropus* were used in classical Greek theory (as they were by later writers such as Martianus and Cassiodorus) to mean the various transpositions of the double-octave scale, which, if one took an octave cross-section at any constant point, would display the different octave species. The different transpositions bore the Greek names Hypodorian, Hypophrygian, and so on. Boethius' preferred term for the transposition scales, however, was *modus*.

It is not surprising, therefore, that ninth-century writers, inheriting a partially conflicting terminology from late antiquity, and also needing to accommodate the new concept of the eight modes of plainchant, should display some inconsistency. In the later part of *Musica enchiriadis*, principally concerned with vertical intervals, *tonus* is restricted to the meaning of an interval of a whole tone. Hence the establishment of *modus* as the term for the plainchant modes. In the first part of his treatise, however, the *Enchiriadis* author has himself used *tonus* for plainchant mode, and is obliged to change course in midstream (actually, in ch. 8).

(v) *Greek Names and Octave Species*

In ch. 9 of *Musica enchiriadis* occurs one of the earliest applications of the names of Greek peoples to the modes:

> Modes or tropes [the uncertainty of terminology is still evident!] are the species of melody [*species modulationum*] which were discussed above, such as the protos autentus or plagis, deuteros autentus or plagis, or the Dorius, Frigius, Lidius mode, etc.

It is the treatise known as *Alia musica*, however, which gives the first comprehensive exposition of this terminology. The author, or editor, appears to have combined material from two previous writers with his own observations (see Mühlmann 1914, Chailley 1965, and Heard 1966), as the title perhaps indicates: 'Alia musica' may be not Latin for 'another treatise on music' but Greek (*halia*) for 'a compilation about music' (Heard 1966). As far as modal terminology is concerned, the author simply assigns the Greek names to the eight plainchant modes without regard to their former nature. In Boethius the names denoted transposition scales, through which the seven species of octave could be displayed; no internal hierarchy of notes was implied. The plainchant modes, on the other hand, cannot be equated simply with scales: a number of notes are more prominent than others, and melodies in a particular mode may be related to each other by melodic charateristics.

Table V.4.1 shows how the Greek names were reordered to provide the medieval modal names. The eighth mode, the Hypermixolydian (literally 'beyond the Mixolydian', rather than Hypomixolydian, 'below the Mixolydian') is derived from Ptolemy, as reported by Boethius. Ptolemy did indeed advocate an eighth transposition scale an octave above the first, hence higher than all the others ('Hyper-'). Writers on the medieval plainchant modes simply changed the name to conform with the other plagal modes. Referring to the pitch-letters derived from Boethius' divisions of the monochord, the author of *Alia musica* justified the distinction between Dorian and Hypermixolydian by invoking the different 'median note' of each mode (*mediam chordam*) which is 'guardian of its quality' (*qualitatis custodem*). For the Hypermixolydian this is M (modern *G*), for the Dorian O (modern *a*).

Not until the humanist rediscovery of classical texts in the sixteenth century were attempts made to restore the Greek names to their original octave species.

In view of the borrowing of the Greek names for octave species in *Alia musica*, it is not surprising that one of the authors should also have taken the radical step of specifying octave ranges for the plainchant modes. One of his successors in the same treatise specified the same ranges split into a fourth plus a fifth, positing the idea of a median tone where the division is to be found (hence the distinction between mode 1 and mode 8). The plagal modes have the modal final as their median note, that is a fourth above the start of the range, whereas the authentic modes have the fifth as their median. The nomenclature *modus* is retained for octave species, while the plainchant mode is *tropus*.

Range is indeed a major preoccupation in the treatise. Even the ranges of the

Table V.4.1. *Greek and medieval modal names*

	Boethius	*Alia musica*	Tonaries, Hucbald, Aurelian, etc.		
	Phrygian	Dorian			Authentus
				Protus	
	Hypodorian	Hypodorian			Plagalis
	Dorian	Phrygian			Authentus
				Deuterus	
	Mixolydian	Hypophrygian			Plagalis
	Hypolydian	Lydian			Authentus
				Tritus	
	Lydian	Hypolydian			Plagalis
	Hypophrygian	Mixolydian			Authentus
				Tetrardus	
		Hypermixolydian			Plagalis

↓ = final of plainchant mode

Noeanne formulas for each mode are stated, although these by no means correspond with the scale-segments defined elsewhere. One senses here a certain tension between melodic pattern (the Noeanne melodies) and abstract concept (scale-segments) in the definition mode.

There is a trace of the same dichotomy in the way in which the chant examples in *Musica enchiriadis* are presented. Each is prefaced, as we have seen, with a phrase which descends through the fifth above the four finals in turn. Although the author does not say so, these are in fact the four possible species of fifth (cf. Ex. V.2.2 and Ex. V.3.5).

(vi) *Italian Theory: The* Dialogus de musica

GS i; Strunk 1950; Oesch 1954; Huglo 1969, 1971 'Prolog'.

The first efflorescence of music theory in the Carolingian century took place primarily in north-east France and the Moselle area, in the late ninth century. From the next few decades less material has come down to us, until at the end of the tenth century and in the early eleventh two new groups of writings appear, from Italian and south German writers respectively.

The Italian texts are characterized by a strongly practical purpose, and are evidently motivated by the need to train the ability to relate heard and sung melodies to written pitches. In some respects they are thus the successors to Hucbald's writings.

The earliest of these texts is an anonymous treatise written in dialogue form probably in Lombardy in the closing years of the tenth century. It was for a long time attributed to the famous abbot Odo of Cluny (d. 942), former pupil of Remigius of Auxerre, who is known to have composed a number of office chants, but who cannot be shown to have left any writings on music theory. The grounds for the attribution are simply that the *Dialogus* mentions the modal corrections made to the antiphon *O beatum pontificem* by a certain abbot Odo. This is naturally of no consequence for the authorship of the *Dialogus* itself, but medieval copies, and later Gerbert, placed Odo's name at the head of the treatise and assumed Odo of Cluny to be the abbot in question. The interrelationship of the writings involved in this complex of material has been clarified by Huglo (1969, 1971, 'Prolog'; see also 1971, *Tonaires*, 183–224 and 'Odo', *NG*):

(i) Prologue 'Formulas quas vobis' (GS i. 248) and tonary (CS ii. 117, from a source rather distant from the archetype); the prologue was written by a Benedictine abbot, the tonary agrees with the liturgical use of Arezzo, so Huglo attributes them to 'abbot Odo of Arezzo', about whom, however, nothing more is known.

(ii) Prologue 'Petistis obnixe' (GS i. 251, Huglo 1971, 'Prolog'); probably conceived as prologue for an antiphoner in alphabetic notation, later used as an introduction to (iii); composed by a monk, possibly Camaldolese or Vallombrosan.

(iii) The dialogue treatise (GS i. 252), in which the Odo of (i) is mentioned.

The first chapters of the *Dialogus* describes the monochord. The crucial remark is that one may mark up an antiphon in a chant-book with the same alphabetic letters that one uses for finding the notes on the monochord. The author's division of the monochord is given in Ex. V.4.2 (cf. that of Boethius in Ex. V.2.1 above).

Subsequent chapters emphasize the difference between whole tone and semitone, and describe the perfect consonances fourth, fifth, and octave. The intervals which

Ex. V.4.2. Division of the monochord in *Dialogus de musica*

	Γ	A	B	C	D	E	F	G	a	b	♭	c	d	e	f	g	a/a
	1	2	3	4	5	6	7	8	9	17	10	11	12	13	14	15	16

1. Take a length of string ΓX

2. divide ΓX by 9: A   ΓX:AX = 9:8
3. divide AX by 9: B   AX :BX = 9:8

4. divide LX by 4: C   LX:CX = 4:3
5. divide AX by 4: D   AX:CX = 4:3
6. divide BX by 4: E   BX:EX = 4:3
7. divide CX by 4: F   CX:FX = 4:3

8. divide ΓX by 2: G   ΓX:GX = 2:1
9. divide AX by 2: a   AX:aX = 2:1
10. divide BX by 2: ♭   BX:♭X = 2:1
11. divide CX by 2: c   CX:cX = 2:1
12. divide DX by 2: d   DX:dX = 2:1
13. divide EX by 2: e   EX:eX = 2:1
14. divide FX by 2: f   FX:fX = 2:1
15. divide GX by 2: g   GX:gX = 2:1

16. divide aX by 2: a/a   aX:a/aX = 2:1

17. divide FX by 4: b   FX:bX = 4:3

occur in chant melodies are then illustrated with chant examples, very much as Hucbald had done, though neither the tritone nor any sixth is admitted here.

It is at this point (ch. 6) that the author discusses two problematic antiphons. *O beatum pontificem* (AS 586) lies mostly in the middle of the D-mode range and resembles mode 2 antiphons like the O-antiphons of Advent; because of two high-lying phrases it must, however, be assigned to mode 1 (cf. prologue to Odo's tonary, GS i. 249, also CS ii. 121). The problem is relatively minor, and is perhaps mentioned only because of the danger of attaching undue importance to the opening alone in deciding modality. The other antiphon, *Domine qui operati*, is more difficult, for it seems to call for both *b♮* and *bb*. It begins like many mode 6 antiphons (cf. *O admirabile commercium*), and we might therefore expect it to be notated with final on *F*. But at a later point, *E♭* would then be required, an impossibility in the notation of the tenth to eleventh century. An answer to the problem would be to notate the antiphon a fifth higher, ending on *c*, so that both *b♮* and *bb* could be used, as is done, for instance, in the Worcester antiphoner (PalMus 12). Another solution would be to rewrite the passages with *E♭/bb*, as is done in the Lucca antiphoner (PalMus 9). The author of the *Dialogus* recommends a different procedure: the opening should be altered to resemble mode 8 antiphons such as *Amen [amen] dico vobis* (AM 489). This would eliminate the semitone step under the final. The antiphon would then presumably be notated in *G*. Strunk (1950, 112) gives the various openings involved (Odo's preferred opening should be emended, however). The scale-segments involved are given in Ex. V.4.3.

Ex. V.4.3. Notes required to notate antiphon *Domine qui operati*, various versions

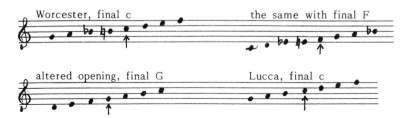

Chapter 8 begins with the well-known definition of 'mode' quoted at the beginning of this section: the final determines the mode. Not only this, it also governs starting-note and range. The rest of the chapter amplifies this—one recollects the examples at the end of Hucbald's treatise—by explaining that the relationship of starting-note to final should be that of the six intervals enumerated previously (minor and major second, minor and major third, fourth, fifth) or the unison. (The only exception is that mode 3 melodies often begin on *c*, a minor sixth above the final *E*.) Individual phrases within the chant (*distinctiones*) had best end on the final, which leads to the rule: 'A chant belongs most to that mode towards which its distinctions most frequently lead.' Chapter 9 defines the range of chants in the various modes (Ex. V.4.4). Eight chapters, one for each mode, provide examples to illustrate the above prescriptions.

Ex. V.4.4. Ranges of chants in the eight modes, *Dialogus de musica*

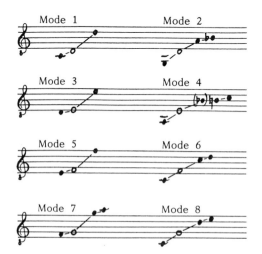

It is clear from the remarks in ch. 6 that the author is perfectly ready to 'improve' chants whose modality is uncertain, and it is not surprising to read in the last chapter of the dialogue that chants which do not obey the rules have been composed by 'presumptuous and corrupt cantors' ('furtivam singularitatem a presumptoribus vitiatisque cantoribus factam esse non dubito'.) It is in precisely this climate that the revisions of the chant repertory undertaken by the Cistercians should be understood.

The last part of the dialogue develops the interesting idea that by virtue of the intervals on either side of it, each note has a certain modal quality. This follows from the concept that the four finals, *D*, *E*, *F*, and *G*, define their respective modes in various ways. Notes which occupy similar scale-segments to one of the four finals therefore share their modal character. For example below *A* there is a tone, above it there is a tone, a semitone and two whole tones; that is also true of *D*. So *A* 'preserves the rule of the first mode: therefore not without reason is it called first'. The 'similitudes' between notes and modes which the author enumerates are given in Ex. V.4.5.

Ex. V.4.5. 'Similitudes' of notes to the various modes, *Dialogus de musica*

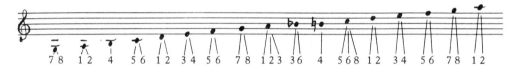

(vii) *Guido of Arezzo*

GS ii; CSM 4; Babb 1978; DMA A.III, IV; Strunk 1950; Oesch 1954; Smits van Waesberghe 1951, 'Guido', 1953, 1957, *Expositiones*, 1974; Waeltner and Bernhard 1976.

Guido of Arezzo was intimately acquainted with the *Dialogus* and even recommends its use on some occasions. But his own teaching goes well beyond it in scope and clarity. Guido was born probably around 990, and was educated in the Benedictine monastery at Pomposa (in the Po delta north of Ravenna), but became unpopular, apparently as a direct result of his music teaching. It was at Pomposa that he invented his system of staff-notation (see above, IV.6). His *Aliae regulae*, conceived as prologue to an antiphoner with the new notation, describes the system. Guido moved to Arezzo around 1025; his most important treatise, *Micrologus*, completed around 1030, was dedicated to Bishop Theodald of Arezzo. Pope John XIX (1024–33) called Guido to Rome to show him the new antiphoner, as Guido reports in his *Epistola ad Michaelem* or *Epistola de ignoto cantu*. His whereabouts in later years is unknown.

Guido's new notation has already been described. As so often, it is the lack of agreement between singers that has caused him to invent the system. Each cantor makes his own antiphoner, he complains, so that instead of the 'Antiphonarium Gregorii' the talk is of Leo's or Albert's antiphoner. Guido sees his system not simply as a better method than the alphabetic way of notating chant. It has implications for the whole way in which chant can be grasped as a tonal construct. This is made clear above all in the *Epistola*.

The *Epistola* introduces the hymn *Ut queant laxis*. The melody Guido uses may well have been composed by him for the purpose in hand, for it is unknown elsewhere at this time. Its well-known peculiarity is that each line begins with a successively higher note of a hexachordal series *C–D–E–F–G–a* (Ex. V.4.6). The reason why the learner should fix the hexachord and its six syllables in the mind is that he will have a point of reference for any chant. Guido says that learning melodies by rote from a master (the traditional method) or by picking them out on the monochord (by means of alphabetic notation, the method of the *Dialogus*) is adequate for beginners but bad for more advanced students. His new learning method reduces the time for learning the chant repertory to two years (previously ten were required). A boy can learn a new chant in three days. This is because the new system helps one to read music previously unknown; and it helps one to notate a melody one knows. The point of the exercise is summed up as follows: 'Ergo ut inauditos cantus, mox ut descriptos videris, competenter enunties, aut indescriptos audiens cito describendos bene possis discernere, optime te iuvabit haec regula.' ('Therefore this rule will be of greatest help to you so that you may sing competently an unknown chant as soon as you see it written, or so that you may be able to perceive straightaway how to write down an unwritten chant that you hear.') The singer is instructed to compare the ends of phrases, in whatever chant is being dealt with, with the notes of the hexachord. This is so that he will be able to gain a proper tonal orientation. The advantages of using a single hexachord as point of reference are many. The hexachord encompasses the cadential formulas of most chants, certainly in the simpler antiphons which were presumably taught first in Guido's song-school; one hexachord is better than four species of fifth or seven species of octave.

Ex. V.4.6. Hymn *Ut queant laxis*, Guido of Arezzo

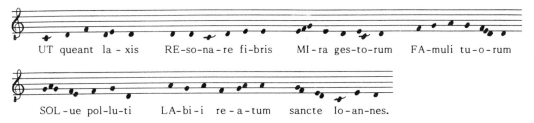

Perhaps we could imagine a choirboy learning the antiphon shown in Ex. V.4.7. He can see the red *F*-line, knows that it marks the semitone step, mi–fa on his hexachord. He can quickly locate the final of the chant in his hexachord, which means he can hear it in his inner ear, or sing it out loud, with the correct constellation of tones and semitones on either side. He can then go on to locate the finals of each phrase, indeed all the notes of the piece, with relative ease. This antiphon is of course a very simple

Ex.V.4.7. Antiphon *Diligite Dominum* (Lucca, Bibl. Cap. 601, PalMus 9, p. 456)

example, but shortly the student will be used to making the connection between clefs, coloured lines, and particular pitches, and be able to read at sight from the stave, using the hexachord only occasionally for reference. This would have been far quicker than learning by rote from a teacher, or from notes picked out on the monochord. (Although he might have attained fluency in reading alphabetic notation at sight, that would not have taken a shorter time than reading staff-notation.) The reverse process, notating *Diligite dominum* when it was already known from memory, could proceed equally quickly through the intermediary of the hexachord. The first phrase would be matched to the syllables *re–re–ut–re–fa–mi–re*, which would then be translated into *D–D–C–D–F–E–D* in staff-notation.

Guido does not mention the steps described above in every detail. I have tried to imagine how it was done. Nor does he say how this learning method was applied to chants with a wide range. Soon after Guido hexachords were imagined on other notes of the scale, on *F* (with b♭, or round b, hence 'hexachordum molle') and on *G* (with b♮, or square b, hence 'hexachordum durum'); the hexachord on *C* was the 'hexachordum naturale'. One other pedagogical device subsequently attributed to Guido is not mentioned anywhere in his writings: the 'Guidonian hand', where each note of the scale has a position on the fingers of the (left) hand, with its correspoding hexachord syllable. The earliest source so far known, Montecassino, Archivio della Badia 318, of the late eleventh century, displays the range Γ–$\overset{a}{a}$, without hexachord syllables, but with the letters T and S indicating the intervals between pitches. The earliest hands with hexachord syllables seem to be somewhat later. Fig. V.4.1 shows the hand in a Bavarian manuscript of the twelfth century, Munich, Bayerische Staatsbibliothek clm 14965b, and the series of hexachords indicated on it. The hand also indicates whole tone and semitone steps, and against certain of the letters are given the modes in which those pitches are the dominant (reciting) note (circled on my copy for ease of identification). The direction of reading is given in a smaller accompanying diagram. (See Smits van Waesberghe 1969, 129 for facsimile, 120–43 for hands of all sorts; also Russell 1981.)

Both the *Regulae* and the *Micrologus* assign modal qualities to the notes of the scale, as the *Dialogus* had done (Guido's term for the correspondences between notes, the *similitudo* of the *Dialogus*, is *affinitas*). The longer discussion in the *Micrologus* (ch. 7) introduces the concept of four *modi vocum* (literally, 'the modes of the notes',

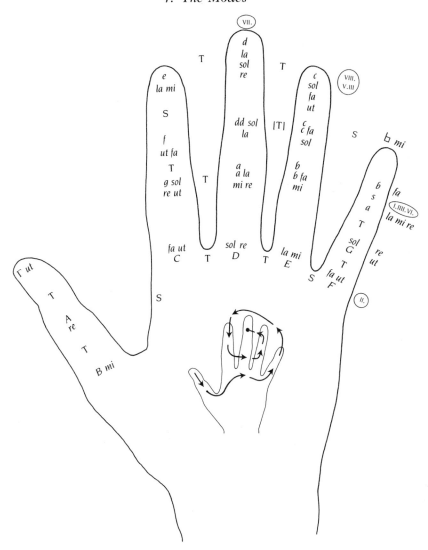

Fig. V.4.1. Guidonian hand after Munich, Bayerische Staatsbibliothek, clm 14965b

perhaps better, 'the modal qualities of the notes'). This describes quite simply the constellation of tones and semitones around each note of the scales. Significantly, each note except *G* is placed in a hexachord, a clear link with the hexachord and its *ut–re–mi* syllables in the (later) *Epistola*. Guido enumerates four patterns (see Ex. V.4.8). The intention is obvious: if we can locate a note in one of these scale-segments, then we can notate it, or sing it, as the case requires.

The *Micrologus* also treats other topics covered by the *Dialogus*: the division of the monochord, the intervals found between notes in plainchant (the same as those enumerated in the *Dialogus*), the names of the modes, their range, the importance of

Ex. V.4.8. The *modi vocum*, Guido of Arezzo

the final, and the 'correction' of wrong melodies (ch. 10: 'De modis, et falsi meli agnitione et correctione'). Two chapters concern organum, and two chapters, on the ethos of the modes and on Pythagoras' discovery of music (!), constitute rare references back to classical theory (there were none in the *Dialogus*). Guido's comparison of the elements of music with the parts of language has already been mentioned (IV.1). Although its roots lie in a tradition of conceiving music in grammatical terms (see Phillips and Huglo 1985, Bower 1989), these chapters (15–16) develop in an original way, with numerous observations on the relationships between musical phrases. Chapter 17 describes the way in which a melody may be composed mechanically by assigning notes according to the vowel in each syllable of the text: let the vowel 'a' be given the note *C*, 'e' be given *D*, 'i' *E*, and so on; several different arrangements are suggested. Guido nevertheless departs from the method at the end of the chant in order to make a satisfactory cadence (see plate 11).

As far as the modality of chant is concerned, Guido's main concern is not an abstract definition. His two primary aims seem to be: (i) to enhance our understanding of chant in ways that will facilitate the reading of notation and, conversely, the notating of chants; and (ii) to provide guidelines for the composition of new chants and the correction of poor ones.

(viii) *South German Writers*

In bulk the Italian theoretical literature of the tenth to eleventh centuries is easily overshadowed by that written in south Germany. From France rather little new writing is known, save for the treatise by Odorannus of Sens (985–1046: see Bautier *et al.* 1972, also Villetard 1912, 1956). Mention should also be made of two letters commenting on Boethius' *De institutione musica*, and a treatise on the measurement of organ-pipes and the division of the monochord, by Gerbert of Aurillac (*c*.940–1003), later Pope Sylvester II (see Sachs 1972 and Huglo, 'Gerbert d'Aurillac', *NG*). The generally meagre quantity of material does not mean, however, that French musicians no longer studied music theory. Proof that they did can be found in manuscripts like Munich, Bayerische Staatsbibliothek, clm 14272. This was mostly copied at Chartres by the monk Hartvic of the Benedictine abbey of St Emmeram in Regensburg, and contains, among other things, Boethius' *De institutione musica*, glossed partly by Fulbert of Chartres (bishop 1007–28), a tonary, writings on the trivium, the three main treatises of the *Enchiriadis* group, a monochord tract, and the *Alia musica* (see RISM B/III/3, 110). In Regensburg it contributed significantly to the studies in music theory which flourished there. Yet, on the whole, works of the originality of those mentioned here do not seem to have been written in France, or have been lost. Smits van Waesberghe repeatedly asserted the existence of a school of music theory in Liège and neighbouring areas, related to the south German tradition, though the true picture remains somewhat unclear. The importance of the south German area can be judged from Table V.4.2, a list of the more important writers and works of the eleventh century.

There is clearly no space here to discuss all members of this colourful group. It contains several very eminent churchmen: Berno, monk of Prüm, called in 1008 by the Emperor Henry II to be abbot of the famous abbey on the Reichenau in Lake Constance; William of Hirsau, monk of St Emmeram in Regensburg, from 1069 abbot of Hirsau in the Black Forest, fountainhead of a great monastic reforming movement, the 'German Cluny'; Theogerus, monk of Hirsau, appointed in 1088 by William to be abbot of St Georgen in the Black Forest, elected bishop of Metz in 1117, though unable to occupy the see; Honorius of Augsburg, perhaps the most widely read popular theologian of his time. Other figures remain somewhat mysterious. Johannes has been identified implausibly as an English monk, or as abbot of the abbey of Afflighem near Brussels; he was almost certainly Bavarian, and possibly wrote his treatise for Bishop Engelbrecht of Passau (1045–60: see Hohler 1980, 57–8). Of Aribo nothing certain is known.

The contributions naturally vary widely in scope and interest. The material connected with St Emmeram in Regensburg exemplifies this (it has been elucidated principally by Bischoff). Otker is known only as the author of a diagram called the *Quadripartita figura* (see Bronarski 1926, also Smits van Waesberghe 1969, 90), apparently for identifying on the monochord the species of fourth and fifth, and the 'chordae constitutivae' of each mode, that is, the notes whose modal character identifies them with each mode, a concept perhaps borrowed by Aribo in a diagram of his own, nicknamed the 'Caprea'. Otloh's musical activities are overshadowed by his many other interests, but he appears as William of Hirsau's partner in two dialogue treatises, on music and on astronomy; he may be author of a Kyrie trope; and he copied two music theory manuscripts, Munich, Bayerische Staatsbibliothek, clm 14523 (which includes all Guido's writings) and clm 18937. William of Hirsau's connection with St Emmeram has already been mentioned. Henricus/Honorius of Augsburg also spent several years as monk in Regensburg. The presence of the *Enchiriadis* treatises at St Emmeram, in a source copied at Chartres, and Otloh's copy of Guido's writings, show how cosmopolitan might be the cultural links of an important Benedictine monastery in its heyday.

Table V.4.2. *Writers on music theory in south Germany in the eleventh century*

Reichenau:
  Pseudo-Bernelinus (before Berno)
    Musica (GS i)
  Berno (d. 1048) (see Oesch 1961)
    Prologus in tonarium (GS ii)
    Tonarius
    De consona tonorum diversitate (DMA A.VI*b*)
    De mensura monochordi (?) (DMA A.VI*a*)
  Hermannus Contractus (1012–54) (see Oesch 1961)
    De musica (GS ii, Ellinwood 1936)
    Explicatio litterarum et signorum (GS ii, Ellinwood 1936)
    Versus ad discernendum cantuum (GS ii, Ellinwood 1936)

Bavarian and related centres:
  Otker of St Emmeram
    Mensura quadripartitae figurae (*c.* 1050?) (Mettenleiter 1865; Bronarski 1926)
  Otloh of St Emmeram (1032–70, older contemporary of William of Hirsau) (Bischoff 1955)
  William of Hirsau (d. 1091)
    Musica (1070–80) (GS ii; Müller 1883; CSM 23)
  Johannes 'Cotto' or 'of Afflighem' (from Passau? *c.* 1060?)
    De musica cum tonario (GS ii; CSM 1; Babb 1978)
  Aribo (of Freising?)
    De musica (1068–78) (GS ii; CSM 2)
  Anon.
    Commentarius in Micrologum (*c.* 1075?) (Smits van Waesberghe 1957, *Expositiones*)
  Frutolf of Michelsberg (Bamberg) (d. 1103)
    Breviarium de musica et tonarius (Vivell 1919)
  Henricus/Honorius of Augsburg ('Augustodunensis', *c.* 1100)
    Dialogus de musica (DMA A.VII; Huglo 1967, 'Henri'; Flint 1982)
  Theogerus of Metz (*c.* 1050–1120, pupil of William of Hirsau at Hirsau)
    Musica (GS ii)

Liège?:
  'Wolf-Anonymous' (*c.* 1060)
    De musica (Wolf 1893)
  Franciscus of Liège (1047–83) or Rudolf of Saint-Trond (*c.* 1070–1132)
    Quaestiones in musica (Steglich 1911)

(ix) *Modal Theory in South Germany*

Guido's influence eventually became very strong in south Germany, as elsewhere, but Berno was writing before it had effect. Nor does Berno use any pitch-notation, where necessary referring instead to the Greek pitch-names, proslambanomenos and so on. There is no explicit connection with sight-singing or notating music. (All passages referring to letter-notation in Gerbert's edition are interpolations, principally from Gerbert's Anonymous I—GS i. 330: see Oesch 1961, 84–90; on the other hand, Smits van Waesberghe in DMA A.VI referred to this author as Berno I.) The use of the monochord is envisaged, as embodiment of the classical scale system, for example when Berno recommends that certain antiphons be transposed up a fifth, for otherwise they will be impossible to locate on the monochord ('Nisi enim hae antiphonae, *Alias oves habeo, Domine qui operati sunt* [mentioned in the *Dialogus de musica*], quae sexti toni sunt, in quintum transponantur locum, hoc est, a parypate meson in trite diezeugmenon, nequaquam in regulari monochordo servare poterunt ordinem suum.')

Berno's principal treatise, conceived as prologue to a tonary, is, according to the author himself, largely a summary of older rules, as is made clear by frequent citations or paraphrases of Hucbald and, to a lesser extent, Pseudo-Bernelinus (GS i. 313). Apart from ch. 4, a speculation on the significance of the numbers 1, 2, 3, 4, and their sum 10, it stays within the confines of basic rudiments, as a brief account of his discussion of modality will make clear. From Pseudo-Bernelinus comes a description of the modes in terms of scale-segments, in fact as species of fourth plus species of fifth, just as in Guido's exposition (see Ex. V.4.8).

There are obvious similarities with the *Alia musica* writings. Berno (GS ii. 69), in the words of Pseudo-Bernelius, says that the modes 'consist of' these tetra- and pentachords ('Protus constat ex prima specie diapente . . .'). His ninth chapter nevertheless sets out guide-lines for range which are more flexible than the octaves just outlined. We should therefore understand the scale-segments as loci within which the chant will typically move. After dealing with chants whose modality is for one reason or another unclear, Berno, like his Italian contemporaries, discusses the relationship between the final and the beginnings and endings of the component phrases of a chant. He also speaks of the *socialitatis foedus* which exists between the finals and notes a fifth higher, and sometimes a fourth higher. A host of examples are adduced to show how some chants are best imagined as being placed in a different part of the two-octave system, to take advantage of the *sociales*, rather than ending on the normal finals. The communion *De fructu operum tuorum* is an example of this (discussed exhaustively by Jacobsthal 1897, 52–62, 136–78, and by Bomm 1929, 60–2, 97–103). Berno regards it as a mode 8 chant, but says that if one starts on the final (our *G*, his lichanos meson) a problem involving the semitone arises in the middle of the chant; for this reason the chant should be assigned to the position a fourth higher, our *c*.

The communion as it is found in the modern Solesmes/Vatican books has our modern major tonality. If one wanted to notate such a piece in the Middle Ages ending with one of the four regular finals, it would have to be placed on *F* and use *b*♭, in other words like our modern F major. The chant lies in the lower register, and is thus assigned to mode 6 (F plagal) in modern books. The reason for Berno's remarks can be seen if the version in the Dijon tonary, Montpellier, Faculté de Médecine H. 159, is inspected, for here the chant is indeed notated with final on *c*, and makes use of both $b\natural$ and $b\flat$ in alternating phrases. The Montpellier manuscript actually assigns the communion to mode 6, whereas for Berno it is a mode 8 piece. Whether one regards the piece as F-mode (with hypothetical $E\natural$ and $E\flat$) or G-mode (hypothetical $F\natural$ and $F\sharp$) appears to be a matter of taste, normally to be decided with reference to the ending: is the note under the final in the last phrase a whole tone (G-mode) or a semitone (F-mode)? (Although by this reasoning Montpellier should have chosen mode 8, it nevertheless assigns the piece to mode 6: is this because in Guido's theory of affinities G has no co-final?) The scale-segments used in the Vatican/ Solesmes version and in Montpellier H. 159 are given in Ex. V.4.9. (The Cistercians had a different solution. *De fructu* has at least one D cadence, and its opening resembles D-mode communions such as *Posuerunt mortalia*. The Cistercian revision eliminated the passages with the $E\flat$/$b\flat$ step, and strengthened the D-mode element, ending the piece on D and creating a mode 1 communion.)

Ex. V.4.9. Notes required to notate communion *De fructu*, two versions

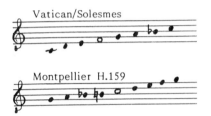

Between Berno and the younger Reichenau monk Hermannus Contractus ('Herman the Cripple'—see the moving biographical sketch by his pupil Berthold, MGH (Scriptores) *Annales et chronica aevi Salici*, 267, PL 143, 25, Ger. trans. Oesch 1961, 117) came the impact of Guido's writings. That is not to say that Hermannus' treatise is in any way like, say, the *Micrologus* in character. (For a German work of that nature one must turn to Johannes 'Cotton'.) But Hermannus uses pitch-letters, and some of his topics reflect ideas in Guido's writings. Of prime interest is Hermannus' response to Guido's hexachordal *modi vocum*, those scale-segments which are decisive for understanding the intervallic relation of the final to the surrounding notes. Hermannus' equivalent are the *sedes troporum*, or 'seats of the modes'. These are determined by taking each of the four tetrachords which make up the diatonic scale (to which William of Hirsau added the synemmenon tetrachord), and adding a tone on either side (see Ex. V.4.10).

Another difference from Guido is Hermannus' insistence on a dual modal function for D (and d). For the author of the *Dialogus* and Guido D was plainly protus, and only A had the same disposition of whole tones and semitones on either side. But Hermannus assigns modal quality on the basis of the position of the note in the four tetrachords. Thus A, D, a, and d have protus quality, B, E, b, and e have deuterus quality, and so on. D, which appears as first note of the *finales* and fourth note of the *graves*, therefore has dual quality.

Ex. V.4.10. Modal qualities of notes, Hermannus Contractus

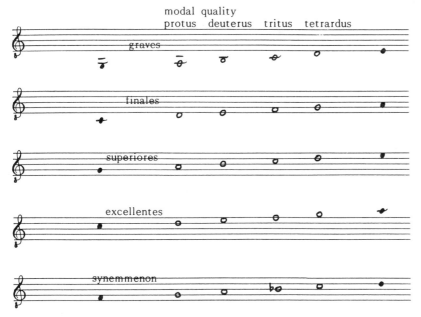

It is therefore not surprising to find Hermannus asserting that the tetrachords *A–D* and *D–G* are different species of fourth. The *A–D* tetrachord starts with the protus note of the *graves* and ends with the protus note of the *finales*, whereas the *D–G* tetrachord starts with the tetrardus note of the *graves* and ends with the tetrardus note of the *finales*. The tetrachords have quite different modal quality and thus constitute the first and fourth species of fourth respectively. By a similar process of matching a protus step with another protus step, Hermannus defines four species of fifth and four species of octave. The fifths run between the *finales* tetrachord and the *superiores* (*D–a, E–b, F–c, G–d*), the octaves between *finales* and *excellentes* (*D–d, E–e, F–f, G–g*).

Although this reasoning has a certain logic, it is clearly at variance with classical theory. And while previous deviations from classical theory—such as, for example, Hucbald's redefinition of the tetrachords—usually had a practical purpose, Hermannus' ideas are not obviously the product of experience in learning and teaching chant. The examples of chants in the different modes which he casually cites near the end of his treatise have no essential connection with what has preceded: 'It also helps greatly in recognizing the quality of the modes (*troporum qualitatem*) if you meditate assiduously upon the customary cadences and *diffinitiones*, many versicles in their modes, and extended chants most of all . . .'. The balance struck between abstract theory and practical application which characterizes the writings of, say, Hucbald and Berno does not obtain in Hermannus' treatise. His theorizing is nevertheless of an original kind, not a reversion to the abundant theoretical knowledge to be found in Boethius.

(x) *Later Syntheses*

During the thirteenth century polyphonic music assumed an importance in the work of progressive theorists at least equal to that of plainchant. It is true that some writers continued to interest themselves in plainchant only, at least judging by the writings that have come down to us: Engelbert of Admont (d. 1331; GS ii. 287–369) is one such. On the other hand we have many short texts, and some complete treatises, which are almost entirely preoccupied with polyphony, for example, the well-known treatise of Anonymous IV (CS i. 327–65; Reckow 1967). But for those writers who attempted to cover theoretical aspects of all branches of practical music, plainchant and polyphony are of equal interest. The transition is unmistakable. Polyphony occupies two of the twenty chapters in Guido's *Micrologus* (*c.*1030). In the twelfth century, polyphony is discussed only in the last two of twenty-nine chapters of the Schneider/London/Naples anonymous edited by Pannain (1920), and in only two of the eighteen by the so-called 'St Martial' or 'De La Fage' anonymous (ed. Seay 1957; see Sarah Fuller 1977, who shows that it is dependent on Cistercian plainchant teaching, for example in its using the term *maneriae* for the four modes D, E, F, and G). Against this we can set the twin treatises by Johannes de Garlandia of about 1240, *De plana musica* and *De mensurabili musica* (see Baltzer, 'Johannes de Garlandia', *NG*). From the early fourteenth century we have the *Lucidarium* (concerning plainchant; ed. Herlinger 1985) and the *Pomerium* (concerning polyphony) of Marchettus of Padua; and of the seven books of Jacobus of Liège's massive *Speculum musicae*, Book 6 concerns plainchant, Book 7 polyphony (CSM 3). (The above-mentioned writers naturally form a mere selection. Much work still remains to be done even on some of the better-known named authors of chant theory; for anonymous ones see the bibliographical information in Palisca, 'Anonymous theoretical writings', *NG*.)

Whatever the intellectual attractions of Hermannus' 'seats of the modes', the practical advantages of Guido's hexachords ensured them universal popularity. By the thirteenth century it was customary to identify pitches not only by their letter but also by their solmization syllables: thus middle *c*, which falls in three hexachords, is *c sol fa ut*: no other pitch has this combination of syllables. The choice between syllables made possible a change from one hexachord to another, a process known as mutation (*mutatio*). If a chant went beyond the range of a single hexachord, it would have to be assigned syllables ('solmized') from an adjacent hexachord. As Johannes de Garlandia explains:

> One undertakes [a mutation] because of ascent or descent, as for example in *C fa ut*, because, if for this note one were to take *fa*, one could ascend as far as the third note. If one wished to reach the fourth note, however, one would have to take *ut* in this same *C fa ut*, which is a mutation of *fa* into *ut*.

The practical application of Guido's *modi vocum* and of Hermannus' 'seats of the modes' (although Hermannus does not explain it) is that chants with chromatic notes can be envisaged and notated at a pitch complementary to the usual one, ending with one of the alternative finals (the 'affinals' in Guido's terminology, usually now 'co-' or 'confinals'). But since both the 'modi vocum' and the 'sedes troporum' were conceived as hexachords, it is clear that the solmization syllables will fit them exactly. Whether the chant closes on a co-final or not, it will be solmized in the same way. So Jacobus of Liège can quite simply say: 'chants which are ended in *re* are of the first and second mode . . . chants which are ended in *mi* are of the third and fourth mode . . . chants ending in *fa* are of the fifth and sixth mode . . . chants which are ended in *sol* are of the seventh and eighth mode . . .' (6. 217). The next chapter (Book 6, ch. 76) deals with chants ending on *ut* and *la*. Those ending on *ut* are to be assigned to mode 5 or 6 if they end on *C*, *F*, or *c*, because of the coincidence and affinity with *fa* at those pitches. (There is a coincidence with *sol* on the higher *c*, but no affinity, because of the whole tone step beneath *sol*.) If *ut* as final falls on *Γ*, *G*, or *g*, then the mode is 7 or 8. If a chant ends on *la*, it may be assigned to modes 1 to 4: if it ends on *a* or *ă*, there is both coincidence and affinity with both *re* and *mi*, so modes 1 to 4 are all possible; if it ends on *E* then *mi* is the only alternative solmization, and the chant should be assigned to mode 3 or 4.

Jacobus can be seen here working out with great thoroughness a technique of modal identification which was implicit in Guido's teaching from the beginning, even if not explained in Guido's own writings. The essence of the technique is the use of an invariable scale-segment—the hexachord—for tonal orientation. It is not basically different from the technique of using the Noeanne formulas to fix in the mind the modality of a chant. The fact that it was an unvarying, abstract pattern made it applicable, through transposition, to any situation likely to arise in the chant repertory, and it was keyed securely into the system of staff-notation. Its success is shown by the fact that even in the much more complicated tonal designs of fourteenth- and fifteenth-century polyphonic music it continued to provide a means of tonal orientation.

VI

Plainchant up to the Eighth Century

VI.1. INTRODUCTION

The rest of this book is a survey which aims to provide a historical background to the previous parts. It is necessarily sketchy, both because of the nature of the evidence and because of the compression of material involved. Particularly for the early centuries, before music was notated, great care is required in considering the nature of whatever plainchant might have been sung. The issues are unusually complicated, much often depending on the interpretation of very brief references, or even single words. There is no space here for a detailed examination, and consequently no room for more than a rough indication of the directions in which recent scholarship has been moving.

The evidence from the centuries before musical notation was developed is perforce more useful for the history of liturgy than of music. While we can learn much about the liturgical function of particular chants, we must continue to speculate about their musical character. This period will be treated only lightly. A special section is nevertheless given to the role of Gregory the Great (d. 604), whose name, more than any other, is indissolubly associated with plainchant.

An important way of filling gaps in our knowledge of early forms of liturgy in the various churches of Christendom has been to compare them with each other and note points of similarity. What is common to them is often the result of a common ancestry. (The classic study of this type is Baumstark 1958.) In a similar fashion—though to a far lesser degree—the veil covering music before the ninth century can be pierced by comparative study of parallel chant repertories. In theory such a study could include all known Christian chant repertories. No attempt is made in this direction, however, beyond a listing of those repertories in VI.2. Later in the book (VIII) the chant which we know as 'Gregorian', the subject of Chapter II above, is placed side by side with the other principal Western chant repertories which have come down to us: the old Italian chant of Milan, Rome, and Benevento, and Old Spanish (or Mozarabic) chant. It is generally agreed, as a very broad hypothesis, that where these repertories show any sort of formal or musical similarity, and yet have developed more or less independently of each other, the similarities may indicate a state of the repertory earlier than the one that has come down to us. If we have three or four similar versions of chant, the common basis of those versions may go back earlier than the ninth century, beyond the time when the first notated books appeared. Yet the fact that these west European repertories diverge as much as they do suggests that extreme caution must be exercised in assessing the possible interrelationships between Western and Eastern chant repertories. And in the East, too, the absence of notation before the ninth century (or later—some repertories were not notated before the present century) raises a formidable barrier.

The first codification in a form of musical notation of the chant we know as 'Gregorian' is the result of the Frankish endeavours to establish Roman liturgy and its chant as normative in their land. There is nevertheless plenty of evidence that the Franks did not simply reproduce the Roman liturgy and its music exactly. Extensive liturgical modifications were necessary to accommodate local requirements. At the same time, the Franks expanded the chant repertory in radically new ways: the repertories of sequences and tropes are the best-known examples of this. Two other achievements of the ninth and tenth centuries are placed in context immediately after the chapter on compositional activity: the codification of the chant repertory with musical notation, and the development of a body of writings on music theory. These have to be understood in the light of both the liturgical renewal and the general intellectual revival of the ninth century, the so-called 'Carolingian renaissance' under Charlemagne and also the 'second Carolingian renaissance' under Charles the Bald. The significance of the 'Carolingian century' is discussed in Chapter VII.

For many centuries both before and after the time when the Franks were learning and codifying the Roman repertory, the Byzantine Empire and church offered a political and liturgical example which might from time to time have been emulated in the West. At a very basic level, Byzantine chant also has roots in the music of the early church which it shared with the Western chant repertories, perhaps discernible in certain musical features common to both Western and Eastern chant. On a more detailed level, we know of Byzantine chants which were translated and sung in the West, and it is probable that the eight-mode system of Gregorian chant is derived from Byzantium. The possible influence of Byzantine chant in the West is considered in VIII.2.

The principal features of the Old Italian chant repertories are discussed in VIII.3–5. As already remarked, the 'Gregorian' repertory was first codified in Francia. Unfortunately, it seems thereby to have displaced local repertories, commonly referred to as 'Gallican', before they achieved written form. The scant remains of Gallican chant are discussed after the Old Italian repertories, in VIII.6. Old Spanish chant was indeed codified, but at a time when diastematic notation (indicating pitch precisely) was not yet used. Only a handful of Old Spanish chants can be transcribed today. Nevertheless, the shape of the melodies codified in adiastematic notation can be recognized sufficiently well to permit a fair degree of certainty about some aspects of Old Spanish chant (VIII.7).

Subsequent chapters pursue the history of Western chant after the Carolingian settlement.

VI.2. THE CHURCHES OF CHRISTENDOM

Baumstark 1958.

The subject of this book is plainchant in the West, meaning the chant sung in the European Latin liturgies used as far east as the areas which correspond roughly to modern Poland, Czechoslovakia, and Hungary. To understand the origins of Western plainchant, a knowledge of the early history of the liturgy is important. This implies consideration not only of the Roman liturgy but also of others, Eastern as well as Western. Ultimately it is also possible to compare the chant repertories of these other rites with Western ones, though in case of many rites the written tradition is very late. Studies of this type have, however, hardly begun, and even questions of basic methodology remain as yet unanswered. It must also be admitted that contact between the different Christian communities was interrupted at an early date by theological differences, for example the separation of communion caused by the Monophysite and other disputes, and by political divisions after the fall of the Roman empire, so that the prospect of meaningful musical comparisons is faint.

The earliest Christian community, in Jerusalem, was presided over by St James after the other Apostles spread abroad, and was the place of the first church council of about the year 49. But after the Jewish revolt of 66 and consequent Roman destruction of the city in 70 Jerusalem had little importance. Then in about 326 Helena, mother of the emperor Constantine, came to venerate the holy places of the city, which began an important fashion of pilgrimage and seems ultimately to have contributed to making the liturgy more topical, commemorative, bound up with the remembrance of persons, places, and events. Uniquely valuable and detailed descriptions of some of its ceremonies have survived, written by the pilgrim Egeria or Etheria, thought to have been a Spanish nun or abbess, who visited many places in Eastern Christendom at the end of the fourth century (*Corpus Christianorum Series Latina*, 175; Eng. trans. Gingras 1970, Wilkinson 1971). At the Council of Chalcedon in 451 the patriarchate of Jerusalem was created. After the Islamic conquest of the Holy Land in the 630s the patriarch resided mostly in Constantinople. Although a fair amount is known about the rite of Jerusalem (not least from Armenian and Georgian borrowings), there are of course no records of its music (see Baldovin 1987, Leeb 1979).

The two most important Middle Eastern Christian centres in the early centuries were not Jerusalem but Antioch and Alexandria.

The church at Antioch in Syria, traditionally founded by St Peter, was by the fourth century third in rank of the patriarchal sees (after Rome and Alexandria). In the next century it was split by christological controversies. The Greek church of Constantinople and the Latin church of Rome have held to the doctrine that Christ has two natures, human and divine, within a single person, affirmed by the Council of Chalcedon. The belief that Christ has two separate persons as well as natures,

Diphysitism, condemned at the Council of Ephesus in 431, was supported by part of the Antiochene church, usually called 'Nestorian' ('East Syrian', 'Church of the East', or 'Assyrian'), which by the end of the Middle Ages had actually spread as far as south India and China. Monophysitism, the doctrine that Christ has one nature and person, condemned at Chalcedon, was supported by the 'Syrian Orthodox' ('Jacobite' or 'West Syrian') party of the church at Antioch as well by the Alexandrine and Armenian churches. The liturgy of the Jacobite church, known as the 'Liturgy of St James' (reflecting its supposed derivation from Jerusalem), adopted Syriac as its language. The Antiochene party which rejected both heresies was called 'Melkite' and retained Greek in its liturgy.

Books of the Syrian rites were rarely notated, though some notated ones, principally Melkite, have survived. Research on the music of the various Syrian liturgies has therefore relied on transcriptions made from modern performance. (See Husmann, 'Syrian Church Music', *NG*; by far the greater part of recent research was accomplished by Husmann.)

The patriarchate of Alexandria, second in rank to Rome before the rise of Constantinople, was traditionally held to have been founded by St Mark, hence the name 'Liturgy of St Mark' for its rite. The church opted for Monophysitism. It became known as the Coptic church after the use of the vernacular (Coptic, descended from ancient Egyptian) in the liturgy. The church became relatively isolated after the conquest of Egypt by Islam in 641. Although a form of ekphonetic notation may have been used in the tenth and eleventh centuries, Coptic chant has been transmitted to the present only orally. (See Borsai, 'Coptic Rite, Music of the', *NG*.)

The Ethiopian church is a daughter of the Syrian Jacobite and Coptic churches, and was subject to the latter until this century. The oldest notated sources of its chant date from the sixteenth century. (See Hannick, 'Ethiopian Rite, Music of the', *NG*.)

The Armenian church was recognized by the country's ruler, Tiridates III, in 301, even before the Roman emperor Constantine's toleration of Christianity. It became schismatic half a century after Chalcedon. Elements of its liturgy are a faithful reproduction of the Jerusalem liturgy of the fifth century (Renoux 1961–2, 1969–71). Both an ekphonetic and an indigenous chant notation are known from Armenian liturgical books, said to date from as early as the ninth century. (See Hannick, 'Armenian Rite, Music of the', *NG*.)

The church of Georgia dates from the fourth century. Monophysite from the Council of Chalcedon onwards, it returned to orthodoxy at the end of the sixth century. Since the development of the Georgian rite was much influenced by the Georgian monasteries in the Holy Land (St Sabas near Jerusalem and on Mount Sinai), it may incorporate some elements of the chant of Jerusalem (see Tarchnischvili 1959–60). Georgian lectionaries with Byzantine ekphonetic notation survive from as early as the tenth century, and chant-books in an indigenous notation are apparently equally ancient. (See Hannick, 'Georgian Rite, Music of the', *NG*.)

In 330 Constantine refounded Constantinople and made it his capital, on the site of

Greek Byzantium, which appears to have had a Christian community from the second century. At a council in the city in 381 the bishop was given precedence second only to Rome, and became patriarch in 451. Much of its early liturgy appears to have derived from Antioch, whence St John Chrysostom came in 398 to be its bishop. The liturgy in general use still bears his name. That reserved for the Sundays of Lent (except Palm Sunday), Maundy Thursday, the Eves of Easter, Christmas, and Epiphany, and the Feast of St Basil (1 January) is known as the Liturgy of St Basil (d. 379).

The political supremacy of Byzantium eventually lent great influence to its liturgy as well. It was translated into Old Slavonic by SS Cyril and Methodius upon the conversion of the Slavs of Moravia in the mid-ninth century, and when Vladimir, prince of Kiev, was baptized in 988, he also introduced the Byzantine liturgy in Old Slavonic. While the Greek expansion into Moravia was short-lived—it clashed with German political interests—the Slavonic liturgy was permanently established in Bulgaria (from the mid-ninth centry) and Serbia (from the tenth). We have musical documentation of Byzantine chant from the ninth century (ekphonetic notation possibly from even earlier), of Russian chant from the late eleventh, of Bulgarian chant only from the thirteenth, and Serbian from the fifteenth. (Byzantine chant is discussed briefly below, VIII.2; on the Slavonic families of chant see Velimirović, 'Russian and Slavonic Church Music', NG.)

The patriarchate of Rome coincided more or less with the western half of the Roman empire, comprising Italia, Illyricum, Africa, Hispania, Gallia, and Britannia.

The bishops of Rome in the first two centuries were all Greek-speaking. Greek remained the language of the Roman liturgy until the fourth century (Klauser 1946). We should in any case expect a strong similarity to have existed between the rites of the various Christian churches at this early period. This means that, with due caution, evidence about one church can be considered for its relevance to another. The later the period, the less safe does this become, for it is by then a matter of assessing influences upon established liturgies, rather than the development of similar usages from common origins. The general picture appears to be one of great conservatism at Rome in face of new liturgical developments. Rome's political importance dwindled after the establishment of Constantinople; it was replaced at times by Milan and Ravenna as imperial residence, and it suffered from barbarian invasion and from wars between the Byzantine empire and Goths and Lombards. Yet this did not weaken its constancy to liturgical tradition. On the other hand, the considerable independence of other Western uses from those of Rome may be attributed to its political weakness. Since it is by no means easy to trace the development of Roman liturgical use through the vicissitudes of the early centuries until the time from which we first have comprehensive liturgical books (the eighth and ninth centuries), attempts to understand its chant in any detail are extremely difficult.

Aquileia in north Italy assumed the rank of patriarchate in the sixth century, from the seventh century being split into two seats, one at Grado, the other at Cividale. Despite speculation about an independent rite and even an independent Old Aquileian chant repertory, practically nothing is known about its early liturgy.

Ravenna and Milan both developed local rites. Ravenna was seat of the imperial residence in the fifth century, and capital of the Byzantine Exarchate from 540 to 751. Again, very little is known of an independent musical repertory. Milan, on the other hand, has retained its liturgical independence until the present day, and its chant repertory, first codified in the twelfth century, constitutes prime evidence of the degree to which local uses varied in the early Middle Ages. It has been argued that the political link between Milan and Benevento in south Italy, first under the Ostrogoths, then the Lombards, provided the basis of a similar chant repertory in the two cities. At any rate, by the time of the writing down of chant in the Beneventan region, another local repertory different from the Roman had obviously developed. (On the local Italian chant repertories see below, VIII.3–5.)

The church in North Africa, centred on Carthage, used a Latin Bible and liturgy even before the church in Rome, but little is known in detail of its rite. Weakened by the Vandal conquest (429)—the Vandals were Arian Christians—restored by the reconquest under Justinian (534), the church faded after the Arab conquest at the end of the seventh century. Nothing of its musical repertory is known.

The term 'Gallican' has been used to indicate all the various north Italian rites independent of Rome and those of Gaul and Spain as well. Alternatively, the liturgies of Spain, Gaul, and the Celtic lands (primarily Ireland) have been grouped together. The term is usually now restricted to the liturgies of Gaul alone. The evidence for these is incomplete and there was doubtless considerable variation from area to area and from century to century. (We are best informed about the church in the Rhône valley.) No written record of Gallican chant was made before the establishment of the Frankish-Roman liturgy in the eighth and ninth centuries. A certain amount of Gallican material survives, however, in 'Gregorian' books. The nature of Gallican chant and the way it differed from Roman is nevertheless not easy to define, not least because of the differences between the two versions of the Roman chant repertory which have come down to us: the 'Gregorian', first codified by the Franks in lands where the Gallican rite had been used, and 'Old Roman', the chant found in books from Rome itself. It is possible to argue that 'Gregorian' chant is at least in some respect a 'Gallicanized' version of the Roman repertory (see VIII.6).

By the expression 'Celtic church' is usually meant the church in Britain and Ireland until the time when Augustine of Canterbury (d. 604 or 605) and later churchmen brought Roman usages, which gradually supplanted Celtic ones. For most of England this happened during the sixth and seventh centuries, while Ireland remained unaffected until considerably later. The process of conversion to Roman customs forms a parallel, some two centuries earlier, with the change in Francia during the eighth and ninth centuries. It should be emphasized that, even less than in Gaul, there is little evidence of a centrally organized church with a 'use' prescriptive within a diocese. Almost no chant which can be considered authentically Celtic has survived: the change to Roman use in England occurred long before musical notaton was used, and no notated books survive from Ireland until those of the thirteenth century, based on English (Sarum) use.

The other branch of the Roman church for whose rite not only the texts but also the music achieved codification was that of Spain. Various terms have been used to designate this rite and its chant: 'Visigothic' (Spain was ruled by the Visigoths from the fifth century until the Islamic invasion of the eighth), 'Old Spanish', and 'Mozarabic' (a term properly used to designate the Christian population living under Arab dominion). Although much can be learnt about its music, practically all surviving chants are recorded in adiastematic notation. The coming of pitch-specific notation coincided with the reconquest of Spain from the Moors and the importation of the Frankish-Roman rite and Gregorian chant. We have almost no manuscripts with Old Spanish chant after this event, and consequently almost no records in pitch-specific notation (see VIII.7).

It will be clear from this brief outline that reconstruction of the chant of the early church is extremely hazardous. By the time written records of the chant repertory of any of the branches of the Christian church were made, they had had ample time to develop along independent lines, through centuries of often radically changing circumstances which cannot fail to have had a considerable impact on the liturgy and its music.

In the next two sections the history of the most important musical forms of the Roman chant are outlined and parallels are drawn with chants of other rites where appropriate. The emphasis is upon the other Western rites. Since almost nothing of the chant of the Celtic lands survives I have largely disregarded Celtic practices (see Warren 1881/1987; Curran 1984). Despite the lean evidence regarding Gallican practices, I have drawn upon them more regularly to point up similarities with other rites.

VI.3. THE EARLY CHURCH

Stäblein, 'Frühchristliche Musik', *MGG*; Hannick, 'Christian Church, Music of the Early', *NG*; Smith 1984; McKinnon 1986, 1987, *Music*.

Until the so-called Edict of Milan of 313, by which Constantine and Licinius sanctioned the freedom of worship of Christians (and adherents of any other religions), Christians were often persecuted savagely, religious meetings took place in private, often in secret, and public worship was impossible. In these circumstances the order of worship remained simple and was in many respects likely to have been unformalized.

Until recently it had been assumed that the Jewish Book of Psalms constituted a chief source of texts for singing in the early church, and that continuity must have existed between Jewish and early Christian worship. This view is no longer accepted. Hannick ('Christian Church, Music of the Early', *NG*) has summarized the large number of witnesses to the singing of non-psalmodic hymns, and both Smith (1984) and McKinnon (1986) have provided ample evidence that Jewish public worship could have given to Christian worship neither its form nor its content. Two types of Jewish worship enter the matter, that of the Temple in Jerusalem, and that of the synagogues (*synagoge* is the Greek translation of Hebrew *beth ha-knesset*: 'place of assembly). In the grand setting of the Temple a different psalm for each day of the week was sung, with instrumental accompaniment, at the daily sacrifice; the Hallel psalms, Pss. 113–18, which have the refrain 'Alleluia', were also sung on important days of the Jewish year. Both because of their special associations and their special manner of performance, it seems unlikely that these psalms would have been adopted by the small, private groups of Christians. The Temple was destroyed by the Romans in the year 70. The evidence from the synagogue is equally unpromising, for a different reason. There is no evidence that in New Testament times meetings in the synagogue were for worship in the normal sense of the word. The synagogue was a secular rather than an exclusively religious meeting place. While passages of scripture (including passages from the psalms) were undoubtedly read there and expounded, it was not normally a place of prayer or psalm-singing. In the decades after the destruction of the Temple, which removed at a stroke the whole focus of Israel's religious life, something like an ordered service of worship became established in the synagogues, a partial substitute for what had been lost. But psalm-singing, or more specifically the singing of the daily psalms once used in the Temple, was one of the last elements of Temple worship to be taken up in the synagogue, to judge by its absence from documents of Jewish religious teaching before the sixth century.

More promising is the evidence linking Christian religious meetings with private Jewish religious exercises. The prayers *shema* and *tefillah*, both part of the Temple liturgy, were to be recited privately, and the Hallel psalms were sung at the ceremonial meal, the Seder, at Passover. Indeed, if the Last Supper of Christ and the disciples took place on the first night of Passover, the likelihood is that the 'hymn' they sang (Matt. 26: 30, Mark 14: 26) was the Hallel.

Most early references to singing during Christian worship are too vaguely expressed to be very enlightening. The words 'psalm' and 'hymn' usually have only the general meaning of 'song of praise', and cannot unconditionally be connected with the Old Testament psalms, or with any particular poetic or musical form.

It has been argued that some early Christian songs are quoted in passages such as 1 Tim. 3: 16, or Rev. 4: 8, but neither constitutes conclusive evidence (McKinnon 1987, *Music*, 16–17). Few newly composed texts have come down to us. The forty-two 'Odes of Solomon', perhaps written towards the end of the first century by a Syrian Christian, are compositions like the Old Testament psalms, and may be representative of new Christian composition (McKinnon 1987, *Music*, 23–4; Charlesworth 1977). Much has been lost, not least the heretical hymns of Bardesanes (Bar-Daisan) of Antioch (154–222) and Paul of Samosata, bishop of Antioch in the third century. Their popularity and consequent danger to orthodox Christians seems to have aroused suspicion against all non-biblical chants, which were then banned in

272 at the second Synod of Antioch and in the fourth-century Canons of Laodicea (the reference is to *idiotikous psalmous*: McKinnon 1987, *Music*, 119).

Something of the nature of Bardesanes' songs can perhaps be gleaned from the songs of Ephrem of Edessa (d. 373), who is supposed to have composed orthodox contrafacta to the melodies which Harmonios, son of Bardesanes, had written for his father's heretical poems. Prominent among them are the *madrasha*, a strophic solo song with choral refrain, and the *sogitha*, a strophic song in dialogue form. Their nearest equivalents in the medieval Western liturgy are perhaps the processional hymns of Passiontide: *Gloria laus et honor*, *Audi iudex mortuorum*, and *Pange lingua*.

Much of the musical component of early Christian worship must have been improvised or at least performed as the occasion demanded, according to no fixed order of service. Something of the character of the services may be gleaned from 1 Cor. 14: 26–7, where Paul comments on what seems to be a largely improvised type of worship, probably prevalent in other places as well as Corinth: 'When ye come together, every one of you hath a psalm, hath a doctrine, hath a tongue, hath a revelation, hath an interpretation.' It has also been supposed that early Christian worship might occasionally have resembled that of other sects, for example the pre-Christian monastic Therapeutae from near Alexandria, whose worship is described by Philo in the early first century (ed. Conybeare 1895, ch. 11). Here the believers rise after their Sabbath meal, form a choir of men and a choir of women, with a musically experienced leader. They sing hymns of varied types, some together, some alternatim (*antiphonois*). On the other hand, in one important description of the eucharist, celebrated on Sunday morning, written by Justin Martyr (*c*100–*c*.165) in Rome, there is no reference to singing of any sort (McKinnon 1987, *Music*, 20).

The distinction between eucharist and other forms of worship is important. In the first three centuries the eucharist was commonly the culmination of a common religious meal, the agape or love-feast. Singing clearly took place during the agape. Tertullian, writing at Carthage in the closing years of the second century, mentions both singing 'from the sacred scriptures' and improvised singing (McKinnon 1987, *Music*, 43, Taft 1985, 26–7). A document known as the *Apostolic Tradition*, attributed to the Roman priest Hippolytus, written in the early third century, describes an agape where the Hallel psalms are sung by various persons, with 'Alleluia' as the refrain (McKinnon, 47). If not by the agape (which gradually fell out of use), then the eucharist might be preceded by other customary observances. Thus in commenting on a charismatic sister of the Montanist movement, Tertullian says she is inspired by the various components of the synaxis which precedes the eucharist: scriptural readings, psalms, a homily, and prayers.

The usual time for the agape and/or eucharist was the evening. Of the possible musical component, if any, of religious observances at other times of the day we know little. The *Apostolic Tradition* describes a number of occasions for private prayer during the day, and a morning communal assembly for instruction as well as the common evening meal.

Although we can speculate, on the basis of later evidence, as to which texts might have been sung during these religious observances, we have almost no precise evidence. The same is naturally true of their music. The single piece which has survived (fragmentarily), the hymn recorded on the Oxyrhynchus papyrus of the late third century, in Egypt, raises as many questions as it answers (facsimile and various transcriptions in Stäblein, 'Frühchristliche Musik', *MGG*). The Greek vocal notation with rhythmic signs allows a relatively secure transcription. It appears to be a solo song in the authentic G-mode (or, less anachronistically stated, using the diatonic Hypolydian scale), with texts in anapaestic dimeter. Although the text is sacred, it is impossible to say whether the music is typical of its time and place, either in its rhythm or its melodic style. The piece was recorded on the back of an (earlier) cereals account, thus for private purposes. Its function (if any) in contemporary Christian worship remains unclear.

VI.4. OFFICE CHANTS BEFORE THE EIGHTH CENTURY

Stäblein, 'Frühchristliche Musik', *MGG*; Hannick, 'Christian Church, Music of the Early', *NG*; Bäumer 1905; Hucke 1953, 'Entwicklung', 1973; Salmon 1959, 1967; Jungmann 1960; Heiming 1961; Leeb 1967; Winkler 1974; Cheslyn Jones *et al.* 1978, ch. V; Bradshaw 1981; Taft 1985; McKinnon 1987, *Music*; Dyer 1989, 'Monastic'.

The official toleration of Christianity and the end of persecution must have had wide-ranging consequences for the form and content of worship. It is no accident that information about the liturgy is much more plentiful from the fourth century onwards. It is often no easier to interpret than that from the previous period because of the differences of observance between different areas. But from the fourth century it is possible to see clearly many of the chief characteristics of the liturgy familiar to us from the Middle Ages. Two developments are particularly important.

First, public worship became a matter of more organized ceremonial conducted by trained personnel in churches built for the purpose. By the end of the fourth century we have ample evidence of formal liturgies of various kinds. This means, for example, that songs were performed by cantors specially detailed to carry out this duty, rather than by any person present inspired to sing (see Hucke 1953, 'Entwicklung', 177–85, with information on the formation of the trained choir; also Foley 1982, Fassler 1985, McKinnon 1987, *Music*, 109). The ceremonial aspects of worship were elaborated in order to add mystery and inspire awe. This can be seen, for example, not only by the liturgy described by Theodore, bishop of Mopsuestia (392–428), north-west of Antioch, but also by his commentary on it (Dix 1945, 282–8). It has been supposed that more extended liturgical actions may have provided the occasion for new accompanying chants.

Secondly, there was a dramatic movement—already under way in Egypt towards the end of the third century—by many religious persons to cut themselves off from the world and lead a solitary life, alone as hermits, as hermits answering occasional calls to meetings, or as communities of monks and nuns. The monastic communities developed their own forms of communal worship, which occasionally interacted with their 'secular' counterparts but remained essentially distinct. Many such communities engaged in more or less continuous worship, usually organized in a system of office hours different from the prayer hours which were observed in the previous centuries. The chief purpose of the religious exercise was continual prayer and praise, particularly through the complete recitation of the Old Testament psalms. This ultimately led, among others, to the type of medieval office known to us in the Roman and Benedictine uses. The secular office, by contrast, was naturally selective in its texts. In recent times it has therefore become usual to distinguish a 'cathedral' or 'clerical' office, the non-monastic use, from a 'monastic' office. We are better informed about monastic practices than cathedral or secular liturgies, because of the survival of the 'rules' of several founding fathers of monastic orders.

Thus from the churches of the western half of the Roman empire there is only fragmentary evidence from before about 600 to inform us about cathedral liturgies, while from the same area we have the *Ordo Monasterii* (part of the so-called *Rule of St Augustine*), probably written in North Africa by a follower of St Augustine (ed. Verheijen 1967), the rules for the monks and nuns of St Caesarius of Arles and his successor Aurelian, bishop of Arles 546–53 (ed. Morin 1942, McCarthy 1960), the late sixth-century rule of St Columbanus for the Celtic church (ed. Walker 1957), and then, for early sixth-century Italy, the *Regula Magistri* (ed. de Vogüé 1964, trans. Eberle 1977) and the Rule of St Benedict (ed. Hanslik 1960, de Vogüé and Neufville 1971–7). With the aid of the last two it has been possible to reconstruct much of the contemporary monasticism in the convents which were attached to the basilical churches in Rome. These are by no means the only uses which can be reconstructed, or about which something is known, as a glance through the accounts of Bradshaw (1981) and Taft (1985) will show.

Common to the cathedral or secular uses was a basic system of daily morning and evening prayer and Sunday eucharist, already established by the end of the fourth century. Many components of morning and evening prayer were characteristic of the time of celebration: Ps. 62 (*Deus Deus meus ad te de luce vigilo*) for morning prayer and Ps. 140 (*Domine clamavi ad te . . . elevatio manuum mearum sacrificium vespertinum*) in the evening. Morning prayer also nearly always included Pss. 148–50, and evening prayer often began with the ceremony of lamplighting. In addition to all these, an all-night vigil service was occasionally performed on the eve of important feast-days. (Those known to Gregory of Tours are listed by Taft 1985, 182.) Here too psalmody took its place alongside readings and prayers, but generally speaking there is little evidence of the recitation of the psalms in cycles.

In these morning and evening services the bulk of the music consists of Old Testament psalms. That is something of a contrast from the early centuries. Other items certainly had a place: the Gloria in excelsis in the morning, and the hymns composed by Hilary of Poitiers and St Ambrose of Milan, for instance. But the psalms are ubiquitous.

In the monastic office the psalter was recited continuously, at first informally and then in groups of psalms distributed among the fixed hours, so arranged that the whole psalter could be performed within a fixed period, usually a week (see the summaries in Bradshaw 1981 and Taft 1985). Cathedral uses with especially numerous psalms, often to be found in the West, seem likely to have been influenced by monastic practice. For example, the Second Council of Tours in 567 specifies what is practically a monastic round: twelve psalms at Lauds, six with 'Alleluia' at Sext, twelve at 'duodecima' (an evening service distinct from the lamplighting ceremony), and a number during the Night Office which varied according to the month—as many as thirty in December (Taft 1985, 149). Ambrosian hymns are permitted, and also hymns by other authors, whose names should be written in the margin (implying a written collection of hymn texts).

We can only speculate on the basis of later evidence what the music during these services was like (for example, four of Ambrose's hymns are thought to have survived with something like their original melodies). Even in important matters of performance practice—the division between soloist and choir, the role of refrain verses, and the use of alternating groups of singers or choirs—there is considerable difficulty in interpreting the early evidence, no doubt because of the considerable variety that existed. Important distinctions which have to be borne in mind are those between monastic and secular practice, and between the performance of the Old Testament psalms and the singing of newly composed verses or songs. The greatest difficulties arise in understanding what is meant by 'response' and 'antiphon' and the concepts 'responsorial' and 'antiphonal psalmody'.

In one sense 'response' and 'antiphon' might seem to be identical, with the meaning of 'refrain', a verse or verses sung after each section of another chant, for example after each verse of a psalm. A possible distinction here would be that the response takes its text from the psalm or other piece being sung (usually its first verse, sung first by the leader and repeated by the chorus), whereas the antiphon is an independently composed song. If that seems a trifling difference, we should remember the sensitivity displayed particularly in the third and fourth centuries in the face of non-biblical songs: those of Bardesanes in Antioch, or those of the Arians in Constantinople described by Socrates and Sozomen (McKinnon 1987, *Music*, 101–4). General conservatism with regard to non-scriptural lyric compositions is one of the things which distinguishes the medieval Roman liturgy most strongly from the Byzantine, where hymn genres such as the kontakion, kanon, and sticheron flourished in great abundance. (The evolution of these is revealing: see Wellesz 1961, chs. VI and VIII–IX.)

The concept of the antiphon and the antiphonal manner of performance has, however, an ancillary meaning of 'alternating' between groups. It has been argued that 'responsorial' relates primarily to the alternation between soloist and chorus, 'antiphonal' to that between two choirs, say the people and the trained choir, or the two halves of a choir. We should also consider the possibility that the two meanings of 'antiphon' may not be mutually exclusive: the antiphon to be both newly composed, non-scriptural, and performed by alternating choirs. Isidore of Seville (*c*.560–633), for example, says that antiphons are chanted by two choirs alternately (*Etymologies* 6. xix, *De ecclesiasticis officiis* 1. vii and ix). He may have been thinking of alternatim performance of the psalm verses to which the antiphons are a counterpart. But in view of the usual solo singing of the psalms themselves at this date, it seems more likely that the antiphon itself is to be chanted first by one side of the choir, then by the other. That is what we find, much later, described by Amalarius of Metz in his *Liber officialis* of *c*.823 (Hanssens, ii. 433, quoted by Dyer 1989, 'Monastic', 68). First the antiphon is begun by a soloist from one choir, and seems to be completed (or repeated) by both choirs together. The performance apparently continues:

> Psalm verse: soloist
> Antiphon: first choir
> Psalm verse: soloist
> Antiphon: second choir
> Psalm verse: soloist
> Antiphon: first choir
> Psalm verse: soloist
> Antiphon: second choir

(and so on)

The great attraction of 'antiphons'—whatever they may have been—is difficult to understand when so little is known about their character and content. There is no space here to examine each reference in order to distinguish between possible meanings (Hucke 1953, 'Entwicklung' provides the most comprehensive and judicious review). I shall extract only a few salient points which are particularly relevant to the medieval usage in the West.

Early reports connect the first singing of antiphons with Antioch, and, lest their non-biblical nature and their use by heretical groups like the Arians should taint their image, a legend grew up that Bishop Ignatius of Antioch (martyred *c*.107) had seen a vision of angels singing 'antiphonal hymns', and introduced the same sort of songs into his own church (McKinnon 1987, *Music*, 102). The fourth century is, as with so many things, the period when the fashion is first widely recorded, and support for its Eastern origin comes from both St Augustine and the biographer of St Ambrose. Augustine says that during the time when Ambrose's church in Milan was under threat from the Arians, in 386–7, 'hymns and psalms' were first sung 'after the manner of the oriental regions' to raise the spirits of those under siege (*Confessions* 9.7.15; McKinnon 1987, *Music*, 154). This has been connected with a remark by Paulinus, former secretary of St Ambrose, writing in 422, but describing the same events as Augustine, that at this time 'antiphons, hymns, and vigils' were first performed in Milan. As so often, there is no precise evidence here of what 'antiphon' means, though its connection with inspiriting congregational singing is clear. (On the other hand, Leeb 1967 pointed out that Ambrose himself only mentions solo psalms with congregational responses in his sermons: see the discussion of the gradual below, VI.5.)

The recent study of Dyer has made it clear that for early monastic psalmody solo performance was the rule. Each monk in turn played his part in singing while his brethren meditated on what was being sung. We do not know what the music for this singing was like. Something akin to the familiar tones may have been used, though solo delivery naturally allowed greater freedom of musical expression.

The meditative role of the monks not taking their turn to sing a psalm was, however, occasionally varied by their adding responses or antiphons to the solo psalm. In John Cassian's *Institutes*, written about 415 for the bishop of Apt, drawing on Cassian's experience of Egyptian monasticism, he says that monks in Palestine first stand to sing three 'antiphona' (usually taken to mean psalms with antiphon refrains), then sit while three monks in turn sing a psalm solo, the other brethren 'responding' (presumably singing short responses). Various arrangements of psalms with antiphons and psalms with responses are to be found among early monastic uses. In some, the balance is weighted heavily in favour of psalms with antiphons. This is the case in the Rule of the Master, for example (ed. de Vogüé 1964; see also Bradshaw 1981, 140, Taft 1985, 122–3), where the proportion is over three to one.

Several items in the Rule of the Master are labelled *responsoria*, and the question arises as to whether these are in fact the same as the old psalms with responses. (See Hucke 1953, 'Entwicklung', 172–3, with reference to the Rule of Benedict, slightly later than the Rule of the Master and drawing heavily upon it.) They appear only as single items, not in the sets typical for the recitation of the whole psalter. In fact, they stand outside the cycles of psalms. We may surmise that pieces like the short responsories known to us from medieval sources may have featured among them, but a more ornate manner of delivery cannot be ruled out.

It is evidently necessary to make a clear distinction between the singing of short responses (such as 'Alleluia') during the cyclic singing of psalms, and the singing of responsories independently of the psalm cycle. It is the latter which we see in the Rule of the Master, and also the Rule of St Benedict, where responsories are performed during the Night Office in alternation with lessons, a new arrangement with which we are familiar from the Middle Ages: the texts are specially selected for their relevance to the lesson. No doubt this could have been done *ad libitum*. If the performance resembled that of medieval short responsories, the singer of the responsory would have selected a passage and sung an opening verse, which would be repeated by the others present (either choir or congregation is conceivable). The soloist would have continued with further verses, the others responding after each verse with the same verse as they had sung at the start. In Benedict's Rule we see this systematically organized for the first time, and may assume that the medieval cycles of lessons and complementary responsories had their birth here.

When perusing old orders of service (such as those gathered together by Bradshaw 1981 or Taft 1985), it is of course not always possible to know whether the responsorial chant is part of a psalm cycle or the complement to a lesson. The sixth-century 'Psalter of St Germain-des-Prés' (Paris, Bibliothèque Nationale, lat. 11947; see Huglo 1982, 'Répons-Graduel'), a psalter with responsory texts, seems to be designed for cyclic performance of the psalter. On the other hand, two lectionary fragments have survived with psalm verses apparently intended for responsorial performance: Wolfenbüttel, Herzog-August-Bibliothek, Weissenburg 76 from the fifth to sixth century (ed. Dold 1936) and Mount Sinai, St Catherine's monastery Gr. 567, possibly a Spanish manuscript of the seventh century (Lowe 1964).

We have two reports of the selection of responsory texts as early as the fifth century (Vogel 1986, 302–3). Gennadius relates that the priest Musaeus of Marseilles made a selection of readings for feast-days and passages for responsories from the psalms (*responsoria psalmorum capitula*) to go with them (McKinnon 1987, *Music*, 170). Sidonius Apollinaris credits Claudianus of Vienne with something similar (cited by Jeffery 1984, 161). With the Wolfenbüttel and Mount Sinai fragments, and with the reports about Musaeus and Claudianus, it is difficult to know if we are dealing with the office responsory or the gradual of mass (the Wolfenbüttel fragment also has what may be offertory texts), but the overall nature and purpose of the extracts is clear.

The early history of the chief forms of office music, the antiphon and the responsory, is therefore by no means clear in detail. Without knowing more about the precise texts sung and their manner of performance, we cannot say very much about their music. The same two forms, or something related to them, were used in the mass liturgy, the subject of the next section, which may help our understanding of the office chants.

Hymns are distinct from the forms discussed so far by reason of their poetic text and their independence from psalmody or readings (although we cannot be sure that all early references to 'hymns' would fit this definition). Being non-scriptural, they were occasionally regarded with the same suspicion as antiphons. They were not part of the early Roman monastic office, nor of the Rule of the Master, but Benedict included them. On the other hand, they are recorded for several other early Western liturgies, particularly secular ones. It has been suggested that a statement by Jerome may refer to the singing of hymns at morning prayer in Roman secular use (Taft 1985, 143). They are known from Milan in St Ambrose's time and north Africa in St Augustine's. For Gaul we have the hymns of St Hilary of Poitiers (*c*.315–67), and they are mentioned at several Gallican synods. The full set of daily hymns of the monastic rule of Arles of the 530s is known (Taft 1985, 103–4). Huglo has reconstructed the whole Gallican hymnal ('Gallican Rite, Music of the', *NG*). Canon 13 of the Fourth Council of Toledo (633), presided over by St Isidore of Seville, allowed the singing during the office of the Gloria in excelsis and other non-biblical hymns such as those of Hilary and Ambrose. That hymns, like antiphons, were valued for the opportunity they gave for popular participation is indicated by the following passage from the Life of St Caesarius of Arles (*c*.470–542):

> He added also and provided that the lay inhabitants memorize psalms and hymns, and chant, some in Greek, others in Latin, proses (*prosas*) and antiphons with a high and melodious voice, like the clergy, so as not to have any time to waste on gossip while in church. (MGH, Passiones vitaeque sanctorum, 463–4; Taft 1985, 151.)

The nature of the 'proses' is uncertain.

The texts of numerous Celtic ecclesiastical hymns have survived, though their liturgical function and position are not often clear (Warren ed. Stevenson 1987, pp. lxxxiii–lxxxix).

The sense of contrasting monastic and non-monastic offices has been dulled somewhat since this early period because of the prevalence of a monastic type since the Carolingian renaissance. The differences between secular and monastic use in the Middle Ages are relatively minor when set beside the adoption by secular uses of the monastic daily Night Office and Little Hours. The distinction between the two types of office was already blurred in Rome from the fifth century onward, when monastic communities became attached to the basilical churches such as the Lateran and St Peter's. Benedict's Rule appears to have been influenced by the practice of these communities. After the reforms of the eighth century in Francia, clerics were increasingly expected to perform a full cycle of office hours after the Roman or a quasi-monastic pattern.

The prevalance of the full monastic-style office is not restricted to Roman/ Benedictine use. We have both secular and monastic Old Spanish sources of office chants; it is clear that the secular services were simply built out into a monastic cycle (see W. C. Bishop 1924). Rather than outline the whole of the Old Spanish monastic liturgy (see Taft 1985, 115–20), I summarize the chief forms of chant, which may be compared with the Roman/Benedictine categories of antiphon, Venite antiphon, great responsory, little responsory, and hymn.

Antiphons for psalms and canticles. In the Night Office and Lauds, two 'normal' antiphons and an alleluiatic antiphon (that is, one with 'alleluia' exclamations in the text) were grouped with a responsory to form the unit known as a 'missa'. (The 'alleluiatici' often served as introit antiphons at mass.) Another type of antiphon distinguished only by the topicality of its text was the 'matutinarium' for Lauds. The 'Benedictiones' were antiphons to be sung with the Benedictus canticle (Dan. 3: 52) of the Night Office. Antiphons were also sung during the other hours.

'*Laudes*', like alleluiatic antiphons without a psalm, were sung in the little hours and Vespers.

Great responsories were sung in each 'missa' of the Night Office, and singly in the other hours.

The 'Lucernarium' or 'Vespertinum' is a responsorial chant (varying numbers of verses are found, and some pieces lack a verse) which begins Vespers.

The 'Sonus' of Lauds appears to resemble the Sacrificium (offertory chant) of mass in form and musical style, and some pieces serve in both functions. They occasionally display long sequence-like melismas with repeat structures.

The Milanese office, like the Roman, betrays a heavy infiltration of monastic forms (see again Bishop 1924). In the time of St Ambrose (bishop of Milan 339–79) a daily morning and evening service and occasional vigil services are documented, though their detailed content is unclear. The medieval order has been edited by Magistretti

(1904–5) and summarized by Lejay ('Ambrosien (rit)' *DACL*) and Borella (1964, ch. 11); it can also be followed through the introduction and transcription in PalMus 6.

The repertory of Milanese office chants consists not only of the expected simpler antiphons to accompany psalms and canticles, and responsories for the readings, but also a number of other items, usually as a part of ceremonial lacking in the Roman/ Benedictine office. The following may be mentioned.

For the morning service (a form of Night Office and Lauds following one another with little division) another responsory was sung after the opening hymn. After the main sequence of Night Office items, concluded outside Advent and Lent by the Te Deum, a special 'antiphona ante crucem', with doxology, was sung at the start of Lauds. The Gloria in excelsis deo was sung after the psalmody. Towards the end of the service a 'psallenda', a processional antiphon with doxology, was sung, as the choir proceeded to the women's baptistery. There a responsory 'in baptisterio' was sung, then another psallenda on the way to the men's baptistery.

The same stational ceremony concluded Vespers. Vespers began with the 'Lucernarium', an elaborate antiphon with verse (cf. the Old Spanish lucernarium). There followed the 'Antiphona in choro' without psalm verse, and the 'Responsorium in choro' sung by a deacon. On important feasts the last two chants were replaced by the 'Responsorium cum infantibus', with its lengthy sequence-like melismas (cf. the Old Spanish sonus).

Thus both Old Spanish and Milanese offices include a number of substantial items not found in Roman use, some of them chants which do not fall into the categories of either responsory or antiphon for the psalms and canticles. Bailey and Merkley (1989, 16) have pointed out a considerable overlapping of use between the very rich Milanese repertory of processional antiphons and office antiphons (usually under the name 'psallenda'). We cannot always know how much of this material was present in the non-Roman rites before the earliest chant-books, written after the eighth century. But we should be aware of the alternative lines of liturgical development which took place in parallel with the Roman one.

The next section pursues some of the same themes as this with regard to the chants of mass.

VI.5. MASS CHANTS BEFORE THE EIGHTH CENTURY

   (i) The Gradual
  (ii) The Introit and Other Chants at the Start of Mass
 (iii) The Chants at Communion
 (iv) The Offertory
  (v) The Chants beside the Lessons

Stäblein, 'Frühchristliche Musik', *MGG*; Hannick, 'Christian Church, Music of the Early', *NG*; Duchesne 1919; Dix 1945; Jungmann 1948 etc. (1962 edn. cited here);

Cheslyn Jones *et al.* 1978, ch. III; Dyer 1982; Jeffery 1984; McKinnon 1987, 'Gradual'; Bailey 1983, *Alleluias*, 1987; McKinnon 1987, *Music*.

(i) *The Gradual*

The presence of responsories in the office, performed singly in response, as it were, to a reading, suggests a parallel with the gradual of mass. McKinnon (1987, 'Gradual') has shown that the evidence for the presence of a distinct chant among the lessons of mass first accumulates towards the end of the fourth century. While in the early centuries the eucharist had been preceded by readings, ending with a homily, there is no definite information about a chant at this point in the service. One arrangement is documented by St Augustine, who refers to the sequence epistle–psalm–gospel as a set of three readings (McKinnon 1987, *Music*, 161). It appears that the psalm may have changed from reading to chant, the gradual in fact (but see section v below for the possibility that the ancestor of the tract was involved in this development).

McKinnon sees the establishment of the gradual as part of the general enthusiasm for the Old Testament psalms which took hold of Christianity in the later fourth century. Whereas an Old Testament psalm might occasionally have been declaimed among the lessons, it now became the norm, and a special manner of performing it arose, whereby the congregation or choir punctuated its performance with a refrain, the respond, a verse taken from the psalm itself. The 'Apostolic Constitutions', which describe the mass liturgy of Antioch in the later fourth century, says that after readings from books of the Old Testament someone should sing the psalms, and the people should respond with 'akrostichia' (McKinnon 1987, *Music*, 109). It was then customary for a homily to be preached on the same respond verse. St Ambrose (d. 397), extolling the virtues of psalm-singing generally, remarks that congregations are noisy when a lesson is being read, but that singing the responses to a psalm involves them in the service (McKinnon 1987, *Music*, 127); while he may not necessary have had the gradual of mass in mind, what he says seems applicable to that chant.

Quite how the chant evolved from what was presumably a fairly simple chant with congregational participation to the elaborate chants we know from many churches of Christendom (Byzantine prokeimenon, Milanese psalmellus, Mozarabic psalmo) is not clear. As early as the late fourth century, however, Augustine refers both to a plainer type of performance of the psalm (note the singular, possibly implying the gradual) which Bishop Athanasius of Alexandria (d. 373) had preferred, and to the entrancingly melodious psalmody of Milan in St Ambrose's day (*Confessions*, 10. 33. 49–50; McKinnon 1987, *Music*, 155).

Jeffery (1984) has provided confirmation of McKinnon's thesis in an unexpected way (his article actually preceded McKinnon's but deals primarily with the next stage in the history of the gradual). A passage in the *Liber pontificalis* which can probably be dated to the early sixth century states that Pope Celestine I (422–32) decreed the 150 psalms should be sung 'before the sacrifice', whereas previously only the epistle and gospel had been recited. In the early ninth century Amalarius of Metz interpreted this to mean that Celestine had introduced the singing of the introit at the start of the Roman mass, an interpretation followed by practically all subsequent writers. But, as Jeffery explains, the passage makes much better sense if Celestine had introduced the singing of a responsorial psalm between epistle and gospel, the gradual in fact. The explanation is especially convincing because Celestine seems to have had experience of the liturgy and its music in Milan at exactly the same time as St Augustine, and might well have wished to introduce the same customs that Ambrose had received from the East and which Augustine said were being taken up all over.

A slightly later interpolation into the *Liber pontificalis* says that the psalms introduced by Celestine were 'antephanatim ex omnibus', which Jeffery explains by the responsory texts having been extracted 'as excerpts' from the parent psalms. The passage reminds one of the later attribution to Gregory the Great of an 'antiphonarius cento', a patchwork of antiphons. Reports, and actual sources, of such excerpts exist from as early as the fifth century (see above, VI.4); some of these may refer to the office responsory, whose history must have run partly parallel to the gradual.

We shall return to the gradual in a discussion of the chants between the lessons as a whole—gradual, alleluia, and tract—at the end of this chapter (section v below).

(ii) *The Introit and Other Chants at the Start of Mass*

Where does Jeffery's new interpretation of Celestine's contribution to the Roman mass leave the introit? The earliest direct evidence for Rome is the Ordo Romanus I, from the early eighth century, where it is described as a psalm with antiphon: as many psalm verses are sung as required for the entrance of pontiff and curia (Andrieu 1931–61, ii. 83). The Gallican antiphona ad praelegendum had psalm verses, as also the Old Spanish praelegendum, and some of the latter were sung both as entrance chants for mass and office antiphons (the 'alleluiatici'), with the same psalm tones. On the other hand, the Milanese entrance song, the ingressa, has no psalm and is therefore by nature more like the large non-Roman processional antiphons. In all these cases, it would seem that ceremonial considerations stimulated the development of the repertory, but evidence for their origins and development is sparse.

Gregorian sources usually indicate a single psalm verse for the introit, but some early books (particularly Paris, Bibliothèque Nationale, lat. 17436) also have a 'Versus ad repetendum'. This corresponds to the instruction in Ordo Romanus XV (third quarter of the eighth century; although it was compiled by an admirer of the Roman liturgy, he is thought to have worked in Burgundy or Austrasia). The following order of performance is indicated (Andrieu 1931–61, iii. 120; cf. Jungmann 1962, i. 417, 420; Dyer 1982, 15):

  introit antiphon
  psalm verse 1
  introit antiphon
  [possibly further psalm verses each followed by the introit antiphon, as time allows]

Gloria patri part 1
introit antiphon
Gloria patri part 2
introit antiphon
Versus ad repetendum
introit antiphon

It has been suggested that the versus ad repetendum is a relic of Gallican practice (Huglo, 'Gallican Rite, Music of the', *NG*; cf. Froger 1948, 'Introït'), but it is found in Old Roman sources.

The introit forms a group with Kyrie and Gloria at the start of the Roman mass which is partly matched by the opening chants in other Western rites. Before proceeding to a brief discussion of the communion chant, which has much in common with the introit, a few words about these other chants are in order.

The first chants in the Gallican, Old Spanish, and Milanese liturgies, the antiphona ad praelegendum and the Milanese ingressa, have already been mentioned. The main chants up to the first lesson are listed in Table VI.5.1.

Table VI.5.1. *Chants at the start of mass in four rites*

Roman	Gallican	Old Spanish	Milanese
Introit	ad praelegendum	ad praelegendum Gloria	ingressa Gloria +
Kyrie			Kyrie
	Trisagion Kyrie ?	Trisagion	
Gloria			
	Benedictus/Sanctus Deus		

The supplicatory verses Kyrie eleison were used on numerous occasions in various rites in various parts of the liturgy, both office and mass, in litanies and processions. The time when litanies with Kyrie eleison became a part of the opening ceremonies of mass is not clear. Possibly Pope Gelasius (492–6) was responsible as far as Rome as concerned (Capelle 1934). Gregory the Great (d. 604) appears to have extended their use to non-festal days. In Milan three simple Kyrie invocations form a mere pendant to the Gloria in excelsis. Porter (1958, 22–3) doubted whether the brief Kyrie chant indicated in the 'Expositio' of the Gallican liturgy was firmly entrenched. Both the Gallican and Old Spanish liturgies had an extended litany chant, the preces, after the lessons.

The Gloria was sung in the morning service as early as the fourth century. Pope Symmachus (d. 514) is said to have introduced or at least extended its use in the Roman mass.

The related Gallican and Old Spanish rites both included the Trisagion *Agios o theos* (sung both in Latin and Greek), like the Gloria a 'song of the angels'. This corresponds to the use of Byzantium and the Liturgy of St James.

Like the Gloria in excelsis, the Benedictus or Canticle of Zacharias (Luke 1: 68–79), has links with morning prayer. The *Expositio antiquae liturgiae Gallicanae* says that the Benedictus was performed 'alternis vocibus' (Ratcliff 1971, 5). In Lent the antiphon *Sanctus Deus angelorum* was sung instead.

The author of the note on Celestine I mentioned above, concerning the gradual, remarked that previous to his introduction of psalmody only the epistle and gospel had been performed before the sacrifice. Possibly he has in mind the cluster of items epistle–gradual–gospel, and his remark has no relevance to what was performed before the epistle. It is also possible that the singing of introit, Kyrie, and Gloria at the start of Roman mass all postdate Celestine (422–32).

(iii) *The Chants at Communion*

Versus ad repetendum are also to be found in the same early sources for the communion chant, and for its Gallican equivalent, the trecanum. In the early eighth-century *Expositio antiquae liturgiae Gallicanae* the 'threefoldness' of the trecanum is described as a reflection of the Holy Trinity (Ratcliff 1971, 16), in a way which Jungmann (1962, ii. 490) interprets as follows:

Father	*Son*	*Holy Spirit*	*Son*	*Father*
antiphon	psalm verse	Gloria	psalm verse	antiphon

The word 'trecanum' has also been explained as a transliteration of the Greek *trikanon*, the name for Ps. 33. No text of a Gallican trecanum actually survives to support either of these explanations.

Ordo Romanus I (Andrieu 1931–61, ii. 105) states that the psalm verses are to be continued until the pontiff gives a sign, whereupon the Gloria and the 'verse' are to be sung. By 'verse' either the communion antiphon (previously called 'antiphona') or a versus ad repetendum might be meant. Once again Ordo Romanus XV (Andrieu 1931–61, iii. 124) is more explicit, and indicates a performance exactly like that of the introit (see above). One branch of the manuscript tradition of Ordo Romanus I indicates that the schola and the subdeacons alternated in the performance, though what portions they sang is not clear.

As might be expected, the singing of a chant at the fraction of the bread, or the distribution of the communion, is attested at an early date. The singing of Ps. 33 during the distribution is recorded both by Cyril, bishop of Jerusalem 349–87, and in the 'Apostolic Constitutions' (liturgy of Antioch, later fourth century; McKinnon 1987, *Music*, 77 and 109). Verse 8 of the psalm, 'O taste and see that the Lord is good', no doubt inspired this choice. From both accounts it appears that a soloist performs the psalm, but there is also early evidence of a choral response (see McKinnon 1987, *Music*, 82, 120, 144.) Indeed when Cyril encourages the people to listen to the cantor singing 'O taste and see', it may partly be because that is the very verse they will themselves take up as a choral response.

Early evidence for Rome is lacking. And the fact that the Western rites differ from one another in the placing and character of their chants in the eucharist as a whole suggests that the tradition of these chants is not as old as that of the gradual. Although Gregory the Great placed the Paternoster before the fraction (he defends the practice in a letter to Bishop John of Syracuse), there is no mention of it in most early ordines (it is absent from Ordo Romanus I, for example, but appears in V, X, and XV). In Ordo Romanus I the Agnus Dei (introduced by Sergius I, 687–701) is sung during the fraction, and the communion chant during the communion of those present.

Gallican usage had the fraction, and an accompanying fraction antiphon (without psalm verse), before the Lord's Prayer. The Agnus Dei was not sung. The trecanum, equivalent in function to the Roman communion chant, followed. This order is also that of the Old Spanish rite: a chant 'ad confractionem' (without psalm verse), then the Paternoster, then a chant 'ad accedentes' (with psalm verses). Only the Old Spanish rite had an antiphon a little earlier in the eucharist, the chant 'ad pacem' for the kiss of peace: this was an antiphon with psalm verse. Both antiphon 'ad accedentes' and 'ad pacem' have the same psalm tones as office antiphons, which is the case with the Roman communion as well.

The Milanese chants follow the same order again: confractorium—Paternoster—transitorium. Neither the fraction nor the communion chant has psalm verses. Oddly, such musical concordances as exist link the Roman communion not with the Milanese transitorium but with the confractorium, while the Milanese transitoria show links with Gallican or even Eastern fraction chants.

(iv) *The Offertory*

The origins of the offertory chant as it appears in the Gregorian tradition of the late eighth century onward are obscure. It was long believed to have been an antiphon with psalm verses chanted in a way similar to the introit and the communion, an 'action chant' like them. Dyer (1982) has shown conclusively that no firm evidence exists for the offertory as an antiphon with psalm verses. Hucke's observations on the responsorial nature of many texts are significant (Hucke 1970). Ordo Romanus I says that it should be terminated when the pontiff gives the choir the nod; but nothing is said about the nature of the chant.

The earliest reference to a chant appears to be in St Augustine's writings, where he reports he was criticized for introducing singing from the psalms both 'ante oblationem' and during communion (*Retractationes* ii. 11 or alternative numbering 37; McKinnon 1987, *Music*, 166). 'Ante oblationem', before the sacrifice, may refer to the gradual rather than the offertory. In any case there is no clear indication as to the manner of performance of these psalms (which has not prevented scholars from assuming that both were done with antiphons). The date of the introduction of an offertory chant into Roman use is unknown.

Chants with an equivalent function in other Western rites are the Old Spanish sacrificium and the Milanese offerenda. Both of these resemble the Roman chant in having one or a few verses, sung in a predominantly (often highly) melismatic manner throughout. The Old Spanish sacrificia are sometimes found as office chants, the soni of the Night Office and Vespers.

The Gallican chant sung during the offertory procession was the sonus. The identity of terminology with the Old Spanish sonus is presumably not accidental; the chant was apparently somewhat different, however. In the description which turns up in Ordo Romanus XV (Andrieu, iii. 122–3 and 74 ff.; Dyer 1982, 15), *Laudate dominum de celis* (called an antiphon) is sung three times during an extended procession. In the *Expositio antiquae liturgiae Gallicanae* (Ratcliff 1971, 10–13) the sonus concludes with a threefold alleluia, the laudes. (The Old Spanish laudes, on the other hand, are the equivalent of the Roman alleluias of mass.) Details of the procession and the concluding threefold alleluia correspond to one of the most splendid parts of the Byzantine mass liturgy, the 'Great Entrance', where the Cheroubikon hymn is sung with a triple alleluia.

(v) *The Chants beside the Lessons*

The early history of the gradual has been touched upon above in a brief discussion of 'antiphonal' and 'responsorial' psalmody. It remains to place it in context in the set of chants sung in close proximity to the lessons of mass. The various Western rites have the arrangements set out in Table VI.5.2.

Table VI.5.2. *Chants beside the lessons in four rites*

Roman	Gallican	Old Spanish	Milanese
	Prophecy	Prophecy	Prophecy
	Benedictiones	Benedictiones	
		Psalmo: Clamor/	Psalmellus
		Threnos	
		Preces	
Epistle	Epistle	Epistle	Epistle
Gradual	Responsorium		
Alleluia/			Alleluia/Cantus
Tract			
	Antiphona ante evangelium		Antiphona ante evangelium
Gospel	Gospel	Gospel	Gospel
	Sanctus post evangelium	Laudes	Antiphona post evangelium
	Preces		

The Milanese antiphons before the gospel appear to have been sung only at Christmas, Epiphany, and Easter, but those after the gospel were sung all the year. These are processional antiphons without verse, to accompany the procession of the deacon to and from the ambo. The Gallican rite had a similar custom, but the return after the gospel was accompanied by singing the Sanctus (as well as in its normal position during the consecration).

The benedictiones in the Gallican rite as described in the 'Expositio' followed one of the readings, presumably the first. It is not clear how much of the canticle, the 'Hymn of the Three Children' (Dan. 3: 52–90) was sung. In the Old Spanish use a variety of related chants were sung, with texts drawn from the canticle, only on the highest feasts.

The Gallican responsorium was sung by boys; its musical character is unknown. The equivalent chants, the Old Spanish psalmo and Milanese psalmellus, like the Roman gradual, are of course ornate. These three are all responsorial in text selection and musical form. The Old Spanish psalmo (*psalmi pulpitales* in the León antiphoner) was on some important feasts followed by a 'clamor', after which the repeat section of the psalmo was sung once again.

It is not clear whether the Gallican rite knew an alleluia approaching the dimensions found in Rome, Milan, and Spain. The threefold alleluia which concluded the offertory chant, the sonus, may have been such a piece. The Old Spanish alleluia chant, the laudes, follows the gospel.

A special chant can be found for the Lenten season in all except Gallican use: the Roman tract replacing the alleluia, the Milanese cantus with a similar function, and the Old Spanish threnos replacing the psalmo.

A number of Gallican preces have survived with music in south French manuscripts. Two can be reckoned concordances with Old Spanish preces. The latter were sung on some Lenten Sundays, but also in the office.

The three most extended chants among these are the gradual, alleluia, and tract, and their equivalents in the non-Roman rites.

Bailey (1987, 35–9) has cautioned against the assumption that all early references to a chant between the lessons concern the gradual, for a congregational response is not always specified. Solo chants without refrain, such as the medieval tract, might also have been sung. If we accept the idea that the chant between epistle and gospel had previously been read as a psalm, then the tract, whose text is psalmic in the medieval sources, is as likely a candidate as the gradual.

The introduction into the Roman mass of the alleluia as an ecstatic melismatic chant between the lessons seems to have gone through progressive stages. The early documentation of its use is difficult to interpret, not least because of possible confusion with simple congregational responses of 'alleluia' and the addition of alleluia extension phrases to many chants when sung during Eastertide. One possible sequence of events is as follows, but other interpretations have been proposed. (See Froger 1948, 'Alléluia', Blanchard 1949, Wellesz 1947 and 1954, Martimort 1970, Jammers 1973, to cite only recent studies.)

In the fourth century Rome may have sung a special, independent alleluia chant on Easter Day only. By the time of John the Deacon (later John I, d. 526) it was sung on other Sundays during Eastertide (John's letter to Senarius; Isidore, *De ecclesiasticis officiis*, speaking of African customs). The singing of the alleluia may have continued to spread through the church year, for Gregory the Great, corresponding with Bishop John of Syracuse in 598, implies that when he allows its use beyond Eastertide this actually represents a cutting back of an even more widespread practice 'brought here by the Greeks' (Gregory's letter). I have here chosen to believe that it was the extent of the alleluia's use that Gregory cut back (the word is *amputavimus*), though other explanations have been offered: that Gregory amputated the melodies we know as sequences, or an extended jubilus for the repeat like the Milanese or Old Roman *melodiae* (Jammers), or introduced verses instead of the jubilus (Wagner II, 92, Apel 1958, 377). All of this is beside the point, however, if Gregory was speaking of alleluia refrains and end-phrases, as is possible. The lack of unanimity in the medieval tradition certainly speaks against any early establishment of the repertory as we know it from the later chant-books.

Special to Rome was the performance of Greek and Latin alleluias during papal vespers on Easter Eve. This and other features of the Old Roman alleluia repertory has led to speculation about possible Byzantine influence. Although the alleluia was sung before the gospel at mass in the Byzantine rite, the musical setting of the word *alleluia* itself remained brief, in contrast to the Western traditions.

Despite the evidence of the early Würzburg lectionary (see above, III. 3. ii), two lessons, rather than three, are to be found in the Roman liturgy (as in the Byzantine). The Old Spanish, Gallican, and Milanese rites have three. Martimort (1984) has pointed out that the fifth-century Jerusalem liturgy, as preserved in the Armenian lectionary, had a varying number of lessons, and he concludes that the number of lessons was fixed in the different rites relatively recently.

The following possible developments in the Western rites may therefore be suggested. If Bailey is right, the tract in Roman use is the descendant of a reading between the epistle and gospel. The gradual was added as a post-epistle responsory. The singing of the alleluia gradually expanded through the church year, replacing the tract in all seasons but Lent. The theory that a chant might once have been a reading appears to require that there were originally four lessons in the Old Spanish and Milanese rites, an arrangement likewise with ancient testimony (Baumstark 1958, 44). At Milan one lesson became the cantus, eventually replaced outside Lent by the alleluia, while the psalmellus entered the liturgy after the prophecy. There is no trace of a tract-like chant in the (admittedly imperfectly known) Gallican rite, but in the Old Spanish rite one might suppose that the threnos suffered the fate of the Roman tract and Milanese cantus. The threnos would have been replaced by the psalmo. The alleluia chant, the laudes, was placed after the gospel; but it too was replaced in Lent by another type of responsorial chant, still called laudes but resembling rather the psalmo.

Explanations of a similar sort can be devised if we follow McKinnon and treat the gradual/psalmo/psalmellus as the successor to a reading. To follow a lesson by a responsorial chant with related text is an ordering well known from the office. But, as Martimort (1970) has pointed out, it is by no means clearly demonstrated in the mass (see also Hucke 1973, 163–71). While the Milanese psalmelli do bear this relationship, other chants do not; the alleluia is linked rather to the gospel, which in most rites follows it. Randel ('Mozarabic Rite, Music of the', *NG*) has argued that the arrangement of psalmo and clamor in the Old Spanish liturgy cannot be the remnant of the singing of a complete psalm.

Gregory the Great (PL 77. 956 ff.) once remarked that mass could last anywhere up to three hours. We should not underestimate the amount of ceremonial and solemn repetition of chants which may regularly have been practised.

VI.6. GREGORY THE GREAT

Hucke 1955, 'Entstehung', 1958, 'Problemen'; Stäblein, 'Gregor I.', *MGG*, 1968, MMMA 2; Steiner, 'Gregorian Chant', *NG*; Hucke, 'Gregory the Great', *NG*.

(i) *Introduction*

The 'Gregorian' chant repertory bears the name of St Gregory the Great, pope from 590 to 604. Gregory's actual share in its composition, or the regulation of its liturgical use, has been much debated. Since there is a gap of nearly three centuries between Gregory's life and the appearance of the first completely notated chant-books, it is unlikely that what finally entered the written musical record is what Gregory knew, even if we subtract the chants for those days whose liturgies were added to the calendar after Gregory's time. It is nevertheless possible to argue that an ancient core of the repertory, in something very much like the state in which we find it in the late ninth century, might date back to Gregory. The arguments are circumstantial, for we have no reliable contemporary witness to any musical activity on Gregory's part, and must rely on inferences drawn from the date of the texts and the style of the music. More difficulty arises from the fact that the very idea of Gregory as a composer cannot be traced back earlier than the eighth century, although there are pieces of evidence that some chants may very well be as old as Gregory's time.

This section therefore examines those of Gregory's writings which touch on musical matters, then the biographical writings of the seventh to the ninth century, by which time the Gregorian legend was fully established. Some parts of the chant repertory which Gregory is alleged to have composed are briefly inspected, and a summary is attempted.

Gregory's own authentic writings do not allude to the composition of music on his part, but nor, in the main, are they of the type that would do so, consisting largely of homilies and commentaries on sacred texts. He was an energetic leader in both civil and ecclesiastical affairs, and a prolific writer, the last of the four great doctors of the medieval church (after Augustine, Jerome, and Ambrose), so that activity at least in the sphere of liturgical organization has never been denied to him: it would be normal on the part of such a church leader. But the only authentic references by him to matters affecting music concern (i) the singing of chant by deacons; and (ii) the occasions when the alleluia and Kyrie might be sung: that is, they are liturgical prescriptions rather than purely musical ones, at least at first sight.

(ii) *Gregory and the Deacons*

At a Council held in 595 (Mansi 1757–98, x. 434), Gregory proposed a number of reforms touching upon the lives of ecclesiastics (though nothing is said of any body which might be called a 'schola cantorum'). One of these arose from the fact that clerics were apparently being promoted to the rank of deacon on account of the beauty of their singing (see Duchesne 1919, 169–70 for epitaphs praising their singing skill). But such an officer should be occupied more in preaching and the care of souls than in 'charming the people with his singing'. Deacons shall therefore henceforth not assume the functions of cantor, but only read the gospel on solemn feasts. 'As to the singing of the psalms and the other lessons, I decree they shall be executed by subdeacons, and, in case of necessity, by clerks in lower orders.'

(iii) *Gregory and the Alleluia*

Part of a letter written in October 598 by Gregory to John, bishop of Syracuse, about Roman usage concerning the singing of the alleluia, has occasioned much dispute (PL 77. 956). The passage on question was brought together with a wealth of other early evidence about the alleluia by Froger (1948, 'Alléluia'). I have expressed reservations above (VI.5.v), as did Martimort (1970), about the usual assumption that Gregory was discussing the alleluia chant of mass.

The same letter touches upon Gregory's provision for singing the Kyrie sometimes with 'other things' (presumably other Latin verses) and on weekdays simply with 'Kyrie eleison' and 'Christe eleison' (see above, II.17).

In another passage Gregory defends placing the Lord's Prayer straight after the Canon of mass, where it is recited by the priest alone.

(iv) *Isidore of Seville; the* Liber pontificalis; *the List of Chant 'Editors' in Ordo Romanus XIX*

As Solange Corbin pointed out (1960, *Église*, 173), the oldest 'life' of Gregory is that of Isidore of Seville (*De illustribus scriptoribus ecclesiasticis*, ch. 27, PL 75. 490), who has nothing to say of liturgical work on Gregory's part. Isidore (*c*.560–636) knew of Gregory from his brother Leander, and was also very familiar with his writings. He was presumably the source of some of the information in the life by his pupil Ildefonsus of Toledo (*c*.607–67) (*De viris illustribus*, ch. 2, PL 96), likewise unhelpful in the present case.

The *Liber pontificalis* (editions by Duchesne and in MGH) is a composite collection of biographies of the popes, which began in the sixth century and was gradually built up, with contributions of increasing length, down to the fifteenth century. The information on Gregory, very scanty, was probably recorded in 638. It says nothing of the composition of chants, and is in fact silent about liturgical activity altogether. Nor does it mention the existence of a schola cantorum, with an interest in which Gregory was also to be credited. Not until the biography of Sergius I (687–701), who is said to have been educated under the care of a 'prior cantorum', does the existence of a school seem to be implied.

An interesting document entitled 'De convivio sive prandio atque cenis monachorum' has survived, which can be dated to the late seventh century (the last pope on it is Martin I, 649–55). It ends with a remarkable list of popes and Roman abbots who, its author claims, made important contributions to liturgy, chant, and other matters. At one time it was claimed to be the composition of John, abbot of the monastery of St Martin and Archicantor of St Peter's, who came to England in Pope Agatho's time (678–81) to teach chant in Northumbria (as recounted by Bede). This attribution has been rejected since Andrieu's edition of the text in no. XIX of his Ordines Romani (Andrieu 1931–61, iii. 223–4), but that the document may stem from late seventh-century Rome (albeit in the ninth-century manuscript St Gall, Stiftsbibliothek 349), is not now seriously questioned. The evaluation of its contents, however, is not easy, particularly as regards the reasons for its selection of personages.

Eight popes, and the three abbots, are mentioned as having 'edited' or 'ordered' (*edidit*; *ordinavit*) the cycle of chants for the year (*omnem annalem cantum*; *anni circoli cantum*). Thus of Boniface II (530–2) it is said that 'inspired by the Holy Spirit, he assembled the rule and arranged the cycle of chants for the year' ('regolam conscripsit et cantilena anni circoli ordinavit'). The remarks are quite formal, and no one personage stands out as more important for music than any other. Knowledge of the liturgy and competence in the celebration of the divine service were of course among the prime requirements of any high-ranking Roman prelate in these times. The progression from monk in one of the Roman monasteries to abbot and at the same time leader of the schola cantorum was normal. The list is nevertheless selective in that not every pope is mentioned. Next after Boniface comes Gregory. Relatively much is made of him, and his commentaries and homilies are all cited in reverent terms; then comes the formal close 'et cantum anni circoli nobili edidit'.

(v) *The Anglo-Saxon Tradition*

An important impulse towards the creation of a Gregorian legend must have come from England. The first reason for this was Gregory's action in sending Augustine to England as a missionary in 597, the first step in the reclamation of England for the Roman church. The second reason was that English monasticism in the later seventh century entered a period of splendid achievement, so that it was eventually able to influence Frankish thinking on liturgy and learning, not least the Frank's notion of Gregory, held in such high esteem in England.

There are two important English sources of information about Gregory. The first is an anonymous life written by a monk of Whitby some time between 704 and 714 (ed. Colgrave 1968); the other is Bede's history of the English church and people, completed in 731 (ed. Plummer, trans. Sherley-Price). The achievements of Gregory here recounted do not include musical ones, and hardly even touch upon liturgy. Of the many charming episodes later to be recounted by John Hymmonides—Gregory's attention to the schola, his compilation of an antiphoner, and so on—the Whitby monk says nothing. We do find for the first time the story of the dove which descends upon Gregory; but he is writing not neumes, as depicted in 'Hartker's Antiphoner', but a commentary on Ezekiel. The author had access to the information about Gregory in the *Liber pontificalis*, and in return his life was known later in Rome, to John Hymmonides. Its only surviving source is an early ninth-century copy, made probably at St Gall and now part of St Gall, Stiftsbibliothek 567.

Bede's history is of special interest as much for what it says about the various cantors who came from Rome in the seventh century as for its information on Gregory. Gregory's answers to questions Augustine sent back regarding problems he was facing are justly regarded as crucial to an understanding of the significance of the mission. Gregory recommends (i. 27) that Augustine make use of whatever liturgical customs seem best for the new English church. (The passage was quoted in the ninth century by Amalarius (ed. Hanssens, i. 363) in justification of his revision of the antiphoner.) This is consistent with the attitude displayed in the abovementioned letter to John of Syracuse, to whom he wrote:

> If this church [i.e. that of Constantinople] or any other has some good thing, then, just as I reprimand my inferiors when they commit a crime, so am I ready to imitate them when they have something of worth. It would be folly so to set up the primacy as to disdain the learning of what is better.

Exactly what Augustine managed to introduce in the way of liturgical customs—Roman, Gallican—is debatable (see Deanesly 1964). But contacts with Rome were maintained. The importance of proper liturgical observance to the life of the church is underlined by the fact that in 680 John, archcantor of St Peter's and abbot of the monastery of St Martin, came from Rome to teach in Northumbria. At Monkwearmouth, 'in accordance with the instructions of the Pope [Agatho], Abbot John taught the cantors of the monastery the theory and practice of singing and reading aloud . . .' (Bede, iv. 18). Liturgical books were produced for reference, cantors from elsewhere came to learn, and John himself travelled about. Bede also says that both in the seventh century and in his own time there came north singers 'trained in vocal music by the successors of blessed Pope Gregory's disciples in Kent'—in this instance Maban, *c*.710 (v. 21); previously, and the first to do so, came Eddi, or Stephen, *c*.670 (iv. 22). The proud link with Gregory was still present in the minds of eighth-century English monks; indeed it was to remain so for many centuries (for the example of eleventh-century Glastonbury see Hiley 1986, 'Thurstan').

A member of the Northumbrian royal house at the time Bede was writing, Egbert, provides the link with the Carolingian era. Ordained deacon at Rome, he was appointed bishop of York about 732. In a letter of 735 which still survives, Bede advised him to apply for the pallium, which he obtained. His brother Eadberht became king in 738, and with his support Egbert founded a cathedral school in York where, among others, Alcuin was a pupil and eventually master. The vitality of the Gregorian tradition was maintained. Indeed, whatever state of codification the liturgy had reached at York in Egbert's time (he died in 766), the composition of the books was attributed to Gregory himself. What is more, these were not books of prayers, which one might expect, but chant-books. In two places in his *Dialogus ecclesiasticae institutionis* (PL. 89, 377–451), a treatise on discipline, reference is made to

> beatus Gregorius in suo antiphonario et missali libro per paedagogum nostrum beatum Augustinum transmisit ordinatum et rescriptum . . .
>
> beatus Gregorius per praefatum legatum, in antiphonario suo et missali . . . non solum nostra testantur antiphonaria, sed et ipsa quae cum missalibus conspeximus ad apostolorum Petri et Pauli limina. (PL 89, 440–2, cited Gevaert 1890, 80–1 and Ashworth 1958.)

Although Gevaert thought the text unauthentic, modern historians seem not to dispute Egbert's authorship (see, for example, Haddan and Stubbs 1869–71, iii. 413–16, and Ashworth 1959, despite the argument that the books cannot have been compiled originally by Gregory). It seems almost certain that these were books with texts only, not notation. Egbert's testimony antedates by only half a century such Italian documents as the Roman antiphoner of Hadrian I (772–95) seen by Amalarius at Corbie, the late eighth-century antiphoner fragments from Lucca (Lucca, Biblioteca Capitolare 490), with the 'Gregorius praesul meritis' prologue (Froger 1979, 'Fragment'), and the eighth-century Italian missal fragments (chant texts in smaller hand) Rome, Biblioteca Apostolica Vaticana, Vat. lat. 10644—see Gamber 1968, no. 1401, also facs. Gamber 1962, 336). The Corbie antiphoner is lost; the others are unnotated.

(vi) The Biographies by Paul Warnefrid and John Hymmonides

Two biographies of Gregory, widely separated in character and date, illustrate the enormous growth of interest in Gregory during the eighth to ninth centuries, that is, at the time during which Frankish rulers made such efforts to 'Romanize' the liturgy in their lands.

Paul Warnefrid ('the Deacon') wrote a brief biography in the second half of the eighth century. It exists in two versions, the original one (ed. Grisar 1887) being shorter and the other (in PL 75. 1–59) interpolated. Even in the PL edition it takes less than twenty columns. John Hymmonides, writing in 872–3, left a text of nearly 200 columns in the modern edition (PL 75. 58–242).

Grisar's edition is based principally on three Montecassino manuscripts of the eleventh century (145, 146, and 110) and sixteen other Italian sources. Other, non-Italian sources are variously interpolated with extra stories and other material. Paul Warnefrid gives the 'Lombard' (south Italian) version of Gregory's work. John Hymmonides (also 'the Deacon') knew of two lives, a Lombard one (Paul's) and an English one (the monk of Whitby's), and at the behest of Pope John VIII (872–82) now provided a Roman vita. Since even his ardent collector's zeal missed some of the stories in the interpolated Paul text, it may be assumed he knew only the original.

Paul does not give the story of the dove which descends upon Gregory, although the homilies are mentioned. Nor does he say anything of liturgical, let alone musical, activity on Gregory's part. The interpolated versions (PL 75. 57–8) add the dove, but during Gregory's work on Ezekiel, not on an antiphoner: a familiar of Gregory's notices that Gregory, dictating in the next room, is making pauses, and, spying through the keyhole, sees that this happens whenever a wondrous dove on his head speaks into his ear. The servant is forbidden to tell of the events, but does so when he hears ill being spoken of Gregory after his death. Also among the interpolations is the information that Gregory arranged the entire population of Rome in seven groups, to sing the Letania septiformis after a visitation of the plague (PL 75. 59).

The great litany, sung as the penitential procession goes from church to church, is described in greater detail by John (PL 75. 80). He knows and describes pictures of Gregory, with a dove, but music does not enter into his explanation (PL 75. 222). After Gregory's death, Peter the Deacon, Gregory's familiar, is himself on his deathbed: seeing that some of Gregory's books are about to be burnt, he forbids the sacrilegious act, citing the dove's miraculous appearance as evidence of the books' holy inspiration.

The main interest, however, of John's book lies in sections ii. 6–10 (PL 75. 90–2), where Gregory's liturgical and musical activity is described in colourful prose (printed as verse by Stäblein, MMMA 2. 143*). In ch. 6 we are told that Gregory 'antiphonarium centonem cantorum studiosissimus nimis utiliter compilavit', 'constituted' the song-school with two dwellings, by St Peter's and the Lateran palace respectively, where his couch, cat-o'-nine-tails for helping the choirboys learn their chant, and the 'authentic antiphoner' are still on display for the reverent visitor.

Although it is easy to dismiss this as pious fiction, John cannot be presumed to have invented wilfully. It is likely that he recounted what was believed in his day, and one must explain how these beliefs arose. The 'authentic antiphoner' may be one of those seen by Egbert, or a successor. The decision to place Gregory's name in writing at its head—this happened to the sacramentary sent by Hadrian to Charlemagne in the late 780s and the only slightly later 'Blandiniensis'—is difficult to pin down chronologically. It may have been contemporaneous with its appearance on the sacramentary. It may have happened in France rather than Rome. But even before the 780s both Rome and its clients evidently believed in Gregory's authorship. Upon the book could easily have followed the couch, the whip, and a belief that Gregory was the originator of arrangements in the field of musical administration as well as composition. Let it be stressed again that we need not mistrust John's veracity, even though we doubt the antiquity of the things he described.

The other chapters in John's narrative speak of the reception of Gregory's chant in other lands. The Germans and Gauls cannot sing it properly (ch. 7); John the Cantor takes it to Britain (ch. 8); Charlemagne tries to ascertain why the Franks are not singing as the Romans do, and Metz is established as a chant centre (chs. 9–10). Intriguingly, John says that when the Roman singers were challenged, 'they probably showed the authentic antiphoner' to the Franks. Does this mean that it had musical notation, in John's day at least? But how old was the book he knew?

As in the case of the sacramentary, whose ancestry scholars do indeed trace back to a (no longer extant) redaction by Gregory I, it is likely that Frankish demands for 'authentic' chant books stimulated their redaction and production in Rome. It is also likely that the Anglo-Saxon tradition was a decisive factor: witness their interest both in Gregory personally and in acquiring authentic books (for example, the booty brought back from six journeys to Rome by Benedict Biscop, abbot of SS Peter and Paul's, Canterbury and founder of the double monastery of Monkwearmouth and Jarrow (d. 689)—related by Bede in his *Historia Abbatum*).

(vii) *The 'Gregorian' Sacramentary*

As already recounted (see above, III.3), Gregory also lent his name to the 'Gregorian sacramentary', a prayer-book of the Roman church which in one form at least—the so-called Hadrianum—was sent to Charlemagne by Pope Hadrian at some time between 784 and 791 at the request of the Frankish ruler. (Charlemagne's request had been conveyed to Hadrian by Paul Warnefrid.) In the letter which accompanied it, and in the earliest surviving copy, Cambrai, Bibliothèque Municipale 164, dated 811–12, it is attributed to Gregory. It seems possible that manuscripts of chant texts with a similar attribution also circulated, and that is the gist of the verse prologue (in hexameters) *Gregorius praesul* (see below, section viii). A Saint-Martial writer of the eleventh century actually said that the *Gregorius praesul* prologue itself was composed by Hadrian.

The Hadrianum represents not a modern book of the late eighth century designed to meet Frankish requirements, but a book of masses as normally celebrated by a Roman pope, the stational liturgy when he visited the various churches of the Eternal City, in a recension probably little different from the state in which it was left by Gregory II, who added the formularies for the Thursdays of Lent. About eighty prayers in the 'Gregorian' sacramentary are reckoned to be by Gregory the Great (Ashworth 1959, 1960), their identification depending on parallels with his authenticated homilies and other writings. But the title of the sacramentary, 'In nomine domini incipit sacramentorium, de circulo anni expositum, a sancto Gregorio papa Romano editum, ex authentico libro bibliothecae cubiculi scriptum', may well refer as much to Gregory II as to Gregory I.

(viii) *The Prologue* Gregorius praesul

Perhaps by analogy with the sacramentary sent to Charlemagne, perhaps because Roman antiphoners already had such a title, the earliest extant 'chant' book bears a title ascribing its contents to Gregory. Brussels, Bibliothèque Royale 10127–10144, the so-called 'Blandiniensis', written somewhere in north France in the last years of the eighth century, begins with the following:

IN DEI NOMEN INCIPIT ANTEFONARIUS

ORDINATUS A SANCTO GREGORIO PER CIRCULUM ANNI

Perhaps just as old, however, is the Lucca source of the *Liber pontificalis*, Lucca, Biblioteca Capitolare 490, possibly once part of a chant-book, which already has one of the longest versions of a complete poem beginning 'Gregorius praesul meritis' (facs. PalMus 2, pl. 3). It is not clear whether its Italian rather than northern provenance is significant. This and other versions of the prologue have been discussed by Stäblein (1968), who also reported an even older source, now lost, from the diocese of Vercelli. Two of the sources edited by Hesbert in AMS (see 2–3) may be cited here: Monza, Basilica S. Giovanni CIX (written in north-east France in the middle of the ninth century), and Paris, Bibliothèque Nationale, lat. 17436 (Soissons, slightly later):

Monza CIX	Paris 17436
GREGORIUS PRAESUL	GREGORIUS PRAESUL
MERITIS ET NOMINE DIGNUS	MERITIS ET NOMINE DIGNUS
UNDE GENUS DUCIT	
SUMMUM CONSCENDIT HONOREM	SUMMUM CONDESCENDENS HONOREM
QUI RENOVANS MONUMENTA	RENOVAVIT MONIMENTA
PATRUMQUE PRIORUM	PATRUM PRIORUM
TUM CONPOSUIT HUNC	ET COMPOSUIT HUNC
LIBELLUM MUSICAE ARTIS	LIBELLUM MUSICAE ARTIS
SCOLAE CANTORUM	SCOLAE CANTORUM
IN NOMINE DEI	PER ANNI CIRCULUM

Stäblein also reports the remarkable fact that the Metz patterning of its liturgy and music on a Roman model went so far as the adaptation of the Gregorius text to

'Anchilramnus presul dum summum conscendit honorem . . .', though without mention of a 'book of musical art' (Stäblein 1968, 543 n. 23, after Andrieu 1930, 350).

(ix) *The Reception of the Gregory Legend in the Ninth Century and Later; the Dove; the Modes*

The dove is not seen in connection with the dictation of music until the frontispiece of the late tenth-century antiphoner of Hartker of St Gall (PalMus II/1), where neumes are actually being written. A chant-book and a representation of Gregory are brought together for the first time in the Monza cantatorium of the mid-ninth century, Monza, Basilica S. Giovanni CIX (north-east French; ed. Hesbert in *AMS*). Here an ivory diptych of the late antiquity depicting two consuls was altered to show Gregory and King David (Stäblein, 'Gregory I.', *MGG*, 773; for fuller iconographical information see Stäblein 1968, n. 39).

Also legendary, it seems, is the attribution to Gregory of the creation of the four plagal modes in response to Ambrose's four authentic ones, a story first put about, as far as I am aware, by Aribo in the eleventh century (GS i. 210, CSM 2).

Gregory's authorship of sacramentary, antiphoner, or missal was variously accepted in the ninth century and later. Amalarius of Metz recorded attributions to Hadrian at the head and tail of the Roman antiphoner (Hanssens, iii. 14; cited Hucke 1955, 'Entstehung', 261 n. 21), but it is clear that he and others of that time had doubts over its authenticity (Hucke, 261–2).

Elsewhere the name of Gregory was plainly used as a means of establishing the authority of the chant. Pope Leo IV (847–55), writing to Abbot Honoratus of (?) Farfa, uses the words 'Gregoriana carmina', perhaps for the first time, while ordering the chant to be sung and the Roman liturgy observed on pain of excommunication (Hucke, 264).

In a short life of Hadrian II which Adhémar of Chabannes appended to a summary of the *Liber pontificalis* (Paris, Bibliothèque Nationale, lat. 2400, Limoges, eleventh century; ed. Duchesne; pp. clxxxii–iv), the verse prologue is attributed to Hadrian I (772–95). This is the remarkable document which attributes the composition of tropes to Gregory I, sequences and more tropes to Hadrian I, their introduction in France to Hadrian I and Charlemagne, and their revivification after a period of negligence to Hadrian II (Gautier 1886, 38). All that was best in medieval chant therefore went back to Gregory.

(x) *Gregory and the Lenten Communions*

Some peculiarities in the choice and the transmission of communion antiphons in the Lenten period have suggested that developments as old as Gregory are at issue. (See Cagin in PalMus 3, Hesbert 1934, 198 and *AMS*, p. xlviii, several articles by Callewaert reprinted 1940, especially Callewaert 1939, Hesbert in PalMus 14, p. 225, and finally Hucke and Huglo, 'Communion', *NG*). The matter is not susceptible to incontrovertible proof and is outlined briefly here only in order to demonstrate the sort of arguments which have been adduced in order to reconstruct an early chapter in chant history.

The masses for the Thursdays of Lent were arranged during the pontificate of Gregory II (715–31). We may assume that the chants for the other twenty-six weekdays were already in existence before then. Nearly all the twenty-six communions for those days have texts taken from Pss. 1 to 26, in numerical order, whereas the added Thursday communions fall outside this scheme. Yet the numerical arrangement is itself probably not ancient, since the other chants for those days are not ordered numerically. The numerically ordered communions nevertheless display a stable musical transmission when finally codified in the late ninth century.

In among the numerically ordered pieces there are, however, five non-psalmodic communions, replacements, as it seems, for communions drawn from Pss. 12, 16–17, 20–1. These are *Oportet te*, *Qui biberit*, *Nemo te condemnavit*, *Lutum fecit* and *Videns dominus*.

Three of the five 'odd' communions quote from the gospel of the mass. Morin thought that the three masses in question were originally masses for the scrutiny of catechumens. Callewaert argued that one of the three, *Videns dominus*, bore the stylistic hallmarks of an authentic text by Gregory, and that *Lutum fecit* and *Qui biberit* might also be his. On this basis, Callewaert suggested that it was Gregory the Great who might well have been responsible for the reorganization of the scrutiny masses, which were shifted from Sundays to weekdays because of their diminishing importance in a completely Christian society (Chavasse 1948). Musicologists are clearly not the only scholars eager to see Gregory as an active reformer of the liturgy.

The five 'odd' communions appear in musical sources with considerable inconsistency, amounting to different melodies. What are thought to be the original melodies, the melodies which appear in the earliest sources, are simple and syllabic, distinctly plainer in musical style than most communions. The other melodies for the five which appear in divergent sources seem to be attempts to provide more 'communion-like' substitutes. But why should the five communions have such a special musical character? Does it mean that the sixth- or seventh-century redactor who was responsible for putting them into place borrowed them from another context, one where their musical style was the appropriate one? That would imply that both sets of melodies were already in existence when the scrutiny masses were reorganized. The sequence of events might have been as follows:

1. The ancient scrutiny masses have old and simple communion antiphon melodies.

2. The communions for the weekdays of Lent have more ornate melodies, of the type more recently in fashion, when the masses for those days (always omitting the Thursdays, however) were compiled.

3. The old scrutiny masses are placed (by Gregory?) into new weekday positions, still retaining their ancient melodies, displacing what are actually newer antiphons.

All this assumes, of course, that we still have something very like the original melodies in our ninth-century musical sources. As I see it, Gregory could not have been responsible both for the liturgical change and for composing text and melody of the 'odd' communions.

Callewaert also suggested, again on grounds of the style and content of their texts, that Gregory was the author of antiphons of Sexagesima and Quinquagesima Sundays. Their music has not, so far as I know, been investigated with this possibility in mind.

(xi) *Conclusions*

It is no doubt spiritually and morally reassuring, for us as for the Carolingians, to be able to attach the name of a canonized author to the body of chant we know as Gregorian. But the available evidence does not permit this. Apart from the insufficient documentary tradition, liturgical and musical analysis has found no strong evidence that the melodies first recorded in the ninth century could be as old as Gregory. No doubt many of the same texts were sung with the same liturgical assignment in Gregory's time as later, but we must reckon with both more subtle and more radical changes in singing practice in the interim, changes which call into question any notion of 'sameness' between the sixth and the ninth century. In this context, the very idea of a 'composer' of chant is anachronistic, and we should do better to uncouple Gregory as a person from the chant repertory.

His name retains its usefulness, in the sense that 'Gregorian' chant is neither of one specific time, nor wholly Roman, nor wholly anything else. A legendary label is as good as any.

VII

The Carolingian Century

VII.1. INTRODUCTION

The events which led to the establishment of the chant repertory we know as 'Gregorian' are inextricably bound up with political developments of the eighth century, principally the establishment of a strong and extensive Frankish kingdom, which by the end of the reign of Charlemagne (768–814) covered an area roughly corresponding to that of modern France, West Germany, and north and central Italy. This is the 'Francia', the empire of the Franks, referred to in this book. Charlemagne was deeply concerned that the church in his domain should follow the liturgy according to Roman practice, as far as possible, which naturally included the plainchant of the liturgy.

The next section recounts some of the stages in the establishment by the Franks of a Roman chant repertory in their empire. Since the needs of the Frankish church did not correspond in every detail with Roman practice, adjustments were necessary. But the Franks also enriched the music of the liturgy in ways which had no precedent in Roman practice (VII.3). A special part in the establishment of a corpus of liturgical chant in Francia was played by notated music-books, which as far as can be seen at the moment became increasingly important at the end of the ninth century (though some scholars would place this development as much as a century earlier). VII.4 comments on the cultural and historical role of the earliest notated books. A final section discusses the place of music in the general cultural awakening popularly known as the 'Carolingian renaissance'.

VII.2. THE ESTABLISHMENT OF ROMAN CHANT IN FRANCIA

Klauser 1933; Hucke 1954, 'Einführung'; Vogel 1960, 1965 'Cultuelle', 1965 'Liturgique'; Stäblein in MMMA 2; Wallace-Hadrill 1983.

Rome in the early centuries had neither the desire nor the power to regulate details of liturgical practice in all parts of Western Europe. The example of Gregory the Great shows this clearly: his recommendation that Augustine use whatever liturgical materials he found good, and his correspondence in liturgical matters with other prelates, where he is at pains to explain and defend what he has done but does not seek to impose it on others (see above, VI.6).

Nor did the other churches of Western Europe expect Rome to be able to lay down a pattern of worship, or try to copy Roman use. The import of the various councils which laid down liturgical prescriptions is not that Roman practice shall be followed but that the liturgy shall be performed in a way that is seemly and proper. The prescriptions naturally took into account local traditions and conditions. This is what we find at the Council of Agde (506), where bishops from the whole of south-west France and Spain gathered, or the Fourth Council of Toledo (633), held under the presidency of St Isidore, which was attended by no less than sixty-two bishops from Spain and Septimania (their liturgical prescriptions are summarized by Taft 1985, 147–8 and 159–60 respectively).

The change in outlook which led to wholesale adoption of Roman practice may well be due to the influence of the Anglo-Saxon missionaries who in the late seventh and early eighth centuries brought new German territories into the Christian fold: Willibrord in Frisia from 690, and Wynfrith, who took the name of Boniface, in central Germany and Bavaria in the first half of the eighth century. After organizing the German church, Boniface turned his attention to the Frankish congregation, reformed through a series of synods in the 740s.

Boniface had been called to Rome in 722, where Gregory II (715–31) gave his blessing to his work. To Anglo-Saxons such as Boniface and his disciples, several of whom became bishops in Germany, the traditional connection of their church with Rome was vital. Their work was accomplished in a spirit of obedience to and in frequent consultation with Rome. Newly founded monasteries in Germany followed the Benedictine rule.

The Frankish church under the newly founded Carolingian dynasty of Pippin III (751–68) was consequently far more Rome-oriented than previously. In 750 Pippin had sent the Anglo-Saxon Burghard of Eichstätt and the Frank Fulrad of Saint-Denis to Rome to gain approval, in effect, for the deposition of the Merovingian king and his own succession. In 754 Pope Stephen II (752–7) visited Pippin to seek military aid against the Lombards, and Pippin was consecrated king at Saint-Denis.

There is no need here to recapitulate subsequent political interchange between Francia and Rome. It constitutes, however, the essential background to the adoption of the Roman liturgy in Francia. The decisions of the numerous Frankish church councils of the mid-eighth century are concerned not with liturgy but with fundamental church discipline, so we cannot trace the details of liturgical interchange very clearly, but its pace unquestionably began to quicken. In 760 Pippin's brother, Bishop Remigius (or Remedius) of Rouen, went to Rome to ask that a Roman teacher of chant be allowed to come north, while monks from Rouen learnt the chant in Rome under George, the *primus scholae*. Simeon, *secundus* of the Roman *schola cantorum*, was sent by Paul I (757–68) to teach Remigius' clerics. After a while George died, and Paul had to recall Simeon to succeed him, while assuring Remigius that the Rouen singers in Rome would be brought to perfection under Simeon's instruction.

Already the records of the Council of Cloveshoe in England (747) speak of a chant-book sent from Rome (VI.6.v). Paul I sent not only Simeon but also books to Rouen: an 'antiphonale', a 'responsale' and books on orthography and the seven liberal arts, all in Greek. (Vogel 1960, 242, MMMA 2, 148*). This evidence that books with chant texts occasionally found their way north from Rome is very important. (I have assumed that at this time they would not be notated.) But on the whole we are much better informed about the Ordines Romani and sacramentaries which were prepared from or under the influence of Roman models than about Roman or Romanized books with chant texts, for we have no chant-books of the eighth century which would reveal the state of the repertory before and during the wave of Roman influence in the second half of the century.

It was Metz, rather than Rouen, which became renowned as a centre of Roman chant in Francia. Chrodegang of Metz (d. 766), the leader of the Frankish church after Boniface was killed on a mission to Frisia in 754, composed a rule for the canons of his city which borrows heavily from Benedictine practice and refers frequently to Roman customs (Mansi xiv. 313; PL 89. 1059–1120). It marks a crucial stage in the 'monasticization' of the office for the secular clergy. The life of Chrodegang by Paul Warnefrid (d. 799) attributes to him the introduction of the 'cantilena Romana' in Metz (MGH (Scriptores) *Scriptores rerum Sangallensium*, etc., 260–70, PL 95. 720; see Buchner 1927). Even a stational liturgy of the papal type, whereby the bishop celebrated mass in different churches on different days, was introduced in Metz (Klauser 1930, Andrieu 1930).

Pepin as the prime mover is then repeatedly mentioned in the numerous documents issued under Charlemagne which concern the singing of chant (Vogel 1960, 230–1, 265–6). What his father had begun became a prime element of Charlemagne's governance. The injunction to sing according to Roman use is repeated time and time again in the numerous capitularies issued under his instructions. The examination of priests included the question as to whether they sang the 'cantus Romanus' in the Night Office and at mass (MGH (Leges) *Capitularia regum Francorum*, i. 235; see also 110 for the required adherence to Roman liturgy).

Charlemagne's promulgation of a uniform Roman liturgy in Francia has been attributed to political motives, as a means of buttressing the political unity imposed in many regions by force of Frankish arms. This view is too cynical. At the beginning of the famous 'Admonitio generalis' of 789, Charlemagne's extensive proposals for ecclesiastical and educational reform, he recollects the work of King Josiah in reforming the worship of Israel (2 Kgs. 22–3; MGH *Capitularia regum Francorum*, i. 54, cited by McKitterick 1977, 2). The good king is careful of the religious observances of his people, suppresses idolatry, and observes the (written) law. Charlemagne's interest could also take a personal turn. If Notker's report is to be believed, he himself listened carefully in order to hear if the singing was correct. His biographer Einhard says he attended church four times a day, and could have intoned the psalms and lessons publicly had he wished to do so. Nor was he without experience of the Roman liturgy, being in Rome four times in all, including the Easter of 787, which he spent with Hadrian I, and Christmas 800, when he was crowned emperor.

If the second half of the eighth century saw the most vigorous contacts between Roman and Frankish singers, there seems to have been a decline in the ninth. Around 820 Helisachar, the chancellor of Louis the Pious (814–40) and abbot of Saint-Riquier and of Saint-Aubin at Angers, wrote to Bishop Nidibrius of Narbonne in terms which suggest that the Roman way of singing responsories was no longer known (see Huglo 1979, 'Remaniements'). At first Metz provided an alternative. The Capitulary of the missi dominici (Charlemagne's emissaries who carried out civil and ecclesiastical visitations) at Thionville in 805 says that for a knowledge of chant one must go to Metz (MGH *Capitularia regum Francorum*, i. 121). Archbishop Leidrad of Lyon (799–814) reformed the liturgy of Lyon according to the use of the imperial palace itself, with the help of a cleric of Metz (MGH (Epistolae) *Epistolae Karolini aevi*, ii. 542, Hucke 1954, 'Einführung', 185, MMA 2, 149*).

The famous story of the Roman and Frankish cantors also gives Metz a place in chant history. The affair, seen through Roman eyes, is told by John Hymmonides (PL 75. 91, Hucke 1954, 'Einführung', 180, MMMA 2 144*). He says that Charlemagne noticed the lack of harmony between Roman and Frankish singing and sent two clerics to Hadrian to learn Roman chant; they came back to Metz and through them the singing in all Gaul was corrected. After they died the situation deteriorated, so Hadrian sent two of his singers to Gaul. The singing in Metz had deteriorated much less than elsewhere.

The life of Charlemagne written by Notker Balbulus of St Gall gives the Frankish side of the story (Hucke 1954, 'Einführung', 178–9). Here the astonishing accusation is made that twelve Roman singers, having been sent to Francia by Pope Stephen II (according to Notker; it would actually have been Stephen III) at Charlemagne's request, set out to confuse the Franks by deliberately singing wrongly, being envious of the glory of the Franks. To circumvent this animosity, an apologetic Pope Leo, Stephen's successor (it would actually have been Hadrian I), suggested that singers be sent disguised to listen to the services sung properly in Rome itself. They came back to Metz and the royal palace at Aachen respectively. Thus the 'ecclesiastica cantilena' is called 'Messine', of Metz.

Even if deliberate misinformation be ruled out, the establishment of Roman chant in Francia was a hazardous undertaking, relying principally on human memory for the learning and promulgation of a vast repertory of music. The process seems to have lasted roughly half a century, coinciding with the reigns of Pepin and Charlemagne. The celebration of mass was altered from whatever Gallican forms were used in Francia to coincide with the Roman liturgy, and the Roman chants of mass had to be learned. We could imagine that the Anglo-Saxon churchmen would have contributed their knowledge of Roman chant, but direct contact with Rome itself was essential. A first generation of singers learned in Rome or from Romans, then taught their art to the next generation. The celebration of the office hours presumably followed a similar course; perhaps Roman/Benedictine practice was established earlier and more firmly, at least in German lands, through the work of Boniface and his fellow Anglo-Saxons.

In the absence of Anglo-Saxon chant-books this remains speculation, but the prestige of Anglo-Saxon learning and liturgical practice cannot have been diminished by the presence at the imperial palace of Alcuin of York as Charlemagne's chief liturgical adviser.

VII.3. THE FRANKISH EXPANSION OF THE CHANT REPERTORY

No book with chant texts sent from Rome to Francia has survived, nor anything that can be called a direct copy of one. The result of the process of establishment of Roman chant in Francia can nevertheless be discerned in the earliest surviving Frankish chant-books, the unnotated graduals edited by Hesbert in *AMS*. And an idea of how they may have differed from Roman use can be gained if we compare them with the earliest surviving books from Rome itself, the Old Roman chant-books of the eleventh century. We can compare, that is, a Frankish version of the Roman liturgy with a Roman one. A list of the principal differences has been drawn up by Huglo (1954, 'Vieux-romain', 108–9). Some concern details of the text: on some occasions the two versions have quite different texts. Quite often this is a case of selecting a different set of mass chants from the common of saints for a particular individual saint. Other differences touch on the selection of the alleluia, a notoriously unstable genre. Whatever the date when these differences of practice arose, it is clear from the history of the various Roman and Frankish sacramentary types that a certain amount of accommodation between different versions of the chant repertory must have gone on. We can get an idea of what was happening also through the work of Amalarius of Metz (see IX.2).

The ninth century saw, however, not just a consolidation of the Roman chant repertory in Francia but also vigorous enrichment. The chronology of this expansion is not always clear, for notated sources appear first in the tenth century. We cannot always be sure whether we are dealing with relatively new items, or compositions of, say, the second Carolingian renaissance of the time of Charles the Bald (king 840–77, emperor 875–7), or of Charlemagne's time, or perhaps even relics of older non-Roman practice which the Franks were unwilling to relinquish.

The main areas of expansion were the following. I give a reference to the chapter earlier in this book where each genre is discussed in more detail.

1. Whereas the repertory of introits, graduals, and communions appears to have remained relatively stable after the Carolingian settlement, the number of alleluias increased dramatically (II.14). The number of frequently sung alleluia melodies in the Old Roman sources (only three common ones) and Milan (one!) may indicate how many the Franks are likely to have known from their Roman teachers. The repertory of melodies by the end of the ninth century was already about sixty (many used for several different texts) and continued to expand rapidly.

Just as dramatic in its way—although the opportunities for singing it were more limited—was the increase in the number of tracts: from twenty-one in the earliest sources to twice that number in the average of the twelfth to thirteenth centuries. Some possible instances of Frankish expansion can actually be seen in the earliest sources. The Old Roman chant-books have Gregorian melodies for the tracts of the Easter Vigil, which should mean that they were adopted from Frankish practice; anciently no tracts were sung here. The tract *Eripe me* for Good Friday is in neither the Old Roman nor all of the earliest Frankish sources, and was still referred to as 'new' in the tenth century (see Hucke in Hermann Schmidt 1956–7, 932). *Domine non secundum*, the tract for Monday, Wednesday, and Friday in Lent, is also a latecomer to the repertory.

The choice of offertory verses was also not fixed at the time of the Frankish settlement. Not only this, recent studies have revealed the existence of non-Roman offertories in the Gregorian repertory which may have come from Gallican use (Baroffio and Steiner, 'Offertory', *NG*, and especially Levy 1984).

2. The origin of the sequences which were sung after the alleluia on feast-days in the Middle Ages is controversial (see the fuller discussion II.22.vi). Whatever the truth of the matter, the repertory at the end of the ninth century probably consisted of only about thirty widely known melodies. Most new compositions circulated only locally. But, as with the alleluia, new compositions were constantly being produced. The total number of melodies known from the Middle Ages is very large, the number of texts vast.

3. The repertory of ordinary-of-mass chants known in Rome was small, and only a few were widely known in Francia. Rapid expansion took place, different pieces being composed in different places. Particularly notable was the composition of trope verses of various kinds. (On the Kyrie repertory and its tropes, see II.17 and II.23.viii, on the Gloria II.18 and II.23.iii, vii, and xi, the Sanctus II.18 and II.23.xi, and on the Agnus II.20 and II.23.xi.) The Nicene Creed was not sung at mass in Rome in Carolingian times, and the Franks sang it here in continuation of older Gallican practice (II.21.).

4. The provision of complementary trope verses, especially for the proper chants of mass, is one of the most striking developments of the new period. Again, the time and place of origin of the practice are obscure (II.23).

5. The famous *Quem queritis* dialogue may well have been written in the ninth century, for it was known all over Europe by the end of the millenium. Its composition has nevertheless been placed in the early tenth century by some. The chief compositional activity in the area of liturgical drama took place, however, after the end of the millenium (II.25).

6. It is still not possible to gain an idea easily of the amount of new composition for the office hours accomplished by the Franks. The sources edited by Hesbert in *CAO* often show striking lack of unanimity in choice of chant, and this variety is matched in the transmission of many musical items. It is presumably the result of conflicting, if already 'Romanized', practices during the eighth century, and the lack of a Roman example against which to control variants. There was naturally a lot of writing for the

offices of local patron saints. The chief musical interest of the latter lies in the frequent adoption of a modal order for the chants, a purely artificial arrangement which reflects the Frankish interest in speculative music theory. (See Crocker 1986 for a musical study of such 'numerical' offices.) The writing of verse texts for such offices is another Carolingian and post-Carolingian development (II.26).

VII.4. THE CODIFICATION OF PLAINCHANT

Treitler 1984; Levy 1987 'Archetype', 1987 'Origin'; David Hughes 1987.

The existence of books containing chant texts has been mentioned repeatedly in previous chapters. The acquiring and copying of service-books and the compilation of new ones was undoubtedly one of the chief concerns of Charlemagne and his advisers. It was part of the great programme of ecclesiastical and educational reform which went to make up the 'Carolingian renaissance' of literacy and learning. The high priest Hilkiah had produced a book of law from which King Josiah learned the ways of righteousness. The written word as a means of ensuring the right performance of the liturgy was obviously of great importance to the Franks. Did their chant-books contain not just chant texts but music-writing? Did the various types of Caroline minuscule script, pioneered chiefly at Corbie in the second half of the eighth century and perfected in Charlemagne's palace school, have a musical parallel or parallels? Where and when was the attempt first made to notate the chant repertory *in toto*? (At least for the mass: all the indications are that the office chants were not codified until later.)

I have discussed above (IV.4) the evidence for and against the notating of chant-books in Charlemagne's time. Neither surviving documentary evidence nor contemporary report allows us to state unequivocally that complete chant-books were prepared for Charlemagne. We could wish for more manuscripts of chant texts from the ninth century, but those that have survived are all unnotated. *Argumenta ex silentio* are never wholly satisfactory, but the silence is in this case almost overwhelming. The graphic materials out of which a notation could emerge were already present: punctuation and accent-signs familiar to all who had received an education in grammar. From the dates of surviving fragments one might even suggest that a system of music-writing was invented at the palace school in Aachen, or Metz, or Tours in Alcuin's time. The ultimate proof is lacking. And this still does not constitute evidence that the newly learned chant repertory was transmitted as a whole in musical notation.

The chant repertory, even after musical notation became usual in the tenth century, had to be learned. The notation controls (in rehearsal, though presumably not during the service itself) the detail of melodies already known by heart. This implies, of course, that when the melody is well enough known, notation is unnecessary for performance. Necessity being the mother of invention, the question is: when did notation become necessary?

The obvious answer is: when the repertory had expanded to a point where it was no longer possible to memorize it all successfully. And this point, I would suggest, was reached later rather than earlier in the ninth century, close to the date of the earliest surviving notated graduals. The problem lay not so much with the 'Gregorian' repertory of proper-of-mass chants as with the sequences, tropes, and ordinary-of-mass melodies being produced in ever-increasing numbers and in musical styles quite different from the formulaic chants learned from Rome. This would also explain why there is no unequivocal sign of a notational archetype from which those manuscripts were descended. Different notational styles had already evolved in different areas. They were now applied to the notating of a repertory still well known but in danger of expanding out of control.

There is a parallel with the situation in the eleventh century, when staff-notation first became widely used for learning (not performing) the repertory (see IV.6 and V.3). The earliest notated chant-books, of about 900, are for the mass. If the date of the Hartker Antiphoner is representative, office-books were not notated for another century. Is it coincidence that the author of the *Dialogus de musica* and Guido of Arezzo speak of pitch-notation primarily in conjunction with the antiphoner? Once again it is possible to envisage the expansion of a repertory to the point of unmanageability, necessitating the invention of radically new pedagogical methods.

I therefore hold to the view that musical notation, while it may possibly have been invented as early as Charlemagne's time, was not used to facilitate learning and teaching during the first heroic phase of establishing the Roman chant repertory in Francia.

VII.5. THE PLACE OF MUSIC IN THE CAROLINGIAN RENAISSANCE

Smits van Waesberghe 1952, 1970; Manfred Schuler 1970; Huglo 1975.

While I have expressed doubt that musical notation was used by the Franks as a principal means of mastering the Roman chant repertory, they were certainly concerned with a musical-technical matter of a different nature, namely, the modal system. Aurelian of Réôme, writing in the decade before 850, says that Charlemagne added four modes to the usual eight. Byzantine musicians, who had prided themselves on the invention of the original eight, thereupon added another four themselves. While Charlemagne's role is doubtless a pious fiction, the earliest surviving tonary— Paris, Bibliothèque Nationale, lat. 13159, from Saint-Riquier, possibly written as early as late eighth century—shows that the eight-mode system was already understood in Charlemagne's time (see III.14). It may indeed have come to the West from Byzantium (see VIII.2).

As I have argued in V.4, the reasons the Franks were interested in the modes were partly practical and partly intellectual. If one understood the mode of an antiphon, an introit, or a communion, one could choose the appropriate tone for the psalm verse(s) which these chants accompanied. The same is true for office responsories. But the presence of other mass chants—graduals, alleluias, and offertories—in the Saint-Riquier tonary (and some later tonaries) suggests that the classification was important for its own sake, a link with the corpus of classical music theory which was being unearthed by the Franks along with so much of the heritage of antiquity, even though it was by no means easy to reconcile with the Roman chant repertory they were trying to master.

Despite the place of music among the seven liberal arts, the 'seven pillars or steps' of Alcuin's educational programme (PL 101. 853), there is little evidence of original thought about music theory at Charlemagne's court. Beside 'David' (Charlemagne), 'Flaccus' (Alcuin), 'Timotheus' (Paulinus of Aquileia), and 'Homer' (Angilbert, later abbot of Saint-Riquier), 'Idithun' had his place as the instructor of the boys in both practical and theoretical music:

> Instituit pueros Idithun modulamine sacro,
> Utque sonos dulces decantent voce sonora.
> Quot pedibus, numeris, rithmo stat musica discant.

(MGH (Antiquitates) *Poetae Latini aevi Carolini*, i. 246; cited by Manfred Schuler 1970 and Huglo 1975). Idithun (Jeduthun) was a musician in the temple of King David (1 Chr. 16: 41–2 and 25: 1; also Pss. 39, 62, 77), and he makes a reappearance in one of those modern psalms, Notker's sequence *Summi triumphum regis* (von den Steinen 1948, i. 240). His identity at Charlemagne's court is unknown. The 'musica' he taught is evidently the knowledge of proportions as exemplified in metrical poetry, as treated in St Augustine's *De musica*. But he must have been a personage of consequence. Several singers of the court chapel are known to have gone to to high ecclesiastical office: Anstrannus, cantor at Charlemagne's court, became bishop of Verdun in 800; Hucbert, praecantor under Louis the Pious, became bishop of Meaux in 823; Johannes, cantor at the court of Lothar II, became bishop of Cambrai in 766 (Manfred Schuler 1970, 26–7).

Charlemagne's own teacher in the quadrivium was Alcuin. A treatise on music which Alcuin is reported to have composed (Manfred Schuler 1970, 34) appears to have been lost, unless it survives in the short text *De octo tonis in musica* (GS i. 26–7), taken up in Aurelian of Réôme's *Musica disciplina* (Gushee 1963 and CSM 21).

Whatever the beginnings under Charlemagne may have been, it remains true that the first flowering of medieval music theory belongs to the second half of the ninth century. Aurelian of Réôme is a somewhat isolated figure, but the writings of Johannes Scotus Eriugena (d. *c*.877), Remigius of Auxerre (d. *c*.900), Hucbald of Saint-Amand (d. 930), and Regino of Prüm are all directly linked to one another and, directly or indirectly, with the entourage of Charles the Bald. Yet the trend which seems to have set in under Louis the Pious, a transference of the main intellectual activity away from the court to the great monasteries and episcopal schools of north France and Germany, continued. The place of origin of some important writings—the *Enchiriadis* group of treatises foremost among them—is still unknown. But it is no accident that Hucbald, Regino, and Berno of Reichenau should all have been abbots, distinguished for much more than their contributions to music theory.

Despite the apparent lack of new writing on music theory in the time of Charlemagne himself, it is there that the foundations were laid for a new understanding of what a 'musician' should be. For Boethius there was an unbridgeable gulf between the connoisseur of 'musica' (the science of music theory), the composer, whose understanding of music is by instinct rather than by the power of reason, and the performer, the mechanic (Bk. 1, ch. 34). Although this doctrine was reproduced in modified form by Aurelian, for example (ch. 7), and in Guido of Arezzo's catchy rhyme 'Musicorum et cantorum', numerous musicians combined all three types in themselves with honour. Perhaps the single most important motive for writing about music among medieval theorists was the desire to improve the performance of the liturgy. (Chapter V above therefore concentrates on writing of a practical intent.) Aurelian and the substantial figure of Regino are early examples of practising church musicians who contributed to the theoretical literature. Hucbald was simultaneously singer in the performance of the liturgy, composer, and theorist. These have their forerunner in the unknown 'Idithun', the instructor in both chant and music theory, the paradigm of the new type of medieval church musician.

VIII

Gregorian Chant and Other Chant Repertories

VIII.1. INTRODUCTION

In this chapter I shall attempt to summarize the relationship of Gregorian chant to other types of early chant in the West, so that Gregorian chant itself may be understood in a rather broader context than would otherwise be possible. Something of the similarities and differences in liturgy between Roman, Milanese, Old Spanish (Mozarabic), and Gallican uses has been indicated in previous chapters (VI.4–5). Now the musical aspects of this relationship are sketched in.

One point of the chapter is to help define the nature of what the Franks learned from Rome and what they recorded in the first music manuscripts, that is, to answer the question, what is Gregorian chant? As is well known, the earliest music manuscripts from Rome itself contain quite different music for the same texts as Gregorian chant, music which is commonly called Old Roman chant. If this is Roman, but differs from Gregorian, in what sense is Gregorian chant 'Roman'?

Old Roman chant is not the only old chant repertory of Italy, and it bears similarities of musical style to the two other extant repertories, namely Milanese and Old Beneventan chant. These provide the musical context for understanding Old Roman chant. The three 'Old Italian' repertories are discussed in VIII.3–5. To understand Gregorian chant, on the other hand, we should at least consider the possibility that the Franks were influenced by the traditional chant of Gaul, that is, Gallican chant. There is an obvious obstacle to this, in that the Gallican rite and its chant were suppressed in favour of Roman use, and practically nothing of the chant repertory survives. But a sister repertory, that of the Old Spanish rite, does survive, albeit in staffless neumatic notation. On this basis we can make at least a tentative guess at the nature of what the Franks suppressed and what may have influenced their understanding of the Roman repertory.

The reader may justifiably have expected greater space to have been devoted to these non-Gregorian chant repertories. There is no doubt that, despite a number of fine studies, they are still not generally well known. The reason for this is the preoccupation of early studies with the music still sung in the Roman church, Gregorian chant. But not only has historical circumstance contributed to their neglect, some of the non-Gregorian repertories have been denigrated as aesthetically inferior to the Gregorian, an unfortunate and baseless justification of something which had quite different causes. The situation at the moment is that while there are excellent studies of individual genres of Old Spanish, Old Roman, and Milanese chant (such as those of Bailey), which are in many respects superior to anything yet available for Gregorian chant genres, and while the whole corpus of Old Beneventan chant has been discussed in exemplary fashion by Kelly, a balanced survey of all of these is not yet possible in the way that it is possible for Gregorian chant, certainly not in a book of this length. In particular, there is a need to return to Gregorian chant in the light of the new knowledge about the other repertories. This is work that will occupy scholars for many years to come.

During much of the period from the fourth to the ninth century, the eastern Roman empire and its capital Byzantium enjoyed greater material wealth, political power, and ecclesiastical prestige than the kingdoms of the West, and often influenced Western affairs directly, most obviously in Italy, parts of which were long under Byzantine rule. The possibility has often been discussed that the Byzantine liturgy and its chant could have influenced Roman use. A preliminary section (VIII.2) is devoted to this question.

VIII.2. THE INFLUENCE OF BYZANTIUM

- (i) Introduction
- (ii) The System of Eight Modes
- (iii) Antiphons for the Adoration of the Cross and Other Chants in Old Italian Repertories
- (iv) The Trisagion
- (v) The Frankish 'Missa graeca'
- (vi) The Communion *Omnes qui in Christo*
- (vii) The Byzantine Alleluias
- (viii) Textual Concordances without Musical Similarity
- (ix) The *Veterem hominem* Antiphons for the Octave of Epiphany

PalMus 5; PalMus 14; Brou 1938–9, 1948–52; Wellesz 1947; Baumstark 1958; Jammers 1962, *Musik*; Levy 1958–63; Huglo 1966; Jammers 1969; Levy 1970; Strunk 1977, 297–330; Kelly 1989, 203–18; Levy, 'Byzantine Rite, Music of the', *NG*.

(i) *Introduction*

At first sight it might seem that a good case could be made out for the likelihood of Byzantine influence on Roman chant. After the fall of the Western Roman empire, Italy was for long periods a territory disputed between Byzantium and the invading Germanic peoples. The reconquest of Italy by Justinian from 533, temporary though it proved, introduced what has been called the 'Byzantine period' in the history of the papacy, lasting over two centuries. Eleven out of thirteen popes between 678 and 752 were Greek or Syrian by birth. The outbreak of the iconoclastic persecution in Byzantium in 726 resulted in the movement of many Greek churchmen to the West. If there were evidence of widespread Byzantine influence in Roman chant it would be easy to explain. On the other hand, the recurrent political and theological differences which separated Rome and Byzantium may as easily have contributed to the separate development of their chant repertories. Some fundamental differences cannot be overlooked. The Western rites never followed Byzantium in the composition of the vast number of kontakia and kanones which constitute one of the glories of Byzantine music. On the other hand, Byzantium remained rigidly economical where Rome developed a cycle of proper-of-mass chants (Strunk 1977, 316): basically only one entrance chant corresponding to the introit, two offertories, two dozen koinonika (communion chants).

The history of the Roman liturgy—and in the absence of musical documents this is all we can argue from—is one of general independence from the Byzantine rite. The influence of Byzantium is seen rather in individual details rather than overall form and content. It may be that some aspects of Roman and Byzantine chant resemble each other because they derive from a common source—though rarely has a serious attempt been made to demonstrate what this might mean in terms of actual pieces of music (see the articles by Levy). Yet the available musical evidence, like the liturgical, relates to isolated items. Most of these are to be found, moreover, not in the Gregorian repertory but in the Old Italian traditions: Old Roman (rarely), Milanese, and Old Beneventan.

Jammers (1962, 183), in a study that influenced the thinking of several other scholars (principally Stäblein: cf. Stäblein, MMMA 2, 58*) argued that Gregorian chant was the result of Old Roman chant carried out at papal behest (specifically, that of Vitalian, 657–72, but on the basis of sources no older than the thirteenth century!) to make it more compatible with the Byzantine practice of ison singing, a type of chanting with vocal drone unfortunately not documented before the fifteenth century. (The obvious deficiencies in Jammers's arguments were stated succinctly by Nowacki 1985, 260–1.) Even if there were better documentary evidence for the 'ison' theory, it seems unwise to propose such all-embracing hypotheses about the style of Byzantine and Roman music in the seventh century on the basis of musical sources of the ninth to tenth centuries (Gregorian), the eleventh (Old Roman), and later (Byzantine).

More significant, it appears to me, is Jammers's observation that chant is bound to a text and often shapes itself according to the needs of text declamation. Differences of language have far-reaching musical consequences.

I have already discussed the possible influence of Byzantine on Western notation (IV.4).

Needless to say, in what follows I have not attempted to list each instance of possible Byzantine influence or Western borrowing. (For further references, see the bibliography in Levy, 'Byzantine Rite, Music of the', *NG*; Brou lists over forty possible borrowings, many of which have not yet been subjected to musical analysis; Jammers 1969 povides another list.) It seems possible to distinguish the following categories: (*a*) musical similarities which may go back to the common roots of Christian music in the early church; that this is by no means unlikely might be argued from the analogy of Greek readings in the Roman mass, which persisted into the ninth century (Vogel 1986, 296–7); (*b*) Western borrowing of texts where no corresponding musical sharing has been established; (*c*) Western versions of Byzantine chants; (*d*) new Western compositions with Greek texts for which no Byzantine model appears to have existed.

Byzantine musical influence can be seen to reduce itself largely to a number of individual instances. The overwhelming impression is that Roman chant developed largely independently of Greek models. There is a somewhat larger number of borrowings in the non-Roman Old Italian liturgies, the Milanese and Old Beneventan.

(ii) *The System of Eight Modes*

Raasted 1966; Bailey 1974; Hucke 1975.

The eight-mode system may have been taken up by the Franks after the Byzantine example. It does not seem to have been part of what they learned from Rome, for it plays no part in the organization of the Old Roman chant repertory. The 'noeanne' formulas of Frankish tonaries, which helped remind singers of the pitch-constellations of various groups of chants, appear to have been borrowed from the intonation formulas of Byzantine chant (see III.14). It is true that the modal system appears to have exerted influence on the shape of certain chants which otherwise had peculiarities consorting uneasily with it. Yet the Gregorian repertory cannot be said to have been formed from the beginning in conformity with the system. The Byzantine influence is therefore indirect and secondary.

(iii) *Antiphons for the Adoration of the Cross and Other Chants in Old Italian Repertories*

The antiphons *O quando in cruce*, *Adoramus crucem tuam*, *Crucem tuam adoremus* for the Adoration of the Cross, and some other Holy Week items, were sung both in Greek and Latin in Benevento and some other Italian centres (Wellesz 1947, 21–2, 68–77; PalMus 14, 305–8; Drumbl 1976; Kelly 1989, 207–18). Some have Byzantine forebears. For example, *O quando in cruce*, found in Beneventan and Ravenna sources, is a version of a Byzantine troparion which can be followed back to the rite of Jerusalem in the seventh century. Its presence in Ravenna should mean that it was already used in the liturgy there before the fall of the Byzantine Exarchate of Ravenna to the Lombards in 752.

These are examples of chants whose sources are principally Italian. Others are the Ambrosian ingressae *Coenae tuae mirabili* and *Videsne Elisabeth*, which both appear to be modelled on Byzantine pieces (PalMus 5, 9–13; Levy 1958–63).

(iv) *The Trisagion*

Another chant from the ceremony of the Adoration of the Cross, the Trisagion *Agios o theos*, is sung in the Roman rite in Greek and Latin as part of the improperia. In the Byzantine rite the Trisagion was sung several times every day in various services. Levy (1972) has argued that both Eastern and Western versions rely on a common modal tradition, and work within the same set of notes, without the one being derived directly from the other.

(v) *The Frankish 'Missa graeca'*

Levy 1958–63; Huglo 1966; Atkinson 1981, 1982 'Missa Graeca', 1989 'Doxa'.

A set of mass chants in Greek appear somewhat sporadically in Western manuscripts of the ninth century onward. Atkinson's careful investigation of the origin and transmission of those for the ordinary of mass (the others are much less widely disseminated) has led him to propose that they were put together between 827 and 835, after the Eastern emperor Michael II had sent a copy of the works of the Pseudo-Dionysius the Areopagite to Louis the Pious. These writings were translated under the direction of Abbot Hilduin of Saint-Denis. Atkinson believes the texts of the 'missa graeca' could have originated at the same time, which coincided with the special concern for the unity of Christendom detectable in the works of such writers as Amalarius of Metz and Agobard of Lyons.

Atkinson argues persuasively that the melody for the Greek Agnus Dei (*O amnos tu theu*), was composed expressly to complete the set. On the other hand, the melodies of the Gloria (*Doxa en ipsistis*), Credo (*Pisteuo eis ena theon*), and Sanctus (*Agios*) may well be earlier borrowings from the East. In fact Levy believes that the Sanctus represents an ancient recitation formula which underlies the whole Anaphora of mass, the Lord's Prayer, and the Te Deum as well (Levy 1958–63; see II.18), comparable in this respect with the Trisagion just mentioned.

The rest of the 'missa graeca'—introit, Kyrie, offertory, and communion—appears to be adapted from Latin chants. The mass was usually sung at Pentecost, when two alleluias would normally have been performed, but no special provision for them appears to have been made. Some other alleluias with Greek texts are known, however, from a small number of mostly north French and English manuscripts: *Dies sanctificatus/Ymera agiasmeni* (Brou 1938–9) and a set in Cambrai, Bibliothèque Municipale 75 (Saint-Vaast at Arras, early eleventh century; Brou 1948[–52], 172–6). None of these appears to be based on Eastern originals, but they are certainly evidence of a lively interest in Greek learning, typical particularly of northern French centres in the second part of the ninth century. The 'missa graeca' was later rounded out at Saint-Denis to make a complete set of prayers for the octave of the feast of the abbey's patron saint.

(vi) *The Communion* Omnes qui in Christo

Levy 1970.

The mass of Easter Eve where the neophytes were baptized has very few chants in the Roman rite and consequently in Gregorian chant-books. Levy has uncovered a number of proper chants for the mass in Old Italian sources. One of these, the offertory chant *Omnes qui in Christo*, can be found in Roman sources as the communion on the Saturday of Easter week. What is more, it appears to be a translation of the Byzantine baptismal chant *Hosoi eis Christon*, with essentially similar music. Other aspects of this set of chants, particularly their modal unity, lead Levy to suspect roots in the early Christian period before the division of the empire and the cleavage between the main Christian churches.

(vii) *The Byzantine Alleluias*

Thodberg 1966.

Like the Roman and other Western repertories, Byzantine chant has a cycle of proper alleluias. They have a standard set of short 'alleluia' openings, one for each of six modes (the two F modes are not used), with florid verses. Thodberg (1966, 168–95) believes that melodic similarities can be demonstrated between three of these melodies and three of the Greek alleluias sung at Roman Easter Vespers and in Easter Week, whose chants are found in the Old Roman chant manuscripts. Since Latin versions of these alleluias are also known, Thodberg's analysis embraces the Gregorian and Milanese traditions as well (*O kyrios = Dominus regnavit decorem*, *Oty theos = Quoniam deus magnus*, *Epy si kyrie = In te domine speravi*; Snow had already noticed the melodic strangeness of these alleluias in the Old Roman repertory: Apel 1958, 499).

(viii) *Textual Concordances without Musical Similarity*

Texts common to Byzantine and Western rites have occasionally been identified, without the melodies being recognizably similar. An example is the text of the Holy Week chant *Vadit propitiator*, which has been identified as part of a kontakion by Romanos the Melode (first half of sixth century). It appears in various liturgical functions with various melodies in the three Old Italian chant repertories: Milanese, Old Roman, and Beneventan. (PalMus 5, 6–9, Hesbert 1938–47: 1945, 73–8, Kelly 1989, 206, 295–6).

(ix) *The* Veterem hominem *Antiphons for the Octave of Epiphany*

Handschin 1954; Lemarié 1958; Strunk 1964 = 1977, 208–19.

Notker Balbulus' life of Charlemagne, written for Charles the Fat in 883–4, relates that the emperor heard members of a Byzantine legation—according to Handschin the embassy sent to Aachen in 802 by the empress Irene—singing in a service on the Octave of Epiphany, and ordered a Latin version of the chants to be made, matching the original syllable for syllable: this was the series of antiphons starting with *Veterem hominem* for Lauds on the Octave Epiphany. The Byzantine originals have been identified, although, as Strunk has stated, without Notker's story one would not have guessed that the pieces were the same. At least the mode is the same in both traditions. If such borrowings are to be recognized on musical grounds alone, closer stylistic analysis of the Western antiphon repertory is clearly required.

VIII.3. OLD ITALIAN CHANT I: ROME

   (i)   Introduction
  (ii)   Sources and Studies of Old Roman Chant
 (iii)   Examples: Communions
 (iv)   Graduals
  (v)   Antiphons
 (vi)   Offertories
 (vii)   Alleluias
(viii)   Oral Tradition

(i) *Introduction*

Three Italian chant repertories have survived, from Rome, Milan, and Benevento respectively, which in many respects are quite different from Gregorian chant. There are faint traces of other repertories as well, but their remains are so slight that I have to pass them by. (See Levy, 'Ravenna Rite, Music of the', *NG*, and Bryant, 'Aquileia', *NG*.)

The Roman chant is for the Roman liturgy, just as is Gregorian chant, the great majority of their texts being the same. The relationship between these two sorts of chant for the one liturgy is the subject of a special section in this chapter. Since the Roman chant was superseded in Rome by Gregorian chant in the thirteenth century, I shall follow the practice of some other writers and call it 'Old Roman'.

The Milanese liturgy is different from the Roman, as I have already indicated above (VI.4–5). Its chant has survived in use until this day.

Although the sources for the Beneventan liturgy before its replacement by the Roman are few, enough is known to allow us to give it a place equivalent to the Milanese. The chant created for this old liturgy may be called 'Old Benevantan', for it was superseded by the Gregorian.

The stylistic similarities in the music of these three chant repertories allow us, with due caution, to group them under the designation 'Old Italian', whose idiom differs in certain fundamental ways from Gregorian chant.

(ii) *Sources and Studies of Old Roman Chant*

Old Roman chant, as already stated, serves the same liturgy as Gregorian chant, but without the liturgical modifications made by the Franks in the eighth and ninth centuries. That means that for the majority of chants of the Roman liturgy, both an Old Roman and a Gregorian version exist. Apart from the musical differences, there are a number of cases where the Old Roman and Gregorian books have a different chant for a particular liturgical occasion: several cases for the mass chants, and many more for the office, where variety between uses is always much greater. Other differences concern the wording of chant texts. Some of the liturgical and textual details peculiar to Old Roman books have occasionally been found in manuscripts of Gregorian chant (Huglo 1954, 'Vieux-romain', Frénaud 1959).

The principal sources of the Old Roman melodies are three graduals (Rome, Biblioteca Apostolica Vaticana, Vat. lat. 5319, San Pietro F. 22, and the Martin Bodmer private collection Bodmer C. 74, formerly Phillipps 16069) and two antiphoners (Rome, Biblioteca Apostolica Vaticana, San Pietro B. 79 and London, British Library, Add. 29988). A facsimile of Bodmer C. 74 has been published (Lütolf 1987) and a transcription of Vat. lat. 5319 (by Landwehr-Melnicki in MMMA 2). Cutter 1979 has edited the texts of Vat. lat. 5319 and San Pietro F.22, together with Georgii's (inaccurate) copy of the texts of Bodmer C. 74.

The earliest of these manuscripts is Bodmer C. 74, dated 1071; the latest is San Pietro F. 22, from the thirteenth century.

The first survey of each of the major Old Roman chant genres was made by Snow (in Apel 1958, 484–505). The best recent survey of this type is Hucke, 'Gregorian and Old Roman Chant', *NG*. Extended studies of most categories have been completed:

introits: Connolly 1972, 'Introits and Communions'; see also Connolly 1972, 'Archetypes', Connolly 1975, Connolly 1980

graduals: Van Deusen 1972; see also Hucke 1955, 'Gregorianischer', 1956

tracts: Hans Schmidt 1955, 1957, 1958

alleluias: Stäblein in MMMA 2, 119*–140*

offertories: Dyer 1971, Kähmer 1971; see also Hucke 1980, 'Aufzeichnung'

communions: Murphy 1977

office antiphons: Nowacki 1980; see also Nowacki 1985

great responsories of mode 2: Cutter 1969; see also Cutter 1967, 'Oral', 1970, 1976

psalm tones: Dyer 1989, 'Singing'

ordinary-of-mass chants: Boe 1982

(iii) *Examples: Communions*

Stated very broadly, and allowing for numerous exceptions, Old Roman chant displays many of the same formal characteristics as Gregorian chant; the text is treated in the same way. The two settings agree on where phrases begin and end, and the settings of those phrases usually have the same shape and tonal character. One finds the same structural elements in both types of chant: intonation figures, recitation passages, cadence figures. The main difference between the Gregorian and Old Roman chant concerns surface detail. Old Roman chant is more ornate. A few examples will illustrate some of the similarities and differences.

Ex. VIII.3.1 is the communion for the mass on Christmas Eve in its Old Roman and Gregorian versions. Most of the Gregorian version can be seen as a recitation around F, with most cadences on D, and the more ornate Old Roman version seems to have the same centre of gravity. In both versions the second phrase rises to a, the third starts on D and ends on F. The last phrase is more adventurous, rising to high c and falling back to D, then up to a again before the final cadence. Some of this is

Ex. VIII.3.1. Old Roman and Gregorian versions of communion *Revelabitur gloria Domini* (Rome, Bibl. Ap. Vat. Vat. lat. 5319, fo. 10ᵛ; Montpellier, Faculté de Médecine H. 159, p. 26)

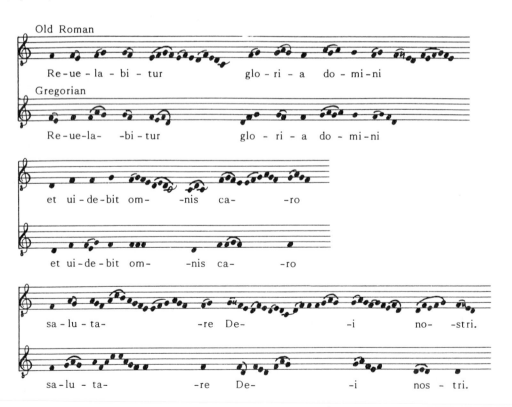

entirely typical for a communion in this mode, but other features, particularly the handling of the last line, show a more than casual resemblance between the two versions.

The Old Roman version has many more notes than the Gregorian. It does not have as many *D–F* or *F–D* leaps, and they tend to occur in the longer melismas. The whole is less 'gapped' than the Gregorian: it is typical that the Old Roman version descends from the high *c* by step. The musician who made this version understands the phrase ends to be in the same place, on the same notes, but marks them by melismas, standard for Old Roman communions. (For a parallel example, see Hucke, 'Gregorian and Old Roman Chant', *NG*, Ex. 1, whose end-phrase 'salutare dei nostri' is identical to Ex. VIII.3.1 in both versions.)

In many places the two versions are almost identical, and there is evidently a close relationship between them. Despite the greater plainness of the one, the verbosity of the other, one would hesitate to say the Old Roman was a decoration of the Gregorian, or the Gregorian a simplification of the other. They are rather two realizations of the same basic idea, one in a rather restrained, the other in a more florid idiom.

(iv) *Graduals*

The standard phrases seen in Gregorian graduals, tracts, and responsories have counterparts in the Old Roman versions. Rather than cite what would be a lengthy example in both versions, I give the Old Roman versions of two verses from graduals of the so-called 'Iustus ut palma' type. In fact the group would have to have another name in the Old Roman repertory, because *Iustus ut palma* was not sung as a gradual in Rome. In Ex. VIII.3.2 I give the verses *Celi enarrant* and *In manibus* for the graduals *A summo celo* and *Angelis suis* respectively. These are three-phrase verses. A comparison can be made with the five-phrase verse in Ex. II.5.1: the three phrases in Ex. VIII.3.2 correspond to the first, second, and fifth phrases in Ex. II.5.1, which end *d, a, a*, as here.

(v) *Antiphons*

The standard Gregorian melodies of the office antiphon repertory have their Old Roman counterparts, and the versions are often closely similar—not surprisingly, in view of the simplicity of the melodies in question. Here Nowacki has made the important observation that the two repertories do not always have the same standard melody (see Nowacki 1985, 246–7, Exx. 4–6). Ex. VIII.3.3 is an example of this: *Valde honorandus est* and *Si vere fratres* are two members of a large family of antiphons using the melody which Nowacki has labelled D.2.i. The equivalent melody in the Gregorian repertory is Ia in Frere's discussion (*AS* 65), Gevaert's theme 6. In the Gregorian tradition, however, *Si vere fratres* has a different melody, a popular G-mode melody, VIIc in Frere, Gevaert's theme 23. This too is well known

Ex. VIII.3.2. Gradual verses *Celi enarrant* and *In manibus portabunt te* (Rome, Bibl. Ap. Vat. Vat. lat. 5319, fos. 7ʳ, 41ᵛ)

in the Old Roman repertory, Type G. 1 in Nowacki's analysis. *Specie tua* is an antiphon with the G-mode melody in both traditions.

(vi) *Offertories*

The Old Roman offertories use two standard types of material (Dyer 'Formula' A and B, Kähmer 'Singweise' 1 and 2) in a way which is not matched in the Gregorian repertory. Formula A is a monotone or decorated recitation which hovers around the semitone step, found in twenty-seven pieces, usually more than once in the same piece. Formula B is a more mobile melodic phrase, and can likewise be used more

Ex. VIII.3.3. Antiphons in Old Roman and Gregorian versions (Rome, Bibl. Ap. Vat. San Pietro B. 79, fos. 34ᵛ, 69ʳ, 139ᵛ; *AS*, 62, 142, 664)

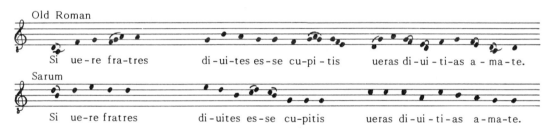

than once in the same offertory; in fact, some offertories consist of little more than repetitions of this same basic phrase. Both formulas appear in pieces in any tonality, though one notices that twelve of the twenty-nine offertories with Formula B have a final on *F*, two more on *C*. This is because Formula B itself is strongly F-orientated. Not surprisingly, some of the Gregorian versions of Old Roman offertories with Formula B have an *F* final.

Ex. VIII.3.4 is the verse *Qui propitiatur* from the offertory *Benedic anima mea* in its Old Roman and Gregorian versions. The offertory respond ends on *D* in the Old

Ex. VIII.3.4. Old Roman and Gregorian versions of offertory verse *Qui propitiatur* (Rome Bibl. Ap. Vat. Vat. lat. 5319, fo. 47ᵛ; Montpellier, Faculté de Médecine H. 159, p. 261)

Roman, on *F* in the Gregorian version, while both versions of the verse end with an imperfect cadence, as it were, on *G*, after being strongly F-orientated. The formulaic make-up of the Old Roman version is quite clear (marked in the example). The two phrases which usually make up the formula appear in the order ab–ab–a–ab. As usual, the accentuation of the text governs the way the formula is used. In this case it makes no sense to align the two versions. The Gregorian version is set out with the same line divisions as the Old Roman, but there is no corresponding repetition of formulas. The melody pushes constantly up to *c*; repetition of small motifs is there, but the vocabulary as well as the form is different from the Old Roman.

Ex. VIII.3.5 is the verse *Quoniam angelis suis* from the offertory *Scapulis suis* (ending *E* in the Old Roman, *G* in the Gregorian version). In this case the Old Roman version ends on *a*, the Gregorian on *G*, just as in Ex. VIII.3.4. The ornate recitation of the Old Roman version is to some extent matched in the Gregorian, in the repeated *c*s or *c–a* oscillations.

The Old Roman offertories, like the Gregorian, occasionally have long melismas. Here the two versions again diverge strongly.

Despite the difference in the text of these versions, and the frequent lack of unanimity in choice of verse(s) (which is characteristic of the Gregorian tradition as a whole) the text tradition is in one important respect largely unanimous: the unusual feature of text repetition is common to both Old Roman and Gregorian offertories

Ex. VIII.3.5. Old Roman and Gregorian versions of verse *Quoniam angelis* from offertory *Scapulis suis* (Rome Bibl. Ap. Vat. Vat. lat. 5319, fo. 43ʳ; Montepellier, Faculté de Médecine H. 159, p. 295)

(though not found in the above examples), and occurs nearly always in the same places.

(vii) *Alleluias*

The ornamented recitation figure of the offertories ('Formula A', Ex. VIII.3.5) is also to be found in some Old Roman alleluias. There are very few Old Roman alleluia melodies—only four are used with any frequency: one in D, one in E, and two in G, of which one is the melody of the alleluias sung at Vespers in Easter Week. This grand series of Easter alleluias, some with Greek texts and some with Latin, has attracted considerable comment, both from Amalarius of Metz in the ninth century and from modern writers (most recently Smits van Waesberghe 1966, Stäblein in MMMA 2, 84*–140*, van Dijk 1969–70). Some of the verses are divided into two parts. In the first part, a recitation in sung on *c* with frequent torculi *cdc*. Then the second half is 'announced' before being sung to a different recitation with torculi *bca*, the torculus we have just seen in the offertory (Ex. VIII.3.5). Ex. VIII.3.6 gives the first verse of *Alleluia Paratum cor meum*.

Ex. VIII.3.6. First verse of *Alleluia Paratum cor meum* (Rome, Bibl. Ap. Vat. Vat. lat. 5319, fo. 89ᵛ)

There are meagre performance rubrics in the only two musical sources, Vatican 5319 and London 29988 (see the facsimiles in *NG* i. 488 and this book, Plate 13). These indicate that the primicerii (leaders of the choir) sang the 'announcements' and the schola took up the rest of the verse, presumably singing the first part. The lengthy description of the ceremony in Ordo Romanus XXVII, in particular of the first alleluia, *Alleluia Dominus regnavit decorem* (Andrieu 1931–61, iii. 363), allowed Smits van Waesberghe to reconstruct a more imposing type of performance, as might have been witnessed in late eighth-century Rome. *Alleluia Dominus regnavit decorem* shares one more peculiarity with other Old Roman alleluias: an extended repeat of the alleluia after the verse(s), called in Ordo Romanus XXVII the 'alleluia secunda', equivalent to the *melodiae* of the Milanese alleluia repertory but only rarely found in the Gregorian. The full performance appears to have been something like this:

Alleluia (+ jubilus?)	primus scholae cum parafonistis infantibus
Alleluia (+ jubilus?)	parafonistae viriles
Alleluia + jubilus	subdiaconus cum infantibus

Verse 1*	subdiaconus cum infantibus
Alleluia (+ jubilus?)	parafonistae viriles
Anouncement	parafonistae viriles
Verse 2, part 1	parafonistae infantes
Announcement	parafonistae viriles
Verse 2, part 2	parafonistae infantes
Alleluia (+ jubilus?)	parafonistae viriles
Announcement	parafonistae viriles
Verse 3, part 1	parafonistae infantes
Announcement	parafonistae viriles
Verse 3, part 2	parafonistae infantes
Request to archdeacon**	primus scholae
Alleluia secunda	primus scholae cum parafonistis infantibus
Alleluia + jubilus	parafonistae viriles

*not sung with the recitation formulas and announcements of verses 2 and 3
**same music as announcement

(viii) *Oral Tradition*

Several writers have pointed out that the Old Roman sources are often in disagreement over melodic details. In particular Cutter (1967, 'Oral') has shown numerous cases of divergence. Nowacki's demonstration (1985) of how changes may have come about in the assignment of typical melodies to particular antiphon texts is another aspect of the continuing oral tradition in Rome. Hucke (1955, 'Gregorianischer') had already realized that in some graduals—a generally very systematically organized group—the Gregorian sources transmitted a frozen, archaic state while the Old Roman ones showed evidence of further stylization.

More surprising is the wide disagreement which Dyer (1989, 'Singing') has shown to have existed between the two Old Roman antiphoners in their repertory and assignment of psalm tones. They have over 100 between them—a prodigality which Dyer associates with the survival of solo psalmody in Rome—but only half are found in both sources. (There is no obvious eight-tone, eight-mode system.)

All this disagreement suggests that the sources were notated in the midst of a continuing tradition of oral transmission and performance. In many respects the Old Roman tradition, even on the eve of its suppression in the thirteenth century, never reached the degree of written fixity indicated in sources of Gregorian chant. This is important for our assessment of the relationship between the Old Roman music as we first see it in the eleventh century, Gregorian chant as we first find it in the sources from around 900, and Roman chant as it might have been known to the Franks in the later eighth century.

Much more could (and has been) said about the interrelationship between particular items and between parts of the repertory in their Old Roman and Gregorian versions (the reader is referred especially to the penetrating essays by Hucke). Before a summing up of the matter is attempted, the other Old Italian repertories will briefly be discussed.

VIII.4. OLD ITALIAN CHANT II: MILAN

PalMus 5–6; Gatard, 'Ambrosien (chant)' (*DACL*); Cattaneo 1950; Hucke 1956; Jesson in Apel 1958, 465–83; Weakland, 'Milanese Rite, Chants of', *NCE*; Baroffio in Fellerer 1972, 191–204; Baroffio 'Ambrosian Rite, Music of the', *NG*; Heiming 1978.

The Milanese (or Ambrosian) rite is related to, but in many respects fundamentally different from, the Roman. The chief differences in the number and nature of important chants have already been noted (VI.4–5). Despite the frequent incompatability of the two rites, some Roman chants were adopted in Milan (see Hucke 1956), and this allows a certain amount of comparative musical analysis. It hardly needs saying that the liturgy and its chant in our medieval sources cannot be exactly that known to St Ambrose (d. 397), whose name has been given to both rite and chant. It seems possible, however, that four hymns by him have survived (see II.15; also Stäblein, 'Ambrosius', *MGG*, Cattaneo 1950).

Our knowledge of the Milanese liturgy in its full medieval splendour depends largely on two sources edited by Magistretti: a manual of the eleventh century containing the texts of the complete liturgy (Magistretti 1904–5), and an ordinal compiled about 1125 by the clerk Beroldus (Magistretti 1894). There are summaries by Lejay ('Ambrosien (rit)', *DACL*), King (1957, *Primatial*, 287–456, for the mass), Paredi ('Milanese rite', *NCE*), and Borella (1964).

The chant of the Milanese rite is the only Latin non-Roman medieval repertory to survive in reasonably complete state in transcribable notation. Reports of attempts to suppress it at various times in the Middle Ages appear to be legendary. For example, in the eleventh century the Milanese chronicler Landulphus (*De ritibus ecclesiasticis* 2. x–xii, PL 147. 855) said that Charlemagne had had an Ambrosian and a Gregorian sacramentary placed side by side on the altar. By the will of God, both opened simultaneously, proving that both were equally authoritative.

Milanese chant survives in a large number of manuscripts (described by Huglo *et al.* 1956), of which the earliest date from the eleventh century. The earliest comprehensive sources are of the twelfth century:

1. three chant-books for the winter season (Milanese books combine mass and office and divide winter from summer services):
   London, British Library, Add. 34209 (facsimile and transcription PalMus 5–6);
   Milan, Biblioteca capitolare F. 2. 2;
   the 'Varese–Eredi Bianchi' manuscript, privately owned;
2. one for the summer services:
   Bedero di Val Travaglia, San Vittore B.

The written musical tradition of Milanese chant, like that of Old Roman chant, therefore begins considerably later than that of Gregorian. An interesting guide to the Milanese chant of the seventeenth century, reflecting the reforms of St Charles Borromeo, was written by Perego (1622).

Two modern service-books edited by Suñol contain editions of the music for mass and vespers respectively, the *Antiphonale missarum* (1935) and *Liber vesperalis* (1939). These do not always reflect medieval practice. An equivalent volume for the morning service did not appear. Critical editions of the alleluias and cantus (tracts) have been made by Bailey (1983, *Alleluias*, 1987) and the complete antiphon repertory has been indexed, edited, and classified according to melodic type by Bailey and Merkley (1989, 1990). A survey of the mass chants as a whole was made by Jesson (1955), and summary accounts of the complete repertory have been made by Jesson (in Apel 1958, 465–83) and Baroffio ('Ambrosian Rite, Music of the', *NG*). Studies of the psalm tones have been made by Bailey (1977, 1978), of the offertories by Baroffio (1964), and Hucke has compared Gregorian, Old Roman, and Milanese graduals (1956). There are studies of the 'responsoria cum infantibus' by Moneta Caglio (1957) and Bailey (1988). Multiple versions of the antiphons of the ferial office were published in parallel by Claire (1975). (For further bibliography see Baroffio, 'Ambrosian Rite, Music of the', *NG*; Heiming 1978 surveys recent liturgical research.)

The simple psalmody of the Milanese office was not organized according to an eight-mode/eight-tone system. The complicated system in the modern *Liber vesperalis*, with over 150 different cadences, is not authentic. Bailey found only sixty-nine in the early sources, of which only twenty-seven were commonly used, and these corresponded with relatively few exceptions to the thirty chief antiphon melodies used. This suggests a possible model for the Roman system before it was organized on modal lines by the Franks.

Until reliable editions from the early sources and analyses of the different genres have been carried out, only general remarks about much of the office repertory are possible. None of the larger antiphon types has been investigated, such as the long morning 'antiphona ad crucem', the evening 'antiphona in choro', and the numerous simpler psallendae. Discussion of the responsory repertory has concentrated on the elaborate melismas in repeat form of the 'responsoria cum infantibus' or 'cum pueris'. Much fundamental work remains to be done. It is clear, for example, that the various types of responsory ('post hymnum' and 'ad lectionem' in the morning, 'in choro' and 'in baptisterio' in the evening) use typical melodies and verse formulas, within each genre and as a larger group, though less unvaryingly than the Gregorian office responsories, and with a number of irregularities in the matching of a particular verse tone to a particular final in the responsory. But no systematic analyses have been published. Jesson has noted instances of recurrent formulas in the psalmelli (graduals). Bailey has edited and analysed all the cantus (tracts) and shown that all are elaborations of the same type-melody, related to the Old Roman and Old Beneventan melodies, which indicates a common parentage.

A formal peculiarity of many Milanese chants—we have seen something like it in the Old Roman offertories—is the repetition of musical phrases or formulas to generate the music for longer verses. This is to be found, as Jesson points out, in numerous non-responsorial chants of mass: ingressae (introits), the 'post evangelium' antiphons, and especially the transitoria (communions: see table in Jesson 1955, 104). The transitorium *Te laudamus*, for example, uses the same music six times over (Jesson 1958, 478). There is great variety in the degree of ornateness in these pieces. The transitorium *Gaude et letare* (PalMus 5, 63, PalMus 6, 72), for example—in the simple form AABAABC—is largely syllabic, with a coating of the little curlicue ornaments (often a pes subbipunctus) beloved of the Milanese singers. The splendid transitorium for the Septuagesima Sunday *Convertimini omnes*, on the other hand, is heavily decorated and includes a lengthy melisma in the final phrase of text. In Ex. VIII.4.1 it is set out in a way which makes clear the repeat structure. As usual, ornamental figures are omnipresent: the torculus subpunctis (pes subbipunctis) appears three times in the A-phrase, twice in C, twice in D, and so on.

Melismas with repeat structure are to be found in many genres of Milanese chant: the psalmellus (gradual), alleluia, and offerenda at mass, the office responsory. Perhaps the best known are those to be found in some office responsories, usually the responsories 'ad lectionem, and 'in baptisterio' of the morning service. These elaborate chants, of which Moneta Caglio (1957) listed forty, often carry the rubric 'cum pueris' or 'cum infantibus'. Some have one long melisma, or *melodiae*, some have two, the *melodiae primae* and *melodiae secundae*. The performance scheme appears to be that usual in responsorial practice in the Roman rite (see II.4), where the responsory is sung twice at the beginning and again complete at the end, but it is conceivable that the repetitions were more numerous on the highest feasts. A melisma may appear within the verse (usually on one of the last syllables). Its primary purpose was to enhance the frequent repetitions of the responsory. The usual place for the *melodiae primae* was the partial repeat of the responsory. The second melisma is for the repeat of the whole responsory. It is not always clear who should sing the various sections, for the early manuals do not specify the arrangement in full. By analogy with other such pieces one might suggest the following (the practice no doubt varied from feast to feast):

responsory	soloist
responsory	magister and boys
responsory	schola
verse	soloist
responsory second part with *melodiae primae*	magister and boys
responsory	schola
responsory with *melodiae secundae*	magister and boys

Ex. VIII.4.1. Transitorium *Convertimini omnes* (London, Brit. Lib. Add. 34209, p. 136)

If this seems excessively elaborate, it is relatively modest in comparison with the singing of the alleluia at mass on Quadragesima Sunday, the day of the 'farewell to the alleluia' before Lent (reconstructed by Bailey 1983, *Alleluias*, 25).

alleluia (twice)	four boys (in ambo)
alleluia (twice)	lectors (in choir)
verse	lectors
alleluia	four boys (in ambo)
alleluia with *melodiae primae*	lectors
alleluia with *melodiae secundae*	four boys (altar steps)
alleluia with *francigenae*	four boys (in choir)

Moneta Caglio listed forty 'responsoria cum infantibus', of which four are reassignments of mass alleluias. Among these he identified some fifteen sets of *melodiae*. Bailey's study (1988) makes it clear that they were frequently re-employed for chants other than their original 'mother' chant. The putative origin can be rediscovered by analysing the *melodiae*, which are usually built up through a process of elaborated repetition of the section of mother chant which they replace and enhance. When *melodiae secundae* are present, they will be a further elaboration of the germ cell. Six such sets of *melodiae* can be discerned. Ex. VIII.4.2 shows one of the shorter examples, the responsory 'in baptisterio' at the morning service of the fourth Sunday in Advent, *Sperent in te* (for a different way of setting out the melisma see Bailey 1988). The second phrase of the respond has the same music as the second phrase of the verse, so that a repeat of the respond from the third phrase seems logical. It is in this phrase that the *melodiae* are sung. The *melodiae* begin with 'Querentes te' (A) and end with 'domine' (XYZ), as in the respond. In between come four phrases. I have marked them with letters to show their relationship to each other and to the mother phrase 'Querentes in te':

A QR A–OPQR–W A–PQR–Q A–PQR A XYZ

The A phrases are ways of getting up to c, the tonal centre of the melisma. The repeats within the melisma are clear enough. W is a pendant to the body of those repeated phrases. If Bailey is right, OPQR should therefore derive from XY, Z being a final cadential flourish. And indeed this seems to be the case, except that O is a preliminary phrase, Q an internal extension. Of course, there is no one right way to analyse such a freely flowing melody, only several plausible ones.

The alleluias show improvisatory melodic generation of a different kind, where, as Bailey has shown, the longer repeated sections can themselves be broken down into small cells often following a repeat scheme: aabbcc, and so on. The Milanese alleluia repertory is small. There are ten alleluia melodies and ten verse types. But since two of the ten melodies are musically related to others, there may once have been only eight. Most of the verse types are related to the alleluia melodies.

Each of the alleluias has at least one *melodiae* for the extended repeat after the verse. Three of them have *melodiae primae* and *melodiae secundae*. *Melodiae tertiae* exist in the shape of the '[melodiae] Francigenae', sometimes supposed to be of

Ex. VIII.4.2. Responsorium cum infantibus *Sperent in te* (London, Brit. Lib. Add. 34209, p. 22)

Gallican origin, though there is no proof of this. It appears to be a nickname like those of sequence melodies: 'Occidentana', 'Metensis maior', 'Bavverisca'. All the *melodiae* can be derived from the mother alleluia, the 'Francigenae' from Alleluia VI in Bailey's edition.

The 'Francigenae' are about 240 notes long; Alleluia V is about 110 notes long, its *melodiae primae* about 240, its *melodiae secundae* over 320 notes long. But there is space here to show only a much more modest example. Ex. VIII.4.3 gives the melodiae of Alleluia VI (originally an Advent alleluia with verse *Venite*). The notes derived from the mother alleluia are marked with an asterisk; brackets between the asterisks show where phrases from the other alleluia are left out.

Ex. VIII.4.3. *Melodiae* of Ambrosian Alleluia VI (London, Brit. Lib. Add. 34209 p. 269)

We do not know how old the Milanese *melodiae* in their present form may be, just as we do not know the age of the Old Roman *melodiae*. Is the habit of singing extended repeats after the alleluia verse in some way Old Italian? But it is found in Old Spanish chant as well. Is it old enough to be a possible model for the Frankish sequence? Did the Frankish sequence derive from (now lost) Gallican alleluias of this sort?

There is common melodic ground between the Milanese alleluias, their Old Italian counterparts in Rome and Benevento, and, presumably through Rome, the Gregorian melodies. Bailey believes all ten Milanese alleluias are related to Old Roman and Gregorian ones, and that the two completely surviving Old Beneventan melodies can be matched as well.

Milanese chant, as will have become clear, is often extremely ornate. This often means that possible cases of identity between Milanese and other chants are difficult to assess, especially when comparing the other ornate Old Italian repertories. Is one comparing 'essential' notes or picking out details of little real significance? A basis for comparison does exist in a number of chants which Milan appears to have borrowed from Rome and which can sometimes be found in as many as four versions: Old Roman and Gregorian, Old Beneventan and Milanese. (Quite often the same piece will appear with different liturgical functions in the different traditions. Well over 100 mass chants with at least partial melodic correspondence between Gregorian and Milanese traditions are listed by Jesson 1955.)

Peculiarities in the handling of formulas, and some cases of very closely similar melodic readings, have led Hucke (1956) to believe that for those Milanese graduals of mode 2 that have Gregorian and Old Roman parallels the Milanese version derives from the Gregorian, not the Old Roman version. It seems likely that this is the case with other borrowings.

Ex. VIII.4.4 gives parallel versions of the introit/ingressa *Invocavit me* for Quadragesima Sunday, from the Gregorian, Old Roman, and Milanese tradition. The Milanese version has an 'alleluia' phrase at the end not reproduced here. Despite the varied ornamental clothing applied to it, the same melody appears to lie under all three versions. Not that one can identify any precise 'Urmelodie', but a common understanding of tonality and melodic shape is present. We can easily understand some of the surface differences as different conventional ways of making identical gestures. The same conventional cadence appears in the Gregorian version at the end of phrases C, E, and F. The Old Roman version has a different way of making a cadence, used in its fullest form at the end of phrases C and E, in a shorter form for phrase B. The Milanese version has yet another way of doing it, seen for phrases C and F. Some other figures seem to be common ornaments: the figure *cbabc* in Milan (3, 10, 16), for example, where the Gregorian version has *cbc*, *ccc*, *cc*, respectively. The opposite figure *aGFGa* appears in the Milanese version at 12, 15, 16, and 18. The Old Roman uses the figure *G(a)bcb*, *a(b)cdc*, or *b(c)ded* six times (7, 10, 12, 19, 25, 35), though this appears in Milan as well (7, 23, 25). If we can see past such conventional figures, the common elements and the differences appear more clearly. The Old Roman version differs from the other two by ending the A phrase on *a*, descending to *G* at 9, and starting the D and F phrases on *c*. The Milanese version stands alone at 12 when it goes for *F* instead of *c*. To this extent the Milanese and Gregorian versions form the closer pair.

It hardly needs saying that such melodic analysis must be carried out on a wide and systematic basis. Arguing from single pieces is of little value. But common sense tells us in this and many other cases that we are dealing with simple and ornate versions of the same melody. Since it seems likely that Milan took the introit over from the Gregorian rather than the Roman tradition, we have here a case of a relatively simple melody being transformed when sung in a more decorative idiom. If we believe that the Franks heard the piece in something like the Old Roman version given here, we might consider their Gregorian version to be a less accomplished, 'ersatz' version of something they found rather difficult to manage in its full splendour. There is, however, another possibility. Both the Milanese and Old Roman versions are recorded in sources no older than the eleventh century. Should we reckon with a

Ex. VIII.4.4. Gregorian, Old Roman, and Ambrosian versions of introit/ingressa *Invocavit me* (Montpellier, Faculté de Médecine H. 159, p. 86; Rome, Bibl. Ap. Vat. Vat. lat. 5319, fo. 41ᵛ; London, Brit. Lib. Add. 34209, p. 147)

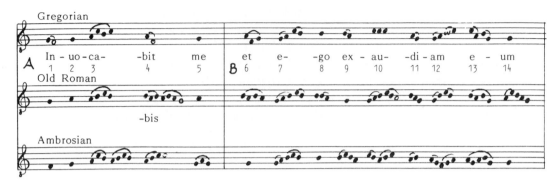

process of accretion of decorative detail between the ninth and eleventh centuries? Did the Romans sing something like the Gregorian version to the Franks, then through further embellishment develop the ornate singing tradition represented by the Old Roman copy of the eleventh century? How old is the ornate style of singing recorded in the Old Roman and Milanese chant books?

Comparison of different Milanese manuscript sources does occasionally reveal small differences of detail, extra passing notes, an extra flourish, the extra repetition of a phrase in a melisma. But the overall level of decoration is hardly altered thereby. Of course, the creation of a written tradition in Milan may well have stopped further embellishment (in the written sources themselves: what was sung may have been another story). But the impression one gains is that the elaboration is not particularly recent. The same is true of Old Roman chant. The differences between the sources are different versions within the same musical style.

Consideration of the third surviving Old Italian chant repertory, the Old Beneventan, also suggests that the high level of decorative detail is not a modern accretion but part and parcel of a general conception of how chant in general should be sung, a conception prevalent in Italy but seemingly foreign to the Franks.

VIII.5. OLD ITALIAN CHANT III: BENEVENTO

Andoyer 1911–21; Hesbert 1938–47; Bailey 1983; Huglo 1985, 'Bénéventain'; Kelly 1985, 'Montecassino', 1987, 1989; PalMus 14, 15, 20.

Despite the modest amount of music surviving from the Old Beneventan repertory, it has been well served by scholarship, principally by Hesbert (especially 1938–47 and in PalMus 14) and recently by Kelly, who has now provided a comprehensive survey of all aspects of the repertory (1989). A set of facsimiles will shortly appear as PalMus 21.

Of the more than eighty sources containing pieces of Old Beneventan chant, none is a complete chant-book for this repertory, and very many have but a handful of items or duplicate each other. The repertory therefore survives in a very incomplete state. It suffered the same fate as all others in Western Europe save the Milanese, giving way to Gregorian chant by the twelfth century, though not before it was recorded in diastematic notation. Much of it survives in the form of items of special local interest or veneration in the midst of Gregorian chant-books. The two principal sources are Benevento, Archivio Capitolare 38 and 40, in both of which a number of Old Beneventan masses (eight and thirteen respectively) are simply copied after the corresponding Gregorian ones.

Much about the Beneventan liturgy remains unknown, but in some respects—chiefly affecting the arrangement of prayers and lessons of mass—it differs from the Roman, often agreeing with Milanese use instead. For example, the introit chant is called the 'ingressa' and has no psalm verse, as at Milan. There are points of musical correspondence with Milan as well, and Beneventan musicians referred to their chant as 'Ambrosian'. It has been suggested that these reflect a period of closer cultural contact between Milan and Benevento, when both were under Lombard domination, although the evidence is not strong enough to speak of a common 'Lombard' liturgical use uniting Milan and south Italy. The Lombard invaders of Italy of the sixth century, gradually converted to Catholicism during the seventh, did not at first conquer Ravenna (Byzantine) or Rome itself. Soon after the Lombard seizure of Ravenna in 751, Charlemagne added the whole of north and central Italy to the Frankish kingdom in 773–4, leaving Benevento as an independent duchy. This survived in one form or another, in the teeth of both Frankish and Byzantine incursions, until its extinction in the mid-eleventh century at the hands of the Normans. As well as Benevento itself, especially the ducal church of Santa Sofia, Montecassino was an important centre of Old Beneventan practice.

The contrast, even conflict, between 'Ambrosian' (Beneventan) and Gregorian chant is illustrated in a poem preserved in Montecassino, Archivio della Badia 318 (Amelli 1913, Cattaneo 1950, 23–6). Charlemagne has ordered that 'Roman' chant be sung in every church. A trial is arranged between two choirboys, singing Roman and Ambrosian chant respectively. After a while the boy singing Ambrosian falls while the other sings on.

Altogether Kelly (1988, 164–5) was able to list sixty-four texts found at least partially in both the Old Beneventan and at least one of the Gregorian, Old Roman, and Milanese repertories. The Milanese repertory, with forty-four concordances, stands the closest, while thirty-nine texts are shared with the Gregorian and thirty-two with the Old Roman. Convincing musical concordances with Old Beneventan chants are fewer: twenty-seven Milanese but only six Gregorian and the same six Old Roman. There is strikingly little uniformity of liturgical assignment, only six items.

Some aspects of Old Beneventan practice stand in contradiction to Roman and Milanese; for example, the Old Beneventan offertories are short, simple pieces without verses (there is some crossing of liturgical boundaries with communions and office antiphons). Then again, in some instances we appear to be dealing with a general Old Italian practice which contradicts Gregorian; for example, the same melody is used for practically all Old Beneventan alleluias.

Like some lengthy Milanese antiphons (an example of a transitorium was given above, Ex. VIII.4.1), some Old Beneventan pieces are composed by the simple repetition of a melodic idea. Kelly gives as examples the especially long Easter ingressa and communion (112–14, 128–9), also three ingressae which share melodic material (117–18). There are seven melodic phrases (A–G) in all, employed in the three pieces as follows:

```
                   ABCDEFG   ABCDEFG   ABCDEFG   ABCDEFG   ABCDEFG
Petrus dormiebat   * * ***   * ** **   * ** **
Surge propera      * * ***   * *    *    *          *      *       * ***
Gaudeamus omnes    **  ***   * *      *      *          *
```

The last phrase ('G') is the most consistently used.

Kelly has discussed in satisfying detail numerous items where the Old Italian and Gregorian repertories have the same basic melody in different stylistic manifestations. If in need of a blunt generalization to counterpoint Kelly's fine discrimination, one might fairly say that the Old Beneventan versions are at least as ornate as any other, and often more ornate. They support the broad hypothesis that Old Italian chant favoured a more elaborate style than is found in the Gregorian tradition. What is more, the Beneventan melodies have an older manuscript tradition than the others. Rome, Biblioteca Apostolica Vaticana, Vat. lat. 10673 (PalMus 14) and Benevento, Archivio Capitolare 33 (PalMus 20), which both contain chants of the Old Beneventan Holy Week liturgy in significant quantities, both date from the early eleventh century.

One example must suffice to show two Old Italian ways of embellishing what is basically a simple recitation around the semitone step. The communion chant *Hymnum canite* is common to the Milanese and the Old Beneventan traditions but is not found in the Roman. In Ex. VIII.5.1 it is transposed up a fifth from the edition in

Ex. VIII.5.1. Ambrosian and Old Beneventan versions of communion *Hymnum canite* (after *Antiphonale . . . Mediolanensis*, 225; Benevento, Arch. Cap. 38, fo. 47ʳ)

Suñol's Milanese *Antiphonale*; the source of the Beneventan version, Benevento 38, has no clefs, so its pitches are also chosen to match it with the common Beneventan recitation pattern (see Kelly 1989, 107 for an extract and other examples). The alleluias at the end of the Old Beneventan piece are not part of the recitation pattern but a typical ending attached to several chants in the Easter season (for other examples see Kelly 1989, 129 and the table on 115).

VIII.6. GALLICAN CHANT

Leclercq, 'Gallicane (liturgie)', *DACL*; Gastoué 1939 (1937–9); Porter 1958; Stäblein, 'Gallikanische Liturgie', *MGG*; Huglo in Fellerer 1972 219–33, 'Gallican Rite, Music of the', *NG*.

The development of liturgies in Gaul was obviously affected by the invasion of Germanic tribes in the fifth century, that is, at a time when forms of public worship all over Christendom were taking on their varied shapes. Visigothic hegemony in the early fifth century was gradually replaced by that of the Franks, who extended their control as far as the Pyrenees in the early sixth century. Their king Clovis was baptized in Reims on Christmas Day 497 or 498. By the mid-sixth century they already ruled an area equivalent to modern France and much of West Germany. Throughout this period liturgical practices developed largely, though not completely, in isolation from Rome, to a certain extent in parallel with those of the Visigothic kingdom in Spain. The principal similarities and differences between the various rites, as far as they affect chants, have been summarized above (VI.4–5). The subsequent adoption of Roman use under Pippin and Charlemagne has also already been outlined (VII.2).

Although we are used to the concept of 'a rite', uniform in form and content wherever it was observed, it is inappropriate for the church in Gaul before the adoption of the Roman rite in the Frankish kingdom. The surviving service-books, rules, synodal prescriptions, descriptions such as the *Expositio antiquae liturgiae gallicanae* (ed. Ratcliff 1971), and passing comments on the liturgy such as those in the *Historia Francorum* by Bishop Gregory of Tours (d. 594) make it clear that there were many variations within a general pattern.

A handful of early Gallican service-books have survived (see Vogel 1986, 108, 275–7, 320–6) and these reflect the diversity of usage. Thus the lectionary in Paris, Bibliothèque Nationale, lat. 10863, sixth to seventh century, and the 'Lectionary of Luxeuil', Paris, Bibliothèque Nationale, lat. 9427, seventh to eighth century, have the same system of three readings at mass, but almost never the same choice of lessons. The sets of prayers for mass first published by F. J. Mone (1850), then edited by Mohlberg *et al.* (1958), differ considerably from those of the so-called 'Missale Gothicum' (ed. Mohlberg 1961), though both follow the same order of service.

By exceptional good fortune some of the surviving books include chant texts:

sixteen for chants after the lessons in Wolfenbüttel, Herzog-August-Bibliothek 4160 (Weissenburg 76), a lectionary of around 500 from southern Gaul (ed. Dold 1936; see also Morin 1937), and the respond texts marked in the so-called 'Psalter of Saint-Germain-des-Prés', Paris, Bibliothèque Nationale, lat. 11947, of the sixth century (Huglo 1982, 'Répons-Graduel'). What may well be Gallican chant texts have also survived alongside Celtic material, for example in the so-called 'Bangor Antiphonary' of the late seventh century (ed. Warren 1893–5; see Curran 1984). The fragments edited by Morin (1905) may be Celtic or Gallican. Several items, distinctive by virtue of their enthusiastic literary style, appear to have survived in Old Spanish or Milanese books. The greater part of what survives with music is, however, to be found in Gregorian chant-books from France. For one reason or another, items from the older indigenous use were preserved alongside the newer Roman repertory.

The remains of the chant repertory of the Gallican rite are so meagre that it is practically impossible to generalize about its musical style. After the pioneering work of Gastoué (1937–9), comprehensive summaries of our present knowledge have been made by Stäblein and Huglo.

A further difficulty exists in that the Gallican chants which did survive in late copies may have been edited to make them conform to the 'Gregorian' style. Abbot Hilduin of Saint-Denis, writing to the emperor Louis the Pious (MGH (Epistolae) *Epistolae Karolini aevi*, iii. 330), said that the office of St Denis included Gallican chants and therefore had to be 'Gregorianized'. When we inspect the music of those chants which seem to be of Gallican origin, it is often difficult, if not impossible, to say what is specifically Gallican about it. That it did indeed sound different in some way is stated by Walahfrid Strabo in *De rebus ecclesiasticis*, ch. 22, written in the 820s (PL 114. 946). He says it can be distinguished by both its texts and its music ('verbis et sono'). He mentions Rome's 'more perfect science of chant' and the 'inelegance' and 'lack of correctness' in Gallican use. The stylistic contrast between the language of Roman and Gallican prayer texts has often been discussed: the concision of the Roman as against the flowery and colourful Gallican. It is naturally more difficult to translate Walahfrid's remarks into concrete musical terms. In practice one has to look for texts with Gallican hallmarks, preferably with a concordance in another non-Roman use, and musical form or melodic style which differs from Gregorian practice. Analytical techniques are not yet well enough developed for distinctions always to be possible. The so-called 'Gallican cadence' (*CDD*, *CEE*, or *DEE*, *FGG*, *Gaa*, etc., often notated with a pes stratus) is not a sure indicator, for although it is not found in the layer of chant taken from Roman use, that is Gregorian chant, it was used very widely in items composed by the Franks from the eighth century onward.

An example of a possible Gallican item is the Palm Sunday processional antiphon *Cum audisset*. This includes a sentence, 'Quantus est iste', which recurs in an antiphon, *Curvati sunt*, known from the Old Spanish and Milanese liturgies. Since *Cum audisset* is not found in Roman books, the probability exists that it is a Gallican piece, like several other processional antiphons (cf. II.9). The recurrence of a number of openings and cadential figures is notable (Ex. VIII.6.1: the scribe has given an alternative text on lines 5–6, presumably not for singing).

The majority of surviving Gallican chants are to be found for occasions when Roman use provided no model. Chants for the votive masses which were added by the Franks to the Roman mass-books may provide the occasion, as in the introit *Prosperum iter faciat* (ed. Stäblein in *MGG*), found in the Aquitanian gradual Paris, Bibliothèque Nationale, lat. 776 (11th c.) for the mass for those setting out on a journey. Another group is constituted by the 'preces', originally a litany at the end of the first part of the Gallican mass, reassigned in Gregorian books to the Minor Litanies of the days before Ascension. This custom of singing litanies in procession was taken over by the Roman church from the Gallican. Huglo (*NG*) has listed twenty-four possibly Gallican items; examples have been edited by Stäblein (*MGG*) and Huglo (1955, 'Preces'). The relationship between these litanies and the Kyries with Latin verses has already been discussed (II.17).

On the other hand, not all additions to the imported Roman repertory can be assumed to be Gallican. Gastoué (1938, 10–11), Stäblein (*MGG*, 1307) and Wright (1989, 47–8) thought that the E-mode melody for the Gradual *Angelis suis* must be Gallican. While the Franks may well have added this gradual to the mass on Quadragesima Sunday (see *AMS*, L–LII), they simply used the usual Gregorian formulas for E-mode graduals.

Some Gallican items appear to have been preserved as an enrichment of Roman use. The Exultet and hymn *Inventor rutili* for the Easter Eve liturgy are examples of this. Others turn up in the liturgies for local French patron saints: St Denis, St Maurice, St Remigius. For example, the mass for St Dagobert, Merovingian king of the Franks, preserved in books from Saint-Denis, includes an antiphon 'ante evangelium', an item of the Gallican mass but not the Roman mass (facsimile and transcription by Stäblein in *MGG*). The presumption is that the music is Gallican, though the precise ways in which it differs from, say, many longer office antiphons in Gregorian books have not yet been distinguished sufficiently clearly.

An important group of possibly Gallican pieces is being uncovered in the offertory repertory. The Gallican sonus accompanied a procession of some splendour, apparently modelled on Byzantine use. Chiefly on grounds of musical style, Gastoué had already believed the offertory *Memor sit dominus* to be Gallican (ed. Stäblein in *MGG*). This piece appears without verse in the Old Roman tradition. The non-Roman offertory *Elegerunt apostoli* survived in Gregorian books for St Stephen's Day in the face of competition from the Roman *In virtute*. On grounds of textual and in part musical concordances with Old Spanish and Milanese chants, Levy (1984) has added more offertories to this group. In most of these cases there are versions in the Old Roman repertory. Acceptance of a non-Roman origin appears to mean that their musical style was 'Romanized' to make the Old Roman versions we know.

Ex. VIII.6.2 gives part of the offertory *Sanctificavit Moyses* (see also the discussion by Baroffio 1967) in Gregorian, Old Spanish (with staffless notation only),

Ex. VIII.6.1. Processional antiphon *Cum audisset populus* (Paris, Bibl. Nat. lat 776, fo. 54ᵛ)

Ambrosian, and Old Roman versions. The text is a free compilation of parts of the book of Exodus. The Gregorian and Old Spanish versions have the respond *Sanctificavit*; after the verses there is a repeat from 'Fecit sacrificium', introduced by a special new phrase 'Tunc Moyses'; there follow the (very long) verses *Locutus est* and *Oravit Moyses* (Ott 1935, 114; Montpellier, Faculté de Médecine H. 159, ed. Hansen, no. 918). The Old Roman version in Rome, Biblioteca Apostolica Vaticana, Vat. lat. 5319 (MMMA 2, 350) splits the verses into four verses each, eight in all. Only the last phrase of the respond 'In conspectu' is given for the repeat. In Milan the first part of the verse *Oravit* is the respond, the verse being formed by the rest of *Oravit* plus what was the repeat in the Gregorian version, 'Tunc Moyses fecit sacrificium'.

If the offertory had come to Rome from elsewhere, one would have expected the Roman musicians to apply their usual formulaic method for providing music (see VIII.3). To some extent in the verses they did so, and for these parts of the verses there is no relation with the Gregorian version. The Milanese version is generally not close to the Gregorian (Baroffio 1967 interprets it as using a basic melody several times over in varied fashion). In so far as the Old Spanish neumes can be properly understood, they appear to indicate a version much closer to the Gregorian, at least in places. Ex. VIII.6.2 shows the close of the second verse, where both Gregorian and Old Spanish versions have a melisma in AAB form (the Old Spanish actually AABB'BB'CCD). The Old Roman version uses the end of a recitation formula for 'Quia ego sum Deus' and the full formula for the last phrase, perhaps evidence of Roman moderation as opposed to Gallican rhetorical display.

Ex. VIII.6.2. Part of offertory *Sanctificavit Moyses* in Old Spanish, Gregorian, Ambrosian, and Old Roman versions (*Antifonario visigotico*, fo. 305ʳ; Montpellier, Faculté de Médecine H. 159, p. 256; *Antiphonale . . . Mediolanensis*, 333; Rome, Bibl. Ap. Vat. Vat. lat. 5319, fo. 130ᵛ)

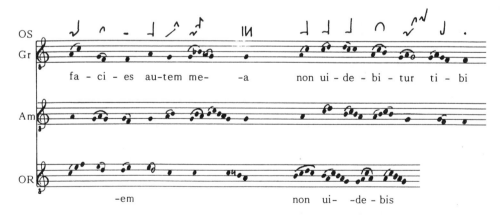

VIII.7. OLD SPANISH (MOZARABIC) CHANT

Liturgy: W. C. Bishop 1924; Prado 1927, 1928; Cabrol, 'Mozarabe (La liturgie)', *DACL*; King 1957, *Primatial*; Rivera Recio 1965.

Chant: Wagner 1928, 1930; Rojo and Prado 1929; Brou 1947, 'Psallendum' and 'Études', 1948, 1950, 1951, 'Alléluia', 1951, 'Séquences'; Husmann 1958–61; Huglo

1964; Pinell 1965; Brockett 1968; Randel 1969, 'Psalmody', 1969, *Psalm Tones*, 1973, 1977, 1985, 'Mozarabic Rite, Music of the', *NG*; Levy 1985.

    Facsimile: *Antifonario visigótico* (León, Archivo Catedral 8); *Antiphonale Hispaniae* (Saragossa).

I have called 'Old Spanish' the chant sung in what is now central and northern Spain and Portugal up until (and in some instances beyond) the imposition of the Roman liturgy and concomitant Gregorian chant in the eleventh century. It is often called 'Visigothic' or 'Hispanic', but mostly 'Mozarabic'. The Visigoths, who ruled most of south-west France in the fifth century before being ousted by the Franks, brought all of Spain under their control in the sixth. Our chief source of knowledge of the liturgy in the Visigothic kingdom is the *De ecclesiasticis officiis* of Isidore of Seville (d. 636), with additional information in his *Etymologiae* and the records of the Fourth Council of Toledo (633) held under his leadership. In 711 the Visigothic kingdom, with the exception of the extreme north-west, was conquered by the Omayyad Arabs. Christian worship nevertheless continued; the word 'Mozarabic' refers to Christians living under Muslim rule. Although a 'Spanish mark' was established under Charlemagne, the Christian reconquest of the peninsula did not get under way until the eleventh century, a major landmark being the conquest of Toledo in 1085. Northern liturgical practices followed northern arms into Spain, not least through Cluniac influence over the monastic houses of northern Spain. Pope Alexander II (1061–73) took a strong interest in the reconquest, and his agent Hugo Candidus was energetic in the replacement of the Old Spanish (Mozarabic) rite. Alexander's policy was continued by Gregory VII (1073–85), prime advocate of papal supremacy. Bernard, appointed archbishop of the newly regained Toledo, was French: he had been abbot of Cluniac Sahagun. Some churches in Toledo were nevertheless permitted to continue with the old rite. (As usual, reports of various legendary 'trials' of the rival rites—in this case by duel and by fire—have come down to us: see King 1957, *Primatial*, 510.)

How much of the chant for this rite survived thenceforth in its early medieval form is not clear. In 1500 and 1502 respectively Cardinal Jiménez de Cisneros published a missal and breviary of a revived Old Spanish rite. The chants differ considerably from any known early sources. There was a further revival, with new chant-books, at the turn of the eighth century under Cardinal Lorenzana, and the Old Spanish rite is still celebrated in one of the chapels of Toledo Cathedral.

Over twenty major sources of Old Spanish chant exist (see Pinell 1965; forty-six sources altogether are listed by Randel 1973, though some contain only one or two pieces). Yet barely two dozen melodies were copied in the transcribable Aquitanian notation imported from the north (written over erasures in Madrid, Academia de la Historia, Aemil. 56). The significance of the various types of Spanish neumes is generally clear, but the pitches of the chants are irrecoverably lost. Despite these obstacles, a surprising amount is known about the repertory, chiefly through the work of Brou and of Randel, whose catalogue of 1973 is a complete list of all known chants, indexed according to liturgical tradition and genre. Randel has also completed penetrating studies of the office responsories and psalm tones. This brief account is chiefly based on his writings.

Like the Gallican rite (in so far as it can be known in any detail), the Old Spanish rite was host to a variety of liturgical traditions, of which four are generally distinguished. They observed the same structure of office and mass but often choose different chants or, in the case of concordances, have different musical variants. One tradition centres on León, one on the upper Ebro valley, the area known as the Rioja, and northern Castile, including the abbey of Santo Domingo de Silos. The Rioja tradition, albeit with variants, was also known in Toledo, where it is known as Toledo 'tradition A'. The Toledan 'tradition B' is restricted to only three sources, but it is this that provided the basis of Cardinal Cisneros's revision in the fifteenth to sixteenth century.

As well as the liturgical differences, two different styles of notation were used, the one upright, used in León and Castile, the other heavily slanted, used in Toledo and perhaps some other centres (see IV.2). The Toledan notation can itself be divided into two traditions. One employs more rounded forms, and was used for the liturgical tradition 'A'. More jagged shapes were used in conjunction with tradition 'B'.

Apart from the documents of St Isidore's time, two early service-books of the period before 711 have survived, a 'Liber ordinum' containing the order of service (ed. Férotin 1904), and the collection of office prayers known as the Verona 'Orationale' (Verona, Biblioteca Capitolare LXXXIX, ed. Vives 1946), particularly important because chant texts are cued in the margin. Since most of these chants appear in later sources, it is assumed that many of the texts of the liturgy were already established before the Moslem conquest. The earliest books with notation are usually dated in the tenth century, but most come from the eleventh. The early books combine office and mass formularies, like the Milanese books. One of the earliest, from León, has been published in facsimile (*Antifonario visigótico*). It contains two prologues, the second of which describe the practice of responsorial and antiphonal psalmody (see Randel 1969, 'Psalmody').

In two important respects Old Spanish chant can be shown to ally itself with the other non-Gregorian repertories and against Gregorian practice. Randel (1977, 1985) has shown that for the antiphonal psalmody of mass and office (the two seem indistinguishable) four tones account for the great majority of surviving examples (admittedly meagre in number). His study of responsory tones in the various traditions also demonstrates that the Gregorian system of eight modes and eight tones did not exist. The León tradition has seven, with four being particularly common, and the Rioja tradition also knew these. Toledo tradition A, although apparently transmitting the same responsory melodies, had a different system of assigning verse tones, and the tones themselves are different, two being particularly common. Toledo tradition B has four tones, one common.

Randel has also discerned the presence of standard phrases in responsorial chants, in the vespertini at the start of Vespers. Type-melodies, on the other hand, underlie the repertory of threni in Lent.

Several different types of chant sometimes include long melismas with repeat structure: the soni of the morning and evening services, the clamor, laudes (alleluia), and sacrificium (offertory) of mass (facsimiles and copies in Brou 1951, 'Alléluia'; see also Husmann 1958–61).

It is frustrating that the two chant repertories which were presumably closest to what the Franks knew before the adoption of Roman practice, their own Gallican chant and that of the sister liturgy in Spain, should both be inaccessible to us. Possible direct links between both of them and the Gregorian repertory were mentioned in the last section, where Gallican offertories were discussed. There is one other way, however, in which the Old Spanish chant repertory might shed light on the origins of Gregorian chant. I have drawn attention to the generally ornate style of the Old Italian chant traditions, which sets them apart from the Gregorian. In this respect Old Spanish chant is more akin to Gregorian. Perhaps that is another small indication that Gregorian chant is the product of non-Italian musicians.

VIII.8. CONCLUSIONS

van Dijk 1961, 1963, 'Papal, 1963, 'Gregory', 1966; Stäblein in MMMA 2; Hucke 1955, 'Gregorianischer', 1956, 'Gregorian and Old Roman Chant', *NG*, 1980, 'View'; Nowacki 1985.

In the last section it was again remarked that a system of eight modes and eight tones which complement them for the singing of office psalms and responsory verses was unique in the West to Gregorian chant. The eight-tone arrangement, and the smaller number of psalm-tone endings by comparison with Old Roman chant, have suggested to Dyer a connection with the change from solo to choral psalm-singing, a novelty of Carolingian practice. This is one among many radical changes brought about by the Frankish church. Others are the obligation of all clerics to celebrate the full round of office hours, the filling out of the Temporale and Sanctorale, and the enlargement of the chant repertory with sequences and tropes. The need for written records to support the huge edifice thus constructed led to the compilation of tonaries and the use of musical notation. All this is Frankish innovation.

It still remains, however, to try and sum up the relationship between the two chants for the Roman liturgy, that is Gregorian and Old Roman chant. On the one hand, the repertory codified by the Franks at the end of the ninth century (earlier according to some) is the Gregorian. On the other hand, the only chant-books written in Rome itself, from the eleventh to the thirteenth century, have Old Roman chant.

Several scholars (notably Smits van Waesberghe, Jammers, and Stäblein) have argued that Gregorian chant was created in Rome itself and coexisted there with Old

Roman. The liturgist Stephan van Dijk argued in a series of articles that Gregorian chant was that sung by the papal schola for the special liturgical use of the papal chapel. Old Roman chant was that of the city churches. But, while the special practice of the papal chapel is well documented, it is by no means clear that a different way of singing should have been created to accompany it. While some chants would be unique to one liturgical use or another, the great majority of chants required would be common to all churches of Rome. We are speaking of variants upon the one Roman liturgy, not differences of the sort that divide Roman from Milanese or Old Spanish use. Nor does it seem likely, on the face of it, that the Gregorian repertory, which is rather sober and simple when compared with Old Roman, should have been created as part of an attempt to imitate 'the majestic court ceremonial of Byzantium', as van Dijk put it (1966, 302). The example of Benevento is important here. Old Beneventan chant and Gregorian did indeed exist side by side, but as representatives of two different rites, the aboriginal Beneventan and the invading Roman respectively. Given the eventual dominance of the Roman rite, Old Beneventan chant was bound to disappear.

Much work has still to be done to understand the musical relationships between Gregorian and Old Roman chant. This has to proceed at two levels. Where there is no demonstrable musical relationship, as for example in many offertories, one needs to ask why this should be so, which involves the history of the liturgical practice of that genre of chant and the particular service of which it is part. Where a musical similarity exists, what is common and what is dissimilar has to be explained, involving minute analysis of idiom and form. In VIII.3 above a number of instances were cited or exemplified, where the same musical idea was realized in different ways, each with its own characteristic turns of phrase. Hucke (*NG*), speaking of graduals, and seeing the Gregorian versions as 'consistent adaptations of the Old Roman ones', has called this 'a translation from a foreign musical language'. Behind this lies the conviction that two different peoples were concerned, the Franks and the Romans, not two groups of musicians within Rome itself. By extension, one should then ask into which language was Roman translated? (To which the only answer is a type of Gallican chant). The evidence to prove or disprove this thesis is, however, lacking. While analysis can establish what constitutes a Gregorian idiom, parallel Gallican examples are lacking, and it seems unlikely that Old Spanish chant will provide an adequate substitute.

Since the codification of the Gregorian repertory preceded that of the Old Roman by some 200 years, it was possible for the Gregorian tradition to influence the Old Roman at the time of its writing down. There are signs that this happened. Not only were Gregorian pieces taken up in Rome itself (for example, the cantica of Holy Saturday, numerous alleluias, not to mention Frankish sequences, alleluia prosulas, ordinary-of-mass chants and the like). There are peculiarities in the readings of some Old Roman chants which suggest 'contamination' by Gregorian versions (Hucke 1980, 'Offertorien'), as if the Old Roman notator had a Gregorian chant-book to hand. (As Hucke points out, notation must have first become known in Rome through the medium of Gregorian books.)

Given the evidence of a continuing oral tradition in Rome even past the time of the earliest notated books, it seems to me impossible that they should represent in all details the melodies sung to the Franks in the eighth century. Since Beneventan and Milanese chant is generally as ornate in idiom as Old Roman, I think it highly likely that the Roman chant of the eighth century already had an ornate idiom, in other words, that Old Roman chant preserves the spirit, if not always the letter, of the eighth-century state. To this extent, the oft-quoted remarks of Johannes Hymmonides (John the Deacon: PL 75. 90, Stäblein in MMMA 2 143*, van Doren 1925, 53) carry conviction: the clumsy Frankish singers were too inexpert to manage the fine detail of the Roman melodies. The theme was later taken up by Adhémar of Chabannes (d. 1034) who refers to the 'tremulas vel vinnolas sive collisibiles vel secabiles voces' of the Romans, in so far as one can understand these terms (*Chronicon*, II. 8, ed. Chavanon 1897, 81; cited van Doren 1925, 53, Hucke 1954, 'Einführung', 181). One would not call the Gregorian melodies clumsy; but it seems conceivable that the Franks sang the Roman chants in a more straightforward idiom than the Romans themselves, something perhaps more akin to that of their native traditions. That is a matter of surface detail. As regards learning the standard formulas, type-melodies and the methods of using them, the Frankish singers were evidently very successful.

IX

Persons and Places

IX.1. INTRODUCTION

This chapter will inevitably seem incomplete to those seeking a full and balanced account of the history of plainchant in all the various churches of Western Europe during the Middle Ages. At present, however, it is not possible to write such an account, for investigation of the various chant traditions has so far proceeded very unevenly, and full coverage would in any case exceed the proper limits of this book. I have therefore ventured a few remarks only on those persons and places, those musicians, churchmen, and churches, that have attracted particular attention so far, or that I myself have found particularly interesting. The purpose of the chapter is partly to point readers in the direction of information they may be seeking, but partly also to give an idea of the sort of research which has been carried out, its aims and methods.

Individual creation, original composition, calls for individual appreciation. But before one can decide what is individual in a particular manuscript, a great deal of patient comparison of sources has to be carried out. Scholars have developed a number of short cuts, as it were, to facilitate comparisons and highlight similarities and differences between sources, and some of these are repeatedly mentioned in what follows. Some are repertorial: the choice of alleluias at mass for the Sundays after Pentecost and of responsories for the Sundays of Advent; others concern musical differences, variants in the detail of melodies (these methods are outlined briefly above, III.15). Such comparisons are naturally only a first step in the evaluation of sources, a means of finding one's bearings in an ocean of evidence, on a veritable mountain of manuscripts. Yet, for many sources little progress beyond these first steps has been made. I hope, therefore, that the reader will forgive the piecemeal approach and gain some benefit from these brief visits to some of the more interesting corners of chant history in the West.

From time to time readers may well find themselves asking, what was chant like at St Gall, at Cluny, at Montecassino? The answer is that in most respects it was like chant anywhere else. Rarely does the repertory of a church stand out by reason of items which are unique in style, or even items which are unique but similar to others. And the small differences in detail which help distinguish one manuscript from another are of bibliographical rather than purely musical interest. No doubt, if one had been able to visit the places in question, one would have been able to distinguish the singing of one medieval schola from another. But it would probably be by virtue of aspects of performance such as voice production, tempo, and dynamic, none of which is reflected in the manuscript sources that have come down to us.

The first section in this part of the historical survey is about Amalarius and Metz in the second quarter of the ninth century, which continues the story from Ch. VII and illuminates some of the consequences of the Carolingian settlement. St Gall (IX.3), Benevento and Montecassino (IX.8), and Limoges (IX.10) are mentioned as three examples of places which made particularly significant contributions to the chant repertory.

These sections concern important historical developments in the chant repertory. But William of Dijon's bringing of chant to a previously devastated Normandy (IX.5), and the chapters on chant in England, Spain, and north and east European countries (IX.6, 11–12), concern the implantation of a previously formed body of music in new territory. Only in the case of Normandy and England have I entered into detail, in order to explain a type of development which was repeated, with local variations, in many other areas. For the last two sections I have limited myself largely to mentioning recent research.

The bringing of a chant repertory to any of these new lands mirrors the Frankish importation of Roman chant. There was a political dimension to almost all these enterprises, in that the local ruler practically always took the initiative in deciding who should come and initiate, restore, or reform worship in his domain. The initiative of Pippin and Charlemagne is echoed later by Richard of Normandy, who brought William north from Dijon, by Edgar of England, and so on.

It may surprise some readers that some obviously important centres such as Chartres and Paris, or Weingarten and Regensburg, have not been singled out for special mention. The liturgy and chant of Paris in the twelfth to thirteenth centuries, for example, has been the subject of much intensive study (most recently Wright 1989), largely because of the polyphony sung there. Such studies could be and in some cases have been made for dozens of churches across Europe. There is, alas, simply no space to report them all here, though I have indicated the more important editions of liturgical books or particular repertories where possible in previous chapters.

Map IX.1 shows the archdioceses of Latin Christianity towards the end of the Middle Ages. Because of numerous reorganizations of church administration it is strictly accurate for no particular date. To some extent the boundaries reflect the national divisions of the early Middle Ages. Thus the archdiocese of Lund included Denmark and what is now southern Sweden, like the medieval kingdom of Denmark. The diocese of Riga is a product of the settlements of the German orders of knights. The Holy Roman or German Empire included in the west the dioceses of Cologne, Trier, Besançon, Vienne, and Arles, in the east Magdeburg, Prague (for the kingdom of Bohemia), and Salzburg. Further east lay the Polish and Hungarian archdioceses. During the Middle Ages Latin Christianity was still expanding eastwards, and Spain

Map IX.1.1. The archdioceses of late medieval Europe

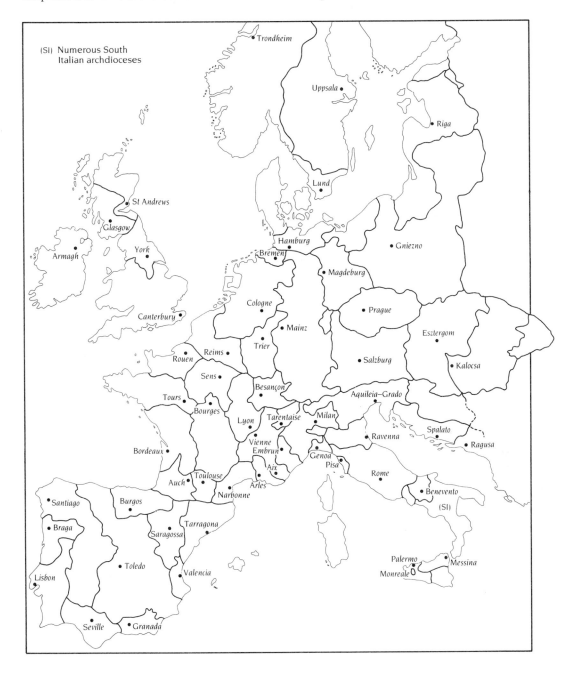

(SI) Numerous South Italian archdioceses

Trondheim

Uppsala

Riga

Lund

St Andrews

Glasgow

York

Armagh

Hamburg

Bremen

Gniezno

Magdeburg

Canterbury

Cologne

Prague

Mainz

Esztergom

Trier

Reims

Salzburg

Rouen

Kalocsa

Sens

Besançon

Aquileia–Grado

Tours

Milan

Bourges

Lyon

Tarentaise

Spalato

Vienne

Ravenna

Ragusa

Bordeaux

Embrun

Genoa

Pisa

Aix

Rome

Auch

Toulouse

Arles

Narbonne

Benevento

Santiago

Burgos

(SI)

Braga

Tarragona

Saragossa

Toledo

Valencia

Palermo

Messina

Lisbon

Monreale

Seville

Granada

was being reconquered from the Moors, so that the Portuguese diocese of Lisbon, and the Spanish ones of Santiago and Toledo, moved correspondingly south. The archdiocese of Seville was only established in 1248, that of Granada only in 1492.

Maps IX.2, 3, and 4 indicate some of the more important churches from which chant sources have survived, in Italy, Germany, and France, respectively. The maps are intended to provide some orientation during the rapid course of this chapter.

Map IX.1.2. Ecclesiastical map of Italy

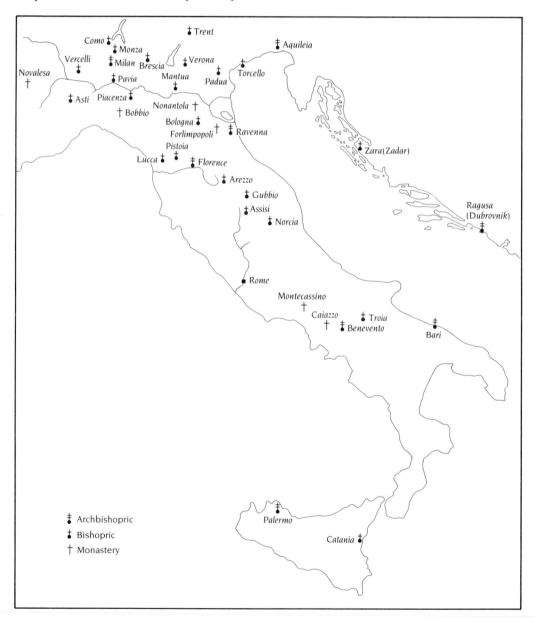

Map IX.1.3. Ecclesiastical map of Germany

Map IX.1.4. Ecclesiastical map of France

IX.2. AMALARIUS OF METZ

Hanssens 1948–50, Cabaniss 1954, Lipphardt 1965, Roger Evans 1977, Huglo 1979, 'Remaniements', Hesbert 1980, 'Amalar'.

Despite the efforts of Pippin and Charlemagne to achieve uniformity of practice, much of the detail of liturgical observance was never standardized. The Roman mass replaced the Gallican, the Benedictine and Roman forms of the office were established. But the texts and chants performed during these services could not be made the same everywhere. Roman exemplars were lacking, inadequate, or self-contradictory, and had to be supplemented for Frankish use. This can be documented for the numerous sacramentaries (collections of prayers for mass), and there is no reason to suppose that things were much different for chant texts. The earliest sources of chant texts for the proper of mass (those edited by Hesbert in *AMS*) indicate that the basic corpus was established in Charlemagne's time, if not earlier. The way differences could arise in the choice and arrangement of office texts is illustrated by the activities of Amalarius of Metz (*c*.775–*c*.850). As to the music for these texts, while we can marvel at the relative uniformity of the earliest preserved copies, there are numerous differences of detail between them.

Amalarius may have been educated under Alcuin at Tours or Aachen, and he is thought to have been archbishop of Trier between 809 and 814. He knew the Byzantine liturgy from an imperial mission to Constantinople in 813–14. He returned via Rome, and had another opportunity to learn Roman practice at first hand on a mission in 830–1. He may have been director of the palace school at Aachen in the 830s (Hanssens, i. 68 ff.). He is famous chiefly for two works on the liturgy. The *Liber officialis* (Hanssens, ii) explains the theological and spiritual significance of each part of the liturgy, texts through the church year, and how it is performed. Though its content is chiefly hermeneutical, it conveys a lot of detail about early ninth-century practice. The *Prologus de ordine antiphonarii* (*c*.831) and *Liber de ordine antiphonarii* (*c*.840) (Hanssens, i. 361–3, iii. 9–109 respectively; trans. Roger Evans 1977) refer to an antiphoner that Amalarius proposed to introduce, though its destination is unclear. Its contents are noted in some detail. Hanssens, in a magnificent appendix (iii. 110–224), sets out its contents, as far as they can be reconstructed, in parallel with the Old Roman antiphoner Rome, Biblioteca Apostolica Vaticana, San Pietro B. 79, one of the earliest Gregorian antiphoners, Paris, Bibliothèque Nationale, lat. 17436 (texts only), and one of the earliest notated antiphoners (St Gall, Stiftsbibliothek 390–391).

The differences between Amalarius and the other sources are very interesting and important, but it is the spirit behind them which interests us here. Amalarius was concerned, like Helisachar before him (see VII.2) and Aurelian after him (V.4), with the literal sense of responsories and their repeat structure. But he tells us that many series of pieces in Rome were simply given as sets from which a choice could be made (this is still the case in the ninth-century antiphoner Paris 17436). Amalarius would prefer a logical order, such as the order of verses as they appear in the Bible. And there are indeed many examples of such numerical ordering in the Gregorian gradual, though not necessarily due to Amalarius' prompting; it may be that he wished to apply to office chants a principle already well established at mass. He tells us that in his antiphoner he marks chants according to the usage he prefers: 'R' for Rome, 'M' for Metz and 'IC' for the 'indulgentia' and 'caritas' which he requests to be extended to his own individual choice.

Amalarius was appointed to the see of Lyon in 835 after its previous archbishop Agobard (814–35/838–40) had been exiled. But it was not difficult for Agobard and his archdeacon Florus to stir up trouble about Amalarius' writings, which were condemned at a synod in Quierzy in 838, whereupon Agobard regained his see. A justificatory letter of Agobard of 838, presented as a 'Praefatio antiphonarii ecclesiae Lugdunensis' in one source, generally called the 'Liber de correctione antiphonarii' (MGH *Epistolae Karolini aevi*, iii. 238; PL 104. 329–50), explains the reasons for Agobard's own revision of the antiphoner, the removal of what was 'superfluous, mendacious, or blasphemous'. Florus' writings (PL 119) also condemn Amalarius' innovations. Agobard's revision chiefly consisted in removing non-biblical texts. Huglo (1979, 'Remaniements') gives numerous examples of Agobard's revisions and others carried out in like spirit in the books of Lyon and neighbouring dioceses. Hymns were not sung in Lyon before the twelfth century, and Agobard may have inspired distrust of troping in some churches (Gy 1983). The Carthusians, too, subsequently eliminated non-scriptural texts from their liturgy. (In Plate 14, a Carthusian manuscript, the scriptural sources of chants are noted in the margin.)

In Rome, Amalarius was unable to obtain a Roman antiphoner for the Emperor Louis, for they had been given to Abbot Wala of Corbie (in 825). Travelling to Corbie, therefore, he was finally able to consult a 'responsoriale' bearing an attribution to Hadrian I (772–95) (Hanssens, i. 361, iii. 14).

Whatever the destination of Amalarius' antiphoner, it does not seem to have been taken up in Metz, in or near which he lived in the latter part of his life. This is clear from a comparison with the ninth-century tonary of Metz (Metz, Bibliothèque Municipale 351, ed. Lipphardt 1965; see 202–11, also Hesbert 1980, 'Amalar'). But the tonary does not agree with what Amalarius himself says was Metz practice ('M'). These complications are witness to a period of fluidity in liturgical repertories. We may surmise that the liturgical reforms of Benedict of Aniane (*c*.750–821), officially promulgated at the Council of Aachen in 817, occasioned both revision and codification, at least in monastic repertories. But their exact workings out in extant liturgical books is unclear.

Careful comparison of their repertories has enabled Lipphardt and Huglo to trace relatives of the Metz tonary in a number of later sources: from Reichenau (Bamberg, Staatsbibliothek lit. 5, of the year 1001), from Quedlinburg (Berlin, Staatsbibliothek, Mus. 40047, *c*.1020), and from Nonantola (Rome, Biblioteca Casanatense 54, early eleventh century). Lipphardt connects this with the effects of the monastic reform movement of Gorze, which lies a few miles west of Metz. Gorze was founded by Chrodegang of Metz, and revived after a period of decline by Bishop Adalbero of Metz (919–62). It and its associated house St Maximin at Trier became widely influential in the second half of the tenth century, introducing reforms of monastic observance into numerous houses at the behest of ecclesiastical or lay patrons. The reform came to Reichenau in the time of Abbot Ruodmann (972–85) in the person of Sandrat of St Maximin in Trier. The effect of such reforms as that of Gorze (see Hallinger 1950–1) and later Hirsau in the Black Forest (Jakobs 1961) upon monastic discipline is well documented, but it is fair to say that their consequences for the chant repertory have not been investigated in sufficient detail. The example of Cluny is discussed below.

IX.3. ST GALL

Schubiger 1858; Marxer 1908; van Doren 1925; Clark 1926; Scheiwiler 1938; von den Steinen 1948; Labhardt 1959–63.

It is unusual for a chant manuscript as old as Metz, Bibliothèque Municipale 351 to survive from a cathedral church. Most of the history of chant in the ninth and tenth centuries has to be reconstructed from monastic books. The monasteries were vital for securing the foundations of the Carolingian renaissance. The earliest surviving tonary, Paris, Bibliothèque Nationale, lat. 13159, is from Saint-Riquier; one of Saint-Riquier's abbots, Angilbert, was Charlemagne's son-in-law. A valuable description of the monastery, its library, and some of its liturgical ceremonial during Angilbert's time survives (ed. Lot 1894), which stops just short of providing the detail about performance practice useful for the musicologist (see Heitz 1963). Wala of Corbie, just mentioned as having carried off an antiphoner during a mission to Rome, was also related to the royal family. Corbie played a leading role in the development of Caroline minuscule script (the ancestor of the type used in this book). Saint-Denis, where Stephen II and his chapel stayed while visiting Pepin in 752, has also been mentioned (VIII.2.iv). Its abbot, Hilduin, responsible for the 'missa graeca', was Louis the Pious's archchancellor. Curiously, as well as the tonary from Saint-Riquier, we have graduals with tonal indications in early sources from both Corbie and Saint-Denis: Paris, Bibliothèque Nationale, lat. 12050, Laon, Bibliothèque Municipale 118, and the 'Mont-Renaud' manuscript, all of the late ninth or early tenth century (III.14.iii). None were originally notated, though the last was provided with notation around the end of the tenth century. The chant tradition of Corbie and Saint-Denis, as revealed both in its selection of chants and the melodic detail of those chants, was closely similar, and one would like to know if this reflects the period when both monasteries were so intimately linked to the Carolingian liturgical reforms.

One of the first and one of the only monasteries before the end of the millennium from which notated manuscripts have been preserved in any numbers is St Gall, south-east of Lake Constance. The earliest, the cantatorium St Gall, Stiftsbibliothek

359 (PalMus II/2), is usually dated around 900. Later in the century appear the gradual St Gall 339 (PalMus 1) and two sources of sequences and tropes, St Gall 484 and 381 (facsimile in preparation by Arlt and Rankin). If the dating of St Gall 359 is correct, the manuscript must have been known by Notker Balbulus ('the Stammerer', d. 912), the famous author of a life of Charlemagne and numerous sequence texts, and Tuotilo (d. 915), who is reputed to have composed tropes while playing on his rotta (a type of triangular psaltery).

Notker's account of the difficulties of Frankish singers learning Roman chant says that of the two Franks who learned in Rome itself, one went to Aachen, the other to Metz. Later musicians were evidently concerned to establish St Gall's claim to a share in the authentic Roman tradition, for Ekkehard IV (c.980–c.1060), in his *Casus Sancti Galli* (probably of the 1040s), says that while one singer did indeed go to Aachen, the other fell ill on the way and stayed at St Gall, teaching the monks from the Roman antiphoner he had with him, and instructing Notker in the meaning of the significative letters used in its notation (MGH *Scriptores rerum Sangallensium*, 102, cited van Doren 1925, 128). There survive several copies of a letter attributed to Notker, written to an otherwise unknown monk Lantpertus, which explains these letters (critical edition by Froger 1962.)

Although Ekkehard IV's chronology has been shown to be shaky, and a certain amount of legend has crept into his narrative, he paints a fascinating and charming picture of life and learning in the monastery, of the activities of Notker and Tuotilo, Ratpert and Hartmann, and Notker's teachers, Iso and the Irishman Moengal or Marcellus (MGH *Scriptores rerum Sangellensium*; the parts relating to music are reproduced by Gautier 1886; see also Schubiger, etc. above). Of the composers of the rich trope repertory in the St Gall manuscripts (the early St Gall 484 and 381, then 376, 378, 380, and 382 from the eleventh century), we know nothing, except for Ekkehard's ascription of two pieces to Tuotilo. This is regrettable, because the early tropers in particular contain many pieces not known elsewhere, including the collection of melismatic extensions to introits, often with prosula texts. On the other hand, the sequence texts which can be more or less reliably attributed to Notker, the famous *Liber hymnorum*, have been identified on stylistic grounds (von den Steinen 1948). The difficulty of matching the neumatic notation of the early sources with appropriate later copies in staff-notation means that no critical musical edition of Notker's book exists, though transcriptions made with reference to early French sources have been published by Crocker (1977). Schubiger edited many of the sequences from later German sources, including St Gall 546, a massive compilation of sequences, ordinary-of-mass chants, and some other items copied by the St Gall cantor Joachim Cuontz in preparation for the canonization of Notker in 1512, 600 years after his death. By this time several of Notker's pieces had fallen out of the repertory, and Cuontz did not know their melodies. (On the ordinary-of-mass chants see Marxer 1908, on the sequences see Labhardt 1959–63.)

No less important than its trope and sequence manuscripts are the St Gall books with proper of mass and office chants, in particular St Gall 359, 339, and the antiphoner compiled around 1000 by Hartker (d. 1011), St Gall 390–1 (PalMus II/1).

The latter is one of the earliest of all notated antiphoners, St Gall 359 one of earliest of all fully notated chant-books. Their importance was recognized by the chant reformers of the nineteenth century, and it was believed for a time that St Gall 359 was the very manuscript brought from Rome by Romanus, hence the publication of a facsimile of it by Lambillotte in 1855. The fine detail of their notation has made the early St Gall books particularly important for the investigation of chant performance. As to the liturgical and melodic tradition they represent, while neither they nor any other books can be shown to transmit any particularly authoritative archetype, Carolingian or other, they belong to a particularly closely knit group of German sources. This is clear from the Solesmes comparison of melodic variants in proper-of-mass chants (*Graduel romain* IV) and also Hesbert's work on Advent responsory series (*CAO* 5–6). It has yet to be shown, however, that St Gall was the *fons et origo* of these traditions in eastern lands.

In Table IX.3.1 I have summarized some of the results of the Solesmes survey. The method of investigation involved identifying 100 'lieux variants' in the proper chants of mass where sources tended to differ from each other. Agreements and disagreements between all the sources were then counted, and groups of sources in total or near agreement were identified. Within a limit of twenty-five disagreements out of 100 (75% agreement), the sources divided into two main blocs, eastern and western, with sources from Vercelli and Le Mans, respectively, forming two eccentric blocs. Within the eastern bloc, however, the great majority of sources were in agreement to within six differences (94% agreement), whereas at that level of agreement the western sources could only form mostly very small groups, representing a single church or religious order. The list omits all single sources which can also be assigned to the main blocs, and those for which insufficient readings were available.

The great majority of the readings on which manuscripts differ are hardly perceptible to modern ears. But there is a quite audible distinction in the treatment of the notes at the semitone steps in the diatonic scale. Where many southern and western sources (Aquitanian and Beneventan ones seem to be most consistent) have a reading such as *ded*, or *aba* at the top of a phrase (or *ab♭a*, if the scribe uses the flat sign at all), northern and especially eastern manuscripts prefer *dfd* and *aca*. Intimately related to this phenomenon is the question of which reciting note, *b* or *c*, is appropriate for the third psalm tone and recitation passages in compositions in modes three and four. The eastern propensity to choose the higher note is often referred to as constituting the 'German chant dialect', though it would seem more appropriate to call it a regional accent, since nothing in the basic vocabulary or grammar of the chants is affected. Numerous examples were given by Wagner (1930–2; see also Wagner 1925), and Heisler (1985, 1987) has recently considerably broadened the scope of the inquiry. (See also Stäblein, 'Deutschland. B. Mittelalter. I. Der römische Choral im Norden', *MGG*, and Agustoni 1987.) Gajard (1954) and Cardine (1954) demonstrated cogently that early St Gall manuscripts used *b*, not *c*, as the reciting note, and this would seem to imply that the preference for the higher note at the semitone step is a relatively later development.

Table IX.3.1. *Sources grouped according to their musical variants*
showing groups of sources with no more than 6 out of 100 disagreements
(number of sources in each group given in parentheses)

Western bloc	Eastern bloc
25 groups, 83 sources	5 groups, 98 sources

Western bloc	Eastern bloc
1. Saint-Denis (earlier), Corbie, Worcester, Downpatrick (5)	1. St Gall, Einsiedeln, Pfävers, Rheinau, Schaffhausen, Augsburg, St Georgen,
2. Saint-Denis (later), Saint-Corneille at Compiègne (3)	St Blasien, Weingarten, Zwiefalten, St Peter at Bantz, Minden, Bamberg,
3. Saint-Maur-des-Fossés (2)	St Emmeram at Regensburg, Prüfening,
4. Paris, Jerusalem (7)	Passau, Salzburg, Admont, Seckau,
5. Paris (another group) (2)	St Lambrecht, Moggio, Kremsmünster,
6. Cambrai (2)	Garsten, St Florian, Seittenstetten, Melk,
7. Arras (2)	Blaubeuern, Stockerau, Klosterneuburg
8. Anchin, Lille, Saint-Amand (3)	(90)
9. Andenne (2)	2. Seeon, Schäftlarn (2)
10. Sens (3)	3. Trier (2)
11. Chartres (2)	4. Monza (2)
12. Lyon (2)	5. Aquileia (2)
13. Autun (2)	
14. Saint-Bénigne at Dijon (2)	
15. Bec (2)	
16. Sarum (3)	
17. Toulouse, Albi, San Millán de la Cogolla (5)	
18. Valence (2)	
19. Ivrea or Pavia (2)	
20. Forlimpopoli, Ravenna (2)	
21. Piacenza (2)	
22. Montecassino, Benevento (11)	
23. Cistercian, Dominican (7)	
24. Carthusians (5)	
25. Franciscans (3)	

IX.4. CLUNY

General: Sackur 1892–4; Joan Evans 1931; Congress Cluny 1949; Congress Todi 1958; Hunt 1967; Wilmart 1970; Valous 1970; Cowdrey 1970.

Customs: PL 149. 633–778; CM 2; CCM 10.

Chant: Leroquais 1935; Hourlier 1951, 'Clunisienne', 1959; Huglo 1957; Renaudin 1972; Garand 1976; Huglo, 'Cluniac Monks', *NG*; Davril 1983; Huglo 1983; Steiner 1984, 'Music', 1987; Holder 1985; Lamothe and Constantine 1986; Hiley 1990.

Perhaps the most famous of all Benedictine monasteries, Cluny, near Mâcon in Burgundy, was founded in 909, long after such Merovingian foundations as Corbie, or Irish ones such as St Gall. It owed obedience to no diocesan authority but only to the pope. This arrangement, and the quality of its monastic observance, inspired other Benedictine houses to reform themselves after the Cluniac model. Others were obliged to do so by their lay or ecclesiastical patrons, or were founded as priories over which Cluny had authority. Under a series of outstanding abbots—Odo (927–42), Mayeul (948–94), Odilo (994–1049), Hugh (1049–1109), and Peter the Venerable (1122–57)—Cluny rose to a position of unprecedented ecclesiastical and temporal power, commanding the obedience of over 1,000 houses, mostly in Burgundy and the rest of France, but also in northern Spain, Italy, and England.

After the twelfth century its influence declined in the face of competition from the newly fashionable ascetic orders, especially the Cistercians, who were responsible for much intemperate polemic against what they saw as a betrayal of the simple ideals of early monasticism. Among other things, the complexity and length of the liturgy was criticized. From the extant evidence it does not, however, seem that Cluny's liturgy was exceptional in its own day. It is now seen not as boundless prolixity for the sake of ritualism alone, but as the expression of a vision of the endless hymn of praise of heaven itself (Waddell 1982, Leclercq 1974, 308).

The surviving chant sources with music from Cluny itself are pitifully few. The earliest are two from about 1075, that is, from the time of Abbot Hugh: a gradual, Paris, Bibliothèque Nationale, lat. 1087, and a noted breviary, Paris 12601 (Hourlier 1959). Unfortunately, later musical manuscripts are almost totally lacking. Only the noted breviary of 1317 of Saint-Victor-sur-Rhins, near Roanne, can be cited, a manuscript which was prepared at Cluny (Davril 1983, Huglo 1983). Recourse is therefore necessary to those from affiliated monastic houses. For manuscripts in staff-notation one may consult, for example, for the mass, a gradual of the early twelfth century from Souvigny (Bourbonnais, compiled for Sauxillanges, Auvergne), Brussels, Bibliothèque Royale II. 3823 (Huglo 1957); for the office, a twelfth-century antiphoner from Saint-Maur-des-Fossés, Paris 12044 (Renaudin 1972); and for mass and office combined, the breviary–missal of Lewes of the late thirteenth century, Cambridge, Fitzwilliam Museum 369 (Leroquais 1935, Holder 1985). (The text of the twelfth-century Cluniac antiphoner of Saint-Maur-des-Fossés, Paris 12584, has been edited by Hesbert in *CAO* 2.)

Presumably because its extant sources, and indeed its time of foundation, were too late in date to interest the Benedictine restorers of plainchant, partly perhaps also because some of the Cistercian criticism has coloured modern attitudes, Cluny's liturgy and chant have not received detailed attention in proportion to their fame. The tendency has been rather to assume Cluniac influence on liturgical music simply because it 'must have been influential'. Thus Abbot Odo, known to have composed three hymns and twelve antiphons for the monastic office of St Martin, was also credited with the compilation of a number of theoretical works, all of which can now be assigned to other authors: the treatise known as the *Dialogus de musica*, which originated in northern Italy in the late tenth century; a tonary and associated treatise likewise of the late tenth century, compiled by Abbot Odo of Arezzo; a tonary of the fourteenth century which is clearly of Franciscan origin (see Huglo, 'Odo', *NG*).

More bizarre is the case of the famous 'Codex Calixtinus' or 'Liber Sancti Jacobi', renowned for the unique collection of conductus (rhymed Latin songs) and polyphonic music which it contains (ed. Whitehill *et al.*, with facsimile of the offices; edition of music also by Wagner 1931). The manuscript was copied in the 1160s, for the great pilgrimage church of Santiago de Compostela, where it still resides. That was not, however, the original destination of the material. The manuscript was written in central France, possibly at Vézelay. It contains texts and music for the festal liturgy of St James, an account of St James's translation to Galicia, his miracles, a book about the pilgrimage routes to Santiago, and a book of the legendary exploits of Archbishop Turpinus, one of the Twelve Peers of Charlemagne. At the beginning of the book its author states that a large proportion of the material was compiled at Cluny, and attempts have therefore been made to understand the book as an instrument of the Cluniac ecclesiastical reconquest of Spain, and the replacement of the Mozarabic rite by the Roman (see Hämel 1950, Stäblein 1963, 'Modale Rhythmen'). This thesis unfortunately takes the nature of the book too seriously. Hohler (1972) has shown that the Pseudo-Turpinus part of the 'Liber Sancti Jacobi' was a text designed to teach Latin to schoolboys, full of deliberate grammatical howlers and malapropisms, not to mention such humorous stories as the one where the Emperor Charlemagne orders his cavalry to charge across rough country with their horses blindfolded. And the St James liturgy cannot be a Cluniac compilation. Among other things, the form of the office hours is not always monastic. And the special Latin songs and polyphony are a type of music which Cluny is unlikely to have cultivated with much vigour.

From the exhaustive description of Cluny's liturgy by Ulrich of Zell (PL 149. 633–778) and the contents of the contemporary gradual Paris 1087, it seems clear that Cluny did not have tropes sung in its liturgy, with the sole exception of a troped Agnus Dei on Easter Day. While the *argumentum ex silentio* is never reliable, Ulrich's habit of dwelling on each unusual detail of the liturgy, not least its more festal music, makes his mention of but one trope telling. Furthermore, sequences were sung with texts at Cluny only on the highest feasts; on other feast-days only the melodies were sung, as wordless melismas. Does this denote a deliberately reserved attitude towards 'non-canonical' material, a later manifestation of the ideas of Agobard of Lyons? (Cf. Gy 1983.) At any rate, with respect to these pieces at least, Cluny appears to have steered a moderate course. Here is none of the inflated liturgical splendour attacked by Cluny's opponents. (See Hiley 1990.)

Cluny's influence upon monastic observance and discipline and relations with ecclesiastical authority has been studied more than its liturgy. When a monastery submitted to Cluniac authority, did it necessarily observe the same liturgy as Cluny? In the case of many houses no service-books survive to give us an answer. The standard tests on the chant repertory provide at least some answers, however. Map IX.4.1 shows the provenance of sources which agree with Cluny in their choice of

Map IX.4.1. Monasteries following the use of Cluny, Saint-Bénigne at Dijon, or Bec

post-Pentecost alleluias and Advent responsories (cf. *CAO* 5–6), and in the melodic readings of their proper-of-mass chants, the Solesmes 'lieux variants' (*Graduel romain* IV). Similar information for two other monastic families is registered on the map, which will be referred to in two subsequent sections.

Although work on many sources remains to be done, it is clear that when a monastery adopted the order of service and selection of chants performed at Cluny, it might continue to sing those chants in the old way. This is the case at Saint-Martial of Limoges (Paris, Bibliothèque Nationale, lat. 1132), Marchiennes (Douai, Bibliothèque Municipale 114), Anchin (Douai 90) and Saint-Amand (Valenciennes, Bibliothèque Municipale 121), all of which have the Cluniac alleluia series but non-Cluniac melodic variants.

Ulrich of Zell (1029–93), the author of the customary just referred to, grew up in Regensburg, where he was educated alongside the great William, future abbot of Hirsau in the Black Forest. Later he entered Cluny, where he wrote the *Consuetudines* to inform William about Cluniac observance. But William did not introduce the Cluny liturgy in the houses which were reformed at first or second hand from Hirsau. Although practically none of Hirsau's liturgical books has survived, we have a good idea of what they contained from manuscripts of monasteries which were reformed from Hirsau. In Hesbert's survey of Advent responsory series, a group of mostly south German monasteries emerged which evidently reflected their common link with Hirsau: Prüfening, Zwiefalten, and St Georgen, Schwarzwald, were founded or reformed from Hirsau itself; SS Ulrich and Afra, Augsburg, and Gengenbach from St Georgen; St Emmeram, Regensburg, from Admont; Rheinau from Petershausen. Assessing such relationships with the aid of the Solesmes 'lieux variants' in proper-of-mass chants is, however, much more problematical, for the observed difference between eastern sources is so slight. It appears that musicologists will have to develop other means of distinguishing between such books.

Within Benedictine monasticism as a whole many monasteries became grouped under the aegis of a founding or reforming centre whose discipline or liturgy they adopted. Cluny, Gorze, and Hirsau, Grandmont, Camaldoli, and Vallombrosa are just a few of the better-known names. Much work remains to be done to discover the significance of these groupings for the chant repertory. The sort of chant propagated in all of them was traditional. Only the Cistercians reformed the actual melodies to be sung (see below, X.2).

IX.5. WILLIAM OF DIJON

Chomton 1900; PalMus 8; Huglo 1956, 'Tonaire', 1971, *Tonaires*; R. Le Roux 1967; Bulst 1973, 1974; Hansen 1974.

William of Dijon was born of a noble Piedmontese family in Volpiano in Lombardy in 961. When already a monk he met Mayeul of Cluny and entered that house, eventually attaining to such esteem that he was entrusted with the reform of Saint-Bénigne at Dijon in 989. He made this the springboard for the reform of other monasteries, particularly in Piedmont. (The customary of one of these survives, from San Benigno at Fruttuaria, close to his birthplace: ed. in CCM 12 and by Albers in CM 2 and 4). In 1001 he was invited by Duke Richard II of Normandy to revive the monastic houses of his duchy, only then beginning to recover from years of devastation and neglect at the hands of Richard's forebears. From Fécamp on the Channel coast he and his successors, in particular his nephew John of Ravenna, who succeeded him as abbot of Fécamp in 1028, and Thierry, abbot of Jumièges and then Mont-Saint-Michel, transformed monastic life in Normandy. On his death in 1031 he was head over more than forty houses in France and Italy. A pious picture of the great man was written by his disciple Rodulphus Glaber, who described his wide-ranging interests, including medical and musical ones. The latter are, however, mentioned in purely general terms—he 'emended' the singing of antiphons, responsories, and hymns so that they were sung better than anywhere else (PL 142. 715)—from which nothing specific can be learned.

While much is known about his reforming work, and the customary he composed for Fruttuaria is extant, it is impossible to know just how much musical expertise William had. The question is not a purely academic one because of the existence of the famous tonary with dual notation, in neumes and letters, Montpellier, Faculté de Médecine H. 159, which may have been composed by him or under his direction (facs. PalMus 8, edn. Hansen 1974).

That the tonary comes from Saint-Bénigne or one of its sister houses in Normandy is beyond dispute. The offertory *Laudate dominum quoniam* was given the prosula *In Hierusalem* in honour of St Benignus (PalMus 8, 217–18, Hansen no. 880–1). And by choice of pieces and by its melodic variants Montpellier H. 159 agrees with other books from Saint-Bénigne, from Jumièges, and Mont-Saint-Michel. By all the usual tests these and other Norman houses form a particularly cohesive group: Fécamp and Saint-Évroult are two other notable members of it, together with some English houses. (As well as *CAO* 5–6 and *Graduel romain* IV, see Le Roux 1967, Michel Robert 1967, Hiley 1980–1 and 1986, 'Thurstan'). These monasteries are represented on Map IX.4.1 above, in company with those following Cluniac use and that of Bec, whose influence is discussed in the next section.

The main part of Montpellier H. 159 is a copy of all the proper chants of mass in tonal, rather than liturgical order. But on its flyleaves is a fragmentary tonary of the conventional kind, that is with the incipits of office antiphons only, which matches a tonary of the same type in the Fécamp antiphoner Rouen, Bibliothèque Municipale 245 (A. 190) (see Huglo, 1956 'Tonaire', 1971, *Tonaires*, 328–33).

The main tonary in Montpellier H. 159 has been described briefly above (III.14), and also its alphabetic notation (IV.6). Given the Italian extraction of William and John of Ravenna, it is perhaps not surprising to find alphabetic notation disseminated among the Norman monasteries, for the anonymous *Dialogus de musica*, written in north Italy in the late tenth century, describes just such a system (albeit without the special signs for liquescence, quilisma, and oriscus in the Montpellier manuscript).

The liturgical ordering of chants which unites books from Saint-Bénigne and those from Norman monasteries is not that of Cluny. The 'Cluniac' nature of William's reforms lies elsewhere. Perhaps it is significant that we do not find tropes for the proper of mass in books from the revived monasteries—a sign of Cluniac 'abstinence'? They do, however, have full complements of sequences, and ordinary-of-mass chants: the twelfth-century troper from Saint-Évroult, for example, Paris, Bibliothèque Nationale, lat. 10508, has one of the biggest collections of ordinary-of-mass chants and tropes of its time, surpassed only by the collections in Norman-Sicilian manuscripts, to which it is in any case related (listed in Hiley 1986, 'Ordinary'). And there is ample evidence of compositional activity in honour of the local patron saints, for all of whom new offices would have been written as soon as musical practice was firmly established. Huglo has speculated that the office of St Benignus, also celebrated in Normandy, might have been written by William himself, though we have no proof of this. At first, as one might expect, outside help was occasionally needed: Ordericus Vitalis, the chronicler of Saint-Évroult, reports that the office of the monastery's patron saint Évroult (Ebrulphus) was first composed according to the secular (Roman) cursus at the request of Abbot Robert de Grantmesnil (1059–61) by Arnulf, the precentor of Chartres cathedral and a pupil of the famous Fulbert. Two young monks, Hubert and Ralph, went to Chartres to hear it sung (in the days before staff-notation was used at Saint-Évroult). The office was given its final monastic form by the addition of nine antiphons and three responsories by Guitmund, who had come to Saint-Évroult with the next abbot, Osbern (1061–6), from the monastery of Sainte-Catherine-du-Mont at Rouen (Chibnall 1969–80, ii. 108–9). Such a pattern of events must have been repeated very often—and of course not in Normandy alone. On numerous occasions the pious composer's name is recorded for posterity by an equally pious chronicler, bypassing for us the anonymity of the liturgical books themselves. The results can easily be seen by turning the pages of, for example, Hesbert's catalogue of manuscripts from Jumièges (MMS 2): Rouen, Bibliothèque Municipale 248 (A. 339), an antiphoner of the thirteenth century, contains the complete proper offices for Philibertus and Aycadrius, the abbey's patron saints, Audoenus (Ouen) of Rouen, and Benignus (Bénigne). Plates LXXVI–LXXVII in MMS 2 give the start of the Philibert office, with the antiphons of the first nocturn of the Night Office in modal order 1–6. The office of St Audoenus, patron of the diocese, was more adventurous, containing numerous prosulas (see Kelly 1977).

IX.6. ENGLAND BEFORE AND AFTER THE NORMAN CONQUEST

Graduale Sarisburiense; *AS*; PalMus 12; Frere 1894, 1898–1901; Eeles 1916; David and Handschin 1935–8; Knowles 1963; Holschneider 1968; Hartzell 1975, 1989; Planchart 1977; Gjerløw 1979; Hiley 1980–1, 1986, 'Thurstan', 1987; Hesbert 1982, 'Antiphonaires'; Underwood 1982; Droste 1983; Rankin 1984, 1987.

After the heroic age of Anglo-Saxon monasticism in the time of Bede and Benedict Biscop, Boniface, and Alcuin, the church in England was practically destroyed by the Scandinavian invaders of the ninth century. It is not at all certain that even a single monastery succeeded in continuing observance of the office. Since, not surprisingly, no service-books have survived this period, the contribution of Anglo-Saxon practice to the introduction of Roman chant under Pippin and Charlemagne is very difficult to assess. If anything did survive into the earliest English chant-books of the late tenth century, we cannot yet recognize it. The evidence seems to tell against such a possibility, for what we see in the later manuscripts is clearly based on continental traditions, principally north French ones.

The English revival, like the Carolingian, depended upon the co-operation of civil and ecclesiastical authority, in the English case between King Edgar (959–75) and three great bishops: Dunstan, Oswald, and Ethelwold. Dunstan (*c*.909–88) was abbot of Glastonbury from about 940, then had to leave England during a period of anti-monastic government in the 950s, spending a few years at the monastery of St Peter, Ghent. In 960 Edgar made him archbishop of Canterbury. Although he counts as an inspiring force behind the monastic revival, it is difficult to see tangible results in the history of liturgical music. He played the harp and made instruments: he gave an organ to Malmesbury, bells to Abingdon and Canterbury. The stories that he composed the Kyrie *Rex splendens* and the antiphon *Gaudent in celis* are, however, false. (For these anecdotes see Stubbs, 1874). Oswald (d. 992) was a monk at Canterbury, then at Fleury (Saint-Benoît-sur-Loire); he returned to England in 959 and was ordained bishop of Worcester by Dunstan in 962. Ten years later he was made archbishop of York. He had followed the usual practice in this period of monastic revival by replacing the clerics of Worcester Cathedral by monks, and he, like his colleagues, was responsible for the restoration and foundation of other monasteries, notably Ramsey, where he installed as first head Germanus, who had studied with him at Fleury. As far as we are concerned, Ethelwold was undoubtedly the most important of these men. He was a monk at Glastonbury, abbot of Abingdon, and from 963 bishop of Winchester, the chief city of Wessex and the main seat of the royal court, where he expelled the clerics from the Old Minster and replaced them with monks. It is from Winchester that most of the earliest English notated liturgical books come. About 972 Ethelwold and Edgar called together an ecclesiastical council to lay down guide-lines for the conduct of the English church. Monks from Ghent and Fleury were there. The document which resulted, the *Regularis concordia* (ed. Symons 1953), describes many interesting details of liturgical practice, including the performance of the *Visitatio Sepulchri* on Easter morning.

Ghent and Fleury were not the only foreign sources of influence. Already at Abingdon Ethelwold had been concerned for unanimity of chant practice and invited monks from Corbie to provide an example in reading and singing (Stevenson 1858, i. 129, cited by Knowles 1963, 552).

After the Norman conquest William I (1066–87) replaced practically the complete Anglo-Saxon church hierarchy, abbots as well as bishops, by Normans. Norman influence had already been felt during the reign of Edward the Confessor. Two more Italians who, like William of Dijon, had been important in Norman monasticism led the reorganized Anglo-Norman church. These were Lanfranc, monk of Bec (not part of William of Dijon's family) from 1042, first prior of St Stephen's, Caen, founded by Duke William, and first archbishop of Canterbury after the Conquest, consecrated by one of his former pupils at Bec, Pope Alexander II. He was succeeded by another Italian, Anselm, prior of Bec from 1063, who in turn became archbishop of Canterbury (1093–1109).

These are the figureheads. We do not know how many English cantors, politically a less important company, were replaced by Normans. But when the English chant-books from before and after the Conquest are compared with their continental counterparts, the various lines of influence stand clearly revealed, first from Corbie, then from monasteries of the Dijon family and from Bec (see for example the demonstrations in Hartzell 1975, Hiley 1980–1, Hesbert 1982, 'Antiphonaires', Hiley 1986, 'Thurstan'). We must of course use these labels with caution, for in practice it is not usually possible to pinpoint the specific source of influence in question. The uses of Corbie and Saint-Denis, for example, are practically indistinguishable according to the rough tests so far developed. Similarly for Fécamp, Jumièges, Mont-Saint-Michel, and so on. But the general pattern is clear. From Winchester we have only pre-Conquest books, which clearly ally themselves with Corbie practice. After the Conquest we can see, for example, the order of service of Bec (judged by the alleluia and responsory test) taken up at Christ Church, Canterbury, Worcester, and Durham, that of Dijon at Westminster, Evesham, and Winchcombe. And study of the melodic variants in their chants shows that Bec versions were taken up at St Albans, Dijon ones at Gloucester. But interestingly, the books of most churches do not show the melodic variants typical of Norman sources. In other words, even when the order of service was revised, the chants were sung as before, something we have already noted in the case of books from Cluniac houses. So we find the Corbie/Winchester melodic tradition perpetuated after the Conquest at Christ Church Canterbury, Crowland, Peterborough, Exeter, Worcester, the Augustinian priory of Guisborough, even the secular use of York, and Downpatrick in northern Ireland. (See Map IX.4.1 above.)

Of the thirty or so liturgical books or fragments with musical notation that have survived from England up to the time of the Conquest (see Rankin 1987) about a third come from Winchester (Rankin 1984 describes the slightly later activity at Exeter). Pride of place is held by the two famous tropers, Cambridge, Corpus Christi College 473, probably written in the closing years of the tenth century, and Oxford, Bodleian Library Bodley 775, from about the middle of the eleventh. At least parts of the Cambridge manuscript may have been written by the cantor of the Old Minster, Winchester, Wulfstan, author of a lost music treatise and of a long poem in honour of St Swithun which contains much invaluable information about the Anglo-Saxon cathedral of Winchester and a celebrated description of its organ. He is generally thought to have been the composer of the organa in the Cambridge manuscript, over

150 second voices to be sung with chants of the mass and office to make two-part polyphony.

Both the tropes and the polyphony in the Winchester manuscripts have been studied in detail (see Planchart 1977, Holschneider 1968). Many of the tropes were known abroad, most in France, some in eastern sources—possible north French intermediaries, such as tropers from Corbie or Saint-Denis, unfortunately do not survive, nor is any troper from Fleury extant. It is a pity that no later English troper of this type exists, one transmitting the pieces in staff-notation, for all the compositions in honour of local English saints, known only from the Winchester sources, remain untranscribable.

It would perhaps be overbold to say that the Norman Conquest, and the new men who took command of the church in England, were directly responsible for a 'purge' of tropes from English books. The evidence is insufficient. But, while the proper tropes are no longer recorded, the advent of the new Norman repertories of ordinary-of-mass chants and their tropes (such as those in the Saint-Évroult troper mentioned above) is clearly visible. The Winchester tropers also contain individual repertories of sequences which do not appear to have survived the Conquest (listed in Frere 1894). About half the seventy-odd texts are unique to Winchester, a strikingly large proportion. Most of the melodies for these texts are known elsewhere. The process must have resembled somewhat that at St Gall, where a century earlier Notker (at least in some cases) provided new texts to replace previous ones which he thought unsatisfactory. Alas, some of the unique Winchester pieces are written to melodies not known elsewhere, and therefore remain lost to us.

The gravity of the changes led to unrest. At Glastonbury the new Norman abbot Thurstan, previously monk at Caen, is reported to have tried to make the monks sing the 'chant of William of Fécamp' (presumably William of Dijon/Volpiano), which, among other things, led to a bloody fight between the recalcitrant monks and Thurstan's retainers. One would like to know more about the 'chant' which caused the trouble: did the monks object to learning new pieces, or to singing old ones in a new way? (On this question, see David and Handschin 1935–8 and Hiley 1986, 'Thurstan').

While the derivation of the monastic uses in England is reasonably clear, the use of the secular cathedrals survives in a frustratingly incomplete state. We have chant-books, both for office and mass, from only three: Salisbury, Hereford, and York.

The foundations of the Salisbury liturgy as we know it appear to have been laid by Bishop (St) Osmund in the late eleventh century. The twelfth-century unnotated gradual Salisbury, Cathedral Chapter Library 149 shows the mass liturgy established before the first notated chant-books appear. These seem to have been due to the initiative of Bishop Richard Poore, who in 1218 changed the site of the cathedral, and who may be the author of the customary and ordinal: these survive in several successive 'editions' (see Frere 1898–1901). Among the earliest chant-books are three notated in very similar hands, of about the second quarter of the thirteenth century:

the graduals London, British Library, Add. 12194 (part facs. in *Graduale Sarisburiense*) and Oxford, Bodleian Library, Rawl. lit. d. 3, and the antiphoner Cambridge, University Library, Mm. 2. 9 (from Barnwell near Cambridge, facs. in *AS*), followed by the noted missal Manchester, John Rylands Library, lat. 24, made in Salisbury around 1260 for Henry of Chichester, canon of Exeter Cathedral, to present to the cathedral there (Hollaender 1942–4).

Salisbury ('Sarum') use became a sort of national rite by the later Middle Ages. The reasons for this are not entirely clear, but among the crucial factors were the following: (i) the seat of the archbishopric, Christ Church Canterbury, was a monastic house whose liturgy could not be automatically used in secular cathedrals and parish churches; (ii) the royal chapel seems to have used the Salisbury liturgy, though from what time is not known; indeed a note in a fourteenth-century Sarum missal states that the use of the royal chapel is actually normative for the rite elsewhere (Hohler 1978, 37); (iii) from the thirteenth century onward, professional workshops in Oxford, London, and later Cambridge appear to have been able to supply books of Salisbury use 'on demand', so to speak, making it easy for churches to acquire books of this type rather than those of their diocesan cathedral (Droste 1983). A good illustration of this is the missal Munich, Bayerische Staatsbibliothek, clm 705, which bears a colophon indicating that it was copied in Oxford, and whose calendar has the dedication feast of the parish church of Lydd (diocese Canterbury).

Early sources from Wales, Ireland, and Scotland are lacking. It was under English influence or in the train of English arms that the chant traditions represented in the earliest extant sources (of the thirteenth century) were formed, with the usual modicum of items in honour of local saints.

Very rarely can we detect 'local flavour' in the chants that were produced locally. There is no fundamental stylistic difference between chant written in England and that written in France. Of the sequences for St Patrick in Irish sources, for example, *Letabundus decantet* is a contrafactum of the popular Christmas sequence (see MMS 4 for a facsimile of Cambridge Add. 710); and while Hesbert comments on the 'Irishness' of *Laeta lux est hodierna* (in the same manuscript), its 'major mode' melody seems to the present author typical of many contemporary compositions. The office of St David in Aberystwyth, National Library 20541 E (see O. T. Edwards 1987) is a contrafactum of the office of St Thomas of Canterbury. (For Scottish remains see McRoberts 1952 etc. and Woods 1987.) The aim is to provide fitting music for the praise of the saint, and what is unusual or radically new will not as a rule be fitting.

IX.7. NORTH ITALIAN TRADITIONS

Planchart 1985.

Apart from the separate rite of Milan, and the scanty remains of other non-Roman uses, chant-books from Italy north of Rome are Gregorian. But they display great variety of content and appearance, some of it the result of establishing local traditions, some due to the importation of foreign practices. A few examples of these overlapping uses may be given here.

The different styles of notation in Italian manuscripts have already been mentioned, and the occurrence of German notation at Monza and Bobbio, Breton notation at Pavia, and Laon notation at Como (IV.2). This accords to some extent with the affiliations revealed by the melodic variants in chants for the proper of mass: Monza (Monza, Basilica S. Giovanni C. 12/75 and C. 13/76) allies itself with the eastern 'bloc'. The nearest relative of Vercelli, Biblioteca Capitolare 186, from Balerna near Lake Como, is the Laon gradual Laon, Bibliothèque Municipale 239. The sources from Pavia (the ascription is provisional), Ivrea, Biblioteca Capitolare 60 and Vercelli, Biblioteca Capitolare 56, are related, though not so closely, to Breton sources. A majority of the tropes in Ivrea 60 appear to have been imported from France, yet its sequence repertory is dominated by German pieces. The latter feature is not unusual in north Italy. The monastery of Nonantola, near Bologna, for example, had a sequence repertory largely based on Notker's collection; one wonders if this fact is connected with the early appearance of a tonary like the Reichenau one in a Nonantola manuscript (Rome, Biblioteca Casanatense 54—see III.14). (Sequences in all Italian sources up to *c*.1200 have been catalogued by Brunner 1985, who is also preparing an edition of the Nonantola sequences.)

Since numerous German clerics were appointed to important positions in the Italian church throughout the early Middle Ages, it is not surprising that the influence of German liturgical practice was strong. It seems certain, for example, that the early appearance of the German or 'Augsburg' form of the *Visitatio Sepulchri*, 'Quem queritis o tremule mulieres', in Aquileia (Udine, Biblioteca Capitolare 234, eleventh to twelfth century) is due to the appointment of Augsburg canons to the patriarchate in the eleventh century: Eberhard in 1042–8 and (more likely) Heinrich in 1077–84. This would also be why the melodic variants in the surviving Aquileian graduals Rome, Biblioteca Apostolica Vaticana, Rossi 76 and Udine 8. 2 (both thirteenth century) agree with sources in the eastern 'bloc'. (On Aquileia see also Huglo 1976).

Apparently ancient non-Roman chants have often turned up in relatively late Italian sources (see Wilmart 1929, Huglo 1952, 'Antiennes', 1954, 'Vestiges', Levy 1970, 1971, Baroffio 1978, Boe 1987). We should not, therefore, automatically assume that items in Italian manuscripts which are also known in France or Germany were necessarily imported into Italy. As a simple illustration of this point one might take the tropes for the introit of mass on Easter Day as they appear in Piacenza, Biblioteca Capitolare 65, the massive compendium of the entire chant repertory of Piacenza cathedral, begun in 1142 not long after the commencement of the new cathedral church. The edition and lists of tropes in CT 3 make comparisons with other repertories easy.

Piacenza 65 begins with a version of the *Quem queritis* dialogue, one where, after the sentences *Quem queritis—Iesum Nazarenum—Non est hic . . . Ite nunciate quia surrexit dicentes*, the piece concludes with the verses *Alleluia resurrexit dominus. Eia carissimi verba canite Christi*, presumably composed to lead into the Easter introit *Resurrexit*. The verse *Eia carissimi* is found otherwise only in a few other Italian sources and in Cambrai, Bibliothèque Municipale 75, from Saint-Vaast at Arras. Where did it originate? That dialogue has been discussed frequently (see Rankin 1985, '*Quem queritis*'). But for the introit itself Piacenza 65 has more trope verses (Ex. IX.7.1).

The trope verses *Qui dicit patri* and *Mirabile laudat* appear elsewhere only in a small number of Italian sources, from Novalesa near Mont Cenis, Pistoia, and Benevento. *Hodie exultent*, however, turns up in a slightly different range of Italian sources and in the two earliest St Gall tropers, St Gall, Stiftsbibliothek 484 and 381. The text was also copied into Munich, Bayerische Staatsbibliothek, clm 14843, which some scholars have dated as early as the ninth century. The second part of the verse, 'resurrexit leo fortis deo gracias', is what usually follows 'Alleluia resurrexit dominus' in the dialogue at the sepulchre. What has happened? Has that sentence been lifted from the dialogue as part of a revision, and if so, was it done at St Gall or in Italy? At St Gall *Hodie exultent iusti* starts a series of verses, rather than finishing one off as in Piacenza and other Italian sources.

In a recent discussion of some other verses beginning 'Hodie . . .' Reier pointed out (*a*) that they had wide distribution in Italian sources, and (*b*) that they all had the same melody (Reier 1981, i. 157). Although Reier's examples were introductory tropes, unlike *Hodie exultent iusti* in Piacenza, *Hodie exultent* has exactly the same melody as the others. That is the same as *Hodie resurrexit leo fortis* in other versions of the Easter dialogue. So we should at least consider the possibility that *Hodie exultent iusti* is an old Italian introductory trope, perhaps displaced by the popular *Quem queritis* dialogue, which in turn in some versions took a verse from *Hodie exultent iusti*. This sort of reasoning carries no conviction if the music is not taken into account. It is the use of the same melody for several different trope verses which is unusual and alerts one to the possibility of an 'original' layer of Italian material.

After these trope verses Piacenza 65 has the Easter *Alleluia Pascha nostrum* with a prosula for the 'Alleluia'. There follows the sequence *Clara gaudia festa paschalia*. The melody of this sequence was widely known, being called 'Angelica' and 'Nobilissima' at Winchester, 'Musa' in Angers, Bibliothèque Municipale 144, and 'Romana' in St Gall (which may not be entirely fanciful). Notker composed two texts for it, *Johannes Jesu Christo* and *Laurenti David magni martyr. Clara gaudia* was known in France, as was another Easter text for this melody, *Dic nobis*, and another

Ex. IX.7.1. Trope verses *Qui dicat patri*, etc. for introit *Resurrexi* (Piacenza, Bibl. Cap. 65, fo. 235ᵛ)

I1   RESUR-RE-XI   ET ADHUC TECUM SUM   AL - LE - LU - IA.

T1   Qui di - cat pa - tri pro-phe-ti-ca uo-ce.

I2   PO-SU-IS-TI   SU - PER ME MANUM TU - AM AL-LE-LU-IA.

T2   Mi-ra-bi-le   lau-dat fi- -li- -us patrem.

I3   MI-RA - BILIS FAC-TA EST SCI-EN-TI-A   TU - A ALLE-LUIA   AL-LE-LU - IA.

Ps   DO-MI-NE PROBASTI ME (rest omitted) ET RESURRECTIO-NEM MEAM.

T3   Ho-di - e ex-ultant ius-ti   re-sur-re-xit le-o for-tis De-o gra-ti-as

di- -ci-te e - ia.   (followed by cues for RESURREXI and GLORIA PATRI)

text for martyrs, *Candida contio*. In the earliest substantial Italian collection of sequences, Monza, Basilica S. Giovanni 13/76, *Clara gaudia* and Notker's two texts all appear. Previous discussion of the sequences has concentrated on Notker's texts and their possible models. Crocker (1977, 146–59) treated *Clara gaudia* and *Dic nobis* as 'West Frankish' but thought neither a wholly convincing starting-point for Notker. He recognized that the melody stood rather outside the mainstream of the sequences he was discussing, but did not pursue the matter further. In fact the melody bears all the hallmarks of an Italian composition (see above, II.22.v): repetitive melody and consequent restricted range, with occasional groups of two or three notes to a syllable (in Italian sources; they are often 'tidied up' in foreign ones). And its prominent place on Easter Day alerts one to the possibility that in Piacenza 65 (as in many other Italian sources: see Brunner 1985, 220) the piece stands in its original place. The text is dramatically colourful, an account of the harrowing of hell, as Crocker observes, with demons howling in line 5, saints acclaiming the Lord in line 6 (Ex. IX.7.2).

Once again it is musical features which point in the direction of what may be the 'original'. In cases like this, Italian chant does not sound like chant everywhere else.

Ex. IX.7.2. Sequence *Clara gaudia festa paschalia* (Piacenza, Bibl. Cap. 65, fo. 236ʳ)

IX.8. BENEVENTO AND MONTECASSINO

The existence of a non-Roman rite and corresponding non-Roman chant in the Lombard duchy of Benevento has already been discussed (VIII.5). The principal sources of Old Beneventan chant are in fact Gregorian books with relics of the displaced repertory still copied alongside the imported one. Of course, these books are also important as sources of Gregorian chant, with the local peculiarities of repertory and melodic readings we should expect of any group of Gregorian sources. Most of the surviving chant manuscripts are from Benevento itself, which has slightly obscured the fact that the great Benedictine monastery of Montecassino lay in the same area and shares many details of chant practice with Benevento (see Planchart 1990). Kelly (1985, 'Montecassino', 1989) has made it clear that Old Beneventan chant was sung at Montecassino as well. Unfortunately, Cassinese sources are fewer and later, and it is therefore difficult to assess its contribution to the establishment of Gregorian chant and the burgeoning repertory of sequences and tropes in the Beneventan region. (The sequences are catalogued by Brunner 1985, the trope repertory is being edited by Boe and Planchart in the series Beneventanum Troporum Corpus.)

The principal sources from the two centres, which date from the early eleventh to the thirteenth century, are the following (in Benevento, Archivio Capitolare, Montecassino, Archivio della Badia, and Rome, Biblioteca Apostolica Vaticana):

from Benevento:
noted missals: Benevento 33 (PalMus 20) and 30 (partially notated)
graduals: Benevento 38, 39, 40, 34 (PalMus 15), and 35
combined missal and breviary: Benevento 19–20
noted breviary (winter part): Benevento 22
antiphoner: Benevento 21 (text edn. by Hesbert in *CAO* 2)
    from Montecassino:
noted missals: Rome, Vat. lat. 6082 (possibly for San Vincenzo al Volturno) and Montecassino 540 (used at Plombariola)
gradual: Montecassino 546 (winter part)
antiphoner: Montecassino 542 (winter part)
troper: Urb. lat. 602 (inventory and partial reconstruction Boe 1985)
theory compendium and tonaries: Montecassino 318 (RISM B/III/2, Huglo 1969, Brunner 1981)
    related sources:
graduals: Rome, Vat. lat. 10673 (only a large fragment remains; PalMus 14) from Apulia or possibly Benevento, and Paris, Bibliothèque Nationale, n. a. 1. 1669 from Gubbio
noted missals: Rome, Barb. 603 from Caiazzo, Baltimore, Walters Art Gallery 6 from Canosa (text ed. Rehle 1972–3), and Rome, Ottob. lat. 576 (partially notated)
troper: Naples, Biblioteca Nazionale VI G. 34 from Troia

(Assignment of these sources to particular institutions and dates is occasionally revised, but the general area is not in doubt. See Boe 1985, Kelly 1989, and the catalogue (in progress) of sources in the Archivio Capitolare at Benevento by Mallet and Thibaut.)

This short list does not mention other types of liturgical books or the numerous fragmentary sources, some from Dalmatia (the Adriatic coast of modern Yugoslavia). In particular, a number of splendid rolls bearing the Exultet chant for the lighting of the paschal candle on Easter Eve have been preserved (Avery 1936, Cavallo 1973). The distinctive script and notation of manuscripts from the area has stimulated scholars to hunt down a great number of sources, and we are relatively well informed about the extent and nature of the surviving corpus (see Loew and Brown 1980).

The story of the supersession of the Old Beneventan rite and chant by the Roman rite and Gregorian chant is a complicated one but very interesting. A great deal of it has been reconstructed by Hesbert (1938–47, PalMus 14, etc.), and Kelly (1989, 18–25) provides a recent summary. The version of the Roman rite as it appears in its earliest Beneventan sources (Benevento 33 from around 1000) has many archaic features which had long been overhauled in the Frankish empire, but which persisted in Rome itself, as may be seen in the Old Roman chant-books of the eleventh century. That the Beneventan books with Roman order of service have Gregorian chant has been taken to denote Frankish, rather than specifically Roman influence, and it has been argued that the Gregorian chant must have arrived in Benevento at the time when the liturgical archaisms were still current in the Frankish empire, that is in the eighth century. It is clear from the music manuscripts themselves that Gregorian and Old Beneventan chant lived side by side, so to speak, for over a century. Could they have coexisted since as far back as the late eighth century? Kelly sees Paul Warnefrid (the Deacon), biographer of Gregory the Great, as a possible agent for Frankish-Gregorian chant in Benevento (Kelly 1989, 22–5). The age of the introduction of Gregorian chant seems to be confirmed by the fact that the Feast of the Twelve Brothers, introduced at Benevento in 760, has Old Beneventan chant, whereas for the Feast of St Mercurius none survives, nor for two saints whose use was taken up at Benevento in the ninth century: Januarius and Bartholomew. Yet for Barbatus, whose cult also developed in the ninth century, some Old Beneventan antiphons do survive.

The role of Montecassino, St Benedict's own monastery, is unclear. Founded in the early sixth century, it was destroyed by the Lombards in 577 and had to be refounded by Petronax of Brescia in 717. In the interim monks from Fleury had carried off the mortal remains of Benedict and his sister Scholastica. So the Frankish 'acquisition' of Roman chant had a tangible monastic counterpart. What the chant of Montecassino may have been like in the eighth century we do not know, which is unfortunate because Boniface sent his disciple Sturmi, future abbot of Fulda, to Rome and Montecassino for two years to study the Benedictine rule. Charlemagne sent there in 787 for a copy of the rule. In 883 it was burned by the Saracens and remained deserted for sixty years. There follows the period for which Old Beneventan chant can be documented at Montecassino, after which we have the following valuable record of moves to introduce Gregorian.

What the non-Italian popes who occupied the Holy See from the end of the tenth century and again from 1046 onward thought of the Old Roman chant they heard in Rome, so different from the Gregorian chant of their native lands, we do not know. It was by the German Gregory V (996–9) that a German type of pontifical was introduced in Rome (see III.12), but we have no specific record of changes in chant practice in Rome until much later. In the mid-eleventh century Montecassino had two German abbots in succession, Richerius and Frederick, the latter becoming Pope Stephen IX in 1057. In 1058, on a visit to Montecassino, he forbade the singing of 'Ambrosianus cantus', by which is meant Old Beneventan chant (MGH (Scriptores) *Die Chronik von Montecassino*, 352–3). The task of suppressing it must have been that of the great abbot Desiderius, the future Victor III, whose abbacy marks a cultural high point in Montecassino's history (Bloch 1985, Cowdrey 1983). It is disappointing that practically no musical source of this time has survived, with the exception of the huge theoretical compendium, Montecassino 318.

The Gregorian tradition established at Benevento and Montecassino and the repertories of tropes and sequences, though naturally containing distinctive local details, were by no means parochial. Many of the tropes and sequences known there appear to have travelled long distances (see Brunner 1981), and the melodic variants in the proper chants of mass, collated by the monks of Solesmes (*Graduel romain* IV), are surprisingly close to those of Aquitanian sources. The reasons for this are not clear.

In the late eleventh century and twelfth, Norman invaders conquered South Italy and Sicily. With them came churchmen and the chant traditions of their native land (Hiley 1981, 1983, 'Quanto', 1988). But this seems to have made little impact upon Beneventan and Cassinese practice. Only one manuscript shows a merging of the different traditions. This is the troper from Troia, Naples VI. G. 34, compiled late in the twelfth century (RISM B/V/1, 175–6; one page facs. Arnese 1967, 146–9 and pl. X, Stäblein 1975, 145).

The manuscript is written in Beneventan script and notation. Even as late as this, there are Old Beneventan remains: the processional antiphon *Deprecamur te domine* in an Old Beneventan version (Levy 1982, 95) and the Old Beneventan *Peccavimus domine* (Kelly 1989, 284), hardly more than a threefold variation of the same simple phrase (Ex. IX.8.1).

But the winds of change have blown strongly. The manuscript has almost nothing in common with the great Beneventan graduals: two of its sequences only are taken from the traditional Beneventan repertory (*Qui purgas animas* and *Precursor Christi*). The Laudes regiae, the acclamatory chant for the crown-wearing of the ruler, sung after the Gloria on the highest feasts of the year, have in this manuscript a version that is peculiar to Norman sources (Kantorowicz 1946). Practically all the Kyrie and Gloria melodies and tropes are taken from the Norman tradition (inventory in Hiley 1986, 'Ordinary'; the Sanctus and Agnus are lost from the end of the manuscript).

Ex. IX.8.1. Rogation antiphon *Peccavimus Domine* (Naples, Bibl. Naz. VI. G. 34, fo. 6ʳ)

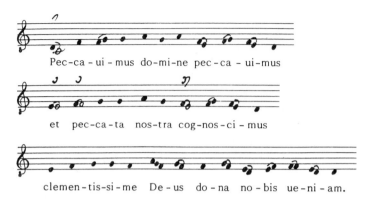

Next come farsed lessons, one of the most extensive collections known (see II.23.xii and Stäblein, 'Epistel', *MGG*), with several unica. Many of these were also known in Sicily and north France, though, since their south Italian sources are equally early, it is not clear where some of these pieces originated. Were they brought by the Normans, or taken back north by them? Ex. IX.8.2 gives the start of the lesson *Lectio libri Sapientiae . . . Qui timet Deum faciet bona* for St John the Evangelist with the farsing verses *Ad laudem regis glorie* unique to this manuscript. The melody of the lesson is largely freely composed; its prose contrasts strongly with the regular, rhyming, farsing verses.

Most of the incomplete collection of alleluias found in Naples VI. G. 34 was not known in Benevento, and the sequences include some new rhymed compositions, typical of north France, particularly Paris, in the later twelfth century: *Potestate non natura*, *Exultemus in hac die*, and so on. The sequence for Troia's patron saints Pontianus and Anastasius is unfortunately incomplete, so that an example of local composition is spoiled for us. St Secundinus has his name inserted in a common sequence borrowed probably from Norman use.

The type of manuscript itself is not one we normally associate with Benevento, where ordinary-of-mass chants, tropes for the proper of mass, and sequences were usually entered in place within each mass, as is often the case in Italian sources. The nearest to it is Vatican Urb. lat. 602, which, however, seems to have had no sequences or farsed lessons. To sum up, Naples VI. G. 34 is a manuscript where the type of notation, traditional Beneventan, gives no clue to the origin of the chants, which range from Old Beneventan through Norman and beyond. The study of overlapping palaeographical and repertorial traditions in manuscripts like this is one of the most fascinating areas of chant research.

Ex. IX.8.2. Start of farsed lesson with verses *Ad laudem regis glorie*, etc. (Naples, Bibl. Naz. VI. G. 34, fo. 46ᵛ)

XI.9. ROME AND THE FRANCISCANS

Hüschen, 'Franziskaner', *MGG*; Abate 1960; van Dijk and Walker 1960; van Dijk 1961, 1963, *Sources*; Salmon 1967; Hesbert 1980, 'Curie'.

I have adopted the view above (see VIII.8) that Old Roman chant was the only chant sung in Rome until about the eleventh century, and that when we meet it in its earliest notated form in 1071 (the date of manuscript Bodmer C. 74) it is a somewhat altered descendent of the chant which the Franks learned from Roman cantors. We do not know how often non-Romans noticed differences from the chant of the Franks. We may assume—though we have no record of it—that the first foreign popes, installed when the Emperor Otto III was trying to reform the papacy, would have been aware of the contrast: the German Gregory V (996–9), and the Frenchman Gerbert of Aurillac, Sylvester II (999–1003). However, as already mentioned (IX.8), it is in the wake of the series of German popes in the mid-eleventh century that we find Stephen IX objecting to 'cantus Ambrosianus' at Montecassino. That is, of course, not Old Roman chant; the point is, it is not Gregorian either, and Gregorian chant is what was sung thereafter at Montecassino. And it may be significant that the earliest surviving Old Roman chant-book dates from a time when the Old Roman repertory may have been coming under pressure, when it may have been felt necessary to take steps for its preservation.

The type of pontifical codified at Mainz in the mid-tenth century, the so-called 'Roman-German' pontifical, was being used in Rome shortly thereafter, brought, it is thought, by Emperor Otto I on one of his many visits. This copious and well-organized collection of pontifical ceremonies is a different matter from a gradual or antiphoner, of course. But it is indicative of the way a non-Roman document could alter Roman liturgical practice. On the other hand, the pontifical compiled for Innocent III (1198–1216) still includes three Old Roman antiphons (Huglo 1954, 'Vieux-romain', 102–4).

Gregorian chant could also have been sung close at hand to the papal chapel from the time when Alexander II (1061–73), a former canon of St Fridian's at Lucca, reformed the monastery attached to the Lateran, whose personnel performed the services in the Lateran basilica (the pope and his chaplains celebrated in the adjacent chapel of St Lawrence). The reform was confirmed in 1105 when the Lateran Congregation of Canons Regular was founded as a descendant of St Fridian's (van Dijk 1961, 424). The ordinal of the canons as codified in the mid-twelfth century is known (ed. Ludwig Fischer 1916), but it contains no chant. (See Gy 1984.)

The first clear evidence that mass and office were performed with Gregorian chant in the papal chapel comes from the thirteenth century, when a series of revisions brought all liturgical practice in Roman churches, papal or civic, into harmony, with which the Franciscan order of friars concorded. Of Innocent III's revision of mass and office of the papal chapel in 1213–16 only the ordinal survives (ed. van Dijk and

Walker 1975). Honorius III (1216–27) made another revision of the breviary for outside use, which the Franciscans adopted in 1230. They adopted a revised missal shortly afterwards. Both breviary and missal were revised by Haymo of Faversham, General of the Franciscan Order from 1240 until his death in 1244, at the request of Innocent IV (1243–54) (ed. van Dijk 1963, *Sources*). From this point papal and Franciscan use were to all intents and purposes the same. The Roman city churches appear to have come into line somewhat gradually. In the 1250s Cardinal John Cajetan Orsini devised a compromise liturgy to combine the use of the Vatican (urban) and the Lateran palace (papal), but on becoming Pope Nicholas III (1277–80) he ordered the destruction of all non-conforming books (van Dijk 1956, 1960).

We can therefore tell what chant was sung in the papal chapel by inspecting notated Franciscan books (papal ones do not survive). The noted missal Naples, Biblioteca Nazionale VI. G. 38 (facs. Arnese 1967, pl. XVI, PalMus 2, pl. 30) has the liturgy of Honorius III's revision (van Dijk 1969–70). The earliest books have central Italian notation (Naples VI. G. 38; see also Hüschen, 'Franziskaner', *MGG*; and *NG* iii. 269 for a breviary of 1224). The preface to the gradual of the 1250s specifically states that square notation should be used (ed. van Dijk 1963, *Sources*, ii. 359; see also i. 110), which by that time was rapidly becoming universally popular.

The Solesmes survey of variant readings on proper-of-mass chants places Naples VI. G. 38 with other Franciscan books in a group showing no strong resemblance to any other. Hesbert's investigation of the responsory series (1980, 'Curie') showed that, once evident innovations have been set aside, an affinity with other central Italian uses became clear—Perugia, Carinola, Florence, and Piacenza. Possibly some such process led to the establishment of the musical readings. The Roman-Franciscan gradual appears to have been adopted by the Celestines, the order of Benedictine monks founded by Celestine V in 1250, to judge from the single source investigated by the monks of Solesmes. On the other hand, Celestine office-books follow the Montecassino order of service. The Olivetans, founded in 1319 by Giovanni Tolomei on Monte Oliveto near Siena, adapted the Roman-Franciscan breviary for monastic use. The same happened in monasteries descended from the reform at Santa Giustina in Padua in the early fifteenth century. Subiaco was reformed from Santa Giustina, and German monks took the reform to the newly founded Melk on the Danube, whence the Roman-Franciscan use in a revised monastic form spread to other Austrian and south German monasteries.

Tonaries survive which were composed by two clerics attached to the curia (Huglo 1971, *Tonaires*, 225–9), though neither has the status of an official document. The treatise *Practica artis musicae* written in 1271 by the English priest Alfred (or Aluredus or Amerus), in the entourage of Cardinal Ottobono Fieschi (the future Hadrian V) contains two tonaries, one compiled according to French and English use, the other according to Roman (ed. Ruini in CSM 25). By contrast, Elias Salomon's *Doctrina scientiae musicae* of 1274 (GS iii), though compiled at the curia, does not reflect its usage.

Churches outside Rome did not stand under any obligation to adopt the Roman-

Franciscan reforms. Even the Lateran basilica celebrated its own form of the office until the late fourteenth century.

The attitude of St Francis and the Franciscans to liturgical music is discussed by D'Angers (1975) and also by Hüschen (MGG). The rhymed office of St Francis was edited by Felder (1901).

IX.10. AQUITAINE AND SAINT-MARTIAL AT LIMOGES

PalMus 13; MMMA 3; Crocker 1957, 1958; Chailley 1957, 1960, *École*; Emerson 1962, 1965; Herzo 1967; Treitler 1967; Steiner 1969; Paul Evans 1970, *Repertory*, 1970, 'Elements'; Brockett 1972; Roederer 1974; Planchart 1977; Planchart and Fuller, 'St Martial' *NG*.

A number of important chant manuscripts have survived from south-west France, that is roughly south of the Loire and as far east as the Saône–Rhône valley, corresponding more or less to the Aquitaine, Gascony, and Septimania of Charlemagne's empire and generally known for convenience as 'Aquitaine'. The manuscripts are usually immediately recognizable by their distinctive notation, one of the earliest whose pitches can be read accurately. They have attracted a great deal of attention, partly because of their legibility, but also because they contain large repertories of tropes and sequences, and because a late sub-group among them contains numbers of the rhythmic, rhyming songs of the twelfth century, some in polyphony. A good number of them passed through the hands of the librarian of the abbey of Saint-Martial at Limoges, Bernard Itier, in the years around 1200, and the label 'Saint-Martial' has sometimes been applied to more or less the whole group. Only a handful can be shown definitely to have been written for Saint-Martial, a few more possibly for other churches in Limoges, while the majority are not from Limoges. The provenance of several is unfortunately not at all clear.

While the sources do indeed share common traditions, to a greater or lesser extent, they are naturally fortuitous survivals, which leave some awkward gaps in our knowledge. The earliest important liturgical source, the unnotated combined gradual and antiphoner Albi, Bibliothèque Municipale 44, dates from the second half of the ninth century. But then there is a complete gap for both types of book for over a century. The earliest collection of tropes and sequences is dated around 930, followed by two particularly large and important collections of a similar nature from the closing decades of the century. We have no notated graduals until well into the next century, and no complete notated office-book until the fourteenth century. (A possible exception is the twelfth-century antiphoner Toledo, Archivo Capitular 44. 2. Its notation is Aquitanian and it is usually said to come from Aquitaine, but its closest 'relatives' in Hesbert's survey—*CAO 5*—are all Spanish.)

The principal manuscripts, in roughly chronological order up to about 1200, are listed in Table IX.10.1. It goes without saying that the label 'troper' covers a great

Table IX.10.1. *Aquitanian sources to* c.*1200*

Manuscript	Type	Provenance	Date
Albi 44	gradual-antiphoner	?	9th c., second half
Paris 1240	troper (p o s t)	Saint-Martial	c.930
Paris 1084	troper (p s t)	?	late 10th c.
Paris 1118	troper (p s t)	Auch ?	late 10th c.
Paris 1085	antiphoner	Saint-Martial	late 10th c.
Paris 1120	troper (p o s f)	Saint-Martial ?	c.1000
Paris 1121	troper (p s f t)	Saint-Martial	c.1000
Paris 887	troper (p o s)	?	early 11th c.
Paris 1138–1338	sequentiary (s)	Limoges	early 11th c.
Paris 909	troper (p s f t)	Saint-Martial	early 11th c.
Paris 1119	troper (p s)	Saint-Martial	c.1030
Paris 1137	sequentiary (o s f)	Limoges	c.1030
Paris 903	gradual (p o s f)	Saint-Yrieix	11th c., first half
Paris 1133	sequentiary (o s f)	Limoges	mid-11th c.
Paris 1135	sequentiary (o s f)	Limoges	mid-11th c.
Paris 1136	sequentiary (o s f)	Limoges	mid-11th c.
London 4951	gradual (f)	Toulouse	mid-11th c.
Paris 776	gradual (f t)	Gaillac near Albi	late 11th c.
Paris 1132	gradual (o s f)	Saint-Martial	late 11th c.
Paris 1134	sequentiary (o s f)	Saint-Martial	late 11th c.
Paris 779	troper (p o)	Limoges?	late 11th c.
Paris 1871	troper (p o s)	?	late 11th c.
Paris 1177	sequentiary (o s f)	?	late 11th c.
Paris 780	gradual (f t)	Narbonne	late 11th c.
Paris 1139	song-book	?	early 12th c.
Paris 3549	song-book	?	12th c.
Paris 3719	song-book	?	12th c.
Paris 778	sequentiary (o s)	Narbonne	late 12th c.
Paris 1086	sequentiary (o s)	Saint-Léonard, Noblat	late 12th c.

p = tropes for proper-of-mass chants
o = ordinary-of-mass chants
s = sequences
f = offertory verses
t = tonary

variety of contents, and I have used it here basically to separate books without the full range of mass-proper chants from graduals. 'Sequentiary' means that the book does not contain tropes for proper-of-mass chants. Two other important sets of chants frequently found are ordinary-of-mass chants and offertory verses. Several of the manuscripts contain tonaries (studied by Huglo 1971, *Tonaires*). But many of them contain much more besides: tracts, alleluias, gradual verses, processional chants, special offices, and so on. (For summaries of the contents of most of these manuscripts, see Emerson, 'Sources, MS, §II', *NG*; *Graduel romain* II; RISM B/V/ 1.) The later books with rhymed versus and Benedicamus songs I have simply called song-books. The dating of many sources is also only approximate; in many cases the manuscripts are composite, containing sections of different sorts written at different periods, brought together for convenience at binding more than for any musical reason.

Two manuscripts now in the chapter library of the cathedral at Apt are also usually considered in conjunction with the Aquitanian sources. Apt 17 is written in Aquitanian notation, while Apt 18 rarely uses it and mostly employs French notation. (This is not in itself surprising, for French notation was used even over the Alps at La Novalesa.) The provenance of Apt 18, the earlier of the two, from *c*.1000, is unclear, while Apt 17, of the mid eleventh century, is from Apt itself. (See Björkvall in CT 5.)

Benedict of Aniane (*c*.750–821), the great monastic reformer and liturgical organizer of Louis the Pious's reign, came of an Aquitanian family and as early as 779 established in his homeland a monastery whose practice was widely imitated. (Late in his life he had the monastery of Inde, or Cornelimünster, near Aachen, built for him by Louis, which was supposed to be a model for other houses.) But it seems unlikely that the later Aquitanian chant-books contain anything that can be identified as springing directly from his work in that region. On the other hand, a number of survivals of Gallican chant have been traced, especially in the graduals Paris 776 and 903 (see VIII.6).

It is, however, for their repertories of sequences and tropes that the majority of the manuscripts are chiefly prized. The literature on their tropes is by now particularly extensive, and only a few points need be made here. Although by repertory the trope collections are related closely to one another, they are by no means uniform, and the interplay of traditions and recasting of material is evident to a considerable degree. This sort of variation between sources characterizes the transmission of tropes right across Europe, of course, and often the Aquitanian sources stand together against others. But within Aquitaine the variations are often great. In Weiss's edition will be found many instances of strongly variant melodies, particularly involving Apt 17. Planchart and others have pointed out that the earliest troper, Paris 1240, often shares variants with north French manuscripts, rather than with later Aquitanian sources. In these cases it may represent an imported early tradition that was subsequently reworked by Aquitanian musicians. (There are outlines of the relationships between sources in Weiss 1964, 'Problem', David Hughes 1966, Hiley 1983, 'Observations'; the easiest way to see the varied selection and arrangement of trope verses is through the volumes of CT and especially Planchart 1977; for further literature see II.23.)

The sequences, too, are found in many and varied traditions (inventories by Crocker 1957, many editions in Crocker 1977). There is a striking contrast between the moderate collection in Paris 1240 and the vast series in Paris 1084 and 1118. The later sources opt for sizes in between those extremes. In Paris 1086 and 778, and in an added section of Paris 1139, the new sequences with regular rhythm and rhyme, associated with Adam of Saint-Victor in Paris, enter the repertory (Husmann 1964).

Two historical events impinged upon the monastery of Saint-Martial and left traces in its musical manuscripts. In the 1020s a movement was set afoot, whose chief apologist was Adhémar of Chabannes, to gain for Martial the rank of apostle. In 1031 the Council of Limoges declared the apostolicity authentic, although the decision had little effect elsewhere (except possibly to stimulate other churches to promote the cause of their own saints). But at Saint-Martial a new mass (beginning with the introit *Probavit*, replacing chants from the common of saints beginning with the introit *Statuit ei*) and a special office were composed. Other saints associated with Martial and with Limoges, Austriclinianus and Valeria, had offices composed at the same time (Emerson 1965). It is likely that Adhémar composed them all. The prime source for these is the manuscript Paris 909, whose physical make-up was extensively rearranged to take in the new material. Ex. IX.10.1 shows the last of the responsories of the Night Office from the office of St Martial. It has two verses, rather than the usual one, both of which, with the Gloria, have the same melody. Two new melismas, both with form AABBC, are provided at the end, to replace the music for 'omnium' when the last part of the responsory is repeated after the verses. The probable order of performance would then have been:

Respond
VI, repetenda with melisma 1
V2, repetenda with melisma 2
Gloria, repetenda

Ex. IX.10.1. Responsory *O sancte Dei apostole* (Paris, Bibl. Nat. lat. 909, fo. 68ʳ)

(Ex. IX.10.1 cont.)

V1.   Ad-   -nun-ci-as – ti   o-pe-ra do – mi – ni

et facta e – ius in-tel-le-   -xis – ti.   Et in – ter...

V2.   Prae-   -fulgens glo-ri – o-sus   in conspectu do-   -mi–ni

tu – o-rum susci-pe   preces ser-uu-lo-   -rum.   Et in-ter...

Gl.   Glo-   -ri-a   patri et fi – li – o   et spi-ri – tu-i sancto

et nunc et per e – on.   Et in-ter...

M1   Om-   -ni – um (populorum)

M2   Om-   -ni – um
                              (populorum)

In 1062 Saint Martial was reformed from Cluny. The change can be seen in the order of chants, for example the post-Pentecost alleluia series in Paris 1132 and 1134. Table IX.10.2 sets out three different series to be found in Limoges books: (a) that of the cathedral of St Stephen, (b) Saint-Martial before the Cluniac reform, (c) the Cluniac series. The Cluniac series of Advent responsories appears in several Limoges books from the same time onward, including the fourteenth-century noted breviaries Paris 783 and 785.

Table IX.10.2. *Post-Pentecost alleluia series in Limoges*

Limoges cathedral		Saint-Martial before 1062	Cluniac Saint-Martial	
Paris 9438	Paris 1137	Paris 909*	Paris 1132	Paris 1134
7: 12	7: 12	7: 12	7: 12	7: 12
17	17		17	17
20	20	20	20	20
30	30		30	30
46	46	46	46	46
64	64		64	64
		70		
77	77	77	77	77
80	80	80	80	80
87	87	87	87	87
89	89	89	89	89
92				
94: 1	94: 1	94: 1	94: 1	94: 1
94: 3	94: 3	94: 3	94: 3	94: 3
		97		
	99			
104	104	104	104	104
107	107	107	107	107
110			110	
		112		
	113: 1	113: 1		113: 1
113B: 11	113B: 11	113B: 11	113B: 11	113B: 11
			116: 1	116: 1
117	117			
121	121			
		124		
129	129	129	129	129
137: 1	137: 1	137: 1	137: 1	137: 1
	145	145	145	145
146	146	146	146	146
		147: 12	147: 12	147: 12
147: 14	147: 14	147: 14	147: 14	147: 14

Highest number of concordances (among these five sources):

Paris 1137	Paris 9438	Paris 1134	Paris 1134	Paris 1132
(21)	(21)	(19)	(22)	(22)

*Paris 1121 has the same series as Paris 909, but there is a lacuna after 104.

Note to Table IX.10.2:

The alleluias are designated by the psalm from which their verse is taken. Where other verses from the psalm were used for medieval alleluias, the one used here is stated.

7: 12	Deus iudex iustus	104	Confitemini domino et invocate
17	Diligam te domine	107	Paratum cor meum
20	Domine in virtute tua	110	Redemptionem misit
30	In te domine speravi . . . accelera	112	Laudate pueri
46	Omnes gentes	113: 1	In exitu Israel
64	Te decet hymnus	113B: 11	Qui timent dominum
70	In te domine speravi . . . eripe me	116: 1	Laudate dominum omnes gentes
77	Attendite popule meus	117	Dextera Dei
80	Exultate Deo	121	Letatus sum
87	Domine Deus salutis mee	124	Qui confidunt
89	Domine refugium	129	De profundis
92	Dominus regnavit decorem	137: 1	Confitebor tibi
94: 1	Venite exultemus	145	Lauda anima mea dominum
94: 3	Quoniam Deus magnus dominus	146	Qui sanat
97	Cantate domino	147: 12	Lauda Iherusalem dominum
99	Iubilate Deo	147: 14	Qui posuit

XI.11. THE HISPANIC PENINSULA AFTER THE RECONQUEST

Sablayrolles 1911–12; Anglès 1922, 1931, 1935, 1938, 1970; Corbin 1952, *Musique*; Donovan 1958; Fernández de la Cuesta 1980, 1985; Huglo 1985, 'Pénétration'; Castro 1990.

Whatever the importance of the Cluniac monks in the 'reconquest' of Spain for Christianity in the latter part of the eleventh century (see VIII.7), little tangible evidence of their liturgical chant has so far been identified in Spain, unless it be negative features such as the failure of the *Quem queritis* and Easter dramatic ceremonies to take root in the western half of the peninsula (Donovan 1958, ch. 6). Studies so far accomplished show that the Gregorian chant-books of north Spain depend heavily, as we should expect, on Aquitanian traditions, just as Aquitanian notation is found in Spanish books of this time (for example, Paris, Bibliothèque Nationale, lat. 742 is an antiphoner of *c*.1200 from Ripoll with Aquitanian notation: Ripoll was reformed from Saint-Victoire at Marseille; see Lemarié 1965). The closest 'relatives' of the following books, as detailed in *Graduel romain* IV, are all Aquitanian: Madrid, Academia de la Historia 18 and 45 (both graduals from San Millán de la Cogolla, early 12th c.), Madrid, Palacio 429 (noted missal, Castilian, late twelfth century), Toledo, Archivo Capitular 35. 10 (gradual, Toledo, thirteenth century). Hesbert's survey of responsory series (*CAO* 5–6) found the Cluniac order in books from Moissac, which was an important source of influence upon Spain, and also in a twelfth-century antiphoner from Santo Domingo de Silos. The eleventh-century Silos manuscripts London, British Library, Add. 30848 and 30850 (facs. in

Antiphonale Silense, text ed. by Hesbert in *CAO* 2, by contrast, share an order unknown elsewhere (the latter, like many other books, includes some survivals of the Old Spanish rite). If the sources now in Toledo were used there, then they reflect a variety of imported uses. The important twelfth-century antiphoner Toledo, Archivo Capitular 44.2, as already mentioned, may be Aquitanian, though all related sources are Spanish; the thirteenth-century antiphoner Toledo 44.3 is Carthusian; the twelfth-century breviary Toledo 48.14 is Camaldolese. Much work remains to be done on the musical traditions of these sources.

A number of important manuscripts containing tropes and sequences have survived, which, while displaying obvious contacts with south French traditions, have numerous individual contributions to the repertory. Among the most important are the Vich tropers, Vich, Museo Episcopal 105 (eleventh to twelfth century) and 106 (twelfth to thirteenth century), and the twelfth-century Huesca troper, which include tropes for the proper chants of mass as well a ordinary-of-mass chants and sequences. Gros (1983) has described and edited the text of the fourteenth-century processional Vich 117 and compared it with its sister manuscript in Erlangen, Institut für Musikwissenschaft (Stäblein 1975, 160–1). There are a number of manuscripts, from Tarragona, Burgos, Gerona, and Tortosa, with ordinary-of-mass chants and sequences (sometimes polyphonic). Of these (brief descriptions and bibliography by Husmann in RISM B/V/1), Barcelona, Biblioteca de Cataluña M. 911, from Gerona in the fifteenth century, may be singled out because of its large provision of responsory prosulas, more numerous than in any other known source (Hofmann-Brandt, 1971, Bonastre 1982)·. All these sources come from Aragon except for the one from Burgos in Castile.

The famous 'Codex Calixtinus', probably copied in central France at a centre such as Vézelay around 1160, was destined for the famous pilgrimage shrine of Santiago da Compostela (ed. Whitehill *et al.* 1944; music alone ed. Wagner 1931; on the nature of the source see Hohler 1972). A copy of it was made at Santiago in 1173 by Arnalt de Munt, a monk of Ripoll (Barcelona, Corona de Aragón 99).

Corbin (1952, *Musique*, with further information in 1960 *Deposition*) has surveyed Portuguese liturgical books and their chant. (See also Bragança 1975, 1985.) It is said that when Lisbon was captured from the Moors in 1147 by a joint Portuguese–English expedition, an Englishman named Gilbert was appointed bishop of Lisbon and introduced the Salisbury missal into use there (Almeida 1910–22, i. 574 n. 1). Queen Philippa of Lancaster is reputed to have heard the canonical hours daily according to Salisbury use (Almeida, i. 627), and may have inspired her sons, Saint Ferdinand (1402–43) and King Duarte (Edward, 1433–8), to do the same. Extant service-books to support these claims have not yet been discovered. Corbin's extensive survey—she was able to list over eighty sources—reveals the expected influence of Aquitanian practice, easily visible in the notation, with other French traditions, including the Cluniac, making a contribution. Numerous Cistercian books survive. Corbin describes a number of local liturgical customs and their music.

IX.12. NORTHERN AND EASTERN EUROPE

(i) Scandinavia and Iceland
(ii) Poland
(iii) Czechoslovakia
(iv) Hungary

The pattern which we have seen when church life was revitalized in Normandy at the beginning of the tenth century was repeated wherever the Roman rite and its plainsong moved into new regions of the West. Much of chant research in these circumstances consists in identifying the starting-point(s) of the 'colonization' and isolating those parts of the repertory composed in the newly established church. The starting-point can often be revealed by palaeographical as well as repertorial studies. The newly composed material was usually called forth by the cult of a local patron saint, which most often entailed a new sequence for the mass and a new office, very possibly modally ordered and with rhyming, rhythmic text. A few remarks have been made about the introduction of Gregorian chant in the Hispanic peninsula; this chapter looks briefly at countries at the opposite end of Europe: Scandinavia, Poland, Czechoslovakia, Hungary, and Dalmatia. I have not attempted to summarize all the results of what in many cases is very distinguished and extensive research, and have simply cited some of the more important studies.

(i) *Scandinavia and Iceland*

The three archdioceses of medieval Scandinavia coincided geographically with medieval, not modern political divisions. Nidaros (Trondheim) covered an area somewhat larger than modern Norway. Lund included Denmark and the south of modern Sweden, where the cathedral was in fact situated. Uppsala took in the rest of modern Sweden and also Finland. The church in Denmark was organized chiefly from north Germany, while that in Norway owed much to England. Liturgical practice in both lands would have been established in the late tenth and eleventh century. At the Reformation liturgical books were systematically destroyed and reused to bind and cover other books: vast numbers of fragments remain.

Numerous fragments from English sources of the eleventh and twelfth centuries have survived in Norwegian collections, many with notation, and these are no doubt indicative of an important source for Norwegian liturgies and their chant (see Toni Schmid 1944, Gjerløw 1961, 1970, 1979). But as Gjerløw has shown, other sources and influences were present: the Nidaros ordinary draws upon both Lanfranc's *Decreta* and the German commentators Bernold of Constance and Honorius of Augsburg (Gjerløw 1968); while much of the order of services can be traced to English uses, no one point of departure emerges. Gjerløw's intensive exploration of the sources of the Nidaros antiphoner (1979) weighs the evidence of all the fragmentary remains, then the choice of chants, and shows once again a strong relationship with the various English traditions as well as some points of contact with German ones. The Nidaros breviary of 1519 has been published in facsimile (*Breviarium Nidrosiense*). The Nidaros sequentiary has been published in facsimile with transcriptions by Eggen (1968).

Remarkable evidence of traditions in Iceland, similar to the Norwegian, and deriving ultimately from the same sources, has been discovered and presented in exemplary fashion once more by Gjerløw (1980). A proper mass and office for Iceland's patron, St Thorlac, have survived (ed. Ottósson 1959).

Apart from around 15,000 fragments with musical notation (Haapanen 1922–32, 1924, Andersen and Raasted 1983), only about a dozen notated Danish liturgical books survive, none of which are graduals or antiphoners (Abrahamsen 1923, Asketorp 1984). Strömberg showed that a pontifical from Reims must have been brought to Lund, interacting with the widespread Roman-German pontifical (Strömberg 1955), and Asketorp was able to support this conclusion to some extent with a study of the melodies (1984). Husmann's comparison of post-Pentecost alleluia series in Danish missals (1962 and 1964–5, 'Studien') shows the expected influence of north German uses. The Lund missal of 1514 has been published in facsimile (*Missale Lundense*).

For Sweden, the fragmentary remains of the printed gradual of Västerås (*c*.1513) have been published in facsimile by Toni Schmid (*Graduale Arosiense*), with sequences for Sweden's patron saints. The Linköping ordinal has been edited by Helander (1957). Kroon (1953) studied Swedish repertories of ordinary-of-mass chants, Moberg (1927, 1947) those of sequences and hymns. Special offices have been studied by Undhagen (1960), Önnerfors (1966), Milveden (1964, 1972), and Lundén (1976). St Bridget (Birgitta: *c*.1303–73) founded the Order of the Brigittines at Vadstena in 1346. A breviary of their house at Syon near London has been edited by Collins (1969).

The thousands of fragments which survive in Finland have been surveyed by Haapanen (1922–32, 1924).

(ii) *Poland*

Morawski 1988.

The unbroken history of Roman Catholicism in Poland since the end of the tenth century has meant that a large number of medieval sources are preserved. Research in the last thirty years has been particularly energetic. Numerous individual sources have been analysed (results published in *Musica medii aevi* and *Muzyka religijna w*

Polsce: Materialy i studia). Most indicators point in the direction of Bavaria, Switzerland, and Austria as the ultimate sources of Poland's chant traditions, apart from books of the religious orders.

We are probably better informed about Polish sequence repertories than those of any other country, thanks to the work of Kowalewicz (1964), Morawski (1973), and Pikulik (1973–6, 1974—101 sources—and 1976). Pikulik (1978, 1990) has catalogued and analysed the ordinary-of-mass repertories (ninety sources), and also Marian alleluias (Pikulik 1984). Polish pontificals have been studied by Miazga (1981).

(iii) *Czechoslovakia*

Konrad 1881; Hutter 1930, 1931; Nejedlý 1904/rev. 1954–6; Schoenbaum 1960; Vanický 1960.

The bishoprics of Prague (Bohemia) and Olomouc (Moravia) had contrasting histories. Until its destruction at the beginning of the ninth century by the Magyars, the church in Moravia was influenced as much by the Greek as by the Roman church. Thereafter it resembled Bohemia in coming under strong German influence. In the tenth century Benedictine monks, from the Reichenau, from Korvey in Saxony, and especially from St Emmeram at Regensburg, played a leading part in the Bohemian church. The results of this in surviving chant manuscripts have yet to be investigated. From the eleventh century collegiate chapters were increasingly important, and the later religious orders made their usual contribution.

Notations in medieval Bohemia have been discussed by Hutter (1930, 1931). Whereas in the twelfth century German notation signs were used, in the thirteenth the Laon type of Gothic notation was taken up. Prague already knew staff-notation in the twelfth century.

Plocek's splendidly detailed catalogue of sources now in Prague (1973), with numerous melodic incipits, gives a good idea of the range of surviving material (see also Podlaha 1910). Among the earliest and most interesting sources is Prague, Archív Metropolitní Kapituly, cim. 4, a collection of the twelfth to thirteenth century from Prague Cathedral of offertory verses and ordinary-of-mass chants (Spunar 1957), many of the latter being 'up-to-date' north French compositions. Numerous manuscripts from Czechoslovakian libraries were filmed by Stäblein and thus available for his work on hymns and the studies of Melnicki, Bosse, Thannabaur, Schildbach, and Hofmann-Brandt. The link with south German and Austrian traditions is clear, while for the repertories of sequences and ordinary-of-mass chants contact is evident with the newer traditions of the twelfth to thirteenth centuries coming from centres such as Paris. The Latin songs usually called *cantiones* were also written in large numbers in Bohemia and Moravia (Orel 1922). The offices of St Procopius and St Ludmila have been studied and edited by Patier (1970, 1986).

(iv) *Hungary*

Szigeti 1963; Dobszay 1985, 1990 'Plainchant'.

The Hungarian church was formally constituted under King (St) Stephen I in 1001, when ten bishoprics under Esztergom (Gran) were established. The Hungarian bishops are reported to have decided in 1100 to normalize their liturgy according to the order of service in Bernold of Constance's *Micrologus de ecclesiasticis officiis* (Kennedy 1956, 234). Repertorial studies have not yet confirmed this.

The earliest surviving sources with notation are from the late eleventh century, the first of many manuscripts taken to Zagreb after its diocese had been founded under Hungarian rule in 1094 (Vidaković 1960). A fully notated antiphoner of the early twelfth century, the so-called 'Codex Albensis' from Székesfehérvár (Stuhlweissenburg, Alba Regia), has been published in facsimile (*Codex Albensis*), as has an early fourteenth-century noted missal from Esztergom (*Missale notatum Strigoniense*).

Chant research has been pursued in Hungary in recent years with exceptional energy and authority. In virtually all areas substantial and systematic studies have been made which often surpass those available for other countries. Radó's catalogue (1973) of liturgical books in Hungarian and neighbouring libraries has provided the foundation for further basic work, in particular the catalogue, analysis, and classification by Szendrei (1981) of over 850 complete and fragmentary sources, with over 100 facsimiles. Equally impressive is Szendrei's account of Hungarian chant notations (Szendrei 1983, with copious facsimiles; revised German version 1988, 'Geschichte'). The dominant type was that developed at Esztergom when staff-notation was taken up at the beginning of the twelfth century, resembling south German staff-notation in many respects, but seemingly a consciously evolved new variety.

Offices for Hungarian saints have been edited by Falvy (1968). The corpus of sequences in Hungarian sources has been edited by Rajecky and Radó (1956), and an edition of the complete repertory of office antiphons is in preparation.

Repertorial studies have concentrated on office-books, a number of which have been completely inventoried and compared with numerous extra-Hungarian traditions as well (see Ullmann 1985, Dobszay 1988, Dobszay and Prószéky 1988).

Hungarian experts distinguish a Zagreb liturgical tradition from the dominant Esztergom one. Sources have been studied by Höfler (1967) and Hudovský (1967, 1971).

The Dalmatian coast area, part of Croatia in the eleventh century and variously under Hungarian, Bosnian, and Serbian control, received chant-books from south Italy and made its own with Beneventan notation (Birkner 1963, Hudovský 1965, Bujić 1968; see also the Dalmatian sources in Loew-Brown 1980 and the items studied by Gyug 1990).

X

Reformations of Gregorian Chant

X.1. INTRODUCTION

Fellerer, 'Choralreform', *MGG*.

The Roman-Franciscan chant repertory does not mark a radically new beginning in the history of Western chant, except in so far as it meant the end of Old Roman chant in the Eternal City itself. Whatever musical reforms it may have involved, they embody a concept of Gregorian chant not in any substantial way different from that found in most churches in most lands. The Cistercian order of monks, however, revised its chant repertory according to fixed principles which entailed quite drastic alterations in the melodies they had inherited. Therefore their chant comes first in this chapter on reforms of the chant repertory.

Since no early medieval chant-books have identical musical readings throughout, they must all, to a greater or lesser extent, have been made by, or according to the directions of, cantors with their own individual ideas about how the melodies should be sung. Medieval writings on chant are liberally strewn with references to cantors who 'emended' the melodies, and books that are 'bene emendati'. The more specific the notation, the more obvious are the differences between the books, as for example in the case of the 'rhythmic' and other signs in the early St Gall and Laon manuscripts. The necessity of specifying pitch relationships added another area of variance between sources. Sometimes it is clear that tonal preconceptions entered into the matter, so that chants were variously notated according to different notions of tonality: less straightforward passages involving semitone steps were made more obvious, tonally eccentric passages were recast in the interests of modal unity, and so on. Some manuscripts stand out as relatively heavily 'edited' in these respects, for example the Saint-Yrieix gradual, Paris, Bibliothèque Nationale, lat. 903 (PalMus 13; see Stuart 1979).

The Cistercians went well beyond this degree of editorial interference, and, while previous versions can all to some extent be called little 'reforms' (but 'editions' would be better), theirs was a major recasting of the repertory. Their revision was adopted in large measure by the Dominicans. After these two I discuss the chant of other religious orders, although these do not encompass major musical departures comparable with that of the Cistercians.

The next major reform of the chant repertory took place in the aftermath of the Council of Trent and can be regarded as part of the Counter-Reformation. This radical overhaul of the Roman chant repertory was but the beginning of a series of recastings of the repertory in other lands, particularly in France, where so-called 'Neo-Gallican' chant was written as a substitute for the medieval or the newer Roman melodies.

X.2. THE CISTERCIANS

Delalande 1949; Marosszéki 1952; Cocheril 1956, 1962; Gümpel 1959; Canal 1960; Hüschen, 'Zisterzienser', *MGG*; Berry, 'Cistercian Monks', *NG*; Huglo 1971, *Tonaires*, 357–67; Waddell 1970, 'Prologue', 1970 'Origin', 1971, 1976, 1980, 1984, 1985; Choisselet and Vernet 1989.

The Cistercian order takes its name from the monastery of Cîteaux, near Dijon, founded in 1098 by the Benedictine monk Robert of Molesme and like-minded brethren, who wished to recover a simpler and stricter form of Benedictine monasticism than that generally practised in their time. It rose to Europe-wide fame and influence through the example and activity of St Bernard of Clairvaux (1090–1153). Daughter houses observed the same liturgy as Cîteaux. Such uniformity was not entirely new, for although Benedictine houses were bound to no one liturgy, identity of liturgical practice between groups of houses can occasionally be found, as when they stand in a mother–daughter relationship. Thus the Norman monasteries refounded from Saint-Bénigne of Dijon all reflect, more or less exactly, the liturgy of Saint-Bénigne. The importance attached to this uniformity and the scale on which it was established in the Cistercian order was, however, unprecedented.

The Cistercian musicians took the performance of the liturgy in its 'pure' and 'original' form very seriously. We know little of the earliest musical practice at Cîteaux, though the office was much shortened in an effort to recover the state prescribed in the Rule of St Benedict, often interpreted in a very literal way. It appears, however, that the ancient reputation of Metz as guardian of the best chant tradition was known at Cîteaux, and monks were sent to Metz to make a copy of the 'authentic' antiphoner there. The thirteenth-century Metz antiphoners Metz, Bibliothèque Municipale 83 and 461 may well represent the tradition which the monks would have encountered (Waddell 1970, 'Origin'). The Cistercian copies are lost. The Messine chant was in any case eventually pronounced unsatisfactory and St Bernard was authorized to oversee a new revision. By about 1190 a complete copy of the Cistercian liturgy had been compiled, to be used as standard for all other manuscripts, now Dijon, Bibliothèque Municipale 114 (facs. Hüschen *MGG*, Choisselet and Vernet 1989); the last five parts of this massive book, including hymnal, gradual, and antiphoner, have unfortunately been missing since the sixteenth century.

The modal considerations which exercised the Cistercians, like other editors, were mentioned in the previous section. To these were added quite new guide-lines. In notating the chants B-flats were to be avoided, by transposition if necessary. Chants were edited to lie predominantly within a single octave, ten notes being the limit. Long melismas were cut back. The last prescription is a clear reflection of a more austere aesthetic, as is the one that repetition of words should be avoided.

The history leading up to these changes and the main idea behind them are described in a short prologue 'Bernardus humilis abbas Clarevallis' written by Bernard and commonly attached to Cistercian chant-books (PL 182, 1122–3, CSM 24. 21–2). The principles of the musical reform are described in a longer tract, regarded as a preface to the chant-books, beginning 'Cantum quem Cisterciensis ordinis ecclesiae' (PL 182. 1122–32, CSM 24. 23–41). This is thought to have been put together by Guy de Cherlieu, monk of Clairvaux and from 1131 to 1157 abbot of Cherlieu. The theoretical part of the tract, which is supposed to justify the emendations, is in fact taken from an earlier treatise, the 'Regula de arte musica' (CS ii. 150–91) of another Guy—Guy d'Eu (Guido Augensis; not Guy de Châlis, as Coussemaker had it). This Guy was also a monk at Clairvaux, apparently from 1132 at Longpont, and later abbot at an unknown monastery. His treatise, beginning 'Premunitos autem esse volumus', was written for his pupil William, first abbot of Rievaulx in Yorkshire, in 1132. Furthermore, the so-called 'Tonale sancti Bernardi' at the end of early Cistercian antiphoners is also from Guy d'Eu's treatise (GS ii. 265–77, PL 182. 1153–66). (This complicated situation would be considerably simplified if these were the same Guys, but the required proof has not yet turned up.)

Marosszéki carried out a number of tests to establish the parentage of the Cistercian chant tradition, and these have to some extent been refined in *Graduel romain* IV and by Hesbert in *CAO* 5. The order of Advent office responsories does indeed agree with the series in the thirteenth-century antiphoner of Saint-Arnoul at Metz, Metz 83. The melodic variants in the gradual point more in the direction of Dijon, situated only 20 km. north of Cîteaux (whereas Metz is over 250 km. further), but no gradual from Metz survives for comparison.

In view of what has been said, it may seem otiose to compare Cistercian books with traditional ones. But the alterations mentioned in the famous preface are not encountered very often. One has only to look at the pages from Cistercian graduals reproduced in PalMus 2–3 and Marosszéki to see how resolutely they keep to the usual path, B-flats, melismas, and all. Marosszéki and Delalande (1949) certainly provide many examples of alterations, but they do not, by and large, leap to the eye in medieval manuscripts.

The notation in Dijon 114 is a variety of French notation on lines with a number of Laon (Messine) features, the type which eventually developed into what is usually called 'Gothic' notation, and in various manifestations this was the notation used in very many subsequent Cistercian books. It has been suggested that Cistercian books were the prime agent for the propagation of this notation in Germany and lands east, in other words of the change from staffless to staff-notation and from German to

Gothic signs (Hourlier 1951, 'Messine', 153–7). Szendrei (1985) has pointed to the lack of exactitude in use of the terms 'Messine' and 'Gothic' and shown that the Cistercians' notation was only one of several varieties involved in the new developments in south Germany and Austria.

On four occasions the Benedictine Rule calls a hymn an 'Ambrosianum'. The Cistercians therefore tried to find the hymns of St Ambrose for their revised office (or rather, office restored to its 'authentic' form and content), travelling to Milan for the purpose (at a time when the Milanese were probably not yet using notated hymnals). A hymnal with thirty-four texts, to nineteen melodies, was compiled, many of them heretofore quite unknown outside Milan. This was used among other places at the monastery of the Paraclete founded by Abelard for Héloïse, and two interesting letters by Abelard survive, commenting on the general strangeness and literary peculiarities of the hymnal (Waddell 1976), one of them Abelard's only letter to St Bernard. The hymnal was indeed eventually found generally unsatisfactory, and a new one prepared with sixty texts to thirty-seven melodies (seven of the previous melodies were dropped). The non-'Ambrosian' hymns were assigned to hours other than Vespers, the Night Office, and Lauds, when the Rule of St Benedict does not use the specific term 'Ambrosianum'. Both the first and second hymnal have been edited by Waddell (1984) with definitive commentary.

X.3. THE DOMINICANS

Bonniwell 1944; Delalande 1949; Hüschen, 'Dominikaner', *MGG*; Gleeson 1972; Berry, 'Dominican Friars', *NG*.

The Dominican order of friars (Ordo Praedicatorum, Order of Friars Preachers), founded by St Dominic, was formally constituted in 1220. Its organization was strongly centralized, like that of the Cistercians, and its chant practice also followed the Cistercian in many respects. The office is, however, celebrated according to the Roman or secular cursus, not the monastic.

The first moves to formulate a standard liturgy appear to have been taken by the second Master-General, Jordan of Saxony (1221–37). A new revision was set in motion in 1244 by John the Teuton (1241–52). At the General Chapter of 1245 a commission of four brothers was appointed, one from each of the four most important provinces of the order, France, England, Lombardy, and Germany, who were to pool their knowledge and materials. (Hesbert 1980, 'Sarum' shows how the Sarum office may have played a part in the final recension.) Working sessions in Angers produced no generally acceptable result, and in 1250 the brothers were commissioned to visit Metz, as the Cistercians had done. After further disagreement the matter was placed in the hands of the newly elected Master-General Humbert of Romans (1254–63), previously Provincial of France and probably already influential in the work of revision. His proposals were approved at the General Chapters of 1255 and 1256. His master copy of the entire liturgy and its music still exists, the 'Correctorium Humberti', or 'Correctorium Sancti Jacobi' (after the Dominican headquarters at Saint-Jacques in Paris where it was made), or 'Le Gros Livre' (now Rome, Santa Sabina XIV. lit. 1). A portable copy for the Master-General's use also survives (London, British Library Add. 23935), and another complete copy of the fourteenth century in Salamanca. A set of rules for the copying of chant-books, based on the Franciscan one, was composed (Van Dijk 1963, *Sources*, 118, Huglo 1967, 'Règlement').

The Cistercian and Dominican series of alleluias for the Sundays after Pentecost are identical, and, as Delalande has shown, the differences during the rest of year are very few. Many of the bowdlerizations of the chant repertory carried out by the Cistercians were perpetuated in the chant of the Dominicans, though in many instances there was a reversion to the norm. Delalande (1949) has provided numerous examples of complete or partial agreement of Cistercian and Dominican books against the general tradition. Occasionally the Dominicans operated where the Cistercians had declined to do so.

Ex. X.3.1 (adapted from Delalande 1949, Tab. IXbis) shows part of the gradual *Exsurge domine fer opem nobis* V. *Deus auribus nostris* (discussed also by Bomm

Ex. X.3.1. From gradual *Exsurge Domine* in Gregorian and Cistercian–Dominican versions (Piacenza, Bibl. Cap. 65; St Petersburg, Saltykov-Shchedrin Public Lib.; Delalande 1949)

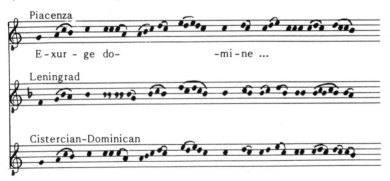

1929, 116–20). This piece is tonally problematic, and several different ways of notating it are to be found. The Cistercians, followed exactly by the Dominicans, devised a unique way which entails a quite drastic revision.

The differences between notated versions concern the relationship between the main reciting note—bb or c, even d near the end of the verse—and the final, either E or D. The situation is complicated by the fact that the reciting note appears to change during the piece and the verse may begin at a level higher than the respond did. All possible combinations of these are to be found:

Respond	reciting note:	c or bb . . . c
	cadence:	E or D
Verse	reciting note:	c or c . . . d or bb . . . c
	cadence:	E or D

To this knot the Cistercians took a sharp sword indeed, staying firmly with c-recitation throughout and then, presumably feeling that the usual descent to the cadence was not adequately prepared, eliminating it altogether by excising a short phrase and ending on a instead of E or D. Ex. X.3.1 gives the beginning of the gradual and the end of the verse in two of the more traditional versions, followed by the Cistercian/Dominican one.

X.4. CHANT IN OTHER RELIGIOUS ORDERS

(i) The Carthusians
(ii) The Premonstratensians

Unlike the Cistercians, other monastic orders did not make a radical revision of the chant repertory but adapted what lay to hand in a fairly straightforward way, as monasteries had traditionally done. While each order was concerned for correctness and uniformity of practice throughout the order, their traditions do not contain many surprises.

(i) *The Carthusians*

Hüschen, 'Kartäuser', *MGG*; Lambres 1970, 1973; Becker 1971, 1975; Huglo 1977, 'Livres'.

The Carthusian order was founded by St Bruno of Cologne at the Grande Chartreuse (Charterhouse) in the Alps a little north of Grenoble in 1084. It was conceived as a contemplative order with a strongly eremitical character. A maximum of twelve monks for each monastery was fixed, and rarely relaxed. The monks lived alone in their cells, coming together only for worship and for meals on feast-days. The Little Hours were said privately. The order had no codified rule until the fifth prior of the Grande Chartreuse, Guigo (Guigues de Châtel), drew up a customary, completed in

1127 (PL 153, 635–760). This included regulations not just for the monastic discipline and way of life but also for the liturgy and its chant, with a tonary (ed. Becker 1975). While the office until then had followed the Roman or secular cursus, Guigo's liturgy had the monastic cursus. Some simplifications of liturgical and musical practice seem to have been introduced: non-biblical texts were avoided, as in the liturgy of Lyon (for that reason hymns were sparingly used); melismas in office responsories were avoided; all those parts of the repertory where expansion can often be found in medieval sources—for example, alleluias, ordinary-of-mass melodies— were kept small. The order had no interest in musical splendour.

Huglo has shown that the post-Pentecost alleluias of Carthusian books correspond with those of Clermont, the Advent responsories with those of Clermont and Aurillac. The melodic variants in proper-of-mass chants agree most often with those of Valence, one of the Aquitanian group of sources. Early Carthusian books have Aquitanian notation, but a very straightforward type of square notation was adopted at the end of the twelfth century which lacks an upward stroke at the start of the clivis (see Plate 14).

(ii) *The Premonstratensians*

Hüschen, 'Augustiner', *MGG*, 'Prämonstratenser', *MGG*; Lefèvre 1933, etc.; Weyns 1967, 1973.

The rule of St Augustine was a document probably drawn up by one of St Augustine's followers and partly on the basis of his writings, which specified a number of particular observances and offered more general comment about the nature of a common religious life. It was little used in the early Middle Ages, but was taken up by the communities of clerics who increasingly from the mid-eleventh century and the period of the 'Gregorian reform'—so called after the reforming pope Gregory VII, 1073–85—formed religious groups devoted to a life of poverty and celibacy in imitation of the first Christians. They became known also as 'Canons Regular' because they followed a Rule (*Regula*). They used the liturgy of whatever diocese they inhabited, except in the case of congregations grouped under the leadership of a particularly influential house. Such were the Augustinian canons of Saint-Victor, Paris, whose version of the Paris liturgy was used by other Victorine Augustinian canons (for example at Bristol in England). The most influential were the Premonstratensians, founded by St Norbert at Prémontré, a little west of Laon, in 1120. They were particularly active in east Germany and Hungary.

The liturgy of the Premonstratensians has been described in various works by Lefèvre, who has edited several ordinals. The text of the mass has been studied by Weyns (1967, 1973). It is not clear to what extent the chant was unified among Premonstratensian houses. The monks of Solesmes tested the melodic variants of one source from Prémontré itself, and other Premonstratensian manuscripts from the Low Countries, Germany, and Austria. The gradual from Prémontré (Soissons,

Bibliothèque Municipale 85) and two of the German sources were closely similar, and similar to books from Liège and Andenne (not Laon). Others were closer to other north French uses, and one was grouped within the eastern European 'bloc'. It is possible that some houses entered the Premonstratensian congregation after their chant tradition was established.

The various congregations of Augustinian Canons Regular are quite distinct from those of the Augustinian hermits or friars. This order was formed by Pope Alexander IV in 1256 from various eremitical communities. They became particularly numerous in German lands. They had no centralized chant tradition.

Little attention has so far been paid to the chant of later religious orders. The Carmelite Order, founded in Palestine in the mid-twelfth century and transferred to Italy a century later, observed the liturgy of an ordinal by Sibert de Beka from 1312, but the melodies were not prescribed. Office sources have been surveyed by Boyce (1990), who has also edited two special offices (1988, 1989). A number of sources with music have survived from the Brigittine Order, founded by St Bridget of Sweden about 1346 at Vadstena. This was originally a double order of both monks and nuns; while the brothers followed the liturgical use of the local diocese, the sisters celebrated an individual Marian office liturgy, referred to as the 'cantus sororum', attributed to Petrus Olavi of Skänninge. Its text has been edited by Collins (1969) and Lundén (1976); Servatius (1990) has made a musical edition of the antiphons. Of the chant of the German reforms of Bursfeld, Kastl, and Melk, relatively little is known, except through the work of Angerer (1974, 1979) on Melk. The Melk reform, in essence yet another return to a 'purer' Benedictine monasticism, was promoted by Albert V of Austria, who founded the famous monastery overlooking the Danube in 1412.

X.5. THE 'MEDICEAN' GRADUAL (1614–15)

Molitor 1901–2; Weinmann 1919.

The threat to the Roman church from Protestantism led to repeated calls for a universal council, which eventually took place in Trent in northern Italy intermittently between 1545 and 1563. One of the minor matters discussed was the music of the church, decrees affecting which were issued at the twenty-second to twenth-fourth sessions. As far as plainchant was concerned, these were couched in general terms. In all music, intelligibility of the text was important. An important provision was made that provincial synods should establish musical practice with respect to local traditions and circumstances.

It was in the Roman chant-books compiled in the aftermath of the Council that concrete shape was given to the broad instructions of the Tridentine decrees. A reformed breviary had been published in 1568, a reformed missal in 1570. In 1577 Giovanni Pierluigi da Palestrina and Annibale Zoilo were commissioned to prepare the music for them, but did not complete the task. Not until 1614 and 1615, in fact, did the two volumes of a new gradual appear, its editors being Felice Anerio and Francesco Soriano, remnants of a six-man committee which had begun work in 1608. The work was published by the Medici Press in Rome, hence its usual appellation of the 'Medicaea' or 'Medicean edition'.

An antiphoner along the same lines was never planned. But already between 1582 and 1588 Giovanni Guidetti, the papal chaplain, singer in the papal chapel and pupil of Palestrina, had published a series of handbooks in small format with music not in the gradual, the most important of which was the *Directorium chori* of 1582, containing recitation tones for prayers, lessons, and psalms, with hymns, versicles, and responses, and short responsories. An important innovation in Guidetti's *Directorium* was the use of proportional note values:

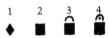

Since the third value only occurred when followed by the first, any piece was exactly measured in units of the second value.

It is important to remember that the use of these books was never officially binding upon the church. It was the example they set that was most telling. The way towards further rewriting of the ancient melodies was open for anyone with the will and ability to follow it.

The drastic nature of the Anerio–Soriano revision cannot be understated. It was a complete recasting of the repertory according to the aesthetic principles of the humanist age, carried out by composers whose musical sensibilities were formed by the polyphonic church music of Palestrina and his contemporaries (see Meier 1969). (Exactly what Palestrina may himself have contributed to the revision is unclear. He was occupied with it in 1578, and his work and ideas would have been known to Soriano, who was his and Zoilo's pupil, and to Anerio, who was Soriano's pupil and Palestrina's successor as composer to the papal chapel.) The corner-stones of the revision were modernization of the tonality and rearrangement of the notes to reflect correct declamatory principles, so that unaccented syllables should not have more notes than accented ones. Every type of chant was affected, the more ornate ones most of all.

Although the Medicaea has not been published in facsimile, it was accurately reproduced in the *Graduale Romanum* published by the firm of Pustet in Regensburg (Ratisbon) in 1871, edited by Franz Xaver Haberl. For thirty years this edition enjoyed papal authority as the approved form of chants for mass, and its contents are widely available for study. Ex. X.5.1 shows the type of revision carried out in the Medicaea. *Omnes de Saba* is the mode 5 gradual for Epiphany. In its Gregorian version, when the melody hovers around *c*, *b*♮ is sung. In the verse the final *F* exerts a stronger tonal attraction and *b*♭s are frequent. The Medicaea avoids *b*♮ altogether. The fluttering repetitions of *c* and *b*♭ are eliminated; notes are not repeated at all

Ex. X.5.1. Medieval and Medicean versions of gradual *Omnes de Saba* (Piacenza, Bibl. Cap. 65; *Graduale*, Rome, 1898)

without a change of syllable (see the final cadences of both parts). Each melisma on the final syllable of a phrase is cut, and the balance between accented and unaccented syllables is altered: for example, *deferentes* is changed from 1–2–2–5 notes to 1–1–6–1, *Saba venient* from 3–13–1–3–1 to 5–2–2–1–2. The accented syllables are marked in the Pustet edition, though not in the Medicaea.

As a piece of early seventeenth-century plainchant the new version is admirably moderate, mellifluous, and consonant. But none of the ecstatic, other-worldly quality of the medieval version remains.

X.6. NEO-GALLICAN CHANT

Bäumer 1905; Henri Leclercq, 'Liturgies néo-gallicanes', *DACL*; [Hiley], 'Neo-Gallican Chant', *NG*; Fontaine 1980; Brovelli 1982.

New Roman liturgical books with texts revised on humanist lines were briefly fashionable, of which the Breviary of the Holy Cross by Cardinal Quiñónez (1535) is an example. But the Council of Trent (1545–63) did not perpetuate such initiatives. The new Roman breviary and missal of Pius V (1568 and 1570 respectively) were by no means drastic revisions of the traditional texts, and would have been found quite normal by the average late-medieval cantor. Those cantors obliged to provide music for services where they were used would have had no difficulty in continuing to employ the traditional chant-books, perhaps one of the many printed editions which appeared in the sixteenth century. They would in any case have had to do precisely this for the office. Popular Italian books included those printed by Junta in Venice: the *Graduale Romanum* of 1606 and 1611 and the *Antiphonarium* of 1603. Another is the *Antiphonarium* published in Antwerp in 1611 for the diocese of Mechelen (Malines), which was used by Haberl for the Pustet antiphoner of 1878. (See Tack 1960 for numerous facsimiles.)

But the seeds of literary and musical revolt against the medieval heritage had been sown, and the next two centuries were to see large numbers of quite new books where texts and melodies were revised or completely replaced. To the extent that they mark a complete break with the medieval chant which is the main subject of this book, they could be omitted from further consideration here. And research on what is in fact a vast heritage of such service-books has hardly begun. A brief statement of the general situation must therefore suffice.

The making of new chant-books was particularly energetic in France, where the church's relationship with Rome was a matter of political and doctrinal controversy. The decisions of the Council of Trent regarding the constitutional structure of the Roman church were not accepted in France, where a strong school of thought supporting freedom from papal authority had existed ever since the thirteenth century. These beliefs were enshrined in the Four Gallican Articles of 1682, drawn up by Bossuet, and although they were formally withdrawn a decade later, their essence was propagated throughout the next century. The tide of opinion did not begin to turn back in an ultramontanist direction until after the restoration of the French monarchy in 1814. The later seventeenth and eighteenth centuries therefore mark the period of composition of what is commonly known as 'Neo-Gallican' chant.

Of the many liturgical books produced in many French dioceses during this period, those for the office are more radical revisions than those for mass. Particularly numerous were the new hymns, by such authors as Santeuil and Coffin (see Pocknee 1954). The most important new liturgies were those of Paris, where Archbishop François de Harlay issued a new breviary in 1680 and Archbishop Charles de Vintimille both a new breviary and a new missal, in 1636 and 1638 respectively. The breviary of Cluny of 1686 was also influential. It has been estimated that ninety out of 139 French dioceses had non-Roman books by the end of the eighteenth century (Fontaine 1980), the Paris books being adopted in over fifty of them.

In the 'Neo-Gallican' chant-books both revision of older chants and composition of quite new ones is to be found. The revision is generally more radical than that of the Medicaea, particularly from the tonal point of view. Accidentals were introduced and whole sections shifted to make more obvious tonal sense (from the Baroque standpoint). Ex. X.6.1 shows two chants for mass at Epiphany from the *Graduel de Paris* of 1754.

Ex. X.6.1. Extracts from *Graduel de Paris*, 1754 (pp. 110, 112)

In the sequence *Ad Jesum accurite* the note values are usually pairs ♩ ♦ , sometimes ♩ ■ . I have assumed that triple time was intended throughout.

In the offertory *Reges Tharsis* I have transcribed note values as follows: ♩ = ♪

■ = ●

♦ = ·

PROSE. Du 1. (Ton)

(Ex.X.6.1 cont.)

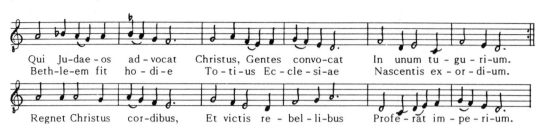

Qui Ju-dae-os ad-vocat Christus, Gentes convo-cat In unum tu-gu-ri-um.
Beth-le-em fit ho-di-e To-ti-us Ec-cle-si-ae Nascentis ex-or-di-um.

Regnet Christus cor-dibus, Et victis re-bel-li-bus Profe-rat im-pe-ri-um.

A- -men.

OFFERTOIRE. Du 8. (Ton)

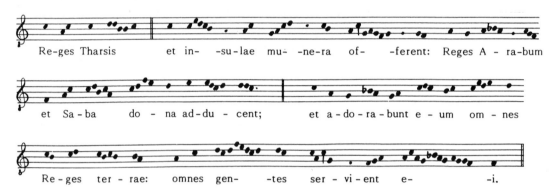

Re-ges Tharsis et in--su-lae mu--ne-ra of--ferent: Reges A-ra-bum

et Sa-ba do-na ad-du-cent; et a-do-ra-bunt e-um om-nes

Re-ges ter-rae: omnes gen--tes ser-vi-ent e- -i.

Apart from this type of chant, two others were developed which have as much in common with Baroque song as with plainchant. These were (i) *chant figuré* or *plainchant*, *plainchant musical*, and (ii) *chant sur le livre* or *fleuretis*. In the first (Pineau 1955, Fuller, 'Plainchant musical', *NG*), a measured and ornamented type of chant was sung by a soloist, choir sections being accompanied by organ harmonies or by a serpent. It is known from treatises by Nivers, Poisson, Lebeuf, and La Feillée, and from the *Cinq messes en plain-chant* (1669) by Henri Du Mont, of which the *Messe royale* was especially popular. In the second (Prim 1961) a soloist improvised florid counterpoint over a chant performed by a choir in rhythm and reinforced again by the serpent.

Neo-Gallican chant in France was not the only repertory of non-Roman and non-medieval chant. Numerous other books of the eighteenth and early nineteenth centuries, especially those published in Germany, are likewise witness to attempts to reform plainchant in accordance with contemporary taste. Those produced by Reiner Kirchrat in Bonn (*Theatrum musicae choralis*, Cologne 1782) and Caspar Ett in Munich (*Cantica sacra*, 1827, with organ accompaniment) are examples. (Tack 1960 gives several facsimiles from German publications.) The French *plainchant musical* is clearly derived from Italian *canto fratto* (as opposed to *canto fermo puro*), exemplified in Viadana's '24 Credo a canto fermo' of 1619. The accompaniment of chant, increasingly common in the seventeenth and eighteenth centuries, was itself an inducement to recompose chants in a style easier to harmonize and more in the style of contemporary solo and concertante motets. (The history of plainchant accompaniment is a subject unfortunately too large for discussion here: see Söhner 1931 and Wagener 1964.)

Contemporary works which offer discussion of or instruction in plainchant are essential for an understanding of the attitudes current in those times. Those written in the seventeenth and eighteenth centuries have been listed and discussed by Fellerer, both briefly (1972, 'Choralpflege') and more comprehensively (1985).

XI

The Restoration of Medieval Chant

XI.1. THE RETURN TO THE SOURCE

Fellerer, 'Choralreform', *MGG*, 1974–5, 1985.

After the period of perpetual revision and new composition it was perhaps inevitable that the idea should have gained ascendancy of returning to the plainchant as it had 'originally' been. Several factors determined the course that events would take.

First, liberalism in the church as a whole was discredited, not least by the French Revolution, which, it was felt, had not been hindered by the Jansenist and Gallican movements in the French church. There was a similar reaction to the ecclesiastical reforms of Holy Roman Emperor Joseph II (1765–90), the general tenor of which had been to increase religious tolerance and limitation of papal authority to spiritual matters. The reaction implied a return to Rome in liturgical as in other matters. The call to action was sounded as early as 1811 by Alexandre-Étienne Choron in his *Considérations sur la nécessité de rétablir le chant de l'Église de Rome dans toutes les églises de l'Empire français*. (Choron had already distinguished himself as an editor of Josquin, Goudimel, Palestrina, and Baroque masters.)

Secondly, the perennial tendency of the church to renew itself by returning to an earlier and supposedly 'purer' state coincided with the wider romantic tendency to idealize the past. When the Age of Reason challenged the old certainties, disaster had ensued. Churchmen yearned for the Age of Faith. To the 'Gothic revival' in church architecture, a comparable revival in church music would be joined. Its musical ideals—held by a remarkably wide body of opinion in Italy, France, and Germany—were embodied in traditional plainchant and the polyphonic music of Palestrina (see Fellerer, 'Caecilianismus', *MGG*).

Thirdly, the techniques of manuscript study which were essential for revealing the secrets of medieval sources had already been developed, particularly by the Benedictine monks of the Congregation of Saint-Maur during the late seventeenth and eighteenth centuries. It would take until the end of the nineteenth century for similar expertise to accumulate in the handling of the first notations, but the whereabouts of many important sources was already known from the Maurists' work, for example, through Mabillon's study of the Ordines Romani, published as early as 1689.

The particular part played by the Benedictine abbey of Solesmes near Le Mans after its restoration by Dom Prosper Guéranger in the 1830s will be discussed in the next section. Solesmes was of course not alone in studying ancient sources as a means of renewing the quality of the church's music.

Some of those who favoured a return to the Roman liturgy were not interested in the medieval melodies, and wished to reintroduce the Medicean and other editions of that period. The Medicaea plus the *Antiphonale* of Peter Liechtenstein, Venice, 1580, formed the basis of the several books produced in the 1840s and 1850s by Duval, de Vogt, and Bogaerts under the auspices of Cardinal Sterckx in Mechelen (Malines). As long as Rome itself remained inactive, this was also the position of the Cecilian movement in Germany, founded in 1868 in Regensburg by Franz Xaver Witt (see Lickleder 1988), and the justification for the editorial line taken by Haberl in the publications by the firm of Pustet in the same city.

Some of the landmarks in recovering the medieval melodies are as follows. In 1846 J.-L.-F. Danjou discovered the eleventh-century manuscript Montpellier, Faculté de Médecine H. 159, with alphabetic letters and neumes in parallel. In 1851 Louis Lambillotte published a hand-drawn facsimile of St Gall, Stiftsbibliothek 359, unfortunately not accurately done, believing that this was the manuscript which, according to St Gall legend, was sent from Rome itself by Pope Hadrian I (IX.3). Lambillotte's *Graduale Romanum* and an *Antiphonarium* were published in 1855 in Paris after his death by Dufour, but did not follow St Gall 359—not surprisingly, since the necessary work of comparing it with other sources to decide on pitches to interpret its neumes would only be accomplished later.

Several editions with partially restored melodies were prominent in the supersession of Neo-Gallican books in France, of which the gradual promoted by the bishops of Reims and Cambrai (1851) is the most significant. It was reputedly based, at least in part, on the Montpellier manuscript and other medieval codices, but the restoration was still half-hearted. The other books took a good deal of account of the Reims–Cambrai edition, but were still more timid, especially in melismatic chants.

Discussion of the merits and demerits of these and other French editions forms a lively chapter in that country's scholarly music criticism, and very numerous were the memoirs published, the committees and societies formed (see Fellerer 1974–5, esp. 143–5 on the Reims–Cambrai commission). In Germany a pioneering role was played by Michael Hermesdorff (1833–85). In 1869 he founded a branch of the Allgemeiner Cäcilienverein for the Trier diocese and from 1872 to 1878 edited the journal *Cäcilia* (founded in 1862, later the *Kirchenmusikalisches Jahrbuch*), which from time to time included facsimiles of medieval chant sources. In 1872 he founded a Verein zur Erforschung alter Choral-Handschriften (Society for the Investigation of Old Chant Manuscripts), and he was Peter Wagner's choirmaster and teacher. He published several chant editions, including a *Graduale . . . Trevirensis* (1863), an *Antiphonale* (1864), *Kyriale* (1869), and a second *Graduale ad normam cantus S. Gregorii* (Trier, 1876–82). The latter was a remarkable publication, which not only restored practically the entire medieval shape of the melodies, but also included small printed neumes, imitating the shapes of later medieval Trier manuscripts, over the music on the staff.

As a means of gauging the relative degree of restoration effected in some of the above-mentioned editions, Ex. XI.1.1 gives the opening of the gradual *Exurge domine non prevaleat homo*. For ease of comparison I have transcribed all the versions on to the modern stave, but reproduced the note-shapes of all the nineteenth-century editions.

The editions of the *Graduale* (1871 and 1873) and *Antiphonale* (1878) published in Regensburg by the firm of Pustet were edited by Franz Xaver Haberl. The gradual was a faithful revival of the Medicaea, the antiphoner was based on the printed editions of Venice 1585 and Antwerp 1611. Chants for new feasts or anything else required by liturgical changes since the seventeenth century were provided by Haberl and Witt. In 1871 Pius IX declared the Regensburg editions to be the only versions officially recognized by the Roman Church, an astonishing monopoly enjoyed by Pustet for thirty years, and the first of its kind, for the Medicean gradual had never had this status.

XI.2. SOLESMES AND THE VATICAN EDITION

Rousseau 1945; Combe 1969; Bescond 1972.

The Benedictine monastery of Saint-Pierre de Solesmes dates back to the early eleventh century, but after the ruin of the French Revolution it had to be founded anew. Dom Prosper Guéranger (1805–75) bought the property in 1832, and settled there with five priests a year later. In 1837 it was constituted an abbey, with Guéranger as its first abbot, and head of the French Benedictine Congregation, by Gregory XVI. (From 1880 to 1895 and again from 1901 to 1922 the community lived in exile, for some years in the latter period at Quarr Abbey on the Isle of Wight.) Guéranger was passionately committed to the restoration of Roman use in France and of authentic melodies in the chant, ideas set out at length in his *Institutions liturgiques* (1840–51).

It was apparently in 1856 that the first work with a medieval chant-book was carried out, when Dom Paul Jausions transcribed the 'Rollington Processional', from Wilton Abbey (now lost: see Combe 1969, 31–2, Benoît-Castelli 1961). In 1860 Dom Joseph Pothier (1835–1923) joined Jausions in the work of transcription. Two years later the transcription on to lines of the first manuscripts in staffless neumes were tackled, Angers, Bibliothèque Municipale 91 by Jausions, and St Gall Stiftsbibliothek 359 by Pothier. In 1864 a *Directorium chori* was compiled. Already by 1867 Pothier felt able to write the first draft of a 'method', of the type which abounded at the time. And by 1868 a new gradual had also been prepared. These eventually became the *Mélodies grégoriennes* of 1880 and the *Liber gradualis* of 1883. Meanwhile a lithographed *Processionale* had been published in 1873.

Ex. XI.1.1. Opening of gradual *Exurge Domine non prevaleat homo*: Montpellier, Faculté de Médecine H. 159, p. 160, *Graduale* (Rome, 1898), p. 79 (Pustet edition), *Graduel romain* (Paris, 1874), p. 205 (Reims–Cambrai edition), *Graduel romain* (Marseilles, 1872), p. 92 (Digne Commission edition), *Graduale . . . Trevirensis* (Trier, 1863), p. 124 (Hermesdorff's first edition), *Graduale ad normam cantus S. Gregorii* (Trier, 1876) (Hermesdorff's second edition)

Pothier was but one star in a brilliant constellation of scholars of liturgy and church history at Solesmes: Pitra (later cardinal, whose aid was later to be invaluable in gaining permission for manuscripts to be brought to Solesmes and photographed), Cagin, Cabrol, and later Quentin and Wilmart, to mention only a few. In 1875 André Mocquereau (1849–1930) joined the community. It was he who conceived the idea of the series of facsimiles *Paléographie musicale*, whose indisputable evidence should support Pothier's publications. The first volume (St Gall 339) appeared in 1889. Particularly significant was the presentation in volumes 2 and 3 of nearly 200 medieval manuscript versions of the same piece, the gradual *Iustus ut palma*. At the international congress on Gregorian chant in New York in 1920 Mocquereau described the *Iustus ut palma* volumes as a 'war machine' ('engin de guerre'), 'une sorte de "tank" scientifique, puissant, invulnérable, capable d'enfoncer tous les raisonnements ennemis'. (Mocquereau 1920–1, 9, cited by Combe 1969, 126.)

In the aftermath of the Franco-Prussian War (and the above was pronounced after the First World War) it was perhaps not only for musicological reasons that French scholars disputed the privilege enjoyed by the Pustet edition. The argument was often acrimonious. The considered views of Haberl, editor of the Regensburg books, laid stress on the papal approbation the Medicaea enjoyed, the necessity of obedience to Rome, the musical value of an edition prepared by such great musicians as Palestrina, Anerio, and Soriano, and the practical merits of the melodies as against the much more difficult versions restored by Solesmes (Haberl 1902).

An international congress held at Arezzo in 1882 vindicated the work of Solesmes, only to have its resolutions rejected by Rome. In 1884 Pitra presented the *Liber gradualis* to Leo XIII, who praised its scientific worth but made it clear that it would not be adopted by the Vatican as normative. An important point was gained when the Jesuit Angelo de Santi, editor of the journal *Civiltà cattolica* in Rome, was won round by Mocquereau, and the students of the French seminary in Rome also adopted the Solesmes melodies. Other Roman choirs, including the Sistine choir itself under Antonio Rella, gradually followed suit. The decree of 1883 confirming support for the Pustet edition was withdrawn by Leo XIII in 1899, and when Pustet's monopoly expired in 1901 it was not renewed. Leo XIII died in 1903 and his successor Pius X acted quickly to sanction the restored chant. The famous *motu proprio* 'Tra le sollecitudini', dated St Cecilia's Day (22 November) 1903, appears to have been drafted by De Santi. A triumphant congress held in Rome in 1904 to mark the thirteenth centenary of St Gregory gave rapturous acclaim to the restorers and the choirs singing the medieval melodies. (Many of the speeches and performances were actually recorded by the Gramophone Company: see Berry 1979.)

Meanwhile a string of important publications had appeared from Solesmes, mostly Pothier's work. A selection of sequences, votive antiphons, and other pieces, the *Variae preces*, was published in 1888. The first edition of the *Liber antiphonarius* was published in 1891. Pothier's old *Processionale* was issued in 1893 as a *Processionale monasticum*, and in 1895 there appeared a *Liber responsorialis*. (These two contain almost the only editions of the great responsories of the Night Office to have been issued, for the various *Antiphonale*s which have appeared are for the day hours only.) A revised *Liber Gradualis* was issued in 1895, and the first edition of the compendium *Liber usualis*.

There remained the task of preparing official Vatican editions of the *Graduale* and *Antiphonale*. In 1904 a commission under Pothier was appointed by the Vatican to supervise the edition, which was to be prepared by the monks of Solesmes. Pothier was no longer resident at Solesmes, having been appointed prior of Ligugé in 1893 and abbot of Saint-Wandrille in 1898. Perhaps inevitably in a project as complex as this, differences between members of the overseeing commission and the executive arose. For example, Pothier favoured *c* rather than *b* as the reciting note in mode 3 and mode 8 chants and recitation formulas. He and others (Wagner, Gastoué) were readier than Mocquereau to admit the validity of later medieval sources, representatives of 'the living tradition'. Bescond (1972, pl. XXIX) gives a facsimile of a page from the preparatory drafts of the *Kyriale*. Here we see the Solesmes proposal for the Kyrie *Lux et origo* and Gloria 'I', with the commission's amendments to Solesmes proposals. The beginnings run thus:

Solesmes: *Gab a ab cbbG*
                Ky- ri-e   e-
Commission: *Gac a accbG*
                Ky- ri-e
Solesmes: *b   aG G G ab b* ab aG G*
                Glo-ri- a in ex-cel-sis de-o
Commission: *b   aG G G ab b* abcb aG G*
                Glo-ri- a in ex-cel-sis   de-o

* = liquescent

The amendments were chiefly those of Pothier, Wagner, Gastoué, and Grospellier. It is clear that no sensible amendments could be made without recourse to manuscript sources, available to very few of the commission. The Solesmes draft was naturally presented without critical apparatus.

Disagreement reached such a pitch that after 1905 Solesmes did not take further part in the editorial work. The *Graduale Romanum* of 1908 was in effect a new edition of Pothier's *Liber gradualis*, re-edited in the light of the Solesmes *Liber usualis* of 1903. The *Antiphonale Romanum* followed in 1912, based on Pothier's *Liber antiphonarius*.

In the same year as the publication of the Vatican *Kyriale* (1905), an alternative Solesmes version was issued by Desclée with rhythmic indications: bars and dots for longer notes, accents to indicate the 'ictus', for that note in a group which the singer should understand to be the most important, without stressing it dynamically. These remained a characteristic of Solesmes editions of the official books, and of course of their own house productions, notably the *Liber usualis*.

XI.3. PRACTICAL EDITIONS AND SCHOLARLY RESEARCH

Stäblein, 'Gregorianik', *MGG*; Hucke 1988, 'Choralforschung'.

After the bruising experience of the Vatican edition, Solesmes remained charged with the preparation of chant editions for the Roman church, a task it continues to perform. During the present century, however, circumstances have changed, in some ways dramatically. Mocquereau worked in the belief that an 'original' form of the earliest melodies could be established by comparison of the oldest sources, in the way that Guéranger had asserted years before: 'When manuscripts of different periods and countries agree upon a particular reading, we can safely assert that we have discovered the Gregorian phrase.' As this book has tried to make clear, it is not at all certain that an 'original' form of this type ever existed. From the Old Roman manuscripts, for example, we can see some of the Roman ways of singing melodies in the eleventh to thirteenth centuries (more than one, for the manuscripts do not agree). We can see several Frankish ways of singing the same melodies in the ninth century, the tenth century, and so on. But the manuscript tradition is too variable for a single 'authentic' reading to be deduced even from a small group of the earliest sources. In *Le Graduel Romain* IV the monks of Solesmes showed how a number of early traditions could be identified, and proposed a comparative edition where the divergent readings would be recorded (see also Froger 1978, 'Edition'). The work is not complete, and a corresponding edition of office chants is also only in the preparatory stages.

It is not only the magnitude of the task which has delayed its completion. The changes in the liturgy announced at the Second Vatican Council (1962–5) have necessitated the preparation of other new service-books, of which the *Psalterium monasticum* (1981) and the *Liber hymnarius* (1983) have so far appeared. These, like all the previous Solesmes editions, are service-books for practical use.

In the history of plainchant research, the practical and the abstract have been inextricably entwined. Scholarship devoted to finding out and understanding the music of the past and present has sometimes been almost inseparable from a religious belief in a 'pure' tradition and the necessity of providing the 'best' attainable form of the melodies for contemporary use. And contemporary use does not stand still. The goal of more recent liturgical reform is no longer the 'Age of Faith' which inspired the efforts of the chant restorers, but rather still earlier centuries from which no liturgical music will ever be recoverable. Christians have not usually found it difficult to worship in churches containing architectural elements of widely differing age, and the liturgy itself is an amalgam of compositions and ceremonial of diverse origins. The plainchant repertory, just like the repertories of polyphonic music which developed alongside it, contains a multiplicity of forms and styles, again of widely differing origin—another point which I hope has become clear in the course of this book. As far as its use in modern worship is concerned there is much to be said for the view of St Gregory, giving advice to his missionary Augustine:

My brother, you are familiar with the usage of the Roman Church, in which you were brought up. But if you have found customs, whether in the Church of Rome or of Gaul or any other that may be more aceptable to God, I wish you to make a careful selection of them, and teach the Church of the English, which is still young in the Faith, whatever you have been able to learn with profit from the various Churches. For things should not be loved for the sake of places, but places for the sake of good things. Therefore select from each of the Churches whatever things are devout, religious, and right; and when you have bound them, as it were, into a Sheaf, let the minds of the English grow accustomed to it. (Bede, *A History of the English Church and People*, i. 27, trans. Sherley-Price, 73.)

That would mean that Carolingian chant of the ninth century could take its place beside rhymed songs of the twelfth or a hymn of the seventeenth century; and, since the 'authenticity' of the chant is no longer an issue, the selection may be made on grounds of religious quality alone.

These are, however, problems of liturgical practice, not musicology, which is surely mature enough as a discipline to distinguish between scholarly and practical concerns in the recovery of old music. The problems facing chant scholarship are great enough without the added complications of liturgical controversy. They concern above all the enormous quantity of potential evidence. What are the right questions to ask when confronted by the huge numbers of liturgical music-books from the ninth century to the present day? How is the material to be controlled and investigated? Exactly what is there in these books, only a tiny fraction of which have been inventoried? What does it tell us about man as a religious being and as a creator of music? Compared with these problems, the tasks which the restorers of the nineteenth century set themselves were essentially simple, concerned with but a part of the chant repertory, mighty though the labours were that were needed to carry out those tasks. It is my hope that this book will have conveyed a sense not only of past achievements but also of the problems that scholarship still faces. But above all, I hope it has helped create an awareness of the inexhaustible musical riches of Western plainchant.

INDEX OF TEXT AND MUSIC INCIPITS

Citations in **bold** indicate a musical example.

INDEX OF MANUSCRIPT AND PRINTED SOURCES

Citations in **bold** indicate facsimile and partial transcription.

INDEX OF NAMES AND TERMS

Citations in **bold** indicate a musical example.